Electrical Engineering
Concepts and Applications

Second Edition

Electrical Engineering
Concepts and Applications

Second Edition

A. Bruce Carlson
David G. Gisser
Rensselaer Polytechnic Institute

 Addison-Wesley Publishing Company
Reading, Massachusetts   Menlo Park, California   New York
Don Mills, Ontario   Wokingham, England   Amsterdam   Bonn
Sydney   Singapore   Tokyo   Madrid   San Juan

This book is in the **Addison-Wesley Series in Electrical and Computer Engineering**

Eileen B. Moran, *Sponsoring Editor*
Bette J. Aaronson, *Production Supervisor*
Sheila Bendikian, *Production Coordinator*
C. J. Petlick of Hunter Graphics, *Designer*
Patricia Steele, *Copy Editor*
Joe Vetere, *Technical Art Consultant*
Scientific Illustrators, *Illustrator*
Dick Hannus, *Cover Designer*
Hugh Crawford, *Manufacturing Supervisor*

Library of Congress Cataloging in Publication Data

Carlson, A. Bruce, 1937–
    Electrical engineering : concepts and applications / A. Bruce
Carlson, David G. Gisser, — 2nd ed.
      p. cm. — (Addison-Wesley series in electrical and computer
    engineering)
    Bibliography: p.
    Includes index.
    ISBN 0-201-14429-8
    1. Electric engineering   I. Gisser, David G. II. Title.
III. Series.
    TK145.C35   1990
   62 1.3 — dc20   89-14937
                                                    CIP

Reprinted with corrections February, 1990

BCDEFGHIJ–HA–9543210

Preface

This text introduces basic concepts of electrical engineering in four general areas: circuits, electronics, information systems, and energy systems. Concepts are related to practical applications to stimulate interest and acquaint students with useful design ideas. We give primary emphasis to elementary analysis and design methods using simplified models of electrical, electronic, and electromechanical devices. But we also alert students to situations where more sophisticated techniques might be needed by discussing the limitations of those methods and models.

The subject matter is arranged in a modular structure, allowing the selection of topics to fit various curriculum settings and courses of different lengths. For beginning students in electrical engineering, we present a motivating and unifying overview of the profession. For students in other engineering disciplines or in the sciences, we provide a working vocabulary of circuits, electronics, instrumentation, and machinery. For all students, we illustrate the significance of integrated circuits and information-processing systems in modern technology.

Prerequisite Background

The text is written at a level suitable for students who have completed at least one term of college physics and mathematics. Most portions have been class-tested with freshmen and sophomore EE majors, with junior "non-EEs," and with technology students.

The prerequisite physics consists of elementary mechanics, dimensional analysis, and some familiarity with electric charge and fields, atomic structure, and sinusoidal waves. Electrical energy, conduction processes, and magnetic phenomena are discussed when introduced.

The prerequisite mathematics consists of trigonometric and exponential functions, quadratic equations, simultaneous linear equations, and elementary calculus. For the most part, the use of calculus is restricted to the slope or rate interpretation of differentiation and the area interpretation of integration. A few results and problems involve integration of simple functions or draw upon differentiation to find a maximum value. An appendix at the back reviews the use of Cramer's rule for the solution of two or three simultaneous linear equations.

Contents and Organization

We have organized the subject matter into four major parts. Part I, Linear Circuits, concerns steady-state circuit analysis and simple transients. Also introduced are controlled sources, op-amps, ideal transformers, and filters. Part II, Electronic Devices and Circuits, emphasizes the external characteristics and applications of diodes, transistors, and selected IC units. FETs and BJTs receive equal attention, as do linear and nonlinear electronic circuits. Part III, Information Systems, develops the concepts of analog and digital information processing as applied to communication, digital instrumentation, and microprocessors. Part IV, Energy Systems, deals with AC power, magnetics, electromechanics, and rotating machines.

All four parts can be covered in a year-long course of about 90 one-hour class sessions. However, we have provided flexible options for shorter courses. In particular, Part I establishes the essential background for each of the other three parts, which are otherwise self-contained. Furthermore, several sections and chapters in Parts II–IV may be taken up out of context if desired.

A one-term survey course, for instance, might cover most of Part I and selected sections from later chapters. An introductory circuits and electronics course could use Parts I and II, perhaps augmented with topics from Part III, while Parts I and IV could be used for a course on circuits and machines. An instrumentation-oriented course might consist of Parts I and III along with sections from Part II. Text material not covered in shorter courses may still be of value to students for reference purposes.

To facilitate course planning, we have indicated in the table of contents the minimum prerequisite for each section. In addition, the symbol † marks optional topics that may be omitted. A few of the optional topics are descriptive in nature, intended for enrichment. Others involve quantitative material that goes somewhat beyond the level of a first course.

Changes in the Second Edition

Although this edition follows the general structure of the first one, the text has been extensively rewritten. Problems at the end of the chapters have been revised, and many new problems are provided. There are also several major changes.

In Part I, systematic node and loop analysis is developed with resistive circuits, a new section relates differential equations to the behavior of dynamic circuits, the introduction of ideal transformers is deferred to AC circuits, and frequency response and transients have been combined into one chapter. Part II contains separate chapters on nonlinear electronic circuits, transistor amplifiers, and power electronics, with more attention to design. Part III has expanded coverage of combinational and sequential logic, including standard logic modules. The chapter on feedback and control systems has been eliminated, but feedback concepts are treated in Part II. Part IV now begins with AC power systems, and mutual inductance in transformers has been added to the chapter on magnetics and electromechanics.

Teaching and Learning Aids

Each chapter starts with a list of learning objectives identified by section. Illustrative examples and exercises are included to help students master the objectives. Answers to the exercises are at the back of the book. There are more than 120 examples and 180 exercises.

Examples and exercises frequently involve numerical computations with realistic parameter values. We assume that students will use calculators for these exercises, especially when performing polar/rectangular vector conversions. (Our numerical values in the examples are rounded off to three significant figures; consequently, some third-digit discrepancies appear between intermediate and final results.)

Besides the exercises, we have provided more than 900 end-of-chapter problems. These problems include simple drills, analysis and design calculations, derivations, and occasional extensions of text topics. Special effort has been made to create problems that are educational, interesting, and "doable."

We expect students to refer to the tables at the back of the book for the mathematical relations needed in certain problems. A table of standard component values is also given at the back for use in electronic design problems.

Instructors should note that the problems are organized by section, in correspondence with the order of text topics and not necessarily in order of increasing difficulty. A solutions manual for instructors is available from the publisher.

Symbols and Units

Pedagogical considerations have led us to adopt certain symbols that, although not unprecedented, may be new to some instructors. Reference voltage polarities are shown by arrows rather than by plus and minus signs, except in the case of voltage sources. We find that voltage arrows make it easier for students to apply Kirchhoff's voltage law. The arrow symbol also goes well with electronic circuit topology and node equations. Different arrowheads distinguish between voltages and currents.

We have purposely avoided boldface and other symbols that would be difficult for instructors to reproduce by hand. Phasors and vectors are indicated by an underbar, such as $\underline{V}$, while absolute-value signs denote magnitudes, such as $|V|$. Script letters are limited to $\mathscr{E}$ (electric field), $\mathscr{R}$ (reluctance), and $\mathscr{F}$ (magnetomotive force). The "loop ell" (ℓ) is used to prevent possible confusion with the numeral 1. We emphasize the special nature of controlled sources with a diamond-shaped symbol.

SI units are employed throughout, of course. However, horsepower (hp) and revolutions per minute (rpm) also appear in conjunction with rotating machinery.

Acknowledgments

We gratefully acknowledge the support of the administration and our colleagues at Rensselaer Polytechnic Institute. Special thanks goes to Professor Charles M. Close, Curriculum Chairman of the Electrical, Computer, and Systems Engineering Department. Many thoughtful comments and suggestions came from the reviewers who read portions of the manuscript, and we extend our appreciation to:

> Professor Ravel F. Ammerman, Colorado School of Mines
> Commander H. David Brown, U.S. Naval Academy
> Professor Karl William Carlson, Mississippi State University
> Professor W. Russell Callen, Jr., Georgia Institute of Technology
> Professor Derald O. Cummings, The Pennsylvania State University and Staff Scientist, Locus, Inc.
> Professor M. Yousif El-Ibiary, University of Oklahoma
> Professor Charles A. Gross, Auburn University
> Professor Herbert Hacker, Jr., Duke University
> Professor E. P. Hamilton III, P.E., University of Texas, Austin
> Professor Nigel T. Middleton, West Virginia University
> Professor William Potter, University of Washington
> Professor James W. Robinson, The Pennsylvania State University
> Professor Ian B. Thomas, University of Texas, Austin

The people at Addison-Wesley have been very helpful to us, especially Bette Aaronson, Production Supervisor, and Tom Robbins, Publisher. Finally, we express thanks to our families for their patience and understanding.

Troy, New York   *A. B. C.*
                                                                    D. G. G.

Contents in Brief

Contents

The numbers in parentheses after each section name identify the previous sections that constitute the minimum prerequisite material. The symbol † identifies optional material that may be omitted without loss of continuity.

Chapter

1 Introduction

Stop for a moment and look around. Close at hand you'll probably find several products of electrical engineering: a lamp that illuminates this page, the alarm clock that awakened you, a stereo system, TV set and radio for entertainment and news, the telephone for communication, and of course your calculator. Surely ours is a "wired world" in which the quality of daily life depends to a significant degree upon the use of electrical phenomena.

This textbook addresses the underlying concepts and methods of electrical engineering behind applications ranging from consumer products and biomedical electronics to instrumentation systems, computers, and machinery. Obviously, we cannot deal with all aspects of these areas. Instead, we emphasize general ideas and techniques that, we hope, will provide the necessary background for you to pursue specific topics in more detail. Consistent with that philosophy, this introductory chapter presents an overview of electrical engineering and a prospectus of the chapters that follow.

1.1 THE SCOPE OF ELECTRICAL ENGINEERING

It seems appropriate to begin with a definition of electrical engineering. In the words of Professor H. H. Woodson:*

Electrical engineering is a profession whose practitioners exploit electromagnetic phenomena and electrical and magnetic properties (and sometimes mechanical, thermal, chemical, and other properties) of matter to do useful things, i.e., someone will pay the engineers to do them. The "useful things" usually involve the processing of information or energy and sometimes both,

---

* From "What is Electrical Engineering?," *IEEE Transactions on Education,* vol. E-22, May 1979, p. 35. (Incidentally, this is the only page cluttered by a footnote. All other references will be found under the heading Supplementary Reading.)

as in the processing of information about energy in the relay protection of a power system; and the engineer's activities generally, but not always, involve equipment.

This definition quite properly focuses on *information* and *energy,* two crucial commodities of modern society. They are central to electrical engineering by virtue of the fact that electrical embodiments of information and energy generally result in more effective processing, control, transmission, and distribution. Looking to the future, those advantages loom with even more importance as we seek greater efficiency in energy utilization and the substitution of low-energy activities (such as long-distance communication) for high-energy activities (such as long-distance travel).

Our definition also emphasizes doing "useful things" with electromagnetic phenomena and, possibly, other properties of matter. Certainly, all engineers should do useful things and, as technology advances, we find increasing interaction between the traditional branches of engineering and science. On the one hand, the fabrication of integrated circuits, for instance, involves chemical, mechanical, metallurgical, and thermal considerations along with electronic expertise. On the other hand, electronic instrumentation and microprocessors have become vital parts of the measurement and control systems employed by aeronautical, biomedical, chemical, civil, mechanical, and nuclear engineers and scientists.

Without belaboring the point, modern technology demands a team approach in which "EEs" and "non-EEs" work together and share a common vocabulary. Accordingly, the goals of this book include sensitizing electrical engineering students to the relevance of nonelectrical topics, and introducing other students to the language of electrical engineers. Both goals will be furthered by considering the scope of electrical engineering.

Perhaps the best indication of that scope is the variety of periodicals published by The Institute of Electrical and Electronics Engineers. The IEEE (I-triple-E) has more than 300,000 members, making it the largest engineering society in the world. Its publications include three general magazines *(IEEE Spectrum, IEEE Potentials,* and *Proceedings of the IEEE)* and 46 special-interest transactions and journals listed in Table 1.1 – 1. The listing has, somewhat arbitrarily, been divided into four categories.

The category headed "Devices" consists of specializations that deal primarily with electromagnetic fields and charged particles and their interaction with matter, often at the microscopic level. The "Circuits and Electronics" category includes several traditional EE activities and some newer ones, essentially characterized by the task of connecting individual devices to achieve a desired circuit behavior. The "Computers and Systems" category reflects the fact that many applications are now so complex that the designer must concentrate on the overall functions rather than the specific details of the constituent circuits that make up a large system.

Table 1.1−1
Transactions and journals of the IEEE.

Devices	Circuits and Electronics	Computers and Systems
Antennas and Propagation	Circuits and Systems	Aerospace and Electronic Systems
Electrical Insulation	Computer-Aided Design of Integrated Circuits and Systems	Automatic Control
Electron Devices		Broadcasting
Electronic Materials	Consumer Electronics	Communications
Lightwave Technology	Industrial Electronics	Computers
Magnetics	Instrumentation and Measurement	Information Theory
Microwave Theory and Techniques		Knowledge and Data Engineering
Nuclear Science	Power Electronics	Pattern Analysis and Machine Intelligence
Plasma Science	Solid-State Circuits	Power Systems
Quantum Electronics		Robotics and Automation
		Software Engineering

Cross-Disciplinary		
Acoustics, Speech, and Signal Processing	Energy Conversion	Professional Communication
Biomedical Engineering	Engineering Management	Reliability
Components, Hybrids, and Manufacturing Technology	Geoscience and Remote Sensing	Semiconductor Manufacturing
Education	Industry Applications	Systems, Man, and Cybernetics
Electromagnetic Compatibility	Medical Imaging	Ultrasonics, Ferroelectrics, and Frequency Control
	Oceanic Engineering	Vehicular Technology
	Power Delivery	

Roughly speaking, the headings at the top of Table 1.1−1 cover a spectrum from the applied physics of device development to the applied mathematics of computers and systems engineering. But all areas of electrical engineering require a working knowledge of physics and mathematics along with engineering methodology and supporting skills in communication and human relations. Thus, the fourth category at the bottom of the table includes specializations that cut across the other categories in one way or another, and several extend into territory not ordinarily associated with electrical engineering.

1.2 HISTORICAL PERSPECTIVE

Having described the present scope of electrical engineering, we should give at least some attention to its historical evolution. For that purpose, the following chronology lists selected discoveries, inventions, and theoretical developments over a span of more than 200 years. Many of the names and events are well known, while others will be unfamiliar to you. Their importance should emerge from the pertinent sections of the text, so you may find it useful to return here on occasion.

Year	Event
1750–1831	*The Beginnings:* Franklin and Coulomb study electric charge; Volta discovers the battery; Fourier and Laplace develop mathematical theories; Ampere, Weber, and Henry conduct experiments on current and magnetism; Ohm's law is stated; Faraday publishes his theory of induction.
1838–1866	*Telegraphy:* Morse perfects his telegraph, and commercial service begins; multiplexing is invented; the first transatlantic cables are laid.
1845	Kirchhoff's circuit laws are stated.
1864	Maxwell's equations predict electromagnetic radiation.
1876–1887	*Telephony:* Bell devises an acoustic transducer; the first telephone exchange, which has eight lines, is constructed in New Haven, Conn.; Strowger invents the automatic switch.
1879–1882	*DC Power Systems:* Edison finds a suitable lamp filament; he establishes in New York City the first electric utility with 59 customers.
1885–1895	*AC Power Systems:* Stanley develops a practical transformer; Steinmetz conceives of phasors for AC circuit analysis; Westinghouse promotes AC systems; Tesla builds an induction motor; in Germany, the first long-distance, three-phase power line is constructed; generators are installed at Niagara Falls.
1887–1897	*Radio:* Hertz verifies Maxwell's prediction; Marconi patents a complete wireless telegraph system.
1904–1920	*Vacuum-Tube Electronics:* Fleming and DeForest build vacuum tubes; electronic amplifiers make possible a transcontinental telephone line; the superheterodyne radio receiver is perfected by Armstrong; KDKA in Pittsburgh is the first AM broadcasting station.
1923–1938	*Television:* Farnsworth and Zworykin devise electronic image formation; DuMont markets cathode-ray tubes; experimental broadcasting begins.
1934	Black invents the negative-feedback amplifier.
1938	Shannon applies Boolean algebra to switching circuits.
1938–1945	*World War II:* Major advances take place in electronics, instrumentation, and theory; radar and microwave systems are developed; operational amplifiers are incorporated in analog computers; FM communication systems are perfected for military applications.
1945–1948	*Systems Theory:* Papers by Bode, Shannon, Wiener, and others establish the basis for systems engineering.
1946	The ENIAC vacuum-tube digital computer is constructed at the University of Pennsylvania.
1948	Long-playing microgroove records are introduced.
1948–1955	*Transistor Electronics:* Schockley, Bardeen, and Bráttain invent the point-contact and junction transistor (Bell Laboratories); the development of the surface-barrier transistor improves manufacturing techniques; transistor radios go into mass production.
1951–1958	*Digital Computers:* UNIVAC I is installed at the US Census Bureau; IBM markets the popular Model 650; the programming language FORTRAN is developed to facilitate scientific use of computers; the transistorized Philco 2000 marks the "second generation" of computing equipment.
1957	The first commercial nuclear power plant becomes operational at Shippingport, Penna.
1958–1961	*Microelectronics:* Hoerni invents the planar transistor (Fairchild Semiconductor); integrated circuits are developed by Kilby (Texas Instruments) and others; commercial production begins.
1960	Laser demonstrations by Maiman.
1962	Telstar I is launched as the first communications satellite.
1962–1966	*Digital Communication:* Commercial data transmission service begins; experiments prove the feasibility of pulse-code modulation for voice and TV; electronic switching systems become operational; practical implementations of error-control coding are devised.
1969	765,000-volt AC power lines are constructed.
1971–1975	*Microcomputers:* MOS technology permits large-scale integration; Hewlett-Packard markets the HP-35 calculator; Intel introduces the 8080 microprocessor chip; semiconductor devices are used for memory.

Year	Event
1975–?	*CAD/CAM:* Computer-aided design and manufacturing come into fruition with advances in interactive computer graphics, software engineering, and parallel processing; artificial intelligence and robotics emerge as new disciplines.
1975–?	*Lightwave Technology:* Fiber optics and optoelectronics go into large-scale use; lasers are employed for office printers, biomedical instruments, and industrial processes.
1980–?	*Digitized Consumer Products:* Digital electronics invade the home in watches, video games, compact-disk players, and kitchen appliances; microprocessors improve automobile performance and diagnostics.
1985–?	*State of the Art:* Recent promising developments include high-power MOS devices, ceramic superconductors, neural networks, expert systems, high-density memory chips, and integrated-services digital networks.

1.3 PROSPECTUS

The text of this book is divided into four major parts: linear circuits; electronic devices and circuits; information systems; and energy systems.

Part I covers the basic concepts and tools needed for the remaining parts, starting in Chapter 2 with a review of electrical quantities and circuit laws. Chapter 3 develops important modeling and analysis techniques in the context of resistive circuits, and Chapter 4 extends those techniques to include amplifiers and the versatile "op-amp." Capacitance and inductance are introduced in Chapter 5, along with the related concept of impedance. Chapters 6 and 7 apply the impedance concept to the study of AC circuits, frequency response, and transients.

Part II begins with the external characteristics of diodes and transistors in Chapters 8 and 9, which also take a brief look at semiconductor physics and fabrication. Chapters 10 through 12 then present the operating principles and design methods for nonlinear electronic circuits, transistor amplifiers, and power electronics.

Part III shifts the emphasis from circuit diagrams to the block diagrams of systems engineering. Analog signal processing and communication systems are discussed in Chapter 13. Chapters 14 and 15 focus on digital systems, including logic design, instrumentation, and computers.

Part IV reconsiders circuits from the energy viewpoint. Chapter 16 discusses AC power and three-phase systems. Magnetic circuits and electromechanical devices are introduced in Chapter 17, while Chapter 18 deals entirely with rotating machines.

Parts II, III, and IV are each relatively self-contained, and some chapters and sections may be taken out of sequence if desired. In this regard, the table of contents indicates the minimum prerequisite material for each section. The symbol † identifies optional material that can be omitted without loss of continuity.

Every chapter has two features intended to guide your study: a list of objectives stated at the beginning, and practice exercises that should help you master the objectives. We recommend that you read the objectives and work the exercises as you come to them, then reread the objectives by way of review. Answers to the exercises are given at the back of the book, along with tables of useful mathematical relationships. Additional problems at the end of each chapter are arranged by section.

If you need further help with a particular topic, you may wish to consult one of the other textbooks listed as Supplementary Reading. Under that heading you will also find more advanced references on the subjects introduced here.

I Linear Circuits

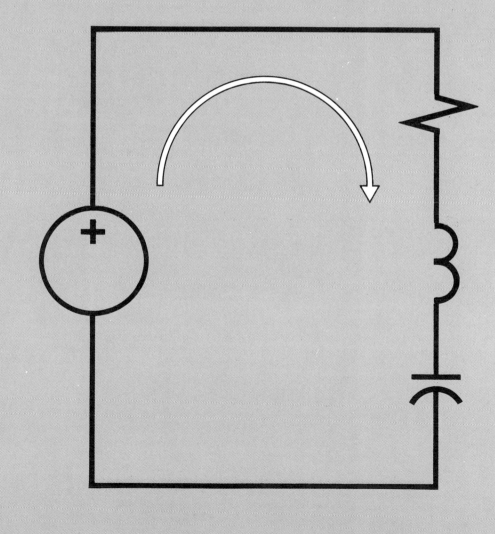

Chapter
2 Circuit Concepts

Whether a simple flashlight or part of a large digital computer, all electrical circuits involve at least four things: energy sources, current, voltage, and resistance. Energy sources provide the driving force that produces currents and voltages in the circuit. Current and voltage, in turn, are the dependent variables that carry out whatever task the circuit has been designed to do. The cause-and-effect relationship between sources and variables is dictated by the devices or elements that make up the circuit and the way they are connected. Of the many different kinds of circuit elements, resistance is the most common.

So we begin by considering these simple but basic circuit concepts. We'll define current and voltage and ideal sources, and develop their relationship to energy and power. We then introduce three important "laws": Ohm's law, which describes resistive circuit elements, and Kirchhoff's current and voltage laws, which govern the interconnection of circuit elements. These three laws are sufficient for analyzing and designing simple but illustrative practical circuits.

Objectives

After studying this chapter and working the exercises, you should be able to do each of the following:

- Define and give the units for current, voltage, and power (Section 2.1).

- Draw the symbols and current-voltage curves for ideal sources and resistance (Sections 2.2 and 2.3).

- Use Ohm's law to calculate current, voltage, resistance, and power, expressing the units with appropriate magnitude prefixes (Sections 2.1 and 2.3).

- Identify loops and nodes in a given circuit, and apply Kirchhoff's laws to them (Section 2.4).
- Solve for the voltages and currents in a simple circuit, using the branch-current or node-voltage method (Sections 2.3 and 2.4).

2.1 CURRENT, VOLTAGE, AND POWER

Figure 2.1–1
Flashlight circuit.

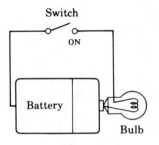

Figure 2.1–1 represents the essential parts of a flashlight — a battery, bulb, switch, and connecting wires. Each element has two electrical contact points or terminals, and turning the switch to the ON position forms a complete loop or closed circuit. The battery's voltage then causes current to go from one terminal, through the bulb and wires, and back to the other battery terminal. As current passes through the bulb, electrical power is converted to heat and light. The current thus serves as a vehicle of power transfer from battery to bulb. Actually, we know that the current consists of moving charges. These charges acquire energy from the battery and deliver energy to the bulb.

This section reviews the fundamental relationships between charge, current, energy, voltage, and electrical power. Thereafter, we will use current, voltage, and power as our primary electrical quantities.

Charge and Current

Electrical charge has two characteristics: The amount or magnitude measured in **coulombs** (C), and the sign or polarity. The charge carried by an electron, for instance, is

$$q_e = -1.60 \times 10^{-19} \text{ C} \qquad \textbf{(1)}$$

which has *negative* polarity. A proton has the same amount of charge but with *positive* polarity. The polarity names were assigned by early experimenters and bear no particular significance. What is significant is the fact that charges of opposite sign tend to neutralize each other. Thus the net charge on a larger particle, such as an ion, is the difference between the amounts of positive and negative charge.

Current exists whenever there is net transfer of charge through a given area in a given time. To illustrate, suppose a positive charge q_1 and a negative charge q_2 both pass from left to right through the shaded area in Fig. 2.1–2. (Situations like this actually occur in a semiconductor diode.) The net charge transfer from left to right will be

$$\Delta q = q_1 + q_2 = q_1 - |q_2|$$

Figure 2.1–2
Charge transfer through an area.

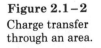
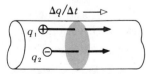

where the absolute-value notation emphasizes that the negative quantity q_2 going from left to right is equivalent to the positive quantity $|q_2|$ going the other way. If this charge transfer takes place in a time interval Δt, the average current is

$$I = \frac{\Delta q}{\Delta t}$$

The current direction is shown by the triangular-headed arrow assuming Δq is positive.

Generalizing using differential quantities, we define **instantaneous current** as

$$i \triangleq \frac{dq}{dt} \tag{2}$$

in which the symbol $\triangleq$ identifies a definition. We measure current in **amperes** (A), called **"amps"** for short, and one ampere equals the transfer of one coulomb per second (C/s). The unit equation is then $1 \text{ A} = 1 \text{ C/s}$.

Frequently, current consists of a large number of identically charged particles, each carrying charge q_0 and moving in the same direction at *average velocity u,* as represented in Fig. 2.1–3. The charge flow through the cross-sectional area A may then be written as

$$i = \frac{dq}{dt} = \frac{dq}{d\ell}\frac{d\ell}{dt} = \frac{dq}{d\ell}\, u \tag{3a}$$

Here, $d\ell = u\, dt$ is the distance traveled by a carrier in time dt and dq equals the total amount of charge in length $d\ell$. If the *carrier density* is n particles per unit volume, then the volume $A\, d\ell$ contains $nA\, d\ell$ carriers at any instant of time so $dq = q_0 nA\, d\ell$. Hence, $dq/d\ell = q_0 nA$ and

$$i = q_0 nuA \tag{3b}$$

which expresses the current in terms of the carriers' charge, volume density, and velocity.

Figure 2.1–3

Current composed of like charge carriers q_0 with density n and average velocity u.

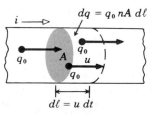

Note that current i has a negative value when q_0 is negative. In such cases we can view it as positive current going in the opposite direction. This happens to be an important observation, because electrons (with negative charge) constitute most electrical currents.

Figure 2.1–4
Charge transferred
when (a) $i = I$,
(b) $i_{av} = I$.

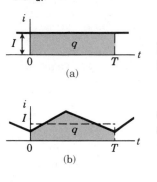

(a)

(b)

Occasionally we will be interested in the *total charge* found by integrating current. In particular, since $dq = i\,dt$, the total charge transported in some time interval, say 0 to T, is

$$q = \int_0^T i\,dt \tag{4a}$$

Equation (4a) says that q equals the *area* under the curve of i versus time from $t = 0$ to $t = T$. If i has the constant value I, then

$$q = IT \tag{4b}$$

as shown in Fig. 2.1–4a. This expression also holds when I stands for the *average value* of i over the time interval in question, defined by

$$I = i_{av} = \frac{1}{T}\int_0^T i\,dt$$

Figure 2.1–4b illustrates the concept of average current.

Energy and Voltage

Returning to the flashlight in Fig. 2.1–1, we noted that the moving charges gave up energy to the bulb. Hence, each charge undergoes a change in potential energy. The electrical variable related to energy change is called **potential difference** or **voltage,** measured in **volts** (V). Specifically, if charge dq gives up energy dw when going from point a to point b, then the voltage across those points is defined to be

Figure 2.1–5
Symbols for voltage
polarity across an
element assumed to be
receiving energy.

$$v \triangleq \frac{dw}{dq} \tag{5}$$

Point a is at the higher potential if dw/dq is positive. Expressing energy in **joules** (J), we have the voltage unit equation $1\ \text{V} = 1\ \text{J/C}$.

Polarity again enters the picture in view of the fact that either dq or dw can be negative quantities. If dq and dw have the same sign, then energy is *delivered* by a positive charge going from a to b or a negative charge going the other way. Conversely, charged particles *gain* energy inside a source where dq and dw have opposite polarities.

Figure 2.1–5 shows an arbitrary circuit element with two symbolic conventions used to indicate voltage polarity. The open-headed arrow in Fig. 2.1–5a points to the end of the element assumed to be at the higher potential, while Fig. 2.1–5b has a plus sign at the higher-potential end and a minus sign at the other end. If both i and v are positive quantities, this element is *receiving* energy because a positive charge goes in the direction of the current arrow from higher to lower potential. If either i or v has a negative value — implying its actual polarity is the reverse of the symbol — then the element must be *supplying* energy.

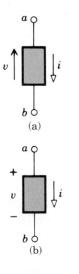

(a)

(b)

Generally, we use the arrow notation for voltage polarity (Fig. 2.1–5a). The type of arrowhead distinguishes voltage and current. Thus an energy-absorbing element has the arrows pointed in *opposite* directions. The other polarity notation will be reserved for voltage sources, in which case we can omit the redundant minus sign at the lower-potential end. A source voltage is sometimes called an **electromotive force** or **emf** to convey the notion that it is the "force" that "drives" the current through the circuit.

Electric Power

Instantaneous power p is defined as the rate of doing work or the rate of change of energy, dw/dt. The electric power in **watts** (W) consumed or produced by a circuit element simply equals its voltage-current product, namely,

$$p = vi \tag{6}$$

so $1\text{ W} = 1\text{ V}\cdot\text{A}$. This equation follows from the fact that voltage is work per unit charge while current is charge transfer per unit time. Thus $vi = (dw/dq)(dq/dt) = dw/dt = p$.

Figure 2.1–6

Measuring power $p = vi$ with an ammeter (AM) and voltmeter (VM).

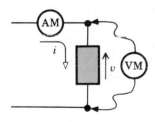

One way of measuring the power consumed by a circuit element is diagrammed in Fig. 2.1–6. We connect a current-sensing instrument called an **ammeter** (AM) so that the current i flows through both the element and the meter. We also connect a **voltmeter** (VM) to read the voltage v across the element. The product of the two meter readings then equals the instantaneous power — providing the voltmeter is ideal so that all of the current i measured by the ammeter passes through the element.

Electric power is the stock-and-trade of electric utility companies, and many electrical devices are characterized by their power rating. But your service bill is for *total energy* consumed, not power. Total energy over a time interval is found by integrating power

$$w = \int_0^T p\, dt \tag{7a}$$

Then, if p has the constant or average value P,

$$w = PT \tag{7b}$$

whose units are watt-seconds or joules. Utility bills are commonly expressed in terms of the **kilowatthour** (kWh),

$$1\text{ kWh} = 3.6 \times 10^6\text{ J} \tag{8}$$

which equals the total energy delivered in one hour when $P = 1000$ W.

Example 2.1–1

Capacity of a battery.

A typical 12-V automobile battery stores about 5×10^6 J of energy, or slightly over one kilowatthour. If the battery is connected to a 4-A headlight, the power delivered to the bulb will be

$$p = 12\text{ V} \times 4\text{ A} = 48\text{ W}$$

Assuming v and i remain constant, the energy consumed in one minute of operation is

$$w = 48 \text{ W} \times 60 \text{ s} = 2880 \text{ J}$$

The total charge that has passed through the bulb during this period is

$$q = 4 \text{ A} \times 60 \text{ s} = 240 \text{ C}$$

equivalent to $240/(1.6 \times 10^{-19}) = 1.5 \times 10^{21}$ electrons.

Incidentally, the maximum charge storage or "capacity" of auto batteries is often rated in ampere-hours (Ah), one ampere-hour being $1 \text{ C/s} \times 3600 \text{ s} = 3600 \text{ C}$. For the battery in question, the capacity equals $5 \times 10^6 \text{ J}/12 \text{ V} = 4.17 \times 10^5 \text{ C} = 116 \text{ Ah}$.

Exercise 2.1 – 1   Find the current when the battery in Example 2.1 – 1 is connected to a 50-W headlight. Assuming v and i remain constant and there is lossless energy transfer from battery to bulb, how long can the headlight be operated before the battery is completely discharged?

Magnitude Prefixes

Current, voltage, power, and other electrical quantities come in very small to very large values. An electronic device, for instance, might have a current of 10^{-6} A and dissipate 10^{-4} W. At the other extreme, a high-voltage transmission system might handle 10^8 W at 10^5 V. Instead of writing powers of 10 all the time, certain **magnitude prefixes** have been adopted to represent them. Table 2.1 – 1 lists the common prefixes we will be using. We then write 2×10^3 W as 2 kW (kilowatts) while 80 μA (micro-amps) stands for 80×10^{-6} A.

Table 2.1 – 1
Magnitude prefixes.

Prefix	Abbreviation	Magnitude
giga-	G	10^{+9}
mega-	M	10^{+6}
kilo-	k	10^{+3}
milli-	m	10^{-3}
micro-	μ	10^{-6}
nano-	n	10^{-9}
pico-	p	10^{-12}

Example 2.1 – 2   Suppose a 50-kV source is rated at a maximum of 20 W. The corresponding current will be

$$i = 20 \text{ W}/50 \text{ kV} = 0.4 \times 10^{-3} \text{ A} = 0.4 \text{ mA} = 400 \ \mu\text{A}$$

We could have obtained this result with fewer intermediate steps by noting that, as far as magnitudes are concerned, $1/1 \text{ k} = 10^{-3} = 1 \text{ m} = 1000 \ \mu$.

Exercise 2.1 – 2   Calculate p when a 50-kV source produces $i = 0.3 \ \mu\text{A}$. Express your result using an appropriate magnitude prefix.

2.2 SOURCES AND LOADS

Figure 2.2 – 1a shows another way of drawing the flashlight circuit, with the battery represented by a box marked "source" and the bulb by a box marked "load," meaning energy sink or consumer of power. Positive current i passes through the load from a to b, and point a is at a higher potential than b. We say that a voltage "drop" v exists across the load from a to b because the potential decreases in the direction of current flow. The power consumed by the load is $p = vi$. On the other side of the circuit, the source provides power, and a voltage "rise" exists across the source from b to a because the potential increases in the direction of current flow.

Figure 2.2 – 1
(a) A source-load circuit. (b) Analogous hydraulic system.

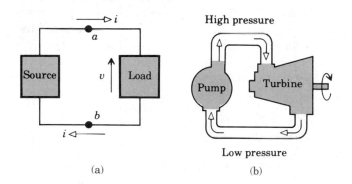

(a)   (b)

Two important observations should be drawn from this discussion. First, a voltage *rise* indicates an electrical *source,* with the charge being raised to a higher potential, while a voltage *drop* indicates a *load,* with charge going to a lower potential. Second, voltage can be thought of as an "across" variable, since we speak of the voltage *across* a circuit element. And current, by the same token, can be thought of as a "through" variable, since current flows *through* a circuit element.

A little confusing? Perhaps a rough analogy will help: We'll briefly describe the operation of the hydraulic system of Fig. 2.2 – 1b, indicating the analogous circuit concepts in parentheses. A pump (the source) forces water flow (the current) through pipes (connecting wires) to drive a turbine (the load). The water pressure (potential) is higher at the inlet port of the turbine than at the output, so a pressure drop (voltage drop) exists across the turbine in the direction the water flows. The pump, on the other hand, raises the pressure (voltage rise). Water flow (current) is a "through" variable, while pressure difference (voltage) is an "across" variable.

In this section we further develop the distinction between sources and loads through consideration of their current-voltage relationships. We also introduce the concept of ideal sources and discuss the significance of idealized device models.

i-v Curves

We have seen that the values and relative polarities of current and voltage characterize the power produced or consumed by a circuit element. Many types of two-terminal elements have a direct relationship between current and voltage that can be expressed as an equation or plotted as a graph called the i-v curve. These curves provide useful information about the nature and behavior of the device. They can be determined experimentally with the help of an adjustable source and the meter arrangement previously given in Fig. 2.1–6.

By way of illustration, consider a typical flashlight bulb and its i-v curve, Fig. 2.2–2. At every point on the curve, i and v have the same sign — both positive or both negative. Hence, the bulb is an electrical load that consumes but never produces power. (We already knew this, of course.) Such devices are said to be *passive,* meaning they have no energy supply of their own.

Figure 2.2–2

Flashlight bulb and its i-v curve.

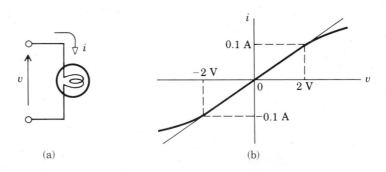

(a)   (b)

The curve also reveals an almost straight-line or *linear* characteristic near the origin, with slope $\Delta i/\Delta v = 0.1$ A/2 V $= 0.05$ A/V. Accordingly, for small values of voltage and current we have the simple approximation

$$i \approx 0.05\, v$$

which holds when -2 V $\leq v \leq 2$ V and -0.1 A $\leq i \leq 0.1$ A. At larger current or voltage values, we would have to take account of the curvature or *nonlinearity* of the i-v plot. The result would be a more complicated expression.

A more intriguing example is the **photodiode** in Fig. 2.2–3. The i-v equation has the form

$$i = I_0(e^{40v} - 1) - I_p$$

where I_0 and I_p are constants, the latter depending upon the light intensity falling on the device. We are not concerned here with the physics underlying this somewhat awesome expression, but rather with its interpretation from the i-v curve. Actually, we must examine two curves, corresponding to the presence or absence of light.

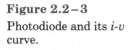

Figure 2.2–3
Photodiode and its i-v curve.

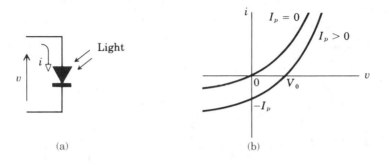

(a)   (b)

With no light on the diode, I_p equals zero and the curve has an asymmetrical shape passing through the origin. Under this condition, the device is passive but draws more current in one direction than the other for the same voltage magnitude. With $I_p > 0$, the curve crosses the vertical axis at $i = -I_p$ and the horizontal axis at the point $v = V_0$. Between these points the device is *active* and *produces power,* since positive current goes from lower to higher potential if $i < 0$. Where does the power come from? From the incident light! The photodiode has become a *solar cell.*

By plotting i versus v in these figures we convey the notion that i is a function of v or that v is the *cause* and i is the *effect.* Sometimes the reverse will be true and it is better to plot v versus i, a v-i curve. Since cause and effect often depend on the specific circuit arrangement, you should be prepared to deal with both i-v and v-i curves.

Ideal Sources

Batteries and AC outlets are familiar electrical sources. Both can be classified as **voltage sources** in the sense that the voltage is essentially independent of the current — though it may vary with time. Formalizing this concept, we say that

> An **ideal voltage source** is one whose terminal voltage v_s is a specified function of time, regardless of the current i through the source.

Figure 2.2–4a shows the symbol we will use for an ideal voltage source with arbitrary time variation.

An ideal **battery** has a constant voltage with respect to time, represented by Fig. 2.2–4b where we use the capitol symbol V_s to emphasize its

Figure 2.2–4

(a) Ideal voltage source. (b) Ideal battery. (c) Ideal sinusoidal (AC) voltage source.

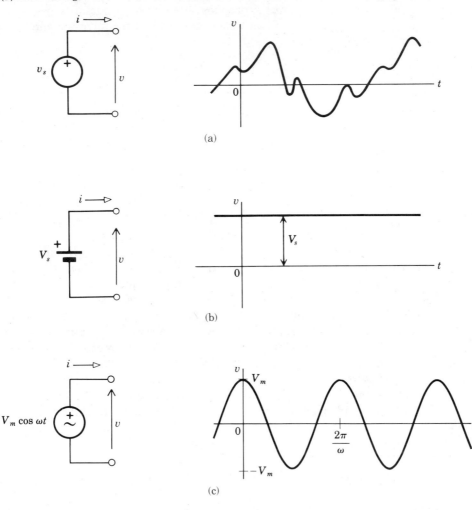

constancy. In normal operation, this source produces a constant current $i = I$ flowing out of the + terminal. The value of I depends on both V_s and the external circuitry connected to the source, whereas V_s remains fixed. Because i is a **direct current,** we call the ideal battery a **DC** source.

Figure 2.2–4c shows the symbol and time variation for a **sinusoidal voltage source** with $v_s = V_m \cos \omega t$. This voltage waveform continuously swings up and down between the peak values $+V_m$ and $-V_m$, as follows from the property of the cosine function that $-1 \le \cos \omega t \le +1$ at any time t. The + sign on the source symbol indicates the terminal at the higher potential whenever $\cos \omega t$ is positive. The rest of the time, when $\cos \omega t$ is

Figure 2.2−5

Ideal current source.

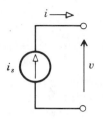

negative, the + terminal is actually at a lower potential than the other terminal. Because this voltage source tends to produce an **alternating current** i, we refer to a sinusoidal source as an **AC source.**

Sometimes it proves convenient to use the concept of an **ideal current source,** defined as one whose current i_s is a specified function of time, regardless of the voltage across its terminals. The circuit symbol is given in Fig. 2.2−5. Although less familiar than voltage sources, current sources play an important role as we'll see when we examine certain electronic devices, notably the transistor.

We underscore the difference between ideal voltage and current sources by plotting their i-v curves in Fig. 2.2−6. At any particular instant of time, a voltage source has a specified voltage value v_s but can supply any amount of current, positive or negative, so its i-v curve is a vertical line intersecting the horizontal axis at $v = v_s$. (The curve for an ideal 9-V battery, for instance, intersects at $v = 9.0$ V.) Conversely, a current source has a specified current value independent of the voltage, so its i-v curve is a horizontal line intersecting the vertical axis at $i = i_s$. (Had we plotted v-i curves, the voltage source would have a horizontal line and the current source a vertical line.)

Figure 2.2−6

i-v curves for (a) an ideal voltage source, (b) an ideal current source.

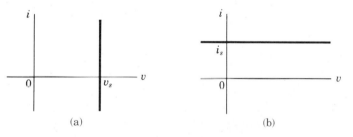

Note that the positive current direction for Fig. 2.2−6 is out of the higher-potential terminal of the source — just the opposite of the curves for passive devices where positive current flows into the higher-potential terminal. This difference reflects the normal mode of operation and allows us to use the i-v curves directly when studying a source-sink combination, as in the example below.

Example 2.2−1

A photodiode.

Suppose a photodiode having $I_0 = 0.1$ μA and $I_p = 0$ is driven by an ideal current source with $i = i_s = 50$ mA, as shown in Fig. 2.2−7a. We can calculate the resulting voltage drop across the photodiode by inserting the values for i, I_0, and I_p into its i-v equation and solving for v, as follows:

$$0.05 = 10^{-7}\,(e^{40v} - 1)$$

$$e^{40v} = \frac{0.05}{10^{-7}} + 1 \approx 5 \times 10^5$$

$$v \approx \frac{1}{40}\,\ln\,(5 \times 10^5) = 0.33 \text{ V}$$

Figure 2.2–7b depicts this solution graphically by plotting the i-v curves for the source and photodiode on the same set of axes; their intersection point gives the value of v, which also equals the voltage rise across the current source.

Figure 2.2–7
Photodiode circuit with an ideal current source.

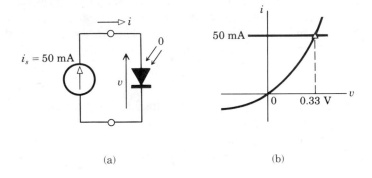

(a)   (b)

Exercise 2.2–1   A device having $i = (0.01\,v)^3$ is connected to a 200-V source. Make a sketch similar to Fig. 2.2–7b, and find the current through the device and the power delivered by the source.

Devices and Models

Theoretically, an *ideal* voltage or current source could produce infinite power $p = vi$, for Fig. 2.2–6 implies apparently unlimited values of v or i. But infinite values are physically impossible so ideal sources cannot exist. Why then, do we bother defining them? The answer to this question is significant, for it relates to one of the most powerful tools in electrical engineering, namely the use of simplified representations or **models** for physical devices.

   Models of natural phenomena — expressed as mathematical relations, curves, etc.— are idealizations representing those aspects of the physical characteristics that are pertinent to a particular application. A good model allows you to predict, with reasonable accuracy, how the device will perform under the expected operating conditions. Thus, for a limited range of current, it might be quite acceptable to pretend that a battery acts like an ideal voltage source. By doing so, we can concentrate on significant factors and effects without getting bogged down in the details of a more accurate but very cumbersome description of a battery. Of course, you must always bear in mind the assumptions and limitations of the model, since predictions that go beyond the model's scope are likely to be invalid — for example, that a battery will produce unlimited power.

Virtually every branch of engineering and science involves mathematical models. (As a case in point, Newton's laws of motion are not absolute laws but models that apply only when velocity is small compared to the speed of light.) Electrical engineering, perhaps, does more with models than some other fields. Actually, every circuit diagram is a model, often a very good one, but still a model. From now on, it should go without saying that the various circuit laws and device representations are approximations of physical reality. Where appropriate, the significant limitations of idealized concepts will be discussed.

Exercise 2.2–2   A certain active device is described by the relationship $(5i)^2 + (0.02v)^2 = 1$ for $i \geq 0$ and $v \geq 0$. Sketch the i-v curve and justify the assertion that this device acts like a current source with $i_s = 200$ mA if $v \leq 5$ V, whereas it acts like a voltage source with $v_s = 50$ V if $i \leq 20$ mA.

Controlled Sources

All of the sources discussed so far fall in the category of **independent sources,** meaning that the source voltage or current is independent of all other voltages and currents. But when we get to the study of transistors, amplifiers, and certain other devices, we'll find it convenient to introduce sources whose voltage or current does depend on the value of some other voltage or current. These models are therefore called **dependent** or **controlled sources.**

By way of example, Fig. 2.2–8a depicts a *voltage amplifier* producing the output voltage $v_{out} = Av_{in}$ where v_{in} is the input voltage and the constant A is the voltage amplification factor. Although the amplifier has very

Figure 2.2–8

(a) Voltage amplifier. (b) Controlled-source model.

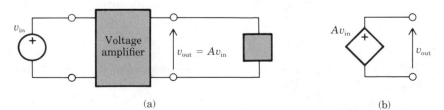

(a)   (b)

complicated internal circuitry, its effect at the output can be represented simply by the source shown in Fig. 2.2–8b, a voltage source controlled by the voltage v_{in}. We use the diamond-shaped symbol here to underscore the special nature of a controlled source.

Conceptually, you may find it helpful to think of this source as an adjustable voltage generator operated by a friendly gnome, as portrayed by Fig. 2.2–9. The gnome has a pair of voltmeters to measure v_{in} and v_{out}, and he continuously readjusts the generator so that the controlled voltage always equals A times the controlling voltage. Thus, for instance, if $A = 20$ and $v_{in} = 5 \cos \omega t$, then our gnome must make sure that v_{out} equals $100 \cos \omega t$ at every instant of time.

Figure 2.2–9
Interpretation of a controlled source.

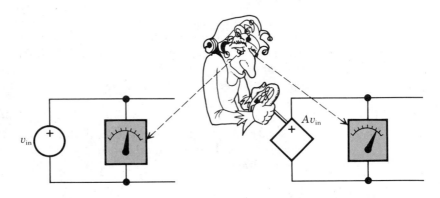

In addition to the voltage-controlled voltage source at hand, we will later encounter current sources controlled by a current or voltage.

2.3   RESISTANCE

A kink in a garden hose impedes the flow of water, producing a pressure drop and conversion of mechanical energy to heat. Similarly, the flow of electric current always encounters some **resistance,** resulting in a voltage drop and the conversion of electric energy to heat. Resistance may be desired in a circuit to produce a voltage drop or energy conversion, or it may be an unwanted but unavoidable part of a device or connecting wire. A **resistor** is a device whose primary electrical characteristic is resistance. The properties of the ideal resistance element are examined here, along with a brief description of electrical conduction in solid materials.

Ohm's Law

An *ideal* or *linear resistance* is an energy-consuming element described by **Ohm's law**

$$v = Ri \tag{1}$$

which means that voltage is directly proportional to current. The proportionality constant R is the value of the resistance, and the units are **ohms** (Ω). Rewriting Ohm's law as

$$R = \frac{v}{i} \tag{2}$$

yields the unit equation

$$1\ \Omega = 1\ \text{V/A}$$

so resistance is the ratio of voltage to current.

Figure 2.3–1 shows the symbol and i-v curve for resistance. Note carefully that the *slope* of the curve is $1/R$ since, from Ohm's law, $i = v/R$. We interpret this curve as follows: If a voltage v is applied across the terminals of a resistance R, the current through the resistance will equal v/R. Resistance resists the flow of current; and the larger R is, the smaller i will be for a given v. By the way, that interpretation should help you keep track of the three different ways of writing Ohm's law — $v = Ri$, $i = v/R$, and $R = v/i$. Had we plotted voltage versus current, its slope would be R rather than $1/R$, meaning that a voltage $v = Ri$ is produced when a current i flows through a resistance R.

Figure 2.3–1

Ideal resistance and its i-v curve.

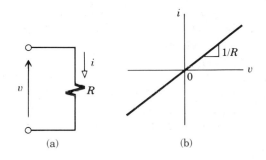

(a)   (b)

For occasional use, it is also convenient to define **conductance** G as the reciprocal of resistance; that is,

$$G \triangleq \frac{1}{R} \tag{3a}$$

so that Ohm's law becomes

$$i = \frac{1}{R} v = Gv \tag{3b}$$

Conductance is measured in inverse ohms or **siemens** (S) in the SI system of units. However, the earlier and rather droll term **mho** ($\mho$) still appears in the literature.

Referring back to Fig. 2.2–2, we find the flashlight bulb had $i \approx 0.05v$ for small voltages and currents. Under this condition, the bulb can be

Figure 2.3–2

ON-OFF switch creates:
(a) a short circuit;
(b) an open circuit.

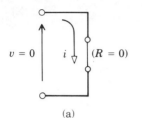

(a)

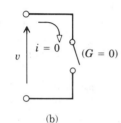

(b)

modeled as a conductance $G = 0.05$ S or resistance $R = 1/G = 20\ \Omega$. Resistance values in typical circuits range from fractions of an ohm to kilohms (kΩ) or even megohms (1 M$\Omega = 10^6\ \Omega$).

A simple ON-OFF switch also may be modeled in terms of resistance, but with very extreme values. Specifically, in the ON or closed position of Fig. 2.3–2a, the switch creates a **short circuit** or zero-resistance path ($R = 0$); then $v = 0 \times i = 0$ for any value of i. But in the OFF or open position of Fig. 2.3–2b, the switch becomes an **open circuit** with zero conductance ($G = 0$); then $i = 0 \times v = 0$ for any value of v.

Power Dissipation and Ohmic Heating

Combining Ohm's law with $p = vi$ gives two equivalent expressions for the power consumed and dissipated by a resistor, namely

$$p = vi = (Ri)i = i^2 R \tag{4a}$$
$$= v(v/R) = v^2/R \tag{4b}$$

Either expression can be used, depending on whether you know the current through the resistor or the voltage across it.

Power dissipation by a resistance element produces heat, a process known as **ohmic heating.** Coils of resistive wire are used precisely for this result in such electrical appliances as ovens, hair dryers, toasters, and so forth. Similarly, an incandescent lightbulb's filament glows when heated by current flow, giving the desired light. Ohmic heating also explains the principle of a simple fuse: when the current reaches the maximum value, heat causes the fuse to melt, thereby "breaking" the circuit.

Unfortunately, however, ohmic heating can cause serious damage to electronic circuits and instrumentation. For this reason, cooling fans are built into some electronic instruments, and large installations such as computer systems require extensive air conditioning.

Example 2.3–1

Figure 2.3–3 shows the circuit and i-v curves for a 9-V battery applied to a 5-kΩ resistance. The current is $i = 9$ V/5 k$\Omega = 1.8$ mA and the power dissipated is $p = (9\ \text{V})^2/5\ \text{k}\Omega = (1.8\ \text{mA})^2 \times 5\ \text{k}\Omega = 16.2$ mW. (Note that power comes out directly in milliwatts when we have voltage in volts,

Figure 2.3–3

Battery-resistance circuit.

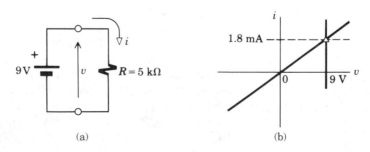

(a)   (b)

resistance in kilohms, and current in milliamps.) Reducing the resistance to 5 Ω gives $i = 1.8$ A and $p = 16.2$ W. In the limit as $R \to 0$, the current and power both become infinite, theoretically, because we have assumed an *ideal* voltage source.

Exercise 2.3 – 1   Find v and p when the battery in Fig. 2.3 – 3 is replaced by a 3-mA current source.

Conduction and Resistivity

Ohm's law, as we stated it, focuses on the external characteristics of a resistive element. But the value of the resistance R depends on the material and shape of the element and its temperature. We'll examine these three factors by taking a brief look at what goes on inside a solid conducting material.

A metallic bar or wire is not just a hollow pipe through which charges may be pumped. Rather, it consists of atoms fixed in an orderly crystal lattice, plus a number of free electrons. These electrons have escaped from their parent atoms, leaving them as positively charged ions, as depicted by Fig. 2.3 – 4a. Both the ions and electrons have kinetic energy that manifests itself in random thermal motion. Binding forces constrain the ions to vibrate about their average lattice positions, while the electrons are free to go flying about erratically, colliding here and there with an ion and changing direction. The net charge flow is zero, however, due to the random electron motion in all directions.

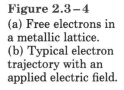

Figure 2.3 – 4
(a) Free electrons in a metallic lattice.
(b) Typical electron trajectory with an applied electric field.

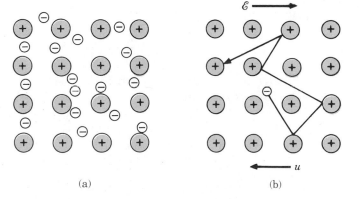

(a)   (b)

Now let an **electric field** $\mathcal{E}$ be applied. The field exerts force on the electrons and they each gain a small velocity component u added to the random motion, so a typical trajectory might look like Fig. 2.3 – 4b. All the electrons thereby drift in the same direction, with average velocity

$$u = -\mu_e \mathcal{E}$$

where μ_e is a proportionality constant called the **mobility.** The minus sign reflects the fact that u is opposite to the direction of $\mathcal{E}$. Since each electron carries negative charge q_e, Eq. (3b) in Section 2.1 tells us that the electron drift constitutes a current through area A given by

$$i = q_e n u A = -|q_e|n(-\mu_e\mathcal{E})A = |q_e|n\mu_e\mathcal{E}A \qquad (5)$$

with n being the material's **electron density.** This current is in the same direction as $\mathcal{E}$ because it consists of negative charges moving in the opposite direction.

Although it may not appear to be so, Eq. (5) is just another statement of Ohm's law. To bring out the connection, we first define the **resistivity** ρ (rho) as

$$\rho \triangleq \frac{\mathcal{E}A}{i} \qquad (6a)$$

$$= \frac{1}{|q_e|n\mu_e} \qquad (6b)$$

Resistivity depends only on the properties of the material in question and has the units of **ohm-meters** ($\Omega \cdot$ m). (We could also define the material's **conductivity** $\sigma = 1/\rho = |q_e|n\mu_e$.) Next, consider the situation in Fig. 2.3–5, where voltage v has been applied to a uniform bar of conducting material having length ℓ and cross-sectional area A. The voltage gives rise to an electric field directed from higher to lower potential and of value

$$\mathcal{E} = \frac{v}{\ell}$$

producing the current

$$i = \frac{\mathcal{E}A}{\rho} = \frac{vA}{\rho\ell} = \frac{v}{(\rho\ell/A)}$$

Finally, observing that this expression is of the form $i = v/R$, we see that the bar has resistance

$$R = \rho\frac{\ell}{A} \qquad (7)$$

Figure 2.3–5
Bar of conducting material with resistivity ρ.

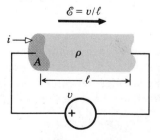

This result also holds for nonmetallic solids with resistivity defined by Eq. (6a).

Equation (7) reveals that resistance is proportional to resistivity and length, but inversely proportional to area. Consequently, a long thin piece of low-resistivity material might have the same resistance as a short, thick piece of high-resistivity material. By the same reasoning, a wire intended to carry large currents should have low resistivity and large area in order to minimize ohmic heating $p = i^2R = i^2\rho\ell/A$. The physical mechanism of this heating in a metal is suggested by Fig. 2.3–4b, namely energy transferred to the lattice when drifting electrons collide with ions.

Table 2.3–1 lists values of ρ for some representative materials. At the one extreme, metals have exceedingly small resistivities due to the large density n of free electrons, and hence are good electrical **conductors.** To illustrate, the 12-gauge copper wire used in many 20-amp household circuits has a radius of about 1 mm, so $A = \pi r^2 \approx 3 \times 10^{-6}$ m^2 and the resistance of a 10-m length will be

$$R = 1.7 \times 10^{-8} \times \frac{10}{3 \times 10^{-6}} \approx 0.06 \ \Omega$$

corresponding to a maximum drop of 1.2 V at the rated current. At the other extreme are the electrical **insulators** (rubber, plastics, etc.) whose lack of free charge carriers yields resistivities so large that ordinary voltages produce virtually no current flow. In between are the **semiconductors,** about which we'll say more in Chapter 8.

Table 2.3–1
Resistivity of various materials at 20°C.

Type	Material	$\rho \ (\Omega \cdot \text{m})$
Conductors	Copper	1.7×10^{-8}
	Aluminum	2.8×10^{-8}
	Nichrome	10^{-6}
	Carbon	3.5×10^{-5}
Semiconductors	Germanium	0.46
	Silicon	2300
Insulators	Rubber	10^{12}
	Polystyrene	10^{15}

The tremendous ratio of available resistivities is one reason why electricity is a convenient method for transporting energy from one place to another. Good insulators keep the energy "contained" within the good conductors, which, in turn, waste little power in ohmic heating. For example, the aforementioned copper wire would dissipate a maximum of about 19 W while delivering approximately 1800 W at 120 V. And a larger wire would further reduce the heat loss.

There is, however, one factor not indicated in Table 2.3–1, namely *temperature dependence.* The given values of ρ are at room temperature (20°C). But resistivity generally increases with temperature for conductors, while it decreases with temperature for insulators. Physically, this

Figure 2.3–6
Strain gauge made with
resistive wire.

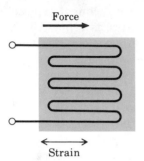

Force

Strain

difference comes about because high temperature tends to liberate charge carriers in an insulator, whereas the increased thermal vibration of the ions in a metal causes more collisions and reduces the electron mobility μ_e. At extremely high temperatures — as might be encountered in an industrial environment — all materials act more or less like semiconductors. Obviously, the circuit designer must take special care in such cases. On the other hand, temperature dependence has practical applications in devices such as the resistance-wire thermometer.

Due, in part, to the temperature dependence of ρ, the v-i curve for many resistive elements becomes *nonlinear* at large values of current. An incandescent lamp filament, for instance, has the characteristic previously seen in Fig. 2.2–2, and its resistance at the operating temperature (around 2000°C) is more than 10 times as large as the "cold" resistance. Consequently, when we first energize a lamp, the initial current flow is substantially higher than the normal operating current. Although its duration is short, this large inrush current may damage mechanical or electronic switches.

Commercially manufactured resistors are designed to be reasonably linear over their intended operating range. They are usually made of carbon, metal film, or very thin wire, and they come in a variety of standard resistance values, precision tolerances, and power ratings. Adjustable resistors are also available; they will be discussed in Section 3.1 under the heading "Potentiometers."

Finally, we should at least mention **superconductors.** These special materials possess the unique property of *zero resistivity* and, hence, they completely eliminate power loss associated with ohmic heating. Metals become superconducting only at temperatures near absolute zero, but some ceramic compounds show promise as superconductors at higher temperatures.

Example 2.3–2
A strain gauge.

A **strain gauge** consists of many loops of thin resistive wire glued to a flexible backing (see Fig. 2.3–6). It is used to measure the fractional elongation or strain $\Delta\ell/\ell$ of a structural member to which it is attached. Straining the wire makes it somewhat longer and thinner, thereby increasing the resistance a small amount, ΔR.

Specifically, if we assume the volume of the wire remains constant, then its area under strained conditions becomes $A' = A\ell/\ell'$ where $\ell' = \ell + \Delta\ell$ is the elongated length. Then, from Eq. (7), the strained resistance R' is

$$R' = \rho\ell'/A' = \rho(\ell + \Delta\ell)^2/A\ell$$
$$= (\rho\ell/A)[1 + 2(\Delta\ell/\ell) + (\Delta\ell/\ell)^2]$$

With $(\Delta\ell/\ell)^2 \ll 1$, we can drop the last term and write $R' \approx R + \Delta R$, where $R = \rho\ell/A$ is the unstrained resistance and

$$\Delta R = 2R(\Delta\ell/\ell)$$

Therefore we can determine mechanical strain by measuring increased resistance.

Exercise 2.3–2   Calculate ΔR when $\Delta\ell/\ell = 10^{-3}$ for a strain gauge made from a 100-cm length of nichrome wire having 0.002-cm radius.

Lumped Parameter Circuits

Let's apply what we now know about resistance to the flashlight circuit of Fig. 2.1–1. Because the bulb is a resistance and there is resistance distributed all along the connecting wires, we can redraw the circuit as in Fig. 2.3–7a, which also includes leakage resistance representing the insulation between the bulb's contacts. If we are concerned only with variables measured at the *terminals* of the various elements, as distinguished from those measured in the interior, we can lump the distributed wire resistance at one point and treat the connecting lines as ideal conductors, as in Fig. 2.3–7b. This is called a **lumped parameter model** in that spatially distributed characteristics — resistance in this case — have been concentrated at one point to simplify analysis. Almost all circuit diagrams employ this approach, thereby focusing on the terminal characteristics of the elements rather than their interior behavior.

Figure 2.3–7

Evolution of a lumped parameter model.

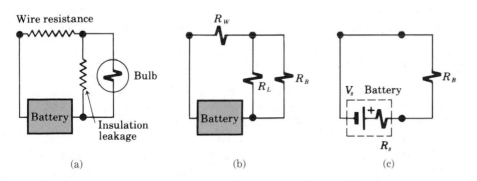

(a)   (b)   (c)

Taking the model one step further, we might ignore the wire resistance entirely since it should be quite small compared to the resistance of the bulb. Similarly, the insulation leakage resistance should be so large that negligible current flows through it. However, the battery also has internal resistance that may cause its terminal behavior to differ significantly from an ideal voltage source. The final diagram then becomes as shown in Fig. 2.3–7c. Hereafter, wire resistance and source resistance will be included when we suspect that they may appreciably affect the circuit's behavior.

2.4 KIRCHHOFF'S LAWS

Figure 2.4–1 illustrates a circuit composed of sources, resistances, and an unspecified element. Circuit analysis is the task of evaluating all voltages and currents, given the source and resistance values and the nature of the other element. Conversely, circuit design involves choosing element values to achieve specified voltages or currents. In either case, one must draw upon **Kirchhoff's laws** (pronounced Kear-koff) as well as Ohm's law.

Figure 2.4–1
Circuit with unknown
voltages and currents.

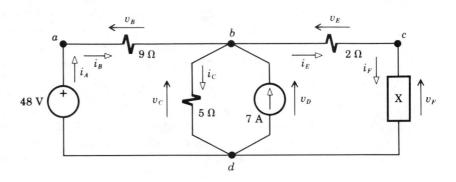

Kirchhoff's two laws are *network laws,* pertaining to the interconnection of elements rather than to individual elements. Ohm's law, in contrast, is an *element law* that describes a particular element irrespective of how it is connected to other elements. After we state Kirchhoff's laws and examine their implication for common circuit configurations, we'll introduce some simple methods of circuit analysis based on those laws.

Kirchhoff's Current Law (KCL)

Kirchhoff's current law describes current relations at the **nodes** in a network — meaning any point where two or more elements are connected, such as points a and b in Fig. 2.4–1. An electrical node is analogous to the junction of two or more water pipes. Clearly, the amount of water coming out of a pipe junction must equal the amount going in, and the same holds for current at a node. Physically, because charge must be conserved and no charge accumulates at a node, the charge flowing out exactly equals the charge flowing in. Expressing charge flow in terms of current, KCL states:

> The net sum of the currents into any node equals zero at each and every instant of time.

We write this symbolically as

$$\sum_{\text{node}} i = 0 \qquad\qquad (1)$$

in which Σ stands for summation.

A key factor in KCL is current *direction,* for we recall that the algebraic sign of a current indicates its direction. Accordingly, a positive current flowing out of a node is equivalent to a negative current flowing into that node. To illustrate, the currents i_C and i_E in Fig. 2.4–2a flow out of node b, while i_B and the source current flow in, so KCL yields

$$i_B - i_C + 7 \text{ A} - i_E = 0$$

Alternatively, we could write

$$i_B + 7 \text{ A} = i_C + i_E$$

which says that the sum of the currents actually flowing in equals the sum of those actually going out.

Figure 2.4–2
Applications of Kirchhoff's current law.

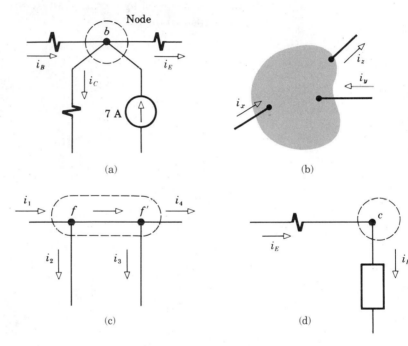

(a)

(b)

(c)

(d)

Moreover, the same statement holds for any portion of a circuit that could be contained within a closed surface, due to the fact that no circuit element accumulates net charge. Thus, in Fig. 2.4–2b, $i_x + i_y = i_z$ regardless of the details of the circuitry inside the enclosure. Similarly, we can write $i_1 = i_2 + i_3 + i_4$ for Fig. 2.4–2c and not even bother with the current going from f to f'; in fact, these two points constitute just one node since they are directly connected by a perfect conductor.

A particularly simple but significant type of node is one where exactly two elements are connected together as in Fig. 2.4–2d. Clearly, $i_E - i_F = 0$ or $i_E = i_F$, and we see that the current must be the same through both elements. This arrangement is called a **series connection.**

Kirchhoff's Voltage Law (KVL)

Kirchhoff's voltage law expresses the principle of conservation of energy in terms of the voltages around a **loop,** a loop being defined as any closed path in a network. Consider Fig. 2.4–3a, which shows one loop from Fig. 2.4–1; if a charge dq goes around this loop, it gains energy from the voltage source and delivers energy to the resistances. Conservation of energy requires that the energy loss equal the energy gain, so $v_B\,dq + v_C\,dq = 48\ \text{V} \times dq$ and hence

$$v_B + v_C - 48\ \text{V} = 0$$

In general, Kirchhoff's voltage law states:

> The net sum of the voltages around any loop equals zero at each and every instant of time.

Using summation notation, KVL is written

$$\sum_{\text{loop}} v = 0 \qquad\qquad \textbf{(2)}$$

in a form analogous to the current law.

Figure 2.4–3
Applications of Kirchhoff's voltage law.

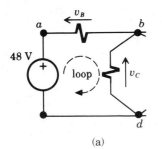 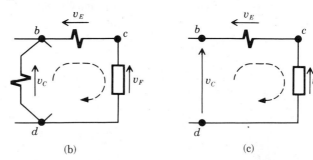

(a)   (b)   (c)

Although we introduced KVL with a loop containing a source, the law also applies to loops such as that in Fig. 2.4–3b. There are no sources here, but the arrow for voltage v_C has the opposite direction of the other two voltage arrows as you travel around the loop. Consequently, KVL requires $v_E + v_F - v_C = 0$ or $v_C = v_E + v_F$. A helpful interpretation of this result comes from the observation that, since v_C is the potential difference between points b and d, it must equal the potential difference between b and c, plus the potential difference between c and d. This interpretation further suggests that KVL applies even to loops that do not have electrical closure. Thus, the voltage across the open circuit from b to d in Fig. 2.4–3c is still given by $v_C = v_E + v_F$.

One other loop from the original circuit deserves special attention, namely the loop in Fig. 2.4–4a formed by two elements connected together at each end, which is called a **parallel connection.** KVL gives us $v_D - v_C = 0$ or $v_D = v_C$; hence, the voltage is the same across both elements. Incidentally, such connections are usually drawn as in Fig. 2.4–4b, where the top and bottom nodes have each been replaced by two points, simply to make life easier for the person drawing the circuit.

Figure 2.4–4
Elements connected in parallel.

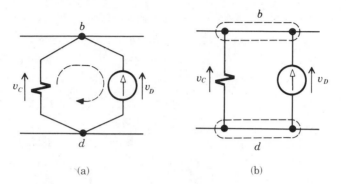

(a)   (b)

Example 2.4–1

Figure 2.4–5a shows a **transistor,** a three-terminal device, connected at nodes B, C, and E. The double-subscript voltage notation here is commonly used for electronic devices, with the first subscript identifying the terminal assumed to be at the higher potential. Thus, v_{CE} stands for the voltage at node C relative to node E. For our present purposes the transistor is taken to have

$$v_{BE} = 0.7 \text{ V} \qquad i_C = 0.9\, i_E$$

Figure 2.4–5
Transistor circuit: (a) schematic diagram; (b) equivalent circuit.

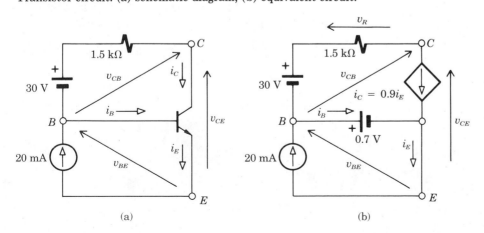

(a)   (b)

This information plus Ohm's and Kirchhoff's laws will allow us to find all other voltages and currents.

First, we incorporate the stated information about the transistor by redrawing the circuit as shown in Fig. 2.4–5b. The 0.7-V battery thus accounts for the fixed value of v_{BE}, while a current-controlled current source represents $i_C = 0.9i_E$. Second, in view of the series connections at nodes E and C, it follows that

$$i_E = 20 \text{ mA} \qquad i_C = 0.9 \times 20 = 18 \text{ mA}$$
$$v_R = 1.5 \text{ k}\Omega \times i_C = 27 \text{ V}$$

Next, using KVL around the outer loop gives

$$v_R + v_{CE} - v_{BE} - 30 = 0$$

so we obtain

$$v_{CE} = 0.7 + 30 - 27 = 3.7 \text{ V}$$
$$v_{CB} = v_{CE} - v_{BE} = 3 \text{ V}$$

Finally, applying KCL to the transistor itself,

$$i_B = i_E - i_C = 2 \text{ mA}$$

which completes the analysis.

Series and Parallel Circuits

Circuits consisting entirely of series or parallel connections are usually the easiest to analyze, because there is only one current value in a series circuit and only one voltage value in a parallel circuit. Despite their simplicity, they play an important role in many applications.

The circuit in Fig. 2.4–6 is a **series circuit** in that each of the three nodes connects exactly two elements. Hence, all three elements carry the same current i, indicated by the single current arrow. Given the source voltages v_1 and v_2 and the value of R, it becomes a simple matter to find v_R from Kirchhoff's voltage law and then solve for i using Ohm's law. Specifically, the voltage across the resistance must satisfy $v_R + v_2 - v_1 = 0$, so

$$v_R = v_1 - v_2 \qquad i = \frac{v_R}{R} = \frac{v_1 - v_2}{R} \tag{3}$$

These seemingly trivial results lead to two important conclusions.

Figure 2.4–6

Series circuit with opposing voltage sources.

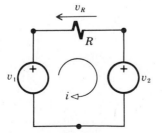

First, the effective source voltage applied to the resistance is the *algebraic sum* of the sources, $v_1 - v_2$. In general, two or more series-connected voltage sources act like one source whose effective voltage equals the algebraic sum of the individual source voltages — as in the common 3-V flashlight with its two 1.5-V batteries.

Second, if v_2 happens to be greater than v_1, then $v_1 - v_2 < 0$ and both v_R and i will be *negative* quantities, meaning that i actually flows the other way. In retrospect, the direction taken for i in the figure was merely a guess based on the assumption that $v_1 > v_2$. Such guesses are not critical for, as we have seen, a "wrong" guess eventually shows up as a negative value.

But the voltage polarity across the resistance must be consistent with Ohm's law and the assumed current direction. Therefore, we will always draw voltage and current arrows in *opposite directions* at each resistance. This **passive convention** will also be used for all other two-terminal elements except sources.

Now consider a **parallel circuit,** Fig. 2.4–7a. Here, the source voltage appears directly across each of the other elements, whose current directions are taken in accordance with the passive convention. Kirchhoff's current law applied at the upper node shows that the total current from the source will be $i = i_R + i_X$ where $i_R = v_s/R$ while i_X depends on the nature of the unspecified element. Residential AC circuits have this type of structure, with many elements in parallel, and the total current equals the sum of the individual device currents.

Figure 2.4–7
(a) Parallel elements with a voltage source. (b) Parallel current sources.

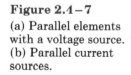 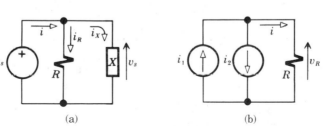

(a)   (b)

If a parallel circuit is driven by one or more current sources, as in Fig. 2.4–7b, then voltage rather than current becomes the unknown that must be found. For the case at hand, KCL and Ohm's law yield

$$i = i_1 - i_2 \qquad v_R = Ri = R(i_1 - i_2) \tag{4}$$

which should be contrasted with the results in Eq. (3).

Exercise 2.4–1

In Fig. 2.4–6, suppose $v_1 = 12$ V and $R = 3\ \Omega$. Redraw the circuit, calculate i, and label all voltage values when the v_2 source is replaced by (a) a closed switch, (b) an open switch, (c) another 3-Ω resistance, (d) a current source with $i_s = 5$ A pointed opposite to i.

Example 2.4–2
Biasing resistors.

Turning from analysis to design, suppose two devices are to be powered by a 12-V battery; device A operates at 4 V and 2.0 A while device B requires

5 V and 1.6 A. Connecting them directly in series or parallel with the battery would not work in view of the differing voltage and current requirements. Those differences are overcome with the help of two resistances arranged as in Fig. 2.4–8.

Figure 2.4–8
Resistors arranged to bias two devices.

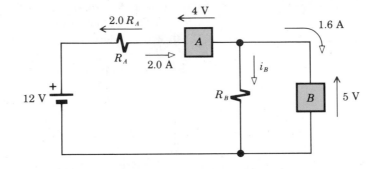

The resistance in series with A provides a voltage drop such that $2.0R_A + 4 + 5 = 12$ to satisfy KVL. Thus $R_A = (12 - 9) \text{ V}/2.0 \text{ A} = 1.5 \text{ }\Omega$. The resistance in parallel with B carries a current $i_B = 2.0 - 1.6$ to satisfy KCL. Thus, $R_B = 5 \text{ V}/0.4 \text{ A} = 12.5 \text{ }\Omega$. Resistors used in this fashion are called **biasing** resistors.

Exercise 2.4–2

Design a circuit with two biasing resistors so that the devices in Example 2.4–2 can be powered by a 6-V battery. Check your work using the appropriate laws.

Branch Currents and Node Voltages

Now that we have Kirchhoff's laws in hand, we can begin a complete analysis of a moderately complicated circuit. Specifically, we will determine all the voltages and currents in Fig. 2.4–1, given that the unspecified element may be modeled as an 8-Ω resistance. Two different methods of attack will be illustrated, one involving branch currents, the other involving node voltages.

For the **branch-current method,** the circuit has been redrawn in Fig. 2.4–9 with three simplifications:

1. Elements in parallel are shown with the same voltage across them.

2. Elements in series are shown with the same current through them.

3. Voltages across resistances are expressed in terms of currents, using Ohm's law and the passive convention.

We have thereby centered attention on three unknowns—labeled i_1, i_2, and i_3—which are the branch currents and have the property that all the unknown voltages are easily found from them.

Figure 2.4–9

Circuit analysis using
branch currents.

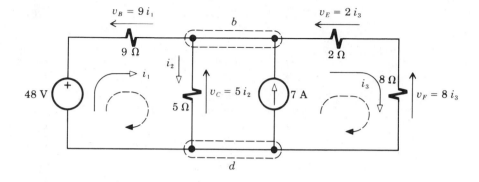

Evaluating the three branch currents requires *three independent equations.* Invoking KCL at node b gives

$$i_1 + 7 = i_2 + i_3$$

and one is tempted to do the same thing at node d. But, as you can check,
KCL at node d yields precisely the same relationship and hence provides
no new information. (Expressed in mathematical parlance, the two node
equations are not independent.) On the other hand, applying KVL around
the left and right loops does give the two additional equations needed

$$9i_1 + 5i_2 = 48 \qquad 5i_2 = 2i_3 + 8i_3$$

A third loop equation can be written for the path around the perimeter, but
it contains the same information as the other two.

We now have a set of three *simultaneous linear equations,* which can be
rewritten as

$$i_1 - i_2 - i_3 = -7$$
$$9i_1 + 5i_2 \qquad = 48 \qquad\qquad \textbf{(5)}$$
$$5i_2 - 10i_3 = 0$$

The corresponding solution for the branch currents is

$$i_1 = 2 \text{ A} \qquad i_2 = 6 \text{ A} \qquad i_3 = 3 \text{ A}$$

Figure 2.4–10

Results of branch-
current analysis.

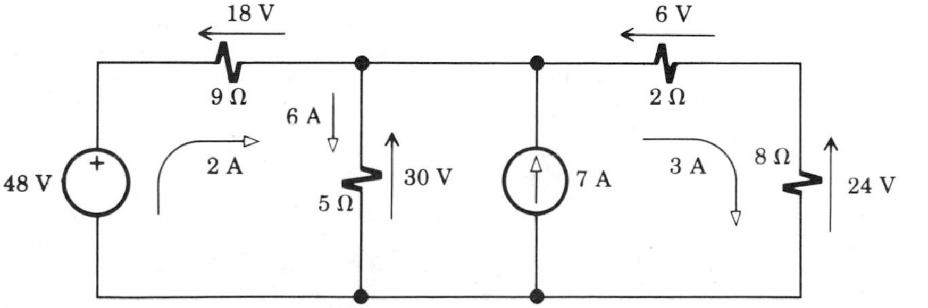

Figure 2.4–10 shows the circuit with all current and voltage values. Testing node and loop sums here provides an easy check on the accuracy of the results. For instance, taking KVL around the perimeter of the circuit confirms that $18 + 6 + 24 = 48$, while KCL used at the bottom node pair confirms that $6 + 3 = 2 + 7$.

For the **node-voltage method** of analysis, our circuit is redrawn in Fig. 2.4–11 and labeled entirely in terms of voltages by a three-step process.

1. Designate one node as the *reference* and draw voltage arrows to all other nodes. These represent the node voltages, of which v_b and v_c are unknown whereas $v_a = 48$ V due to the voltage source between node a and the reference.

2. Invoke KVL to express all remaining voltages as the difference between node voltages.

3. Use Ohm's law to write all currents through resistances in terms of node voltages, again following the passive convention.

Thus, for instance, the voltage across the 5-Ω resistance is the node voltage v_b and the current is $v_b/5$. Similarly, the voltage across the 2-Ω resistance is $v_b - v_c$ and the current is $(v_b - v_c)/2$.

Figure 2.4–11

Circuit analysis using node voltages.

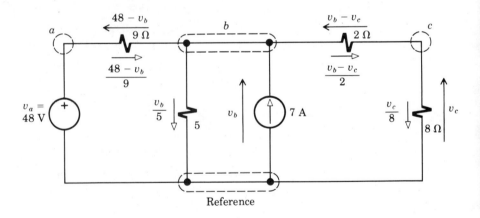

Having gone through these preliminaries, all that remains is to set down KCL equations at each node having an unknown voltage. Since only v_b and v_c are unknown, we write for node b

$$\frac{48 - v_b}{9} + 7 = \frac{v_b}{5} + \frac{v_b - v_c}{2}$$

and for node c

$$\frac{v_b - v_c}{2} = \frac{v_c}{8}$$

Regrouping and clearing fractions yields the pair of linear equations

$$73v_b - 45v_c = 1110$$
$$-4v_b + 5v_c = 0$$

(6)

which can be solved to obtain

$$v_b = 30 \text{ V} \qquad v_c = 24 \text{ V}$$

All the other unknown voltages and currents quickly follow from these voltage values and Fig. 2.4–11.

This node-voltage method has the general advantage of immediately giving us the correct number of independent equations. In addition, it usually involves fewer unknowns than branch-current analysis, thereby reducing algebraic labor and chances of arithmetic errors. The circuit at hand, for instance, has three branch currents but only two unknown node voltages.

One final point ought to be made here. As just demonstrated, circuit analysis frequently calls for the solution of simultaneous linear equations with two or more unknowns. Such sets of equations can always be solved using successive substitution to reduce the problem to one equation with one unknown. However, a more efficient approach is solution by **Cramer's rule,** which involves the evaluation of determinants. If Cramer's rule is new to you, or if you've forgotten how to use it, then you should read the short presentation on the subject found at the back of this book. For practice with Cramer's rule, you may try it on Eqs. (5) and (6) and check your work against the given solutions.

Example 2.4–3

Let's apply node-voltage analysis to the circuit in Fig. 2.4–12. This circuit has a form commonly occuring in electronics, with currents in milliamps and resistances in kilohms.

Figure 2.4–12
Circuit for Example 2.4–3.

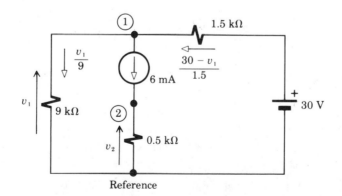

After identifying the unknown node voltages v_1 and v_2, we can label the currents without explicitly showing the difference voltages. The KCL equation at node 1 is then

$$\frac{30 - v_1}{1.5 \text{ k}\Omega} = \frac{v_1}{9 \text{ k}\Omega} + 6 \text{ mA}$$

which yields $v_1 = 18$ V. The voltage v_2 did not enter into this calculation because the current source fixes its value at $v_2 = 0.5 \text{ k}\Omega \times 6 \text{ mA} = 3$ V. Since node 1 is at a higher potential than node 2, the current source happens to be consuming power here, namely $p = (v_1 - v_2) \times 6 \text{ mA} = 90$ mW.

Exercise 2.4–3   Rework Example 2.4–3 using branch-current analysis, and show that $i_1 = v_1/9 \text{ k}\Omega = 2$ mA.

Exercise 2.4–4   Figure 2.4–13 represents a battery-charging circuit. Use the node-voltage method to find all voltages and currents and to show that i_2 is negative.

Figure 2.4–13
Model of a battery-charging circuit.

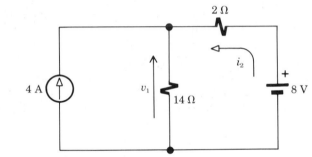

PROBLEMS

2.1–1 A wire with $n = 10^{29}$ electrons/m³ has area $A = 1$ mm² and carries current $i = 40$ mA. Find the number of electrons that pass a given point in one second, and determine their average velocity.

2.1–2 Do Problem 2.1–1 with $A = 0.5$ mm² and $i = 2$ μA.

2.1–3 Do Problem 2.1–1 with $A = 2.5$ cm² and $i = 4$ kA.

2.1–4 Calculate the total charge transferred and the value of i_{av} over $0 \le t \le 1$ min when $i(t) = 10^{-3}t$ A.

2.1–5 Do Problem 2.2–4 with $i(t) = 1 - 0.02t$ A.

2.1–6 Do Problem 2.2–4 with $i(t) = 1 - 0.1t$ A.

2.1–7 A 1.2-V nickel-cadmium battery stores 24 kJ. Find the stored charge and the ampere-hour rating.

2.1–8 A new 9-V alkaline battery stores 900 C. Find the stored energy and the ampere-hour rating.

2.1–9   A 12-V automobile battery is rated at 500 Ah. Find the total charge and energy stored.

2.1–10 If a utility company charges 6 cents/kWh, then what does it cost to operate a 2-W night light continuously for 30 days?

2.1–11 A load operating continuously at 100 V consumes 6 kWh in one day. Find the average power and current.

2.1–12 An automobile clock draws 5 mA from the 12-V battery. How many kilowatt hours does it consume in one year?

2.2–1   Suppose the source in Fig. 2.2–1a has $v = 12 - 3i$ V. Find the supplied power p when $i = 1$, 2, and -1 A. Compare each value with the power supplied by an ideal 12-V source.

2.2–2   Do Problem 2.2–1 for a source with $v = 12 - 6i$ V.

2.2–3   Do Problem 2.2–1 for a source with $v = 12 - 7i$ V.

2.2–4   A certain source produces $i = 5 - 0.01v^2$ A. Justify the assertion that it approximates an ideal current source when $|v| \leq 7$ V.

2.2–5   A certain source produces $i = 4/\sqrt{64 + v^2}$. Justify the assertion that it approximates an ideal current source when $|v| \leq 4$ V.

2.2–6   A certain source produces $i = 3v/\sqrt{4 + v^2}$. Justify the assertion that it approximates an ideal current source when $v \geq 5$ V.

2.3–1   Example 2.3–1 demonstrates that volts, milliamps, kilohms, and milliwatts constitute a **consistent set of units** that does not require conversion of magnitude prefixes. Fill in the blanks in the table below so that each row forms another consistent set.

Voltage	Current	Resistance	Power
V	mA	kΩ	mW
mV	—	kΩ	—
—	kA	—	kW

2.3–2   Add rows to the table in Problem 2.3–1 for:
(a) current in μA and resistance in MΩ;
(b) current in kA and power in MW.

2.3–3   Add rows to the table in Problem 2.3–1 for:
(a) voltage in μV and power in pW;
(b) resistance in MΩ and power in W.

2.3–4   A resistor is fabricated by depositing a thin carbon film on a cylinder 9 mm long and 2 mm in diameter. Estimate the film's thickness t in micrometers for $R = 25$ Ω.

2.3–5   Do Problem 2.3–4 for $R = 10$ Ω.

2.3–6   A 60-W, 120-V lightbulb has a filament with 5 μm radius.
(a) Find the length of the filament in centimeters if $\rho = 10^{-6}$ Ω · m at the operating temperature.
(b) Find the inrush current if $\rho = 5 \times 10^{-8}$ Ω · m at room temperature

2.3–7   Do Problem 2.3–6 for a 150-W lightbulb.

2.3–8   The resistance of a **fuse** increases with temperature such that $R = R_c(1 + \alpha T)$, where R_c is the "cold" resistance, α is the temperature coefficient, and T the temperature rise above 20°C. The temperature rise is given by $T = kp$ where k is a constant and p is the power dissipated by the fuse. Obtain an expression for R in terms of the current i through the fuse, and show that it "blows out" $(R \to \infty)$ at $i = 1/\sqrt{\alpha k R_c}$.

2.4–1   Let $v_s = 12$ V and $R = 3$ kΩ in Fig. 2.4–7a. Find i when $i_X = 4$ mA and $i_X = -4$ mA.

2.4–2   Let $v_s = 15$ V and $R = 5$ kΩ in Fig. 2.4–7a. Find i when $i_X = 5$ mA and $i_X = -5$ mA.

2.4–3   Find the value of i_4 in Fig. P2.4–3, given that $v_2 = 12$ V but R_3 and R_4 are unknown. Hint: Consider Fig. 2.4–2b.

Figure P2.4–3

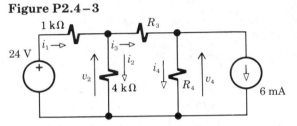

2.4–4   Do Problem 2.4–3, given that $v_2 = 16$ V.

2.4–5   Given that $i_X = 2$ A in Fig. P2.4–5, use KVL at nodes d and c and KCL around loops $bcde$ and ecd to find v_s.

Figure P2.4–5

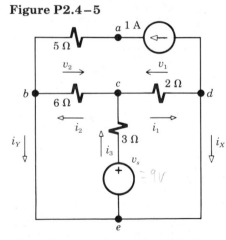

2.4–6   Given that $i_Y = 3$ A in Fig. P2.4–5, use KVL at nodes b and c and KCL around loops $bcde$ and ecb to find v_s.

2.4–7   Find i_E, v_{CE}, and v_{CB} in Fig. P2.4–7 when the transistor has $i_C = 10i_B$ and $v_{BE} = 0.7$ V.

2.4–8   Find i_E, v_{CE}, and v_{CB} in Fig. P2.4–7 when the transistor has $i_C = 20i_B$ and $v_{BE} = 0.6$ V.

Figure P2.4–7

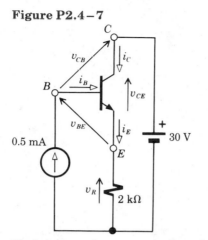

2.4–9 Find the branch currents i_1, i_2, and i_3 in Fig. P2.4–9 when $i_s = 10$ mA and $v_s = 16$ V.

Figure P2.4–9

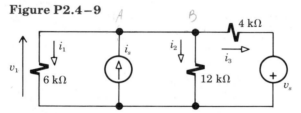

$$\frac{V_1}{6} - 5mA + \frac{V_B}{12} + \frac{V_B - (-V_s)}{4} = 0$$

2.4–10 Find the branch currents i_1, i_2, and i_3 in Fig. P2.4–5 when $v_s = 9$ V.

2.4–11 Let $R_3 = 8\ \Omega$ in Fig. P2.4–11. Use KCL equations at nodes a and b and one KVL equation to find the branch currents i_1, i_2, and i_3.

Figure P2.4–11

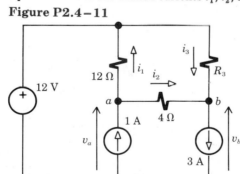

2.4–12 Find all the branch currents in Fig. P2.4–3 when $R_3 = 5\ \Omega$ and $R_4 = 2\ \Omega$. You will need two KCL equations and two KVL equations.

2.4–13 Use the node-voltage method to find v_1 in Fig. P2.4–9 when $i_s = 5$ mA and $v_s = 10$ V.

2.4–14 Find v_a and v_c relative to node e in Fig. P2.4–5 when $v_s = 12$ V.

2.4–15 Find v_2 and v_4 in Fig. P2.4–3 when $R_3 = 18\ \Omega$ and $R_4 = 2\ \Omega$.

2.4–16 Find v_a and v_b in Fig. P2.4–11 when $R_3 = 2\ \Omega$.

Chapter

3 Resistive Circuits

This chapter applies basic circuit concepts to the study of circuits consisting of sources and linear resistance. There are two major reasons why resistive circuits deserve further attention. First, they have many practical applications, either in their own right or as models of more complicated circuits. Second, the analysis techniques developed here will be extended in subsequent chapters when we deal with other elements and electronic devices. Accordingly, a firm grasp of resistive circuits serves as a foundation for more exciting topics to come.

We begin where the last chapter left off, namely with the application of Ohm's and Kirchhoff's laws to resistive circuits. We'll examine series and parallel configurations that occur time and again in practice, including source-load circuits. Three important circuit theorems are presented next for their value in both analysis and design work. Then we develop systematic methods of circuit analysis using node and loop equations. The last section deals with basic DC meters and measurement techniques.

Objectives

After studying this chapter and working the exercises, you should be able to do each of the following:

- Use divider ratios and equivalent resistance to find all the voltages and currents in a network comprised of series and parallel resistors (Section 3.1).

- Calculate the current and power delivered to a load resistance from a real source, and state the condition for maximum power transfer (Section 3.1).

- Identify the form of the equivalent model for a two-terminal resistive network, with or without internal sources (Section 3.2).

- State Thévenin's theorem and use Thévenin and Norton equivalent circuits to simplify circuit analysis (Section 3.2).

- Analyze a circuit containing two sources by applying the superposition theorem (Section 3.2).

- Write and solve node and loop equations for a circuit with up to three unknowns (Section 3.3).†

- Draw and explain the basic circuits used for DC meters and null measurements (Section 3.4)†

3.1 SERIES AND PARALLEL RESISTANCE

Often, the "load" network connected to a source consists of several resistive elements in series and parallel arrangements. Finding all the voltages and currents in such a circuit could entail considerable work. But suppose you are concerned with only a few things — the current drawn from the source, perhaps, or the voltage across a particular element. For those calculations, the technique of series/parallel reduction and the handy voltage and current divider ratios can be invoked to expedite matters.

Our goal in this section is to develop efficient methods for the analysis of circuits with series and parallel resistance. We'll also consider the effect of internal resistance in real sources.

Series Resistance and Potentiometers

When two resistances are in series they must carry the same current. Applying this fact, plus KVL and Ohm's law to Fig. 3.1–1a, we have

$$v_s = v_1 + v_2 = (R_1 i) + (R_2 i) = (R_1 + R_2)\, i$$

which is of the form $v_s = R_{eq} i$ with

$$R_{eq} = R_1 + R_2 \tag{1}$$

The sum $R_{eq} = R_1 + R_2$ is called the **series equivalent resistance.** It is equivalent in the sense that two resistances in series have the same $v\text{-}i$ characteristic as one resistance whose value is the sum of the two. Hence,

Figure 3.1–1
Series equivalent resistance.

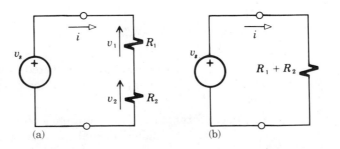

(a)   (b)

replacing R_1 and R_2 by a single resistance of value $R_1 + R_2$, as in Fig. 3.1–1b, produces no changes in the rest of the circuit. While we would seldom make the actual physical replacement, we often do this mentally for calculations. To illustrate, if $v_s = 12$ V, $R_1 = 5\ \Omega$, and $R_2 = 1\ \Omega$, then $i = 12$ V$/(5 + 1)\ \Omega = 2$ A.

Equation (1) readily extrapolates to the case of n resistances in series. The equivalent resistance then becomes

$$R_{eq} = R_1 + R_2 + \cdots + R_n \tag{2}$$

The derivation should be obvious.

Figure 3.1–2

Voltage divider.

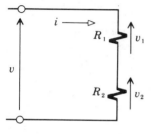

Now consider the portion of a circuit shown in Fig. 3.1–2 where the voltage v is known (not necessarily as a source voltage). We want to find the voltage v_1 across resistance R_1. This happens to be a common task that keeps popping up in circuit analysis and design. Clearly, $v_1 = R_1 i$ and $i = v/(R_1 + R_2)$. Therefore

$$v_1 = \frac{R_1}{R_1 + R_2}\, v \tag{3}$$

which gives us v_1 directly in terms of v without the intermediate calculation of i.

The circuit configuration in Fig. 3.1–2 is called a **voltage divider,** $R_1/(R_1 + R_2)$ being the **voltage-divider ratio.** Similarly, we would use the ratio $R_2/(R_1 + R_2)$ to get v_2. The name comes from the fact that the total voltage v is "divided" between the two resistances and, in fact, $v_1 = v_2 = v/2$ when $R_1 = R_2$. You will find it handy to memorize Eq. (3), noting carefully that v_1 is proportional to R_1. Also note the approximations

$$v_1 \approx \begin{cases} v & R_1 \gg R_2 \\ \dfrac{R_1}{R_2}\, v & R_1 \ll R_2 \end{cases}$$

so most of the voltage appears across R_1 when $R_1 \gg R_2$, and vice versa.

Figure 3.1–3a is the schematic symbol for a resistive device called a **potentiometer,** known informally as a "pot." The third terminal, labeled w, is a movable contact point or **wiper** that intercepts a portion of the total resistance R_{ab}. Hence, for a particular wiper position, the potentiometer acts like the two resistors R_{aw} and R_{wb} shown in Fig. 3.1–3b, where $R_{aw} + R_{wb} = R_{ab}$. Precision potentiometers are available with dials indicating the wiper position as accurately as 0.1%.

Connecting the wiper to terminal a puts a short-circuit path around R_{aw}, thus creating the *adjustable resistance* $R = R_{wb} \le R_{ab}$ depicted by Fig. 3.1–3c. Similarly, we obtain an *adjustable voltage* from a battery using the circuit in Fig. 3.1–3d. The potentiometer forms a voltage divider across V_s to yield

$$v_w = \frac{R_{aw}}{R_{ab}}\, V_s$$

Figure 3.1–3

Potentiometer: (a) symbol; (b) equivalent network; (c) adjustable resistance; (d) adjustable voltage.

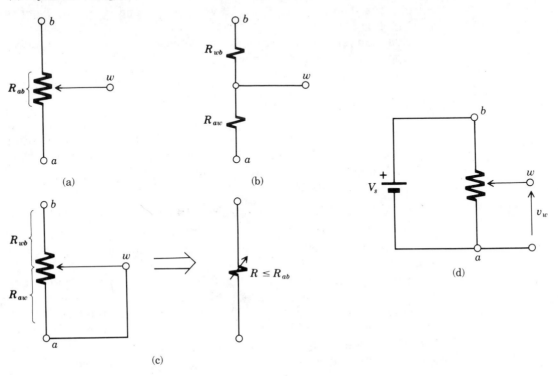

If the battery is replaced by a time-varying source voltage $v_s(t)$, then the output voltage $v_w(t)$ has the same waveshape but reduced in magnitude by the adjustable factor $R_{aw}/R_{ab} \leq 1$. The volume control in a radio and various other electrical controls often employ potentiometers in this manner.

Parallel Resistance

When two resistances are in parallel the same voltage must appear across both of them. Thus, in Fig. 3.1–4a,

$$i = i_1 + i_2 = \frac{v_s}{R_1} + \frac{v_s}{R_2} = v_s\left(\frac{1}{R_1} + \frac{1}{R_2}\right)$$

which has the form $i = v_s/R_{eq}$ with

$$R_{eq} = \left(\frac{1}{R_1} + \frac{1}{R_2}\right)^{-1} = \frac{R_1 R_2}{R_1 + R_2} \tag{4}$$

Equation (4) is the equivalent resistance for two parallel resistances, and both circuits in Fig. 3.1–4 would draw the same current from the source.

Figure 3.1–4
Parallel equivalent
resistance.

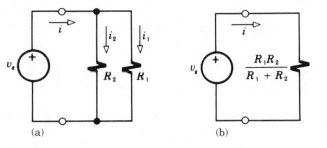

(a)　　　　　　　　　　(b)

The right-hand side of Eq. (4) occurs so often in circuit work that we will give it the special notational symbol

$$R_1 \| R_2 \triangleq \frac{R_1 R_2}{R_1 + R_2} \tag{5}$$

where $R_1 \| R_2$ is read as "R_1 in parallel with R_2." Taking $R_1 = 4\ \Omega$ and $R_2 = 6\ \Omega$ for a simple example, the parallel equivalent value is $4\|6 = (4 \times 6)/(4 + 6) = 24/10 = 2.4\ \Omega$.

Observe that the parallel equivalent value is always smaller than either term, and $R_1 \| R_2 = R_1/2$ when $R_2 = R_1$. This should be contrasted with the series case where $R_1 + R_2$ is always larger than either term and $R_1 + R_2 = 2R_1$ when $R_2 = R_1$. Moreover, if $R_1 \gg R_2$ then $R_1 \| R_2 \approx R_2$, whereas $R_1 + R_2 \approx R_1$.

The extension to n parallel resistances is not as obvious as the series case. To derive it we use *conductance* $G_1 = 1/R_1$ and Ohm's law written as $i_1 = G_1 v$, etc. Then

$$i = i_1 + i_2 + \cdots + i_n$$
$$= G_1 v + G_2 v + \cdots + G_n v$$

which has the form $i = G_{eq} v$ with the equivalent conductance being

$$G_{eq} = G_1 + G_2 + \cdots + G_n \tag{6a}$$

or

$$\frac{1}{R_{eq}} = \frac{1}{R_1} + \frac{1}{R_2} + \cdots + \frac{1}{R_n} \tag{6b}$$

Taking the reciprocal of $1/R_{eq}$ finally gives the parallel equivalent resistance, a routine computation with a calculator. You should prove to yourself that Eq. (6b) leads to Eq. (4) when $n = 2$. (But don't try to use the product-over-sum expression for $n > 2$.)

Occasionally, you may encounter a situation in which the resistance between two nodes must be reduced from its present value R_1 to a new value $R_{eq} < R_1$. The desired result is easily obtained by paralleling R_1 with

$$R_2 = \frac{R_1 R_{eq}}{R_1 - R_{eq}} \tag{7}$$

This "reverse parallel" formula follows directly from Eq. (4).

Figure 3.1–5
Current divider.

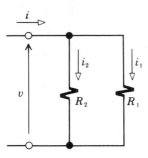

Just as series resistors form a voltage divider, parallel resistors act as a **current divider.** Specifically, if we know the total current i in Fig. 3.1–5, then the current i_1 through R_1 is given by

$$i_1 = \frac{R_2}{R_1 + R_2}\, i \tag{8}$$

$$\approx \begin{cases} i & R_1 \ll R_2 \\ \dfrac{R_2}{R_1}\, i & R_1 \gg R_2 \end{cases}$$

Equation (8) should be compared with Eq. (3), paying special attention to the fact that i_1 is proportional to R_2, so a large R_2 "forces" more current through R_1.

Example 3.1–1
An electric range.

The surface cooking units of an electric range sometimes have two resistance elements and a special switch that connects them to the voltage source individually, in series, or in parallel, as represented by Fig. 3.1–6. If $R_1 < R_2$, then there are four increasing resistance values

$$R_1 \| R_2 \qquad R_1 \qquad R_2 \qquad R_1 + R_2$$

Since the voltage is constant, we have $p = v^2/R$ and the lowest "heat" is $p_{\min} = v^2/R_{\max} = v^2/(R_1 + R_2)$. Similarly, $p_{\max} = v^2/R_{\min} = v^2/(R_1 \| R_2)$. Connecting the elements individually gives the two intermediate "heats" v^2/R_2 and v^2/R_1. Four more heat settings are possible when v has two values, such as 120 V and 240 V.

Figure 3.1–6
Surface unit of an
electric range.

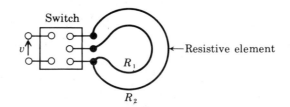

Exercise 3.1–1

Let $v = 120$ V and $R_1 = 60\ \Omega$ in Fig. 3.1–6. Find R_2 such that $p_{\min} = 80$ W. Then calculate the other three values of p.

Series/Parallel Reduction

Now we're prepared to tackle problems in which a source is connected to a network consisting entirely of series and parallel resistors. These circuits can be analyzed by the series/parallel reduction method, as follows:

1. Repeatedly apply the series and parallel formulas until the entire network has been reduced to a *single equivalent resistance*.

2. Use the equivalent resistance to calculate the voltage or current at the network's terminals.

3. Work forward through the network to find any other voltages or currents of interest from appropriate divider ratios.

An example should help clarify this valuable and simple method.

Example 3.1–2

Consider the circuit in Fig. 3.1–7a, where all resistances are in kilohms. The load network contains two 20-kΩ resistors in parallel and a series connection of 4 kΩ, 5 kΩ, and 6 kΩ. We can therefore replace these five elements by the equivalent resistances

$$20\|20 = 10 \text{ k}\Omega \qquad 4+5+6 = 15 \text{ k}\Omega$$

The resulting partially reduced network shown in Fig. 3.1–7b has 2 kΩ in series with $10\|15 = 6$ kΩ. Hence, $R_{eq} = 2 + 6 = 8$ kΩ for the entire load. It then follows from Fig. 3.1–7c that the load draws $i = 40$ V/8 kΩ = 5 mA and dissipates $p = i^2 R_{eq} = 200$ mW.

"Unfolding" some of the equivalent resistances now takes us back to Fig. 3.1–7b. Since i divides between the 10-kΩ and 15-kΩ equivalent

Figure 3.1–7

Circuit analysis using series/parallel reduction.

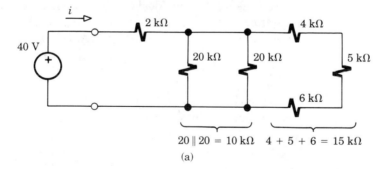

(a)

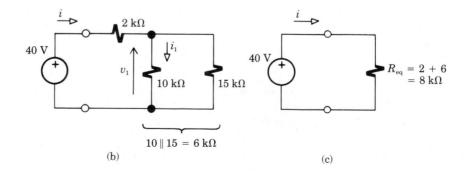

(b)   (c)

resistances, the current labeled i_1 is easily found via

$$i_1 = \frac{15}{15 + 10} \times 5 \text{ mA} = 3 \text{ mA}$$

This result clearly agrees with $v_1 = 40 - 2i = 30$ V.

Exercise 3.1–2   Find the equivalent resistance of the load network in Fig. 3.1–8. Then calculate i, p, and v_2.

Figure 3.1–8
Circuit for Exercise
3.1–2.

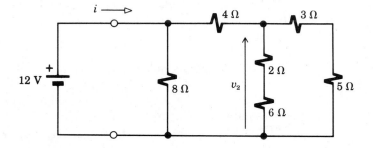

Source Loading and Power Transfer

Up to this point we have assumed all sources to be *ideal*. But a real voltage source differs from the ideal by the presence of *internal resistance*. This resistance causes the terminal voltage to decrease as increasing current is drawn from the source. Hence, the v–i relationship has the form

$$v = v_s - R_s i \tag{9}$$

The corresponding circuit model of a real voltage source is given by Fig. 3.1–9a, including an arbitrary load. The model itself consists of an ideal voltage source v_s in series with a resistance R_s.

Figure 3.1–9
(a) Model of a real voltage source. (b) Open-circuit voltage. (c) Short-circuit current.

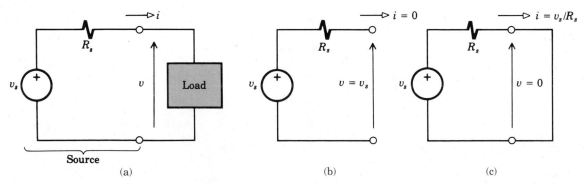

We will call v_s the **source voltage** and R_s the **source resistance**. Actually, it would be more correct to say that v_s is the *open-circuit source voltage* because the terminal voltage v equals v_s only under the open-circuit condition in Fig. 3.1–9b, where $i = 0$, so $v = v_s - R_s i = v_s$. Figure 3.1–9c shows the other extreme condition in which a perfect conductor across the source forces $v = 0$ and draws the *short-circuit current v_s/R_s*. (Of course, short-circuit current is primarily a theoretical concept since a real voltage source could be damaged in the short-circuit condition.)

A real current source also has internal resistance, resulting in an i–v relationship of the form

$$i = i_s - \frac{v}{R_s} \tag{10}$$

Figure 3.1–10 is the corresponding circuit model using an ideal current source in parallel with R_s. Here, the **source current** i_s is identical to the short-circuit current, while the open-circuit voltage equals $R_s i_s$.

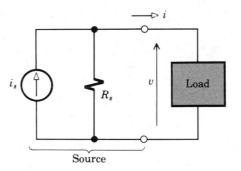

A closer examination of Figs. 3.1–9a and 3.1–10 reveals that the load has no way of knowing whether it's connected to a voltage source or a current source, since neither v nor i are held fixed by the source. Consequently, we could represent a real voltage source by the model for a real current source, and vice versa! In fact, both source models have exactly the same terminal relationships when $R_s = v_s/i_s$, a property easily confirmed from Eqs. (9) and (10). The practical distinction between real voltage and current sources becomes apparent when we examine loading.

Let a load resistance R_L be connected to a real voltage source, as in Fig. 3.1–11a. The terminal voltage v will be

$$v = v_s - R_s i = \frac{R_L}{R_L + R_s} v_s \tag{11}$$

which is less than the open-circuit source voltage v_s due to the voltage drop $R_s i$ across the internal source resistance. (A familiar demonstration of that voltage drop occurs when you start an automobile engine with the headlights on; the large current drawn by the starter decreases the voltage

Figure 3.1 – 11
Loading effect caused
by source resistance.

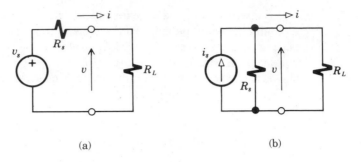

(a)   (b)

across the battery terminals and the headlights momentarily dim.) This loading effect is negligible when $R_s \ll R_L$, and thus $R_s i \ll v_s$.

Similarly, with R_L connected to the current source in Fig. 3.1–11b, i_s divides between R_s and R_L, and

$$i = i_s - \frac{v}{R_s} = \frac{R_s}{R_L + R_s} \, i_s \tag{12}$$

so that the load current is less than the short-circuit source current because of current v/R_s bypassed through the internal source resistance. The loading effect here will be negligible when $R_s \gg R_L$, and thus $v/R_s \ll i_s$.

We now see that the key difference between a practical voltage source and a practical current source is the value of R_s relative to the operating conditions. A "good" voltage source — that is, one that acts almost like an ideal voltage source — has a *small* internal resistance so that $v \approx v_s$ over the expected current range. Conversely, a "good" current source has a *large* internal resistance so that $i \approx i_s$ over the expected voltage range.

Besides causing loading effect, the internal resistance of a source dissipates power whenever we attempt to transfer power from the source to a load. We'll investigate this situation using the voltage-source model in Fig. 3.1–11a. Power p_L is delivered to R_L via the current $i = v_s/(R_s + R_L)$, which also produces ohmic heating power p_s in R_s. Thus,

$$p_L = R_L i^2 = \frac{R_L}{(R_L + R_s)^2} \, v_s^2$$

$$p_s = R_s i^2 = \frac{R_s}{(R_L + R_s)^2} \, v_s^2 \tag{13}$$

A comparison of these expressions reveals that $p_L = (R_L/R_s)p_s$ so the ratio R_L/R_s plays a dominant role here.

Figure 3.1–12 plots p_L and p_s versus R_L/R_s. We see that p_s steadily decreases from its maximum value v_s^2/R_s at $R_L/R_s = 0$, approaching zero as R_L/R_s increases. Therefore, wasted power in the source is minimized by making R_L/R_s as large as possible. On the other hand, the curve for p_L has a distinct *maximum* value

$$p_{\max} = \frac{v_s^2}{4R_s} \tag{14}$$

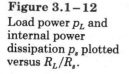

Figure 3.1–12
Load power p_L and internal power dissipation p_s plotted versus R_L/R_s.

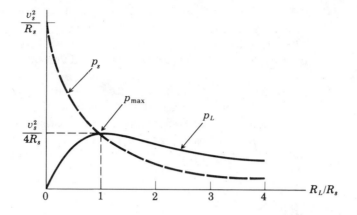

which occurs when $R_L/R_s = 1$. These results are derived from Eq. (13) by solving $dp_L/dR_L = 0$. We therefore state that:

> When a source has fixed internal resistance R_s, maximum power transfer to a load requires $R_L = R_s$.

Under this condition the load is said to be **matched** to the source.

You should carefully observe that maximum power transfer differs from maximum **efficiency**. Efficiency (Eff) is the ratio of load power to total power generated by the source. Thus,

$$\text{Eff} = \frac{p_L}{p_L + p_s} \tag{15}$$

usually expressed as a percentage. But p_L equals p_s when the load is matched (see Fig. 3.1–12), so maximum power transfer corresponds to 50% efficiency, with equal power dissipation in source and load. Moreover, the terminal voltage drops to $v = v_s/2$ when $R_L = R_s$. Clearly, electric utility companies would not, nor should not, strive for maximum power transfer. They seek instead maximum efficiency by making p_s as small as possible.

When do we want maximum power transfer? Primarily in those applications where voltage or current signals convey information — communication systems and computers being common examples. In such applications the information-bearing signals account for only a small fraction of the total power consumed by the system, so 50% efficiency is not a serious drawback.

Example 3.1–3

A certain automobile battery has $v_s = 12$ V and $R_s = 0.02 \ \Omega$ when freshly charged. Hence, the maximum power available from the battery is

$$p_{\text{max}} = \frac{(12 \text{ V})^2}{4 \times 0.02 \ \Omega} = 1800 \text{ W}$$

This maximum power is transferred to $R_L = R_s = 0.02 \ \Omega$, so $v = v_s/2 =$

6 V and $i = v_s/2R_s = 300$ A. A more realistic load current would be $i = 10$ A, in which case $v = 11.8$ V, $p_L = 118$ W, $p_s = 2$ W, and Eff $= 118/120 \approx 98\%$. As the battery "runs down" its internal resistance increases, thereby decreasing the terminal voltage and efficiency.

Exercise 3.1–3   An electric lawnmower is connected to an outlet with a long extension cord, so the equivalent source applied to the motor has $v_s = 120$ V and $R_s = 5$ Ω. Electrically, the motor acts like a load resistance whose value depends on the amount of work being done. Draw the circuit diagram and tabulate the values of i, v, p_L, p_s, and Eff for $R_L = 5$ Ω and 35 Ω.

3.2 CIRCUIT THEOREMS

This section presents three theorems that often simplify the tasks of circuit analysis or design. Accordingly, you should view them as valuable working tools rather than abstract formalisms. Although introduced here in the context of resistive circuits, these theorems will later take on added meaning for circuits containing nonresistive elements and electronic devices.

Equivalence and Equivalent Resistance

We previously learned how to calculate the equivalent resistance of series and parallel resistors. Now we'll extend that technique with the help of the generalized concept of equivalence for one-port networks.

A **one-port network,** or "one-port" for short, may contain any number of elements but it has just *one pair of terminals* for connection to the world outside. Since the only accessible quantities of a one-port are the voltage and current at its terminals, we say that:

> Any two one-port networks are *equivalent* if they have exactly the same terminal i–v characteristics.

In other words, equivalent one-ports always result in the same terminal voltage and current when they are connected to identical external circuitry, regardless of what that circuitry may be. Equivalence holds only with respect to the *terminal* behavior, not to what goes on *inside* the networks. Thus, we use equivalent networks to deal exclusively with terminal characteristics, temporarily ignoring internal conditions.

To bring out the meaning and value of equivalence, consider the two one-port networks on the left-hand side of Fig. 3.2–1. We will eventually show that these are equivalent one-ports. Accordingly, we can use the lower one to represent the external behavior of the more complicated upper one. For instance, to find the resulting current when the load in Fig. 3.2–1a is a 4-kΩ resistor, we refer instead to Fig. 3.2–1b and immediately see that $i = 18$ V$/(5 + 4)$ kΩ $= 2$ mA.

Figure 3.2–1

Two equivalent
one-port networks.

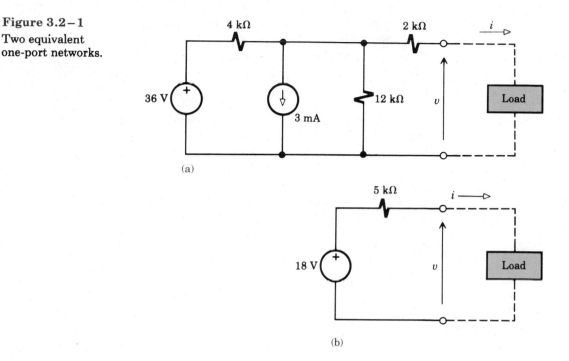

(a)

(b)

Turning specifically to equivalent resistance, suppose the load net-work in Fig. 3.2–2a contains only series and parallel resistors. Then we know that the terminal voltage and current are related by $v = R_{eq}i$, with

$$R_{eq} \triangleq v/i \qquad \qquad (1)$$

Figure 3.2–2b diagrams the corresponding equivalent one-port. The theorem of **equivalent resistance** goes even further along this line, stat-ing that:

> If a one-port network consists entirely of linear resistances or linear resistances and controlled sources, but no independent sources, then the terminal voltage and current are related by $v = R_{eq}i$ where R_{eq} is a constant.

Hence, for purposes of analysis, any such network can be replaced by a single resistor whose value is given by Eq. (1).

Figure 3.2–2

(a) Arbitrary one-port
load. (b) Equivalent
resistance.

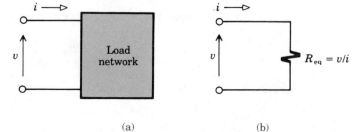

(a)   (b)

The following two examples illustrate equivalent resistance calculations when series/parallel reduction does not work, and when the network
includes a controlled source. The fact that the terminal v/i ratio is a
constant in both cases agrees with the prediction of the equivalent resistance theorem.

Example 3.2–1   Suppose we want to find the equivalent resistance of the *unbalanced bridge
network* in Fig. 3.2–3a, which cannot be simplified by series/parallel reduction. Instead, we mentally apply an arbitrary source voltage v, and we
calculate $R_{eq} = v/i$ by finding the resulting current i in terms of v. To this
end, the circuit is redrawn as Fig. 3.2–3b and labeled with the node voltages and corresponding branch currents.

Figure 3.2–3
Unbalanced bridge network.

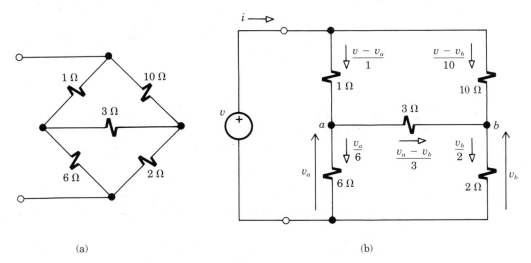

(a)   (b)

There are five branch currents here but only two unknown node voltages, so we'll first use the node-voltage method to solve for v_a and v_b in
terms of v. Summing the currents into nodes a and b gives

$$\frac{v - v_a}{1} - \frac{v_a - v_b}{3} - \frac{v_a}{6} = 0 \qquad \frac{v - v_b}{10} + \frac{v_a - v_b}{3} - \frac{v_b}{2} = 0$$

After regrouping terms and clearing of fractions we have the pair of equations

$$9v_a - 2v_b = 6v$$
$$-10v_a + 28v_b = 3v$$

Simultaneous solution then yields

$$v_a = 3v/4 \qquad v_b = 3v/8$$

Next, since we want i in terms of v, we use KCL at the bottom node to obtain

$$i = \frac{v_a}{6} + \frac{v_b}{2} = \frac{(3v/4)}{6} + \frac{(3v/8)}{2} = \frac{5}{16}v$$

Therefore,

$$R_{eq} = v/i = 16/5 = 3.2\ \Omega$$

which is the desired result.

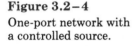

Example 3.2–2

Now consider the one-port in Fig. 3.2–4. This network contains a voltage-controlled current source producing the controlled current $i_c = gv$. The proportionality constant g has the units of conductance, and the control voltage v happens to be the terminal voltage in this case. Similar situations occur in practice as models of transistors. Here, we'll find and interpret $R_{eq} = v/i$.

Figure 3.2–4

One-port network with a controlled source.

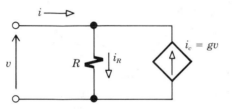

Applying KCL at the upper node gives $i + i_c = i_R$ with $i_R = v/R$, so

$$i = \frac{v}{R} - i_c = \frac{v}{R} - gv = \frac{1 - gR}{R}v$$

Hence, we see that

$$R_{eq} = \frac{v}{i} = \frac{R}{1 - gR}$$

which is, indeed, a constant. However, the value of R_{eq} depends on the value of the product gR in the denominator. If $gR = 1/2$, for instance, then $1 - gR = 1/2$ and $R_{eq} = 2R$. Or if $gR = 1$, then $1 - gR = 0$ and $R_{eq} = \infty$. The equivalent resistance is greater than R in these cases because the internal controlled source provides part or all of the current through R, thereby reducing the input current i for a given value of v. With $i < v/R$, the equivalent resistance must be greater than R.

But suppose that $gR = 2$ so $1 - gR = -1$ and we get $R_{eq} = -R$, a *negative* equivalent resistance. Negative equivalent resistance means that the controlled source pumps current *out* of the upper terminal when $v > 0$. Nonetheless, the network still has $v = R_{eq}i$ at its terminals, and we can still use the model in Fig. 3.2–2b to predict the terminal behavior, whether R_{eq} is positive or negative.

Exercise 3.2–1   Obtain an expression for R_{eq} in Fig. 3.2–4 when $i_c = \beta i$, where i is the input current and β is a constant.

Exercise 3.2–2   Suppose you know that $v = 12$ V in Fig. 3.2–1a. Use Fig. 3.2–1b to find the equivalent resistance of the load.

Thévenin's Theorem

Obtaining full advantage from the equivalence concept requires methods for finding a simple one-port network equivalent to a more complicated network. Thévenin's theorem has special significance in this regard, for it leads to two basic equivalent structures when the one-port in question contains *independent sources.*

Figure 3.2–5a represents an arbitrary linear network consisting of at least one independent source along with resistances and, perhaps, controlled sources. Let v_{oc} be the network's open-circuit voltage, and let i_{sc} be the short-circuit current. The **Thévenin resistance** of the network is defined by the ratio

$$R_o \triangleq \frac{v_{oc}}{i_{sc}} \tag{2}$$

We write the Thévenin parameters as v_{oc}, i_{sc}, and R_o, rather than v_s, i_s, and R_s previously used to model a single source, because the networks under consideration here may include more than one source.

Figure 3.2–5
(a) Arbitrary one-port source. (b) Thévenin equivalent circuit. (c) Norton equivalent circuit.

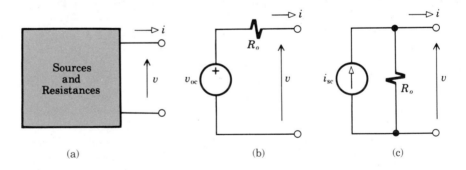

(a)   (b)   (c)

For purposes of modeling the network in Fig. 3.2–5a, **Thévenin's theorem** states that:

> Any one-port network composed entirely of sources and resistances is equivalent to a voltage source v_{oc} in series with R_o or a current source i_{sc} in parallel with R_o.

The model with the voltage source diagrammed in Fig. 3.2–5b is known as the **Thévenin equivalent circuit,** while the model with the current source in Fig. 3.2–5c is known as the **Norton equivalent circuit.**

Any two of the three Thévenin parameters fully characterize the terminal behavior of the network, and we obtain an equivalent circuit by determining the open-circuit voltage and short-circuit current. Alternatively, having found either v_{oc} or i_{sc}, we can get R_o by invoking the following corollary to Thévenin's theorem:

> The Thévenin resistance equals the equivalent resistance seen between a one-port's terminals when all independent sources are suppressed.

Sources are *suppressed* by making the replacements

$$\text{voltage source} \rightarrow \text{short circuit}$$
$$\text{current source} \rightarrow \text{open circuit}$$

These replacements ensure that no energy flows from the sources. A network with all independent sources suppressed is said to be "dead."

Generally, the corollary provides a simpler way of computing R_o when no controlled sources are present. Then, given R_o and either v_{oc} or i_{sc}, we obtain the third parameter via $i_{sc} = v_{oc}/R_o$ or $v_{oc} = R_o i_{sc}$.

Besides representing complete one-ports, Thévenin's theorem may be applied to portions of a network to simplify intermediate calculations. Moreover, successive conversions back and forth between Thévenin and Norton circuits often saves considerable labor in circuit analysis, especially when there are two or more sources present. As a rule, the Thévenin circuit works best when dealing with a series connection, whereas the Norton circuit is used with parallel connections.

Example 3.2–3
Calculating Thévenin parameters.

Suppose you want an equivalent circuit for the one-port in Fig. 3.2–6a. For this task, you need to find any two of the three Thévenin parameters.

First, you might redraw the circuit as shown in Fig. 3.2–6b with $i = 0$ for the open-circuit condition. Then v_{oc} follows from the voltage-divider ratio

$$v_{oc} = \frac{10 \text{ k}\Omega}{25 \text{ k}\Omega} \times 75 \text{ V} = 30 \text{ V}$$

Next, you might draw the short-circuit condition in Fig. 3.2–6c. No current flows through the 10-kΩ resistor here, so

$$i_{sc} = 75 \text{ V}/15 \text{ k}\Omega = 5 \text{ mA}$$

Hence, from Eq. (2),

$$R_o = 30 \text{ V}/5 \text{ mA} = 6 \text{ k}\Omega$$

Or you could replace the voltage source with a short circuit to get the "dead" network in Fig. 3.2–6d. The equivalent resistance looking back into the terminals is

$$R_o = 15 \text{ k}\Omega \| 10 \text{ k}\Omega = 6 \text{ k}\Omega$$

in agreement with the previous calculation.

Figure 3.2–6
Circuits for Example
3.2–3.

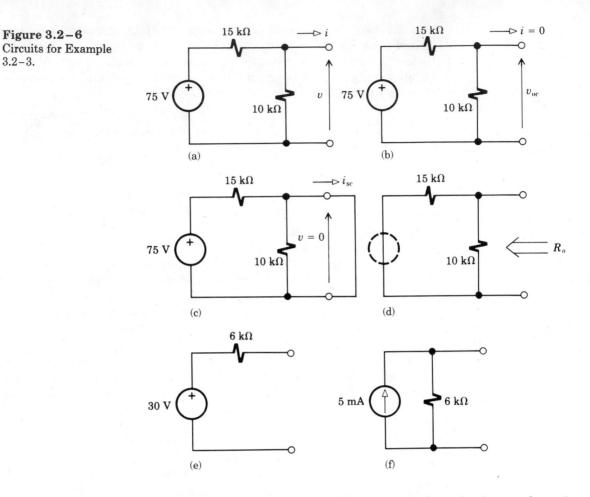

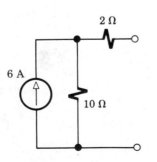

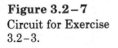

The resulting Thévenin and Norton equivalent circuits are shown in parts (e) and (f) of the figure.

Exercise 3.2–3   Obtain the Thévenin and Norton equivalent circuits for the network in Fig. 3.2–7.

Figure 3.2–7
Circuit for Exercise
3.2–3.

Example 3.2−4
Thévenin/Norton
conversions.

Now consider again the battery-charging circuit from Fig. 2.4−13, re-peated here as Fig. 3.2−8a. If you want to compute the charging current i without solving the entire circuit, you can simply "Thévenize" the left-hand portion of the network. This yields Fig. 3.2−8b, from which $i = (56 − 8)$ V$/(14 + 2)$ $\Omega = 3$ A. Similarly, "Nortonizing" the right-hand portion leads to Fig. 3.2−8c and facilitates the node-voltage calculation: $v = (14 \| 2)$ $\Omega \times (4 + 4)$ A $= 14$ V. Little tricks such as these come easily with practice, especially if you watch for Thévenin and Norton structures.

Figure 3.2−8
Circuit analysis using
Thévenin/Norton
conversion.

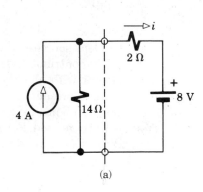

(a)

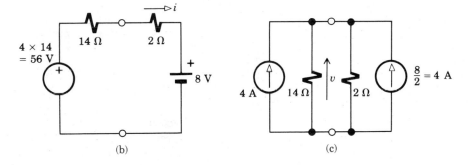

(b)   (c)

Exercise 3.2−4

Use Thévenin/Norton conversions and the results of Example 3.2−3 to find the current i in Fig. 3.2−9.

Figure 3.2−9
Circuit for Exercise
3.2−4.

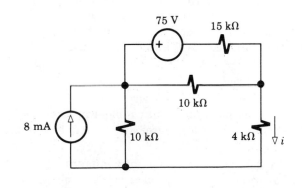

Superposition

Another helpful technique for dealing with networks containing two or more sources stems from the principle of superposition or **superposition theorem,** as follows:

> When a current or voltage in a linear network is the result of several sources acting together, its value is the algebraic sum of the individual contributions from each independent source acting alone.

Thus, a complicated problem reduces to several easier problems — one for each independent source — whose answers are added together to get the final result. The contribution from any one source is determined by analyzing the network with all other independent sources suppressed. (Controlled sources, if present, are never suppressed.)

Figure 3.2–10
Circuit analysis using superposition.

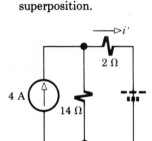

(a)

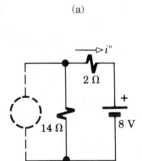

(b)

Superposition holds for any type of physical system where cause and effect are linearly related. Formally speaking, let x be the cause and y the corresponding effect related to x by

$$y = f(x) \tag{3a}$$

where $f(x)$ is a function of x. The function is *linear* if replacing x by a sum $x' + x''$ produces

$$y = f(x') + f(x'') \tag{3b}$$

which requires that $f(x' + x'') = f(x') + f(x'')$.

Ohm's law, $v = Ri$, is a linear relationship, for if $i = i' + i''$ then $v = R(i' + i'') = Ri' + Ri''$, and hence the voltage is a linear function of current. We have repeatedly used the phrase *linear resistance* to underscore this fact. As a counter example, the power dissipated by a resistance is *not* a linear function of current, since $p = Ri^2$ and $R(i' + i'')^2/ \neq Ri'^2 + Ri''^2$. Superposition therefore applies to voltage and current calculations in a network consisting of linear resistances, but it does not apply to power calculations directly.

The battery-charging circuit of Fig. 3.2–8a provides a good illustration of how to use superposition. It is redrawn in Fig. 3.2–10a with the battery suppressed (short-circuited) and in Fig. 3.2–10b with the current source suppressed (open-circuited). We compute the charging current i by finding its two components i' and i''. Specifically, $i' = (14/16) \times 4 = 3.5$ A and $i'' = -8/16 = -0.5$ A, so $i = i' = 3.5 - 0.5 = 3.0$ A, which agrees with our previous result.

Exercise 3.2–5

Apply superposition to the problem in Exercise 3.2–4. You may wish to "Thévenize" or "Nortonize" a portion of the circuit as a preliminary step.

Example 3.2–5
A voltage adder.

Figure 3.2–11a shows a circuit configuration called a **voltage adder:** v_1 and v_2 are *input* signals while v is the resulting *output*. The drawing is in a kind of short-hand, with all voltages measured with respect to the common bottom terminal identified by a **ground** symbol ($\doteq$). Thus, the actual circuit configuration looks like Fig. 3.2–11b.

Figure 3.2–11
Voltage adder.

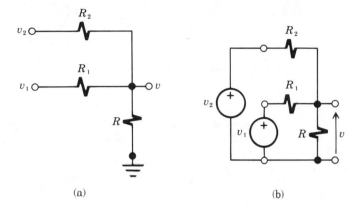

(a) (b)

We apply the superposition principle by mentally suppressing v_2, which puts R_2 directly in parallel with R. (Redraw the circuit with v_2 replaced by a short circuit if you do not visualize this immediately.) The output voltage contribution from v_1 is then

$$v' = \frac{R\|R_2}{R_1 + (R\|R_2)}\, v_1$$

By symmetry, the contribution v'' due to v_2 has the same form, but with interchanged subscripts. Therefore, in the common case where $R_2 = R_1$,

$$v = v' + v'' = \frac{R}{2R + R_1}\,(v_1 + v_2)$$

This justifies the name "voltage adder" since the output voltage v is proportional to $v_1 + v_2$.

Example 3.2–6

We finally return to the network at the beginning of this section (Fig. 3.2–1a) to determine its equivalent circuits by using all the tools developed here.

First, invoking the corollary to Thévenin's theorem, we find R_o by suppressing all sources as in Fig. 3.2–12a. This diagram shows that $R_o = 2 + (4\|12) = 5$ kΩ, the resistance value previously given in Fig. 3.2–1b.

Second, we need either v_{oc} or i_{sc}. The former will be easier to find in this case because no current flows through the 2-kΩ resistance under open-circuit conditions, and we can ignore it completely.

Applying superposition to the task, Fig. 3.2–12b immediately gives us $v'_{oc} = (12/16) \times 36 = 27$ V for the voltage source's contribution. As for the contribution from the current source, we start with Fig. 3.2–12c and then Thévenize the portion to the left of the dashed line, being careful with the

Figure 3.2–12
Circuits for Example
3.2–6.

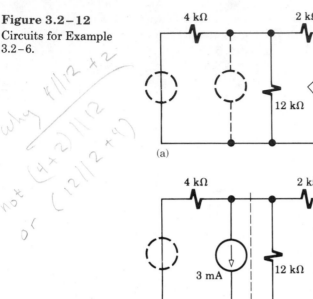

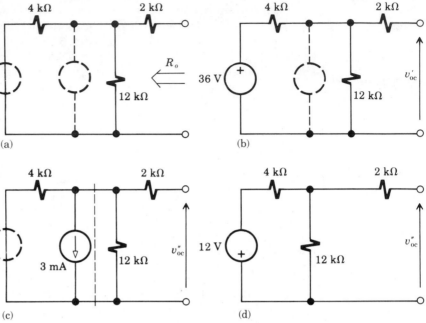

why $1/12 + 2$
not $(4+2)||12$
or $(12||2 + 4)$

(a)   (b)

(c)   (d)

polarity. Figure 3.2–12d is the result, and gives $v''_{oc} = (12/16) \times (-12) = -9$ V. Hence $v_{oc} = 27 - 9 = 18$ V, and Fig. 3.2–1b is, indeed, the Thévenin equivalent circuit. The Norton circuit would have $i_{sc} = 18$ V/5 k$\Omega = 3.6$ mA.

Had there been a resistance directly across the terminals — as in Fig. 3.2–6c, for instance — i_{sc} would be easier to find than v_{oc} because the short circuit bypasses current around that resistance and it can be ignored in the calculation.

3.3 NODE AND LOOP ANALYSIS[†]

Any resistive circuit can be analyzed using the techniques previously developed. However, the two methods presented here provide more efficient ways of translating circuit diagrams into network equations. Node analysis is a systematic application of the node-voltage concept, while loop analysis involves the new concept of loop currents.

Both methods lead to equations in standard form for solution by Cramer's rule or digital computer, the latter being necessary when there are many unknowns. In fact, special computer programs have been written for network analysis based on node voltages or loop currents. Effective use of these programs therefore requires familiarity with the underlying methods.

Node Analysis

The circuit in Fig. 3.3–1 has three unknown node voltages labeled v_1, v_2, and v_3 — all taken to be positive with respect to the reference node at the bottom. (In this regard, you may wish to review the discussion of node voltages in Section 2.4.) If we can determine these three voltages, then all other unknowns are easily found from them and the circuit diagram. For instance, the voltage across the 6-Ω resistor equals $v_1 - v_2$ so the branch current going from node 1 to node 2 will be $(v_1 - v_2)/6$.

Figure 3.3–1
Circuit with three
unknown node voltages.

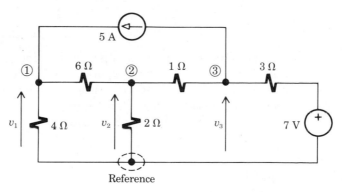

Applying KCL to sum the currents into each of the three nodes, and using KVL and Ohm's law, we get the node equations

$$5 - \frac{v_1}{4} - \frac{v_1 - v_2}{6} = 0$$

$$\frac{v_1 - v_2}{6} - \frac{v_2}{2} - \frac{v_2 - v_3}{1} = 0$$

$$\frac{v_2 - v_3}{1} - 5 - \frac{v_3 - 7}{3} = 0$$

Combining and regrouping terms then yields

$$\left(\frac{1}{4} + \frac{1}{6}\right) v_1 \qquad\qquad - \frac{1}{6} v_2 \qquad\qquad\qquad = 5 \qquad\qquad \textbf{(1a)}$$

$$- \frac{1}{6} v_1 + \left(\frac{1}{6} + \frac{1}{2} + \frac{1}{1}\right) v_2 \qquad - \frac{1}{1} v_3 = 0 \qquad\qquad \textbf{(1b)}$$

$$- \frac{1}{1} v_2 + \left(\frac{1}{1} + \frac{1}{3}\right) v_3 = \frac{7}{3} - 5 \qquad \textbf{(1c)}$$

We thus have the three equations needed to solve for v_1, v_2, and v_3.

The set of simultaneous equations in Eq. (1) are written in **standard form,** with the unknowns on the left and the source terms on the right. Putting simultaneous equations in standard form is a necessary step prior to solving them by Cramer's rule or digital computer. The following interpretation of these equations will show how they can be obtained more directly from the circuit diagram.

Consider Eq. (1a), which expresses KCL at node 1. Comparing terms with the circuit diagram reveals the following properties:

- The coefficient of v_1 equals the sum of all *conductances* connected at node 1.

- The coefficient of v_2 is the negative of the conductance connecting node 1 to node 2, and v_3 does not appear here simply because zero conductance connects nodes 1 and 3.

- The term on the right-hand side equals the *source current* into node 1.

Equation (1b) has these same properties since node 2 is connected by conductances to nodes 1 and 3, and no source current goes into node 2. Equation (1c) also has the same properties since the 7-V source in series with the 3-Ω resistor connected to node 3 can be replaced by the Norton equivalent circuit in Fig. 3.3–2. Hence, the net source current into node 3 equals $(7/3) - 5$.

Figure 3.3–2
Equivalent source currents into node 3.

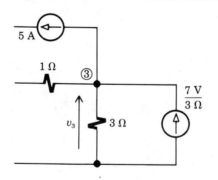

As a generalization of the foregoing observations, suppose a circuit has N unknown node voltages, all taken to be positive with respect to a reference node. The set of standard-form node equations for most (but not all) such circuits will look like

$$
\begin{aligned}
G_{11}v_1 - G_{12}v_2 - \cdots - G_{1N}v_N &= i_{s1} \\
-G_{21}v_1 + G_{22}v_2 - \cdots - G_{2N}v_N &= i_{s2} \\
&\ \ \vdots \\
-G_{N1}v_1 - G_{N2}v_2 - \cdots + G_{NN}v_N &= i_{sN}
\end{aligned}
\tag{2}
$$

The symbols used here are interpreted as follows:

$$v_k = \text{Unknown voltage at node } k$$
$$G_{kk} = \text{Sum of all conductances connected to node } k$$
$$G_{km} = G_{mk} = \text{Equivalent conductance connected directly}$$
$$\text{between nodes } k \text{ and } m$$
$$i_{sk} = \text{Net equivalent source current into node } k$$

Note on the left-hand side of Eq. (2) that the coefficients along the main diagonal ($G_{11}, G_{22}, \ldots$) are positive, whereas the off-diagonal coefficients are negative and symmetrical ($-G_{21} = -G_{12}$, etc.). The equivalent source currents on the right-hand side may be positive, negative, or zero.

Based on this interpretation of the conductance and current terms, you should be able to find their values by inspection of a circuit diagram together with a few side calculations. Then you just insert those values into Eq. (2) with the appropriate value of N. Having thus obtained the node equations in standard form, you can proceed with the task of solving them.

There is, however, one restriction on the validity of Eq. (2) and our inspection technique, namely:

Any *ideal voltage sources* (without series resistance) must have one terminal connected to the reference node so that each source establishes a known node voltage.

This restriction seldom causes problems in view of the fact that most practical circuits with multiple voltage sources usually do have a common ground that can be chosen as the reference node.

Example 3.3–1

The circuit in Fig. 3.3–3a is to be analyzed using node equations. The bottom node is an appropriate reference, and v_1 and v_2 are the only unknown node voltages since the voltage source puts node 3 at -20 V relative to the reference. We'll write the node equations in a consistent set of units with voltages in volts, currents in milliamps, and conductances in millisiemens (the reciprocal of kilohms).

The conductance terms for the standard-form equations are seen by inspection to be

$$G_{11} = \frac{1}{5+7} + \frac{1}{6} = \frac{1}{4} \qquad G_{22} = \frac{1}{5+7} + \frac{1}{3} + \frac{1}{5} + \frac{1}{20} = \frac{2}{3}$$

$$G_{12} = G_{21} = \frac{1}{5+7} = \frac{1}{12}$$

However, the equivalent source currents are not immediately obvious because two resistors connect directly to the voltage source at node 3. Replacing this source with two identical sources in parallel leaves the circuit unchanged, and it allows us to split node 3 as shown in Fig. 3.3–3b. Hence,

$$i_{s1} = 8 - \frac{20}{6} = \frac{14}{3} \qquad i_{s2} = -\frac{20}{3}$$

Figure 3.3–3
(a) Circuit for Example 3.3–1. (b) Using two voltage sources to find equivalent
source currents.

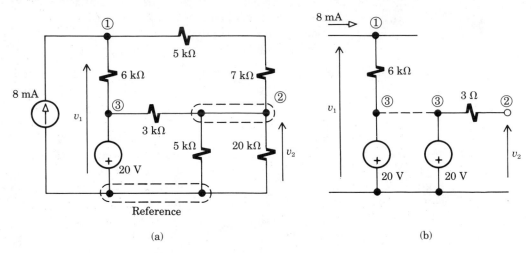

(a)   (b)

where $-20/6$ and $-20/3$ are the Norton equivalent currents into nodes 1
and 2, respectively.

Substituting our conductance and current terms into Eq. (2) with
$N = 2$ gives

$$\frac{1}{4} v_1 - \frac{1}{12} v_2 = \frac{14}{3}$$

$$-\frac{1}{12} v_1 + \frac{2}{3} v_2 = -\frac{20}{3}$$

Then, after clearing the fractions, we have

$$3v_1 - v_2 = 56$$
$$-v_1 + 8v_2 = -80$$

and simultaneous solution yields

$$v_1 = 16 \text{ V} \qquad v_2 = -8 \text{ V}$$

This completes the node analysis unless you want to use the results to find
some other voltage or a branch current.

Figure 3.3–4
Circuit for Exercise
3.3–1 and 3.3–2.

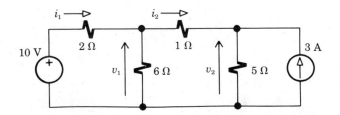

Exercise 3.3–1   Use node analysis to find v_1, v_2, i_1, and i_2 in Fig. 3.3–4.

Loop Analysis

If a circuit has more than three unknown node voltages, or if the quantities of interest are currents rather than voltages, then loop equations may be easier to write and solve. The essential concept for loop analysis is the loop current.

Consider the circuit in Fig. 3.3–5, where each "hole" or "window pane" is called a **mesh** and has been labeled with a **loop current**. These currents are visualized as traveling completely around the closed path of elements that frame the meshes. Thus, the branch current through an element on the perimeter will be the same as a loop current, whereas the branch current through any inner element consists of two loop currents. For instance, loop current i_1 is the branch current through the 7-Ω resistor, but the branch current through the 2-Ω resistor is $i_1 - i_2$. If we can find all the unknown loop currents, then we can determine all the branch currents and the resulting voltages from the circuit diagram.

Figure 3.3–5
Circuit with three unknown loop currents.

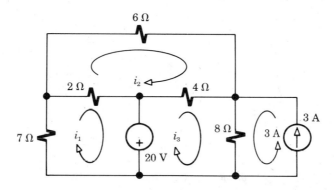

The circuit at hand happens to have three unknown loop currents, i_1, i_2, and i_3, since the remaining loop current passes through the current source and must equal 3 A. The current source establishes the direction of the known loop current. The unknown loop currents may rotate either clockwise or counter-clockwise, but we choose to draw them so that they all have the *same rotational direction*.

To find the unknown currents, we apply KVL together with Ohm's law to sum the voltage drops around each of the three loops. Thus,

$$7i_1 + 2(i_1 - i_2) - 20 = 0$$
$$6i_2 + 4(i_2 - i_3) + 2(i_2 - i_1) = 0$$
$$4(i_3 - i_2) + 8(i_3 + 3) + 20 = 0$$

Combining and regrouping terms then yields the standard-form equations

$$(7 + 2)i_1 \qquad\qquad -2i_2 \qquad\qquad\qquad = 20 \qquad\qquad \textbf{(3a)}$$
$$-2i_1 + (6 + 4 + 2)i_2 \qquad -4i_3 = 0 \qquad\qquad \textbf{(3b)}$$
$$-4i_2 + (4 + 8)i_3 = -20 - (8 \times 3) \qquad \textbf{(3c)}$$

We'll interpret these equations to show how they can be obtained more directly from the circuit diagram.

Consider Eq. (3a), which states KVL for loop 1. Comparing terms with the circuit diagram reveals the following properties:

- The coefficient of i_1 equals the sum of all *resistances* around loop 1.
- The coefficient of i_2 is the negative of the resistance common to loops 1 and 2, and i_3 does not appear here because loops 1 and 3 have no common resistance.
- The term on the right-hand side equals the *source voltage* that tends to drive i_1 around loop 1.

Equation (3b) has these same properties since loop 2 has branch resistance in common with loops 1 and 3, and there is no voltage source in loop 2. Equation (3c) also has the same properties since the 3-A source in parallel with the 8-Ω resistor can be replaced by the Thévenin equivalent circuit in Fig. 3.3–6. Hence, the net source voltage opposes i_3 and equals $-20 - (8 \times 3)$.

Figure 3.3–6
Equivalent source voltages around loop 3.

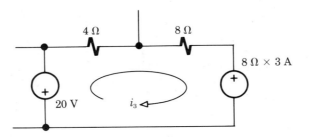

As a generalization of the foregoing observations, consider a circuit having N unknown loop currents, all circulating in the same clockwise or counter-clockwise direction. The standard-form node equations for most (but not all) such circuits will look like

$$
\begin{aligned}
R_{11}i_1 \ - R_{12}i_2 \ - \cdots - R_{1N}i_N &= v_{s1} \\
-R_{21}i_1 + R_{22}i_2 \ - \cdots - R_{2N}i_N &= v_{s2} \\
&\vdots \\
-R_{N1}i_1 - R_{N2}i_2 - \cdots + R_{NN}i_N &= v_{sN}
\end{aligned}
\tag{4}
$$

The symbols used here are interpreted as follows:

$$
\begin{aligned}
i_k &= \text{Unknown current around loop } k \\
R_{kk} &= \text{Sum of all resistances around loop } k \\
R_{km} = R_{mk} &= \text{Equivalent resistance common to loops } k \text{ and } m \\
v_{sk} &= \text{Net equivalent source voltage around loop } n
\end{aligned}
$$

Note that Eq. (4) has the same structure and symmetry as our previous node equations.

Based on this interpretation of the resistance and voltage terms, you should be able to find their values by inspection of a circuit diagram together with a few side calculations. Then you just insert those values into Eq. (4), with the appropriate value of N, and proceed with the solution.

However, there are two restrictions on the validity of Eq. (4) and the inspection technique for loop equations:

1. The circuit must be *planar* so that no branch crosses over another without a node connection.

2. Any *ideal current sources* (without parallel resistance) must be on the perimeter of the circuit so that each source establishes a known loop current.

When a circuit cannot be redrawn to satisfy these restrictions, node analysis may be easier to carry out. If the circuit is planar but has a current source in an interior branch, then loop equations can be written from scratch as we first did for Fig. 3.3–5.

Example 3.3–2

The circuit previously studied using node analysis in Example 3.3–1 has been redrawn in Fig. 3.3–7a for loop analysis. Although this circuit has four meshes, the parallel 5-kΩ and 20-kΩ resistors can be replaced by a 4-kΩ equivalent resistance to eliminate one mesh, and the current source establishes a known loop current. Hence, we have just two unknown loop currents to find. As before, we'll work with current in milliamps and resistance in kilohms.

Figure 3.3–7
(a) Circuit for Example 3.3–2. (b) Using two current sources to find equivalent source voltages.

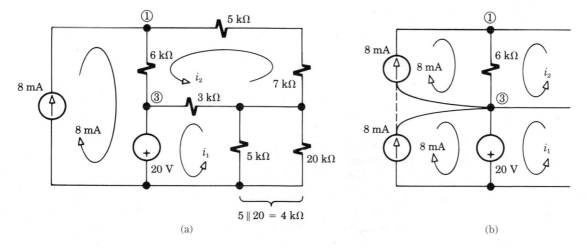

(a)   (b)

The resistance terms for the standard-form loop equations are seen by inspection to be

$$R_{11} = (5\|20) + 3 = 7 \qquad R_{22} = 7 + 5 + 6 + 3 = 21$$
$$R_{12} = R_{21} = 3$$

But determining the net equivalent source voltages is complicated by the fact that the current source cannot immediately be converted to Thévenin form. To get around this problem, without affecting the circuit, we put two identical current sources in series and tie their junction to node 3, as drawn in Fig. 3.3–7b. Now we see that the current source produces no voltage in loop 1. Hence,

$$v_{s1} = 20 \qquad v_{s2} = -6 \times 8 = -48$$

since the Thévenin equivalent voltage in loop 2 opposes i_2 and equals $6 \text{ k}\Omega \times 8 \text{ mA} = 48 \text{ V}$.

Substituting our resistance and voltage terms into Eq. (4) with $N = 2$ gives

$$7i_1 - 3i_2 = 20$$
$$-3i_1 + 21i_2 = -48$$

Simultaneous solution then yields

$$i_1 = 2 \text{ mA} \qquad i_2 = -2 \text{ mA}$$

Using these results, we find that the branch current going down through the 6-kΩ resistor is $8 \text{ mA} + i_2 = 6 \text{ mA}$, so the voltage at node 1 relative to the bottom node will be

$$v_1 = (6 \text{ k}\Omega \times 6 \text{ mA}) - 20 \text{ V} = 16 \text{ V}$$

in agreement with the value previously obtained. Other branch currents and voltages can be determined from similar calculations.

Exercise 3.3–2   Rework Exercise 3.3–1 using loop analysis.

Circuits with Controlled Sources

Node or loop analysis by inspection is still possible when the circuit contains a *controlled source*. However, the method requires three extra steps, as follows:

1. Write the standard-form node or loop equations treating the controlled source as an independent source whose value is represented symbolically.

2. Write an additional equation, called the **constraint equation,** to relate the value of the controlled voltage or current to the unknown node voltages or loop currents.

3. Insert the constraint equation into the node or loop equations, and then solve for the unknowns.

An example should clarify this process.

Example 3.3–3

The circuit in Fig. 3.3–8 contains a source voltage v_c controlled by the voltage v_x. We'll formulate the equations required to find the node voltages v_1 and v_2.

Figure 3.3–8
Circuit for Example 3.3–3.

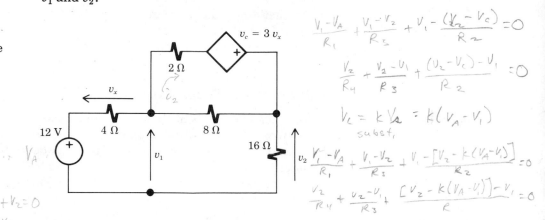

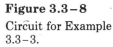

The standard-form node equations are easily obtained by Thévenin-to-Norton conversion of both sources, treating the controlled voltage as a source voltage of value v_c. The Norton equivalent current from node 1 to node 2 is then $v_c/(2 \ \Omega)$, so we get

$$\left(\frac{1}{4}+\frac{1}{2}+\frac{1}{8}\right) v_1 - \left(\frac{1}{2}+\frac{1}{8}\right) v_2 = \frac{12}{4} - \frac{v_c}{2}$$

$$-\left(\frac{1}{2}+\frac{1}{8}\right) v_1 + \left(\frac{1}{2}+\frac{1}{8}+\frac{1}{16}\right) v_2 = \frac{v_c}{2}$$

Next, we need the constraint equation relating v_c to v_1 and/or v_2. Since $v_c = 3v_x$ and $v_x + v_1 = 12$ V, our constraint equation is

$$v_c = 3(12 - v_1) = 36 - 3v_1$$

After substituting for v_c and rearranging terms, the set of equations becomes

$$-\frac{5}{8} v_1 - \frac{5}{8} v_2 = -15$$

$$\frac{7}{8} v_1 + \frac{11}{16} v_2 = 18$$

which can now be solved for v_1 and v_2. Observe that inserting the constraint equation has destroyed the symmetry properties of the standard-form equations.

Exercise 3.3–3   Apply loop analysis to find the current going left to right through the 2-Ω resistor in Fig. 3.3–8.

3.4 DC METERS AND MEASUREMENTS †

We conclude this chapter with an introduction to DC measuring instruments that can be described in terms of resistive circuits. The capitalized symbols V and I will be used throughout as a reminder that we are dealing with constant or slowly varying voltage and current.

Voltmeters and Ammeters

Routine measurements of DC current and voltage are usually made with **direct-reading meters** incorporating a **d'Arsonval moving-coil mechanism.** The operating principles of this device will be covered in Chapter 17; here it suffices to say that the meter has an indicating pointer whose angular deflection depends on the amount and direction of the average current I through the meter, as illustrated by Fig. 3.4–1a. The maximum or **full-scale deflection** corresponds to a constant current I_{fs} flowing in the proper direction, and pointer deflection is proportional to DC current values between zero and I_{fs} with an accuracy of 2–5% of the full-scale value.

Regardless of the specific details of the internal configuration, the equivalent circuit seen at the meter's terminals is nothing more than a resistance R_m representing the fact that the meter consumes some energy, as Fig. 3.4–1b shows. Therefore, the voltage required to produce full-scale deflection is

$$V_{fs} = R_m \, I_{fs} \qquad (1)$$

The characteristics of a given meter are fully specified by any two of the three parameters I_{fs}, R_m, and V_{fs}. Typical values for a low-current meter are $I_{fs} = 50 \ \mu A$ and $R_m = 3 \ k\Omega$, so $V_{fs} = 150 \ mV$, while a high-current meter might have $I_{fs} = 10 \ mA$ and $R_m = 5 \ \Omega$. Low-current meters are said to have greater **sensitivity** because full-scale deflection requires less current.

With the aid of additional resistors and switches, a simple meter can measure voltage or current in various ranges. Figure 3.4–2, for example, shows the arrangement for voltage measurement with two ranges. The unknown voltage to be measured is V_u and the range-setting resistors R_1 and R_2 are called **multipliers.**

To analyze this **voltmeter** circuit, assume the switch to be in the upper position, putting R_1 in series with R_m. From the voltage-divider ratio, the voltage across the meter element is $V = R_m V_u/(R_m + R_1)$, so that

Figure 3.4–1
A DC meter and its equivalent circuit.

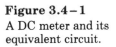

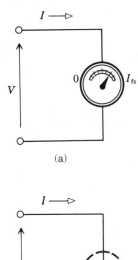

(a)

(b)

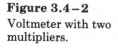

Figure 3.4−2
Voltmeter with two
multipliers.

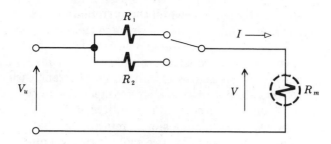

full-scale deflection corresponds to

$$V_{u_{\text{fs}}} = \frac{R_m + R_1}{R_m} V_{\text{fs}} = \left(1 + \frac{R_1}{R_m}\right) V_{\text{fs}} \tag{2}$$

The same equation applies for the other switch position, with R_2 in place of R_1.

Equation (2) shows that $V_{u_{\text{fs}}} \geq V_{\text{fs}}$; the multiplier accounts for the difference between the unknown voltage and the meter voltage, as brought out by the alternative expression $V_{u_{\text{fs}}} = V_{\text{fs}} + R_1 I_{\text{fs}}$. To illustrate, take a high-current meter with $I_{\text{fs}} = 10$ mA, $R_m = 5\ \Omega$, $R_1 = 95\ \Omega$, and $R_2 = 495\ \Omega$. The voltage ranges then will be $(1 + 95/5) \times V_{\text{fs}} = 20 \times 50$ mV $= 1$ V and $(1 + 495/5) \times 50$ mV $= 5$ V.

Putting resistors in parallel with the meter element produces the current-measuring **ammeter** circuit of Fig. 3.4−3. The resistors R_1 and R_2 are known as **shunts,** and full-scale deflection corresponds to $I_{u_{\text{fs}}} \geq I_{\text{fs}}$ with the shunt carrying the excess current.

Figure 3.4−3
Ammeter with two
shunts.

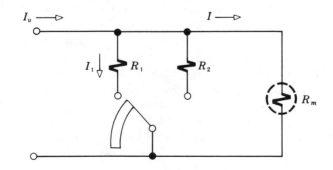

The switch in a multirange ammeter must be a special type with a *make-before-break* design. As suggested by the extended contact in Fig. 3.4−3, this switch "makes" the connection to R_2 before it "breaks" contact with R_1, so the meter element always has shunt resistance. Were this not the case, the entire current I_u would flow through the meter and might destroy the mechanism by excessive ohmic heating. Voltmeters do not require this special switch (Why?), but all meter elements should be protected by a fuse to prevent damage from excessive current.

Since a simple meter can be used for voltage or current measurement, it is possible to construct an instrument called a **multimeter** that measures multiple ranges of voltage and current with one meter element. A rotary switch with several sets of mechanically coupled contacts allows for the proper connection of multipliers and shunts. To illustrate, Fig. 3.4–4 shows the circuit for a simple multimeter that has one voltage range and one current range. The two switches are coupled to move together and the terminal marked "NC" has no connection to it. Redraw the circuit, if necessary, to convince yourself that R_v is a multiplier when the switches are in the upper position while R_a is a shunt when they are in the lower position.

Figure 3.4–4
Simple multimeter.

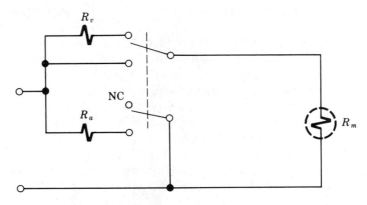

Multimeters are portable and very handy but — like the separate ammeter and voltmeter circuits — they do consume power from the quantity being measured and thereby may significantly alter the value. You must be particularly alert to the possibility of loading effect in voltage measurements. The input resistance of a multirange voltmeter is usually stated in *ohms per volt* of full-scale deflection. For instance, when set on the 5-V scale, a typical 20 kΩ/V meter has an input resistance of 20 kΩ/V × 5 V = 100 kΩ and would draw 50 μA when reading 5 V. The higher the input resistance, the less loading there will be. Other methods for minimizing or eliminating the loading effect will be covered later.

Exercise 3.4–1   Show that the ammeter circuit in Fig. 3.4–3 has

$$I_{u_{fs}} = \left(1 + \frac{R_m}{R_1}\right) I_{fs}$$

Exercise 3.4–2   If the meter in Fig. 3.4–4 has $I_{fs} = 50$ μA and $R_m = 3$ kΩ, then what values of R_v and R_a yield $V_{u_{fs}} = 30$ V and $I_{u_{fs}} = 2$ A?

Ohmmeters

Figure 3.4–5a diagrams a simplified **ohmmeter** circuit for DC resistance measurement with a d'Arsonval (or equivalent) meter movement. The resistance R is chosen to yield full-scale deflection with the instrument's terminals short-circuited so $R_u = 0$. Thus

$$\frac{V_s}{R + R_m} = I_{\text{fs}}$$

Then, with the unknown resistance R_u in the circuit

$$I = \frac{V_s}{R_u + R + R_m} = \frac{I_{\text{fs}}}{1 + R_u/(R + R_m)} \tag{3}$$

which shows that I is inversely proportional to R_u. Specifically, half-scale deflection ($I = 0.5 I_{\text{fs}}$) corresponds to $R_u = R + R_m$, whereas $I = 0$ when $R_u = \infty$ (open-circuited terminals). Accordingly, the ohms scale is *backwards* and *nonlinear*, as illustrated in Fig. 3.4–5b taking $R + R_m = 300\ \Omega$.

Figure 3.4–5
Ohmmeter: (a) circuit; (b) nonlinear scale.

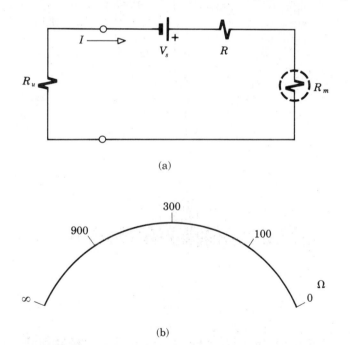

(a)

(b)

Most multimeters include an ohmmeter circuit similar to Fig. 3.4–5, but in a slightly more complicated form to provide multiple ranges with different half-scale values. A variable resistance, usually labeled "Ohms adjust," is included to compensate for the changing value of V_s as the multimeter's battery ages. A multimeter with scales for volts, ohms, and milliamps is known by the acronym VOM.

Pay careful attention to the fact that the simple ohmmeter can only be used to measure isolated resistors or the equivalent resistance of a connection of resistors. It should not be used to measure the resistance of an electronic component that might be damaged by the sensing current I, nor can it measure the value of a resistance embedded in a network.

Bridges and Null Measurements

Null measurements are made with bridge circuits and related configurations. They differ from direct measurements in that the quantity being measured is compared with a known reference quantity, analogous to weight measurement with a balancing scale. The balancing strategy avoids unwanted interaction effects and usually results in greater precision than direct measurement, which depends on the accuracy of a meter movement.

To introduce the null-measurement principle, consider the **potentiometric voltage measurement** circuit in Fig. 3.4–6. This circuit allows you to measure an unknown source voltage V_u by comparison with the adjustable voltage V_w derived from a standard reference voltage V_{ref}. You simply adjust the potentiometer's wiper until the zero-center meter indicates that

$$V_{uw} = V_u - V_w = 0$$

Hence, $V_u = V_w = (R_{aw}/R_{ab})V_{ref}$, and the value of V_u can be determined by reading the calibrated dial of the potentiometer. No current flows between nodes u and w in the balanced condition, making the null method ideal for measuring sensitive electrochemical processes and the like.

Figure 3.4–6
Potentiometric voltage measurement.

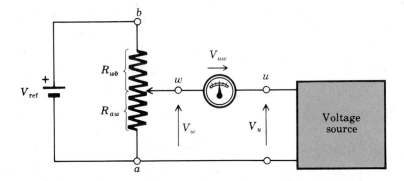

An automated **self-balancing potentiometer** incorporates a small motor driven by the difference voltage $V_u - V_w$. The motor moves the wiper until balance is achieved and $V_u - V_w = 0$. The self-balancing system will continuously follow a time-varying voltage, if not too rapid, and is often used for strip-chart and X-Y recorders.

Another null-measurement instrument called the **Wheatstone bridge** is diagrammed in Fig. 3.4–7a. This circuit is designed for precise

Figure 3.4−7

Wheatstone bridge: (a) diagram; (b) equivalent circuit when balanced.

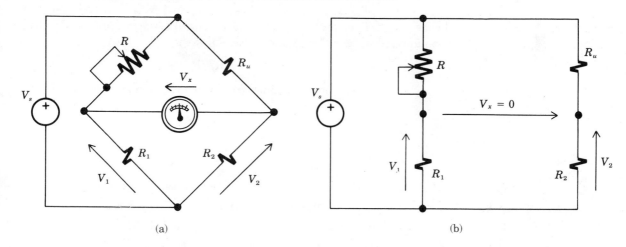

(a)   (b)

measurement of an unknown resistance R_u with the help of a calibrated potentiometer R plus two known resistances R_1 and R_2. We adjust R until the bridge is balanced with $V_x = 0$. Since no current flows through the meter under balanced conditions, we can redraw the circuit as shown in Fig. 3.4−7b where

$$V_1 = \frac{R_1}{R_1 + R} V_s \qquad V_2 = \frac{R_2}{R_2 + R_u} V_s$$

But $V_x = V_2 - V_1 = 0$, so $V_2 = V_1$ and hence

$$R_u = (R_2/R_1)R \qquad\qquad (4)$$

which is independent of the source voltage V_s.

Exercise 3.4−3

Figure 3.4−8 is a Wheatstone bridge equipped with two potentiometers and connected to a strain gauge of the type discussed in Example 2.3−2. A

Figure 3.4−8

Strain measurement with a Wheatstone bridge.

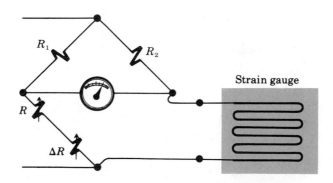

Strain gauge

small mechanical strain $\Delta \ell / \ell$ increases the resistance of the strain gauge from its unstrained value R_u to $R_u + \Delta R_u$, where $\Delta R_u = 2R_u(\Delta \ell / \ell)$.

Let the bridge be initially balanced using R with $\Delta R = 0$ and no strain on the gauge. Then let the bridge be rebalanced using ΔR after strain has been applied. Under these conditions, show that

$$\Delta \ell / \ell = \Delta R / 2R$$

Thus, the strain can be determined from the potentiometer settings R and ΔR.

PROBLEMS

3.1-1   Generalize Fig. 3.1-2 and Eq. (3) for the case of n resistors in series.

3.1-2   Show that putting n equal resistors in parallel yields $R_{eq} = R/n$.

3.1-3   Obtain an expression similar to Eq. (4) for three parallel resistors.

3.1-4   Obtain an expression similar to Eq. (4) for four parallel resistors.

3.1-5   Obtain the condition on R_1 and R_2 in Fig. 3.1-6 such that $p_{max} = 4p_{min}$. Why would this be undesirable?

3.1-6   Show how you would connect four 3-Ω resistors to get: (a) $R_{eq} = 4\ \Omega$; (b) $R_{eq} = 5\ \Omega$.

3.1-7   Show how you would connect four 4-Ω resistors to get: (a) $R_{eq} = 10\ \Omega$; (b) $R_{eq} = 3\ \Omega$.

3.1-8   Show how you would connect five 5-Ω resistors to get: (a) $R_{eq} = 4\ \Omega$; (b) $R_{eq} = 6\ \Omega$.

3.1-9   Let $R_3 = 6\ \Omega$ in Fig. P3.1-9. Find v_1, i_2, i_3, and the power supplied by the source.

Figure P3.1-9

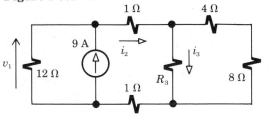

3.1-10   Do Problem 3.1-9 with $R_3 = 60\ \Omega$.

3.1-11   Find i, v_1, and v_2 in Fig. P3.1-11 when both wipers are set at the middle of the potentiometers.

Figure P3.1-11

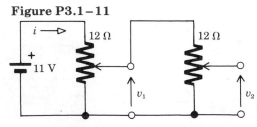

3.1–12 Find i, v_1, and v_2 in Fig. P3.1–11 when both wipers are set one-third from the bottom of the potentiometers.

3.1–13 Find i, v_1, and v_2 in Fig. P3.1–11 when both wipers are set one-third from the top of the potentiometers.

3.1–14 Suppose the source modeled in Fig. 3.1–11 has $v_s = 10$ V and $i_s = 5$ A.

(a) Calculate R_s and p_{max}.
(b) What's the condition on R_L such that $v \geq 0.95v_s$?

3.1–15 Suppose the source modeled in Fig. 3.1–11 has $i_s = 2$ A and $R_s = 80$ Ω.

(a) Calculate v_s and p_{max}.
(b) What's the condition on R_L such that $i \geq 0.9i_s$?

3.1–16 Suppose the source modeled in Fig. 3.1–11 has $i_s = 5$ mA and $R_s = 20$ kΩ.

(a) Calculate v_s and p_{max}.
(b) What's the condition on R_L such that $v \geq 0.9v_s$?

3.1–17 Derive Eq. (14) by solving $dp_L/dR_L = 0$ for R_L.

3.2–1 Find $R_{eq} = v/i$ for Fig. P3.2–1 when $R_x = 20$ Ω and $v_c = 2v$.

Figure P3.2–1

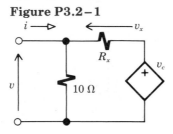

3.2–2 Find $R_{eq} = v/i$ for Fig. P3.2–1 when $R_x = 20$ Ω and $v_c = 5v$.

3.2–3 Find $R_{eq} = v/i$ for Fig. P3.2–1 when $R_x = 5$ Ω and $v_c = 2v_x$.

3.2–4 Find $R_{eq} = v/i$ for Fig. P3.2–1 when $R_x = 5$ Ω and $v_c = -2v_x$.

3.2–5 Perform Thévenin/Norton conversions to obtain the Thévenin equivalent circuit for Fig. P3.2–5 when $R_x = 0$.

Figure P3.2–5

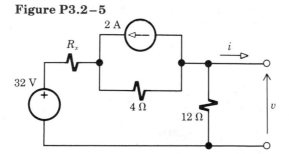

3.2–6 Perform Thévenin/Norton conversions to obtain the Thévenin equivalent circuit for Fig. P3.2–6 when $R_x = 0$.

3.2–7 Perform Thévenin/Norton conversions to obtain the Norton equivalent circuit for Fig. P3.2–5 when $R_x = 2$ Ω.

Figure P3.2−6

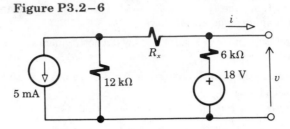

3.2−8   Perform Thévenin/Norton conversions to obtain the Norton equivalent circuit for Fig. P3.2−6 when $R_x = 18$ kΩ.

3.2−9   Let $i_s = 5$ A and $v_s = 10$ V in Fig. P2.4−9. Use Thévenin/Norton conversion to find i_3.

3.2−10   Let $i_s = 3$ A and $v_s = 32$ V in Fig. P2.4−9. Use Thévenin/Norton conversion to find v_1.

3.2−11   Let $R_1 = 10$ kΩ in Fig. P3.2−11. Use Thévenin/Norton conversion to find v_2.

Figure P3.2−11

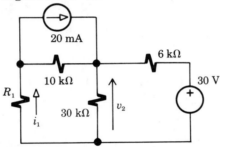

3.2−12   Use Thévenin/Norton conversion to determine the value of R_1 in Fig. P3.2−11 such that $i_1 = 5$ mA.

3.2−13   Let $i_s = 3$ mA in Fig. P2.4−9. Apply superposition to find v_s such that $i_3 = 4$ mA.

3.2−14   Let $v_s = 10$ V in Fig. P2.4−9. Apply superposition to find i_s such that $v_1 = 9$ V.

3.2−15   Find v_a in Fig. P3.2−15 using superposition.

Figure P3.2−15

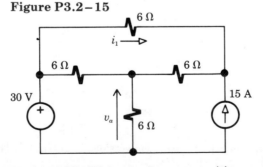

3.2−16   Find i_1 in Fig. P3.2−15 using superposition.

3.2−17   Let $R_3 = 2$ Ω in Fig. P2.4−11. Apply superposition to find i_1.

3.2−18   Let $R_3 = 8$ Ω in Fig. P2.4−11. Apply superposition to find v_b.

3.3–1 Write and solve node equations for v_2 and v_4 in Fig. P2.4–3 with $R_3 = R_4 = 1\ \Omega$.

3.3–2 Write and solve node equations for v_2 and v_4 in Fig. P2.4–3 with $R_3 = 6\ \Omega$ and $R_4 = 4\ \Omega$.

3.3–3 Use node analysis to find v in Fig. P3.2–6 when $R_x = 18\ \text{k}\Omega$ and a 20-kΩ resistor is connected across the output.

3.3–4 Solve node equations for v_2 in Fig. P3.2–11 when $R_1 = 10\ \text{k}\Omega$.

3.3–5 Suppose an external source provides $i_1 = 5\ \text{A}$ in Fig. P3.3–5. Solve three node equations to calculate $R_{eq} = v_1/i_1$.

Figure P3.3–5

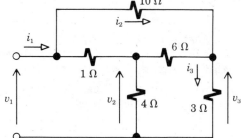
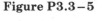

3.3–6 Solve node equations for v_a in Fig. P3.2–15.

3.3–7 Find v_2 and v_3 in Fig. P3.3–5 when an external source provides $v_1 = 8\ \text{V}$.

3.3–8 Find v_a and v_b in Fig. P2.4–11 when $R_3 = 20\ \Omega$.

3.3–9 Write and solve loop equations for i_1 and i_3 in Fig. P2.4–3 with $R_3 = R_4 = 1\ \Omega$.

3.3–10 Write and solve loop equations for i_1 and i_3 in Fig. P2.4–3 with $R_3 = 6\ \Omega$ and $R_4 = 4\ \Omega$.

3.3–11 Use loop analysis to find i in Fig. P3.2–6 when $R_x = 0$ and a 12-kΩ resistor is connected across the output.

3.3–12 Solve loop equations for i_1 in Fig. P3.2–11 when $R_1 = 20\ \text{k}\Omega$.

3.3–13 Suppose an external source provides $v_1 = 32\ \text{V}$ in Fig. P3.3–5. Solve three loop equations to calculate $R_{eq} = v_1/i_1$.

3.3–14 Find i_1 and i_3 in Fig. P3.3–5 when an external voltage source provides $v_1 = 16\ \text{V}$ and the 10-Ω resistor is replaced by a current source such that $i_2 = 1\ \text{A}$.

3.3–15 Find i_2 and i_3 in Fig. P3.3–5 when an external source provides $i_1 = 2\ \text{A}$.

3.3–16 Solve loop equations for i_1 in Fig. P3.2–15.

3.3–17 Write a node equation to find v_a in Fig. P3.3–17 when $v_c = 12i_1$.

Figure P3.3–17

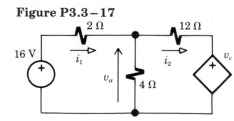

3.3 – 18   Write a loop equation to find i_a in Fig. P3.3–18 when $i_c = v_1/2$.

Figure P3.3 – 18

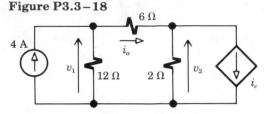

3.3 – 19   Find v_1 and v_2 in Fig. P3.3–18 when $i_c = 2i_a$.

3.3 – 20   Find i_1 and i_2 in Fig. P3.3–17 when $v_c = 2v_a$.

3.4 – 1   Let Fig. 3.1–3d have $R_{aw} = R_{wb} = 500\ \Omega$ and $V_s = 9$ V. What value of v_w would be measured by a voltmeter with $R_1 + R_m = 2$ kΩ?

3.4 – 2   A certain voltmeter reads $v_w = 4$ V in Fig. 3.1–3d when $R_{aw} = R_{wb} = 300\ \Omega$ and $V_s = 10$ V. What is the value of the voltmeter resistance $R_1 + R_m$?

3.4 – 3   A multimeter circuit like Fig. 3.4–4 has a meter with $I_{fs} = 1$ mA and $R_m = 20\ \Omega$. Specify the shunt and multiplier resistors needed to get full-scale ranges of 1 mA, 1 A, 0.2 V, and 20 V.

3.4 – 4   Do Problem 3.4–3 for a meter with $I_{fs} = 100\ \mu$A and $R_m = 2$ kΩ.

3.4 – 5   Do Problem 3.4–3 for full-scale ranges of 5 mA, 500 mA, 0.5 V, and 50 V.

3.4 – 6   The movement in a certain multimeter has $I_{fs} = 100\ \mu$A and $R_m = 200\ \Omega$. Find the meter's input resistance for the 5-V scale and the 10-mA scale.

3.4 – 7   Do Problem 3.4–6 for a movement with $V_{fs} = 0.1$ V and $R_m = 50\ \Omega$.

3.4 – 8   Suppose the element in Fig. 2.1–6 is a resistor R, and the voltmeter has a resistance of 5 kΩ/V. What is the value of R if the ammeter reads 0.9 mA and the voltmeter reads 1.8 V on the 2-V scale?

3.4 – 9   Do Problem 3.4–8 for an ammeter reading of 0.62 mA and a voltmeter reading of 6 V on the 10-V scale.

3.4 – 10   Let the ohmmeter in Fig. 3.4–5a have $V_s = 3$ V and $I_{fs} = 50\ \mu$A. What values of R_u correspond to readings of 20%, 50%, and 80% of full scale?

3.4 – 11   Do Problem 3.4–10 with $V_s = 9$ V and $I_{fs} = 4.5$ mA.

3.4 – 12   Suppose the values of R_1 and R_2 in the Wheatstone bridge in Fig. 3.4–7 are known to an accuracy of $\pm 0.5\%$ and the potentiometer dial can be read with an accuracy of $\pm 1\%$. Find the maximum and minimum possible values of R_u to three significant figures when the bridge is balanced with the following nominal settings: $R_1 \approx 400\ \Omega$, $R_2 \approx 10\ \Omega$, $R \approx 200\ \Omega$.

3.4 – 13   Do Problem 3.4–12 with the following nominal settings: $R_1 \approx 100\ \Omega$, $R_2 \approx 6000\ \Omega$, $R \approx 40\ \Omega$.

Chapter

4 Amplifiers and Op-Amps

A typical electronic system, say a stereo amplifier or TV set, contains hundreds of elements and numerous interconnected circuits. Trying to understand the operation of such a system would be very difficult if we had to deal with each component individually. And designing a system part-by-part would be impossible. Obviously, that's not the way it's done. Instead, the designer first works with functional building blocks to obtain a system-block diagram with the desired characteristics. The system diagram then serves as the basis for designing the circuit blocks.

This chapter develops the concepts, modeling, and analysis of amplifiers as electrical building blocks. We'll start with ideal amplifiers modeled by controlled sources, and we'll add resistances to represent and study the external behavior of real electronic amplifiers. Then we introduce the versatile operational amplifier, or "op-amp," along with some of its practical applications. Later chapters will consider electronic components and the implementation of amplifier circuits.

Objectives

After studying this chapter and working the exercises, you should be able to do each of the following:

- Represent the behavior of amplifiers with controlled sources and resistances (Section 4.1).
- Use circuit models to find the gain of an amplifier or cascaded amplifiers, including loading effects (Section 4.1).
- Express amplifier gain in decibels (Section 4.1).
- State the major features of an operational amplifier (Section 4.2).
- Apply the concept of the virtual short to analyze simple circuits with op-amps (Section 4.2).

■   Design an op-amp system with noninverting, inverting, and/or summing amplifiers (Section 4.2).

4.1 AMPLIFIER CONCEPTS

Amplifiers are networks containing electronic devices capable of "enlarging" or "magnifying" the variations of an electrical signal represented by a voltage or current waveform. As such, amplifiers play essential roles in multitudinous applications — radio and TV sets, biomedical instrumentation, industrial process control, and so on, almost without end. This section introduces the general concepts of amplifiers and amplification systems.

Figure 4.1–1a shows a familiar amplification system consisting of an electric guitar, amplifier, and loudspeaker. Figure 4.1–1b represents those same units in a generalized *block diagram*. The *input signal,* denoted $x(t)$, is the voltage or current waveform produced by some *source* — the pickup on the guitar, for instance, or perhaps an electrode sensing brain-wave activity. In any case, we desire an amplified *output signal* $y(t)$ to drive an energy-consuming *load* such as a loudspeaker or strip-chart recorder.

Figure 4.1–1

(a) Amplification system. (b) Block diagram.

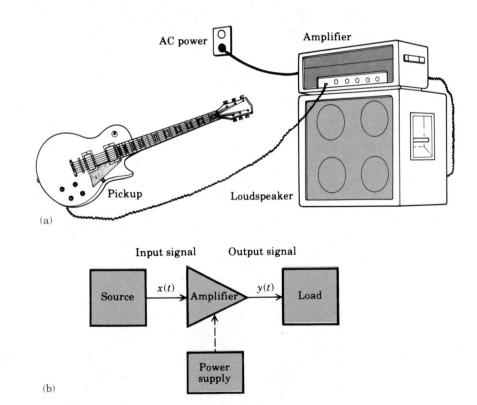

Amplification is achieved with the help of electronic devices (usually transistors) that have the ability to control a large voltage or current in proportion to the smaller input voltage or current. Of course, the large voltage or current must come from something other than the input source, so every amplifier has an associated power supply — a battery or AC-to-DC converter.

We are not concerned here with the internal workings of amplifiers and power supplies. Rather, we focus on the input-output relationship, the corresponding equivalent circuits using controlled sources, and the differences between ideal and practical amplifiers.

Ideal Amplifiers

Mathematically, the action of an **ideal amplifier** is expressed by the simple equation

$$y(t) = Ax(t) \tag{1}$$

where A is a constant called the **amplification.** Since $y(t)/x(t) = A$ at any instant of time, plotting y versus x yields a straight line through the origin with *slope A*. Figure 4.1–2a shows this plot, known as the amplifier's **transfer curve,** along with typical waveforms for the case when $A > 1$. We see that the output signal $y(t)$ has the same relative variations as the input signal $x(t)$, but with bigger excursions in the positive and negative directions. Figure 4.1–2b shows the transfer curve and typical waveforms for a *negative* amplification, $A < -1$. The amplifier still produces an enlarged output signal but with a sign inversion or polarity reversal. Many amplifiers have negative amplification, which may or may not pose a problem in a particular application.

Figure 4.1–2

Transfer curves and illustrative waveforms for an amplifier: (a) $A > 1$; (b) $A < -1$.

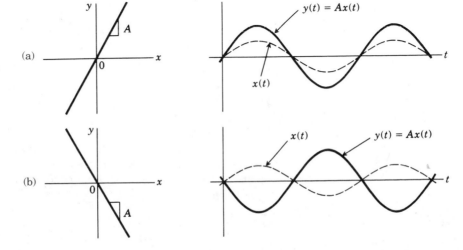

The transfer curve of a *real* amplifier usually approximates a straight line for limited input values, so $y(t) \approx Ax(t)$. But real amplifiers cannot produce output voltage or current beyond certain bounds dictated by the electronic devices and the power supply. Consequently, the transfer curve eventually becomes *nonlinear* for extreme input values. This nonlinearity causes *distortion* of the output signal, and we say that the amplifier is being *overdriven*. Hereafter, unless otherwise indicated, we'll focus on amplifiers operating in their linear region.

Figure 4.1–3a gives the circuit model of an **ideal voltage amplifier.** The amplifier has two pairs of terminals, an input pair to which we connect the source voltage being amplified, and an output pair connected to a load resistance. The input terminals ideally lead to an open circuit, so that $i_{in} = 0$ and $v_{in} = v_s$. Amplification of the input voltage is represented by a voltage-controlled voltage source $v_a = \mu v_{in}$ at the output, resulting in $v_{out} = v_a = \mu v_{in}$. Hence, the source-to-output voltage amplification or **gain** will be

$$A_v = \frac{v_{out}}{v_s} = \frac{\mu v_{in}}{v_{in}} = \mu \tag{2}$$

The symbol μ (mu) traces historically to the voltage amplification factor of a triode vacuum tube.

Figure 4.1–3

(a) Ideal voltage amplifier. (b) Ideal current amplifier.

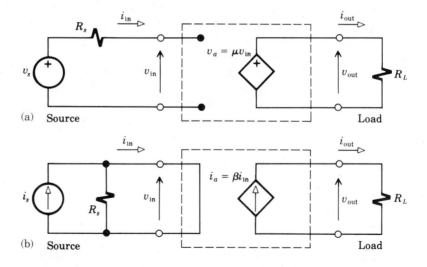

(a)   Source   Load

(b)   Source   Load

The circuit model of an **ideal current amplifier** in Fig. 4.1–3b also has two pairs of terminals. But the input terminals here lead to a short circuit so $v_{in} = 0$ and $i_{in} = i_s$, where i_s is the source current to be amplified. Amplification is then represented by a current-controlled current source $i_a = \beta i_{in}$ at the output, producing $i_{out} = i_a = \beta i_{in}$. Hence, the source-to-output current gain will be

$$A_i = \frac{i_{out}}{i_s} = \frac{\beta i_{in}}{i_{in}} = \beta \tag{3}$$

The symbol β (beta) comes from the current amplification factor of a bipolar junction transistor.

Observe that the performance of these ideal amplifiers is independent of the source and load resistances, R_s and R_L. Hence, the gains A_v and A_i equal the controlled-source amplification factors μ and β, respectively. Moreover, ideal amplifiers draw no power from the signal source since either $i_{in} = 0$ or $v_{in} = 0$. The output power delivered to the load comes entirely from the amplifier's power supply which, by convention, was not included in the circuit model.

Amplifier Models and Analysis

Unlike ideal voltage and current amplifiers, the performance of real amplifiers depends on the source and load circuits. We represent that dependence here by adding resistances to our ideal amplifier models. Then we investigate the implications of the resulting loading effects at input and output.

Figure 4.1–4 gives two possible models for a real amplifier. Both of these models have an **input resistance** R_i that equals the equivalent resistance seen looking into the input terminals, so $R_i = v_{in}/i_{in}$. It is used in the circuit model to represent the fact that real amplifiers always draw some power from the input source. Both circuits also include an **output resistance** R_o to represent internal resistance associated with the controlled source.

Figure 4.1–4
Models of a real amplifier.

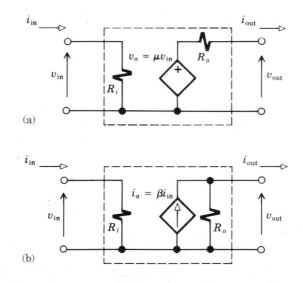

(a)

(b)

The Thévenin output circuit (Fig. 4.1–4a) has R_o in series with the controlled voltage source v_a, and we interpret v_a as the open-circuit output voltage. The Norton output circuit (Fig. 4.1–4b) has R_o in parallel with the controlled current source i_a, which we correspondingly interpret as the short-circuit output current. The wire connecting the lower input and output terminals carries no current but reflects the fact that these two terminals are at the same potential, usually the common ground point, with respect to signal variations.

It must be emphasized that Figs. 4.1–4a and 4.1–4b are different but *equivalent* models for a given practical amplifier. To convert from one to the other we use our previous relationship between the Thévenin and Norton parameters, namely $v_{oc} = R_o i_{sc}$, which becomes $v_a = R_o i_a$ in our present notation. Although these models omit certain other phenomena present in actual amplifiers, they still provide us with valuable approximations for analysis purposes.

The impact of input and output resistance on amplification is brought out by Fig. 4.1–5, where an input voltage source and output load have been connected to the model from Fig. 4.1–4a. Since $v_{in} \neq v_s$ and $v_{out} \neq v_a$ due to loading effect, we must proceed with care when calculating $A_v = v_{out}/v_s$. To this end, we first expand v_{out}/v_s as a "chain" expression

$$A_v = \frac{v_{in}}{v_s} \times \frac{v_a}{v_{in}} \times \frac{v_{out}}{v_a} \tag{4}$$

This expression permits us to examine each of the voltage ratios.

Figure 4.1–5
Amplifier model with source and load.

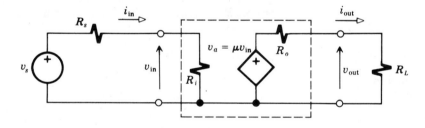

Recalling the voltage-divider relation, the input and output ratios are

$$\frac{v_{in}}{v_s} = \frac{R_i}{R_s + R_i} \qquad \frac{v_{out}}{v_a} = \frac{R_L}{R_o + R_L}$$

and, by definition of the controlled source,

$$\frac{v_a}{v_{in}} = \mu$$

Substituting these into Eq. (4) gives

$$A_v = \left(\frac{R_i}{R_s + R_i}\right) \mu \left(\frac{R_L}{R_o + R_L}\right) \tag{5}$$

Clearly, $|A_v| < |\mu|$ and its actual value depends on all four resistances because of loading at the input and output.

Input loading is negligible if the input resistance is large enough, so $v_{in} \approx v_s$ if $R_i \gg R_s$. Likewise, $v_{out} \approx v_a$ if the amplifier has small output resistance, $R_o \ll R_L$. Therefore, when

$$R_i \gg R_s \qquad R_o \ll R_L \qquad \qquad \textbf{(6a)}$$

we have

$$A_v \approx \mu \qquad \qquad \textbf{(6b)}$$

and a practical amplifier closely approximates an ideal voltage amplifier.

If we are concerned with current amplification, we use Norton circuits for both the source and amplifier and write $A_i = i_{out}/i_s = (i_{in}/i_s)(i_a/i_{in})(i_{out}/i_a)$. The details are left as an exercise.

Example 4.1–1

A certain medical transducer produces the oscillating voltage signal $v_s = 0.5 \cos \omega t$ V. The transducer has $R_s = 2$ kΩ and a maximum current limitation of ± 0.1 mA. We want to amplify v_s for plotting on a strip-chart recorder having a ± 10-V scale and 75-Ω input resistance. We propose to use an amplifier with $R_i = 8$ kΩ, $\mu = -20$, and $R_o = 25$ Ω.

Figure 4.1–6 shows the complete equivalent circuit, including polarity inversion to account for the negative amplification factor. Substituting values into Eq. (5) gives the voltage gain

$$A_v = \frac{8}{2+8} \times (-20) \times \frac{75}{25+75} = 0.8 \times (-20) \times 0.75 = -12$$

Note that input and output loading reduce the gain magnitude by $0.8 \times 0.75 = 0.6$. Hence, $v_{out} = A_v v_s = -6 \cos \omega t$ V, which oscillates over ± 6 V and is appropriate for the recorder's range. The resulting input current is $i_{in} = v_s/(R_s + R_i) = 0.05 \cos \omega t$ mA, and falls within the transducer's limitation.

Figure 4.1–6
Equivalent amplifier circuit for Example 4.1–1.

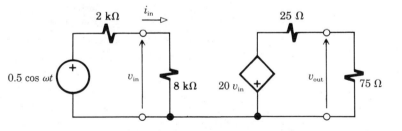

Exercise 4.1–1

By comparing the open-circuit output voltage and short-circuit output current, show that the two models in Fig. 4.1–4 are equivalent if $\mu/\beta = R_o/R_i$.

Exercise 4.1–2

Let a current source i_s with parallel resistance R_s be connected to the input of Fig. 4.1–4b, and let a load resistor R_L be connected at the output. Show that

$$A_i = \frac{i_{\text{out}}}{i_s} = \frac{R_s \beta R_o}{(R_s + R_i)(R_o + R_L)}$$

What are the conditions on the resistances such that $A_i \approx \beta$?

Power Gain and Decibels

In addition to voltage and current amplification, we will be interested in the **power gain** of an amplifier. Referring to Fig. 4.1–7, let P_{in} be the *average input power* to the amplifier from the source. Thus, P_{in} is the average value of the instantaneous power $p_{\text{in}} = v_{\text{in}} i_{\text{in}}$. Similarly, let P_{out} be the *average output power* from the amplifier to the load. The power gain is then defined as the ratio of output to input power,

$$G \triangleq \frac{P_{\text{out}}}{P_{\text{in}}} \tag{7}$$

and, therefore, $P_{\text{out}} = GP_{\text{in}}$. (Note that G defined here has no relationship to the symbol for conductance.)

Figure 4.1–7

Block diagram of power gain.

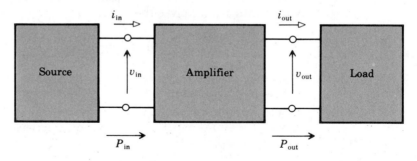

As previously observed, an ideal voltage or current amplifier has $p_{\text{in}} = 0$, so $P_{\text{in}} = 0$, whereas $P_{\text{out}} \neq 0$, meaning that ideal amplifiers have *infinite* power gain. Practical amplifiers always have finite power gain, of course, but the value of G can be very large–so large that a logarithmic measure called the **decibel** (dB) often proves useful. We define the dB power gain to be

$$G_{\text{dB}} \triangleq 10 \log G = 10 \log (P_{\text{out}}/P_{\text{in}}) \tag{8}$$

where log stands for the common logarithm, $\log_{10}$. For example, if a certain amplifier produces $P_{\text{out}} = 5$ W when $P_{\text{in}} = 0.5$ mW, then

$$G = \frac{5\text{ W}}{0.5\text{ mW}} = 10^4 \qquad G_{\text{dB}} = 10 \log 10^4 = 10 \times 4 = 40\text{ dB}$$

We usually drop the dB subscripts when there is no danger of confusion, so writing $G = 40$ dB clearly means that $G_{\text{dB}} = 40$ dB and $G = 10^4$.

Being a logarithmic unit, the decibel converts powers of 10 to products of 10. Hence,

$$G = 10^n \leftrightarrow G_{dB} = 10n \text{ dB}$$

A somewhat strange but logical consequence of this measure is that unit power gain ($G = 1.0 = 10^0$) corresponds to zero dB gain ($G_{dB} = 10 \times 0 = 0$ dB). Furthermore, if $G < 1.0$ then $G_{dB} < 0$ dB. Other dB values are easily computed with the help of a calculator, but you may find it handy to memorize some of the common pairs listed in Table 4.1-1. Conversion from a dB value back to actual power gain is accomplished via

$$\frac{P_{out}}{P_{in}} = G = 10^{0.1 G_{dB}} \tag{9}$$

obtained by inverting Eq. (8).

Table 4.1-1	dB	G	A_v or A_i	dB	G	A_v or A_i
Selected decibel values.	−10	0.1	0.316	3	2	1.414
	−6	0.25	0.5	6	4	2.0
	−3	0.5	0.707	10	10	3.16
	0	1.0	1.0	20	100	10.0

Strictly speaking, the decibel is defined only for power gain or, equivalently, power ratios. Nonetheless, in practice we often express voltage or current gain in decibels, defined by

$$A_{dB} \triangleq 10 \log A^2 = 20 \log |A| \tag{10}$$

where A represents A_v or A_i, as the case may be. This definition is based on the fact that power is proportional to the square of voltage or current. Consequently, power gain is proportional to the square of the voltage or current amplification. Thus, for instance, we can say that A_v equals 40 dB when $A_v = \pm 100 = \pm 10^2$ even if $G \neq 40$ dB. In such cases, a voltage gain of $10n$ dB simply means that the amplification equals $\pm 10^{n/2}$.

Example 4.1-2

Suppose v_s in Fig. 4.1-6 stays constant at +0.5 V, so $i_{in} = 0.05$ mA and $v_{out} = -6$ V. The corresponding signal powers are $P_{in} = i_{in}^2 R_i = 0.02$ mW and $P_{out} = v_{out}^2/R_L = 480$ mW. Thus, the amplifier provides a power gain of $480/0.02 = 24{,}000$ and

$$G_{dB} = 10 \log 24{,}000 = 43.8 \text{ dB}$$

as contrasted with the voltage gain

$$A_{dB} = 10 \log (-12)^2 = 21.6 \text{ dB}$$

In view of the relatively small voltage amplification, most of the power gain comes about by "transferring" the signal voltage from a large input resistance to a much smaller load resistance.

Exercise 4.1–3   A current amplifier having $R_i = 2 \text{ k}\Omega$, $\beta = 100$, and $R_o = 5 \text{ k}\Omega$ is connected between a 50-Ω load and a source with $R_s = 1 \text{ k}\Omega$ and $i_s = 3$ mA. Draw the complete equivalent circuit and label it with all current values. Then calculate A_i and G in dB.

Cascaded Amplifiers

Frequently, we find it necessary to put two or more amplifiers in **cascade** —the output of the first being applied to the input of the second, and so forth. The amplification produced by the cascade is then proportional to the *product* of the individual gains and, as a result, can be much larger than that achieved by a single amplifier.

To demonstrate this cascade multiplication property, consider two cascaded voltage amplifiers with a source and load, as shown in Fig. 4.1–8. Observing that the first output voltage becomes the second input voltage, so, $v_{12} = v_{\text{out}_1} = v_{\text{in}_2}$, the overall voltage amplification is found to be

$$A_v = \frac{v_{\text{out}}}{v_s} = \frac{v_{\text{in}}}{v_s} \times \frac{v_{a1}}{v_{\text{in}}} \times \frac{v_{12}}{v_{a1}} \times \frac{v_{a2}}{v_{12}} \times \frac{v_{\text{out}}}{v_{a2}}$$

$$= \left(\frac{R_{i1}}{R_s + R_{i1}}\right)\mu_1\left(\frac{R_{i2}}{R_{o1} + R_{i2}}\right)\mu_2\left(\frac{R_L}{R_{o2} + R_L}\right)$$

Minimizing loading effects now requires $R_{o1} \ll R_{i2}$, as well as $R_{i1} \gg R_s$ and $R_{o2} \ll R_L$, in which case

$$A_v \approx \mu_1\mu_2$$

We then have the product of the individual amplification factors. Similar expressions hold for two cascaded current amplifiers.

Figure 4.1–8
Cascade of two voltage amplifiers.

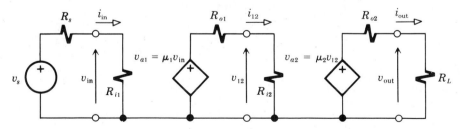

When loading is negligible between cascaded amplifiers and at the input and final output, the individual gains are $A_1 \approx \mu_1$, $A_2 \approx \mu_2$, etc., and n cascaded voltage amplifiers yield

$$A_v \approx A_1 A_2 \cdots A_n \qquad \textbf{(11)}$$

Accordingly, we could get $A_v = 10^4$ from two amplifiers with $A_1 = A_2 = 100$ or from four amplifiers with $A_1 = A_2 = A_3 = A_4 = 10$, and so forth. We can also correct for unwanted polarity inversion by cascading an even number of inverting amplifiers. Thus if $A_1 = -20$ and $A_2 = -5$, then $A_v = (-20) \times (-5) = +100$.

Exercise 4.1–4   Convert Eq. (11) to decibels, thereby demonstrating that the cascade amplification in dB equals the *sum* of the individual dB values.

4.2 OPERATIONAL AMPLIFIERS

Not too long ago, an electrical engineer who needed an amplifier had to sit down and design it completely, using perhaps a dozen or more individual components. Thankfully, all that has been changed by the modern integrated-circuit operational amplifier, commonly called an "op-amp." Now you can buy an op-amp off the shelf and connect two or three elements to it to create a more reliable, less expensive amplifier than one designed from scratch. The versatile op-amp has thus become a major building-block in today's electronic systems, performing a wide variety of signal-processing operations in addition to amplification.

The name "operational amplifier" actually refers to a large family of general-purpose and special-purpose amplifiers having three characteristics: large input resistance, small output resistance, and very large voltage gain. After describing these characteristics, we'll introduce the valuable concept of the ideal op-amp and use it to explain several op-amp circuits. Other applications are explored in later chapters and in the problems at the end of this chapter.

Op-Amp Characteristics

An operational amplifier, symbolized by Fig. 4.2–1a, is a high-gain voltage amplifier that responds to the *difference* between *two input voltages*. The input terminals, labeled $+$ and $-$, are called the **noninverting** and **in-**

Figure 4.2–1

Operational amplifier: (a) symbol; (b) power-supply and load connections.

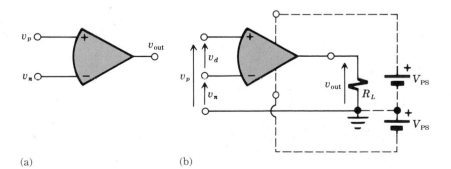

(a)   (b)

verting inputs, respectively. Despite the labels, the applied voltages v_p and v_n may be either polarity with respect to ground. The output v_{out} depends on the **difference voltage**

$$v_d = v_p - v_n$$

which appears across the input terminals. Figure 4.2–1b shows these voltages explicitly and also indicates how the necessary power-supply voltages $+V_{\text{PS}}$ and $-V_{\text{PS}}$ are connected along with the load resistance.

An op-amp's transfer curve, v_{out} versus v_d, ideally has the shape plotted in Fig. 4.2–2. This curve consists of a *linear* middle region bounded by positive and negative *saturation* regions where the output stays pinned at $v_{\text{out}} \approx +V_{\text{PS}}$ or $v_{\text{out}} \approx -V_{\text{PS}}$. The linear region has slope A, so

$$v_{\text{out}} = Av_d = A(v_p - v_n) \tag{1}$$

Accordingly, $v_{\text{out}} = 0$ when $v_p = v_n$. Increasing v_p relative to v_n causes v_{out} to move in the positive direction, whereas increasing v_n causes v_{out} to move in the negative direction—which explains the names "noninverting" and "inverting" for the two input terminals. Aside from the difference-voltage input, the distinctive characteristic of an op-amp is its gigantic gain A, typically 100,000 or more!

Figure 4.2–2
Op-amp transfer curve.

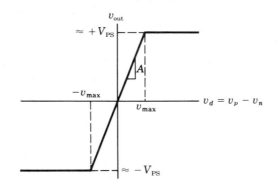

When operating within the linear region, an op-amp behaves more or less like the ideal voltage amplifier diagrammed in Fig. 4.2–3. (The effects of finite input resistance and nonzero output resistance will be considered later.) This idealized circuit model holds only over the output range

$$-V_{\text{PS}} < v_{\text{out}} < +V_{\text{PS}} \tag{2}$$

corresponding to the input range

$$-v_{\text{max}} < v_d < v_{\text{max}} \tag{3a}$$

where

$$v_{\text{max}} \approx V_{\text{PS}}/A \tag{3b}$$

Figure 4.2–3

Simplified op-amp
model for linear
operation.

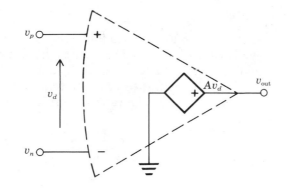

Since the op-amp has gain $A \geq 10^5$, and since the power-supply voltage V_{PS} is usually less than 100 V, it follows from Eq. (3) that linear amplification requires the difference voltage to be restricted such that

$$|v_d| < v_{\max} \approx V_{\mathrm{PS}}/A \leq 0.001 \text{ V}$$

Otherwise, if $|v_d| \geq v_{\max}$, then Fig. 4.2–2 indicates that the output saturates at $v_{\mathrm{out}} \approx \pm V_{\mathrm{PS}}$ and ceases to respond to further changes of v_p or v_n. We then would have extreme nonlinear distortion instead of the linear amplification represented by the model in Fig. 4.2–3.

In view of the restriction on v_d, an op-amp alone is not a very useful amplifier, for it would be driven immediately into saturation by any input voltage v_{in} whose variations exceed about one millivolt. We'll subsequently see that:

> All linear op-amp circuits include a **negative feedback connection,** forming a closed loop from the output terminal back to the inverting terminal.

Properly designed feedback results in a greatly reduced difference voltage, $|v_d| \ll |v_{\mathrm{in}}|$, thereby preventing saturation even though $|v_{\mathrm{in}}| > v_{\max}$.

With the feedback connection in place, the op-amp's input terminals act as a **virtual short** in the sense that $v_d \approx 0$ (like a short circuit) while $i_p = i_n = 0$ (like an open circuit). The concept of a virtual short turns out to be very handy for analyzing and designing linear circuits with op-amps. We'll represent it symbolically by Fig. 4.2–4, where the double-headed arrow denotes the virtual short and implies that $v_p \approx v_n$. Unlike the equivalent circuit in Fig. 4.2–3, this op-amp model does not give an explicit relationship for v_{out} in terms of v_d. Instead, the value of v_{out} must be whatever is needed to satisfy the virtual-short conditions with feedback connected.

The following investigations of specific circuit configurations should clarify the role of feedback and the use of the virtual short. In all cases, we'll assume operation within the linear region where $|v_{\mathrm{out}}| < V_{\mathrm{PS}}$.

Figure 4.2–4

Ideal op-amp with
virtual short.

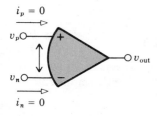

Noninverting Amplifiers

The op-amp circuit in Fig. 4.2–5a functions as a **noninverting voltage amplifier.** The input is applied directly to the noninverting terminal, so $v_{in} = v_p$ and $i_{in} = i_p = 0$. The resistors R_F and R_1 constitute a feedback connection from the output to the inverting terminal. We'll perform a detailed analysis of this configuration to bring out the feedback effects.

Figure 4.2–5
Noninverting amplifier: (a) circuit diagram; (b) simplified equivalent circuit.

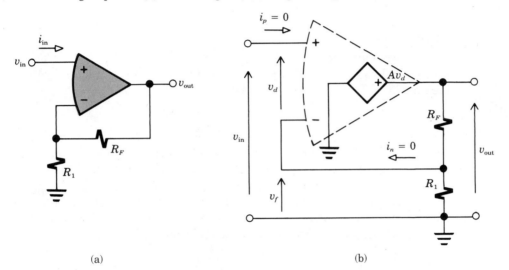

(a)   (b)

Replacing the op-amp symbol with the equivalent circuit from Fig. 4.2–3 gives the diagram in Fig. 4.2–5b. Since $i_n = 0$, the external resistors form a voltage divider producing the **feedback voltage**

$$v_f = Bv_{out} \qquad B = \frac{R_1}{R_F + R_1} \tag{4}$$

Thus, $v_f + v_d = v_{in}$, and the difference voltage becomes

$$v_d = v_{in} - v_f = v_{in} - Bv_{out}$$

where the subtraction signifies *negative* feedback. Substituting $v_{out} = Av_d$ and solving for v_d and v_{out} finally yields

$$v_d = \frac{1}{1 + AB} v_{in} \qquad v_{out} = \frac{A}{1 + AB} v_{in} \tag{5}$$

The product AB appearing here is called the **loop gain,** as distinguished from the **open-loop gain** A of the op-amp alone. The overall amplification v_{out}/v_{in} is known as the **closed-loop gain.**

Most op-amps have sufficiently large gain A that the loop gain satisfies the condition

$$AB \gg 1 \qquad (6)$$

Then $1 + AB \approx AB$ in Eq. (5), and

$$|v_d| \approx \frac{1}{AB} |v_{in}| \ll |v_{in}|$$

which confirms that feedback does, indeed, reduce the difference voltage —as required for linear amplification of v_{in}. Furthermore, the resulting input-output relation simplifies to

$$v_{out} \approx \frac{1}{B} v_{in} = \frac{R_F + R_1}{R_1} v_{in} \qquad (7)$$

Thus, the closed-loop gain depends entirely on the feedback factor B associated with the voltage divider.

Figure 4.2–6 summarizes our results in the form of a circuit model for the noninverting amplifier when $AB \gg 1$. There are no loading effects at input or output, so

$$A_v = \frac{v_{out}}{v_s} = \frac{R_F + R_1}{R_1} v_{in}$$

Accordingly, you can achieve almost any desired A_v by picking appropriate values for R_F and R_1, subject to the constraint that they should not be too small or too large. Resistance values in the range of 1 kΩ to 100 kΩ work well in most cases. If needed, adjustable gain is easily implemented with a potentiometer as the voltage divider. The op-amp's gain is unimportant, provided that $A \gg A_v$ to fulfill the loop-gain condition.

Figure 4.2–6

Circuit model of the noninverting amplifier.

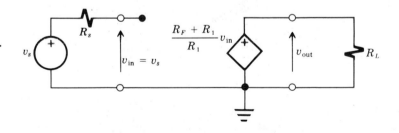

Now let's rework the analysis of the noninverting amplifier using the virtual-short model shown in Fig. 4.2–7. The virtual short requires that

$$v_d = v_{in} - v_f \approx 0$$

so $v_f \approx v_{in}$, where $v_f = B v_{out}$ as before. Thus,

$$v_{out} \approx v_{in}/B$$

Figure 4.2–7

Analysis of a
noninverting amplifier
using the virtual short.

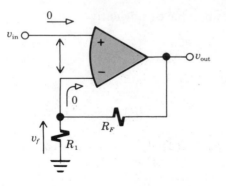

which agrees with Eq. (7). We therefore deduce that the virtual-short
model yields accurate results whenever $AB \gg 1$.

Given that most op-amps have $A \geq 10^5$, we'll hereafter take advantage
of the quicker virtual-short analysis method. We'll also assume that $A \to \infty$
and write $v_d = 0$, rather than $v_d \approx 0$.

Consider, for example, the special type of noninverting amplifier in
Fig. 4.2–8, known as a **voltage follower.** The direct feedback connection
produces $v_f = v_{out}$ and the virtual short forces $v_f = v_{in} - v_d = v_{in}$. Hence,

$$v_{out} = v_{in} \tag{8}$$

which says that the output voltage "follows" the variations of the input
voltage. The same conclusion can be reached starting from Eq. (7) by
letting $R_F \to 0$ and $R_1 \to \infty$ so that Fig. 4.2–5a reduces to Fig. 4.2–8.
Despite the unity voltage gain, voltage followers are capable of large current and power gain. They are used in practice to *buffer* a high-resistance
source and minimize loading effects.

Figure 4.2–8
Voltage follower.

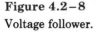

Example 4.2–1

Suppose you have an op-amp with $A = 10^5$ and $V_{PS} = 12$ V, and you want
to amplify an input signal having $|v_{in}(t)| \leq 0.2$ V. Ensuring linear operation
with $|v_{out}| < 12$ V requires that $|v_{out}/v_{in}| < 12/0.2 = 60$, so a closed-loop
gain of 50 would provide an ample safety margin. Consequently, you might
take $R_1 = 1$ kΩ and $R_F = 49$ kΩ in Fig. 4.2–5a to obtain

$$B = \frac{1}{49+1} = \frac{1}{50} \qquad AB = \frac{10^5}{50} = 2{,}000 \gg 1$$

Inserting these values in Eq. (5) gives

$$\frac{v_d}{v_{\text{in}}} = \frac{1}{2{,}001} \approx 0.0005 \qquad \frac{v_{\text{out}}}{v_{\text{in}}} = \frac{10^5}{2{,}001} = 49.975 \approx 50$$

Notice that the difference voltage is only about 0.05% of v_{in} and has a maximum value of $0.0005 \times 0.2 \text{ V} = 0.1 \text{ mV}$, whereas $|v_{\text{out}}|_{\text{max}} \approx 50 \times 0.2 \text{ V} = 10 \text{ V} = 10^5 |v_d|_{\text{max}}$.

Exercise 4.2−1   Use the equivalent circuit from Fig. 4.2−3 to obtain expressions for v_d/v_{in} and $v_{\text{out}}/v_{\text{in}}$ for the voltage follower in Fig. 4.2−8 when the op-amp has finite gain. Evaluate your results taking $A = 10^5$.

Inverting and Summing Amplifiers

The op-amp circuit in Fig. 4.2−9a functions as an **inverting voltage amplifier.** The input v_{in} is applied through R_1 to the inverting terminal, while the noninverting terminal is grounded. The virtual short now provides a *"virtual ground"* with $v_n = 0$, as indicated in Fig. 4.2−9b, and feedback takes the form of the current i_f through R_F. Since $i_{\text{in}} + i_f = i_n = 0$, we must have

$$i_f = -i_{\text{in}}$$

where

$$i_f = \frac{v_{\text{out}} - v_n}{R_F} = \frac{v_{\text{out}}}{R_F} \qquad i_{\text{in}} = \frac{v_{\text{in}} - v_n}{R_1} = \frac{v_{\text{in}}}{R_1}$$

Figure 4.2−9
Inverting amplifier: (a) circuit diagram; (b) analysis using the virtual short; (c) circuit model.

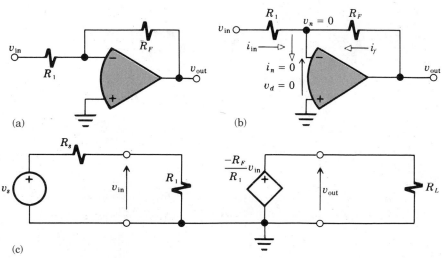

Therefore, $v_{\text{out}}/R_F = -v_{\text{in}}/R_1$ and

$$v_{\text{out}} = -\frac{R_F}{R_1} v_{\text{in}} \qquad \textbf{(9)}$$

This circuit thus *inverts* and *amplifies* the input by the factor R_F/R_1. If $R_F = R_1$, then $v_{\text{out}} = -v_{\text{in}}$ and the circuit becomes a *unity-gain inverter*.

In contrast to the noninverting amplifier, the inverting amplifier draws input current and has the finite input resistance

$$R_i = v_{\text{in}}/i_{\text{in}} = R_1$$

Accordingly, we must give attention to source loading using the circuit model in Fig. 4.2–9c. If the source has open-circuit voltage v_s and internal resistance R_s, then

$$A_v = \frac{v_{\text{out}}}{v_s} = \frac{R_1}{R_s + R_1}\left(-\frac{R_F}{R_1}\right) = -\frac{R_F}{R_s + R_1}$$

Whenever possible, we take $R_1 \gg R_s$ to minimize loading and get $A_v \approx -R_F/R_1$.

Connecting an additional input resistor to the inverting terminal changes the inverting amplifier into the **summing amplifier** shown in Fig. 4.2–10. Routine analysis with the virtual short yields

$$v_{\text{out}} = -\left(\frac{R_F}{R_1} v_1 + \frac{R_F}{R_2} v_2\right) \qquad \textbf{(10a)}$$

$$= -\frac{R_F}{R_1}(v_1 + v_2) \qquad R_2 = R_1 \qquad \textbf{(10b)}$$

Hence, this circuit inverts, amplifies, and sums the two input voltages. The generalization of Eq. (10) for three or more inputs should be obvious.

Figure 4.2–10
Summing amplifier with two inputs.

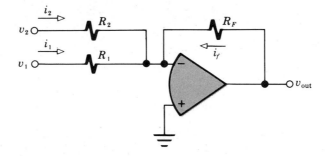

We've now seen how op-amp circuits perform the operations of amplification, inversion, and summation. The next example considers the subtraction operation.

Example 4.2–2

We are given two signal sources generating $|v_a(t)| \le 30$ mV and $|v_b(t)| \le 120$ mV, each having $R_s = 20\ \Omega$. We want to design an op-amp system that produces

$$v_{\text{out}}(t) \approx 200v_a(t) - 25v_b(t)$$

There are a number of different solutions to this problem, but two op-amps seem to be required in view of the subtraction needed to form v_{out}.

One solution is based on rewriting the output as $v_{\text{out}} \approx -25[(-8v_a) + v_b]$. Comparing this expression with Eq. (10b) suggests an inverting summing amplifier with $R_2 = R_1$, $R_F/R_1 = 25$, and inputs $v_1 = -8v_a$ and $v_2 = v_b$. The input $v_1 = -8v_a$ can be obtained by applying v_a to an inverting amplifier with $R_F/R_1 = 8$. Figure 4.2–11 shows the complete system diagram taking $R_1 = 1$ kΩ for both amplifiers so that $R_i = R_1 \gg R_s$.

Figure 4.2–11
Circuit for Example 4.2–2.

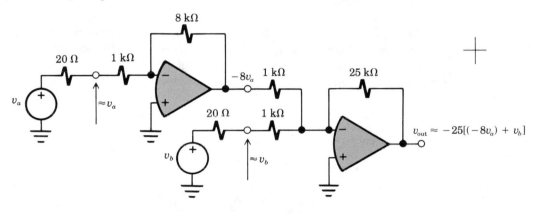

But we should also take account of the maximum output voltage. We compute $|v_{\text{out}}|_{\text{max}}$ by assuming *worst-case conditions*, so v_a hits its positive maximum at the same time that v_b is at its negative maximum, or vice versa. Then

$$|v_{\text{out}}|_{\text{max}} = 200|v_a|_{\text{max}} + 25|v_b|_{\text{max}}$$
$$= 200 \times 30\ \text{mV} + 25 \times 120\ \text{mV} = 9\ \text{V}$$

Therefore, the system must have $V_{\text{PS}} \ge 9$ V to ensure operation within the linear region.

Exercise 4.2–2

Let the circuit in Fig. 4.2–10 have $R_1 = 1$ kΩ and $V_{\text{PS}} = 12$ V. Determine values for R_F and R_2 to get $v_{\text{out}} = -K(3v_1 + v_2)$ where K is as large as possible when $|v_1| \le 0.2$ V and $|v_2| \le 0.6$ V.

Exercise 4.2–3   Show that Fig. 4.2–12 is an *inverting current amplifier* with $A_i = i_{out}/i_s = -(R_F + R_1)/R_1$.

Figure 4.2–12
Inverting current
amplifier.

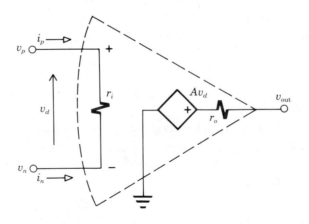

Input and Output Resistance [†]

Up to this point we have ignored the op-amp's internal resistance. A real op-amp always has input resistance $r_i < \infty$ and output resistance $r_o > 0$, which are included in the more accurate circuit model of Fig. 4.2–13. Representative values of these resistances are $r_i \approx 1$ MΩ and $r_o \approx 100$ Ω. However, feedback modifies the equivalent resistance seen at the input and output of a linear op-amp circuit. We'll investigate this feedback effect for the case of the noninverting amplifier. Other configurations can be analyzed in a similar manner.

Figure 4.2–13
Op-amp model with
input and output
resistance.

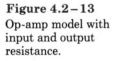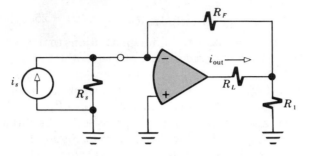

Figure 4.2–14a shows a noninverting amplifier circuit with the op-amp model from Fig. 4.2–13. This arrangement is called a **series-shunt** configuration because the feedback network is connected in series at the input and in parallel or "shunt" at the output. For convenience, we'll assume that R_F draws negligible current i_f and we'll temporarily let $R_L = \infty$ so $i_{out} = 0$. Under these conditions,

$$v_{out} = v_{out\text{-}oc} = Av_d$$

in which A must be interpreted as the op-amp's *open-circuit* voltage gain. The feedback voltage is $v_f = Bv_{out}$, where $B = R_1/(R_F + R_1)$ if $|i_n| \ll |i_f|$.

Figure 4.2–14
Noninverting amplifier with internal op-amp resistance:
(a) equivalent circuit;
(b) circuit model.

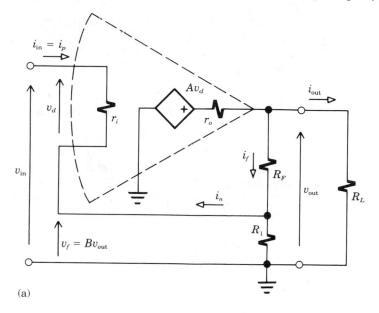

(a)

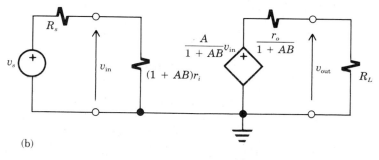

(b)

The input resistance of the amplifier circuit is defined by $R_i = v_{in}/i_{in}$, as distinguished from the op-amp's input resistance r_i. We calculate R_i by noting that $v_d = r_i i_{in}$ and $v_f = Bv_{out} = ABv_d$, so

$$v_{in} = v_d + v_f = r_i i_{in} + ABr_i i_{in} = (1 + AB)r_i i_{in}$$

Therefore,

$$R_i = (1 + AB)r_i \qquad \textbf{(11)}$$

and we see that series-connected feedback at the input *increases* the input resistance.

The output resistance of the amplifier circuit equals the Thévenin resistance seen looking back into the output terminals, which will differ from the op-amp's output resistance r_o. We can easily find the Thévenin resistance by writing $R_o = v_{\text{out-oc}}/i_{\text{out-sc}}$, since we already know that

$$v_{\text{out-oc}} = Av_{\text{in}}/(1 + AB)$$

Setting $R_L = 0$ in Fig. 4.2–14a forces $v_{\text{out}} = 0$ and $v_d = v_{\text{in}} - Bv_{\text{out}} = v_{\text{in}}$, so

$$i_{\text{out-sc}} = Av_d/r_o = Av_{\text{in}}/r_o$$

Hence, upon dividing $v_{\text{out-oc}}$ by $i_{\text{out-sc}}$,

$$R_o = \frac{r_o}{1 + AB} \tag{12}$$

which indicates that feedback connected in shunt across the output *decreases* the output resistance.

The resulting improved circuit model for the noninverting voltage amplifier is diagrammed in Fig. 4.2–14b. Since we usually have $1 + AB \gg 1$, R_i should be large enough and R_o should be small enough to justify ignoring input and output loading, as we did for the simpler model back in Fig. 4.2–6. When $AB = 1000$, for instance, $R_i \approx 1000 \times 1 \text{ M}\Omega = 10^9 \ \Omega$ and $R_o \approx 100 \ \Omega/1000 = 0.1 \ \Omega$. The very large value of R_i also justifies the assumption that $|i_n| \ll |i_f|$ so $B = R_1/(R_F + R_1)$.

PROBLEMS

4.1–1   Obtain an expression for r_m so the controlled source in Fig. 4.1–4a could be labeled $v_a = r_m i_{\text{in}}$.

4.1–2   Obtain an expression for g_m so the controlled source in Fig. 4.1–4b could be labeled $i_a = g_m v_{\text{in}}$.

4.1–3   Consider Fig. 4.1–5 with $R_s = 1 \text{ k}\Omega$, $R_i = 4 \text{ k}\Omega$, $R_o = 100 \ \Omega$, and $R_L = 500 \ \Omega$. Find μ needed to get $A_v = 20$. Then calculate the corresponding values of the current gain $A_i = i_{\text{out}}/(v_s/R_s)$ and the power gain G. Express G in dB.

4.1–4   Consider Fig. 4.1–5 with $R_s = 1 \text{ k}\Omega$, $R_i = 3 \text{ k}\Omega$, $R_o = 4 \ \Omega$, and $R_L = 8 \ \Omega$. Find μ needed to get $A_v = 1$. Then calculate the corresponding values of the current gain $A_i = i_{\text{out}}/(v_s/R_s)$ and the power gain G. Express G in dB.

4.1–5   Let the current amplifier in Fig. 4.1–4b be driven by a source current i_s with resistance $R_s = 300 \ \Omega$, and let $R_L = 1 \text{ k}\Omega$ be connected at the output. Find β needed to get $A_i = 60$ when $R_i = 200 \ \Omega$ and $R_o = 5 \text{ k}\Omega$. Then calculate the corresponding values of the voltage gain $A_v = v_{\text{out}}/(R_s i_s)$ and the power gain G. Express G in dB.

4.1–6   Let the current amplifier in Fig. 4.1–4b be driven by a source current i_s with resistance $R_s = 50 \ \Omega$, and let $R_L = 0.5 \text{ k}\Omega$ be connected at the output. Find β needed to get $A_i = 1$ when $R_i = 10 \ \Omega$ and $R_o = 2.5 \text{ k}\Omega$. Then calculate the corresponding values of the voltage gain $A_v = v_{\text{out}}/(R_s i_s)$ and the power gain G. Express G in dB.

4.1–7   An amplifier with $R_i = 1 \text{ k}\Omega$ provides $G = 33$ dB when connected to a 200-Ω load. What are the values of P_{out} and v_{out} if $v_{\text{in}} = 0.5$ V?

4.1–8 An amplifier with $R_i = 2$ kΩ provides $G = 56$ dB when connected to an 8-Ω load. What are the values of P_{out} and v_{out} if $v_{in} = 0.1$ V?

4.1–9 Obtain a general expression for the power gain in Fig. 4.1–5. Then show that $G = \mu^2 R_s/4R_L$ when the resistances are matched for maximum power transfer at the input and output.

4.1–10 Consider a cascade of two identical current amplifiers like Fig. 4.1–4b with $R_i \ll R_s$, $R_o \gg R_L$, and $R_i = R_L$. Derive approximate expressions for A_i and G.

4.1–11 Let both amplifiers in Fig. 4.1–8 have $R_i = 5$ kΩ, $\mu = -40$, and $R_o = 200$ Ω. What value of R_L yields $A_v = 500$ when $R_s = 3$ kΩ?

4.1–12 Let the first voltage amplifier in Fig. 4.1–8 be replaced by a current amplifier with $R_{i1} = 400$ Ω, $\beta = 25$, and $R_{o1} = 1$ kΩ. Write a chain expression to find the value of μ_2 that yields $A_v = -300$ when $R_{i2} = 1$ kΩ, $R_{o2} = 2$ Ω, $R_s = 100$ Ω, and $R_L = 8$ Ω.

4.1–13 Let the second voltage amplifier in Fig. 4.1–8 be replaced by a current amplifier with $R_{i2} = 400$ Ω, $\beta = -50$, and $R_{o2} = 300$ Ω. Write a chain expression to find the value of μ_1 that yields $A_v = -80$ when $R_{i1} = 20$ kΩ, $R_{o1} = 1$ kΩ, $R_s = 5$ kΩ, and $R_L = 100$ Ω.

4.1–14 Let both voltage amplifiers in Fig. 4.1–8 be replaced by identical current amplifiers with $R_i = 2$ kΩ, $R_o = 3$ kΩ, and current amplification β. Write a chain expression to find the value of β that yields $A_v = 250$ when $R_s = 1$ kΩ and $R_L = 600$ Ω.

4.2–1 Plot v_{out} versus v_{in} for Fig. 4.2–5a when $R_1 = 2$ kΩ, $R_F = 8$ kΩ, and $V_{PS} = 15$ V.

4.2–2 Plot v_{out} versus v_{in} for Fig. 4.2–9a when $R_1 = 2$ kΩ, $R_F = 6$ kΩ, and $V_{PS} = 15$ V.

4.2–3 Derive Eq. (10a) by applying the virtual-short properties to Fig. 4.2–10.

4.2–4 Let a third input v_3 and resistor R_3 be connected at the input in Fig. 4.2–10. Find v_{out} by applying the virtual-short properties.

4.2–5 Obtain an expression for v_{out} in Fig. 4.2–7 when a current source i_s with parallel resistance R_s is connected at the input.

4.2–6 Use a node-voltage equation and the result of Exercise 4.2–3 to find the voltage v_1 across R_1 in Fig. 4.2–12.

4.2–7 Design a system with two ideal op-amps and input v_a that produces $v_{out} = -100v_a$. The input must see $R_i = \infty$, and all resistors must be in the range 1 kΩ–10 kΩ.

4.2–8 Design a system with three ideal op-amps and inputs v_a and v_b that produces $v_{out} = -2v_a - 100v_b$. The two inputs must see $R_i = \infty$, and all resistors must be in the range 1 kΩ–10 kΩ.

4.2–9 Design a system with two ideal op-amps and inputs v_a, v_b, and v_c that produces $v_{out} = 50v_a + 50v_b - 5v_c$. The three inputs must see $R_i = 2$ kΩ, and all resistors must be in the range 1 kΩ–10 kΩ.

4.2–10 Design a system with two ideal op-amps and inputs v_a, v_b, and v_c that produces $v_{out} = 8v_a - 5v_b - 5v_c$. The three inputs must see $R_i = 2$ kΩ, and all resistors must be in the range 1 kΩ–10 kΩ.

4.2–11 Figure P4.2–11 is a **negative-resistance converter**. Show that $v_{in}/i_{in} = -R_L$.

4.2–12 Figure P4.2–12 is an **inverting/noninverting circuit**. Find v_{out} and show that adjusting the potentiometer changes the output from $v_{out} = -v_{in}$ when $\alpha = 0$ to $v_{out} = +v_{in}$ when $\alpha = 1$.

Figure P4.2–11

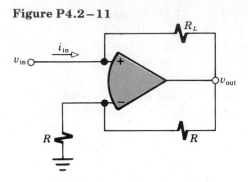

Figure P4.2–12

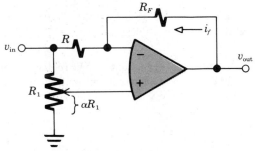

4.2–13   Figure P4.2–13 is a **noninverting summing amplifier** with two inputs. Use superposition to find v_p and v_{out}. Then show that your result reduces to $v_{\text{out}} = (R_2 v_1 + R_1 v_2)/R$ when $R_1 + R_2 = R_F + R$.

Figure P4.2–13

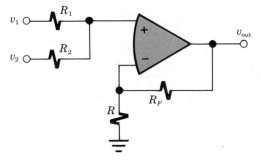

4.2–14   Let a third input v_3 and resistor R_3 be connected at the input in Fig. P4.2–13. Use superposition to find v_p and v_{out}. Then show that your result reduces to $v_{\text{out}} = (R_F + R)(v_1 + v_2 + v_3)/3R$ when $R_1 = R_2 = R_3$.

4.2–15   The circuit in Fig. P4.2–15 becomes a **difference amplifier** when R_4 is removed. Obtain an expression for v_{out}, and show that $v_{\text{out}} = K(v_2 - v_1)$ when $R_F = KR_1$ and $R_3 = KR_2$.

4.2–16   Let R_3 be removed from Fig. P4.2–15, and let $R_1 = R_2 = R_4 = R$. Find v_{out} when

$v_2 = 0$ and when $v_1 = 0$. Then apply superposition to obtain a general expression for v_{out}.

Figure P4.2–15

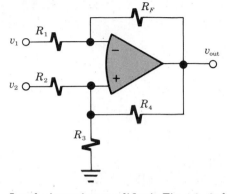

4.2–17   Let the inverting amplifier in Fig. 4.2–9a be built with an op-amp having finite gain A. Use the model from Fig. 4.2–3 to find $v_{\text{out}}/v_{\text{in}}$ and $i_{\text{in}}/v_{\text{in}}$.

4.2–18   Real op-amps may have an **offset voltage** v_{os} and **bias currents** i_{bp} and i_{bn} represented by the model in Fig. P4.2–18. Use this model to show that if $v_{\text{in}} = 0$ and $i_{bp} = i_{bn} = i_b$, then the resulting output of an inverting or noninverting amplifier is $v_{\text{out}} = (1 + R_F/R_1)v_{os} + R_F i_b$.

Figure P4.2–18

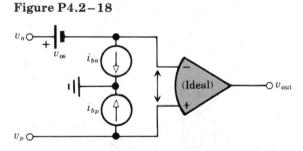

4.2–19   A **compensating resistance** R_c is often put in series with the noninverting terminal of an op-amp to reduce the effect of the bias currents i_{bp} and i_{bn} in Fig. P4.2–18. Using this model with $v_{os} = 0$, obtain an expression for v_{out} when an inverting or noninverting amplifier includes R_c and has $v_{\text{in}} = 0$. Then show that $v_{\text{out}} = R_F(i_{bn} - i_{bp})$ if $R_c = R_1 \| R_F$.

Chapter

5 Capacitance, Inductance, and Impedance

This chapter introduces the two *energy-storage* circuit elements, *capacitance* and *inductance.* These elements possess the distinctive ability to absorb energy in a circuit (acting somewhat like a sink), store the energy temporarily, and return the stored energy to the circuit (acting somewhat like a source).

Circuits containing such elements have electrical *memory* in the sense that energy stored at an earlier time may contribute to the present value of a voltage or current. Consequently, these circuits behave in a markedly different way compared to the behavior of "memory-less" resistive circuits — especially when time-varying sources are involved. The term *dynamic response* is used to describe the behavior of an energy-storage circuit driven by a time-varying source.

We begin with the defining relationships and properties of capacitance and inductance. Then we'll formulate and briefly investigate some differential circuit equations that govern dynamic response. This investigation leads to the important concept of *impedance,* which allows us in many practical situations to treat energy-storage elements in a manner similar to the way we treat resistance.

Objectives

After studying this chapter and working the exercises, you should be able to do each of the following:

- Write expressions for the voltage-current relationships and for the energy stored by capacitance and inductance (Section 5.1).

- Analyze a circuit with stored energy under DC steady-state conditions (Section 5.1).

- Calculate equivalent capacitance and inductance (Section 5.1).

- Distinguish between the natural response, forced response, and complete response of a dynamic circuit (Section 5.2).
- Write impedance expressions for individual circuit elements, and state the conditions under which the impedance concept is applicable (Section 5.3).
- Find the equivalent impedance or admittance of a simple one-port network and the transfer function of a two-port network (Section 5.3).

5.1 CAPACITANCE AND INDUCTANCE

Capacitors are circuit elements that store energy in an electric field, while inductors store energy in a magnetic field. Their characteristics are presented here in preparation for the study of dynamic circuits.

Capacitors and Capacitance

Figure 5.1–1a depicts the essential parts of a **capacitor.** It consists of two metal plates separated by insulating material called a **dielectric.** A source connected across the capacitor will establish an electric field between the plates, which stores energy drawn from the source. The insulating dielectric prevents current flow when the applied voltage is *constant* (DC), but a *time-varying* voltage produces a current proportional to the rate of voltage change, namely

Figure 5.1–1
(a) Capacitor.
(b) Circuit symbol.

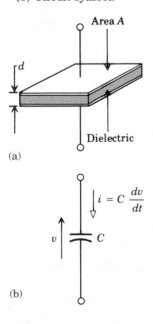

$$i = C \frac{dv}{dt} \tag{1}$$

The proportionality constant C is the **capacitance** measured in **farads** (F), having the unit equation

$$1\ F = \frac{1\ A}{1\ V/s} = 1\ A \cdot s/V$$

since the units of dv/dt are volts per second (V/s). Figure 5.1–1b gives the circuit symbol for capacitance.

Actually, the farad turns out to be a huge quantity, and practical capacitors have values more conveniently expressed in *microfarads* or even *picofarads* ($1\ \mu F = 10^{-6}$ F and $1\ pF = 10^{-12}$ F). For example, the capacitance of the parallel-plate structure in Fig. 5.1–1a is

$$C = \frac{\epsilon A}{d} \tag{2}$$

where A is the area of the plates, d their separation, and ϵ the **permittivity** of the dielectric. A typical dielectric such as mica has $\epsilon \approx 5 \times 10^{-11}$ F/m, so one-meter by one-meter plates, with one millimeter spacing, corresponds to $C \approx 5 \times 10^{-11} \times 1/10^{-3} = 5 \times 10^{-8}$, or only 0.05 μF.

Commonly available capacitors range from a few picofarads to a few thousand microfarads. The larger values are achieved by rolling layers of aluminum foil and dielectric film into a tubular structure. **Electrolytic capacitors** employ a polarized dielectric that restricts operation to relatively low voltages of one polarity. Variable capacitors are easily built, by using movable metal plates with air as the dielectric, but their capacitance values are relatively small due to the wider spacing and the lower permittivity $\epsilon_0 = 10^{-9}/36\pi$ F/m. Capacitance also occurs naturally between any two conducting surfaces. When conductors come into close proximity, like wires in a cable or the leads of an electronic device, the resulting capacitance may be large enough to influence circuit behavior. This is called **stray** or **parasitic capacitance,** and it sometimes causes significant problems.

To obtain the voltage-current relationship for capacitance, we rewrite Eq. (1) as

$$dv = \frac{1}{C} i \, dt$$

Integrating both sides from $t = -\infty$ to an arbitrary time instant t then yields

$$v(t) = \frac{1}{C} \int_{-\infty}^{t} i \, dt \qquad \textbf{(3)}$$

The notation $v(t)$ emphasizes that v is a function of time t. Thus, the voltage across a capacitor depends on the entire past history of the current through it, from $t = -\infty$ to the present. In other words, a capacitor "remembers" past values of current. This *memory* capability of capacitance is directly related to its stored energy, as we'll soon see.

Although we can never know the entire past history of a capacitor, we often do know the voltage at some initial time, say $t = t_0$. To obtain a more useful expression for v given i for $t \geq t_0$, let the *initial voltage* be

$$V_0 = v(t_0) = \frac{1}{C} \int_{-\infty}^{t_0} i \, dt$$

Breaking the integration in Eq. (3) into two parts then gives

$$v(t) = V_0 + \frac{1}{C} \int_{t_0}^{t} i \, dt \qquad t \geq t_0 \qquad \textbf{(4)}$$

The second term here represents the change in voltage after the initial time.

To illustrate Eq. (4), let $v(0) = V_0$ and let a constant current $i = I$ be applied at $t = 0$. Then

$$v(t) = V_0 + \frac{I}{C} t \qquad t \geq 0 \qquad \textbf{(5)}$$

Figure 5.1–2

Constant current and voltage ramp.

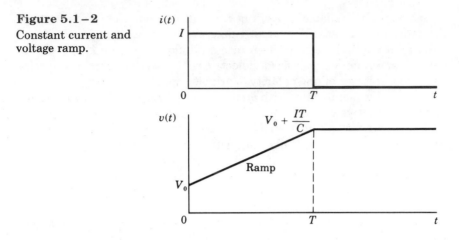

which means that a constant current yields a linearly increasing voltage or *ramp* waveform, like Fig. 5.1–2. If the current stops at time $t = T$, as it does in the figure, the voltage stops increasing but stays constant at $v(t) = V_0 + IT/C$ for $t \geq T$—even though $i = 0$ for $t > T$.

We can further interpret the capacitive memory effect by recalling that $i = dq/dt$, so the charge q stored by a capacitor at any instant of time is

$$q = \int_{-\infty}^{t} i\, dt$$

Thus, rewriting Eq. (3) in terms of q we have

$$q = Cv \qquad (6)$$

which tells us that a one-farad capacitor stores one coulomb of charge per volt. It also tells us that the initial voltage V_0 in Eq. (4) corresponds to an initial charge $q_0 = CV_0$. Therefore, capacitive memory takes the form of *stored charge.*

Now suppose a source is momentarily applied across a capacitor. This produces a charge $+q = Cv$ on the positive terminal plate and an equal but opposite induced charge $-q$ on the other plate (see Fig. 5.1–3). The *work* done by the source to charge the capacitor is

Figure 5.1–3

Charge storage in a capacitor.

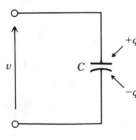

$$w = \frac{1}{2C} q^2 = \tfrac{1}{2} C v^2 \qquad (7)$$

This result follows from Eq. (6), the definition $v = dw/dq$, and the fact that $q\, dq = d(\tfrac{1}{2}q^2)$. Thus,

$$dw = v\, dq = \frac{q}{C}\, dq = \frac{1}{C}\, d(\tfrac{1}{2}q^2)$$

so $w = q^2/2C$ and letting $q = Cv$ gives the second expression in Eq. (7). The work done by the source also must equal the *energy stored* by the capacitor, for there are no losses here. To show that Eq. (7) *is* the stored energy, we consider the instantaneous power $p = dw/dt = vi$. Since $i = C\, dv/dt$,

$$dw = p\, dt = vi\, dt = vC\frac{dv}{dt}\, dt = Cv\, dv = C\, d(\tfrac{1}{2}v^2)$$

and therefore the stored energy is $w = \tfrac{1}{2}Cv^2$ as expected.

After an ideal capacitor has been charged and the source removed, a voltage $v = q/C$ remains across the terminals because there is no conducting path for the charge. This charge is stored indefinitely until an energy-consuming device is connected across the terminals. Capacitors are therefore useful in applications such as spot welders, electronic flash lamps, and pulsed lasers requiring large bursts of energy that can be built-up and stored during the relatively long period between bursts.

Ideal capacitors are said to be **lossless** in the sense that there is no internal resistance present to provide a conduction path between the plates. Real capacitors, however, always have some **leakage** that acts like a large parallel resistance through which the capacitor gradually discharges itself. Discharge time for a good quality capacitor may be hours or days.

Example 5.1–1

Various **waveform generators** take advantage of the voltage ramp produced across a capacitor by a constant current. As an example of this, the *square-wave* current $i(t)$ plotted in Fig. 5.1–4a generates a *triangle-wave* voltage $v(t)$, Fig. 5.1–4b. We analyze such waveforms in a piecewise fashion, separately considering each successive time interval defined by the points at which $i(t)$ changes value.

Assuming that the capacitor is initially uncharged at $t = 0$, Eq. (5) becomes $v(t) = It/C$ over the first interval $0 \le t \le T$, and $v(T) = IT/C$ at $t = T$. For the second interval we use Eq. (4) with $i = -I$, $t_0 = T$, and the new initial value $V_0 = v(T)$, so

$$v(t) = \frac{IT}{C} + \frac{1}{C}\int_T^t (-I)\, dt = \frac{IT}{C} - \frac{I}{C}(t - T) \qquad T \le t \le 2T$$

This expression reveals that $v(t)$ decreases linearly from $v(T)$ with slope $dv/dt = i/C = -I/C$. The third interval is just like the first, since $v(t) = 0$ at $t = 2T$ when the current changes again to $i = I$. Hence, the waveform repeats itself periodically thereafter as shown.

The resulting variation of energy stored by the capacitor is $w(t) = \tfrac{1}{2}Cv^2(t)$, as plotted in Fig. 5.1–4c. Multiplying the voltage and current waveforms gives the instantaneous power $p(t) = v(t)i(t)$ in Fig. 5.1–4d. Observe that $w(t)$ decreases when $p(t) < 0$, meaning that the capacitor is discharging and returning stored energy to the current source. Also note that the average value of $p(t)$ will be zero, reflecting the fact that capacitance does not dissipate power.

Figure 5.1–4
Waveforms for
Example 5.1–1.

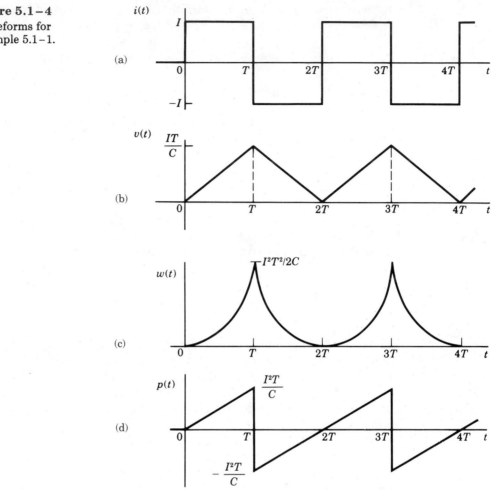

Example 5.1–2
Op-amp integrator.

The integral sign in Eq. (3) suggests that capacitors might be used to construct **integrators** for time-varying signals. Figure 5.1–5 illustrates one possible implementation involving an operational amplifier with capacitance in the feedback path. In view of the virtual short at the input of an op-amp,

$$i_C = i_R = \frac{v_1}{R}$$

and

$$v_2 = -v_C = -\frac{1}{C}\int_{-\infty}^{t} i_C\, dt$$

Figure 5.1–5
Op-amp integrator.

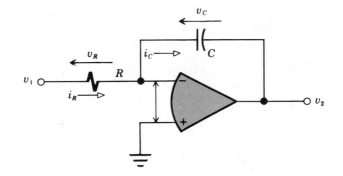

Substituting for i_C then yields

$$v_2 = -\frac{1}{RC}\int_{-\infty}^{t} v_1\, dt \tag{8}$$

so the output voltage v_2 is proportional to the integral of the input v_1. This op-amp integrator has applications in signal-processing systems and as the building block of **analog computers.**

Exercise 5.1–1

The waveform in Fig. 5.1–6 is of the type used to drive the horizontal sweep in a TV set. Note that the time axis is in *microseconds* (μs). Plot the current $i(t)$ that will produce $v(t)$ across $C = 0.02\ \mu$F. Also sketch and label $w(t)$ and $p(t)$.

Figure 5.1–6
TV sweep waveform.

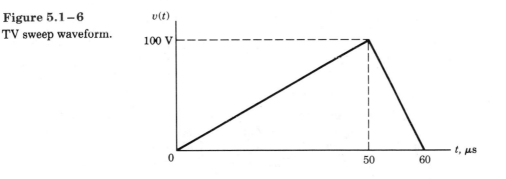

Inductors and Inductance

An **ideal inductor** consists of a length of lossless (resistanceless) wire wound into a coil, like Fig. 5.1–7a. Energy is stored in the magnetic field established around the coil when a current flows through it. A constant current produces no voltage drop across the inductor, but a time-varying current produces the voltage

$$v = L\frac{di}{dt} \tag{9}$$

Figure 5.1 – 7
(a) Inductor. (b) Circuit symbol.

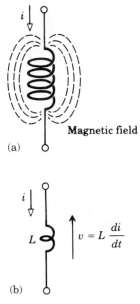

Magnetic field

(a)

(b)

$$v = L \frac{di}{dt}$$

with L being the **inductance** measured in **henrys** (H). The unit equation for inductance is

$$1 \text{ H} = 1 \text{ V} \cdot \text{s/A}$$

Figure 5.1 – 7b is the circuit symbol.

Later in Chapter 17 we will show that a cylindrical coil with length ℓ, area A, and N turns of wire has

$$L \approx \mu \frac{N^2 A}{\ell}$$

where μ is the **permeability** of the core upon which the coil is wound. Practical values of inductance range from microhenrys to a few henrys, the larger values requiring hundreds of turns and a high-μ magnetic core material such as iron. The magnetic core is made movable relative to the coil for a variable inductance. Every circuit also has small amounts of stray inductance due to the fact that current through *any* conductor, even a straight wire, creates a magnetic field around it.

Close examination of the v-i relation in Eq. (9) shows that inductance is the "opposite" of capacitance in that $v = L \ di/dt$ becomes $i = C \ dv/dt$ if i and v are interchanged, and C and L are interchanged. This interchange relationship is called **duality,** and it holds for all the capacitance equations we have previously derived. Thus, replacing C and v by their *duals* (L and i) in $w = \frac{1}{2}Cv^2$, we have the energy stored by an inductor

$$w = \tfrac{1}{2}Li^2 \tag{10}$$

Likewise, the dual of the capacitance voltage integral is

$$i(t) = \frac{1}{L} \int_{-\infty}^{t} v \ dt \tag{11a}$$

$$= I_0 + \frac{1}{L} \int_{t_0}^{t} v \ dt \qquad t \geq t_0 \tag{11b}$$

where I_0 is the initial current. We see from Eq. (11) that inductance "remembers" past values of the applied voltage. We also see that a constant voltage $v = V$ applied at $t = 0$ produces a linearly increasing current

$$i(t) = I_0 + \frac{V}{L} t \qquad t \geq 0$$

Thus, we could relabel Fig. 5.1 – 2 using duality.

Despite the dualism between inductance and capacitance, a real inductor does not hold stored energy as well as a real capacitor. Doing so would require sustaining the current i through a short-circuit across the inductor's terminals, and a real inductor always has some winding resistance in the coil that dissipates the energy rather quickly. Capacitors, therefore, are used almost exclusively for storing energy for any appreciable duration.

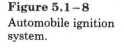

Figure 5.1–8
Automobile ignition
system.

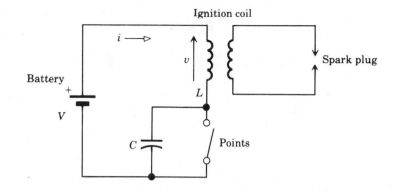

Example 5.1–3
Automobile ignition.

One commonplace application of inductive energy storage is an automobile ignition system, diagrammed in simplified form in Fig. 5.1–8. The switch (or "points") is closed initially to establish a current $i = Vt/L$ through the inductance associated with one winding of a two-winding device called the ignition coil. Thus, after T seconds have elapsed, there will be energy $w = \frac{1}{2}Li^2 = V^2T^2/2L$ stored in the magnetic field.

When the switch is opened, the current rapidly drops to zero and produces a large voltage spike, $v = L\, di/dt$. Magnetic coupling then produces an even larger spike across the other winding. We have now obtained an electrical arc through the gap of the spark plug as the stored energy becomes liberated. A capacitor (known as the "condenser") provides a current path to minimize arcing across the points that would damage the contacts.

Exercise 5.1–2

Confirm the duality result in Eq. (10) by finding w directly from $p = vi = dw/dt$ with $v = L\, di/dt$.

Exercise 5.1–3

A 100-mH inductor is in series with a 250-μF capacitor and an AC source. The voltage across the capacitor is $v_C = 80 \sin 100t$ V. Find the current i, the voltage v_L across the inductor, and the total voltage v across the two elements.

DC Steady-State Behavior

Duality applies, as well, to DC behavior, because an inductor has no voltage drop when the current is constant, whereas a capacitor has no current flow when the voltage is constant. Inductance thus acts like a DC *short circuit*, and capacitance acts like an *open circuit* or DC block. Accordingly, when all voltages and currents are constant in a circuit containing energy-storage elements, we can analyze it by making the following mental replacements:

Capacitance → Open circuit

Inductance → Short circuit

Figure 5.1–9
Circuit in DC steady-state condition.

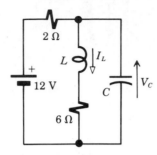

In Fig. 5.1–9, for instance, we easily find that $I_L = 12\text{ V}/8\ \Omega = 1.5$ A and $V_C = (6\ \Omega/8\ \Omega) \times 12$ V $= 9$ V.

This analysis, however, holds only when all voltages and currents are constant and, hence, all time derivatives equal zero. When such is the case, we say that the circuit is in the DC *steady-state* condition. In the next section we'll develop the tools needed to determine when a circuit has reached the steady state.

Exercise 5.1–4

Suppose the battery in Fig. 5.1–9 is replaced by a DC current source with $I_s = 30$ A and $R_s = 16\ \Omega$. Find I_L and V_C, and calculate the total stored energy when $L = 2$ mH and $C = 50\ \mu$F.

Equivalent Inductance and Capacitance

Although the *i-v* relationships for inductance and capacitance look quite different from Ohm's law, they are nonetheless *linear* relationships. Hence, superposition applies to networks containing inductance and capacitance.

The linear property of inductance is established simply by letting $i = i' + i''$ in $v = L\,di/dt$. Then

Figure 5.1–10
Inductors in series.

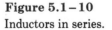

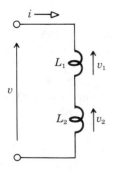

$$v = L\frac{d(i' + i'')}{dt} = L\left(\frac{di'}{dt} + \frac{di''}{dt}\right) = L\frac{di'}{dt} + L\frac{di''}{dt} = v' + v''$$

so the total voltage equals the sum of the voltages produced by the two current components. Proof of linearity for capacitance likewise follows by letting $v = v' + v''$ in $i = C\,dv/dt$.

A direct consequence of linearity is that inductors or capacitors connected in series or parallel have the same effect as one *equivalent* inductance or capacitance. For instance, referring to the series-connected inductors in Fig. 5.1–10, we see that

$$v = v_1 + v_2 = L_1\frac{di}{dt} + L_2\frac{di}{dt} = (L_1 + L_2)\frac{di}{dt}$$

Thus, the equivalent value for series inductance is just the sum

$$L_{eq} = L_1 + L_2 \qquad (12)$$

Similarly, for two inductors connected in parallel, the parallel equivalent inductance is found from

$$\frac{1}{L_{eq}} = \frac{1}{L_1} + \frac{1}{L_2} \qquad (13)$$

Note that equivalent inductance obeys exactly the same rules as equivalent resistance, so these equations can immediately be generalized for three or more inductors.

However, Eqs. (12) and (13) become invalid if interaction exists between the magnetic fields of the inductors. When magnetic fields are in close proximity, their interaction produces a coupling effect known as **mutual inductance.** Mutual inductance is the physical phenomenon underlying the action of a transformer, and it will be discussed further in Chapter 17.

Meanwhile, consider the two parallel-connected capacitors in Fig. 5.1–11, where

$$i = i_1 + i_2 = C_1 \frac{dv}{dt} + C_2 \frac{dv}{dt} = (C_1 + C_2) \frac{dv}{dt}$$

Hence, the equivalent capacitance value is

$$C_{eq} = C_1 + C_2 \qquad (14)$$

Since parallel capacitance adds, like parallel *conductance* rather than resistance, parallel capacitors are often used to build up a large capacitance "bank" for energy storage. For two capacitors connected in series, the equivalent capacitance is found from

$$\frac{1}{C_{eq}} = \frac{1}{C_1} + \frac{1}{C_2} \qquad (15)$$

The extension of Eqs. (14) and (15) for three or more capacitors should be obvious.

Figure 5.1–11
Capacitors in parallel.

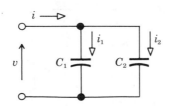

5.2 DYNAMIC CIRCUITS

When a circuit contains resistance plus one or more energy-storage elements, its voltage-current relationship takes the form of a *differential equation* due to the presence of the derivative terms $C\,dv/dt$ or $L\,di/dt$. Such circuits are said to be **dynamic** because the voltage and current waveforms generally differ from the source waveform, in contrast with a resistive circuit whose waveforms always have the same shape.

This section starts with the formulation of differential equations for dynamic circuits. We then briefly consider the task of solving simple differential equations. But our purpose here is not a comprehensive study of differential equations. Instead, by investigating dynamic circuit behavior, we will identify some important characteristics of the *natural response, forced response, steady-state response,* and *transient response.* These response classifications establish a framework for the more efficient analysis methods developed subsequently.

Differential Circuit Equations

Consider a series RL circuit driven by a time-varying voltage $v(t)$, as diagrammed in Fig. 5.2–1. To find the current $i(t)$, we use Kirchhoff's voltage law

$$v_L + v_R = v$$

together with the element relations

$$v_L = L\frac{di}{dt} \qquad v_R = Ri$$

Here the time dependence is implied but not explicitly written, so v_L stands for $v_L(t)$, etc. Inserting the element equations into KVL yields

$$L\frac{di}{dt} + Ri = v \tag{1}$$

which gives us an indirect rather than a direct relationship between the driving voltage $v(t)$ and the resulting current $i(t)$.

Figure 5.2–1
Series RL circuit.

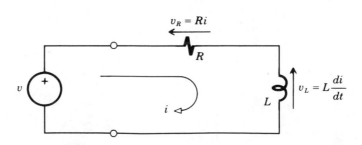

Equation (1) is called a first-order inhomogeneous differential equation — quite a mouthful! It is a **differential** equation because it includes both the unknown current i and its derivative di/dt. The designation **first-order** means that no higher derivatives of the unknown are present, while **inhomogeneous** refers to the *nonzero forcing function* $v = v(t)$ on the right-hand side. It is also a linear equation in the usual sense that superposition applies. Equations such as this occur time and again in conjunction with dynamic circuits.

As a case in point, the voltage across the parallel RC circuit in Fig. 5.2–2a is related to the source current via

$$C\frac{dv}{dt} + \frac{1}{R}v = i \tag{2}$$

This equation has the same mathematical form as Eq. (1) with v being the unknown and i the forcing function. You can easily derive Eq. (2) from KCL and the element equations. Incidentally, the parallel RC circuit may be viewed as the *dual* of the series RL circuit by taking conductance $G = 1/R$ as the dual of resistance and "parallel-connected" as the dual of "series-connected," in addition to the L-C and v-i duality.

Figure 5.2–2
Parallel and series RC circuits.

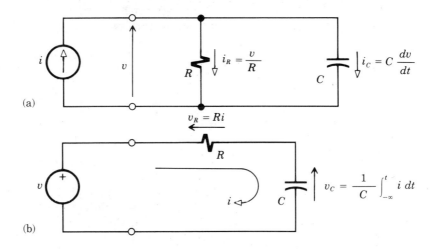

For another example, look at the series RC circuit in Fig. 5.2–2b. Direct application of KVL gives the *integral equation*

$$Ri + \frac{1}{C}\int_{-\infty}^{t} i\, dt = v$$

We convert this to a differential equation by differentiating both sides with respect to time and using the property that

$$\frac{d}{dt}\frac{1}{C}\int_{-\infty}^{t} i\, dt = \frac{1}{C}\frac{d}{dt}\int_{-\infty}^{t} i\, dt = \frac{1}{C}i$$

Thus,

$$R\frac{di}{dt} + \frac{1}{C}i = \frac{dv}{dt} \tag{3}$$

which again has the same form as Eq. (1) with dv/dt as the forcing function.

We summarize the foregoing results by letting $y(t)$ stand for the unknown voltage or current and letting $F(t)$ stand for the forcing function. Then *any* circuit consisting of resistance and one energy-storage element is represented by a differential equation having the generic form

$$a_1\frac{dy}{dt} + a_0 y = F(t) \tag{4}$$

where a_1 and a_0 are constants representing element values. It must be emphasized that such an equation cannot be solved by algebraic manipulations. For instance, if you rewrite Eq. (4) as

$$y = \frac{1}{a_0}F(t) - \frac{a_1}{a_0}\frac{dy}{dt}$$

then you get no closer to the solution for y because its unknown derivative dy/dt now appears on the right-hand side. We'll come back to this problem after considering a circuit with two energy-storage elements.

Figure 5.2–3 diagrams a series *LRC* circuit along with its element voltages. Applying KVL directly yields the *integral-differential equation*

$$L\frac{di}{dt} + Ri + \frac{1}{C}\int_{-\infty}^{t} i\,dt = v$$

After differentiating both sides we get

$$L\frac{d^2i}{dt^2} + R\frac{di}{dt} + \frac{1}{C}i = \frac{dv}{dt} \tag{5}$$

where d^2i/dt^2 is the derivative of di/dt or the *second derivative* of i. Equation (5) is called a *second-order* differential equation and we refer to Fig. 5.2–3 as a **second-order circuit,** as distinguished from the previous *first-order circuits* that led to first-order differential equations.

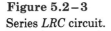

Figure 5.2–3
Series *LRC* circuit.

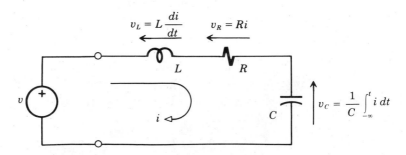

Second-order equations arise whenever a circuit has two energy-storage elements that cannot be combined into one equivalent inductance or capacitance. Any second-order circuit is described by a differential equation in the form

$$a_2 \frac{d^2 y}{dt^2} + a_1 \frac{dy}{dt} + a_0 y = F(t) \tag{6}$$

which is the logical extension of Eq. (4). And, as you might guess, the differential equation of a circuit with n energy-storage elements will generally include all derivatives of the unknown up to the nth-derivative. The circuit would therefore be called an *nth-order circuit.*

Exercise 5.2–1 Obtain a differential equation for the circuit in Fig. 5.2–2b with v_C as the unknown.

Natural Response

The task of solving a differential circuit equation is best accomplished by dealing first with the special case known as the **free** or **natural response.** This term refers to the behavior of the voltages or currents in a dynamic circuit without an applied source, so the *forcing function equals zero.* Lacking an external source, the natural response is driven entirely by *initial stored energy* within the circuit.

Consider the parallel *RC* circuit with zero input current shown in Fig. 5.2–4. If the capacitor has been initially charged to voltage V_0 at $t = 0$, then a current i_R will flow through R for $t \geq 0$ as the capacitor discharges its stored energy. The behavior of the natural-response voltage, denoted $v_N(t)$, obeys the differential equation

$$C \frac{dv_N}{dt} + \frac{1}{R} v_N = 0 \tag{7}$$

which is obtained from Eq. (2) with $i = 0$ and $v = v_N$. Equation (7) is a **homogeneous differential equation,** and mathematicians call $v_N(t)$ the **complementary solution** of the related inhomogeneous equation.

To find the natural response voltage, we rewrite Eq. (7) as

$$\frac{dv_N}{dt} = -\frac{1}{RC} v_N \tag{8}$$

which reveals that $v_N(t)$ must be *proportional to its own derivative.* The only time function possessing this property is the *exponential* waveform Ke^{pt}, where $e = 2.718\ldots$ is the natural logarithm base and K and p are constants. Indeed, if we assume that

$$v_N(t) = Ke^{pt}$$

Figure 5.2–4

Natural response of an *RC* circuit.

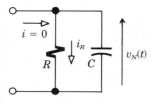

then

$$\frac{dv_N}{dt} = K\frac{d}{dt}(e^{pt}) = Kpe^{pt} = p(Ke^{pt}) = pv_N$$

which agrees with Eq. (8) if $p = -1/1RC$. The constant K is now evaluated by setting $t = 0$ and using the **initial condition** $v_N(0) = V_0$, so

$$v_N(0) = Ke^0 = K = V_0$$

Therefore, taking $K = V_0$ and $p = -1/RC$, we have

$$v_N(t) = V_0 e^{-t/RC} \qquad t \geq 0 \qquad (9)$$

which is our final result.

Equation (9) tells us that the natural response voltage across a parallel RC circuit starts at the initial value V_0 and *decays exponentially* toward zero volts as the capacitor discharges its stored energy through the resistor. This waveform is plotted in Fig. 5.2–5. We see that the natural response drops to about 37% of V_0 at $t = RC$, and it becomes vanishingly small for $t > 5RC$.

Figure 5.2–5
Decaying exponential
waveform $V_0 e^{-t/RC}$.

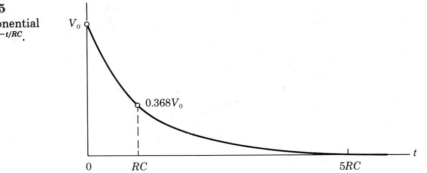

Our results for the parallel RC circuit easily generalize to other first-order circuits as follows. The natural response $y_N(t)$ for *any* first-order circuit obeys the homogeneous equation related to Eq. (4), namely

$$a_1\frac{dy_N}{dt} + a_0 y_N = 0$$

Hence,

$$\frac{dy_N}{dt} = -\frac{a_0}{a_1}y_N$$

whose solution is

$$y_N(t) = Ke^{pt} \qquad p = -a_0/a_1 \qquad (10)$$

This waveform will be a decaying exponential in the usual case where a_1 and a_0 are positive so that p is negative. The constant K is evaluated using the initial condition on $y_N(t)$.

For the natural response of a circuit with two energy-storage elements, the homogeneous equation related to Eq. (6) is

$$a_2 \frac{d^2 y_N}{dt^2} + a_1 \frac{dy_N}{dt} + a_0 y_N = 0 \tag{11}$$

This equation cannot be rewritten to obtain a simple relation for dy_N/dt as before. However, if we again assume the exponential function $y_N = Ke^{pt}$ and insert it into Eq. (11), then we get

$$a_2 p^2 y_N + a_1 p y_N + a_0 y_N = (a_2 p^2 + a_1 p + a_0) y_N = 0$$

Hence, the exponential function is a valid solution when the value of p is such that

$$a_2 p^2 + a_1 p + a_0 = 0 \tag{12}$$

Equation (12) is called the **characteristic equation** of a second-order circuit. And, being a quadratic, it has *two roots*

$$p_1 = -\frac{1}{2a_2}(a_1 - \sqrt{a_1^2 - 4a_2 a_0})$$

$$p_2 = -\frac{1}{2a_2}(a_1 + \sqrt{a_1^2 - 4a_2 a_0})$$

Since p_1 and p_2 both satisfy Eq. (12), and since superposition holds, the natural response consists of a sum of two exponential functions in the form

$$y_N(t) = K_1 e^{p_1 t} + K_2 e^{p_2 t} \tag{13}$$

Evaluating the two constants K_1 and K_2 then requires *two initial conditions,* consistent with the fact that the circuit stores energy in two separate elements.

We'll postpone further analysis of the second-order natural response to Chapter 7. The important point to notice here is that if p_1 and p_2 are both negative, then $y_N(t)$ decays with time and eventually goes to zero — just as it does in a first-order circuit with $p < 0$. The nth-order versions of Eqs. (11) – (13) likewise show that the natural response of an nth-order circuit consists of n decaying exponentials when all n roots of the characteristic equation are negative. The practical significance of decaying exponential functions is underscored in the following example.

Example 5.2 – 1

Suppose the circuit in Fig. 5.2 – 4 actually represents a real capacitor with capacitance $C = 300 \ \mu F$ and leakage resistance $R = 2 \ M\Omega$. The capacitor was initially charged to 1000 V and then discharges through its own leakage resistance. Inserting $RC = (2 \times 10^6)(300 \times 10^{-6}) = 600$ and $V_0 = 1000$ into Eq. (9) gives the natural response voltage

$$v_N(t) = 1000\, e^{-t/600} \qquad t \geq 0$$

Thus, the voltage at $t = 10$ min $= 600$ s has only decayed to $1000\, e^{-1} = 368$ V, and you have to wait until $t > 5RC = 50$ min before $v_N < 1000\, e^{-5} \approx 7$ V. The slow rate of decay is consistent with the very small leakage current $i_R = v_N/R \leq 500\ \mu A$.

These numerical values happen to be typical of a capacitor in a high-voltage power supply, as in a TV set, which explains why you might get a nasty "zap" from the capacitor even when the power has been off for several minutes. You can speed up the decay by putting a small external resistance R_x across the capacitor's terminals so that C discharges through the parallel combination $R_x\|R \approx R_x$. If $R_x = 20\ \Omega$, for instance, then $R_x C = 0.006$ and $v_N \leq 7$ V for $t \geq 5RC = 0.03$ s.

Forced and Complete Response

Now consider a dynamic circuit with an applied time-varying source, creating a nonzero forcing function $F(t)$. The **forced response**, denoted in general by $y_F(t)$, is defined as the solution of the inhomogeneous differential equation obtained without regard to initial conditions. For a first-order circuit described by Eq. (4), $y_F(t)$ is a function containing no unknown constants that satisfies

$$a_1 \frac{dy_F}{dt} + a_0 y_F = F(t) \tag{14}$$

Mathematicians call $y_F(t)$ the **particular solution** of the inhomogeneous differential equation.

Solving Eq. (14) with an arbitrary $F(t)$ is a difficult chore. Fortunately, the problem has received considerable attention, and particular solutions have been found for many forcing functions of practical interest. A few of these known cases are listed in Table 5.2–1. Moreover, the given solutions are valid not only for first-order circuits but also for *any* linear differential equation. To use this table, you substitute the expression for $y_F(t)$ into the differential equation and evaluate the constants A and B by equating the left-hand side of the differential equation to the forcing function in question. We'll illustrate this method shortly.

Table 5.2–1
Particular solutions.

$F(t)$	$y_F(t)$
$\alpha t + \beta$	$At + B$
αe^{st}	Ae^{st}
$\alpha \cos \omega t + \beta \sin \omega t$	$A \cos \omega t + B \sin \omega t$

Although $y_F(t)$ satisfies the inhomogeneous equation with $F(t)$, it does not necessarily represent the *complete* response of a dynamic circuit. The difference between the forced and complete response hinges upon the *initial conditions.* For if a circuit has stored energy at the time the source is applied, then its complete response clearly consists of the natural response along with the forced response. The **complete response** $y(t)$ is therefore given by

$$y(t) = y_F(t) + y_N(t) \tag{15}$$

where $y_N(t)$ includes one or more constants to be evaluated from the initial conditions on $y(t)$. Equation (15) also holds even when the circuit contains no initial stored energy because the energy-storage elements cannot adjust immediately to the conditions required for the forced response. The natural response in Eq. (15) thus accounts for the readjustment process, with or without initial stored energy.

Finally, suppose that all the terms in the natural response decay with time and become vanishingly small. Dynamic circuits with this property are said to be **stable.** The complete response thus eventually simplifies to

$$y(t) \approx y_F(t)$$

We then say that the circuit has reached the **steady-state condition,** meaning that the complete response equals the forced response. The preceding readjustment behavior that involves both the natural and forced response is known as the **transient response.**

Most circuits of practical interest are stable. (In fact, any dynamic circuit with resistance will be stable, provided that it does not contain an internal source of energy.) Consequently, special analysis methods have been developed for the study of steady-state and transient response. Some of these methods will be presented in the next section and in Chapters 6 and 7. We close here with an example of a complete solution using the direct or "brute-force" approach.

Example 5.2–2

The series RL circuit in Fig. 5.2–6 is driven by a sinusoidal voltage $v(t) = 100 \sin 600t$ starting at $t = 0$. We want to find the resulting current $i(t)$, given the initial condition $i(0) = 0$. We'll start with the forced response, then add the natural response and apply the initial condition to obtain the complete response.

Figure 5.2–6
Circuit for Example 5.2–2.

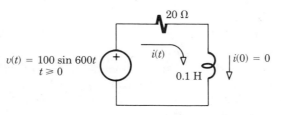

The inhomogeneous equation for the forced response $i_F(t)$ is obtained from Eq. (1) with the stated element values and forcing function $v(t)$, so

$$0.1 \frac{di_F}{dt} + 20i_F = 100 \sin 600t$$

Since the forcing function has the form $\alpha \cos \omega t + \beta \sin \omega t$ with $\alpha = 0$ and $\omega = 600$, the appropriate particular solution from Table 5.2–1 is

$$i_F(t) = A \cos 600t + B \sin 600t$$

Differentiation then yields

$$di_F/dt = 600A \sin 600t - 600B \cos 600t$$

since $d(\cos \omega t)/dt = \omega \sin \omega t$ whereas $d(\sin \omega t)/dt = -\omega \cos \omega t$.

After substituting for i_F and di_F/dt in the inhomogeneous equation and regrouping terms, we obtain

$$(60A + 20B) \sin 600t + (20A - 60B) \cos 600t = 100 \sin 600t$$

The constants A and B must satisfy this relationship for all values of t, which will be true only if

$$60A + 20B = 100 \qquad 20A - 60B = 0$$

We thus find that $A = 1.5$ and $B = 0.5$, giving the forced response

$$i_F(t) = 1.5 \cos 600t + 0.5 \sin 600t$$

This waveform and its two components are plotted in Fig. 5.2–7a. Notice that the sum turns out to be another sinusoidal waveform, a property that we'll examine more closely in Chapter 6. Also notice that $i_F(0) = 1.5$, which means that we need to add the natural response to be in agreement with the initial condition $i(0) = 0$.

The natural response $i_N(t)$ obeys the homogeneous equation $0.1 \ di_N/dt + 20i_N = 0$ or $di_N/dt = -200i_N$, so $p = -200$ and $i_N(t) = Ke^{-200t}$. The complete response is then

$$i(t) = i_F(t) + i_N(t) = 1.5 \cos 600t + 0.5 \sin 600t + Ke^{-200t}$$

We can now evaluate K from the initial condition by setting $t = 0$ to get

$$i(0) = 1.5 + K = 0$$

Thus, $K = -1.5$ and our final result is

$$i(t) = 1.5 \cos 600t + 0.5 \sin 600t - 1.5e^{-200t} \qquad t \geq 0$$

Figure 5.2–7b shows the complete response $i(t)$ along with $i_F(t)$ and $i_N(t)$. The natural-response component dies away rather quickly, leaving the circuit in the steady-state condition with $i(t) \approx i_F(t)$ for $t > 0.025$ s. The transient response corresponds to the interval $0 \leq t \leq 0.025$ where $i_N(t)$ is a significant part of $i(t)$.

Figure 5.2-7
(a) Sinusoidal forced response $i_F(t)$.
(b) Complete response $i(t) = i_F(t) + i_N(t)$.

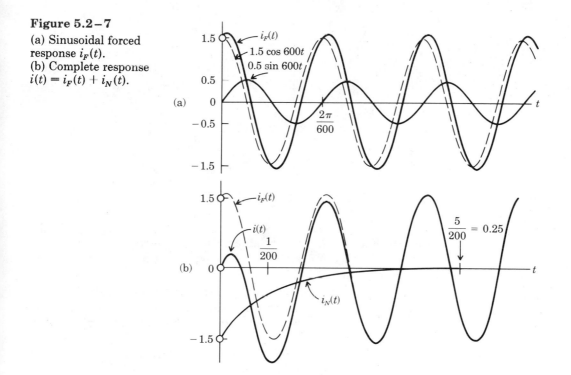

Exercise 5.2-2

Find the complete response $v(t)$ for the circuit in Fig. 5.2-2a when $R = 50\ \Omega$, $C = 0.01$ F, $i(t) = 3t$ for $t \geq 0$, and $v(0) = 0$.

5.3 IMPEDANCE AND TRANSFER FUNCTIONS

As we have seen, the natural response of a dynamic circuit involves exponential time functions. Furthermore, Table 5.2-1 indicates that if the forcing function is exponential, then the forced response is also exponential. The special properties of exponential voltage and current waveforms lead us in this section to the concept of *impedance,* the most valuable and important tool for the analysis of dynamic circuit response.

The Impedance Concept

Let the source voltage or current for a dynamic circuit be an exponential function of the form

$$v = V_0 e^{st} \qquad i = I_0 e^{st}$$

where the constant s has the units of inverse time, $(\text{sec})^{-1}$, so that st and e^{st} are dimensionless quantities. Exponential functions exhibit a decaying, constant, or growing time behavior that depends on the value of s, as

Figure 5.3−1
Exponential waveforms
$v = V_0 e^{st}$ or $i = I_0 e^{st}$.

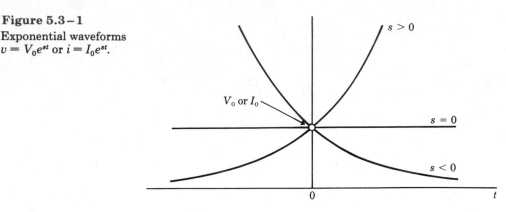

illustrated by Fig. 5.3–1. The special case $s = 0$ corresponds to a DC source with $v = V_0$ or $i = I_0$. Otherwise, v and i vary with time, passing through the value V_0 or I_0 at $t = 0$.

We'll be concerned here entirely with the resulting forced response, which equals the steady-state response when the circuit is stable. Accordingly, we don't need subscripts to distinguish between the natural and forced response.

The key properties of an exponential function pertinent to circuit analysis are, for any constant A,

$$\frac{d}{dt}(Ae^{st}) = sAe^{st} \qquad \int_{-\infty}^{t} Ae^{st}\,dt = \frac{A}{s}e^{st} \tag{1}$$

Thus the derivative or integral of an exponential is *another exponential* having the *same value of s*, but a different *magnitude*. Consequently, if the applied voltage or current has an exponential time variation, then *all* forced voltages and currents will be exponential waveforms, and any derivative or integral terms in the circuit equation reduce to algebraic expressions.

To demonstrate these points, take the voltage applied to a series LRC circuit (Fig. 5.3–2) to be $v = V_0 e^{st}$ with given values of V_0 and s. We then assume that $i = I_0 e^{st}$, where, for the moment, the constant I_0 is unknown.

Figure 5.3−2
Series LRC circuit.

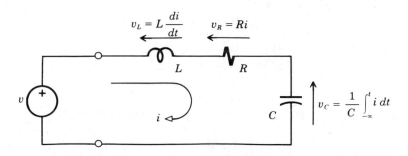

Using Eq. (1) and the element relations, we find the individual voltages are

$$v_L = L\frac{di}{dt} = L(sI_0e^{st}) = (sL)I_0e^{st}$$

$$v_R = Ri = RI_0e^{st}$$

$$v_C = \frac{1}{C}\int_{-\infty}^{t} i\,dt = \frac{1}{C}\left(\frac{I_0}{s}e^{st}\right) = \left(\frac{1}{sC}\right)I_0e^{st}$$

We then define the **element impedances**

$$Z_L \triangleq sL \qquad Z_R \triangleq R \qquad Z_C \triangleq \frac{1}{sC} \tag{2}$$

such that

$$v_L = Z_L I_0 e^{st} \qquad v_R = Z_R I_0 e^{st} \qquad v_C = Z_C I_0 e^{st} \tag{3}$$

Significantly, each of these expressions has the form $v = Zi$, since $i = I_0e^{st}$.

But the applied voltage $v = V_0e^{st}$ equals the sum $v_L + v_R + v_C$, so

$$Z_L I_0 e^{st} + Z_R I_0 e^{st} + Z_C I_0 e^{st} = V_0 e^{st}$$

or, upon factoring,

$$(Z_L + Z_R + Z_C)I_0 e^{st} = V_0 e^{st}$$

The time variation e^{st} now cancels out on both sides, leaving

$$I_0 = \frac{V_0}{Z_L + Z_R + Z_C}$$

which expresses I_0 in terms of the source parameters V_0 and s and the element values as contained in the impedances Z_L, Z_R, and Z_C. We have thereby found the unknown constant in the assumed current $i = I_0e^{st}$, and Eq. (3) gives the individual voltages — all of them being exponential waveforms.

Observe that the previous analysis did not involve a differential circuit equation because algebraic voltage-current relations in terms of impedances have replaced the derivative and integral. Additional consideration of the impedance concept is therefore in order.

The impedance $Z_R = R$ associated with a resistance equals the resistance value itself, independent of s, and clearly has the units of *ohms*. The other two impedances depend on the source parameter s as well as the element value, but dimensional analysis shows that the units of $Z_L = sL$ and $Z_C = 1/sC$ also equal ohms. Accordingly, each of the relations in Eq. (3) may be viewed as modified versions of Ohm's law with *impedance replacing resistance*. Putting this another way, impedance "impedes" the flow of current just as resistance "resists" it.

Further confidence in this interpretation is gained by setting $s = 0$, representing a DC source; then $Z_L = 0$ while $Z_C = \infty$, agreeing with our prior conclusions that inductance acts like a DC short circuit and capacitance acts like a DC open circuit. If $s < 0$, representing a decaying exponential, then Z_L and Z_C will have negative values. *Negative impedance* simply means that the energy-storage element is returning energy to the circuit rather than storing it.

To summarize our results so far: The impedance concept provides a direct method for finding a dynamic circuit's steady-state response caused by an exponential forcing function. Although that might seem to be a rather limited situation, we will see in Chapters 6 and 7 how impedance relates to steady-state AC circuit analysis and to the study of transient behavior. Furthermore, advanced mathematical methods permit one to represent virtually any time function with a combination of exponentials. (These methods are called *transforms* because they transform differential equations into algebraic equations.) The remainder of this chapter, therefore, concentrates on impedance per se, along with the related concepts of admittance and transfer functions.

Example 5.3–1

Let the circuit in Fig. 5.3–2 have $L = 0.25$ H, $R = 400$ Ω, and $C = 100$ μF, and let $v = 30e^{-200t}$. Since $s = -200$, the inductor and capacitor impedances are

$$Z_L = -200 \times 0.25 = -50 \ \Omega$$

$$Z_C = \frac{1}{-200 \times 100 \times 10^{-6}} = -50 \ \Omega$$

Thus, $I_0 = 30/(400 - 50 - 50) = 0.1$, so

$$i = 0.1e^{-200t} \qquad v_R = 40e^{-200t} \qquad v_L = v_C = -5e^{-200t}$$

The fact that v_R is actually greater than the applied voltage means that energy stored in L and C is being returned to the circuit. The steady-state condition therefore requires initial stored energy.

Exercise 5.3–1

Confirm by substitution that the result for i in Example 5.3–1 satisfies Eq. (5) in Section 5.2.

Circuit Analysis Using Impedance and Admittance

Just as we defined the equivalent resistance of a resistive circuit, we can define **equivalent impedance** of a dynamic circuit. Referring to Fig. 5.3–3, let a one-port network consist of linear elements (but no independent sources), and let the applied source be an exponential function, either voltage or current. Then both terminal variables v and i will be exponentials in the steady state, and their ratio is the network impedance

$$Z(s) \triangleq \frac{V_0 e^{st}}{I_0 e^{st}} \tag{4}$$

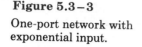

Figure 5.3–3

One-port network with exponential input.

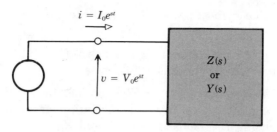

The functional notation $Z(s)$ emphasizes the dependence of impedance on the source parameter s. We also define the **admittance** as the reciprocal of impedance

$$Y(s) \triangleq \frac{1}{Z(s)} = \frac{I_0 e^{st}}{V_0 e^{st}} \tag{5}$$

whose units are siemens (or mhos), the same as for conductance. Table 5.3–1 lists the impedance and admittance of the individual circuit elements.

Table 5.3–1

Element	Impedance	Admittance
Resistance	R	$G = \dfrac{1}{R}$
Inductance	sL	$\dfrac{1}{sL}$
Capacitance	$\dfrac{1}{sC}$	sC

Equivalent impedance (or admittance) is found from a circuit diagram by combining the individual impedances (or admittances) as if they were resistances (or conductances). For instance, the series LRC circuit in Fig. 5.3–2 has

$$Z(s) = Z_L + Z_R + Z_C = sL + R + \frac{1}{sC} = \frac{s^2 LC + sRC + 1}{sC} \tag{6}$$

Similarly, recalling that parallel conductances add, you will find that the admittance of the parallel RC circuit back in Fig. 5.2–2a is

$$Y(s) = Y_C + Y_R = sC + G \tag{7a}$$

and

$$Z(s) = \frac{1}{Y(s)} = \frac{1}{sC + G} = \frac{R}{sCR + 1} \tag{7b}$$

You can also obtain this last result by writing $Z(s) = Z_R \| Z_C$.

Comparing Eqs. (6) and (7) reveals a significant property of impedance or admittance when written in terms of s as a ratio of polynomials:

The highest power of s never exceeds the number of energy-storage elements.

This property provides a quick check for possible manipulation errors. Another simple check is as follows:

The values of $Z(s)$ or $Y(s)$ obtained by setting $s = 0$ and $s = \infty$ should agree with direct calculations from the circuit diagram, using the fact that $Z_L = 0$ and $Z_C = \infty$ when $s = 0$ whereas $Z_L = \infty$ and $Z_C = 0$ when $s = \infty$.

These two tests for accuracy are particularly important when the network involves both series and parallel connections.

Once we know $Z(s)$ or $Y(s)$ for a given network, it becomes a trivial matter to find the steady-state response caused by an exponential forcing function. Specifically, a voltage $v = V_0 e^{st}$ produces the current $i = I_0 e^{st}$ with

$$I_0 = \frac{V_0}{Z(s)} = Y(s)V_0 \tag{8a}$$

while a current $i = I_0 e^{st}$ produces the voltage $v = V_0 e^{st}$ with

$$V_0 = \frac{I_0}{Y(s)} = Z(s)\, I_0 \tag{8b}$$

These equations play the same role in the study of dynamic circuits that Ohm's law did for that of resistive circuits.

If we are concerned with an internal voltage or current, as distinguished from the terminal variables in Fig. 5.3–3, we take the impedance concept one step further and use *impedance in place of resistance* in the various techniques previously developed for resistive circuits:

- Branch-current and node-voltage equations
- Voltage and current dividers
- Thévenin's theorem
- Thévenin/Norton conversion

Furthermore, the *superposition* principle can be invoked to deal with the case of a sum of exponential functions. For instance, if the current through an impedance $Z(s)$ is

$$i = I_1 e^{s_1 t} + I_2 e^{s_2 t} \tag{9a}$$

then the resulting voltage will be

$$v = Z(s_1)I_1 e^{s_1 t} + Z(s_2)I_2 e^{s_2 t} \tag{9b}$$

Note that $Z(s)$ may take on different values when $s = s_1$ and $s = s_2$.

The following examples illustrate some of the techniques of circuit analysis using impedance and admittance.

Figure 5.3–4
(a) Circuit with exponential input. (b) Impedance diagram.

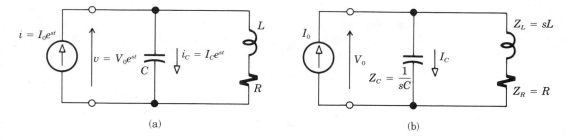

(a) (b)

Example 5.3–2

The circuit in Fig. 5.3–4a is in the steady-state condition driven by $i = I_0 e^{st}$. We are asked to find the resulting terminal voltage v and the current i_C through the capacitor.

Since all waveforms here are proportional to e^{st}, we can suppress the time variation and relabel the circuit with impedances and voltage and current constants, as shown in Fig. 5.3–4b. Combining the series imped-ance $Z_L + Z_R$ in parallel with Z_C gives the terminal equivalent impedance

$$Z(s) = \frac{(Z_L + Z_R)Z_C}{(Z_L + Z_R) + Z_C} = \frac{(sL + R)/sC}{sL + R + 1/sC} = \frac{sL + R}{s^2 LC + sCR + 1}$$

The presence of s^2 but no higher powers of s corresponds to the order of the circuit ($n = 2$ energy-storage elements). Setting $s = 0$ gives $Z(0) = R$, as expected under DC steady-state conditions. Setting $s = \infty$ gives $Z(\infty) = 0$, which agrees with the fact that the capacitor becomes a short circuit with $Z_C = 1/sC \to 0$ as $s \to \infty$.

Having fully tested our expression for $Z(s)$, we use Eq. (8b) to obtain

$$V_0 = Z(s)I_0 = \frac{sL + R}{s^2 LC + sCR + 1} I_0$$

To find the capacitor's current, we'll use the impedance version of the current-divider relationship

$$I_C = \frac{Z_L + Z_R}{Z_L + Z_R + Z_C} I_0 = \frac{sC(sL + R)}{s^2 LC + sCR + 1} I_0$$

The same result is also obtained from $I_C = V_0/Z_C$.

Example 5.3–3
Miller-effect capacitance.

Figure 5.3–5a depicts an inverting voltage amplifier with a capacitor C connecting the input and output terminals. We'll assume exponential steady-state conditions as represented by Fig. 5.3–5b (with e^{st} sup-pressed), and we'll find the resulting equivalent admittances $Y_1(s) = I_1/V_1$ and $Y_2(s) = I_2/V_2$.

Figure 5.3–5

Voltage amplifier with
Miller-effect
capacitance.

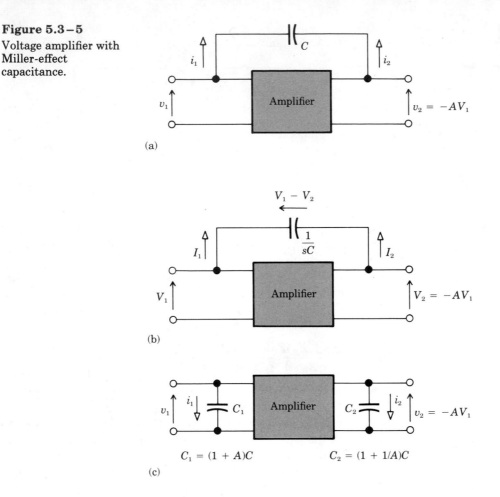

(a)

(b)

(c)

$$C_1 = (1 + A)C \qquad\qquad C_2 = (1 + 1/A)C$$

Since the voltage across the capacitor equals the potential difference $V_1 - V_2$, and since $V_2 = -AV_1$, the current I_1 is

$$I_1 = \frac{V_1 - V_2}{1/sC} = sC(V_1 + AV_1)$$

Hence, $Y_1(s) = I_1/V_1 = sC(1 + A)$, which we rewrite in the form

$$Y_1(s) = sC_1 \qquad C_1 = (1 + A)C$$

Similar analysis using $V_1 = -V_2/A$ yields $I_2 = sC(V_2 + V_2/A)$, so

$$Y_2(s) = sC_2 \qquad C_2 = \left(1 + \frac{1}{A}\right)C$$

The apparent multiplication of the capacitance value by the factors $(1 + A)$ and $(1 + 1/A)$ is known as the **Miller effect.**

Although we derived our results under exponential steady-state conditions, the equivalent capacitance values hold for any time variations. Accordingly, Fig. 5.3–5a can be replaced by Fig. 5.3–5c where C_1 appears directly across the input and C_2 appears directly across the output. This equivalent configuration will be used for the analysis of transistor amplifiers in Chapter 11.

Exercise 5.3–2   Let capacitor C be added in parallel with the resistor in Fig. 5.2–1. Find $Z(s)$ and obtain an expression for the voltage v_R when $v = V_0 e^{st}$. Check your result by setting $s = 0$.

Transfer Functions

Again consider a series LRC circuit, but this time arranged as the *two-port* network in Fig. 5.3–6 with the output voltage taken across the resistance. To characterize the input-output relationship, we use the voltage **transfer function**

$$H(s) = \frac{V_{out}e^{st}}{V_{in}e^{st}} = \frac{Z_R}{Z_L + Z_C + Z_R} = \frac{sCR}{s^2LC + sRC + 1} \tag{10}$$

Note that $H(s)$ has been written as a ratio of polynomials, with the denominator being the same polynomial that appeared in $Z(s)$, Eq. (6). Given the values of R, C, s, and V_{in}, Eq. (10) could be used to compute V_{out}.

Figure 5.3–6

Two-port network with input and output voltages.

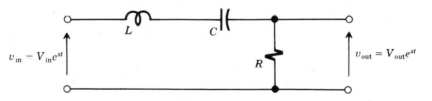

Going from this particular case to an arbitrary linear network, let $x = X_{in}e^{st}$ denote the applied exponential voltage or current and let $y = Y_{out}e^{st}$ be any forced response of interest, either voltage or current. (Be careful not to confuse the scale factor Y_{out} with our symbol for admittance.) We then make the general definition of a transfer function

$$H(s) \triangleq \frac{Y_{out}e^{st}}{X_{in}e^{st}} \tag{11}$$

This definition reduces to our previous definition of impedance or admittance when x and y are taken to be the terminal variables of a one-port network.

If you happen to know the differential equation relating the input x to the variable y of interest, then it becomes a very simple matter to find the transfer function. By way of illustration, the differential equation for many second-order circuits has the general form

$$a_2 \frac{d^2y}{dt^2} + a_1 \frac{dy}{dt} + a_0\, y = b_1 \frac{dx}{dt} + b_0\, x \qquad (12)$$

where the forcing function on the right-hand side may involve x and/or its first derivative. Substituting $x = X_{in}e^{st}$ and $y = Y_{out}e^{st}$ into Eq. (12) and factoring common terms gives

$$(a_2 s^2 + a_1 s + a_0)\, Y_{out}e^{st} = (b_1 s + b_0) X_{in}e^{st}$$

Hence,

$$H(s) = \frac{b_1 s + b_0}{a_2 s^2 + a_1 s + a_0} \qquad (13)$$

which follows from the definition of $H(s)$ in Eq. (11).

After comparing Eqs. (12) and (13) you should be able to write out $H(s)$ directly from inspection of a differential circuit equation of any order n. Conversely, you can apply impedance analysis to find $H(s)$, and then obtain the nth-order differential equation directly from the transfer function.

Now recall that the *characteristic equation* of a second-order circuit described by Eq. (12) is

$$a_2 p^2 + a_1 p + a_0 = 0$$

as previously given in Section 5.2. The polynomial on the left-hand side of the characteristic equation is here seen to be identical to the denominator in Eq. (13) with $s = p$. Hence, the denominator polynomial of $H(s)$ is called the **characteristic polynomial.** Since the roots of the characteristic equation determine the *natural response,* it follows that we can use $H(s)$ instead of a differential equation to investigate a circuit's transient behavior — a topic pursued further in Chapter 7.

Relative to the forced or steady-state response, knowledge of $H(s)$ helps us study how the behavior of the output y depends on the element values and the time variation of the source as represented by s. A case of special interest occurs when $H(s)$ turns out to be constant, independent of s. In particular, if

$$H(s) = K$$

then we know that

$$y(t) = Kx(t)$$

for *any* input waveform $x(t)$.

Example 5.3 – 4
Compensated probe.

Figure 5.3 – 7a depicts a **compensated 10X probe** of the type often used with a cathode-ray oscilloscope (CRO). The CRO has a very large input resistance R and the connecting cable introduces stray capacitance C in parallel with R. Without the probe, the waveform v_{out} displayed by the CRO

Figure 5.3−7

(a) Compensated 10X
probe. (b) Impedance
diagram.

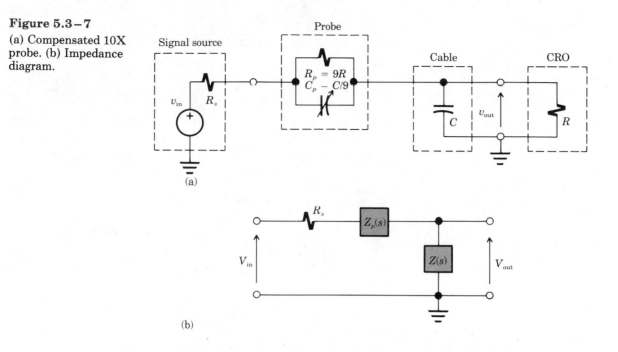

(a)

(b)

would differ from the desired signal waveform v_{in} due to capacitive energy
storage. The probe consists of resistance $R_p = 9R$ in parallel with a compensating capacitance adjusted to get $C_p = C/9$, which results in $v_{out}(t) \approx
K v_{in}(t)$.

To confirm that this compensation method works, we draw the corresponding impedance circuit in Fig. 5.3−7b with

$$Z(s) = \frac{R}{sCR + 1}$$

$$Z_p(s) = \frac{R_p}{sC_p R_p + 1} = \frac{9R}{s(C/9)(9R) + 1} = 9\,\frac{R}{sCR + 1} = 9Z(s)$$

Application of the impedance voltage-divider relation yields

$$H(s) = \frac{Z(s)}{Z(s) + Z_p(s) + R_s} = \frac{Z(s)}{10Z(s) + R_s} = \frac{1}{10 + R_s/Z(s)}$$

where $R_s/Z(s) = (R_s/R)(sCR + 1)$.

Usually, the source resistance R_s is very small compared to the CRO
input resistance R. We therefore assume that $R_s/Z(s) \ll 10$, in which case

$$H(s) \approx 1/10$$

Hence, the transfer function is essentially a constant, as desired. We call
this a "10X" probe because $v_{out} \approx v_{in}/10$ and we must multiply the CRO
scale by 10 to get the actual values of v_{in}.

Exercise 5.3–3   Obtain an expression for $H(s)$ in Fig. 5.3–4, taking the output as the voltage across L and the input as the applied current.

PROBLEMS

5.1–1   A 20-μF capacitor has $v = -50$ V at $t = 0$ and $i = 3$ mA for $t \geq 0$. Find v and w at $t = 1$ sec.

5.1–2   An 8-μF capacitor has $v = 0$ at $t = 0$. What constant current I results in $w = 1$ J at $t = 0.1$ sec?

5.1–3   Sketch and label $v(t)$, $w(t)$, and $p(t)$ for an 8-μF capacitor whose current waveform in milliamps is plotted in Fig. P5.1–3. Assume that $v(0) = 0$.
Figure P5.1–3

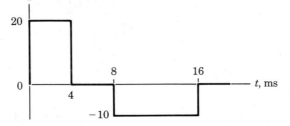

5.1–4   Sketch and label $i(t)$, $w(t)$, and $p(t)$ for a 2-μF capacitor whose voltage waveform is plotted in Fig. P5.1–4.
Figure P5.1–4

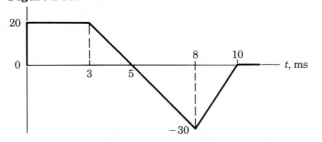

5.1–5   Show that the op-amp circuit in Fig. 5.1–5 becomes a **differentiator** when the resistor and capacitor are interchanged.

5.1–6   The electric field strength between the capacitor plates in Fig. 5.1–1a is $\mathcal{E} = v/d$, and the dielectric breaks down if $\mathcal{E}$ exceeds a certain value. Calculate the maximum possible stored energy when the capacitor's volume is $Ad = 1$ cm^3 and the dielectric is mica, for which $\epsilon \approx 5 \times 10^{-11}$ F/m and $\mathcal{E}_{max} \approx 2 \times 10^8$ V/m.

5.1–7   Do Problem 5.1–6 with an air dielectric, for which $\epsilon \approx 10^{-9}/36\pi$ F/m and $\mathcal{E}_{max} \approx 3 \times 10^6$ V/m.

5.1–8   A 0.8-H inductor has $i = -5$ A at $t = 0$ and $v = 12$V for $t \geq 0$. Find i and w at $t = 1$ sec.

5.1–9   A 0.5-H inductor has $i = 0$ at $t = 0$. How long will it take to get $w = 1$ J if $v = 10$ V for $t \geq 0$?

5.1–10 Sketch and label $i(t)$, $w(t)$, and $p(t)$ for a 40-mH inductor whose voltage waveform is plotted in Fig. P5.1–3. Assume that $i(0) = 0$.

5.1–11 Sketch and label $i(t)$, $w(t)$, and $p(t)$ for a 0.4-H inductor whose current waveform in milliamps is plotted in Fig. P5.1–4

5.1–12 Suppose the op-amp circuit in Fig. 4.2–9b has R_1 replaced by capacitor C and R_F replaced by inductor L. Show that $v_{out} = -LC\, d^2v_{in}/dt^2$.

5.1–13 Find v_{in}, i_1, and i_2 in Fig. P5.1–13 under DC steady-state conditions with $i_{in} = 2$ A. Then calculate the total stored energy when $L = 5$ mH and $C = 25$ μF.

Figure P5.1–13

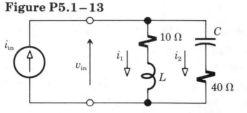

5.1–14 Find i_{in}, v_1, and v_2 in Fig. P5.1–14 under DC steady-state conditions with $v_{in} = 15$ V. If $L = 16$ mH and the total stored energy is 5 mJ, then what is the value of C?

Figure P5.1–14

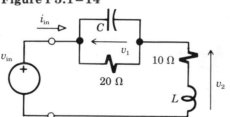

5.1–15 Find v_{in}, i_1, and i_2 in Fig. P5.1–15 under DC steady-state conditions with $i_{in} = 1$ A. If $C = 80$ μF and the total stored energy is 80 mJ, then what is the value of L?

Figure P5.1–15

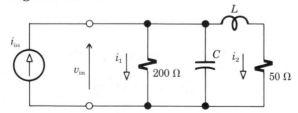

5.1–16 Suppose parallel inductors L_1 and L_2 draw the total current i, and i_1 is the current through L_1. Use Eqs. (11a) and (13) to show that $i_1/i = L_2/(L_1 + L_2)$.

5.1–17 Suppose series capacitors C_1 and C_2 have the total voltage drop v, and v_1 is the voltage across C_1. Use Eqs. (3) and (15) to show that $v_1/v = C_2/(C_1 + C_2)$.

5.1–18 Suppose you have available six 30-μF capacitors rated for up to 800 V. Devise a capacitance bank for maximum possible energy storage when connected to a 1400-V source. How many joules will it store?

5.1–19 Do Problem 5.1–18 with an 1800-V source.

5.2–1   Obtain a differential equation relating i_R to i in Fig. 5.2–2a.

5.2–2   Obtain a differential equation relating v_L to v in Fig. 5.2–1.

5.2–3   Let the op-amp circuit in Fig. 4.2–7b have a capacitor C in parallel with R_F. Obtain a differential equation relating v_{out} to v_{in}.

5.2–4   Let the op-amp circuit in Fig. 4.2–7b have an inductor L in parallel with R_F. Obtain a differential equation relating v_{out} to i_{in}.

5.2–5   Let an inductor L replace R in Fig. 5.2–2a, so $i = i_L + i_C$. Obtain a differential equation relating i_L to i.

5.2–6   Let an inductor L be added in parallel with R and C in Fig. 5.2–2a. Obtain a differential equation relating v to i in Fig. 5.2–2a.

5.2–7   Obtain a differential equation relating v_C to v in Fig. 5.2–2b when an inductor L is added in parallel with C.

5.2–8   Let an inductor L be inserted in series with C in Fig. 5.2–2a, so $v = v_C + v_L$. Obtain a differential equation relating v_C to i.

5.2–9   Solve Eq. (3) for the forced response i_F when $R = 5$, $C = 0.5$, and $v = 3t^2 + 9t$.

5.2–10   Solve Eq. (5) for the forced response i_F when $L = 3$, $R = 0$, $C = 0.02$, and $v = 5 \cos 4t$.

5.2–11   Solve Eq. (5) for the forced response i_F when $L = 3$, $R = 1$, $C = 0.02$, and $v = 5 \sin 4t$.

5.2–12   Find the complete response $i(t)$ from Eq. (1) with $L = 2$, $R = 6$, $v(t) = 12t - 10$ for $t \ge 0$, and $i(0) = 2$.

5.2–13   Find the complete response $v(t)$ from Eq. (2) with $C = 0.05$, $R = 2$, $i(t) = 3e^{5t}$ for $t \ge 0$, and $v(0) = 1$.

5.2–14   Find the complete response $i(t)$ from Eq. (3) with $R = 2$, $C = 0.2$, $v(t) = 10 \sin 5t$ for $t \ge 0$, and $i(0) = 0$.

5.3–1   Use impedance analysis to show that capacitors C_1 and C_2 in series have $C_{eq} = C_1 C_2/(C_1 + C_2)$.

5.3–2   Use impedance analysis to show that inductors L_1 and L_2 in parallel have $L_{eq} = L_1 L_2/(L_1 + L_2)$.

5.3–3   Capacitor C' is connected in series with a parallel RC branch. Find $Y(s)$ as a ratio of polynomials.

5.3–4   Inductor L' is connected in parallel with a series LR branch. Find $Z(s)$ as a ratio of polynomials.

5.3–5   Find $Y(s)$ as a ratio of polynomials when a capacitor is added in parallel with the inductor in Fig. 5.2–1.

5.3–6   Find $Z(s)$ as a ratio of polynomials when an inductor is inserted in series with the capacitor in Fig. 5.2–2a.

5.3–7   Let Fig. P5.1–13 have $L = 300$ mH and $C = 500$ μF. Evaluate the branch impedances to find v_{in} and i_1 when $i_{in} = 0.5e^{100t}$ A.

5.3–8   Let Fig. P5.1–14 have $C = 50$ μF and $L = 120$ mH. Evaluate the branch impedances to find i_{in} and v_2 when $v_{in} = 14e^{250t}$ V.

5.3–9   Do Problem 5.3–8 with $v_{in} = 20e^{-500t}$ V.

5.3–10   Let Fig. P5.1–15 have $C = 50$ μF and $L = 50$ mH. Evaluate the branch impedances to find v_{in} and i_2 when $i_{in} = 2e^{-500t}$ A.

5.3–11   Find $H(s) = I_2/I_{in}$ for Fig. P5.1–13 with $L = 2$ H and $C = 1/1000$ F. Check your result as $s \to 0$ and $s \to \infty$. Then write the differential equation relating i_2 to i_{in}.

5.3–12   Find $H(s) = V_1/V_{in}$ for Fig. P5.1–14 with $C = 1/200$ F and $L = 5$ H. Check your result as $s \to 0$ and $s \to \infty$. Then write the differential equation relating v_1 to v_{in}.

5.3–13   Find $H(s) = V_2/V_{in}$ for Fig. P5.1–14 with $C = 1/40$ F and $L = 2$ H. Check your result as $s \to 0$ and $s \to \infty$. Then write the differential equation relating v_1 to v_{in}.

5.3–14   Find $H(s) = I_1/I_{in}$ for Fig. P5.1–15 with $C = 1/200$ F and $L = 1$ H. Check your result as $s \to 0$ and $s \to \infty$. Then write the differential equation relating i_1 to i_{in}.

5.3–15   Find $H(s) = V_{out}/V_{in}$ for the amplifier circuit model in Fig. P5.3–15 when $Z(s) = 1/sC_L$.

Figure P5.3–15

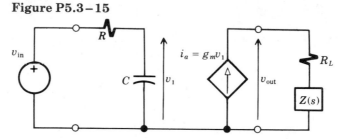

5.3–16   What element should be used for $Z(s)$ in the amplifier circuit model in Fig. P5.3–15 to get $v_{out} = g_m R_L v_{in}$ for any input waveform?

5.3–17   The **gyrator** represented by Fig. P5.3–17 is an electronic network having the properties that $i_1 = -v_2/r_2$ and $i_2 = v_1/r_1$, where r_1 and r_2 are constants with the units in ohms.

(a) Find $H(s) = V_2/V_1$ when a load impedance $Z_2(s)$ is connected across the output terminals. Then show that $Z_1(s) = V_1/I_1 = r_1 r_2/Z_2(s)$.

(b) Obtain and interpret the equivalent circuit for $Z_1(s)$ when $Z_2(s)$ is a parallel RC network.

Figure P5.3–17

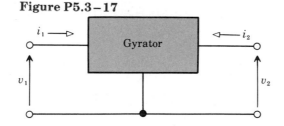

Chapter

6 AC Circuits

AC circuits have long been the bread and butter of electrical engineering — in power transmission, consumer products, lighting, and machinery. Furthermore, an understanding of AC circuit response is an essential prerequisite for topics such as filters, electronic amplifiers, and communication systems. Thus, our work here has both immediate and subsequent applications.

We begin by defining phasor notation and AC impedance as the keys to efficient circuit analysis in the AC steady state. These tools are then used to find the voltage-current relationships in circuits that include capacitance and inductance as well as resistance. We'll also study resonance, average power, and circuits with transformer coupling. An optional section at the end describes residential circuits and wiring. The next chapter will consider frequency-response characteristics of AC circuits, while Chapter 16 deals with AC power systems.

Objectives

After studying this chapter and working the exercises, you should be able to do each of the following:

- Represent a sinusoidal waveform by a phasor, and use phasor addition to sum sinusoids at the same frequency (Section 6.1).

- Calculate the equivalent AC impedance of a network, and find its AC resistance and reactance (Section 6.2).

- Obtain the AC steady-state response of a circuit and calculate the average power dissipation (Sections 6.2 and 6.3).

- Construct phasor diagrams for the voltages and currents in RC, RL, and RLC circuits (Section 6.3).

- Evaluate and interpret the resonant frequency and quality factor of a series or parallel RLC circuit (Section 6.3).

- Apply superposition to a circuit having AC sources at different frequencies (Section 6.3).†

- State the properties of an ideal transformer, and use reflection to analyze an AC circuit with transformer coupling (Section 6.4).

- Design an AC circuit for maximum power transfer (Section 6.4).†

6.1 SINUSOIDS AND PHASORS

Suppose a stable linear circuit is driven by a sinusoidal source applied sufficiently long ago that the natural response has died away to zero. Then, as we saw in Section 5.2, the complete response simply equals the forced response and we say that the circuit has reached **AC steady-state conditions.** Under these conditions, the following two properties hold:

- All voltages and currents within the circuit are sinusoids having the same oscillation frequency as the source.

- The peak values and relative positions of the voltage and current waveforms depend on the element values and the source parameters.

Thus, given the element values and source parameters, AC steady-state circuit analysis reduces to the problem of finding the peak values and relative positions of sinusoidal waveforms.

After a review of sinusoidal waveforms, this section presents important mathematical tools for working with sinusoids. We'll introduce the phasor concept as the vector representation of sinusoids, and we'll develop the algebra of complex numbers as a means of manipulating phasor relationships.

Sinusoidal Waveforms

Consider the sinusoidal voltage plotted in Fig. 6.1–1. We presume that this waveform continues its periodic oscillation for an indefinite amount of time, consistent with AC steady-state conditions. Any such waveform can be written as

$$v = V_m \cos (\omega t + \theta_v) \tag{1}$$

There are three and only three parameters here: the **amplitude** V_m, the **angular frequency** ω, and the **phase angle** θ_v. An AC current $i =$

Figure 6.1–1
Waveform of the AC voltage $V_m \cos (\omega t + \theta_v)$.

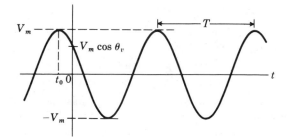

Figure 6.1−2
Rotating vector
representing
$V_m \cos(\omega t + \theta_v)$.

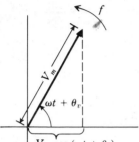

$I_m \cos(\omega t + \theta_i)$ would have the same waveshape, with I_m and θ_i replacing V_m and θ_v.

The amplitude V_m in Eq. (1) equals the *peak value* of v in either the positive or negative direction. The angular frequency ω is the oscillation rate in *radians per second* (rad/s), and the time required for one full oscillation is the **period**

$$T \triangleq 2\pi/\omega \tag{2a}$$

Since the waveform goes through one complete cycle per period, we also define the **cyclical frequency**

$$f \triangleq 1/T = \omega/2\pi \tag{2b}$$

which is measured in cycles per second or **hertz** (Hz). Although we usually speak of frequency in hertz, we use ω to avoid writing $2\pi f$ in expressions like Eq. (1). When $f = 60$ Hz, for instance, $\omega = 2\pi 60 \approx 377$ rad/s.

The phase angle θ_v in Eq. (1) is measured in radians or, more commonly, in degrees. If $\theta_v > 0$, then the central positive peak of the waveform is advanced to the left of the time origin and occurs at time

$$t_0 = -\frac{\theta_v(\text{rad})}{\omega} = -\frac{\theta_v(\text{rad})}{2\pi} T = -\frac{\theta_v(\text{deg})}{360°} T \tag{3}$$

Thus, $v = V_m$ at $t = t_0$ in Fig. 6.1−1, whereas $v = V_m \cos \theta_v$ at $t = 0$. A negative phase angle obviously means a delayed central peak occurring at $t_0 > 0$. Equation (3) is obtained from Eq. (1) by setting $\omega t_0 + \theta_v = 0$ and recalling that 2π rad $= 360°$.

The three sinusoidal parameters in Eq. (1) and Fig. 6.1−1 are displayed in a different way by Fig. 6.1−2, which provides further interpretation of the parameters. For purposes of this drawing, we view the waveform $v = V_m \cos(\omega t + \theta_v)$ as the projection of a *rotating vector*. The vector has length V_m, it rotates counterclockwise at the cyclic rate $f = \omega/2\pi$, and it forms the angle θ_v relative to the horizontal axis at time $t = 0$. Therefore, at any time t, the horizontal projection equals $V_m \cos(\omega t + \theta_v)$. We'll exploit this vector picture for the phasor representation of sinusoids.

Exercise 6.1−1

Let $v = 6 \cos(50\pi t - 90°)$. Calculate t_0 and sketch and label the voltage waveform. Your result should be consistent with the trigonometric identity $\cos(\phi - 90°) = \sin \phi$ for any angle ϕ.

Phasor Representation

In view of the fact that all AC voltages and currents in a given circuit have the same frequency as the source, differing only in amplitude and phase angle, unnecessary pencil-pushing is eliminated if we represent sinusoidal voltage and current by the short-hand **phasor notation**

$$\underline{V} = V_m \underline{/\theta_v} \qquad \underline{I} = I_m \underline{/\theta_i}$$

The symbol $\underline{V}$ stands for a *fixed vector* of length or magnitude V_m at an angle θ_v relative to the horizontal axis, like a snapshot of the rotating vector in Fig. 6.1–2 taken at time $t = 0$. Likewise for the phasor $\underline{I}$.

The term *phasor* designates any vector that represents a sinusoidal time function. Phasor notation allows us to concentrate on the unknown amplitude and angle — V_m and θ_v, or I_m and θ_i — without having to write down the full-blown AC expression at each step of the analysis. Once we find the unknowns, we simply plug their values into $v = V_m \cos (\omega t + \theta_v)$ or $i = I_m \cos (\omega t + \theta_i)$ — and this is the only time we need the entire sinusoidal expression.

Besides notational convenience, phasors readily lend themselves to the task of adding or subtracting sinusoids. In particular, the sum of two or more sinusoidal waveforms having the same frequency can be evaluated by phasor summation. Consequently,

> Kirchhoff's laws hold in phasor form, with each sinusoid replaced by its phasor representation.

This important statement merely reflects the property that the sum of the horizontal projections of two or more vectors equals the horizontal projection of the vector sum.

Consider, for instance, the AC current $i = i_1 + i_2$ in Fig. 6.1–3a. If you know that $i_1 = I_1 \cos (\omega t + \theta_1)$ and $i_2 = I_2 \cos (\omega t + \theta_2)$, then

$$i = I_1 \cos (\omega t + \theta_1) + I_2 \cos (\omega t + \theta_2)$$
$$= I_m \cos (\omega t + \theta_i)$$

where I_m and θ_i could be found by applying trigonometric identities. But the desired result is obtained with much less effort from the phasor version

$$\underline{I} = \underline{I}_1 + \underline{I}_2 = I_1\ \underline{/\theta_1} + I_2\ \underline{/\theta_2}$$

Figure 6.1–3b diagrams this phasor summation taking $\theta_1 > 0$ and $\theta_2 < 0$. The negative value of θ_2 is indicated here by the *clockwise* angle arrow, and we see that θ_i also happens to be negative when $\underline{I}_1$ and $\underline{I}_2$ are as shown. Such

Figure 6.1–3

Phasor summation of AC currents.

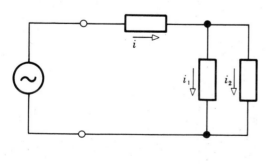

(a)

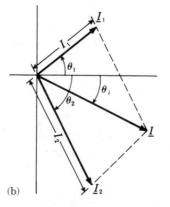

(b)

phasor sums can be evaluated by conventional vector methods or with the help of complex algebra discussed next.

Complex Algebra

Frequently in engineering analysis we need to represent a quantity that has two distinct but related parts. This can be done using a vector or by using a complex number. The advantages of both methods are combined if a complex number is treated as a vector in a *complex plane*.

The **complex plane,** Fig. 6.1–4, is a two-dimensional space wherein any point $\underline{A}$ is uniquely specified by two **rectangular coordinates:** its horizontal coordinate or **real part** $\text{Re}[\underline{A}] = A_r$ and its vertical coordinate or **imaginary part** $\text{Im}[\underline{A}] = A_i$. Of course there's nothing "imaginary" about the quantity $\text{Im}[\underline{A}]$; we simply use the designations "real" and "imaginary" to distinguish between the two directional components. Likewise, the **imaginary unit** $j \triangleq \sqrt{-1}$ merely serves as a way of labeling the vertical component of $\underline{A}$ when it is written as the **complex number**

$$\underline{A} = \text{Re}[\underline{A}] + j\,\text{Im}[\underline{A}] = A_r + jA_i \tag{4}$$

Either part of a complex number may be positive, negative, or zero — but it is always a *real* quantity. The quantity $\underline{A}$ is complex only because its imaginary part is multiplied by j.

Figure 6.1–4
The complex plane and a complex number $\underline{A}$.

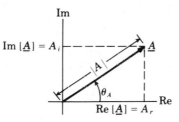

We also identify the point $\underline{A}$ by its **polar coordinates:** the **magnitude** $|\underline{A}| = |A|$, equal to the distance from the origin to the point, and the **angle** $\measuredangle \underline{A} = \theta_A$, measured counterclockwise relative to the positive real axis. We then write $\underline{A} = |A|\,\underline{/\theta_A}$, a complex-plane *vector*.

To convert from the polar coordinates to rectangular coordinates, we see in Fig. 6.1–4 that

$$A_r = \text{Re}[\underline{A}] = |A|\cos\theta_A$$
$$A_i = \text{Im}[\underline{A}] = |A|\sin\theta_A \tag{5}$$

and conversely

$$|A| = |\underline{A}| = \sqrt{A_r^2 + A_i^2} \tag{6a}$$

$$\theta_A = \measuredangle \underline{A} = \arctan\frac{A_i}{A_r} \tag{6b}$$

Equation (6b) assumes $A_r \geq 0$; if $A_r < 0$, the angle can be found from

$$\theta_A = \pm 180° - \arctan\left(\frac{A_i}{-A_r}\right) \qquad \text{(6c)}$$

Electronic calculators are made-to-order for such computations—especially if the calculator happens to have direct rectangular-to-polar and polar-to-rectangular conversion. (But pity the thousands of engineering students in decades past who sweated over these problems with pencil, paper, and slide rule!)

Euler's theorem formally links complex numbers and vectors via the exponential function. It states that, for any angle ϕ,

$$e^{j\phi} = \cos\phi + j\sin\phi \qquad \text{(7)}$$

which is a complex number having

$$\mathrm{Re}[e^{j\phi}] = \cos\phi \qquad \mathrm{Im}[e^{j\phi}] = \sin\phi$$

Comparison with Eq. (5) then reveals that

$$|e^{j\phi}| = 1 \qquad \sphericalangle\, e^{j\phi} = \phi$$

so we can write

$$e^{j\phi} = 1 \,\underline{/\phi}$$

Hence $e^{j\phi}$ is a *unit-length vector* at angle ϕ, as drawn in Fig. 6.1–5. Accordingly, an arbitrary vector with magnitude $|A|$ and angle θ_A can be written as

$$\underline{A} = |A|\, e^{j\theta_A}$$

which is the mathematical expression for the polar form, containing exactly the same information as our informal notation $\underline{A} = |A| \,\underline{/\theta_A}$.

Figure 6.1–5

Euler's theorem in the complex plane.

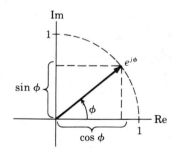

Addition and *subtraction* in the complex plane obey the usual rules of vector arithmetic. You separately add or subtract the horizontal and vertical components. Thus, given the rectangular coordinates of $\underline{A}$ and $\underline{B}$,

$$\underline{A} \pm \underline{B} = (A_r + jA_i) \pm (B_r + jB_i)$$
$$= (A_r \pm B_r) + j(A_i \pm B_i) \qquad \text{(8)}$$

Multiplication and *division*, however, are more easily done with polar coordinates using the exponential properties $e^{\alpha}e^{\beta} = e^{(\alpha+\beta)}$ and $1/(e^{\alpha}) = e^{-\alpha}$ for any α and β. Specifically,

$$\underline{A}\,\underline{B} = (|A|\,e^{j\theta_A})(|B|\,e^{j\theta_B}) = |A||B|\,e^{j(\theta_a+\theta_B)} = |A||B|\,\underline{/\theta_A + \theta_B} \qquad (9)$$

and

$$\underline{A}/\underline{B} = \frac{|A|\,e^{j\theta_A}}{|B|\,e^{j\theta_B}} = \frac{|A|}{|B|}\,e^{j(\theta_A-\theta_B)} = \frac{|A|}{|B|}\,\underline{/\theta_A - \theta_B} \qquad (10)$$

Note carefully that magnitudes multiply or divide, whereas angles add or subtract.

Equation (9) helps explain how the imaginary unit represents the vertical direction of the complex plane. To see this connection, observe that j is a complex number with zero real part and unit imaginary part, so

$$j = 0 + j1 = 1\,\underline{/90°} \qquad (11a)$$

Multiplying any vector A by j thus gives

$$j\underline{A} = (1\,\underline{/90°})(|A|\,\underline{/\theta_A}) = |A|\,\underline{/\theta_A + 90°}$$

which is equivalent to rotating $\underline{A}$ through an angle of 90°. If $\underline{A} = j$ then $j \times j = 1\,\underline{/90° + 90°} = 1\,\underline{/180°}$. Consistency therefore requires the "imaginary" property

$$j^2 = 1\,\underline{/180°} = -1 \qquad (11b)$$

and hence $j = \sqrt{-1}$. A related property is the reciprocal

$$\frac{1}{j} = -j = 1\,\underline{/-90°} \qquad (11c)$$

which is consistent with Eq. (10) when $\underline{A} = 1$ and $\underline{B} = j$. Figure 6.1–6 summarizes these vector characteristics of j.

Figure 6.1–6

Vectors representing j, j^2, $-j$, and $-j^2$.

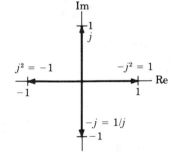

Multiplication can be carried out in rectangular form if you bear in mind that $j^2 = -1$. Thus

$$\underline{A}\,\underline{B} = (A_r + jA_i)(B_r + jB_i)$$
$$= A_rB_r + jA_rB_i + jA_iB_r + j^2A_iB_i$$
$$= (A_rB_r - A_iB_i) + j(A_rB_i + A_iB_r)$$

Division in rectangular form is also possible with the help of the **complex conjugate,** formed by inverting the sign of the imaginary part of a complex number. The complex conjugate of $\underline{A} = A_r + jA_i$ is

$$\underline{A}^* \triangleq A_r - jA_i = |A| \; \underline{/\theta_A} \tag{12}$$

Multiplying $\underline{A}$ by $\underline{A}^*$ then yields a purely real quantity

$$\underline{A}\,\underline{A}^* = |A|^2 = A_r^2 + A_i^2 \tag{13}$$

We apply this property in rectangular division to "rationalize" the denominator, as follows:

$$\underline{A}/\underline{B} = \frac{\underline{A}\,\underline{B}^*}{\underline{B}\,\underline{B}^*} = \frac{1}{|B|^2}\,(A_r + jA_i)(B_r - jB_i)$$

$$= \frac{A_r B_r + A_i B_i}{B_r^2 + B_i^2} + j\,\frac{A_i B_r - A_r B_i}{B_r^2 + B_i^2}$$

Clearly, polar multiplication and division are easier and more direct if we want the results in polar form.

Example 6.1–1

Suppose we need to find $i = i_1 + i_2$ in Fig. 6.1–3a, given that $i_1 = 6 \cos (50t + 30°)$ and $i_2 = 8 \cos (50t - 60°)$. To carry out phasor summation, we first perform the polar-to-rectangular conversions

$$\underline{I}_1 = 6 \; \underline{/30°} = 6 \cos 30° + j6 \sin 30° = 5.20 + j3$$

$$\underline{I}_2 = 8 \; \underline{/-60°} = 8 \cos (-60°) + j \sin (-60°) = 4 - j6.93$$

Adding the real and imaginary parts then gives

$$\underline{I} = \underline{I}_1 + \underline{I}_2 = (5.20 + 4) + j(3 - 6.93) = 9.20 - j3.93$$

Hence, the magnitude and angle of $\underline{I}$ are

$$I_m = |\underline{I}| = \sqrt{9.20^2 + (-3.93)^2} = 10$$

$$\theta_i = \sphericalangle \underline{I} = \arctan (-3.93/9.20) = -23.1°$$

so $\underline{I} = 10 \; \underline{/-23.1°}$ and

$$i = 10 \cos (50t - 23.1°)$$

which is our final result.

Exercise 6.1–2

Draw the phasor diagram representing

$$v = 12 \cos (\omega t - 30°) + 20 \cos (\omega t + 45°)$$

Then find $\underline{V}$ in both rectangular and polar forms and write v as a single cosine function.

Exercise 6.1–3

Let $\underline{A} = 3 - j4$. Find $j\underline{A}$, $\underline{A}\,\underline{A}$, and $1/\underline{A}$ in rectangular and polar forms.

6.2 AC IMPEDANCE

Having introduced phasor notation and the algebra of complex numbers, we're now prepared to examine the role of impedance in steady-state AC circuit analysis. We'll also define the related concepts of AC resistance and reactance, and we'll develop some techniques for impedance calculations.

AC Impedance and Admittance

Let Fig. 6.2–1 represent a stable linear one-port network (containing no independent sources) with an AC excitation at frequency ω. The steady-state phasor voltage and current at the input terminals are, in general,

$$\underline{V} = V_m \underline{/\theta_v} \qquad \underline{I} = I_m \underline{/\theta_i}$$

as indicated on the diagram. The network's **AC impedance** is a complex quantity or vector $\underline{Z}$ obtained by setting $s = j\omega$ in $Z(s)$, so that

$$\underline{Z} \triangleq Z(j\omega) \tag{1}$$

The AC impedance relates the terminal phasors via

$$\underline{V} = \underline{Z}\underline{I} \tag{2}$$

which may be called "Ohm's law for AC circuits."

Equation (2) is similar to our impedance relationship for exponential time variations except that we are now dealing with vector quantities, each having a magnitude and angle. This somewhat complicates the calculations although the principles are simple enough. In particular, if $\underline{Z}$ has the polar form

$$\underline{Z} = |Z| \underline{/\theta_z}$$

then Eq. (2) becomes

$$V_m \underline{/\theta_v} = (|Z| \underline{/\theta_z})(I_m \underline{/\theta_i}) = |Z| I_m \underline{/\theta_z + \theta_i}.$$

Hence,

$$V_m = |Z| I_m \qquad \theta_v = \theta_z + \theta_i \tag{3}$$

and we have separate equations for the magnitudes and angles. The expression $\underline{V} = \underline{Z}\underline{I}$ contains both of these equations in vector notation.

We can also define the **AC admittance** of a network as

$$\underline{Y} \triangleq Y(j\omega) = \frac{1}{Z(j\omega)} \tag{4}$$

Since $\underline{Z} = 1/\underline{Y}$, the magnitudes and angles of the impedance and admittance are related by

$$|Z| = \frac{1}{|Y|} \qquad \sphericalangle\underline{Z} = -\sphericalangle\underline{Y} \tag{5}$$

obtained from Eq. (10), Section 6.1, with $\underline{A} = 1$ and $\underline{B} = \underline{Y}$.

Figure 6.2–1
One-port with AC input.

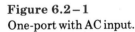

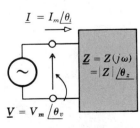

Our next task will be an examination of the characteristics of $\underline{Z}$ (or $\underline{Y}$) for specific circuits. Before doing so, however, we outline the proof of $\underline{V} = \underline{Z}\underline{I}$, which hinges upon Euler's theorem to represent sinusoidal functions as *complex exponentials* of the form $e^{j\omega t}$.

The voltage phasor $\underline{V} = V_m \underline{/\theta_v} = V_m e^{j\theta_v}$ in Fig. 6.2–1 stands for the AC waveform $v = V_m \cos{(\omega t + \theta_v)}$. But consider, instead, the related complex voltage

$$\underline{v} = V_m \cos{(\omega t + \theta_v)} + jV_m \sin{(\omega t + \theta_v)}$$

$$= V_m e^{j(\omega t + \theta_v)} = V_m e^{j\theta_v} e^{j\omega t} = \underline{V}e^{j\omega t}$$

where we have used Euler's theorem with $\phi = \omega t + \theta_v$. We call $\underline{v} = \underline{V}e^{j\omega t}$ a *rotating phasor* since it is a vector of length V_m and instantaneous angle $\omega t + \theta_v$, just like the rotating vector back in Fig. 6.1–2. The horizontal projection of $\underline{v}$ then equals the actual AC voltage since

$$\text{Re}[\underline{v}] = \text{Re}[\underline{V}e^{j\omega t}] = V_m \cos{(\omega t + \theta_v)}$$

Similarly, let

$$\underline{i} = I_m e^{j\theta_i} e^{j\omega t} = \underline{I}e^{j\omega t}$$

so that $\text{Re}[\underline{i}] = I_m \cos{(\omega t + \theta_i)} = i$.

We have thereby represented the AC voltage and current as the real part of exponential time functions with $s = j\omega$. It now follows from Eq. (8b), Section 5.3, that if $\underline{i} = \underline{I}e^{j\omega t}$ and $\underline{v} = \underline{V}e^{j\omega t}$, then

$$\underline{V} = Z(j\omega)\underline{I} = \underline{Z}\underline{I}$$

which holds even when $\underline{V}$, $\underline{I}$, and $s = j\omega$ are complex quantities. For the real sinusoids $i = \text{Re}[\underline{i}]$ and $v = \text{Re}[\underline{v}]$, all imaginary parts disappear from the final result but the magnitudes and angles are still related by $\underline{V} = \underline{Z}\underline{I}$.

Example 6.2–1

Consider a series RL network with $R = 8\ \Omega$ and $L = 0.5$ H driven in the steady state by an AC current $i = 2 \cos 30t$, so $\omega = 30$ rad/s, which corresponds to $f = 30/2\pi \approx 5$ Hz. We are asked to find the resulting steady-state voltage across the terminals of the network.

First, we set $s = j\omega = j30$ in the impedance $Z(s) = R + sL$ to obtain the network's AC impedance

$$\underline{Z} = 8 + j30 \times 0.5 = 8 + j15 = 17\ \underline{/61.9°}$$

Second, we write the current phasor $\underline{I} = 2\ \underline{/0°}$. Third, application of Ohm's law for AC circuits yields the voltage phasor

$$\underline{V} = (17\ \underline{/61.9°})(2\ \underline{/0°}) = 34\ \underline{/61.9°}$$

Thus, we finally have $v = 34 \cos{(30t + 61.9°)}$.

Exercise 6.2–1

Assume a complex current $\underline{i} = 2e^{j30t}$ through the RL network in Example 6.2–1. Find the resulting complex voltage $\underline{v} = R\underline{i} + L\,d\underline{i}/dt$ and confirm that $\text{Re}[\underline{v}]$ equals the stated result.

AC Resistance and Reactance

If a network happens to consist of just one element, then its AC impedance and admittance are found simply by setting $s = j\omega$ in our prior expressions for $Z(s)$ and $Y(s)$. Specifically, drawing upon Table 5.3–1, we have

$$\underline{Z}_R = R \qquad\qquad \underline{Y}_R = G = \frac{1}{R} \qquad\qquad \textbf{(6a)}$$

$$\underline{Z}_L = j\omega L = \omega L \,\underline{/90^\circ} \qquad \underline{Y}_L = \frac{1}{j\omega L} = \frac{1}{\omega L}\,\underline{/-90^\circ} \qquad \textbf{(6b)}$$

$$\underline{Z}_C = \frac{1}{j\omega C} = \frac{1}{\omega C}\,\underline{/-90^\circ} \qquad \underline{Y}_C = j\omega C = \omega C\,\underline{/90^\circ} \qquad \textbf{(6c)}$$

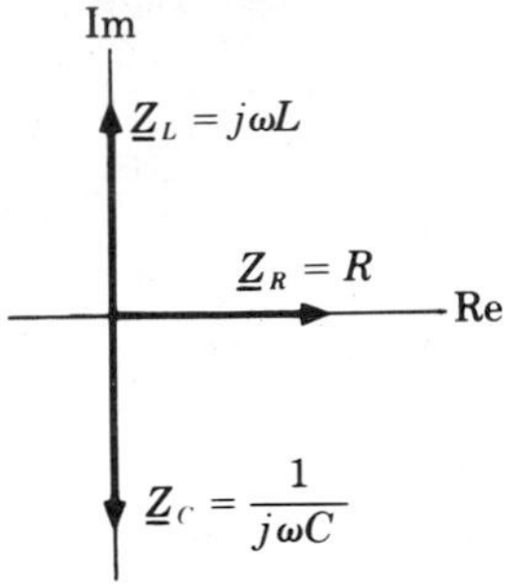

Figure 6.2–2
Impedance vectors.

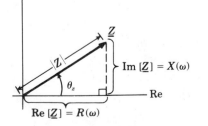

Figure 6.2–2 shows the individual impedance vectors and emphasizes that the energy-storage elements always have an angle of $\pm 90^\circ$ due to the presence of j when we set $s = j\omega$. As before, the magnitude of AC impedance is measured in ohms.

When a network consists of both resistance and energy storage elements, its equivalent AC impedance will be a vector having real (horizontal) and imaginary (vertical) components, written in the general form

$$\underline{Z} = R(\omega) + jX(\omega) \qquad\qquad \textbf{(7a)}$$

with

$$R(\omega) \triangleq \text{Re}[\underline{Z}] \qquad X(\omega) \triangleq \text{Im}[\underline{Z}] \qquad\qquad \textbf{(7b)}$$

The real part $R(\omega)$ is called the **AC resistance;** it may or may not involve ω, depending on the particular circuit. The imaginary part $X(\omega)$ is called the **reactance;** it always involves ω because reactance reflects the presence of energy storage by inductance and/or capacitance. From Eq. (6), the reactances of inductance and capacitance by themselves are

$$X_L = \text{Im}[\underline{Z}_L] = \omega L \qquad X_C = \text{Im}[\underline{Z}_C] = -\frac{1}{\omega C} \qquad\qquad \textbf{(8)}$$

which have opposite signs and reciprocal frequency dependence.

Equation (7) expresses $\underline{Z}$ in rectangular form, with horizontal and vertical components as shown in the **impedance triangle** of Fig. 6.2–3.

Figure 6.2–3
Impedance triangle.

Thus, the corresponding polar coordinates are

$$|Z| = |\underline{Z}| = \sqrt{R^2(\omega) + X^2(\omega)}$$

$$\theta_z = \sphericalangle \underline{Z} = \arctan \frac{X(\omega)}{R(\omega)}$$

(9)

These conversion formulas are important because we often get $\underline{Z}$ in rectangular form from a circuit diagram, whereas the computations represented by $\underline{V} = \underline{Z}\underline{I}$ require $\underline{Z}$ in polar form.

Example 6.2–2
Impedance of a parallel RC circuit.

In Section 5.3, we found that a parallel RC circuit has $Z(s) = R/(sCR + 1)$, so

$$\underline{Z} = Z(j\omega) = \frac{R}{j\omega CR + 1}$$

We put this into rectangular form using the complex conjugate to rationalize the denominator, obtaining

$$\underline{Z} = \frac{R}{1 + j\omega CR} \times \frac{1 - j\omega CR}{1 - j\omega CR} = \frac{R - j\omega CR^2}{1 + (\omega CR)^2}$$

$$= \frac{R}{1 + (\omega CR)^2} + j \frac{-\omega CR^2}{1 + (\omega CR)^2}$$

By comparison with Eq. (7), the AC resistance and reactance are

$$R(\omega) = \frac{R}{1 + (\omega CR)^2} \qquad X(\omega) = -\frac{\omega CR^2}{1 + (\omega CR)^2}$$

The *negative* value of $X(\omega)$ is characteristic of *capacitive reactance*.

Exercise 6.2–2

Obtain expressions for $R(\omega)$ and $X(\omega)$ for a parallel RL circuit, starting from $Z(s)$, and confirm that $X(\omega) > 0$.

Impedance Calculations

At last we put to work the tools that have been developed for AC circuit calculations. In general we proceed as follows:

1. Identify the frequency ω of the source and its amplitude and phase (I_m and θ_i or V_m and θ_v).

2. Find the impedance at the source frequency, and convert it to polar form $\underline{Z} = |Z| \,/\underline{\theta_z}$.

3. Apply Ohm's law for AC circuits, $\underline{V} = \underline{Z}\underline{I}$, to obtain the phasor response $\underline{V} = V_m \,/\underline{\theta_v}$ or $\underline{I} = I_m \,/\underline{\theta_i}$.

4. Substitute the phasor parameters into the corresponding AC waveform $v(t)$ or $i(t)$.

Additional intermediate steps may involve a voltage-divider ratio, Thévenin's theorem, etc., with AC impedance in place of resistance.

For the impedance calculation per se, we obviously could start with $Z(s)$ as we did in Example 6.2–2. This approach yields a rectangular expression that brings out the dependence on the source frequency ω, but conversion to polar form often involves a lot of algebra. An alternative approach having particular value for numerical computations starts with the AC impedance of the individual elements; these are then combined in the usual manner with the help of the rules for vector addition, multiplication, etc. Sometimes it is easier to find the equivalent admittance and get the impedance from $\underline{Z} = 1/\underline{Y}$. The following example illustrates these techniques.

Example 6.2–3

Suppose the circuit in Fig. 6.2–4a has $R_1 = 80\ \Omega$, $L = 0.14$ H, $C = 2.5\ \mu F$, and $R_2 = 800\ \Omega$. If the source voltage is $v(t) = 60 \cos 1000t$, then $\omega = 1000$ rad/s and the impedances of the energy-storage elements are

$$\underline{Z}_L = j1000 \times 0.14 = j140\ \Omega = 140\ \Omega\ \underline{/90°}$$

$$\underline{Z}_C = \frac{1}{j1000 \times 2.5 \times 10^{-6}} = -j400\ \Omega = 400\ \Omega\ \underline{/-90°}$$

Figure 6.2–4
(a) Circuit for Example 6.2–3. (b) Frequency-domain diagram. (c) Equivalent impedance.

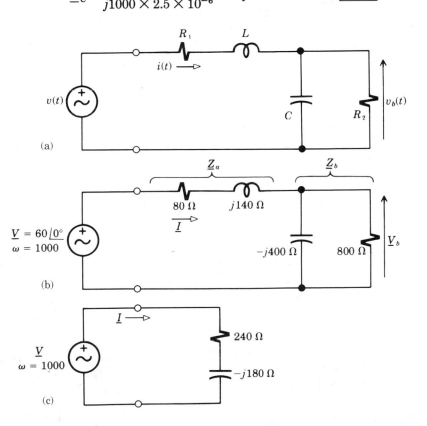

while, of course, $\underline{Z}_{R_1} = 80\ \Omega\ \underline{/0°}$ and $\underline{Z}_{R_2} = 800\ \Omega\ \underline{/0°}$. Thus, for analysis purposes, the circuit becomes as shown in Fig. 6.2–4b. We call this the **frequency-domain circuit** because the impedance values depend on the source frequency ω. (You should always draw a frequency-domain diagram to begin AC circuit analysis.)

To evaluate the network's equivalent impedance $\underline{Z}$, we first observe that the impedance of the series RL section is

$$\underline{Z}_a = 80\ \Omega + j140\ \Omega$$

Next, since parallel admittances add, we find that the parallel RC section has

$$\underline{Y}_b = \frac{1}{800} + \frac{1}{-j400} = \frac{1+j2}{800}$$

and therefore

$$\underline{Z}_b = \frac{1}{\underline{Y}_b} = \frac{800}{1+j2} \times \frac{1-j2}{1-j2}$$

$$= 160\ \Omega - j320\ \Omega = 358\ \Omega\ \underline{/-63.4°}$$

We calculated $\underline{Z}_b$ in both rectangular and polar form in anticipation of subsequent steps.

Adding the real and imaginary parts of $\underline{Z}_a$ and $\underline{Z}_b$ gives the total equivalent impedance

$$\underline{Z} = \underline{Z}_a + \underline{Z}_b = (80 + 160) + j(140 - 320)$$

$$= 240\ \Omega - j180\ \Omega = 300\ \Omega\ \underline{/-36.9°}$$

Thus, at $\omega = 1000$, the equivalent AC resistance and reactance of the network are

$$R(\omega) = \mathrm{Re}[\underline{Z}] = 240\ \Omega$$

$$X(\omega) = \mathrm{Im}[\underline{Z}] = -180\ \Omega$$

The AC resistance in this case is less than the DC resistance (which equals $80 + 800 = 880\ \Omega$), and the reactance is a negative quantity. As far as the source is concerned, it "sees" the equivalent circuit of Fig. 6.2–4c, which includes a capacitance (and no inductance) because of the sign of reactance. This frequency-domain circuit model holds *only* for $\omega = 1000$ (so $f = 159$ Hz).

The resultant current from the source is now easily found with Ohm's law for AC circuits. Specifically,

$$\underline{I} = \underline{V}/\underline{Z} = \frac{60\ \mathrm{V}\ \underline{/0°}}{300\ \Omega\ \underline{/-36.9°}} = 0.2\ \mathrm{A}\ \underline{/36.9°}$$

and the corresponding AC waveform is

$$i(t) = 0.2\cos(1000t + 36.9°)\ \mathrm{A}$$

Having determined $\underline{I}$ and $\underline{Z}_b$, it's an easy task to calculate the voltage $v_b(t)$ across the parallel RC section in Fig. 6.2–4c. We simply write

$$\underline{V}_b = \underline{Z}_b\underline{I} = 358\ \underline{/-63.4^\circ} \times 0.2\ \underline{/36.9^\circ} = 71.6\ \text{V}\ \underline{/-26.5^\circ}$$

Alternatively, had we been concerned only with $v_b(t)$, we could have skipped the calculation of $i(t)$ and gone immediately to the voltage-divider expression

$$\underline{V}_b = \frac{\underline{Z}_a}{\underline{Z}_a + \underline{Z}_b}\underline{V} = \frac{358\ \underline{/-63.4^\circ}}{300\ \underline{/-36.9^\circ}}\ 60\ \underline{/0^\circ} = 71.6\ \text{V}\ \underline{/-26.5^\circ}$$

Either method yields $v_b(t) = 71.6 \cos(1000t - 26.5^\circ)$.

Notice that $v_b(t)$ has a larger amplitude (71.6 V) than the source voltage (60 V). This rather surprising result reflects the fact that $\underline{Z}_b$ has a larger magnitude than the total impedance $\underline{Z}_a + \underline{Z}_b$ due to partial cancellation of the reactances. The related phenomenon of *resonance* will be explored in the next section.

Exercise 6.2–3

Let the circuit in Fig. 6.2–5 have $R_1 = 12\ \Omega$, $C = 250\ \mu\text{F}$, $L = 30\ \text{mH}$, and $R_2 = 24\ \Omega$. Calculate the equivalent AC resistance and reactance for $\omega = 800$, and find $i(t)$ when $v(t) = 100 \cos 800t$ V.

Figure 6.2–5
Circuit for Exercise 6.2–3.

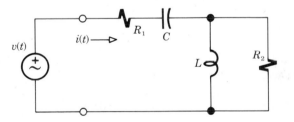

Example 6.2–4
Ladder network.

The network in Fig. 6.2–6a has a "ladder" configuration, the resistors being the "rungs" of the ladder. We are given that $L = 1\ \text{mH}$, $R_1 = 20\ \Omega$, $C = 5\ \mu\text{F}$, $R_2 = 10\ \Omega$, and $\omega = 5000$, so the frequency-domain diagram

Figure 6.2–6
(a) Ladder network. (b) Frequency-domain diagram.

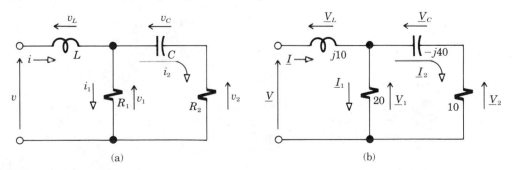

(a) (b)

becomes as shown in Fig. 6.2–6b. We are asked to find the input impedance $\underline{Z} = \underline{V}/\underline{I}$ and the phasor current ratio $\underline{I}_2/\underline{I}$.

Although we might use series/parallel reduction or node-voltage equations here, a more efficient method for circuits of this type is to *assume* a convenient value of the voltage or current farthest from the source and then work backward. All ratios calculated in this manner will be independent of the assumed value since the circuit is *linear*.

Accordingly, let's assume for convenience that

$$\underline{I}_2 = 1 \,\underline{/0^\circ} = 1 + j0$$

in which case

$$\underline{V}_2 = 10\underline{I}_2 = 10 \qquad \underline{V}_1 = \underline{V}_2 + \underline{V}_C = (10 - j40)\underline{I}_2 = 10 - j40$$

Going back toward the source gives

$$\underline{I}_1 = \underline{V}_1/20 = 0.5 - j2 \qquad \underline{I} = \underline{I}_1 + \underline{I}_2 = 1.5 - j2$$

$$\underline{V} = \underline{V}_1 + \underline{V}_L = 10 - j40 + j10(1.5 - j2) = 30 - j25$$

Therefore, upon forming the appropriate ratios, we have

$$\frac{\underline{I}_2}{\underline{I}} = \frac{1}{1.5 - j2} = 0.4 \,\underline{/53.1^\circ}$$

$$\underline{Z} = \frac{\underline{V}}{\underline{I}} = \frac{30 - j25}{1.5 - j2} = 15.6 \ \Omega \,\underline{/13.3^\circ}$$

Since $\sphericalangle \underline{Z} > 0$, we conclude that $X(\omega) > 0$ so the network has an inductive reactance when $\omega = 5000$.

Exercise 6.2–4   Find the ratio $\underline{V}_b/\underline{V}$ for the circuit in Fig. 6.2–4b by assuming that $\underline{V}_b = 800 + j0$.

6.3 AC CIRCUIT ANALYSIS

This section goes into further details of the behavior of circuits under AC steady-state conditions. We'll start with the construction of phasor diagrams, which are then used to examine resonance effects. We'll also study instantaneous and average power, and circuits with sources at different frequencies.

Phasor Diagrams

Phasor diagrams provide an informative picture of the voltage and current relationships in an AC circuit. As a simple but important example, consider the series RC circuit in Fig. 6.3–1a with a voltage source having

Figure 6.3–1
(a) Series RC circuit.
(b) Phasor diagram.
(c) Waveforms.

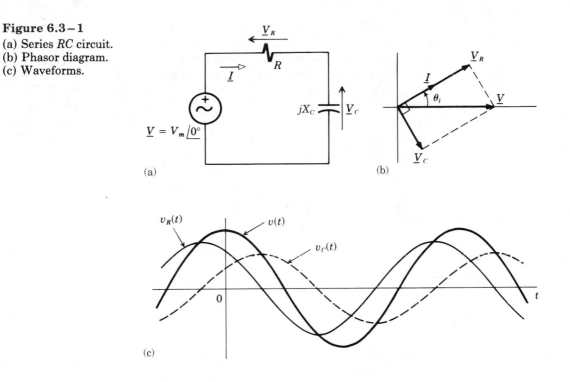

(a)

(b)

(c)

$\underline{V} = V_m \, \underline{/0°}$. (We normally take zero phase angle for an applied source.)
The impedance is

$$\underline{Z} = R + jX_C = R - j/\omega C$$

which has

$$|Z| = \sqrt{R^2 + X_C^2} = \sqrt{R^2 + (1/\omega C)^2} \tag{1}$$

$$\theta_z = \arctan \frac{X_C}{R} = -\arctan \frac{1}{\omega CR}$$

The angle θ_z is negative because X_C is negative and $\arctan(-\phi) = -\arctan \phi$.

The magnitude and angle of the resulting current phasor are found
from $\underline{I} = \underline{V}/\underline{Z}$ with $|\underline{V}| = V_m$ and $\theta_v = 0°$, so

$$I_m = \frac{V_m}{\sqrt{R^2 + (1/\omega C)^2}} \qquad \theta_i = +\arctan \frac{1}{\omega CR} \tag{2}$$

The corresponding phasor voltages across the individual elements are

$$\underline{V}_R = R\underline{I} = RI_m \, \underline{/\theta_i} \qquad \underline{V}_C = jX_C\underline{I} = \frac{I_m}{\omega C} \, \underline{/\theta_i - 90°}$$

where the $-90°$ shift in the phase of $\underline{V}_C$ comes from $jX_C = (1/\omega C) \, \underline{/-90°}$.

Figure 6.3–1b diagrams these phasors along with $\underline{V}$ and $\underline{I}$. The current phasor $\underline{I}$ is said to **lead** the voltage phasor $\underline{V}$ because $\theta_i > \theta_v = 0$. The phasor $\underline{V}_R$ is colinear with $\underline{I}$ and also leads $\underline{V}$, whereas $\underline{V}_C$ is perpendicular to $\underline{I}$ and **lags** $\underline{V}$. If we let these phasors rotate counterclockwise at rate $f = \omega/2\pi$ and take their horizontal projections, we get the actual AC wave-forms $v(t)$, $v_R(t)$, and $v_C(t)$ of Fig. 6.3–1c. The current waveform is not shown, but it has the same shape as $v_R(t)$. You can visually check that $v_R(t) + v_C(t)$ equals $v(t)$ at any instant of time, as required by Kirchhoff's voltage law. The phasor diagram conveys this fact in the vector sum $\underline{V}_R + \underline{V}_C = \underline{V}$.

Now consider a parallel RC circuit driven by a current source with $\underline{I} = I_m \,\underline{/0°}$, as in Fig. 6.3–2a. The parallel structure suggests using the equivalent AC admittance

$$\underline{Y} = \frac{1}{R} + \frac{1}{jX_C} = \frac{1}{R} + j\omega C$$

Solving $\underline{V} = \underline{I}/\underline{Y}$ then yields

$$V_m = \frac{I_m}{\sqrt{(1/R)^2 + (\omega C)^2}} \qquad \theta_v = -\arctan \omega CR \qquad \textbf{(3)}$$

which you should contrast with Eq. (2).

Figure 6.3–2
(a) Parallel RC circuit.
(b) Phasor diagram.

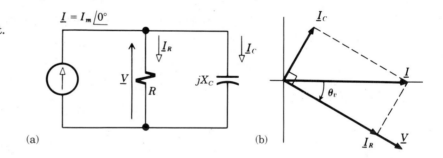

(a)   (b)

The resulting phasor currents through the individual elements are

$$\underline{I}_R = \frac{1}{R}\underline{V} = \frac{V_m}{R}\,\underline{/\theta_v} \qquad \underline{I}_C = \frac{1}{jX_C}\underline{V} = \omega CV_m\,\underline{/\theta_v + 90°}$$

We thereby get the phasor diagram of Fig. 6.3–2b. Here, $\underline{I}_R$ is colinear with $\underline{V}$, and $\underline{I}_C$ is perpendicular to $\underline{I}_R$. The phasor sum $\underline{I}_R + \underline{I}_C = \underline{I}$ corresponds to Kirchhoff's current law, $i_R(t) + i_C(t) = i(t)$, and one could easily sketch waveforms like those in Fig. 6.3–1c.

Note that $\underline{I}$ leads $\underline{V}$ in Fig. 6.3–2b, just as in Fig. 6.3–1b. This reflects the fact that an RC circuit, whether series or parallel, has capacitive reactance $X(\omega) < 0$, so $\theta_z < 0$ and $\theta_i > \theta_v$. The opposite phasor relationship holds for inductive reactance, in which case $\underline{I}$ *lags* $\underline{V}$.

Figure 6.3–3
Circuit and phasor diagram for Example 6.3–1.

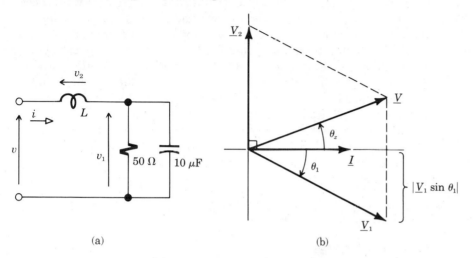

(a)   (b)

Example 6.3–1

Given the network in Fig. 6.3–3a under AC steady-state conditions with $\omega = 1000$, we are asked to determine what value of L results in $\theta_z > 0$. For this task, we'll construct a phasor diagram taking $\underline{I} = I_m \underline{/0°}$ so $\theta_z = \theta_v - \theta_i = \theta_v$.

Since $jX_C = 1/j\omega C = 100/j$, the phasor voltage $\underline{V}_1$ across the parallel RC section is related to the current $\underline{I}$ by

$$\underline{I} = \left(\frac{1}{50} + j\frac{1}{100}\right)\underline{V}_1 = \frac{2+j}{100}\underline{V}_1 = \frac{\sqrt{5}}{100}\underline{/26.6°}\ \underline{V}_1$$

from which we obtain $|\underline{V}_1| = 20\sqrt{5}I_m$ and $\theta_1 = -26.6°$. The phasor voltage across the inductor is

$$\underline{V}_2 = j\omega L\underline{I} = 1000LI_m\ \underline{/90°}$$

Thus, $\underline{I}$ lags $\underline{V}_2$ by 90°.

Figure 6.3–3b shows these phasor relationships along with the terminal voltage $\underline{V} = \underline{V}_1 + \underline{V}_2$. The diagram has been constructed assuming $\theta_z > 0$, and we see that this assumption holds when

$$|\underline{V}_2| > |\underline{V}_1 \sin \theta_1|$$

Inserting numerical values yields the condition $1000LI_m > 20\sqrt{5}I_m \times \sin 26.6°$ or $L > (20\sqrt{5}/1000)\sin 26.6° = 0.02$. Therefore, if $L > 0.02$ H, then the equivalent impedance of the network has $\theta_z > 0$ — signifying *inductive* reactance.

Exercise 6.3–1

Construct phasor diagrams for a series RL circuit and a parallel RL circuit to confirm that inductive reactance results in $\underline{V}$ leading $\underline{I}$ in both cases.

Resonance

Inductance and capacitance have opposite AC properties in two respects: inductive reactance $X_L = \omega L$ is positive and proportional to frequency, while capacitive reactance $X_C = -1/\omega C$ is negative and inversely proportional to frequency. A circuit containing both types of energy-storage elements exhibits a distinctive behavior owing to the joint effect of these properties. Depending on the excitation frequency, the inductance or capacitance may dominate, or the two reactances may cancel out each other and produce the condition known as resonance. Here we will examine the properties of simple series and parallel resonant circuits; the discussion is continued in the next chapter under the heading of bandpass filters.

Consider the series RLC circuit represented by Fig. 6.3–4. Its total impedance is the sum

$$\underline{Z} = R + jX_L + jX_C = R + j\omega L - \frac{j}{\omega C} = R + j\left(\omega L - \frac{1}{\omega C}\right) \qquad \textbf{(4)}$$

which has the *net reactance*

$$X = X_L + X_C = \omega L - \frac{1}{\omega C}$$

Clearly, if $\omega L > 1/\omega C$ then $X > 0$ and the circuit is **inductive** because it has positive reactance like a series RL circuit. Conversely, if $\omega L < 1/\omega C$ then $X < 0$ and the circuit is **capacitive.** The borderline between these two cases occurs when $\omega L = 1/\omega C$, called the *resonance* condition. The **resonant frequency** ω_0 satisfies $\omega_0 L = 1/\omega_0 C$, so

$$\omega_0^2 = \frac{1}{LC} \qquad \omega_0 = \frac{1}{\sqrt{LC}} \qquad \textbf{(5)}$$

The circuit appears to be purely *resistive* at ω_0 since $X = 0$ and $\underline{Z} = R$.

Figure 6.3–4
Series resonant circuit.

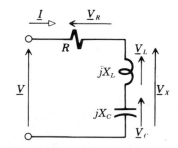

The corresponding phasor diagrams provide further insight about these three conditions. Suppose a circuit is driven by an AC voltage source with $\underline{V} = V_m \underline{/0°}$. The resulting current phasor $\underline{I} = \underline{V}/\underline{Z}$ will have

$$I_m = \frac{V_m}{\sqrt{R^2 + X^2}} = \frac{V_m}{\sqrt{R^2 + \left(\omega L - \dfrac{1}{\omega C}\right)^2}} \quad \textbf{(6a)}$$

$$\theta_i = -\arctan\frac{X}{R} = -\arctan\left(\frac{\omega L}{R} - \frac{1}{\omega CR}\right)$$

and the individual phasor voltage drops will be

$$\underline{V}_R = RI_m \underline{/\theta_i} \qquad \underline{V}_L = \omega L I_m \underline{/\theta_i + 90°} \qquad \underline{V}_C = \frac{I_m}{\omega C} \underline{/\theta_i - 90°} \quad \textbf{(6b)}$$

Since $\underline{V}_L$ and $\underline{V}_C$ are colinear but have opposite directions, it proves convenient to introduce the net reactive voltage phasor

$$\underline{V}_X = \underline{V}_L + \underline{V}_C$$

Kirchhoff's voltage law in phasor form then becomes

$$\underline{V}_R + \underline{V}_X = \underline{V}$$

which will help our phasor constructions.

Figure 6.3–5a shows the phasor diagram below resonance (that is, $\omega < \omega_0$), where $\underline{V}_X$ is in the same direction as $\underline{V}_C$ so $\underline{I}$ leads $\underline{V}$ and the circuit is capacitive (see Fig. 6.3–1b for comparison). The conditions exactly at resonance are shown in Fig. 6.3–5b. Here $\underline{V}_L$ and $\underline{V}_C$ cancel each other leaving $\underline{V}_X = 0$, so $\underline{V}_R = \underline{V}$ and $\underline{I}$ is in phase with $\underline{V}$—just as if the circuit consisted only of resistance. Finally, Fig. 6.3–5c gives the relationships above resonance, where $\underline{V}_X$ is in the same direction as $\underline{V}_L$, so $\underline{I}$ lags $\underline{V}$, and the circuit is inductive.

Figure 6.3–5
Series resonance phasor diagrams: (a) $\omega < \omega_0$; (b) $\omega = \omega_0$; (c) $\omega > \omega_0$.

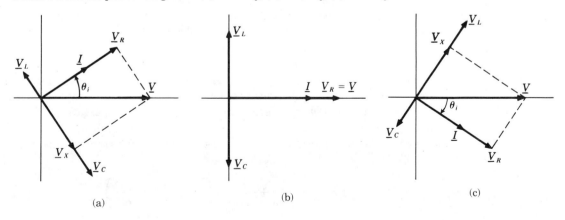

(a) (b) (c)

Although the net reactive voltage $\underline{V}_X$ has zero amplitude at resonance, the individual amplitudes $|\underline{V}_L|$ and $|\underline{V}_C|$ are quite large — possibly even greater than the source amplitude V_m! Specifically, using Eq. (6b) with $\omega = \omega_0$, the voltage amplitudes across the inductor and capacitor are

$$|\underline{V}_L| = \omega_0 L I_m = \omega_0 L \frac{V_m}{R} = \frac{1}{R}\sqrt{\frac{L}{C}}\, V_m$$

$$|\underline{V}_C| = \frac{I_m}{\omega_0 C} = \frac{1}{\omega_0 C}\frac{V_m}{R} = \frac{1}{R}\sqrt{\frac{L}{C}}\, V_m$$

where we have substituted $\omega_0 = 1/\sqrt{LC}$. We now define the **quality factor** of a series resonant circuit to be

$$Q \triangleq \frac{\omega_0 L}{R} = \frac{1}{\omega_0 C R} = \frac{1}{R}\sqrt{\frac{L}{C}} \tag{7}$$

Thus, at resonance,

$$|\underline{V}_L| = |\underline{V}_C| = Q V_m \tag{8}$$

These amplitudes will be greater than V_m if $Q > 1$, a phenomenon known as the *resonant voltage rise*.

When first encountered, this phenomenon seems paradoxical. But bear in mind that $\underline{V} = \underline{V}_R + \underline{V}_L + \underline{V}_C$, and referring to Fig. 6.3–5b shows that the amplitudes $|\underline{V}_L|$ and $|\underline{V}_C|$ can be arbitrarily large compared to V_m since they cancel each other out in the phasor sum. Of course the individual voltages $v_L(t)$ and $v_C(t)$ do exist, and the resonant voltage rise is sometimes used to obtain voltages greater than the available source voltage. (It also may cause a shocking surprise when you're working with a circuit you don't know is resonant.)

The physical explanation of resonance relates to *energy storage,* in that a large amount of energy shuttles to and fro between the inductor and capacitor at resonance, without appearing anywhere else in the circuit. The same effect occurs in mechanical suspension systems with insufficient damping. If you have ever ridden in an automobile with worn-out shock absorbers you probably experienced an unpleasant throbbing at some particular speed — a mechanical resonance condition with energy oscillating between kinetic energy in the vibrating chassis and potential energy in the springs. Replacing the shock absorbers increases the damping and decreases the mechanical Q, thereby decreasing the vibration amplitude.

We've dealt first with series resonance because its properties are probably simpler to grasp than parallel resonance. But we can now apply the series results to the parallel RLC circuit driven by a current source, Fig. 6.3–6, if we use the admittance

$$\underline{Y} = \frac{1}{R} + \frac{1}{jX_C} + \frac{1}{jX_L} = \frac{1}{R} + j\left(\omega C - \frac{1}{\omega L}\right) \tag{9}$$

Figure 6.3–6
Parallel resonant circuit.

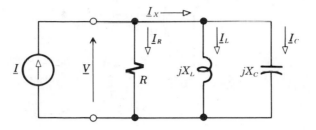

Comparison with Eq. (4) shows that $\underline{Y}$ has the same form as $\underline{Z}$ for a series RLC circuit with L and C interchanged and R replaced by $1/R$ — another example of *duality*. Moreover, the current and voltage phasors are related by $\underline{V} = \underline{I}/\underline{Y}$, which is the dual of $\underline{I} = \underline{V}/\underline{Z}$. Thus our previous equations hold for parallel resonance when $\underline{I}$ and $\underline{V}$ are interchanged, along with the foregoing modifications.

When the parallel RLC circuit in Fig. 6.3–6 operates at resonance, the phasors $\underline{I}_L$ and $\underline{I}_C$ have equal magnitudes and opposite directions, so

$$\underline{I}_X = \underline{I}_L + \underline{I}_C = 0 \qquad \underline{I}_R = \underline{I}$$

However, the circuit exhibits a resonant *current* rise in the sense that

$$|\underline{I}_L| = |\underline{I}_C| = QI_m \tag{10}$$

The quality factor now is given by the dual of Eq. (7), namely

$$Q = \omega_0 CR = \frac{R}{\omega_0 L} = R\sqrt{\frac{C}{L}} \tag{11}$$

Note that a high-Q parallel circuit has large resistance, whereas a high-Q series circuit has small resistance. The resonant frequency itself, however, remains the same since interchanging L and C in Eq. (5) still yields

$$\omega_0 = \frac{1}{\sqrt{LC}}$$

You can easily show from Eq. (9) that the imaginary part of $\underline{Y}$ equals zero when $\omega = \omega_0$.

Example 6.3–2

Figure 6.3–7a represents a parallel RLC circuit driven by an AC voltage source with internal resistance. The capacitor C is to be chosen for parallel resonance at the source frequency $\omega = 5000$, and the resulting amplitudes of the capacitor's current i_C and the terminal voltage v are to be found.

The value of C is determined from Eq. (5) with $\omega_0 = 5000$ and $L = 10$ mH $= 0.01$ H, yielding

$$C = 1/\omega_0^2 L = 4 \times 10^{-6} = 4 \ \mu\text{F}$$

Figure 6.3–7
Circuit and frequency-
domain diagram for
Example 6.3–2.

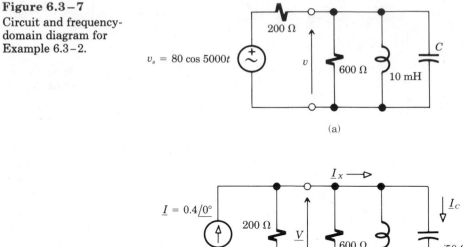

(a)

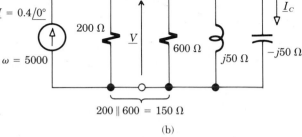

(b)

For the remaining calculations, we convert the voltage source into its
Norton equivalent and draw the frequency-domain circuit in Fig. 6.3–7b
where $\omega_0 L = 1/\omega_0 C = 50\ \Omega$. The equivalent source current has amplitude
$I_m = (80\ \text{V})/(200\ \Omega) = 0.4\ \text{A}$ and the parallel equivalent resistance be-
comes $R = 200\|600 = 150\ \Omega$. Thus, using Eqs. (10) and (11).

$$Q = R/\omega_0 L = 150/50 = 3 \qquad |I_C| = QI_m = 1.2\ \text{A}$$

Finally, the amplitude of v is

$$|\underline{V}| = RI_m = 150 \times 0.4 = 60\ \text{V}$$

which follows from the fact that $\underline{I}_X = 0$ at resonance.

Exercise 6.3–2

Calculate ω_0 and Q for Fig. 6.3–6 with $R = 400\ \Omega$, $L = 1\ \text{mH}$, and $C = 0.1\ \mu\text{F}$. Then construct a phasor diagram showing all the currents when $\underline{V} = 8\ \text{V}\ \underline{/0°}$ and $\omega = 0.8\omega_0$. Use your diagram to find $I_m = |\underline{I}|$.

Instantaneous and Average Power

Under AC steady-state conditions, all voltages and currents in a circuit are
sinusoids. Accordingly, we expect the instantaneous power to oscillate
sinusoidally. Our objectives here are to analyze instantaneous power in an
AC circuit and to find the resulting average power.

Consider a one-port network driven in the steady-state by an AC source, so the terminal voltage and current will be

$$v(t) = V_m \cos(\omega t + \theta_v) \qquad i(t) = I_m \cos(\omega t + \theta_i)$$

Figure 6.3–8 shows these two waveforms taking $\theta_v = 0$ and $\theta_i < 0$. Also plotted is the product $p(t) = v(t)i(t)$, the **instantaneous power** delivered by the source. The oscillatory shape of this waveform means that the source alternately supplies power to the network, when $p(t) > 0$, and receives power back from the network, when $p(t) < 0$. However, the instantaneous positive peak is bigger than the negative peak. We therefore conclude that, on the average, the source delivers more power than it receives. From symmetry, the dashed line halfway between the positive and negative peaks represents the **average power** P or the average rate at which electrical energy is delivered to the network. Thus, the total energy delivered in one period T is $w = PT$.

Figure 6.3–8

AC voltage, current, and power waveforms.

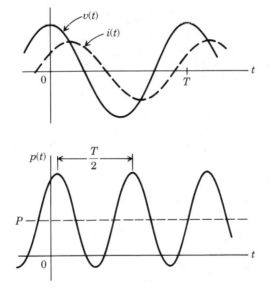

For a quantitative analysis of AC power, we first write out the instantaneous power as

$$p(t) = v(t)i(t) = V_m I_m \cos(\omega t + \theta_v) \cos(\omega t + \theta_i)$$

Next, we invoke the trigonometric identity

$$\cos \alpha \cos \beta = \tfrac{1}{2} \cos(\alpha - \beta) + \tfrac{1}{2} \cos(\alpha + \beta)$$

which leads to

$$p(t) = \tfrac{1}{2} V_m I_m \cos(\theta_v - \theta_i) + \tfrac{1}{2} V_m I_m \cos(2\omega t + \theta_v + \theta_i) \qquad \textbf{(12)}$$

The second term here, being a sinusoid, averages to zero and does not contribute to the average power. Instead, this term represents energy flow back and forth between the source and reactive elements (inductors and capacitors), elements that never dissipate power. The first term of Eq. (12), being a constant, therefore must equal the average power P dissipated by resistance within the network in question.

More informative expressions for the average power are obtained by starting with $P = \frac{1}{2} V_m I_m \cos (\theta_v - \theta_i)$ and recalling that $V_m = |Z| I_m$ and $\theta_i = \theta_z - \theta_v$. Thus, we can write P in terms of the network's impedance as

$$P = \tfrac{1}{2} V_m I_m \cos \theta_z = \tfrac{1}{2} \frac{V_m^2}{|Z|} \cos \theta_z = \tfrac{1}{2} I_m^2 |Z| \cos \theta_z \qquad (13)$$

Alternatively, since $|Z| \cos \theta_z = \mathrm{Re}[\underline{Z}] = R(\omega)$, we have

$$P = \tfrac{1}{2} I_m^2 R(\omega) \qquad (14)$$

This simple result serves as a further interpretation of the AC resistance.

If the network happens to contain only *reactive elements*, then $R(\omega) = 0$ and $P = 0$—consistent with the property that ideal inductors and capacitors never dissipate power. On the other hand, if the network consists entirely of *resistance*, then $R(\omega) = R$ and $V_m = R I_m$ where R denotes the equivalent resistance. Hence, Eq. (14) reduces to

$$P = \tfrac{1}{2} I_m^2 R = \tfrac{1}{2} \frac{V_m^2}{R} \qquad (15)$$

Equation (15) also holds for any resistor R *within* a network, provided that V_m and I_m are interpreted relative to the resistor's terminals.

AC power relationships can be tidied up by introducing the **root-mean-square (rms) value** of voltage or current. Given an arbitrary but periodic voltage waveform $v(t)$, its rms value is defined in general as

$$V_{\mathrm{rms}} \triangleq \sqrt{\frac{1}{T} \int_0^T v^2(t)\, dt} \qquad (16)$$

The terminology "root-mean-square" reflects the fact that V_{rms} equals the square-root of the mean (average) value of the square of $v(t)$. The rms value of a periodic current waveform is defined in the same fashion. For the particular case of sinusoidal voltage and current with amplitudes V_m and I_m, respectively, the corresponding rms values are

$$V_{\mathrm{rms}} = \frac{V_m}{\sqrt{2}} \qquad I_{\mathrm{rms}} = \frac{I_m}{\sqrt{2}} \qquad (17)$$

Standard residential AC circuits in the United States have $V_{\mathrm{rms}} \approx 120$ V, so the peak voltage is actually $V_m = \sqrt{2}\, V_{\mathrm{rms}} \approx 170$ V.

Substituting V_{rms} and I_{rms} into any of our previous power relationships eliminates the factor $\frac{1}{2}$. For instance, Eq. (14) becomes

$$P = I_{\mathrm{rms}}^2 R(\omega).$$

This expression says that an AC current produces the same average power dissipation in the AC resistance $R(\omega)$ as would be produced by a DC current $I = I_{rms}$ through a resistance $R = R(\omega)$. In other words, I_{rms} and V_{rms} are the **effective values** of AC current or voltage insofar as power is concerned. We'll make more use of rms values for the study of AC power systems in Chapter 16.

Example 6.3–3

The circuit analyzed in Example 6.2–3 is redrawn here in Fig. 6.3–9 for purposes of power considerations. We previously found that $R(\omega) = 240 \; \Omega$ and $I_m = 0.2$ A when $V_m = 60$ V. Hence, from Eq. (14), the source delivers the average power

$$P = \tfrac{1}{2}(0.2 \text{ A})^2 \times 240 \; \Omega = 4.8 \text{ W}$$

which is dissipated by the two resistors. Since the peak current through R_1 is $|\underline{I}| = I_m = 0.2$ A, this resistor dissipates

$$P_1 = \tfrac{1}{2}(0.2 \text{ A})^2 \times 80 \; \Omega = 1.6 \text{ W}$$

For the power dissipated by R_2, we observe that $|V_b| = 71.6$ V so Eq. (15) gives

$$P_2 = \tfrac{1}{2} \frac{(71.6 \text{ V})^2}{800 \; \Omega} = 3.2 \text{ W}$$

Thus, $P_1 + P_2 = P$, as expected.

Figure 6.3–9
Circuit for Example 6.3–3.

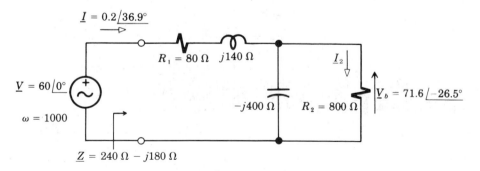

You should note here that $P \neq \tfrac{1}{2}I_m^2(R_1 + R_2)$ because R_1 and R_2 carry different currents. This fact is incorporated in the AC resistance $R(\omega) = 240 \; \Omega \neq R_1 + R_2$.

Exercise 6.3–3

A parallel RL network having $R = 10 \; \Omega$ and $L = 0.1$ H is driven by the AC current $i = 4 \cos 100t$ A. Calculate P from Eq. (14). Then find V_m and check your result using Eq. (15). (Save your results for use in Exercise 6.3–4).

AC Superposition †

Suppose you encounter a circuit having two or more AC sources at *different frequencies,* or one source whose waveform consists of the sum of two or more sinusoids at different frequencies. If the circuit is linear, then you can find any response of interest by applying the principal of superposition.

However, keep in mind that superposition holds for instantaneous voltage or current, not for phasors representing sinusoids at different frequencies. Thus, the proper procedure is as follows:

1. Draw the frequency-domain diagram for one of the frequencies in question, suppressing any other independent sources at different frequencies. Analyze this diagram to find the corresponding phasor response and convert it to a sinusoidal function.

2. Repeat the first step for all other independent sources at different frequencies.

3. Add the sinusoidal responses at different frequencies to obtain the total response due to all sources.

This procedure remains valid when DC and AC sources are present, since a DC source can be treated as producing a cosine wave at $\omega = 0$.

Having found the total response, the instantaneous power could then be determined from the general definition $p(t) = v(t)i(t)$. But we are usually more interested in *average* power, which is easily found with the help of the property that:

> The total average power dissipated in a network driven by AC sources at different frequencies equals the sum of the average powers due to each source acting alone.

In other words, superposition also holds for average AC power — even though it is not valid for instantaneous power.

To prove this superposition property, let the current through a resistance R be the sum of two sinusoids, say

$$i(t) = I_1 \cos (\omega_1 t + \theta_1) + I_2 \cos (\omega_2 t + \theta_2)$$

with $\omega_1 \neq 0$, $\omega_2 \neq 0$, and $\omega_1 \neq \omega_2$. The average power then equals R times the average value of $i^2(t)$, where

$$
\begin{aligned}
i^2(t) = {} & I_1^2 \cos^2 (\omega_1 t + \theta_1) + I_1^2 \cos^2 (\omega_2 t + \theta_2) \\
& + 2I_1 I_2 \cos (\omega_1 t + \theta_1) \cos (\omega_2 t + \theta_2) \\
= {} & \tfrac{1}{2} I_1^2 + \tfrac{1}{2} I_1^2 \cos 2(\omega_1 t + \theta_1) + \tfrac{1}{2} I_2^2 + \tfrac{1}{2} I_2^2 \cos 2(\omega_2 t + \theta_2) \\
& + I_1 I_2 \cos [(\omega_1 - \omega_2)t + \theta_1 - \theta_2] \\
& + I_1 I_2 \cos [(\omega_1 + \omega_2)t + \theta_1 + \theta_2]
\end{aligned}
$$

which has been expanded via standard trigonometric identities. The first and third terms in the expansion are constants, while all other terms average to zero. Therefore,

$$P = (\tfrac{1}{2} I_1^2 + \tfrac{1}{2} I_2^2)R = P_1 + P_2 \tag{18}$$

where

$$P_1 = \tfrac{1}{2} I_1^2 R \qquad P_2 = \tfrac{1}{2} I_2^2 R$$

so the total average power equals the sum of the average powers due to individual components of $i(t)$. We can also write Eq. (18) as $P = I_{rms}^2 R$ with the rms value of $i(t)$ given by

$$I_{rms} = \sqrt{\tfrac{1}{2} I_1^2 + \tfrac{1}{2} I_2^2} = \sqrt{I_{1_{rms}}^2 + I_{2_{rms}}^2} \tag{19}$$

Like results hold when $i(t)$ consists of more than two sinusoids.

Example 6.3 – 4

Figure 6.3–10a shows a series RLC circuit with the applied voltage

$$v = 12 \cos 10{,}000t + 10 \sin 40{,}000t \text{ V}$$

We'll find the total current by writing $i = i' + i''$ and finding the phasor $\underline{I}'$ at $\omega = 10{,}000$ and the phasor $\underline{I}''$ at $\omega = 40{,}000$.

Figure 6.3–10
AC circuit analysis using superposition.

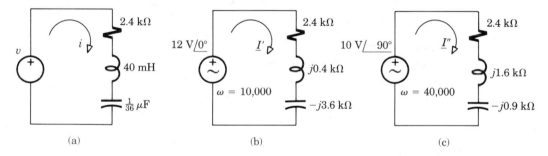

(a)   (b)   (c)

The two frequency-domain diagrams are given in Fig. 6.3–10b, where all impedances are in kilohms and the diagram for $\omega = 40{,}000$ has the voltage phasor at an angle of $-90°$ since $\sin 40{,}000t = \cos(40{,}000t - 90°)$. We now see that

$$\underline{I}' = \frac{12 \; \underline{/0°}}{2.4 + j(0.4 - 3.6)} = 3 \text{ mA} \; \underline{/+53.1°}$$

$$\underline{I}'' = \frac{10 \; \underline{/-90°}}{2.4 + j(1.6 - 0.9)} = 4 \text{ mA} \; \underline{/-106.3°}$$

so the resulting current in milliamps is

$$i = 3 \cos (10{,}000t + 53.1°) + 4 \cos (40{,}000t - 106.3°)$$

The resistor absorbs the average powers

$$P' = \tfrac{1}{2}|I'|^2 R = 10.8 \text{ mW} \qquad P'' = \tfrac{1}{2}|I''|^2 R = 19.2 \text{ mW}$$

for a total of $P = P' + P'' = 30$ mW.

Exercise 6.3–4   Find the total voltage v across the network in Exercise 6.3–3 when $i = 4 \cos 100t - 2 \cos 300t$ A.

6.4 TRANSFORMER CIRCUITS

Many AC circuits involve a source network magnetically coupled to a load network by a *transformer*. The physical principles underlying transformer action will be covered in Chapter 17. Here, we'll state the properties of an *ideal* transformer and then use those properties to develop the method of impedance reflection for the study of AC circuits with ideal transformers. We'll also see how a transformer can be employed to maximize power transfer from an AC source to a given load impedance.

The Ideal Transformer

A transformer consists of two coils of wire with magnetic coupling between them, symbolized by Fig. 6.4–1. The coil on the input side is called the **primary winding**, and the coil on the output side is called the **secondary winding**. Both windings act like short circuits under DC steady-state conditions, and a constant current through one winding produces no effect in the other winding. But the transformer springs to life when either winding has a *time-varying excitation*, which activates the magnetic coupling and produces an induced voltage in the opposite winding.

Figure 6.4–1
Symbol for an ideal transformer.

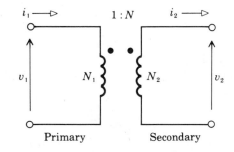

The behavior of an **ideal transformer** with time-varying excitation depends entirely on its **turns ratio**

$$N \triangleq \frac{N_2}{N_1} \tag{1}$$

where N_1 and N_2 are the number of turns in the primary and secondary windings, respectively. The terminal voltages and currents are then related by

$$v_2 = Nv_1 \qquad i_2 = \frac{i_1}{N} \tag{2}$$

Thus, an ideal transformer multiplies the input voltage by the turns ratio and divides the input current by the turns ratio. Many real transformers approximate these characteristics to a greater or lesser degree.

Since no direct electrical connection exists between primary and secondary, **polarity dots** at one end of each winding indicate the relative voltage polarities as follows:

> The potential difference from the dotted to undotted end of the secondary winding has the same polarity as the potential difference from the dotted to undotted end of the primary winding.

Accordingly, we draw the reference voltage arrows in Fig. 6.4–1 taking both dotted ends to be at the higher potential. The reference current arrows then reflect the convention that the primary receives power from a source (not shown), and the secondary delivers power to a load (not shown).

If v_1 varies with time and if $N > 1$, then $|v_2| = N|v_1| > |v_1|$ so we have a **step-up transformer** whose output voltage magnitude will be greater than the input. But note from Eq. (2) that $|i_2|$ will be less than $|i_1|$ when $N > 1$. Conversely, a **step-down transformer** with $N < 1$ produces $|v_2| < |v_1|$ and $|i_2| > |i_1|$. A transformer with $N \neq 1$ can be either step-up or step-down, depending upon which winding is used for the input. A transformer with $N = 1$ also has practical value as an **isolation transformer** that isolates or decouples the DC potential levels on either side without affecting the time-varying quantities.

Regardless of the value of N, an ideal transformer delivers at the output the same amount of power it receives at the input. We prove this property by starting with the instantaneous output power $p_2 = v_2 i_2$ and substituting from Eq. (2) to obtain

$$p_2 = v_2 i_2 = Nv_1 \times \frac{i_1}{N} = v_1 i_1 = p_1$$

where p_1 is the instantaneous input power. Hence, an ideal transformer transfers power without any internal dissipation. Furthermore, even though the winding symbols in Fig. 6.4–1 look like inductors, an ideal transformer never stores any energy.

Sometimes a transformer has three or more windings, or one winding has additional terminals known as **taps.** For instance, Fig. 6.4–2 shows an ideal transformer with a tap dividing the secondary into two segments having N_2 and N_3 turns. Each secondary segment is magnetically coupled

Figure 6.4–2
Transformer with a
tapped secondary.

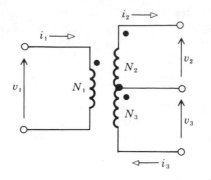

to the primary so that

$$v_2 = \frac{N_2}{N_1} v_1 \qquad v_3 = \frac{N_3}{N_1} v_1 \qquad (3)$$

Such transformers thus provide two or more different output voltages from one source, as needed in an electronic power supply or a dual-voltage AC system.

Exercise 6.4–1   Use the lossless power transfer property to find i_1 in Fig. 6.4–2 in terms of i_2 and i_3.

Impedance Reflection

Far and away the most common type of transformer circuit has the structure of Fig. 6.4–3a, where an ideal transformer serves as an *interface* between an AC source and a load. Since any AC source network can be

Figure 6.4–3
Transformer
interfacing a source
and load.

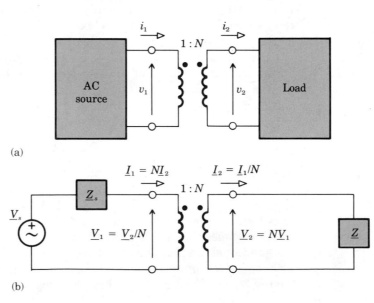

presented by its Thévenin model and any passive load network can be represented by its equivalent impedance, the corresponding frequency-domain diagram becomes as shown in Fig. 6.4–3b. The properties of the ideal transformer are displayed here in phasor form, namely

$$\underline{V}_1 = \underline{V}_2/N \qquad \underline{I}_1 = N\underline{I}_2$$
$$\underline{V}_2 = N\underline{V}_1 \qquad \underline{I}_2 = \underline{I}_1/N \tag{4}$$

We'll draw upon these equations to obtain two other circuits better suited to further analysis.

First, we calculate the equivalent impedance seen looking into the primary. Since $\underline{I}_2 = \underline{V}_2/\underline{Z} = N\underline{V}_1/\underline{Z}$, the primary current is

$$\underline{I}_1 = N\underline{I}_2 = N(N\underline{V}_1/\underline{Z}) = N^2\underline{V}_1/\underline{Z}$$

and therefore

$$\underline{V}_1/\underline{I}_1 = \underline{Z}/N^2 \tag{5}$$

We call $\underline{Z}/N^2$ the **reflected load impedance** seen in the primary. Figure 6.4–4a diagrams the equivalent circuit with the reflected load impedance replacing the transformer and load.

Figure 6.4–4
Equivalent transformer circuits: (a) load reflected into the primary; (b) source reflected into the secondary.

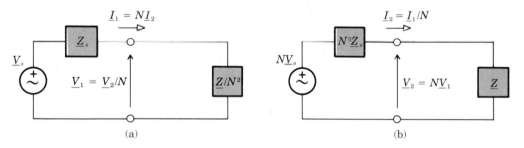

(a)   (b)

To illustrate an important implication of Eq. (5), suppose the load is a series RLC network so $\underline{Z} = R + j\omega L + 1/j\omega C$. The reflected load impedance can then be written as

$$\underline{Z}/N^2 = R/N^2 + j\omega(L/N^2) + 1/j\omega(N^2C)$$

which corresponds to a series network comprised of R/N^2, L/N^2, and N^2C. Furthermore, we state in general that:

> A **reflected load network** has the same structure as the load network with each R and L divided by N^2 and each C multiplied by N^2.

This property allows you to reflect the *entire* load network into the primary, element by element, without having to calculate the equivalent load impedance $\underline{Z}$.

Next, we seek the Thévenin equivalent circuit seen looking back into the secondary. This is easily done in an indirect fashion by observing in Fig. 6.4–3b that

$$\underline{V}_1 = \underline{V}_s - \underline{Z}_s\underline{I}_1 = \underline{V}_s - \underline{Z}_s(N\underline{I}_2)$$

Thus, the secondary voltage and current are related by

$$\underline{V}_2 = N\underline{V}_1 = N[\underline{V}_s - \underline{Z}_s(N\underline{I}_2)]$$
$$= N\underline{V}_s - (N^2\underline{Z}_s)\underline{I}_2 \tag{6}$$

Based on Eq. (6), we can replace the original source model and the transformer with the new equivalent circuit in Fig. 6.4–4b. Here, $N\underline{V}_s$ is the open-circuit voltage reflected across the secondary, while $N^2\underline{Z}_s$ is the reflected source impedance.

Depending upon the quantities of interest, either of these two circuits may be used to analyze the original circuit in Fig. 6.4–3. The primary voltage and current are more easily found by reflecting the load into the primary (Fig. 6.4–4a), while the secondary voltage and current are more easily found by reflecting the source into the secondary (Fig. 6.4–4b). If the load happens to have a parallel configuration, then it may be advantageous to put the reflected source in Norton form.

However, the reflection method does not work in those rare cases where part of the source current goes to the load by a path around the transformer. Figure 6.4–5 illustrates a circuit of this type. Such circuits must be analyzed by inserting the transformer relationships from Eq. (4) into the equations for the external circuitry.

Figure 6.4–5
Transformer circuit that cannot be analyzed by reflection.

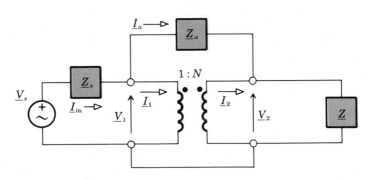

Example 6.4–1

Figure 6.4–6a represents an electronic oscillator that generates a 12-V sinusoid at $\omega = 50{,}000$ and has 1-kΩ source resistance. A 20-V battery powers the oscillator and is connected to it through an ideal transformer,

Figure 6.4–6
(a) Transformer circuit with AC and DC sources. (b) Frequency-domain diagram with reflected AC source.

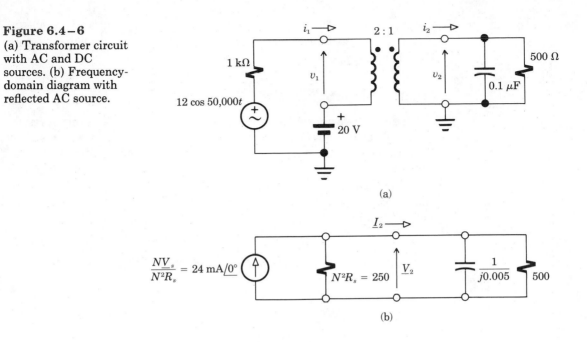

(a)

(b)

which also couples the oscillator to an RC load. The transformer's isolation property has been exploited here to ground the undotted end of the secondary while the grounded battery puts the undotted end of the primary 20 V above ground. The label denotes a 2:1 step-down transformer, so its primary-to-secondary turns ratio is actually $N = \frac{1}{2}$.

By superposition, the primary current i_1 consists of an AC component i_1' plus a DC component $i_1'' = -20 \text{ V}/1 \text{ k}\Omega = 20 \text{ mA}$. The DC component produces no effect in the secondary, so i_2 will be a sinusoid at $\omega = 50,000$. We'll find this current by reflecting the AC source into the secondary. Then we'll find the AC primary component via $i_1' = Ni_2$.

The frequency-domain circuit obtained by reflection is shown in Fig. 6.4–6b, where we have converted the Thévenin source model to Norton form with parallel resistance $N^2 \times 1 \text{ k}\Omega = 250 \text{ }\Omega$ and source current $(N \times 12 \text{ V } \underline{/0°})/(250 \text{ }\Omega) = 24 \text{ mA } \underline{/0°}$. The resulting voltage across the secondary is easily calculated from the node equation

$$\left(\frac{1}{250} + j0.005 + \frac{1}{500}\right)\underline{V}_2 = 24 \times 10^{-3}$$

which yields $\underline{V}_2 \approx 3 \text{ V } \underline{/-40°}$. Now we can find $\underline{I}_2$ via

$$\underline{I}_2 = \left(j0.005 + \frac{1}{500}\right)\underline{V}_2 \approx 16 \text{ mA } \underline{/20°}$$

Hence, $\underline{I}_1' = N\underline{I}_2 \approx 8 \text{ mA } \underline{/20°}$ and the total primary current is $i_1 = i_1' + i_1'' = 8 \cos(50,000t + 20°) - 20 \text{ mA}$.

Exercise 6.4–2   Let the circuit in Fig. 6.4–3 have $\underline{V}_s = 25$ V $\underline{/0°}$, $\underline{Z}_s = (3 + j4)\Omega$, $N = 2$, and $\underline{Z} = (12 - j9)\Omega$. Use the reflection in Fig. 6.4–4b to calculate $\underline{I}_2$ and $\underline{V}_2$. Then find $\underline{I}_1$ and $\underline{V}_1$.

Maximum Power Transfer †

In Section 3.1 we showed that if a voltage source has fixed internal resistance R_s, then maximum power transfer to a load resistance requires $R_L = R_s$. Now we examine a similar problem represented by the frequency-domain circuit in Fig. 6.4–7. Here we want to maximize the average power transferred from an AC voltage source with impedance $\underline{Z}_s$ to a load impedance $\underline{Z}$. Both impedances may consist of AC resistance plus reactance, so we'll write

$$\underline{Z}_s = R_s + jX_s \qquad \underline{Z} = R + jX \tag{7}$$

where the dependence on ω is understood but omitted for brevity.

Figure 6.4–7
AC power transfer
from source to load.

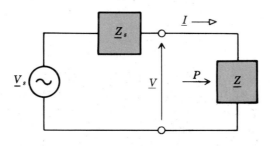

Since average power is proportional to I_m^2, and since $I_m = |\underline{I}| = |\underline{V}_s|/|\underline{Z}_s + \underline{Z}|$, maximum power transfer involves *minimizing* the quantity

$$|\underline{Z}_s + \underline{Z}|^2 = (R_s + R)^2 + (X_s + X)^2$$

Presumably, R_s and either X_s or X are fixed, while the other two components are variable. Of course, the load must have $R > 0$; otherwise, it could not absorb power from the source. Under these conditions, the minimum value of $|\underline{Z}_s + \underline{Z}|^2$ occurs when $X_s + X = 0$ or

$$X = -X_s \tag{8a}$$

This means that X should be inductive if X_s is capacitive, and vice versa. Since the reactive components cancel out, similar to series resonance, the equivalent circuit reduces to a voltage source in series with R_s and R. Hence, maximum power transfer additionally requires that

$$R = R_s \tag{8b}$$

just as we found in the case of a purely resistive circuit.

The two requirements in Eqs. (8a) and (8b) can be combined with the help of complex-conjugate notation by writing

$$\underline{Z} = \underline{Z}_s^* \tag{9}$$

We then say that the impedances are **matched** for maximum power transfer. As was true for purely resistive circuits, maximum transfer of AC power corresponds to 50% efficiency in the sense that equal amounts of power are dissipated in the source and load. We are willing to accept such low efficiency only when it goes along with getting as much power as possible from an information signal source.

When both R_s and R have fixed values, maximum power transfer requires the help of a transformer inserted as shown in back in Fig. 6.4–3. The load impedance reflected into the primary then becomes $\underline{Z}/N^2$, and power transfer is maximized by taking

$$\underline{Z}/N^2 = \underline{Z}_s^*$$

The transformer turns ratio N should thus be chosen to satisfy this condition.

Example 6.4–2

The left side of Fig. 6.4–8a is the frequency-domain model for the output of an amplifier operating at $\omega = 10^5$. The load has a fixed resistance of 50 Ω, and we want to find the load reactance X and the turns ratio N to achieve maximum power transfer.

Figure 6.4–8
Frequency-domain diagrams for Example 6.4–2.

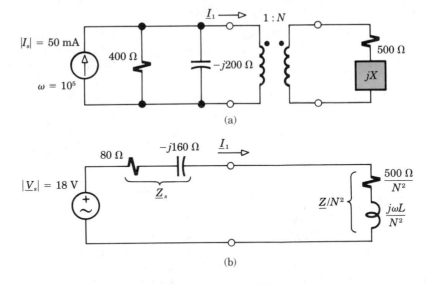

(a)

(b)

Performing a Norton-to-Thévenin conversion on the source yields

$$\underline{Z}_s = 400\|(-j200) = 80\ \Omega - j160\ \Omega$$

$$|\underline{V}_s| = |\underline{Z}_s\underline{I}_s| \approx 18\ \text{V}$$

Figure 6.4–8b gives the equivalent circuit with the load impedance reflected into the primary. Impedance matching therefore requires an inductive load reactance and a turns ratio such that

$$500\ \Omega/N^2 = R_s = 80\ \Omega \qquad \omega L/N^2 = -X_s = +160\ \Omega$$

We therefore need

$$N = \sqrt{500/80} = 2.5$$

and the required inductance is $L = 160N^2/\omega = 10$ mH.

With matched impedances, the primary current from the source has amplitude $|\underline{I}_1| = (18 \text{ V})/(80 \text{ }\Omega + 80 \text{ }\Omega) \approx 113$ mA. Thus, since the transformer is assumed to be lossless, the average power delivered to the load will be

$$P = \tfrac{1}{2}|\underline{I}_1|^2 \times (500 \text{ }\Omega)/N^2 \approx 51 \text{ mW}$$

Without the impedance-matching inductor, the current would drop to $|\underline{I}_1| = 18/|80 + 80 - j160| \approx 80$ mA which reduces the transferred power to $P \approx 25$ mW.

Exercise 6.4 – 3   Let $\underline{Z}_s = 100 \text{ }\Omega \text{ } \underline{/45°}$ and $\omega = 10^4$ in Fig. 6.4 – 7. What two series elements should make up the load impedance for maximum power transfer? Find the resulting value of P when $|\underline{V}_s| = 20$ V.

6.5   RESIDENTIAL CIRCUITS AND WIRING †

No discussion of AC circuits would be complete without at least some coverage of the wiring arrangements that distribute electric power to the outlets of a typical home. This section, therefore, describes several of the major features of residential circuits. As we will be concerned only with voltage and current magnitudes, capital letters will be used throughout to denote rms values.

A word of warning before we begin: This material is intended for illustrative purposes and does not qualify you to be an electrician!

Dual-Voltage Service Entrance

Most homes today have a **dual voltage** AC supply provided by a three-line entrance cable like that of Fig. 6.5 – 1a. One line, called the **neutral,** is connected to an **earth ground.** The other two lines, labeled B and R, are "hot" in the sense that they have 170-V peak sinusoidal potential variations relative to ground potential, with one sinusoid inverted in polarity compared to the other. The two sources shown actually correspond to the secondaries of a center-tapped transformer, like that in Fig. 6.4 – 2, whose primary is connected to the power line at a utility pole.

Formally, we write $v_B = 170 \cos \omega t$ and $v_R = -v_B = -170 \cos \omega t$, with rms values $V_B = V_R = 170/\sqrt{2} = 120$ V. The line-to-line voltage is then $v_{BR} = v_B - v_R = 340 \cos \omega t$, as shown in Fig. 6.5 – 1b, and has the rms value $V_{BR} = 340/\sqrt{2} = 240$ V.

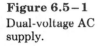

Figure 6.5–1
Dual-voltage AC supply.

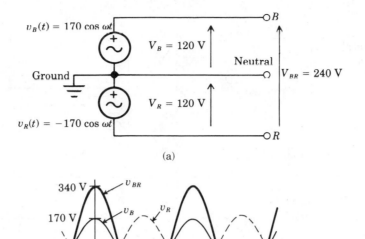

(a)

(b)

After passing through the electric meter that measures energy consumption, the entrance cable terminates at the **main panel.** Here, the hot lines connect to individual circuits for lighting, appliances, and so forth, while the neutral connects to a **busbar** and thence to the local earth ground.

Figure 6.5–2 diagrams a main panel with **breakers** serving the joint role of disconnecting switches and overcurrent protection. The modern thermal-magnetic breaker is a spring-loaded switch having a bimetallic element that opens under small but continuous current overload, and a magnetic coil that trips the switch instantly under heavy overloads. The breaker labeled GFCI has additional features described later. Older designs would have separate switches and fuses instead of breakers.

The figure also shows four different types of circuits going out of the panel. Each circuit has a minimum of three wires, and they have been labeled in accordance with standard insulation color codes as follows:

Hot = B (black) or R (red)
Neutral = W (white)
Ground = G (green) or uninsulated

Every outgoing hot wire must connect to a breaker, whereas every neutral wire and ground wire must be tied directly to earth ground at the neutral busbar. The functional difference between neutral and ground wires will be brought out by considering circuit wiring.

Figure 6.5–2
Main panel with
breakers and typical
circuit connections.

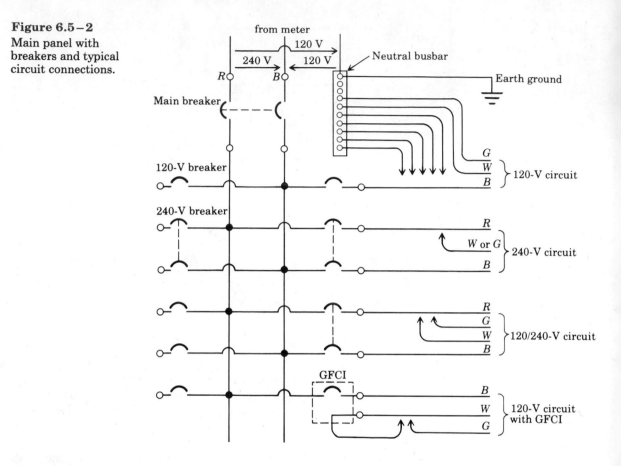

Wiring and Grounding

Figure 6.5–3 shows the wiring and wire resistances from the panel to a
120-V grounded outlet. Several other outlets, lights, etc., would typically be
connected in parallel on the same circuit. Under normal conditions, current flows to the load (not shown) through only hot and neutral wires;
hence the ground terminal at the outlet is at zero volts with respect to earth

Figure 6.5–3
Wiring resistance
between panel and
outlet.

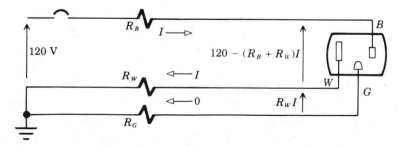

ground, despite the resistance R_G. On the other hand, current I through R_W and R_B causes the neutral terminal to be at $R_W I$ volts with respect to earth ground, and the available load voltage becomes $120 - (R_W + R_B)I$. Resistance in the entrance cable increases this loading effect even further, so home appliances must be designed to operate over a range of $110–120$ V.

From the viewpoint of an electronics engineer, the ground wire provides a valuable *reference potential* for voltage measurements, independent of the neutral and its voltage offset. In addition, electrical interference and ground-loop problems are minimized by connecting all instruments to one ground point. But vastly more important is the ground wire's role in protecting human life against electrical shock.

Table 6.5–1 lists effects of various levels of 60-Hz AC current on the human body. The $100–300$ mA range turns out to be the most dangerous. Larger currents induce a temporary heart contraction that actually protects it from fatal damage. As the table implies, the amount of current rather than voltage is the key factor in electrical shock; voltage enters the picture when we take account of body resistance, which ranges from around 500 kΩ down to 1 kΩ, depending on whether the skin is dry or wet. Thus, a person with wet skin risks electrocution from AC voltages as low as 100 V.

Table 6.5–1
Effects of AC electrical shock.

Current	Effects
1–5 mA	Threshold of sensation.
10–20 mA	Involuntary muscle contractions ("can't-let-go").
20–100 mA	Pain, breathing difficulties.
100–300 mA	Ventricular fibrilation, POSSIBLE DEATH.
>300 mA	Respiratory paralysis, burns, unconsciousness.

Now suppose you touch the metal frame of an ungrounded appliance that has an internal wiring fault between the frame and the hot line as in Fig. 6.5–4a. Your body then provides a possible conducting path to ground, and you may experience a serious shock, especially if you're standing on a damp concrete floor or other grounded surface so that the current passes through your chest. The circuit breaker offers no help in this situation, for it is designed to protect the circuit — not you — against excess currents of 15 A or more.

An internal connection from the frame to neutral significantly improves the situation, but even better is the grounded frame shown in Fig. 6.5–4b. This arrangement keeps the frame at ground potential in absence of a wiring fault or, at worst, a few volts from ground if a fault results in current through the ground wire. But the best possible shock protection is afforded by the **ground-fault circuit interrupter** (GFCI) shown in Fig. 6.5–4c. The GFCI has a sensing coil around the hot and neutral wires, and any ground-fault current — through you *or* the ground wire — that results in an imbalance $|I_B - I_W| > 5$ mA induces a current in the sensing coil and

Figure 6.5–4
Appliance with wiring
fault: (a) ungrounded;
(b) grounded; (c) with
ground-fault circuit
interrupter.

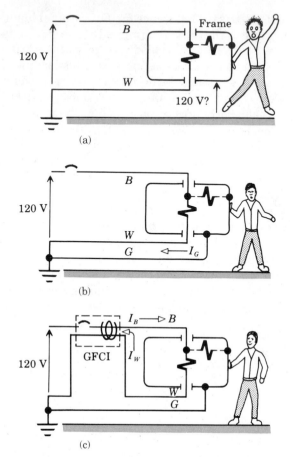

opens the circuit. The GFCI may be located at an outlet or it may be part of a circuit breaker at the main panel, as in Fig. 6.5–2.

Figure 6.5–5 diagrams three wiring patterns involving 240 V: (a) a 240-V load, such as an electric heating unit; (b) a dual-voltage load, such as an electric range with 120-V and 240-V elements; (c) a so-called "three-wire circuit" with two 120-V loads sharing common neutral and ground wires. In this last case (intended solely to reduce wire costs), $I_W = I_B - I_R$ and the neutral current will be zero if the loads are equal.

Finally, Fig. 6.5–6 illustrates how a light or some other device can be controlled independently from two different locations using single-pole double-throw (SPDT) switches commonly known as "three-way" switches. The hot wire is switched between two "travelers" at the first switch and from the travelers to the light at the second; therefore, we have a complete circuit only when both switches are either up or down, and flipping either switch opens the circuit. Control at three or more locations is also possible using "four-way" switches that interchange the travelers. In any case, neutral and ground wires are never switched.

Figure 6.5–5
(a) 240 V load.
(b) Dual-voltage load.
(c) Three-wire circuit for two 120-V loads.

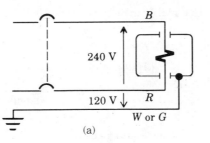

(a)

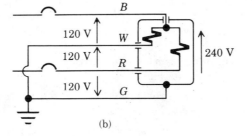

(b)

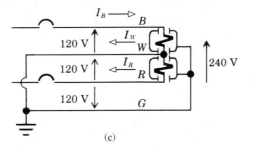

(c)

Figure 6.5–6
Connection of three-way switches.

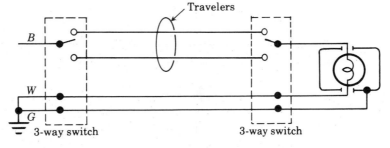

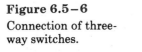

PROBLEMS

6.1–1 Evaluate $\underline{A} + j\underline{B}$ and $\underline{B}/\underline{A}^*$ in rectangular and polar form with $\underline{A} = -6 + j8$ and $\underline{B} = 17\ \underline{/-118.1°}$.

6.1–2 Do Problem 6.1–1 with $\underline{A} = 24 + j7$ and $\underline{B} = -18 + j24$.

6.1–3 Evaluate $\underline{A} - j\underline{B}$ and $\underline{B}\underline{A}^*$ in rectangular and polar form with $\underline{A} = 13\,\underline{/-157.4°}$ and $\underline{B} = -9 + j12$.

6.1–4 Do Problem 6.1–3 with $\underline{A} = 41\,\underline{/167.3°}$ and $\underline{B} = 15\,\underline{/53.1°}$.

6.1–5 Taking $\underline{A} = A_r + jA_i = |\underline{A}|\,\underline{/\theta}$, simplify each of the following expressions: $\underline{A} + \underline{A}^*$, $\underline{A} - \underline{A}^*$, $1/\underline{A}^*$, $\underline{A}^*/\underline{A}$.

6.1–6 Carry out the division $(1 + jx)/(1 + jy)$ in rectangular and polar forms to show that $\arctan[(x - y)/(1 + xy)] = \arctan x - \arctan y$.

6.1–7 Use the series expansions for e^x, $\cos\phi$, and $\sin\phi$ to justify Eq. (7).

6.1–8 Show from Eq. (9) that

$$\frac{1}{2}(e^{j\phi} + e^{-j\phi}) = \cos\phi \qquad \frac{1}{2j}(e^{j\phi} - e^{-j\phi}) = \sin\phi$$

6.1–9 Use the results of Problem 6.1–8 to derive the trigonometric identity $\sin\alpha\sin\beta = \frac{1}{2}\cos(\alpha - \beta) - \frac{1}{2}\cos(\alpha + \beta)$.

6.1–10 Use the results of Problem 6.1–8 to derive the trigonometric identity $\cos(\alpha + \beta) = \cos\alpha\cos\beta - \sin\alpha\sin\beta$.

6.1–11 Draw the corresponding phasor diagrams and express $v_1 + v_2$ and $v_1 - v_2$ in the form of Eq. (1) when $v_1 = 10\cos(\omega t + 135°)$ and $v_2 = 10\cos(\omega t + 45°)$.

6.1–12 Do Problem 6.1–11 with $v_1 = 2\cos\omega t$ and $v_2 = \sqrt{8}\cos(\omega t + 135°)$.

6.1–13 Do Problem 6.1–11 with $v_1 = 2\cos\omega t$ and $v_2 = 2\cos(\omega t - 60°)$.

6.2–1 Let the circuit in Fig. P6.2–1 have $R_1 = 5\,\Omega$, $L = 1$ mH, $R_2 = 0$, $C = 20\,\mu$F, and $i = 4\cos 5000t$ A. Calculate $\underline{Z}$ and v. Then find i_1 and i_2, and use a phasor sum to confirm that $i_1 + i_2 = i$.

Figure P6.2–1

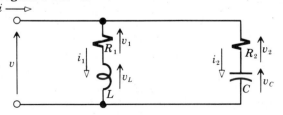

6.2–2 Do Problem 6.2–1 with $R_1 = 0$, $L = 10$ mH, $R_2 = 15\,\Omega$, $C = 25\,\mu$F, and $i = 3\cos 2000t$ A.

6.2–3 Let the circuit in Fig. P6.2–3 have $C = 10\,\mu$F, $R_1 = 75\,\Omega$, $R_2 = 0$, $L = 50$ mH, and $v = 25\cos 2000t$ A. Calculate $\underline{Z}$ and i. Then find v_C and v_1, and use a phasor sum to confirm that $v_C + v_1 = v$.

Figure P6.2–3

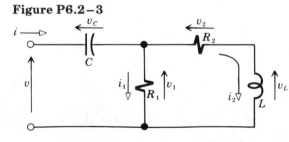

6.2–4   Do Problem 6.2–3 with $C = 100 \ \mu\text{F}$, $R_1 = 20 \ \Omega$, $R_2 = 4 \ \Omega$, $L = 8 \ \text{mH}$, and $v = 10 \cos 1000t$ A.

6.2–5   Assume that $\underline{I}_L = 1$ in Fig. P6.2–5 to find $\underline{I}_L/\underline{I}$ and $\underline{V}/\underline{I}$ when $R_1 = 19$, $R_2 = 20$, and $1/\omega C = \omega L = 10$. Express your results in polar form.

Figure P6.2–5

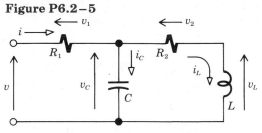

6.2–6   Assume that $\underline{I}_2 = 1$ in Fig. P6.2–3 to find $\underline{V}_1/\underline{V}$ and $\underline{V}/\underline{I}$ when $R_1 = R_2 = 4$, $1/\omega C = 2$, and $\omega L = 8$. Express your results in polar form.

6.2–7   Assume that $\underline{I}_2 = 1$ in Fig. P6.2–1 to find $\underline{I}_2/\underline{I}$ and $\underline{V}/\underline{I}$ when $R_1 = R_2 = 2$ and $1/\omega C = \omega L = 6$. Express your results in polar form.

6.2–8   Figure P6.2–8 represents a source network with $\underline{V}_s = 260 \ \text{mV} \underline{/0°}$ and $\omega = 10^6$. Let $L = 0$.

Figure P6.2–8

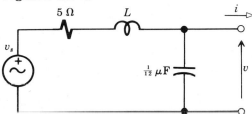

   (a) Obtain the Thévenin equivalent circuit by calculating the phasors $\underline{V}_{oc}$ and $\underline{I}_{sc}$ and $\underline{Z}_o = \underline{V}_{oc}/\underline{I}_{sc}$. Check $\underline{Z}_o$ by direct impedance calculation with $\underline{V}_s = 0$.
   (b) Find the output current i when the source drives 10-Ω resistance.

6.2–9   Do Problem 6.2–8 including $L = 24 \ \mu\text{H}$.

6.2–10 Suppose the network in Fig. 6.2–4a has $X(\omega) = 0$ at $\omega = 2000$ when $R_1 = 14 \ \Omega$, $C = 5 \ \mu\text{F}$, and $R_2 = 75 \ \Omega$. What are the corresponding values of $R(\omega)$ and L?

6.2–11 Suppose the network in Fig. P6.2–3 has $X(\omega) = 0$ at $\omega = 2000$ when $R_1 = 30 \ \Omega$, $R_2 = 10 \ \Omega$, and $L = 10 \ \text{mH}$. What are the corresponding values of $R(\omega)$ and C?

6.2–12 At a certain frequency, a 9-Ω resistor in series with inductance L has the same impedance as a 10-Ω resistor in parallel with a 0.5-H inductor. What are the values of ω and L?

6.2–13 At a certain frequency, a 200-Ω resistor in parallel with capacitance C has the same admittance as a 100-Ω resistor in series with a 10-μF capacitor. What are the values of ω and C?

6.2–14 Let Fig. P6.2–1 have $R_1 = 600 \ \Omega$, $L = 1 \ \text{H}$, $R_2 = 0$, and $C = 1 \ \mu\text{F}$. Find the frequency $\omega > 0$ at which $\text{Im}[Y(j\omega)] = 0$, and determine the corresponding value of $\underline{Z}$.

6.2–15 Do Problem 6.2–14 with $R_1 = 0$, $L = 0.1 \ \text{H}$, $R_2 = 30 \ \Omega$, and $C = 100 \ \mu\text{F}$.

6.2−16 Figure P6.2−16 is a **Maxwell bridge** designed to measure a coil's resistance R_u and inductance L_u by adjusting R_1 and R_2 until $v_a = v_b$. Write expressions for the phasors $\underline{V}_a$ and $\underline{V}_b$ to show that the balanced condition corresponds to $R_u = R_2 R/R_1$ and $L_u = CR_2 R$.

Figure P6.2−16

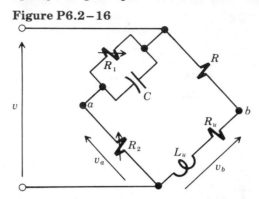

6.2−17 Suppose the bridge in Fig. P6.2−16 is modified by putting R_1 in series with C. Write expressions for the phasors $\underline{V}_a$ and $\underline{V}_b$ to find R_u and L_u in terms of ω and the other elements when $v_a = v_b$. Compare your results with those given in Problem 6.2−16.

6.3−1 Let the network in Fig. P6.2−1 have $\underline{Z} = 2 - j2\ \Omega$, $R_1 = \omega L = 2\ \Omega$, and $R_2 = 0$. Construct a diagram showing all the voltage and current phasors when $\underline{I} = 1 + j0$ A.

6.3−2 Let the network in Fig. P6.2−3 have $\underline{Z} = 2 + j1.2\ \Omega$, $1/\omega C = 0.8\ \Omega$, and $R_2 = 0$. Construct a diagram showing all the voltage and current phasors when $\underline{I} = 1 + j0$ A.

6.3−3 Let the network in Fig. P6.2−3 have $\underline{Z} = 5 + j0\ \Omega$, $1/\omega C = 5\ \Omega$, $R_2 = 4\ \Omega$, and $\omega L = 8\ \Omega$. Construct a diagram showing all the voltage phasors when $\underline{I} = 1 + j0$ A.

6.3−4 Let the network in Fig. P6.2−5 have $\underline{Z} = 8 - j6\ \Omega$, $R_1 = 5\ \Omega$, $R_2 = 12\ \Omega$, and $\omega L = 6\ \Omega$. Construct a diagram showing all the voltage phasors when $\underline{I} = 1 + j0$ A.

6.3−5 Suppose the series resonant circuit in Fig. 6.3−4 has $|\underline{V}| = 8$ V, $R = 40\ \Omega$, and $L = 50$ mH. Find C such that $\omega_0 = 2000$. Then calculate Q and $|\underline{V}_L|$ and $|\underline{V}_R + \underline{V}_L|$ at resonance.

6.3−6 Let the series resonant circuit in Fig. 6.3−4 have $\omega_0 L = 50\ \Omega$ and $Q = 25$. If the circuit is driven at resonance by a 10-V source with internal resistance $R_s = 3\ \Omega$, then what are the values of $|\underline{V}_C|$ and $|\underline{V}|$?

6.3−7 A coil with inductance L and series resistance R is connected to a 1-μF capacitor and a 5-V AC source. Measurements show that $|\underline{V}_C|_{max} = 100$ V at $\omega = 5000$. Find L and R.

6.3−8 The winding resistance of a certain coil is given by $R_w = 100\sqrt{L}\ \Omega$, where L is the coil's inductance in henrys. The coil is connected to a capacitor, forming a series resonant circuit with $R = R_w$. Determine the values of C, L, and R_w if $\omega_0 = 10^6$ and $Q = 500$.

6.3−9 Suppose the parallel resonant circuit in Fig. 6.3−6 has $|\underline{I}| = 2$ mA, $R = 15$ kΩ, and $C = 0.4\ \mu$F. Find L such that $\omega_0 = 50{,}000$. Then calculate Q and $|\underline{V}|$ and $|\underline{I}_L|$ at resonance.

6.3–10   Let the parallel resonant circuit in Fig. 6.3–6 have $\omega_0 L = 200 \ \Omega$ and $Q = 25$. If the current source includes parallel internal resistance $R_s = 15 \ k\Omega$ and it produces $|\underline{I}| = 0.5$ mA, then what are the values of $|\underline{V}_C|$ and $|\underline{V}|$ at resonance?

6.3–11   Stray capacitance between the windings of a 0.1-mH inductor makes it act like a parallel LC circuit resonant at $\omega = 10^6$. What external resistor and capacitor should be connected in parallel to get $\omega_0 = 500{,}000$ and $Q = 80$?

6.3–12   Given a parallel resonant circuit with $L = 20 \ \mu H$, $\omega_0 = 500{,}000$, and $Q = 100$, what external resistor and capacitor should be connected in parallel to get the lower values $\omega_0' = 250{,}000$ and $Q' = 40$?

6.3–13   The circuit in Fig. 6.2–4a may exhibit series resonance. Set $\text{Im}[Z(j\omega_0)] = 0$ to obtain an expression for ω_0^2/ω_{LC}^2 where $\omega_{LC}^2 = 1/LC$. Then determine the condition on L such that $0 < \omega_0^2 < \infty$, so resonance can occur.

6.3–14   Do Problem 6.3–13 for the circuit in Fig. P6.2–3.

6.3–15   The circuit in Fig. P6.2–1 may exhibit parallel resonance with $R_1 = 0$. Set $\text{Im}[Y(j\omega_0)] = 0$ to obtain an expression for ω_0^2/ω_{LC}^2 where $\omega_{LC}^2 = 1/LC$. Then determine the condition on C such that $0 < \omega_0^2 < \infty$, so resonance can occur.

6.3–16   Do Problem 6.3–15 with $R_1 > R_2$.

6.3–17   Use Eq. (14) to find the average power dissipated in Fig. P6.2–1 when $R_1 = 20 \ \Omega$, $L = 10$ mH, $R_2 = 0$, $C = 100 \ \mu F$, and $v = 10 \cos 1000t$ V. Then confirm that $\frac{1}{2}|\underline{I}|^2 R_1 = P$.

6.3–18   Use Eq. (14) to find the average power dissipated in Fig. P6.2–3 when $C = 250 \ \mu F$, $R_1 = 10 \ \Omega$, $R_2 = 0$, $L = 5$ mH, and $v = 10 \cos 2000t$ V. Then confirm that $\frac{1}{2}|\underline{I}|^2 R_1 = P$. (Results from this problem may be used in Problem 6.3–21.)

6.3–19   Use Eq. (14) to find the average power dissipated in Fig. P6.2–5 when $R_1 = 1 \ \Omega$, $C = 250 \ \mu F$, $R_2 = 8 \ \Omega$, $L = 4$ mH, and $v = 10 \cos 1000t$ V. Then calculate the power dissipated in each resistor, and confirm that $P_1 + P_2 = P$.

6.3–20   Use Eq. (14) to find the average power dissipated in Fig. P6.2–3 when $C = 100 \ \mu F$, $R_1 = 20 \ \Omega$, $R_2 = 4 \ \Omega$, $L = 4$ mH, and $v = 10 \cos 2000t$ V. Then calculate the power dissipated in each resistor, and confirm that $P_1 + P_2 = P$. (Results from this problem may be used in Problem 6.3–22.)

6.3–21   Find the total average power dissipated and the total input current i when the circuit in Problem 6.3–18 has $v = 10 \cos 2000t - 4 \cos 1000t$.

6.3–22   Find the total average power dissipated and the total input current i when the circuit in Problem 6.3–20 has $v = 10 \cos 2000t + 12 \cos (1000t - 20°)$.

6.4–1   Show that $i_1 = N^2 C \, dv_1/dt$ when capacitance C is connected across the secondary in Fig. 6.4–1.

6.4–2   Show that $v_1 = (L/N^2) \, di_1/dt$ when inductance L is connected across the secondary in Fig. 6.4–1.

6.4–3   Let the load in Fig. 6.4–3 be a parallel RL network. By considering admittances, confirm that the reflected load network consists of R/N^2 in parallel with L/N^2.

6.4–4   Let the load in Fig. 6.4–3 be a parallel RC network. By considering admittances, confirm that the reflected load network consists of R/N^2 in parallel with $N^2 C$.

6.4–5   Reflect each end into the middle section in Fig. P6.4–5 to find $\underline{I}_2$, $\underline{I}_1$, and all the voltage phasors when $N_2 = 2$, $\underline{Z}_2 = 30 + j30 \ \Omega$, $N_4 = 4$, and $\underline{Z}_4 = 800 - j32 \ \Omega$.

Figure P6.4−5

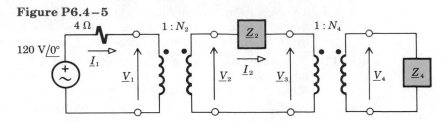

6.4−6 Do Problem 6.4−5 with $N_2 = 5$, $\underline{Z}_2 = 60 + j20\ \Omega$, $N_4 = \frac{1}{4}$, and $\underline{Z}_4 = 15 - j20\ \Omega$.

6.4−7 Do Problem 6.4−5 with $N_2 = \frac{1}{2}$, $\underline{Z}_2 = 5 - j5\ \Omega$, $N_4 = 3$, and $\underline{Z}_4 = 180 + j180\ \Omega$.

6.4−8 Do Problem 6.4−5 with $N_2 = \frac{1}{4}$, $\underline{Z}_2 = 0.75 + j0.5\ \Omega$, $N_4 = \frac{1}{2}$, and $\underline{Z}_4 = 3 - j2\ \Omega$.

6.4−9 Figure P6.4−9 is a parallel resonant circuit. Calculate ω_0 and Q, given that $N = 2$, $R_x = 60\ \text{k}\Omega$, and $C_x = 0.25\ \mu\text{F}$.

Figure P6.4−9

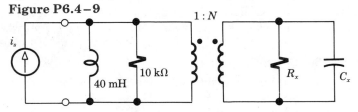

6.4−10 Let the parallel resonant circuit in Fig. P6.4−9 have $R_x = 10\ \text{k}\Omega$ and $C_x = 1\ \mu\text{F}$. Find N needed to get $\omega_0 = 10{,}000$, and determine the resulting Q.

6.4−11 Let the parallel resonant circuit in Fig. P6.4−9 have $R_x = 160\ \text{k}\Omega$. Find N and C_x such that $Q = 50$ and $\omega_0 = 2500$.

6.4−12 Use Eq. (4) to find $\underline{V}_2, \underline{I}_a$, and $\underline{I}_{in}$ in Fig. 6.4−5 when $\underline{V}_s = 24\ \text{V}\ \underline{/0°}$, $\underline{Z}_s = 0$, $N = 3$, $\underline{Z}_a = -j12\ \Omega$, and $\underline{Z} = 36\ \Omega$.

6.4−13 Use Eq. (4) to find $\underline{V}_2, \underline{I}_a$, and $\underline{I}_{in}$ in Fig. 6.4−5 when $\underline{V}_s = 25\ \text{V}\ \underline{/0°}$, $\underline{Z}_s = j3\ \Omega$, $N = 2$, and $\underline{Z}_a = \underline{Z} = 20\ \Omega$. (Hint: First write $\underline{V}_2, \underline{I}_a$, and $\underline{I}_{in}$ in terms of $\underline{V}_1$.)

6.4−14 Show for the circuit in Fig. 6.4−5 that $\underline{I}_{in}/\underline{V}_1 = (N^2/\underline{Z}) + (N - 1)^2/\underline{Z}_a$. (Hint: Note that $\underline{V}_2/\underline{Z} = \underline{I}_2 + \underline{I}_a$.)

6.4−15 Let $\underline{Z}_s = 50\ \Omega\ \underline{/25°}$ and $\omega = 10^4$ in Fig. 6.4−7. By considering the admittances, find the two elements that should be connected in parallel to form $\underline{Z}$ for maximum power transfer.

6.4−16 Do Problem 6.4−15 with $\underline{Z}_s = 250\ \Omega\ \underline{/-40°}$ and $\omega = 10^5$.

6.4−17 Suppose the transformer circuit in Fig. 6.4−3b has $\underline{Z}_s = 300\ \Omega\ \underline{/60°}$ and $\underline{Z} = 6\ \Omega + jX$. Find the values of N and X needed to get maximum power transfer.

6.4−18 Suppose the transformer circuit in Fig. 6.4−3b has $\underline{Z} = 480\ \Omega\ \underline{/-30°}$ and $\underline{Z}_s = 26\ \Omega + jX_s$. Find the values of N and X_s needed to get maximum power transfer.

6.4−19 If the circuit in Fig. P6.4−9 has $R_x = 400\ \Omega$ and operates at $\omega = 50{,}000$, then what values of N and C_x result in maximum power transfer to R_x?

Frequency Response and Transients

This chapter concludes our study of linear circuits with an introduction to frequency response and transient behavior. Both of these topics are important for their practical applications, as well as for a more complete understanding of the effects of energy storage in electrical networks.

Under AC steady-state conditions, circuits containing reactive elements exhibit frequency dependence in the sense that the response varies with the source frequency. We'll examine this concept of *frequency response* and show how simple networks function as *filters*. We'll also develop the Bode plot as a graphical technique for displaying frequency response.

We then turn to the *transient* behavior that takes place before a circuit reaches AC or DC steady-state conditions. After reviewing the natural response of first-order circuits, we'll examine switched DC transients of the type that occur in pulsed operation and electronic waveform generators. Finally, we consider the natural response and transient behavior of second-order circuits containing two energy-storage elements.

Objectives

After studying this chapter and working the exercises, you should be able to do each of the following:

- Sketch the frequency response and calculate the cutoff frequency of a first-order lowpass or highpass filter (Section 7.1).

- Calculate the center frequency and bandwidth of a second-order bandpass filter and sketch its frequency response (Section 7.1).

- Design a simple lowpass, highpass, or bandpass filter using an op-amp (Section 7.1).†

- Construct the Bode plot for a transfer function that consists of a product of first-order functions (Section 7.2).†

- Calculate the time constant of a first-order circuit (Section 7.3).

- Sketch the DC transient response, step response, and pulse response of a first-order circuit (Section 7.3).

- Obtain the characteristic equation of a second-order network, and determine if it is overdamped, underdamped, or critically damped (Section 7.4).†

- Find the natural response and step response of a second-order circuit (Section 7.4).†

7.1 FREQUENCY RESPONSE AND FILTERS

The previous chapter dealt with AC circuits operating at fixed source frequencies, so we paid little attention to the frequency dependence. Here, we consider ω to be variable and we'll formulate the corresponding transfer functions of simple filters in order to focus explicitly on their frequency-response characteristics. Basic lowpass, highpass, and bandpass filters will be examined, along with op-amp circuits that combine amplification and filtering.

Lowpass and Highpass Filters

Figure 7.1–1a represents a series RL circuit driven by an AC source $v = V_m \cos \omega t$ having variable frequency ω. The resulting current for a particular value of ω will be $i = I_m \cos (\omega t + \theta_i)$ with

$$I_m = \frac{V_m}{\sqrt{R^2 + (\omega L)^2}} \qquad \theta_i = -\arctan \frac{\omega L}{R} \tag{1}$$

which are obtained from simple impedance analysis. If all other parame-

Figure 7.1–1
(a) *RL* circuit.
(b) Frequency-response curves.

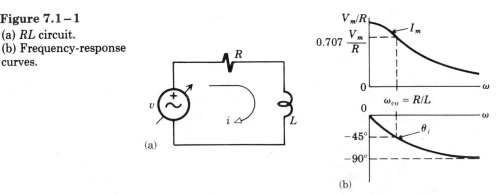

(a)

(b)

ters are kept fixed, then I_m and θ_i vary with ω as shown in Fig. 7.1–1b when the source frequency goes from low to high values. At the low-frequency end, where $\omega \to 0$, we have

$$I_m = \frac{V_m}{\sqrt{R^2 + 0}} = \frac{V_m}{R} \qquad \theta_i = -\arctan 0 = 0°$$

This low-frequency response agrees with our prior conclusion that inductance acts like a *DC short circuit,* since an AC source at $\omega = 0$ would actually be a DC source with $v = V_m \times \cos 0 = V_m$. At the other end of the frequency axis, where $\omega \to \infty$, we see that

$$I_m = 0 \qquad \theta_i = -\arctan \infty = -90°$$

meaning that the inductance effectively becomes an *open-circuit* or "AC choke" at high frequencies.

An easily computed intermediate point on the curves in Fig. 7.1–1b occurs where $\omega = R/L$ so $\omega L = R$ and

$$I_m = \frac{V_m}{\sqrt{R^2 + R^2}} = \frac{V_m}{\sqrt{2}\,R} = 0.707 \frac{V_m}{R}$$

$$\theta_i = -\arctan 1 = -45°$$

This point serves as a rough boundary between the two extremes, and we define the so-called **cutoff frequency**

$$\omega_{co} = R/L \tag{2}$$

The cutoff frequency is also known as the **half-power point** because the average power dissipation is $P = V_m^2/4R$ at ω_{co}, precisely one-half of the maximum value $P = V_m^2/2R$ obtained when $\omega \to 0$.

In view of the shape of I_m versus ω in Fig. 7.1–1b, we say that a series *RL* circuit functions as a **lowpass filter** — meaning that a low-frequency applied voltage produces much more current than a high-frequency source. Thus, if the applied signal actually consists of several different frequency components, then the circuit "passes" all low frequencies at $\omega \ll \omega_{co}$ but rejects or "filters out" all high frequencies at $\omega \gg \omega_{co}$.

We formalize this filtering property by arranging the circuit as the two-port network in Fig. 7.1–2, with input voltage phasor $\underline{V}_{in} = V_{in}\,\underline{/\theta_{in}}$

Figure 7.1–2
RL lowpass filter.

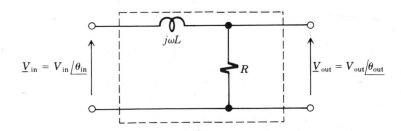

and output $\underline{V}_{out} = V_{out} \, \underline{/\theta_{out}}$. We then introduce the *AC voltage transfer function* as the ratio of voltage phasors at an arbitrary frequency ω. Specifically, let

$$H(j\omega) \triangleq \frac{\underline{V}_{out}}{\underline{V}_{in}} = \frac{V_{out}}{V_{in}} \, \underline{/\theta_{out} - \theta_{in}} \qquad (3)$$

which corresponds to setting $s = j\omega$ in the transfer function $H(s)$ from Section 5.3. In general, $H(j\omega)$ is a complex quantity and can be written in the polar form

$$H(j\omega) = |H(\omega)| \, \underline{/\theta(\omega)}$$

where, by definition,

$$|H(\omega)| = \frac{V_{out}}{V_{in}} \qquad \theta(\omega) = \theta_{out} - \theta_{in} \qquad (4)$$

We call $|H(\omega)|$ the **amplitude ratio** and $\theta(\omega)$ the **phase shift.**

To clarify these terms, note that any AC input voltage $v_{in}(t) = V_{in} \cos (\omega t + \theta_{in})$ produces the steady-state output

$$v_{out}(t) = V_{out} \cos (\omega t + \theta_{out})$$

whose amplitude and phase are found from Eq. (4) as

$$V_{out} = |H(\omega)| V_{in} \qquad \theta_{out} = \theta(\omega) + \theta_{in}$$

Furthermore, if the input consists of two or more sinusoids, say

$$v_{in}(t) = V_1 \cos (\omega_1 t + \theta_1) + V_2 \cos (\omega_2 t + \theta_2) + \cdots$$

then superposition holds and the steady-state output will be

$$v_{out}(t) = |H(\omega_1)| V_1 \cos [\omega_1 t + \theta(\omega_1) + \theta_1]$$
$$+ |H(\omega_2)| V_2 \cos [\omega_2 t + \theta(\omega_2) + \theta_2] + \cdots$$

Thus, by evaluating $|H(\omega)|$ and $\theta(\omega)$ at each value of ω, we can determine the effects of a filter on each frequency component of the input signal.

The transfer function for the filter in Fig. 7.1–2 is easily found from the impedance voltage divider

$$\underline{V}_{out} = \frac{R}{R + j\omega L} \, \underline{V}_{in} = \frac{1}{1 + j(\omega L/R)} \, \underline{V}_{in}$$

Upon inserting the cutoff frequency $\omega_{co} = R/L$ we obtain

$$H(j\omega) = \frac{\omega_{co}}{\omega_{co} + j\omega} = \frac{1}{1 + j(\omega/\omega_{co})} \qquad (5a)$$

which has

$$|H(\omega)| = \frac{1}{\sqrt{1 + (\omega/\omega_{co})^2}} \qquad \theta(\omega) = -\arctan \frac{\omega}{\omega_{co}} \qquad (5b)$$

Figure 7.1–3
Amplitude ratio and phase shift of a lowpass filter.

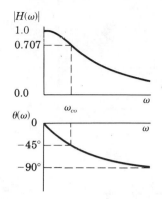

Figure 7.1–3 plots the corresponding frequency-response curves, $|H(\omega)|$ and $\theta(\omega)$ versus ω.

We see that this filter is, indeed, a *lowpass* filter because any input signal components at $\omega \ll \omega_{co}$ are passed to the output with amplitude and phase virtually unchanged, while any components at $\omega \gg \omega_{co}$ have greatly reduced output amplitudes. The cutoff frequency is defined such that

$$|H(\omega_{co})| = 1/\sqrt{2} = 0.707$$

so ω_{co} roughly divides the *passband* from the *stopband*.

An RC circuit arranged as shown in Fig. 7.1–4 also acts as a lowpass filter. The lowpass filtering property here comes from the fact that the capacitor is an open circuit for $\omega = 0$ but becomes a short circuit for $\omega \to \infty$ and "shorts out" any high-frequency voltage components across the output. The transfer function is identical to Eq. (5) with the cutoff frequency now given by

$$\omega_{co} = 1/RC \tag{6}$$

Likewise, the frequency-response curves have the same shape as Fig. 7.1–3.

Figure 7.1–4
RC lowpass filter.

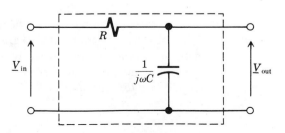

Figure 7.1–5

Highpass filters.

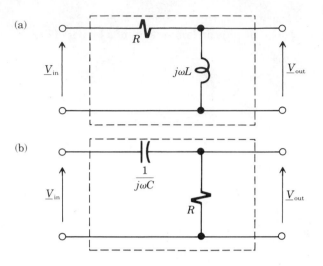

(a)

V_{in} R $j\omega L$ V_{out}

(b)

V_{in} $\dfrac{1}{j\omega C}$ R V_{out}

Figure 7.1–6

Amplitude ratio and phase shift of a highpass filter.

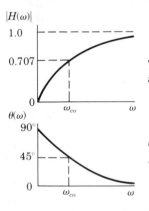

$|H(\omega)|$

1.0

0.707

0

ω_{co} ω

$\theta(\omega)$

90°

45°

0

ω_{co} ω

Interchanging the resistance and reactance in our lowpass filter circuits yields the **highpass filters** diagrammed in Fig. 7.1–5. The highpass transfer function for either of these circuits can be written as

$$H(j\omega) = \frac{j\omega}{\omega_{co} + j\omega} = \frac{j(\omega/\omega_{co})}{1 + j(\omega/\omega_{co})} \qquad \textbf{(7a)}$$

with $\omega_{co} = R/L$ or $1/RC$. Converting Eq. (7a) to polar form yields the amplitude ratio and phase shift

$$|H(\omega)| = \frac{(\omega/\omega_{co})}{\sqrt{1 + (\omega/\omega_{co})^2}} \qquad \theta(\omega) = 90° - \arctan\frac{\omega}{\omega_{co}} \qquad \textbf{(7b)}$$

The resulting frequency-response curves plotted in Fig. 7.1–6 confirm that these filters block low frequencies ($\omega \ll \omega_{co}$) and pass high frequencies ($\omega \gg \omega_{co}$).

Despite the similarity of RL and RC circuits, virtually all simple filter designs employ capacitors rather than inductors due to the greater availability and lower cost of capacitors and the unwanted resistance associated with inductors. More sophisticated filters, using two or more reactive elements, have a "squarer" frequency response of the type sketched in Fig. 7.1–7, where $|H(\omega)|$ is essentially constant or "flat" over the passband and there is a relatively narrow transition between the passband and stopband.

Finally, it should be noted that we usually express values of frequency in *hertz*, meaning cyclical frequency f rather than ω. All of our results here still hold for cyclical frequency, since $\omega = 2\pi f$ so factors of 2π cancel out in any frequency ratios. Thus, $\omega/\omega_{co} = f/f_{co}$, and the cyclical cutoff frequency is $f_{co} = \omega_{co}/2\pi$.

Figure 7.1–7

Amplitude ratio of improved lowpass filter.

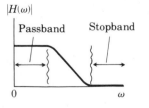

$|H(\omega)|$

Passband   Stopband

0   ω

Example 7.1–1   A home intercom system has developed an annoying 10-kHz "whistle" whose amplitude is about ten-percent of the typical voice signal. Knowing that voice frequency components much above 3 kHz are unimportant for intelligibility, the owner decides to insert an RC lowpass filter with $f_{co} = \omega_{co}/2\pi = 4$ kHz to get rid of the whistle while keeping the voice signal. He chooses $R = 2$ kΩ, arbitrarily, and computes the capacitance $C = 1/R\omega_{co} \approx 0.02\ \mu$F.

To analyze the strategy here, let's model the voice signal as a 3-kHz sinusoid with 5-V amplitude, so the total input signal is the sum

$$v_{in}(t) = 5\cos\omega_1 t + 0.5\cos\omega_2 t$$

where $\omega_1 = 2\pi \times 3$ kHz and $\omega_2 = 2\pi \times 10$ kHz. By superposition, the output will be the sum of the individual outputs determined from the transfer function given in Eq. (5) with $\omega_1/\omega_{co} = 3$ kHz/4 kHz $= 0.75$ and $\omega_2/\omega_{co} = 10$ kHz/4 kHz $= 2.5$, respectively. Inserting numerical values yields

$$H(j\omega_1) = 0.80\ \underline{/-36.9°} \qquad H(j\omega_2) = 0.371\ \underline{/-68.2°}$$

from which

$$v_{out}(t) = 4.0\cos(\omega_1 - 36.9°) + 0.19\cos(\omega_2 - 68.2°)$$

Therefore, the whistle amplitude is now down to roughly five-percent of the signal's slightly reduced amplitude — not much improvement, but a first-order filter can do no better in this particular case.

After installing the filter, the owner is dismayed to find virtually no output signal at all, voice or whistle. What's been overlooked is the *loading effect* due to a 50-Ω source resistance and its matched load resistance, the equivalent circuit with filter being as diagrammed in Fig. 7.1–8a. However, the owner can still achieve the desired filtering by discarding the 2-kΩ resistance entirely and calculating C from the circuit in Fig. 7.1–8b. The Thévenin equivalent circuit seen by the capacitance has $v_{oc} = v_{in}/2$ and $R_{eq} = 50\|50 = 25\ \Omega$, so getting $f_{co} = 4$ kHz requires $C = 1/(25\Omega \times 2\pi \times 4 \times 10^3$ Hz$) \approx 1.6\ \mu$F. The output is then $v_{out} \approx v_{in}/2$ at $\omega \ll \omega_{co}$ due to the voltage divider.

Figure 7.1–8
Circuits for Example 7.1–1.

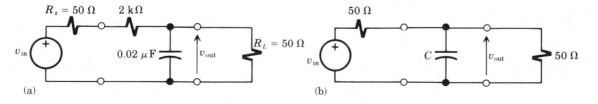

(a)   (b)

Exercise 7.1–1   Confirm that both circuits in Fig. 7.1–5 have $H(j\omega)$ as given by Eq. (7a).

Exercise 7.1–2   A certain telemetry signal consists of two individual signals: $v_a(t)$, with frequencies below 10 kHz, and $v_b(t)$, with frequencies above 50 kHz. The circuit in Fig. 7.1–9 has been proposed as a means of separating the two signals. Determine appropriate values for R_a and R_b. Then calculate the actual output amplitudes when $v(t)$ is a 30-kHz sinusoid with 10-V amplitude.

Figure 7.1–9

Circuit for Exercise 7.1–2.

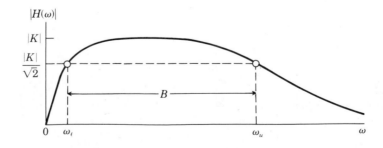

Bandpass Filters

Bandpass filters include at least two reactive elements arranged to achieve a frequency-response characteristic that combines lowpass and highpass filtering. Thus, they pass a frequency band falling between the lower and upper cutoffs. Such filters play important roles in communication electronics (radio, TV, and radar) and in various instrumentation systems.

A simple bandpass filter might have the characteristic sketched in Fig. 7.1–10, where ω_ℓ and ω_u stand for the **lower** and **upper cutoff frequencies,** respectively. The amplitude ratio is relatively "flat" over the passband and has the value

$$|H(\omega)| \approx |K| \qquad \omega_\ell < \omega < \omega_u$$

We call $|K|$ the **midband gain,** and we define the cutoff frequencies by the property that

$$|H(\omega_\ell)| = |H(\omega_u)| = |K|/\sqrt{2} \tag{8a}$$

Figure 7.1–10

Amplitude ratio of a wideband filter.

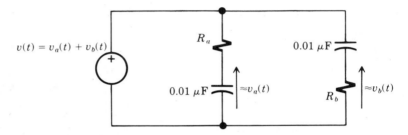

The filter's **bandwidth** is

$$B = \omega_u - \omega_\ell \qquad \textbf{(8b)}$$

and we say we have a **wideband filter** when $\omega_u \gg \omega_\ell$ so $B \approx \omega_u$. The transfer function of a wideband filter is expressed mathematically by

$$H(j\omega) = K \frac{j(\omega/\omega_\ell)}{1 + j(\omega/\omega_\ell)} \frac{1}{1 + j(\omega/\omega_u)} \qquad \textbf{(9)}$$

If $K = 1$, then Eq. (9) is simply the product of first-order highpass and lowpass functions with cutoff frequencies $\omega_\ell \ll \omega_u$. Notice that the *highpass* function accounts for the *lower* cutoff frequency ω_ℓ, and vice versa.

Many applications call for a **narrowband filter** whose passband is centered at some frequency ω_0 and whose bandwidth B is small compared to the center frequency. This type of characteristic can be implemented with a high-Q resonant circuit "tuned" to the desired center frequency. Consider, for instance, the tuned circuit in Fig. 7.1–11a. Qualitatively, we see that the inductor shorts out low-frequency voltage components while the capacitor shorts out high frequencies; consequently, the filter passes only those frequencies in the vicinity of resonance where the LC section has maximum impedance.

Figure 7.1–11
Tuned circuits for narrowband filters.

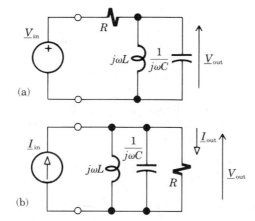

To emphasize that we are dealing with *parallel* resonance, we perform a Thévenin-to-Norton conversion and redraw the circuit per Fig. 7.1–11b. Taking the input and output to be the indicated currents then, relative to Fig. 7.1–11a, $\underline{I}_{in} = \underline{V}_{in}/R$ and $\underline{V}_{out} = R\underline{I}_{out}$, so $\underline{I}_{out}/\underline{I}_{in} = \underline{V}_{out}/\underline{V}_{in}$ and the two circuits have identical transfer functions. In practice, the parallel configuration driven by a current source often occurs as the model of a tuned amplifier.

For a quantitative analysis of tuned circuits, we use the parallel admittance $Y(j\omega) = \underline{I}_{in}/\underline{V}_{out}$ and we write $\underline{I}_{out} = \underline{V}_{out}/R = \underline{I}_{in}/RY(j\omega)$, from which

$$H(j\omega) = \frac{\underline{V}_{out}}{\underline{V}_{in}} = \frac{\underline{I}_{out}}{\underline{I}_{in}} = \frac{1}{RY(j\omega)} = \frac{1}{1 + j(\omega CR - R/\omega L)}$$

This expression can be put in a more convenient form by introducing the resonant frequency and quality factor

$$\omega_0 = 1/\sqrt{LC} \qquad Q = \omega_0 CR = R/\omega_0 L \qquad \text{(10)}$$

so that

$$\omega CR = \omega_0 CR \frac{\omega}{\omega_0} = Q\frac{\omega}{\omega_0} \qquad \frac{R}{\omega L} = \frac{R}{\omega_0 L}\frac{\omega_0}{\omega} = Q\frac{\omega_0}{\omega}$$

Making these substitutions yields

$$H(j\omega) = \frac{1}{1 + jQ\left(\dfrac{\omega}{\omega_0} - \dfrac{\omega_0}{\omega}\right)} \qquad \text{(11a)}$$

with

$$|H(\omega)| = \frac{1}{\sqrt{1 + Q^2\left(\dfrac{\omega}{\omega_0} - \dfrac{\omega_0}{\omega}\right)^2}} \qquad \text{(11b)}$$

$$\theta(\omega) = -\arctan Q\left(\frac{\omega}{\omega_0} - \frac{\omega_0}{\omega}\right)$$

which are plotted in Fig. 7.1–12. The midband "gain" is $|H(\omega_0)| = 1$.

This figure clearly supports our hunch that the tuned circuit performs a bandpass filtering function, passing only those frequencies in the vicinity of ω_0. The two cutoff frequencies are defined from Eq. (8a) to be where $|H(\omega)|^2 = \frac{1}{2}$, which means that

$$Q^2\left(\frac{\omega}{\omega_0} - \frac{\omega_0}{\omega}\right)^2 = 1$$

at $\omega = \omega_\ell$ and ω_u. Solving the resulting quadratic equation yields

$$\omega_\ell = \omega_0\sqrt{1 + \left(\frac{1}{2Q}\right)^2} - \frac{\omega_0}{2Q}$$

$$\omega_u = \omega_0\sqrt{1 + \left(\frac{1}{2Q}\right)^2} + \frac{\omega_0}{2Q} \qquad \text{(12a)}$$

Therefore, the filter's bandwidth is

$$B = \omega_u - \omega_\ell = \frac{\omega_0}{Q} \qquad \text{(12b)}$$

Figure 7.1–12
Amplitude ratio and
phase shift of a
narrowband filter.

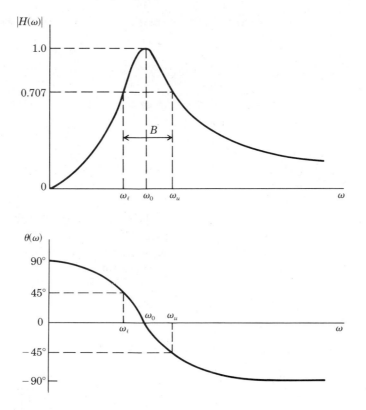

so a high-Q filter has a narrow bandwidth. When $Q \gg 1$, we can use the
approximations

$$\omega_\ell \approx \omega_0 - \frac{\omega_0}{2Q} = \omega_0 - \frac{B}{2} \qquad \omega_u \approx \omega_0 + \frac{\omega_0}{2Q} = \omega_0 + \frac{B}{2}$$

These are derived from Eq. (12a) with $(1/2Q)^2 \ll 1$. In this very common
case, the amplitude-ratio curve is essentially symmetrical around ω_0.

Lastly, we should take account of the inevitable series resistance associated with a real inductor, which leads to the more realistic tuned-circuit
model in Fig. 7.1–13. The admittance of this circuit is

$$Y(j\omega) = \frac{1}{R_1} + j\omega C + \frac{1}{R_2 + j\omega L} = \frac{1}{R_1} + \frac{j\omega C R_2 - \omega^2 L C + 1}{R_2 + j\omega L}$$

Figure 7.1–13
Practical tuned circuit.

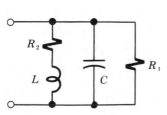

However, if $R_2 \ll \omega_0 L$ then $R_2 + j\omega L \approx j\omega L$ for the frequencies of interest and

$$Y(j\omega) \approx \frac{1}{R_1} + \frac{CR_2}{L} + j\omega C + \frac{1}{j\omega L}$$

which has the same form as that of a parallel RLC circuit with

$$\frac{1}{R} = \frac{1}{R_1} + \frac{CR_2}{L} \tag{13}$$

Our previous results therefore apply to Fig. 7.1–13 by computing R from Eq. (13).

Example 7.1–2
Tuned amplifier.

Figure 7.1–14 represents the output stage of a tuned amplifier with transformer coupling to a 50-Ω load. The amplifier is to operate over a frequency range of 200 ± 5 kHz, delivering maximum signal power to the load. We need to find C, L, and the turns ratio N to suit the specifications. We also want to determine $\underline{V}_{out}$ when $\underline{I}_s = 10$ mA $\underline{/0°}$ and the source frequency is $f = 200$ kHz.

Figure 7.1–14
Output stage of a tuned amplifier.

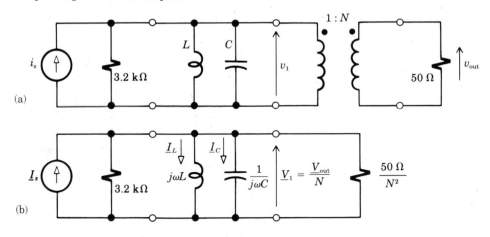

First, we refer the load resistance to the primary side of the transformer, thereby obtaining the frequency-domain equivalent circuit of Fig. 7.1–14b. Since the parallel reactances cancel out at resonance, we get maximum power transfer when the referred load resistance equals the source resistance. Accordingly, we want

$$\frac{50\ \Omega}{N^2} = 3.2\ \mathrm{k}\Omega \qquad N = \sqrt{\frac{50}{3200}} = \frac{1}{8}$$

which calls for a step-down transformer. The equivalent parallel resistance is then $R = 3.2 \text{ k}\Omega/2 = 1.6 \text{ k}\Omega$.

Next, from the frequency specifications we take $\omega_0 = 2\pi \times 200 \text{ kHz}$, $\omega_\ell = 2\pi \times 195 \text{ kHz}$, and $\omega_u = 2\pi \times 205 \text{ kHz}$, so $B = 2\pi \times 10 \text{ kHz}$ and

$$Q = \omega_0/B = 200/10 = 20$$

Inserting values in Eq. (10) now yields

$$C = \frac{Q}{\omega_0 R} = 9.95 \times 10^{-9} \approx 0.01 \ \mu\text{F}$$

$$L = \frac{R}{\omega_0 Q} = 6.37 \times 10^{-5} \approx 64 \ \mu\text{H}$$

Finally, to compute $\underline{V}_{\text{out}}$ when $\underline{I}_s = 10 \text{ mA} \ \underline{/0°}$ and $\omega = \omega_0$ we draw upon the fact that $\underline{I}_L + \underline{I}_C = 0$ at resonance. Hence, $\underline{V}_{\text{out}} = N\underline{V}_1 = N(R\underline{I}_s) = 2 \text{ V} \ \underline{/0°}$.

Exercise 7.1–3   The intermediate-frequency (IF) amplifier in an FM tuner has $\omega_0 = 2\pi \times 10.7 \text{ MHz}$ and $B = 2\pi \times 250 \text{ kHz}$. Find Q, R, and C for the corresponding tuned circuit when $L = 10 \ \mu\text{H}$.

Op-Amp Filters †

Loading effects and other shortcomings of passive filters may be overcome by including an amplifying device to make an **active filter.** We'll briefly introduce this technique using the ideal op-amp circuit in Fig. 7.1–15. If the impedances had been resistances, then you would recognize this configuration as a simple inverting amplifier with $v_{\text{out}}/v_{\text{in}} = -R_F/R_1$. Putting impedances in place of resistances immediately gives us the transfer function

$$H(j\omega) = \underline{V}_{\text{out}}/\underline{V}_{\text{in}} = -Z_F(j\omega)/Z_1(j\omega) \tag{14}$$

The frequency-response characteristics therefore depend on $Z_F(j\omega)$ and $Z_1(j\omega)$.

Figure 7.1–15

Active filter with ideal op-amp.

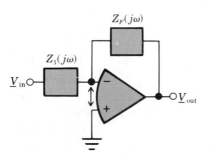

210   Chapter 7   Frequency Response and Transients

Figure 7.1–16
(a) Active lowpass
filter. (b) Active
highpass filter.

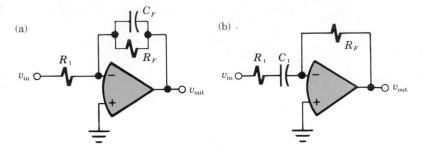

Figure 7.1–16a diagrams a lowpass op-amp filter implemented by taking

$$Z_1(j\omega) = R_1 \qquad Z_F(j\omega) = R_F\|\left(\frac{1}{j\omega C_F}\right) = \frac{R_F}{1 + j\omega C_F R_F}$$

Thus, from Eq. (14),

$$H(j\omega) = -\frac{R_F}{R_1}\frac{1}{1 + j(\omega/\omega_{co})} \qquad \omega_{co} = \frac{1}{R_F C_F} \qquad (15)$$

which is a first-order lowpass function with low-frequency gain $H(0) = -R_F/R_1$. The input impedance equals R_1, which should be large compared to the source impedance to minimize loading effect. Similar analysis of the highpass filter circuit in Fig. 7.1–16b shows that

$$H(j\omega) = -\frac{R_F}{R_1}\frac{j(\omega/\omega_{co})}{1 + j(\omega/\omega_{co})} \qquad \omega_{co} = \frac{1}{R_1 C_1} \qquad (16)$$

so the high-frequency gain is $H(\infty) = -R_F/R_1$.

The lowpass and highpass filtering operations are combined in the circuit of Fig. 7.1–17 to obtain a wideband bandpass filter. The resulting transfer function is given by Eq. (9) with

$$K = -R_F/R_1 \qquad \omega_\ell = 1/R_1 C_1 \qquad \omega_u = 1/R_F C_F \qquad (17)$$

provided that $\omega_u \gg \omega_\ell$. The amplitude-ratio curve was previously plotted in Fig. 7.1–10.

Figure 7.1–17
Active bandpass filter.

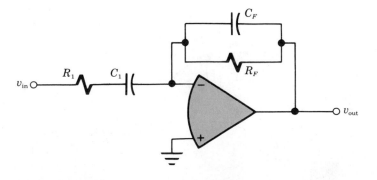
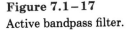

More sophisticated op-amp filters have "squarer" frequency responses, adjustable gain, and large input impedance. Some of these circuits are included in the problems at the end of this chapter.

Example 7.1–3

Suppose that, along with the 10-kHz whistle, the intercom system in Example 7.1–1 also suffers from a low-frequency 60-Hz "hum" and an inadequate signal level. We can cure all these problems in one fell swoop with an op-amp bandpass filter, using element values chosen according to the following line of reasoning.

Since the source has $R_s = 50\ \Omega$ we take $R_1 = 1\ \mathrm{k\Omega} \gg R_s$ to minimize input loading. We then want C_1 such that the lower cutoff frequency from the highpass function falls above 60 Hz to reduce the hum, but below the significant voice frequencies. An appropriate compromise frequency might be 200 Hz, obtained with $C_1 \approx 0.08\ \mu\mathrm{F}$. The midband voltage gain between 200 Hz and 4 kHz equals $-R_F/R_1$, and we can increase the signal level by a factor of three, say, if $R_F = 3\ \mathrm{k\Omega}$. This, in turn, leads to $C_F \approx 0.01\ \mu\mathrm{F}$ for the lowpass function to have an upper cutoff of 4 kHz.

Exercise 7.1–4

Use Eq. (14) to derive $H(j\omega)$ for the op-amp bandpass filter in Fig. 7.1–17.

7.2 BODE PLOTS †

The transfer function $H(j\omega)$ of a two-port network actually represents two related functions, the amplitude ratio or gain $|H(\omega)| = |H(j\omega)|$ and the phase shift $\theta(\omega) = \measuredangle H(j\omega)$. A **Bode plot** (pronounced Bo-dee) displays these functions in a special way: the gain is expressed in *decibels* and plotted along with the phase shift on a *logarithmic frequency axis*. The resulting curves are valuable for studying the frequency-response characteristics of filters, amplifiers, and linear systems in general.

Bode plots can be sketched rapidly by taking advantage of simple asymptotic behavior, rather than making extensive numerical calculations. We'll develop that construction technique here, beginning with factored transfer functions. Then we'll examine the basic first-order and second-order functions obtained by factoring a more complicated expression.

Factored Transfer Functions

Given a transfer function, the starting point for constructing its Bode plot is to decompose $H(j\omega)$ into a *factored* expression in the form

$$H(j\omega) = KH_1(j\omega)H_2(j\omega) \cdots \qquad (1)$$

The constant K is chosen to be a real number, possibly negative, and $H_1(j\omega), H_2(j\omega), \ldots$, are chosen to be simple functions with known Bode plots.

Having factored $H(j\omega)$, the overall gain can be written as the product

$$|H(\omega)| = |K||H_1(\omega)||H_2(\omega)| \cdots$$

which becomes a *sum* when we express gain in **decibels** (dB). Specifically, we define the dB gain by

$$H_{\mathrm{dB}} \triangleq 10 \log |H(\omega)|^2 = 20 \log |H(\omega)| \qquad \textbf{(2)}$$

Then, since the logarithm of a product equals the sum of the logarithms, we obtain the overall dB gain

$$H_{\mathrm{dB}} = K_{\mathrm{dB}} + H_{1_{\mathrm{dB}}} + H_{2_{\mathrm{dB}}} + \cdots \qquad \textbf{(3)}$$

where $K_{\mathrm{dB}} = 20 \log |K|$, $H_{1_{\mathrm{dB}}} = 20 \log |H_1(\omega)|$, etc.

Furthermore, since the angle of a product of complex number equals the sum of the individual angles, the overall phase shift $\theta = \sphericalangle H(j\omega)$ can also be written as a sum. Thus, from Eq. (1),

$$\theta = \theta_K + \theta_1 + \theta_2 + \cdots \qquad \textbf{(4)}$$

where $\theta_1 = \sphericalangle H_1(j\omega)$, etc., while $\theta_K = 0°$ or $\pm 180°$ depending on whether K is positive or negative.

Equations (3) and (4) bring out the underlying strategy here. For if you know the Bode plots of the individual factors of $H(j\omega)$, then you can simply *add them together* to get the complete Bode plot. Accordingly, we next focus attention on the individual factors that might make up a typical transfer function per Eq. (1).

First-Order Functions

Most of the factored functions of interest are first order, consisting of constants and ω but no higher powers of ω. The simplest case is the linear *ramp* function

$$H_r(j\omega) = j\omega/W = (\omega/W) \; \underline{/90°} \qquad \textbf{(5a)}$$

where W stands for any positive real constant. The gain and phase are

$$H_{r_{\mathrm{dB}}} = 20 \log (\omega/W) \qquad \theta_r = 90° \qquad \textbf{(5b)}$$

which are plotted against the *logarithmic* frequency axis in Fig. 7.2–1. Observe that the gain curve appears as a straight line passing through 0 dB at $\omega = W$. This line has a slope of $+20$ dB per **decade,** meaning that $H_{r_{\mathrm{dB}}}$ increases by 20 dB when the frequency increases by a factor of 10. The slope is also equivalent to about $+6$ dB per **octave,** an octave being a frequency increase by a factor of 2.

Next, consider the case of a first-order *highpass* function with $\omega_{\mathrm{co}} = W$, so

$$H_{\mathrm{hp}}(j\omega) = \frac{j(\omega/W)}{1 + j(\omega/W)} = \frac{j\omega}{W + j\omega} \qquad \textbf{(6a)}$$

Figure 7.2–1

Bode plot of the ramp function $H_r(j\omega) = j\omega/W$.

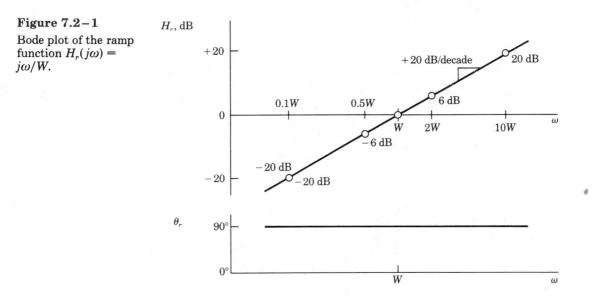

Its Bode plot can be constructed by taking advantage of the low-frequency and high-frequency approximations

$$H_{hp}(j\omega) \approx \begin{cases} j\omega/W = H_r(j\omega) & \omega < 0.1W \\ j\omega/j\omega = 1 \ \underline{/0^\circ} & \omega > 10W \end{cases} \qquad \textbf{(6b)}$$

We also draw upon Fig. 7.1–6 to obtain the midpoint values at $\omega = W$, namely

$$H_{hp_{dB}} = 20 \log (1/\sqrt{2}) \approx -3 \text{ dB} \qquad \theta_{hp} = 45^\circ \qquad \textbf{(6c)}$$

Since $H_{hp} \approx -3$ dB at $\omega = W$, the cutoff frequency of a first-order filter is sometimes referred to as the "3-dB frequency."

The resulting gain and phase curves are plotted in Fig. 7.2–2 along with straight-line **asymptotic approximations.** Being straight lines, these asymptotes are easily drawn on semilogarithmic graph paper. We call W the **corner frequency** or **break frequency** in this context because the gain approximation initially rises with a 20-dB slope and then "breaks" at $\omega = W$ and becomes a horizontal line. However, the phase approximation does not turn a corner at W; instead it breaks at $0.1W$ and $10W$ and has a slope of -45° per decade between the horizontal low-frequency and high-frequency asymptotes.

Often, the asymptotic approximations provide sufficient information about the frequency response. When more precise values are needed in the range $0.1W \leq \omega \leq 10W$, you just plot additional points using the indicated correction terms relative to the asymptotes and draw a smooth curve through them. Figure 7.2–2 therefore deserves careful examination.

Figure 7.2–2
Bode plot of the highpass function $H_{hp}(j\omega) = j\omega/(W+j\omega)$.

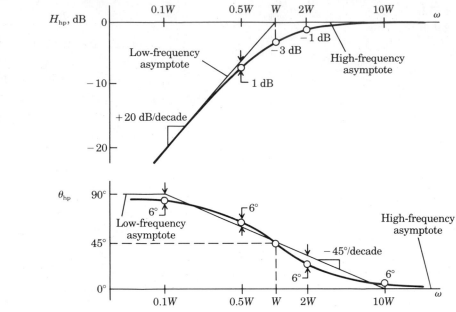

Finally, consider the case of a first-order *lowpass* function with $\omega_{co} = W$, so

$$H_{\ell p}(j\omega) = \frac{1}{1+j(\omega/W)} = \frac{W}{W+j\omega} \tag{7a}$$

The low-frequency and high-frequency asymptotic approximations are

$$H_{\ell p}(j\omega) \approx \begin{cases} W/W = 1 \; \underline{/0^\circ} & \omega < 0.1W \\ W/j\omega = 1/H_r(j\omega) & \omega > 10W \end{cases} \tag{7b}$$

Since the high-frequency approximation equals the *reciprocal* of $H_r(j\omega)$, its plot is obtained from Fig. 7.2–1 simply by *changing the signs* of the gain and phase. Thus, as shown in Fig. 7.2–3, the gain curve starts as a horizontal line at 0 dB but falls off with a slope of -20 dB per decade above the break frequency W. Similarly, the phase curve now goes from 0° to -90°.

Combinations of the three simple functions $H_r(j\omega)$, $H_{hp}(j\omega)$, and $H_{\ell p}(j\omega)$, together with their reciprocals, cover an amazingly wide variety of practical cases. Two examples should help demonstrate this point and illustrate the technique of Bode plotting.

Example 7.2–1

A Bode plot is to be constructed for a filter having the transfer function

$$H(j\omega) = \frac{j1200\omega}{(20+j\omega)(400+j\omega)}$$

Figure 7.2–3
Bode plot of the lowpass function $H_{lp}(j\omega) = W/(W + j\omega)$.

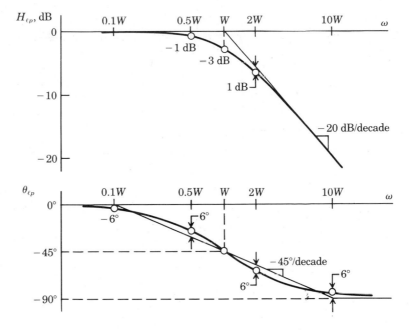

Careful examination of this expression reveals that it consists of a high-pass function and a lowpass function. With a little manipulation, $H(j\omega)$ can be factored to get

$$H(j\omega) = 3 \frac{j\omega}{20 + j\omega} \frac{400}{400 + j\omega}$$

which we recognize as a *bandpass* function with $K = 3$, $\omega_\ell = 20$, and $\omega_u = 400$. When you factor a transfer function, you should always put the terms in order of increasing break frequencies as has been done here.

We construct the Bode plot by graphically adding the asymptotes of highpass and lowpass functions that have break frequencies of 20 and 400, respectively. Then, accounting for K, we shift the entire gain curve up by $K_{dB} = 20 \log 3 \approx 9.5$ dB. Had K been a negative quantity, the phase curve would also be shifted up or down to include $\theta_K = \pm 180°$.

Figure 7.2–4 shows the individual asymptotic approximations from Figs. 7.2–2 and 7.2–3, the sums with K_{dB} added to the gain curve, and the final smooth plots obtained using the correction terms. Observe that the gain correction terms do not "overlap" because the break frequencies are more than a decade apart — consistent with the characteristics of a wide-band filter. However, the phase corrections do have some overlap; thus, for instance, the total phase correction at $\omega = 40$ consists of $-6°$ from the highpass function at $\omega = 2\omega_\ell$ plus another $-6°$ from the lowpass function at $\omega = \omega_u/10$.

Figure 7.2–4
Bode plot for Example
7.2–1: (a) asymptotes;
(b) final curves with
correction terms.

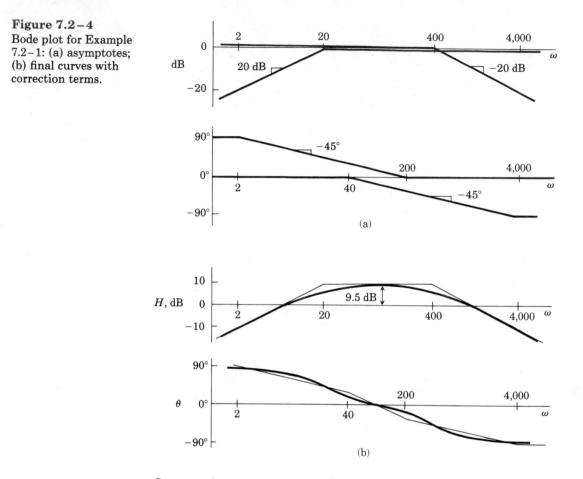

(a)

(b)

Example 7.2–2

Suppose that we have a transfer function given by

$$H(s) = \frac{25{,}000(s + 10)}{s^3 + 550s^2 + 25{,}000s}$$

We'll need to set $s = j\omega$ to obtain $H(j\omega)$. But before taking that step, we should try to rewrite the polynomial in the denominator as a product of first-order functions. This turns out to be an easy task since

$$s^3 + 550s^2 + 25{,}000s = s(s + 50)(s + 500)$$

Thus,

$$H(j\omega) = \frac{25{,}000(j\omega + 10)}{j\omega(j\omega + 50)(j\omega + 500)} = \frac{10 + j\omega}{j\omega} \frac{50}{50 + j\omega} \frac{500}{500 + j\omega}$$

which has the form $H(j\omega) = H_1(j\omega)H_2(j\omega)H_3(j\omega)$.

The first term $H_1(j\omega) = (10 + j\omega)/j\omega$ is the reciprocal of a highpass function with $W = 10$, while $H_2(j\omega)$ and $H_3(j\omega)$ are lowpass functions with break frequencies at 50 and 500. Since $K = 1$ in this case, $K_{dB} = 20 \log 1 = 0$ dB so the gain curve does not require vertical shifting.

The complete Bode plot is drawn in Fig. 7.2–5. The individual asymptotes have been omitted, but they come from Fig. 7.2–2 (with the signs reversed) and from Fig. 7.2–3. Due to the joint effects of $H_2(j\omega)$ and $H_3(j\omega)$, the high-frequency gain has a slope of -40 dB per decade and the phase approaches $-180°$.

Figure 7.2–5
Bode plot for Example 7.2–2.

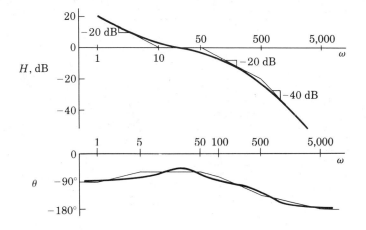

Exercise 7.2–1   Draw the Bode plot of $H(j\omega) = j1000\omega/(100 + j\omega)^2$ and determine the maximum value of H_{dB}. Hint: There are two break frequencies at $\omega = 100$.

Exercise 7.2–2   Use the reciprocal of the lowpass function to obtain the gain and phase asymptotes for $H(j\omega) = -8\,(50 + j\omega)/(400 + j\omega)$.

Second-Order Functions

Second-order functions involve ω^2 as well as ω, and usually occur in a form equivalent to

$$H(j\omega) = \frac{1}{1 + j2\zeta(\omega/\omega_0) - (\omega/\omega_0)^2} \tag{8}$$

The parameter ζ (zeta) is called the **damping ratio**. If $\zeta > 1$, then we have an **overdamped** case and $H(j\omega)$ can be rewritten as a product of two first-order lowpass functions. If $\zeta < 1$, then we have an **underdamped** or **resonant** case with break frequency ω_0 and asymptotes determined from the approximations

$$H_{dB} \approx \begin{cases} 0 & \omega < 0.1\,\omega_0 \\ -40 \log(\omega/\omega_0) & \omega > 10\,\omega_0 \end{cases}$$

$$\theta \approx \begin{cases} 0° & \omega < 0.1\,\omega_0 \\ -180°C & \omega > 10\,\omega_0 \end{cases} \tag{9}$$

Figure 7.2−6
Bode plot of second-order function with damping ratio ζ.

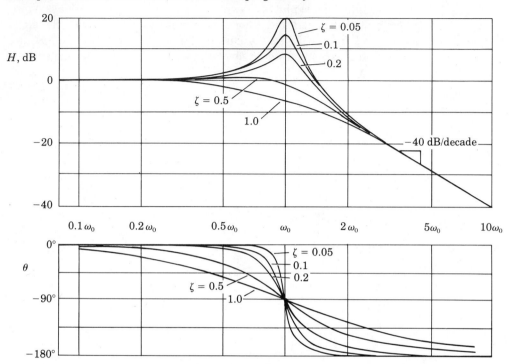

However, the exact behavior in the vicinity of ω_0 depends critically upon the damping ratio.

Figure 7.2−6 shows the Bode plot of Eq. (8) for selected values of ζ. The asymptotes are clearly poor approximations near ω_0 when $\zeta < 0.5$, and the exact curves do not lend themselves to simple correction terms. Consequently, you may need to resort to a calculator or computer program for the detailed calculations if accuracy is at all important.

Exercise 7.2−3   Show that the tuned-circuit transfer function in Eq. (11), Section 7.1, can be written as

$$H(j\omega) = \frac{j2\zeta(\omega/\omega_0)}{1 + j2\zeta(\omega/\omega_0) - (\omega/\omega_0)^2} \tag{10}$$

with $\zeta = 1/2Q$. Then use Figs. 7.2−1 and 7.2−6 to sketch the Bode plot when $Q = 5$.

7.3 FIRST-ORDER TRANSIENTS

The remainder of this chapter is devoted to the transient behavior of dynamic circuits. Transients occur before a circuit reaches steady-state conditions and in some cases, the response consists entirely of transients.

The study of transients is important because, on the one hand, they often turn out to be limiting factors in the performance of electrical systems. On the other hand, they may be put to beneficial use in wave-shaping circuits, timing devices, and various other practical applications.

To divide this topic into manageable parts, we focus here on first-order transients. We'll start with a review and elaboration of the natural response of RC and RL circuits. Then we'll examine transients in first-order circuits with switched sources, including the important step and pulse response.

Natural Response of RC and RL Circuits

The switched RC circuit diagrammed in Fig. 7.3–1a represents a situation frequently encountered in practice. The switch has been in the upper position for a long time prior to $t = 0$, thus charging the capacitor to $v_C = V_0$ volts. The charged capacitor acts as a DC block under steady-state conditions, so $i = 0$ for $t < 0$. The switch is then thrown to the lower position at $t = 0$, which puts a short across the series RC network for $t > 0$ and we have the equivalent circuit of Fig. 7.3–1b.

Figure 7.3–1
(a) Switched RC circuit. (b) Circuit for $t > 0$. (c) Natural-response waveforms.

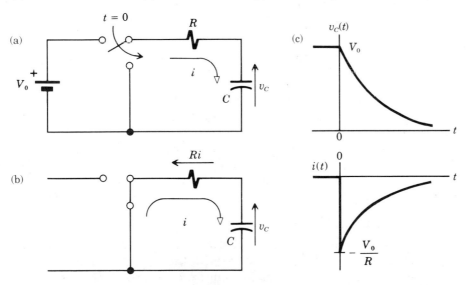

Since the forcing function equals zero for $t > 0$, the resulting behavior of $v_C(t)$ and $i(t)$ will be the *natural response*. We know from Section 5.2 that the natural response of an RC circuit is a decaying exponential $Ke^{-t/RC}$, where K stands for the initial value at $t = 0$. Thus, we only need to find the initial values of v_C and i in Fig. 7.3–1b.

Presumably, the initial voltage across the capacitor equals V_0 because its stored energy is $w = \frac{1}{2}CV_0^2$ for $t < 0$ and energy usually does not change value suddenly. To test this hunch in a general way, we draw upon the integral relation for capacitor voltage written in the form

$$v_C(t^+) = v_C(t^-) + \frac{1}{C} \int_{t^-}^{t^+} i \, dt$$

where t^- and t^+ stand for the instants immediately before and after an arbitrary time t. Clearly, integrating any *finite* current i from t^- to t^+ will yield a vanishingly small result. And *infinite* current, besides being physically impossible, would produce infinite voltage across R in Fig. 7.3–1b and thereby violate KVL. We therefore conclude that

$$v_C(t^+) = v_C(t^-) \tag{1}$$

Equation (1) mathematically expresses the important property:

> The voltage across a capacitor never undergoes a discontinuous change when the current through it remains finite.

This property is the **continuity condition for voltage across capacitance,** analogous to the condition on the velocity of a moving mass whose momentum tends to prevent sudden velocity changes.

For the circuit at hand, we set $t = 0$ in Eq. (1) to obtain $v_C(0^+) = v_C(0^-) = V_0$. Hence, the natural-response voltage is

$$v_C(t) = V_0 \, e^{-t/\tau} \qquad t > 0 \tag{2}$$

where we've introduced the exponential **time constant** τ (tau) defined as

$$\tau = RC \tag{3}$$

The initial current is then easily found by observing in Fig. 7.3–1b that $Ri(0^+) + v_C(0^+) = 0$. Thus, $i(0^+) = -V_0/R$ and

$$i(t) = -\frac{V_0}{R} \, e^{-t/\tau} \qquad t > 0 \tag{4}$$

The negative value of i reflects the fact that the capacitor is now discharging its stored energy through R. The instantaneous power dissipated by the resistance is $p = Ri^2$, and the total energy it absorbs equals the initial stored energy $\frac{1}{2}CV_0^2$.

The waveforms $v_C(t)$ and $i(t)$ are sketched in Fig. 7.3–1c, including the steady-state values for $t < 0$. Unlike $v_C(t)$, the current exhibits a sudden change at the switching instant, jumping from $i(0^-) = 0$ to $i(0^+) = -V_0/R$ before decaying exponentially toward zero. In view of this behavior, we cannot meaningfully speak about the value of $i(t)$ at $t = 0$. However, the continuity condition on $v_C(t)$ allows us to state that $v_C(0) = V_0$.

Now consider the switched RL circuit in Fig. 7.3–2. The constant current source establishes $i_L = I_0$ and $w = \frac{1}{2}LI_0^2$ for $t < 0$, while $v = 0$ under

Figure 7.3–2
Switched *RL* circuit.

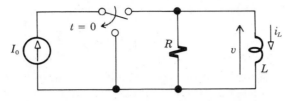

DC steady-state conditions. When the switch is thrown to the lower position at $t = 0$, the energy stored by the inductor produces current through R, and $i_L(t)$ and $v(t)$ will be the natural response for $t > 0$. This circuit is precisely the *dual* of Fig. 7.3–1a, so we can apply our previous results here by interchanging voltage and current.

Accordingly, the initial values must satisfy the **continuity condition for current through inductance,** namely:

> The current through an inductor never undergoes a discontinuous change when the voltage across it remains finite.

This condition is expressed mathematically as

$$i_L(t^+) = i_L(t^-) \tag{5}$$

Thus, $i_L(0^+) = i_L(0^-) = I_0$ and $v(0^+) = -Ri_L(0^+)$, so the natural response current and voltage are

$$i_L(t) = I_0\, e^{-t/\tau} \qquad v(t) = -RI_0\, e^{-t/\tau} \qquad t > 0 \tag{6}$$

Although $i_L(t)$ has continuity at the switching instant, $v(t)$ jumps from $v(0^-) = 0$ to $v(0^+) = -RI_0$. The time constant is obtained by duality, replacing C with L and R with $1/R$ in Eq. (3) to get

$$\tau = L/R \tag{7}$$

Incidentally, note from Eqs. (3) and (7) that the time constant of an RC or RL network equals the reciprocal of the corresponding filter's *cutoff frequency,* i.e., $\tau = 1/\omega_{co}$.

Finally, to generalize our discussion, let $y(t)$ be a natural-response voltage or current in any network consisting of resistance and one energy-storage element — either capacitance or inductance. Viewed from the energy-storage element, the rest of the network acts as one *equivalent resistance* R_{eq}, so the time constant can be calculated from

$$\tau = R_{eq}C \qquad \text{or} \qquad \tau = L/R_{eq} \tag{8}$$

which generalizes Eqs. (3) and (7). If we know from continuity at some instant $t = t_0^+$ that

$$y(t_0^+) = Y_0$$

then the subsequent natural response will be

$$y(t) = Y_0\, e^{-(t-t_0)/\tau} \qquad t > t_0 \tag{9}$$

To find the time t at which $y(t)$ reaches a specified value, we can use

$$t = t_0 + \tau \ln\left[\frac{Y_0}{y(t)}\right] \qquad (10)$$

Equation (10) is derived by taking the natural logarithm of $y(t)$ in Eq. (9).

Since Eq. (9) holds for *any* natural response in *any* first-order circuit, we should give further attention to the corresponding waveform plotted in Fig. 7.3–3. The *initial slope,* indicated by the dashed line, is

$$\frac{dy(t_0^+)}{dt} = \frac{-Y_0}{\tau} \qquad (11a)$$

where the notation $dy(t_0^+)/dt$ stands for dy/dt evaluated at $t = t_0^+$. Hence, the initial decay can be approximated by

$$y(t) \approx y(t_0^+) + \frac{dy(t_0^+)}{dt}(t - t_0) = Y_0\left(1 - \frac{t - t_0}{\tau}\right) \qquad (11b)$$

which holds for $t_0 < t < t_0 + (\tau/10)$. But the slope progressively decreases as time increases, and $y(t)$ still equals about 37% of its initial value after one time constant has elapsed. Although $y(t)$ never completely reaches zero in finite time, it does become negligibly small for most purposes after five constants where $y(t_0 + 5\tau) \approx 0.007Y_0$—less than 1% of the initial value.

Figure 7.3–3
Decaying exponential
$y(t) = Y_0 e^{-(t-t_0)/\tau}$.

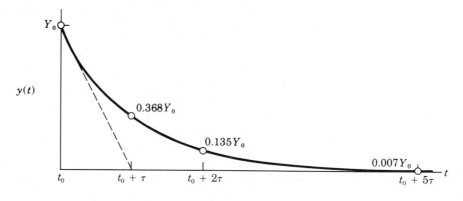

Example 7.3–1

The circuit in Fig. 7.3–4 has reached DC steady-state conditions when the switch is opened at $t = 0$. We want to find $i(t)$ and $v(t)$ for $t > 0$. We also want to evaluate the initial slope of $i(t)$, and to determine when $i(t)$ drops to one-fourth of its initial value.

Since the current i has continuity here, we first use a DC current divider to obtain

$$I_0 = i(0^+) = i(0^-) = \frac{1}{2}\frac{60 \text{ V}}{(5 + 20\|20)\Omega} = 2 \text{ A}$$

Figure 7.3−4
Circuit for Example
7.3−1.

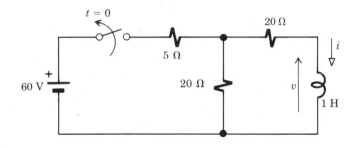

Next, we calculate the time constant by observing that the equivalent resistance seen from the inductor with the switch open is $R_{eq} = 20 + 20 = 40\ \Omega$, so $\tau = 1\ H/40\ \Omega = 25$ ms. Therefore, for $t > 0$,

$$i(t) = I_0\,e^{-t/\tau} = 2\ A\ e^{-t/(25\ ms)}$$

$$v(t) = -R_{eq}I_0\,e^{-t/\tau} = -80\ V\ e^{-t/(25\ ms)}$$

Continuity of current in this circuit thus generates a negative voltage "spike" whose peak value $|v(0^+)| = R_{eq}I_0 = 80$ V exceeds the 60-V source voltage.

The waveform $i(t)$ has the shape of Fig. 7.3−3, and its initial slope is

$$di(0^+)/dt = -I_0/\tau = 0.08\ A/ms = 80\ A/s$$

The time at which $i(t)$ equals $0.25\ I_0 = 0.5$ A is

$$t = 25\ ms \times \ln(2\ A/0.5\ A) \approx 34.7\ ms$$

which is obtained from Eq. (10).

Exercise 7.3−1 Let the inductor in Fig. 7.3−4 be replaced by a 1-μF capacitor. Find $v(t)$ and $i(t)$ for $t < 0$ and $t > 0$.

Switched DC Transients

Switched DC transients occur when the source applied to a network makes a discontinuous change from one constant value to another. Eventually, the circuit reaches DC steady-state conditions forced by the new source value. Our concern here is the resulting transient behavior, which involves both the forced response and the natural response.

Suppose, for example, that the switch in Fig. 7.3−5a has been in the upper position for a long time and is moved to the lower position at $t = 0$. We know that the initial voltage across the capacitor is

$$V_0 = v(0^+) = v(0^-) = \frac{R_2}{R_1 + R_2}\,V_A$$

Figure 7.3–5
(a) RC circuit with
switched voltage.
(b) Equivalent circuit
for $t > 0$.

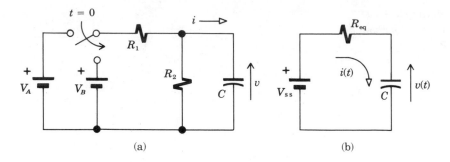

(a)   (b)

and we see that the final steady-state voltage will be

$$V_{ss} = v(\infty) = \frac{R_2}{R_1 + R_2} V_B$$

But how does $v(t)$ make the transition from V_0 to V_{ss} as $t \to \infty$?

To answer that question, we first draw the Thévenin equivalent circuit as viewed by the capacitor for $t > 0$. The corresponding open-circuit voltage and equivalent resistance are $v_{oc} = R_2 V_B/(R_1 + R_2) = V_{ss}$ and $R_{eq} = R_1 \| R_2$, giving us the circuit model in Fig. 7.3–5b. Next, we recall from Section 5.2 that the *complete response* consists of the steady-state forced response $v_F(t) = V_{ss}$ plus a natural-response component $v_N(t)$. Thus,

$$v(t) = v_F(t) + v_N(t) = V_{ss} + K e^{-t/\tau} \qquad t > 0$$

where $\tau = R_{eq} C$ but the constant K depends upon the initial condition. Setting $t = 0^+$ to evaluate K from the known initial value yields

$$v(0^+) = V_{ss} + K = V_0$$

Hence, $K = V_0 - V_{ss}$ and we finally obtain

$$v(t) = V_{ss} + (V_0 - V_{ss})e^{-t/\tau} \qquad \textbf{(12)}$$

which is the complete response for $t > 0$.

The validity of Eq. (12) is quickly checked by noting that $v(0^+) = V_0$ and $v(\infty) = V_{ss}$, as expected. Furthermore, if $V_{ss} = 0$ — meaning no applied source for $t > 0$ — then Eq. (12) reduces to the correct natural response $v(t) = V_0 e^{-t/\tau}$. In any case, the time interval from $t = 0^+$ to $t \approx 5\tau$ comprises the transitional readjustment between the initial and final values.

Equation (12) also readily generalizes to any first-order circuit with a DC source that switches at time $t = t_0$. The resulting transient waveform for any voltage or current $y(t)$ is given by

$$y(t) = Y_{ss} + (Y_0 - Y_{ss})e^{-(t-t_0)/\tau} \qquad t > t_0 \qquad \textbf{(13)}$$

The initial value $Y_0 = y(t_0^+)$ and the final value $Y_{ss} = y(\infty)$ are determined from continuity and DC steady-state analysis. The time constant is found by calculating the Thévenin equivalent resistance seen from the energy-storage element after the switching time.

Example 7.3–2   For a numerical illustration of transient behavior, let the element values in Fig. 7.3–5a be

$$V_A = 12 \text{ V} \quad V_B = -20 \text{ V} \quad R_1 = 4 \text{ k}\Omega \quad R_2 = 12 \text{ k}\Omega \quad C = 100 \ \mu\text{F}$$

Then we have

$$V_0 = \frac{12}{16} (12) = 9 \text{ V} \qquad\qquad R_{\text{eq}} = 4\|12 = 3 \text{ k}\Omega$$

$$V_{\text{ss}} = \frac{12}{16} (-20) = -15 \text{ V} \qquad \tau = 3 \text{ k}\Omega \times 100 \ \mu\text{F} = 0.3 \text{ sec}$$

Since $t_0 = 0$, Eq. (13) gives

$$v(t) = -15 + 24 \ e^{-t/0.3} \qquad t > 0$$

which is plotted in Fig. 7.3–6. The transient interval lasts about $5\tau = 1.5$ sec, after which $v(t) \approx V_{\text{ss}} = -15$ V.

Figure 7.3–6
Waveform for Example
7.3–2.

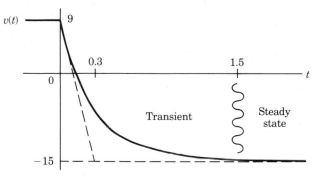

If desired, the current waveform can be found from Fig. 7.3–5b where $i(t) = [V_{\text{ss}} - v(t)]/R_{\text{eq}}$. Thus, for instance, its initial value is $i(0^+) = (-15 - 9)/3 = -8$ mA.

Exercise 7.3–2   Let the capacitor in Fig. 7.3–5a be replaced by a 0.3-H inductor. Find the values of $i(t)$ and $v(t)$ at $t = 0^-$, 0^+, and ∞ when $V_A = 12$ V, $V_B = -20$ V, $R_1 = 4$ kΩ, and $R_2 = 12$ kΩ. Then sketch the waveforms $i(t)$ and $v(t)$.

Step and Pulse Response

Two special transients having particular importance in digital electronics and related applications are the step and pulse response. We'll discuss these waveforms with reference to the RC circuit in Fig. 7.3–7, taking the output to be the voltage across the capacitor.

The **step response** is defined as the output produced by a constant input starting at $t = 0^+$ when the network contains no initial stored energy.

Figure 7.3–7
RC circuit with input
$v_s(t)$.

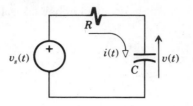

To model this case, we take $v(0^-) = 0$ in Fig. 7.3–7 and we write the source voltage as

$$v_s(t) = \begin{cases} 0 & t < 0 \\ V_s & t > 0 \end{cases}$$

This expression represents an input voltage "step" of "height" V_s occurring at $t = 0$. The corresponding step response follows directly from Eq. (12) with initial value $V_0 = 0$ and final value $V_{ss} = V_s$. Thus,

$$v(t) = V_s - V_s\, e^{-t/\tau} = V_s(1 - e^{-t/\tau}) \qquad t > 0 \qquad \textbf{(14)}$$

which is plotted in Fig. 7.3–8.

Figure 7.3–8
Step response and rise
time t_r.

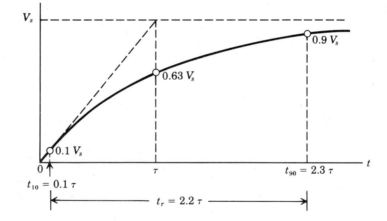

From our previous study of the exponential function, we know that $e^{-t/\tau}$ decreases almost linearly near $t = 0$ so $(1 - e^{-t/\tau})$ begins to *rise* in a nearly linear fashion and

$$v(t) \approx \frac{V_s}{\tau} t \qquad 0 < t < \frac{\tau}{10} \qquad \textbf{(15)}$$

This is a good approximation up to the 10% time $t_{10} \approx 0.1\tau$, where $v(t_{10}) = 0.1\, V_s$. The rise continues less rapidly, passing through the 63% point at $t = \tau$ where $v(\tau) = V_s(1 - e^{-1}) \approx 0.63\, V_s$, and is essentially over at the 90% time $t_{90} \approx 2.3\tau$, where $v(t_{90}) = V_s(1 - e^{-2.3}) = 0.9\, V_s$. The time interval between t_{10} and t_{90} is termed the **rise time**

$$t_r \triangleq t_{90} - t_{10} \approx 2.2\tau \tag{16}$$

Rise time serves as a measure of the "speediness" of the step response.

Rise-time considerations play a major role in the design of high-speed digital electronic circuits. They also figure in more prosaic circumstances such as the simple charging of a capacitor. To clarify this point, observe from Fig. 7.3–8 that v never reaches the full value V_s exactly, but that there will be no measurable difference between v and V_s if the source has been applied for a time interval large compared to t_r. That's what we imply in statements such as "The switch has been closed for a *long* time."

The **pulse response** is defined as the output produced by a *rectangular input pulse* when the circuit contains no initial stored energy. We'll let D stand for the *duration* of the input pulse and we'll model this case by writing

$$v_s(t) = \begin{cases} V_s & 0 < t < D \\ 0 & \text{otherwise} \end{cases}$$

which is shown in Fig. 7.3–9a. For the purpose of analysis, we view this pulse as the *sum of two step functions* — an upward step at $t = 0$ followed by a downward step at $t = D$, as in Fig. 7.3–9b. Invoking superposition now yields a resulting pulse response that is the sum of two step responses — the second step response *delayed D seconds* from the first.

Before proceeding further, we need to take a short digression on the subject of **time delay.** Delaying any time function, say $y(t)$, by D units shifts the entire function to the right along the time axis and produces a new time function written as $y(t - D)$. Conversely, $y(t + D)$ is the time-advanced version of $y(t)$. In short, replacing t with $t \mp D$ shifts a time function to the right by $\pm D$ units. Let's now apply superposition and the time-delay concept to calculate the pulse response of an RC circuit.

Since there is no initial stored energy at $t = 0$, the pulse response starts out exactly like the step response. Hence,

$$v'(t) = V_s(1 - e^{-t/\tau}) \qquad t > 0$$

which holds until the pulse ends at $t = D$. But we have modeled the end of the pulse as a delayed step of height $-V_s$, and the response to that term alone is found from Eq. (14) by replacing V_s with $-V_s$ and substituting $t - D$ for t to account for the delay. Thus,

$$v''(t) = -V_s[1 - e^{(t-D)/\tau}] \qquad t > D$$

The total pulse response is then $v(t) = v'(t) + v''(t)$, in which $v''(t) = 0$ for $t < D$.

Our final result, therefore, must be written as two equations

$$v(t) = V_s(1 - e^{-t/\tau}) \qquad 0 < t < D \tag{17a}$$

Figure 7.3–9

(a) Rectangular pulse.
(b) Decomposition into two step functions.

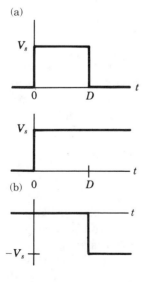

(a)

(b)

and

$$v(t) = V_s(1 - e^{-t/\tau}) - V_s[1 - e^{-(t-D)/\tau}] \tag{17b}$$
$$= V_s[e^{-(t-D)/\tau} - e^{-t/\tau}] \qquad t > D$$

Equation (17a) describes the transient during the applied pulse, and Eq. (17b) the transient after the pulse has ended. Setting $t = D$ in either of these gives the value of the response at the end of the pulse, namely

$$v(D) = V_s(1 - e^{-D/\tau}) \tag{18}$$

Equation (17b) now can be expressed in terms of $v(D)$ as

$$v(t) = v(D)e^{-(t-D)/\tau} \qquad t > D \tag{19}$$

This equation simply represents a decaying *natural response* starting at $t = D$ with initial value $V_0 = v(D)$.

Figure 7.3–10 shows the pulse response for three different values of the time constant τ relative to the pulse duration D. When $\tau \ll D$ the response rises quickly since $t_r = 2.2\tau \ll D$; then it essentially flattens off at the applied pulse height and decays quickly from $v(D) \approx V_s$. When $\tau \approx D$ the rise and decay are less rapid, and $v(D)$ must be computed from Eq. (18). When $\tau \gg D$ the rise approximates a linear ramp and $v(D) \approx V_s D/\tau \ll V_s$.

Figure 7.3–10
Pulse response with:
(a) $\tau \ll D$; (b) $\tau \approx D$;
(c) $\tau \gg D$.

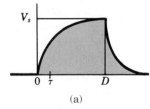
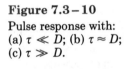

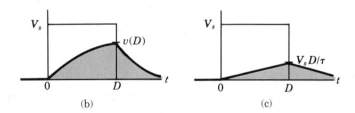

(a)   (b)   (c)

We therefore reach the important conclusion that the shape of the pulse response of an RC (or RL) circuit depends critically on the relative values of τ and D. Nonetheless, the decaying "tail" at the end of the pulse always has the shape of the natural response and falls off at the same speed that the step response rises. Hence, the *rise* time t_r also serves as a measure of the *fall time*. A reasonably faithful reproduction of a rectangular pulse shape requires

$$\tau \leq \frac{D}{5} \tag{20}$$

so that $v(D) \geq 0.99\, V_s$.

Another important conclusion here relates to the *frequency response* of the network. The waveforms in Fig. 7.3–10 correspond to the pulse response of a *lowpass filter*, since we took the output voltage across the capacitor in Fig. 7.3–7. The filter's cutoff frequency is $\omega_{co} = 1/\tau$, so the reproducibility condition in Eq. (20) becomes

$$\omega_{co} \geq \frac{5}{D}$$

Thus, a small pulse duration requires a large cutoff frequency.

Exercise 7.3−3   Let $R = 1$ kΩ in Fig. 7.3−7 and let the input be a rectangular pulse with $V_s = 10$ V and $D = 100$ μs.
(a) Calculate $v(D)$ and sketch $v(t)$ for $0 \leq t \leq 600$ μs when $C = 0.5$ μF.
(b) Evaluate the condition on C for faithful reproduction.

7.4 SECOND-ORDER TRANSIENTS †

When a circuit contains two or more energy-storage elements, its natural response may include time functions other than the simple decaying exponentials that characterized first-order circuits. Of particular interest are networks with at least one capacitor and one inductor, for such networks have three possible types of natural behavior: *overdamped, underdamped,* and *critically damped.*

This section examines the three types of natural response and the resulting transients that occur in second-order circuits. The extension to higher-order circuits is conceptually straightforward, but it often involves laborious computations. Fortunately, much can be learned about higher-order circuits from the study of second-order transients.

Overdamped Natural Response

Let $x(t)$ stand for the voltage or current applied to a second-order network, and let $y(t)$ be any voltage or current of interest within the network. The differential equation relating $y(t)$ to $x(t)$ was discussed in Section 5.2. We also saw that when $x(t) = 0$ for $t > 0$, the natural response produced by initial stored energy is a sum of *two* exponential functions having the general form

$$y(t) = K_1 e^{p_1 t} + K_2 e^{p_2 t} \qquad t > 0 \tag{1}$$

The constants p_1 and p_2 depend entirely upon the element values and are evaluated by solving the characteristic equation derived from the differential equation.

But recall from Section 5.3 that the *transfer function* relating $y(t)$ to $x(t)$ contains exactly the same information as the differential equation. Hence, we can obtain p_1 and p_2 by forming the appropriate transfer function and finding the roots of the characteristic polynomial in the denominator of $H(s)$. To elaborate on this method, let the transfer function of an arbitrary second-order circuit be written as

$$H(s) = \frac{N(s)}{s^2 + 2\alpha s + \beta^2} \tag{2}$$

Here, the element values have been incorporated in the constants α and β and in the numerator polynomial symbolized by $N(s)$. Setting the denominator of Eq. (2) equal to zero immediately gives us the **characteristic equation**

$$s^2 + 2\alpha s + \beta^2 = 0 \tag{3}$$

Each second-order circuit has a unique characteristic equation, since only the numerator polynomial $N(s)$ depends upon which variable is taken to be the response $y(t)$.

The two values of s that satisfy Eq. (3) correspond to p_1 and p_2 in Eq. (1). Thus, by application of the quadratic formula, we get

$$p_1 = -\alpha + \sqrt{\alpha^2 - \beta^2} \qquad p_2 = -\alpha - \sqrt{\alpha^2 - \beta^2} \tag{4}$$

If the element values are such that

$$\alpha > 0 \qquad \beta^2 < \alpha^2$$

then p_1 and p_2 will be *real, negative,* and *unequal.* The natural response $y(t)$ therefore consists of two decaying exponentials with time constants

$$\tau_1 = -1/p_1 \qquad \tau_2 = -1/p_2$$

We say in this case that the network is *overdamped.* Second-order networks containing resistance and either two capacitors or two inductors *always* have an overdamped natural response. Networks containing one capacitor and one inductor may also be overdamped, depending on the element values.

Now we turn our attention to the constants K_1 and K_2 in Eq. (1), which are calculated from the initial conditions on the natural response. Specifically, if we know the *initial value* $y(0^+)$, then it follows from Eq. (1) with $t = 0^+$ that

$$K_1 + K_2 = y(0^+) \tag{5a}$$

However, we still need another equation to evaluate K_1 and K_2 individually. For this purpose, assume that we also know the *initial slope* $dy(0^+)/dt$. Differentiating Eq. (1) gives $dy(t)/dt = K_1 p_1 e^{p_1 t} + K_2 p_2 e^{p_2 t}$, so setting $t = 0^+$ yields

$$p_1 K_1 + p_2 K_2 = \frac{dy(0^+)}{dt} \tag{5b}$$

Equations (5a) and (5b) constitute a pair of simultaneous equations that can be solved for K_1 and K_2. The required initial value and slope are obtained from the continuity conditions on the two energy-stored elements — a problem we'll take up subsequently.

By extension of the preceding analysis, the natural response of a network containing n energy-storage elements has the form

$$y(t) = K_1 e^{p_1 t} + K_2 e^{p_2 t} + \cdots + K_n e^{p_n t}$$

where $p_1, p_2, \ldots, p_n$ are the roots of an nth-order polynomial. Although the characteristic polynomial may be easily found from the transfer function, calculating the n roots is usually a difficult chore. Furthermore, the constants $K_1, K_2, \ldots, K_n$ must be evaluated from a set of n simultaneous equations, after using n continuity conditions to obtain $y(0^+)$, $dy(0^+)/dt$, $d^2 y(0^+)/dt^2$, etc. Consequently, analyzing the natural response of nth-order circuits with $n \geq 3$ generally requires computer-aided numerical methods.

Example 7.4 – 1
Consider the circuit in Fig. 7.4 – 1a whose input voltage $v_S(t)$ has been turned off at $t = 0$. We want to find the resulting natural-response voltage $v(t)$ across the parallel RC section for $t > 0$.

Figure 7.4 – 1
Circuit and impedance diagram for Example 7.4 – 1.

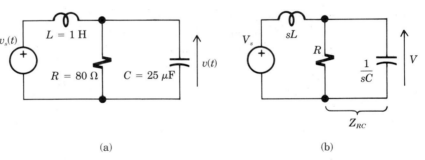

(a)   (b)

Our first step is to formulate the transfer function from the impedance diagram in Fig. 7.4 – 1b. Using a voltage-divider ratio with $Z_{RC} = R \| (1/sC) = R/(sCR + 1)$ we obtain

$$H(s) = \frac{V}{V_s} = \frac{Z_{RC}}{sL + Z_{RC}} = \frac{R}{s^2 LCR + sL + R} = \frac{1/LC}{s^2 + \dfrac{1}{RC} s + \dfrac{1}{LC}}$$

Hence, the characteristic equation is

$$s^2 + 2 \frac{1}{2RC} s + \frac{1}{LC} = 0$$

Comparison with Eq. (3) then shows that

$$\alpha = \frac{1}{2RC} = 250 \qquad \beta^2 = \frac{1}{LC} = 40{,}000$$

so $\alpha > 0$ and $\beta^2 < \alpha^2$. The corresponding roots from Eq. (4) are

$$p_1 = -250 + \sqrt{250^2 - 40{,}000} = -100$$
$$p_2 = -250 - \sqrt{250^2 - 40{,}000} = -400$$

We therefore have an overdamped circuit with time constants $\tau_1 = 1/100 = 10$ ms and $\tau_2 = 1/400 = 2.5$ ms.

Next, suppose that the initial value and slope are known to be

$$v(0^+) = 6 \text{ V} \qquad dv(0^+)/dt = -3000 \text{ V/s}$$

Inserting these along with the roots into Eq. (5) gives the pair of equations

$$K_1 + K_2 = v(0^+) = 6$$

$$(-100)K_1 + (-400)K_2 = \frac{dv(0^+)}{dt} = -3000$$

Simultaneous solution yields $K_1 = -2$ V and $K_2 = 8$ V, so our final result becomes

$$v(t) = -2e^{-100t} + 8e^{-400t} \qquad t > 0$$

This waveform is plotted in Fig. 7.4–2. The negative values of $v(t)$ imply that current driven by stored energy in the inductor temporarily recharges the capacitor in the opposite direction before $v(t)$ decays to zero.

Figure 7.4–2
Waveforms of
overdamped natural
response.

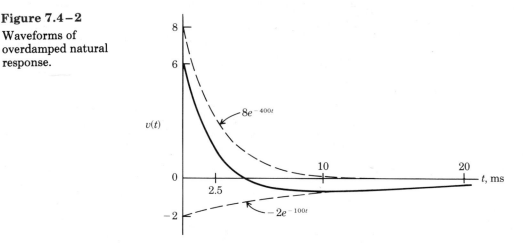

Exercise 7.4–1   The series circuit in Fig. 7.4–3 has $v_s(t) = 0$ for $t > 0$, and the current $i(t)$ is the response under consideration.

(a) Use the terminal impedance $Z(s)$ to obtain $H(s) = I/V_s = 1/Z(s)$, and show therefrom that

$$\alpha = R/2L \qquad \beta^2 = 1/LC$$

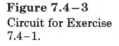

Figure 7.4–3
Circuit for Exercise
7.4–1.

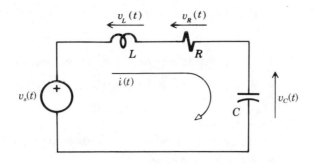

(b) Evaluate p_1 and p_2 with $L = 1$ H, $R = 300\ \Omega$, and $C = 50\ \mu$F. Then find $i(t)$ for $t > 0$, given that $i(0^+) = 0.5$ A and $di(0^+)/dt = 0$.

Underdamped Natural Response

When a second-order circuit contains one capacitor and one inductor, the element values may be such that p_1 and p_2 turn out to be *complex conjugates* with negative real parts. In particular, if the characteristic polynomial $s^2 + 2\alpha s + \beta^2$ has

$$\alpha > 0 \qquad \beta^2 > \alpha^2$$

then its roots are

$$p_1 = -\alpha + j\omega_N \qquad p_2 = p_1^* = -\alpha - j\omega_N \tag{6a}$$

where

$$\omega_N \triangleq \sqrt{\beta^2 - \alpha^2} \tag{6b}$$

We call this the **underdamped** case because the natural response will have an *oscillatory* decaying behavior.

To show that complex roots correspond to oscillating time functions, we first observe from Eq. (5b) that the constants K_1 and K_2 must also be complex conjugates when p_1 and p_2 are complex conjugates. Accordingly, we write K_1 and K_2 in the polar form

$$K_1 = \tfrac{1}{2}Ae^{j\theta} \qquad K_2 = K_1^* = \tfrac{1}{2}Ae^{-j\theta} \tag{7a}$$

in which

$$A \triangleq 2|K_1| \qquad \theta \triangleq \measuredangle K_1 \tag{7b}$$

Substituting Eqs. (6a) and (7a) into Eq. (1) then gives the natural response as

$$y(t) = \tfrac{1}{2}Ae^{j\theta}e^{(-\alpha + j\omega_N)t} + \tfrac{1}{2}Ae^{-j\theta}e^{(-\alpha - j\omega_N)t}$$

$$= \tfrac{1}{2}Ae^{-\alpha t}[e^{j(\omega_N t + \theta)} + e^{-j(\omega_N t + \theta)}]$$

But Euler's theorem says that $e^{j\phi} = \cos\phi + j\sin\phi$ for any ϕ, so $e^{-j\phi} =$

$\cos(-\phi) + j\sin(-\phi) = \cos\phi - j\sin\phi$ and it follows that

$$e^{j\phi} + e^{-j\phi} = 2\cos\phi$$

Therefore, by letting $\phi = \omega_N t + \theta$, our expression for $y(t)$ simplifies to

$$y(t) = Ae^{-\alpha t}\cos(\omega_N t + \theta) \tag{8}$$

All imaginary quantities have thus disappeared, as they must for $y(t)$ to be a real function of time.

Figure 7.4–4 illustrates the underdamped natural response described by Eq. (8). This waveform oscillates at the rate ω_N, called the *natural frequency,* and its amplitude or *envelope* $Ae^{-\alpha t}$ decays with time constant $\tau = 1/\alpha$. Physically, oscillation results from the exchange of stored energy flowing back and forth between the capacitor and inductor. If the circuit were lossless, then $\alpha = 0$ and the oscillations would go on forever without decay. Oscillations do not occur in an overdamped circuit, which dissipates energy at a rate faster than it can be exchanged.

Figure 7.4–4
Underdamped natural response.

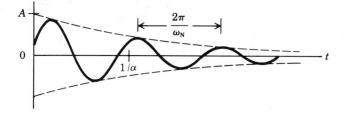

Example 7.4–2

Suppose the inductance in Example 7.4–1 is decreased to $L = 64$ mH, so

$$\beta^2 = 1/LC = 625{,}000 > \alpha^2$$

We now have an underdamped circuit with $\alpha = 250$ and natural frequency

$$\omega_N = \sqrt{625{,}000 - 250^2} = 750 \text{ rad/sec}$$

If the initial value and slope of $v(t)$ remain the same, then the initial-condition equations from Eq. (5) become

$$K_1 + K_2 = 6$$

$$(-250 + j750)K_1 + (-250 - j750)K_2 = -3000$$

After substituting $K_2 = 6 - K_1$ from the first equation, the second equation reduces to $j1500K_1 = -1500 + j4500$, and

$$K_1 = 3 + j1 = \sqrt{10}\,\underline{/18.4°}$$

Equation (7b) then yields

$$A = 2|K_1| = 2\sqrt{10} \qquad \theta = \sphericalangle K_1 = 18.4°$$

We thus obtain the oscillatory natural response

$$v(t) = 2\sqrt{10}\, e^{-250t} \cos{(750t + 18.4°)}$$

Since the oscillation period is $2\pi/750 \approx 8$ ms and $5\tau = 5/250 = 20$ ms, $v(t)$ will go through about 2.5 cycles before dying away.

Exercise 7.4–2   Let the circuits in Example 7.4–1 and Exercise 7.4–1 both have $L = 0.25$ H and $C = 25\ \mu$F. What's the condition on the value of R for an underdamped natural response in each case?

Critical Damping

Critical damping is the dividing line between overdamped and underdamped natural behavior. It occurs when the element values exactly satisfy the condition

$$\alpha^2 = \beta^2$$

and hence

$$p_1 = p_2 = -\alpha \tag{9}$$

We then say that the roots are **repeated.**

Further analysis using the homogeneous differential equation reveals that the special case of repeated roots produces a natural-behavior term proportional to $te^{-\alpha t}$, along with $e^{-\alpha t}$. Thus, instead of Eq. (1), a critically damped natural response takes the form

$$y(t) = K_1 e^{-\alpha t} + K_2 t e^{-\alpha t} \tag{10}$$

The first term of Eq. (10) is just a decaying exponential. However, as seen in Fig. 7.4–5, the second term starts with a linear rise and goes through a maximum value of $K_2/\alpha e$ at $t = 1/\alpha$ before it decays towards zero.

Figure 7.4–5
Critically-damped
component $K_2 e^{-\alpha t}$.

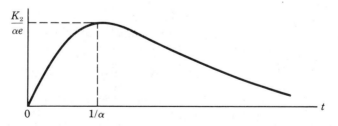

With critical damping, the constants K_1 and K_2 in Eq. (10) are given by

$$K_1 = y(0^+) \qquad K_2 = \frac{dy(0^+)}{dt} + \alpha K_1 \tag{11}$$

These expressions are obtained from $y(t)$ and $dy(t)/dt$ with $t = 0^+$.

Complete Response

The complete transient response of a second-order circuit, like that of a first-order circuit, consists of the forced response plus the natural response. We therefore write the complete response as

$$y(t) = y_F(t) + y_N(t) \qquad t > 0 \tag{12}$$

The forced component $y_F(t)$ is produced by the applied input $x(t)$ and usually can be found via steady-state analysis methods. The natural-behavior component $y_N(t)$ is given by Eq. (1), (8), or (10), depending upon the degree of damping. In all cases, however, $y_N(t)$ contains two constants that must be evaluated from initial conditions.

If the circuit in question is either overdamped or underdamped, so $p_1 \neq p_2$, then we find K_1 and K_2 by solving the pair of equations

$$K_1 + K_2 = y(0^+) - y_F(0^+)$$

$$p_1 K_1 + p_2 K_2 = \frac{dy(0^+)}{dt} - \frac{dy_F(0^+)}{dt} \tag{13}$$

If the circuit is critically damped, so $p_1 = p_2 = -\alpha$, then

$$K_1 = y(0^+) - y_F(0^+)$$

$$K_2 = \frac{dy(0^+)}{dt} - \frac{dy_F(0^+)}{dt} + \alpha K_1 \tag{14}$$

The values of $y_F(0^+)$ and $dy_F(0^+)/dt$ in Eqs. (13) and (14) are determined directly from the forced response, whereas the values of $y(0^+)$ and $dy(0^+)/dt$ must satisfy the continuity conditions and take account of the input $x(t)$ at $t = 0^+$.

A systematic procedure for finding the initial value and slope of any variable $y(t)$ is as follows:

1. Write down the initial input value $x(0^+)$, and evaluate initial capacitor voltages and inductor currents using the continuity relations

$$v_C(0^+) = v_C(0^-) \qquad i_L(0^+) = i_L(0^-) \tag{15}$$

2. Label the circuit diagram with symbols (not values) for all voltages and currents for $t > 0$. Then set $t = 0^+$ and apply Ohm's and Kirchhoff's laws to find the initial values of all other voltages and currents from $x(0^+)$, $v_C(0^+)$, and $i_L(0^+)$.

3. Calculate $dx(0^+)/dt$ from $x(t)$, and use $i_C(0^+)$ and $v_L(0^+)$ to evaluate the slopes

$$\frac{dv_C(0^+)}{dt} = \frac{i_C(0^+)}{C} \qquad \frac{di_L(0^+)}{dt} = \frac{v_L(0^+)}{L} \tag{16}$$

4. Apply Ohm's and Kirchhoff's laws as needed to obtain $dy(0^+)/dt$ from the values of $dx(0^+)/dt$, $dv_C(0^+)/dt$, and $di_L(0^+)/dt$.

This procedure may also be used to find the initial conditions for any *natural* response, in which case $x(0^+) = 0$ and $dx(0^+)/dt = 0$ by definition.

A second-order transient of special importance is the *step response* produced by an input step

$$x(t) = \begin{cases} 0 & t < 0 \\ x_s & t > 0 \end{cases} \tag{17a}$$

It then follows that

$$x(0^+) = X_s \qquad \frac{dx(0^+)}{dt} = 0 \tag{17b}$$

The forced response under DC steady-state conditions is simply a constant, say

$$y_F(t) = Y_{ss} \tag{18a}$$

from which we get

$$y_F(0^+) = Y_{ss} \qquad \frac{dy_F(0^+)}{dt} = 0 \tag{18b}$$

Since the circuit has zero stored energy for $t < 0$, Eq. (15) reduces to

$$v_C(0^+) = v_C(0^-) = 0 \qquad i_L(0^+) = i_L(0^-) = 0 \tag{19}$$

The initial value and slope of $y(t)$ are then easily found using the systematic method, as seen in the following example.

Example 7.4–3

Second-order step response.

The natural response of the circuit in Fig. 7.4–6 was studied in previous examples. Here we'll examine the transient response produced by an input voltage step $v_s(t) = V_s$ for $t > 0$. But before specifying the particular output, let's illustrate our systematic method by finding the initial value and slope of all voltages in the circuit.

Figure 7.4–6

Circuit for Example 7.4–3.

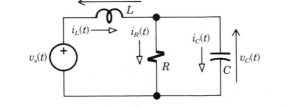

We are given that $v_s(t) = V_s$ for $t > 0$, and that the circuit has no stored energy for $t < 0$. Hence,

$$v_s(0^+) = V_s \qquad v_C(0^+) = 0 \qquad i_L(0^+) = 0$$

Applying Ohm's and Kirchhoff's laws then yields

$$i_R(0^+) = \frac{1}{R} v_C(0^+) = 0 \qquad i_C(0^+) = i_L(0^+) - i_R(0^+) = 0$$

$$v_L(0^+) = v_s(0^+) - v_C(0^+) = V_s$$

Therefore, from Eq. (16),

$$\frac{dv_C(0^+)}{dt} = \frac{i_C(0^+)}{C} = 0 \qquad \frac{di_L(0^+)}{dt} = \frac{v_L(0^+)}{L} = \frac{V_s}{L}$$

The other initial slopes are

$$\frac{di_R(0^+)}{dt} = \frac{1}{R} \frac{dv_C(0^+)}{dt} = 0$$

$$\frac{di_C(0^+)}{dt} = \frac{di_L(0^+)}{dt} - \frac{di_R(0^+)}{dt} = \frac{V_s}{L}$$

$$\frac{dv_L(0^+)}{dt} = \frac{dv_s(0^+)}{dt} - \frac{dv_C(0^+)}{dt} = 0$$

where we have used $dv_s(0^+)/dt = 0$ to evaluate $dv_L(0^+)/dt$.

Now let's specify the output to be the voltage $v_C(t)$ across the parallel RC section. Clearly, the corresponding forced response is the DC steady-state value $v_C = V_s$. Thus, assuming that $\alpha^2 \neq \beta^2$, the step response will be

$$v_C(t) = \begin{cases} V_s + K_1 e^{p_1 t} + K_2 e^{p_2 t} & \alpha^2 > \beta^2 \\ V_s + 2|K_1| e^{-\alpha t} \cos(\omega_N t + \sphericalangle K_1) & \alpha^2 < \beta^2 \end{cases}$$

The constants K_1 and K_2 are given by Eq. (13) with $y(0^+) = v_C(0^+) = 0$, $y_F(0^+) = V_s$, $dy(0^+)/dt = dv_C(0^+)/dt = 0$, and $dy_F(0^+)/dt = 0$, so

$$K_1 + K_2 = -V_s \qquad p_1 K_1 + p_2 K_2 = 0$$

which can be solved for K_1 and K_2, given p_1 and p_2.

If we take $p_1 = -100$ and $p_2 = -400$ (as in Example 7.4–1), then $K_1 = -4V_s/3$ and $K_2 = V_s/3$. Hence, we get the overdamped step response

$$v_C(t) = V_s \left(1 - \frac{4}{3} e^{-100t} + \frac{1}{3} e^{-400t} \right)$$

If we take $p_1 = -250 + j750$ and $p_2 = -250 - j750$ (as in Example 7.4–2), then $K_1 = (-3 + j1)V_s/6 = \sqrt{10}(V_s/6) \underline{/161.6°}$ and

$$v_C(t) = V_s \left[1 + \frac{\sqrt{10}}{3} e^{-250t} \cos(750t + 161.6°) \right]$$

which is an underdamped step response.

Both of these waveforms are plotted in Fig. 7.4–7. We see that the underdamped step response rises more rapidly than the overdamped re-

Figure 7.4−7
Waveforms of
overdamped and
underdamped step
response.

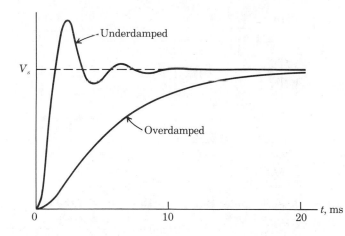

sponse. However, underdamping also produces *overshoot* in the sense that
the transient portion of $v_C(t)$ has oscillatory peaks above the steady-state
value V_s. Critical damping would yield the most rapid initial rise without
overshoot and oscillations. Incidentally, critical damping is just what you
want for an automobile's suspension system, whose dynamic behavior acts
essentially like that of a second-order circuit.

Exercise 7.4−3   Let the circuit in Example 7.4−3 be critically damped, so $p_1 = p_2 = \alpha$.
Show that the step-response voltage across the capacitor is

$$v_C(t) = V_s(1 - e^{-\alpha t} - \alpha t e^{-\alpha t})$$

Exercise 7.4−4   Suppose the output in Example 7.4−3 is specified to be the voltage across
the inductor. Find the step response of $v_L(t)$ when $\alpha = 100$ and $\beta^2 =$
100,000.

PROBLEMS

7.1−1   Suppose a resistor R' is added to Fig. 7.1−5a in parallel with L. Show that the
transfer function has the form $H(j\omega) = (j\omega/a)/[1 + (j\omega/b)]$ with $b < a$. Then
sketch $|H(\omega)|$ versus ω taking $R' = 4R$ and $R/L = 500$.

7.1−2   Suppose a resistor R' is added to Fig. 7.1−2 in parallel with L. Show that the
transfer function has the form $H(j\omega) = (1 + j\omega/a)/(1 + j\omega/b)$ with $b < a$. Then
sketch $|H(\omega)|$ versus ω taking $R' = 4R$ and $R/L = 500$.

7.1−3   Suppose a resistor R' is added to Fig. 7.1−5b in parallel with C. Show that the
transfer function has the form $H(j\omega) = (a + j\omega/b)/(1 + j\omega/b)$ with $a < 1$. Then
sketch $|H(\omega)|$ versus ω taking $R' = 4R$ and $1/RC = 100$.

7.1−4   The ladder network in Fig. P7.1−4 becomes a lowpass filter when $\underline{Z}_1 = \underline{Z}_3 = R$ and
$\underline{Z}_2 = \underline{Z}_4 = 1/j\omega C$. Let $\underline{V}_{out} = 1$ to find $\underline{V}_{in}$ and show that $H(j\omega) = \underline{V}_{out}/\underline{V}_{in} =$
$1/(1 - x^2 + j3x)$ where $x = \omega RC$. Then solve the quadratic resulting from
$|H(j\omega)|^2 = \frac{1}{2}$ to obtain ω_{co} in terms of RC.

Figure P7.1–4

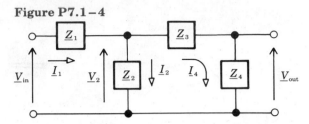

7.1–5   The ladder network in Fig. P7.1–4 becomes a highpass filter when $\underline{Z}_1 = \underline{Z}_3 = 1/j\omega C$ and $\underline{Z}_2 = \underline{Z}_4 = R$. Let $\underline{V}_{out} = 1$ to find $\underline{V}_{in}$ and show that $H(j\omega) = \underline{V}_{out}/\underline{V}_{in} = x^2/(x^2 - 1 - j3x)$ where $x = \omega RC$. Then solve the quadratic resulting from $|H(j\omega)|^2 = \frac{1}{2}$ to obtain ω_{co} in terms of RC.

7.1–6   A second-order **Butterworth lowpass filter** has

$$H(j\omega) = \frac{1}{1 - (\omega/\omega_{co})^2 + j\sqrt{2}\,(\omega/\omega_{co})}$$

(a) Show that $|H(\omega)| = 1/\sqrt{1 + (\omega/\omega_{co})^4}$. Then sketch the amplitude ratio and compare it with the amplitude ratio of a simple RC highpass filter.

(b) Let inductance L be inserted in series with R in Fig. 7.1–4. Obtain expressions for L and R in terms of ω_{co} and C such that this circuit becomes a Butterworth filter.

7.1–7   A second-order **Butterworth highpass filter** has

$$H(j\omega) = \frac{1}{1 - (\omega_{co}/\omega)^2 - j\sqrt{2}\,(\omega_{co}/\omega)}$$

(a) Show that $|H(\omega)| = 1/\sqrt{1 + (\omega_{co}/\omega)^4}$. Then sketch the amplitude ratio and compare it with the amplitude ratio of a simple RC highpass filter.

(b) Let capacitance C be inserted in series with R in Fig. 7.1–5a. Obtain expressions for C and R in terms of ω_{co} and L such that this circuit becomes a Butterworth filter.

7.1–8   Consider Fig. P5.3–15 with $Z(s) = sL$ and resistance $R_o = 4R_L$ added in parallel with the current source. Obtain the transfer function in the form $H(j\omega) = K(1 + j\omega/\omega_2)/(1 + j\omega/\omega_1)(1 + j\omega/\omega_3)$. Then sketch $|H(\omega)|$ versus ω for the case when $L = 0$ and when $L = R_L RC$. What is the purpose of including L?

7.1–9   Carry out the details leading to Eq. (12a).

7.1–10   Let $\omega = \omega_0(1 + \delta)$ in Eq. (11a), and let $\delta \ll 1$ so $(1 + \delta)^{-1} \approx 1 - \delta$. Derive the handy approximation

$$H(j\omega) \approx \frac{1}{1 + j2Q(\omega - \omega_0)/\omega_0} = \frac{1}{1 + j2(\omega - \omega_0)/B}$$

which holds for $|\omega - \omega_0| \ll \omega_0$.

7.1–11   Figure 7.1–4 can act like a bandpass filter when inductance L is added in series with R. Write $H(j\omega)$ in terms of $\omega_0 = 1/\sqrt{LC}$ and $Q_s = 1/\omega_0 CR$. Then sketch $|H(\omega)|/|H(\omega_0)|$ versus ω taking $Q_s \gg 1$.

7.1–12   Figure 7.1–5a can act like a bandpass filter when capacitance C is added in series with R. Write $H(j\omega)$ in terms of $\omega_0 = 1/\sqrt{LC}$ and $Q_s = 1/\omega_0 CR$. Then sketch $|H(\omega)|/|H(\omega_0)|$ versus ω taking $Q_s \gg 1$.

7.1–13 A low-frequency vibration analyzer requires a tuned circuit like Fig. 7.1–11b with $R = 20$ kΩ, $\omega_0 = 50$ rad/s, and $B = 5$ rad/s.

(a) Find the required values of C and L.

(b) Devise a circuit implementation that eliminates the large inductor by using an additional capacitor and the gyrator from Problem 5.3–17 with $r_1 = r_2 = 1000$ Ω.

7.1–14 Do Problem 7.1–13 with $R = 10$ kΩ, $\omega_0 = 25$ rad/s, and $B = 5$ rad/s.

7.1–15 Let the coil in Fig. 7.1–13 have $L = 100$ μH and $R_2 = 0.1$ Ω. Find the values of C and R_1 such that $\omega_0 = 200,000$ rad/s and $B = 2000$ rad/s.

7.1–16 Do Problem 7.1–15 with $\omega_0 = 50,000$ rad/s and $B = 1000$ rad/s.

7.1–17 Suppose the coil in Fig. 7.1–13 has $L = 10$ μH and $R_2 = 0.5$ Ω, and you want to get $\omega_0 = 10^6$ rad/s and $B = 10^4$ rad/s.

(a) Find the required values of C and R_2.

(b) Devise a circuit implementation using the negative-resistance converter from Problem 4.2–11.

7.1–18 Do Problem 7.1–17 with $\omega_0 = 500,000$ rad/s and $B = 25,000$ rad/s.

7.1–19 Let the op-amp in Fig. 7.1–15 have finite gain A, as modeled by Fig. 4.2–3. Show that

$$H(j\omega) = \frac{-A\underline{Z}_F}{(A+1)\underline{Z}_1 + \underline{Z}_F}$$

7.1–20 Use the result given in Problem 7.1–19 to obtain expressions for $H(j\omega)$ and ω_{co} when the lowpass filter in Fig. 7.1–16a has a finite op-amp gain A. Check your result by letting $A \to \infty$.

7.1–21 Use the result given in Problem 7.1–19 to obtain expressions for $II(j\omega)$ and ω_{co} when the highpass filter in Fig. 7.1–16b has a finite op-amp gain A. Check your result by letting $A \to \infty$.

7.1–22 Consider the active filter in Fig. P7.1–22 with $\underline{Z}_a = R$ and $\underline{Z}_b = 1/j\omega C$.

(a) Use the fact that $\underline{V}_{out} = K\underline{V}_p$ and write a node-voltage equation for $\underline{V}_x$ to obtain $H(j\omega) = \underline{V}_{out}/\underline{V}_{in}$.

(b) How should R, C, and K be chosen to get a frequency response like the Butterworth lowpass filter in Problem 7.1–6?

Figure P7.1–22

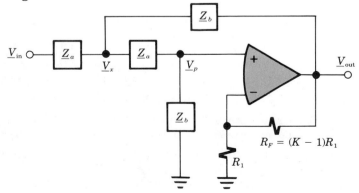

7.1–23 (a) Do Problem 7.1–22(a) with $\underline{Z}_a = 1/j\omega C$ and $\underline{Z}_b = R$.

(b) How should R, C, and K be chosen to get a frequency response like the Butterworth highpass filter in Problem 7.1–7?

7.2–1 Construct the complete Bode plot for $H(j\omega) = -20,000/(20 + j\omega)(200 + j\omega)$.

7.2–2 Construct the complete Bode plot for $H(j\omega) = 8(j\omega)^2/(50 + j\omega)(500 + j\omega)$.

7.2–3 Construct the complete Bode plot for $H(j\omega) = (400 + j\omega)/20(40 + j\omega)$.

7.2–4 Construct the complete Bode plot for $H(j\omega) = -j\omega(100 + j\omega)/200(10 + j\omega)$.

7.2–5 Draw the asymptotic Bode plot for $H(j\omega) = 8000j\omega/(10 + j\omega)(40 + j\omega)(80 + j\omega)$. Then add the dB correction terms to find the maximum gain and the frequencies at which the gain drops 3 dB below the maximum.

7.2–6 Do Problem 7.2–5 for $H(j\omega) = -2000(j\omega)^2/(50 + j\omega)(200 + j\omega)(400 + j\omega)$.

7.2–7 Do Problem 7.2–5 for $H(j\omega) = j\omega(800 + j\omega)/(50 + j\omega)(200 + j\omega)$.

7.2–8 Do Problem 7.2–5 for $H(j\omega) = -4000(10 + j\omega)/(100 + j\omega)(200 + j\omega)$.

7.2–9 Construct the asymptotic Bode plot for $H(j\omega) = 400(10 + j\omega)/(40 + j\omega)^2$. Hint: If $H_2(j\omega) = [H_1(j\omega)]^2$ then the Bode plot of $H_2(j\omega)$ is easily obtained by doubling the decibel and degree values on the plot of $H_1(j\omega)$.

7.2–10 Do Problem 7.2–9 for $H(j\omega) = (80 + j\omega)^2/4(10 + j\omega)(40 + j\omega)$.

7.2–11 Do Problem 7.2–9 for $H(j\omega) = 100(50 + j\omega)(j\omega)^2/(j\omega)(100 + j\omega)^2(400 + j\omega)$.

7.2–12 Use Fig. 7.2–6 to plot H_{dB} for $H(j\omega) = 100j\omega/(10 + j\omega)(100 + j2\omega - \omega^2)$.

7.2–13 Use Fig. 7.2–6 to plot H_{dB} for $H(j\omega) = (j\omega)^2/(100 + j4\omega - \omega^2)$.

7.2–14 Use Fig. 7.2–6 to plot H_{dB} for $H(j\omega) = (100 + j\omega - \omega^2)/(10 + j\omega)^2$, which is the transfer function of a **notch filter.**

7.3–1 The switch in Fig. P7.3–1 has been in the upper position a long time before $t = 0$. Find $v(t)$ and $i(t)$ for $t > 0$ taking element X to be a 5-μF capacitor.

Figure P7.3–1

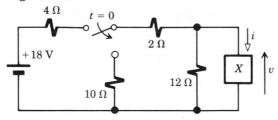

7.3–2 Do Problem 7.3–1 taking element X to be a 0.3-H inductor.

7.3–3 The switch in Fig. P7.3–3 has been in the upper position a long time before $t = 0$. Find $v(t)$, $i(t)$, and $v_1(t)$ for $t > 0$ taking element X to be a 30-μF capacitor.

7.3–4 Do Problem 7.3–1 taking element X to be a 0.2-H inductor.

7.3–5 Given a 12-V battery with $R_s = 0.6$ Ω, devise a circuit using a switch, inductor, and resistor to produce a 1000-V voltage spike with $\tau = 20$ μs across a 100-Ω resistance.

7.3–6 Given a 12-V battery with $R_s = 0.6$ Ω, devise a circuit using a switch, capacitor, and resistor to produce a 30-A current spike with $\tau = 10$ μs through a 0.2-Ω resistance.

Figure P7.3−3

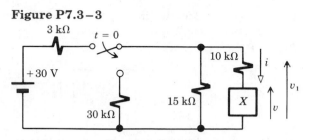

7.3−7 As $t \to \infty$, the total charge that passes to the left through R in Fig. 7.3−1b must equal the initial charge stored by C. Using this fact, show that

$$\int_0^\infty e^{-at}\, dt = \frac{1}{a}$$

for any constant a.

7.3−8 The total energy dissipated by R in Fig. 7.3−2 is $w = \int_0^\infty p\, dt$ for $t > 0$. Use the integral relationship from Problem 7.3−7 to confirm that w equals the initial energy stored by L.

7.3−9 The total energy dissipated by R in Fig. 7.3−1b is $w = \int_0^\infty p\, dt$ for $t > 0$. Use the integral relationship from Problem 7.3−7 to confirm that w equals the initial energy stored by C.

7.3−10 Let Fig. P6.2−1 have $R_1 = 25\ \Omega$, $L = 0.1$ H, $R_2 = 1$ kΩ, and $C = 4\ \mu$F. Find $v_C(t)$ and $v_L(t)$ for $t > 0$ when $v = 10$ V and there is no initial stored energy.

7.3−11 Let Fig. P6.2−1 have $R_1 = R_2 = 2$ kΩ, $L = 80$ mH, and $C = 0.5\ \mu$F. Find $i_1(t)$ and $i_2(t)$ for $t > 0$ when $v = 10$ V and there is no initial stored energy.

7.3−12 Let Fig. P6.2−1 have $R_1 = 500\ \Omega$, $L = 50$ mH, $R_2 = 250\ \Omega$, and $C = 0.4\ \mu$F. Find $i_1(t)$ and $i_2(t)$ for $t > 0$ when $v = 10$ V and there is no initial stored energy.

7.3−13 The switch in Fig. P7.3−13 has been at the upper position a long time before $t = 0$, when it goes to the lower position. The switch returns to the upper position at $t_0 = 1$ sec. The resistance values in are in ohms, and element X is a 2-H inductor.
(a) Find $i(t)$ for $0 < t < t_0$, and determine when $i(t) = 0$.
(b) Find $i(t)$ for $t > t_0$, and sketch the waveform for $0^- \leq t \leq 3$ sec.

Figure P7.3−13

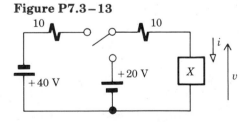

7.3−14 The switch in Fig. P7.3−13 has been at the upper position a long time before $t = 0$, when it goes to the lower position. The switch returns to the upper position at $t_0 = 1$ sec. The resistance values in are in kilohms, and element X is a 20-μF capacitor.
(a) Find $v(t)$ for $0 < t < t_0$, and determine when $v(t) = 0$.
(b) Find $v(t)$ for $t > t_0$, and sketch the waveform for $0^- \leq t \leq 3$ sec.

7.3–15   Do Problem 7.3–13 for the circuit in Fig. P7.3–15, where the resistance values are in ohms and element X is a 4-H inductor.

Figure P7.3–15

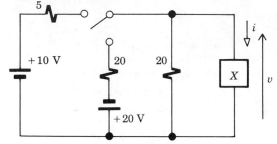

7.3–16   Do Problem 7.3–14 for the circuit in Fig. P7.3–15, where the resistance values are in kilohms and element X is a 50-μF capacitor.

7.3–17   Let the capacitor in Fig. 7.3–7 be replaced by inductor L, and let $v_s(t)$ be the rectangular pulse in Fig. 7.3–9a with $V_s = 10$ V. By considering $i(t)$, find and sketch $v(t)$ for $t \geq 0^-$ when there is no initial stored energy and $L/R = 2D$.

7.3–18   Do Problem 7.3–17 with $L/R = 0.5D$.

7.3–19   Suppose $v_s(t)$ in Fig. 7.3–7 is the double-pulse waveform in Fig. P7.3–19. Sketch and label $v(t)$ with $\tau = 0.5D$.

Figure P7.3–19

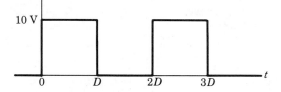

7.3–20   Do Problem 7.3–19 with $\tau = D$.

7.4–1   Suppose the network in Fig. P6.2–1 has $R_1 = R$ and $R_2 = 2R$. Find $H_a(s) = V/I = 1/Y(s)$ and $H_b(s) = V_C/I = (V_C/V)(V/I)$ to show in both cases that $\alpha = 3R_1/2L$ and $\beta^2 = 1/LC$.

7.4–2   Suppose the network in Fig. P6.2–3 has $R_2 = R$ and $R_1 = 4R$. Find $H_a(s) = I/V = 1/Z(s)$ and $H_b(s) = I_2/V = (I_2/I)(I/V)$ to show in both cases that $\alpha = (L + 4CR_2^2)/8LCR_2$ and $\beta^2 = 5/4LC$.

7.4–3   Suppose the network in Fig. P6.2–5 has $R_1 = R$ and $R_2 = 3R$. Find $H_a(s) = I/V = 1/Z(s)$ and $H_b(s) = V_C/V = (V_C/I)(I/V)$ to show in both cases that $\alpha = (3CR_1^2 + L)/2LCR_1$ and $\beta^2 = 4/LC$.

7.4–4   Show that critical damping for the circuit in Exercise 7.4–1 corresponds to $Q = \frac{1}{2}$, where Q is the quality factor for series resonance.

7.4–5   Given $y = Kte^{-\alpha t}$, verify the location and value of the maximum as shown in Fig. 7.4–5.

7.4–6   Consider the homogeneous differential equation $d^2y/dt^2 + \alpha\, dy/dt + \beta^2 y = 0$. Prove by substitution that the critically damped response in Eq. (10) is the solution if and only if $\alpha^2 = \beta^2$.

7.4 – 7   The network in Fig. P6.2–1 has $i = 0.1$ A for all $t < 0$ and $i = 0$ for $t > 0$. Use the results given in Problem 7.4–1 together with Eqs. (15) and (16) to find the natural response $v_C(t)$ for $t > 0$ when $R_1 = 400\ \Omega$, $R_2 = 800\ \Omega$, $L = 2$ H, and $C = 10\ \mu$F.

7.4 – 8   The network in Fig. P6.2–3 has $v = 10$ V for all $t < 0$ and $v = 0$ for $t > 0$. Use the results given in Problem 7.4–2 together with Eqs. (15) and (16) to find the natural response $i_2(t)$ for $t > 0$ when $R_1 = 200\ \Omega$, $R_2 = 50\ \Omega$, $L = 0.5$ H, and $C = 50\ \mu$F.

7.4 – 9   The network in Fig. P6.2–5 has $v = 40$ V for all $t < 0$ and $v = 0$ for $t > 0$. Use the results given in Problem 7.4–3 together with Eqs. (15) and (16) to find the natural response $v_C(t)$ for $t > 0$ when $R_1 = 100\ \Omega$, $R_2 = 300\ \Omega$, $L = 0.25$ H, and $C = 25\ \mu$F.

7.4 – 10   Let the network in Fig. P6.2–1 be driven by current i, and let $R_1 = 20\ \Omega$ and $R_2 = 80\ \Omega$. Show that

$$H(s) = \frac{I_2}{I} = \frac{s^2 + 20s/LC}{s^2 + 100s/L + 1/LC}$$

Then find the step response $i_2(t)$ when $L = 0.1$ H, $C = 40\ \mu$F, and $i = 2$ A for $t > 0$.

7.4 – 11   Let the network in Fig. P5.1–14 be driven by v_{in}. Show that

$$H(s) = \frac{V_1}{V_{\text{in}}} = \frac{1/LC}{s^2 + (L + 200C)s/20LC + 3/2LC}$$

Then find the step response $v_1(t)$ when $L = 0.1$ H, $C = 50\ \mu$F, and $v_{\text{in}} = 6$ V for $t > 0$.

7.4 – 12   Let the network in Fig. P6.2–3 be driven by voltage v, and let $R_1 = 20\ \Omega$ and $R_2 = 30\ \Omega$. Show that

$$H(s) = \frac{V_C}{V} = \frac{(sL + 50)/20LC}{s^2 + (L + 600C)s/20LC + 5/2LC}$$

Then find the step response $v_C(t)$ when $L = 0.1$ H, $C = 500\ \mu$F, and $v = 5$ V for $t > 0$.

Part II Electronic Devices and Circuits

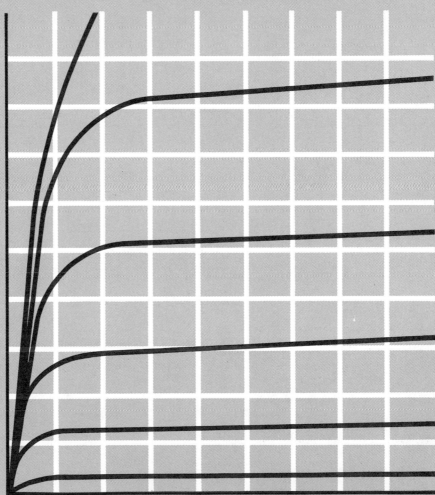

Chapter

8 Diodes and Semiconductors

This chapter launches our investigation of electronic devices and circuits by examining the class of elements called *diodes*. A diode is a two-terminal element that exhibits *nonlinear* electrical behavior, as do most electronic devices. Nonlinearity complicates circuit analysis and calls for new methods of attack. But the richness and variety of practical applications more than compensates for the somewhat increased analytical difficulty.

Our starting point will be the ideal diode concept, followed by the properties of semiconductor diodes and an overview of the underlying physical phenomena. Then we introduce two important methods for dealing with nonlinear electronic devices: load-line analysis using the i-v curve, and piecewise linearization using an idealized model. The chapter closes with a brief survey of some special-purpose two-terminal semiconductor devices. Many of the ideas developed here will be put to further use when we come to transistors in Chapter 9.

Objectives

After studying this chapter and working the exercises, you should be able to do each of the following:

- Sketch the i-v curves for an ideal diode, semiconductor diode, and Zener diode (Sections 8.1 and 8.2).

- Identify the current components in a *pn* junction, and explain the operating principles of a semiconductor diode (Section 8.1).

- Apply load-line analysis to find the operating point of a nonlinear element (Section 8.2).

- Develop an appropriate piecewise-linear model for a given nonlinear element (Section 8.2).

- Analyze a resistive diode circuit using the breakpoint method (Section 8.2).
- Describe the characteristics and applications of special-purpose devices (Section 8.3).†

8.1 IDEAL AND SEMICONDUCTOR DIODES

This section introduces the concept of an ideal diode, whose i-v characteristics are then compared with those of a semiconductor diode. We'll also describe the properties of semiconductors and pn junctions in order to explain the external behavior of semiconductor diodes.

The Ideal Diode

Fundamentally, a diode serves as an electrical *one-way valve*. It conducts current much more readily in one direction, called the **forward direction,** than it does in the other direction, called the **reverse direction.** Most semiconductor diodes conduct so well in the forward direction that they almost seem to be short circuits, while so little current flows in the reverse direction that they seem to be open circuits. Putting this another way, a diode acts like a *switch* that closes to allow current flow in the forward direction but opens to prevent current flow in the reverse direction.

A hypothetical device having precisely those characteristics is called the **ideal diode** and symbolized by Fig. 8.1–1a. With a negative applied voltage, the ideal diode is off and no current flows — like the open switch in Fig. 8.1–1b. With a slightly positive applied voltage, say $v = 0^+$, the ideal diode is on and any amount of current may flow in the forward direction — like the closed switch in Fig. 8.1–1c. The value of i in this case would be limited by external circuitry, and the short-circuit condition of the ON state makes it impossible to have any positive voltage $v > 0$ across the diode.

Figure 8.1–1
Ideal diode: (a) symbol; (b) OFF state; (c) ON state; (d) i-v curve.

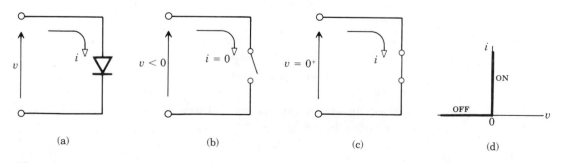

(a)	(b)	(c)	(d)

The arrow-like device symbol indicates the forward direction for current $i \geq 0$, while the bar at the tip of the arrowhead denotes the barrier to conduction in the reverse direction.

We summarize these ideal characteristics by plotting the i-v curve in Fig. 8.1–1d. This nonlinear curve consists simply of two straight lines, corresponding to the diode's OFF and ON states. Curves like this are said to be **piecewise linear.** Despite its idealized nonlinear behavior, the ideal diode turns out to be a practical model for real diodes in many situations — as you'll see when we examine the characteristics of semiconductor diodes.

Example 8.1–1
Half-wave rectifier.

A **half-wave rectifier** consists of an alternating voltage source in series with a diode and a resistor, as diagrammed in Fig. 8.1–2a assuming an ideal diode. The diode voltage v obviously has the same polarity as the source voltage v_s. Thus, when $v_s < 0$, the diode is off and $i = 0$, so $v_R = 0$ and $v = v_s$. Conversely, when $v_s > 0$, the diode is on and $v = 0$, so $v_R = v_s$ and $i = v_s/R$.

Figure 8.1–2b shows the resulting waveforms for a sinusoidal source voltage $v_s = V_m \sin \omega t$. Current now flows only during the positive half

Figure 8.1–2
Half-wave rectifier: (a) circuit; (b) waveforms.

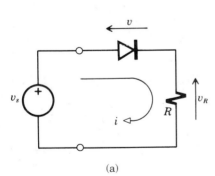

(a)

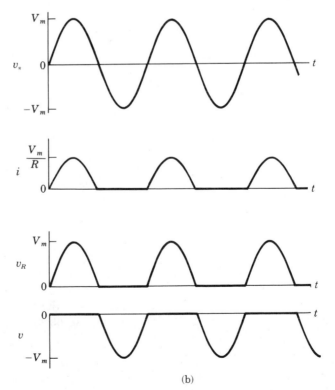

(b)

cycles of v_s. Accordingly, the waveforms i and v_R are zero during the negative half cycles, when $v_s < 0$. We call this circuit a rectifier because it draws *unidirectional* or *nonalternating* current from an AC source. Rectification is the first step for the important task of AC-to-DC conversion in electronic power supplies.

Exercise 8.1–1   Redraw the waveforms in Fig. 8.1–2 for the case where the diode's direction is reversed.

Semiconductor Diodes

Figure 8.1–3 gives the circuit symbol and i-v curve for a typical **semiconductor diode.** The solid arrowhead symbol distinguishes this device from an ideal diode. The i-v curve has three distinct regions marked **forward bias, reverse bias,** and **reverse breakdown,** depending on the value of v. The diode conducts easily in the forward-bias region with $v > 0$, whereas very little negative current flows in the reverse-bias region until we reach reverse breakdown at $v = -V_Z$. Under normal operating conditions we keep $v > -V_Z$ to avoid reverse breakdown, and further discussion of that behavior is deferred to the next section.

Figure 8.1–3
Semiconductor diode:
(a) symbol; (b) i-v curve.

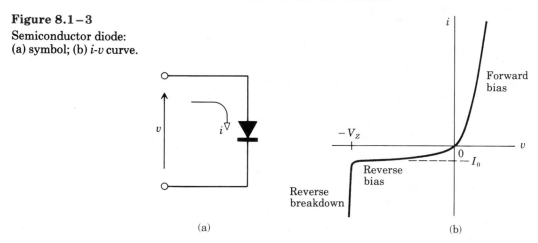

(a)   (b)

The theoretical i-v equation for the forward-bias and reverse-bias regions of a semiconductor diode at room temperature is given by

$$i = I_0(e^{40v} - 1) \tag{1}$$

The physical phenomena underlying this expression will be discussed subsequently. Meanwhile, we observe that $e^4 \gg 1$ and $e^{-4} \ll 1$ so Eq. (1) becomes

$$i \approx \begin{cases} I_0 e^{40v} & v > 0.1 \text{ V} \\ -I_0 & v < -0.1 \text{ V} \end{cases} \tag{2}$$

which emphasizes the difference between forward-bias and reverse-bias behavior. The constant I_0 is called the **reverse saturation current** since $i \approx -I_0$ for $v < -0.1$ V. The value of I_0 varies from diode to diode, typically falling between a few picoamps (10^{-12} A) and a fraction of a milliamp.

In view of the exponential factor in Eq. (1), the apparent shape of the i-v curve depends critically upon the scale of the voltage and current axes. Figure 8.1–4 illustrates this point taking $I_0 = 1$ nA $= 10^{-9}$ A. Part a of the figure brings out the reverse saturation current by restricting the scale to minute values of voltage and current. With larger values included, that same curve appears as shown in part b, where we no longer see the reverse current because $I_0 \ll 1$ μA. Going even further in this direction yields Fig. 8.1–4c in which the i-v curve now looks more like two straight lines.

Figure 8.1–4
Semiconductor diode curves plotted on different scales (omitting reverse breakdown).

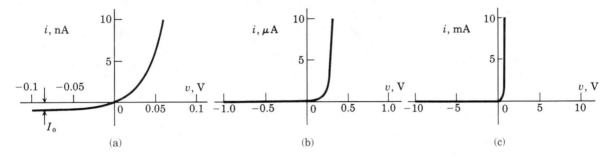

A comparison of Fig. 8.1–4c with the i-v curve for an ideal diode (Fig. 8.1–1d) suggests that we can use the ideal diode as a *model* for a semiconductor diode whenever the forward voltage drop and reverse current of the semiconductor diode are unimportant. By modeling a semiconductor diode as an ideal diode, we avoid the analytical complications that come with the nonlinear i-v relationship in Eq. (1). We'll pursue this point further in Section 8.2.

Example 8.1–2

A semiconductor diode having $I_0 = 10$ μA is connected in series with a 1-kΩ resistor and a voltage source. The source forward biases the diode such that $i = 30$ mA. The resulting diode voltage drop v is found from Eq. (1) written as

$$10^{-5}(e^{40v} - 1) = 30 \times 10^{-3}$$

so $e^{40v} - 1 = 3000$ and

$$v = (\ln 3001)/40 = 0.200 \text{ V}$$

The total source voltage must be $v_s = v_R + v = 30.20$ V.

Exercise 8.1–2   What source voltage would yield $i = -8\,\mu A$ in the circuit in Example
8.1–2?

Semiconductors and Doping

A brief discussion of **semiconductors** is needed at this point to help
explain the internal workings of semiconductor diodes. In their pure crystalline form, semiconductors are materials that have no free electrons at
absolute zero temperature, but they can conduct at higher temperatures
thanks to thermal ionization. Two such materials are silicon and germanium, which come from the fourth column of the table of the elements.
Some alloys of third and fifth column elements also act like fourth-column
elements. We'll focus our attention primarily on silicon, since it is the basis
for most semiconductor technology.

Each atom in the lattice of a silicon crystal has four nearest neighbors
and shares its four valence electrons with them, one electron to each
neighbor. Although the actual lattice structure is three dimensional, the
two-dimensional representation of Fig. 8.1–5 very nicely shows an approximate cross section of the crystal. The shared electrons constitute
covalent bonds, represented by pairs of lines. The label "+4" on the atoms
means a charge of $+4\,|q_e|$ without the valence electrons. The net electric
charge is, of course, zero.

Figure 8.1–5
Silicon crystal lattice
with covalent bonds.

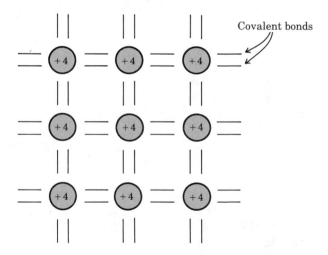

As the temperature increases, the atoms vibrate more and more ener-
getically about their average positions. The valence electrons also gain
energy. At room temperature, some electrons with higher than normal
kinetic energy manage to break loose completely from their bonds and fly
off as **free electrons.** Conduction may now occur thanks to the presence
of this somewhat limited number of free electrons.

There is also another equally important but less obvious mechanism for conduction. When a free electron moves off, it leaves a single electron instead of a pair orbiting around the two parents. The net charge in that vicinity is now positive. Suppose that the electron marked a in Fig. 8.1-6 has become a free electron and disappears from the picture. With an overall electric field $\mathscr{E}$ applied, all orbits are distorted and bulge slightly to the right. Suddenly, electron b leaves its job of bonding the horizontal pair and replaces the vacancy at a. Since b is now a vacant position, the electron at c quickly moves to fill it. Then d fills the c vacancy, and so on. It's easy to see that, on the average, electrons are moving from left to right, even though they are not free in the usual sense.

Figure 8.1-6

Migration of bound electrons from left to right, equivalent to a positive hole going from right to left.

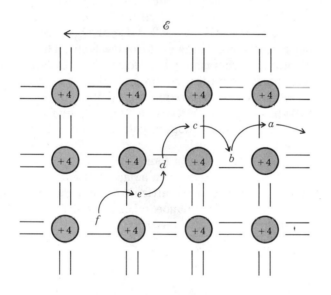

It's more convenient, however, to think of a missing orbital electron as a positive charge at its former location rather than as an absence of a negative charge. We call these fictitious positive charges **holes**, for that's what they really are; and they play a major role in the electrical properties of semiconductors. To clarify, observe from Fig. 8.1-6 that the hole moves from right to left, in the general direction of the field $\mathscr{E}$, whereas the escaped negatively-charged electrons go off to the right. Therefore, both holes and electrons contribute to conduction, although the holes have less mobility due to the covalent bonding.

With or without an applied field, holes exhibit random thermal motion and, from time to time, a hole and an electron in the same vicinity will **recombine** — the electron becoming bound again, and the hole disappearing. But other electron-hole pairs are continually being *generated* by thermal ionizations, and equilibrium conditions require equal rates of generation and recombination. For a pure or *intrinsic* semiconductor, the equilibrium density of free electrons is called the **intrinsic concentration,** denoted by n_i and expressed in electrons per unit volume (m^{-3}). The equilibrium hole concentration p_i must also equal n_i because generation and recombination always involve *electron-hole pairs.* The value of n_i for silicon at room temperature is

$$n_i \approx 1.5 \times 10^{16} \text{ electrons/m}^3 \qquad\qquad (3)$$

Since silicon has about 10^{28} atoms/m³, it follows that only a tiny fraction of the atoms will be ionized at any instant. However, thermal generation increases n_i rapidly when the temperature increases.

Most semiconductor devices are made by adding controlled amounts of impurities to pure crystalline silicon, producing a **doped** semiconductor. Doping elements from the fifth column of the periodic table (phosphorus, arsenic, antimony) are called **donors** because they donate their fifth valence electron to the pool of free electrons. The resulting ***n*-type** semiconductor conducts mostly by electron flow. Doping elements from the third column (aluminum, gallium, indium) are called **acceptors** because they accept a free electron to make up a full set of covalent bonds. The resulting ***p*-type** semiconductor conducts mostly by the motion of holes.

To explain how doping works, consider the region in an *n*-type crystal where a donor atom has replaced the silicon atom, as shown at the center of Fig. 8.1–7. Since only four of the five valence electrons are needed as covalent bonds for the four adjacent atoms, the fifth electron orbiting the donor atom requires so little additional energy to escape that at any reasonable temperature it goes off as a free electron, leaving a net charge of $+1$ at the location of the impurity.

Figure 8.1–7
Silicon crystal with
n-type doping by donor
atom.

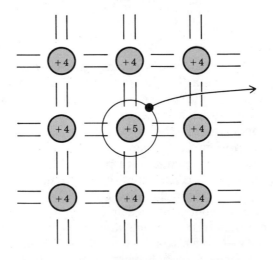

Usually, we make the donor concentration much larger than n_i, which increases the electron concentration n and simultaneously decreases the hole concentration p via recombination. Thus, a typical n-type semiconductor has

$$n \gg n_i \qquad p \ll n_i$$

Accordingly, we say that electrons are the **majority carriers** while the holes are the **minority carriers.** Nonetheless, the concentration product remains constant at

$$np = n_i^2 \tag{4}$$

This relationship reflects the fact that the generation and recombination rates must still be equal in a doped semiconductor.

A similar situation exists for acceptor doping. In this case an impurity atom, while electrically neutral, would have only three valence electrons to share with four neighbors. The result is a vacancy or hole. It takes very little thermal energy for the hole to move off, leaving a fixed negative charge (the fourth and borrowed electron) in the vicinity of the impurity atom. Conduction is, in this instance, almost entirely by holes, which are now the majority carriers since their concentration is about the same as that of the impurity atoms. We thus have $p \gg n_i$ and $n \ll n_i$.

Whether n-type or p-type, a relatively small amount of doping produces a drastic change in the concentration of majority and minority carriers and significantly reduces the resistance of a doped semiconductor. The resistivity is calculated by extension of Eq. (6b), Section 2.3, taking account of the fact that holes are less mobile than electrons but have the same charge magnitude. Thus,

$$\rho = \frac{1}{|q_e|n\mu_e + |q_e|p\mu_p} = \frac{1}{|q_e|(n\mu_e + p\mu_p)} \tag{5}$$

where, for silicon at room temperature,

$$\mu_e \approx 0.13 \qquad \mu_p \approx 0.05$$

which are expressed in SI units.

pn Junctions

There are several methods for producing a semiconductor crystal that has an abrupt change from a predominantly donor or n-type region to a predominantly acceptor or p-type region. The boundary between the two regions is called a **pn junction.** When terminals are attached to the p and n sides of the crystal, we have a **semiconductor diode.**

Figure 8.1–8 is a simplified representation of a pn junction, with p-type material to the left of the dashed line and n-type material to the right. The circles stand for impurity atoms only. The scale of the drawing is quite different from our previous figures, for we have omitted the millions of silicon atoms fixed in the lattice between the impurity atoms. The impurity atoms, of course, are actually distributed at random.

Figure 8.1–8
pn junction showing
depletion region.

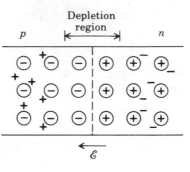

The donor atoms in the *n*-type region are positively ionized, while the acceptor atoms in the *p*-type region are negatively ionized. The resulting free electrons and holes move about randomly, but the net charge equals zero in any area of reasonable size — except in the immediate vicinity of the junction. Here there is a **depletion region** almost totally devoid of mobile charges, because any holes moving from left to right across the junction have recombined with electrons; likewise for electrons moving across the junction from right to left. Thus, the depletion region has an electrostatic field $\mathcal{E}$ pointing from the immobile positive ions on the *n* side to the immobile negative ions on the *p* side. We infer quite correctly from the presence of this electric field that there is an effective potential difference between the *p* and *n* sides of the crystal.

Now let's assume that the right-hand part of the *n*-type region is connected to a metal wire, and the left-hand part of the *p* region is similarly

Figure 8.1–9

(a) Short-circuited *pn* junction. (b) Potential diagram with contact potential V_d.

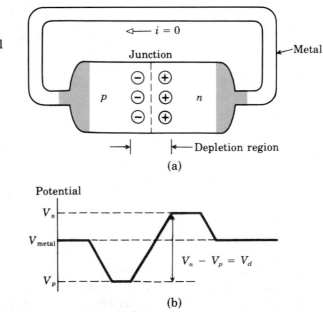

connected to the other end of the same wire. If all parts are at the same temperature, we would expect no net current flow, since the system has no energy source. Nevertheless, the internal potentials of the wire, the p-type material, and the n-type material are all different, as we show in Fig. 8.1–9, where the potential of the metal has been chosen arbitrarily. The potential difference $V_d = V_n - V_p$ is called the **contact potential** of the junction, and typically equals a few tenths of a volt.

Although no current flows through the connecting wire that short-circuits a pn junction, there are actually *two opposing currents* within the diode itself. Figure 8.1–10 depicts these currents, labeled I_0 and I_d, and their constituent carriers. A description of how these currents originate will help explain the i-v characteristics of a semiconductor diode with an external applied voltage.

Figure 8.1–10

Current components in a pn junction.

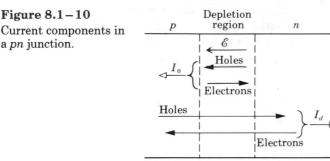

The current I_0 originates in or near the depletion region where thermal ionization of the silicon atoms continually generates new hole-electron pairs, just as in an intrinsic semiconductor. The internal electric field $\mathcal{E}$ separates the holes and electrons and sweeps them in opposite directions, producing equivalent positive charge transfer from the n side to the p side. We call I_0 the *drift current* because it consists of charge carriers that drift under the action of an electric field. Since the carriers in I_0 are thermally generated, this current is quite small at room temperature but it increases rapidly when the temperature goes up. Specifically, if $I_0 = I_0(T_1)$ at temperature T_1, then at a higher temperature T_2,

$$I_0(T_2) \approx I_0(T_1) \times 2^{(T_2 - T_1)/10} \tag{6}$$

where T_1 and T_2 are expressed in *degrees kelvin* (K). Equation (6) says that I_0 approximately doubles when the temperature increases by 10 K.

The current I_d also consists of electrons and holes. However, it originates outside the depletion region and travels *against* the electric field, unlike ordinary currents. To understand this unusual process, recall that electrons and holes have thermal motion with randomly directed velocities. Consequently, a few of the holes on the p side will be injected into the depletion region with sufficient velocity to overcome the potential barrier

$V_n - V_p$ and enter the n-type material—where they are *minority* carriers and quickly recombine with majority-carrier electrons. Likewise, some electrons from the n side will cross the depletion region and recombine in the p-type material. We call I_d the **diffusion current.** It overcomes the potential barrier by virtue of the *diffusion process* that always exists between regions of higher and lower concentrations of particles. Familiar examples of diffusion in action are the evaporation of water and the odor of gasoline, both due to the natural movement of particles to regions of lower concentration.

For the equilibrium condition back in Fig. 8.1–9, with $V_n - V_p = V_d$, the value of V_d must be such that the diffusion current exactly equals the drift current and $i = I_d - I_0 = 0$. But I_d is highly sensitive to the potential barrier, so an external applied voltage upsets the equilibrium and produces a net current—which we know will be much greater in one direction than the other. Let's see how this works in terms of the two current components.

Consider first the situation in Fig. 8.1–11a, where an external source raises the potential of the n side of the diode relative to the p side. This voltage adds to the contact potential, increases the diffusion barrier, and effectively stops I_d if V_s exceeds a few tenths of a volt. The remaining current is $i \approx -I_0$, temperature dependent but relatively small. Thus, the junction is *reverse biased,* and we recognize I_0 as the *reverse saturation current.*

Figure 8.1–11
pn junction with: (a) reverse bias; (b) forward bias.

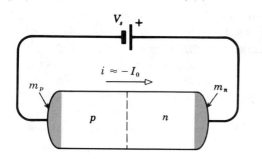

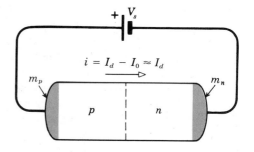

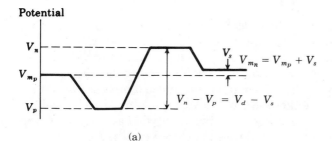

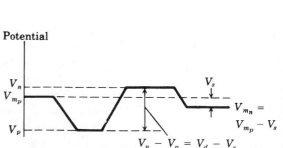

(a)

(b)

Next, let the applied voltage raise the potential of the p side as in Fig. 8.1–11b. This voltage has little effect on I_0, but it lowers the potential barrier and allows the diffusion current to increase markedly. The junction is now *forward biased,* and diffusion accounts for the major portion of the net current $i = I_d - I_0 > 0$.

For an arbitrary applied voltage v, the diffusion current theoretically has an exponential dependence of the form

$$I_d = Ae^{\lambda(v - V_d)} = Ae^{-\lambda V_d} e^{\lambda v}$$

Here, A is a proportionality constant while

$$\lambda = |q_e|/kT = 11,600/T \qquad \textbf{(7)}$$

with k being Boltzmann's constant and T the absolute temperature. The total current is then

$$i = I_d - I_0 = Ae^{-\lambda V_d} e^{\lambda v} - I_0$$

But $i = 0$ when $v = 0$, which requires that $Ae^{-\lambda V_d} = I_0$ and thus

$$i = I_0 e^{\lambda v} - I_0 = I_0(e^{\lambda v} - 1) \qquad \textbf{(8)}$$

If we take room temperature to be $T = 290$ K, then $\lambda = 40$ and Eq. (8) becomes the i-v relationship for a semiconductor diode as previously stated in Eq. (1).

Example 8.1–3

A certain diode at room temperature is found to have $i = 16\ \mu$A when $v = 50$ mV. With $\lambda v = 40 \times 0.05 = 2$ in Eq. (8), we get the drift current

$$I_0 = 16\ \mu\text{A}/(e^2 - 1) = 2.5\ \mu\text{A}$$

Since $i = I_d - I_0$, the corresponding diffusion current must be $I_d = i + I_0 = 18.5\ \mu$A.

Exercise 8.1–3

Use Eqs. (6) and (7) to find the new values of I_0, I_d, and i for the diode in Example 8.1–3 when the temperature increases 20 K to $T = 310$ K, but v is held constant.

8.2 DIODE CIRCUIT ANALYSIS

This section presents two methods for analyzing nonlinear diode circuits. The first method confronts nonlinearity head-on, using a graphical or numerical approach based on the concept of the load line. The second method is a linearization approach, using piecewise-linear models to approximate nonlinear diode behavior. We'll employ ideal diodes to construct models of the forward-bias and reverse-breakdown conditions. Then we'll develop the breakpoint analysis technique for circuit models containing ideal diodes.

Load-Line Analysis

Consider the seemingly trivial circuit in Fig. 8.2–1, where the unspecified element on the right is nonlinear. This element could be a semiconductor diode or any two-terminal device having a nonlinear i-v curve expressed by the function

$$i = f(v) \tag{1}$$

The left-hand side of the circuit represents, in general, the Thévenin model of any DC source network.

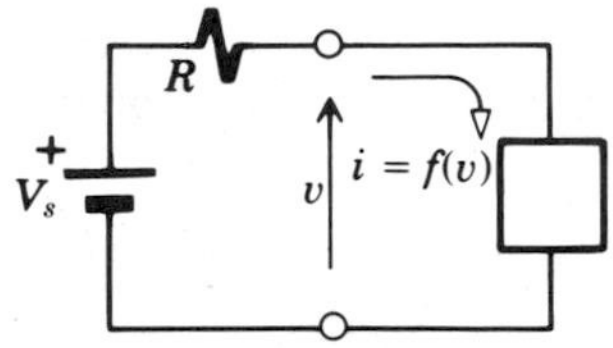

Figure 8.2–1
Circuit with a
nonlinear device.

Despite the presence of the nonlinear element, Kirchhoff's laws still hold here since they pertain to connection relationships independent of the particular elements. And, of course, the linear resistance still obeys Ohm's law. It therefore follows that

$$i = \frac{V_s - v}{R} \tag{2}$$

which is called the **load-line equation.** This equation describes the left-hand side of the circuit, regardless of the nonlinear load element on the right-hand side. Conversely, the device equation $i = f(v)$ describes the nonlinear element itself, regardless of the rest of the circuit.

The device equation and the load-line equation involve two variables, i and v, whose values must simultaneously satisfy both equations. Accordingly, we equate the right-hand sides of Eqs. (1) and (2) to obtain the nonlinear voltage equation

$$\frac{V_s - v}{R} = f(v) \tag{3}$$

The solution of Eq. (3) will be denoted by $v = v_Q$, which is the device's *operating voltage* in this circuit. The corresponding operating current is $i_Q = f(v_Q) = (V_s - v_Q)/R$. But solving Eq. (3) requires the use of graphical or numerical methods.

Graphical solution begins with a plot of the device curve $i = f(v)$. Next, we superimpose the load line on the same set of axes, as illustrated in Fig. 8.2–2. The load line is easily drawn by connecting the point representing the *open-circuit voltage, $v = V_s$* at $i = 0$, to the point representing the *short-circuit current, $i = V_s/R$* at $v = 0$. The intersection of the load line and the

Figure 8.2–2

Load-line analysis showing operating point Q.

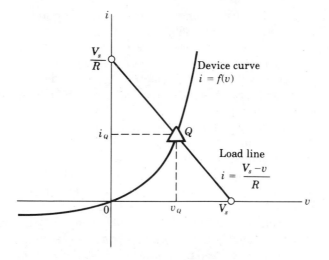

device curve is then the **operating point** Q, since v_Q satisfies Eq. (3) and $i_Q = f(v_Q)$. This method has merit because it works even when the device is described by an experimental i-v curve. However, graphical analysis may entail some rather tedious plotting, and it has limited accuracy.

A more accurate, more modern approach would be a numerical solution arrived at by calculator or computer. To set up the computation, we rewrite Eq. (3) in the form

$$f(v) - \frac{V_s - v}{R} = 0$$

Then we find v_Q using a *root-finding algorithm* such as bisection or Newton's method. The disadvantage of numerical analysis is that it requires an appropriate mathematical expression for $f(v)$.

By way of example, take the circuit in Fig. 8.2–3, where the nonlinear device is a forward-biased semiconductor diode with $I_0 = 1$ nA $= 10^{-6}$ mA. Setting $V_s = 10$ V and $R = 2$ kΩ in Eq. (3), we have the nonlinear equation

$$5 \text{ mA} - \frac{v}{2 \text{ k}\Omega} = 10^{-6} \text{ mA} \ (e^{40v} - 1)$$

Figure 8.2–3

Nonlinear circuit with a diode.

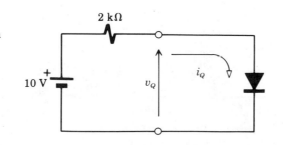

Numerical solution yields the operating-point voltage $v_Q = 0.385$ V, from which $i_Q = 5 - (0.385/2) = 4.81$ mA.

But what do we get if we assume an *ideal* diode in this circuit? We get a much simpler problem whose results are not very far from the mark. Specifically, from the diagram we conclude that current flows in the forward direction, so $v_Q = 0$ V and $i_Q = 10 \text{ V}/2 \text{ k}\Omega = 5$ mA. The latter is off by about four percent, and the small but nonzero voltage drop across the real diode may or may not be significant, depending on the application.

In summary, we've found that analyzing even a simple circuit with a real diode (or other nonlinear element) is a messy problem, at best, whereas the assumption of an ideal diode leads to quick but less accurate results. We'll shortly turn our attention to developing improved models using ideal diodes and other linear elements to represent more accurately the behavior of real diodes. The load-line concept will reappear in the next chapter when we talk about transistors; this concept also helps in the study of waveform distortion caused by nonlinearity, as illustrated in the following example.

Example 8.2–1

Let the battery in Fig. 8.2–3 be replaced by a sinusoidal source $v_s(t) = 0.1 \sin \omega t$ V, and let the rest of the circuit be a 10-MΩ resistance and a semiconductor diode with $I_0 = 1$ nA. We can find the time-varying current $i(t)$ by drawing several load lines corresponding to different values of $v_s(t)$. Four such load lines are shown in Fig. 8.2–4a, representing $v_s/R = \pm 10$ nA when $v_s = \pm 0.1$ V (that is, when $\sin \omega t = \pm 1$) and $v_s/R = \pm 5$ nA when $\sin \omega t = \pm 0.5$ so $v_s = \pm 0.05$ V. Reading off the values of i at the

Figure 8.2–4
Load line and waveforms for Example 8.2–1.

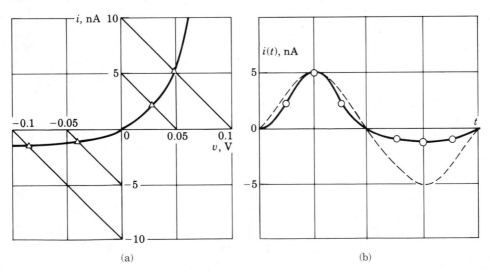

(a)   (b)

intersection points and plotting them versus time gives the waveform $i(t)$ in Fig. 8.2–4b. Figure 8.2–4b also includes a sinusoid with the same positive peak as $i(t)$, which brings out the fact that the current has neither a sinusoidal shape nor a half-wave rectified shape like that of Fig. 8.1–2b.

Exercise 8.2–1   Consider a *square-law diode* described by

$$i = \begin{cases} Kv^2 & v \geq 0 \\ 0 & v < 0 \end{cases}$$

Plot the i-v curve for $0 \leq v \leq 5$ V, taking $K = 0.02$. Draw the load line and find v_Q and i_Q when this diode is inserted in Fig. 8.2–3 with $V_s = 4$ V and $R = 10\ \Omega$. (Save your plot for use in Exercise 8.2–2). Then sketch $i(t)$ for $v_s(t) = 4 \sin \omega t$.

Forward-Bias Models

The most apparent difference between a real diode and the ideal diode is the nonzero voltage drop when a real diode conducts in the forward direction. Figure 8.2–5 shows three ways of modeling this feature with an ideal

Figure 8.2–5
Forward-biased
diode models with:
(a) threshold voltage
V_γ; (b) forward
resistance R_f;
(c) V_γ and R_f.

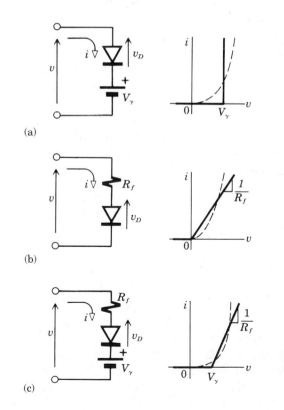

(a)

(b)

(c)

diode and additional elements. The dashed i-v curves represent the actual diode in question, while the solid lines are the piecewise-linear approximations corresponding to each model.

The piecewise-linear curves follow directly from the property that an ideal diode acts like a switch and its voltage v_D can only be negative (when $i = 0$) or zero (when $i > 0$). The corner or intersection of the two straight-line segments is known as the **breakpoint.** It occurs when the diode changes states and, therefore, has both zero voltage drop and zero current.

For the model in Fig. 8.2–5a, the battery has polarity such that $v_D = v - V_\gamma$, where v is the external terminal voltage. If $v < V_\gamma$, then $v_D < 0$ so the ideal diode is off and $i = 0$. The diode goes on when $v = V_\gamma$, and any amount of current may then flow. The resulting i-v curve breaks at the point where $i = 0$ and $v = V_\gamma$. Thus, it looks like the curve of an ideal diode offset to the right by V_γ, which is called the **threshold voltage.**

For the model in Fig. 8.2–5b, the **forward resistance** R_f has no effect when $v < 0$ so the ideal diode is off and $i = 0$. But the diode is on when $v > 0$, so $v_D = 0$ and $i = v/R_f$. The i-v curve therefore breaks at $i = 0$ and $v = 0$, and has slope $1/R_f$ for $v > 0$.

Combining forward resistance and threshold voltage yields the model in Fig. 8.2–5c, a quite satisfactory approximation for most real diodes operating in the forward-bias region. The value of V_γ roughly corresponds to the "knee" of the actual i-v curve which, for silicon diodes, is normally taken to be

$$V_\gamma \approx 0.7 \text{ V} \tag{4}$$

The value of the forward resistance R_f generally falls in the range 5–$50\ \Omega$. Such a small resistance may often be ignored, thereby reducing our model to Fig. 8.2–5a. That same model also holds for the reverse-bias region when the reverse saturation current I_0 is negligible, as is usually the case in practice.

Example 8.2–2

Suppose that the half-wave rectifier back in Fig. 8.1–2 has peak AC source voltage $V_m = 3$ V, load resistance $R = 100\ \Omega$, and a real diode modeled by Fig. 8.2–5c with $V_\gamma = 0.7$ V and $R_f = 10\ \Omega$. When the voltage v across the diode is greater than V_γ, the ideal diode in the model is on, and we have the equivalent circuit of Fig. 8.2–6a. Thus

$$v_R = \frac{100\ \Omega}{110\ \Omega}\ (v_s - 0.7 \text{ V})$$

whose maximum value is 2.1 V when $v_s = 3$ V. The condition $v \geq 0.7$ V corresponds to $v_s \geq 0.7$ V, obtained by solving the circuit for v_s with the breakpoint values $v = 0.7$ and $i = 0$.

When $v_s < 0.7$ V, our model predicts zero current flow, since the ideal diode will be off. The real diode would actually have $|i| \leq I_0$ in the reverse-bias state. If $I_0 = 0.1\ \mu$A, for instance, then $|v_R| \leq 10^{-5}$ V, and we are justified in ignoring the reverse current.

Figure 8.2–6
Circuit and waveforms for Example 8.2–2.

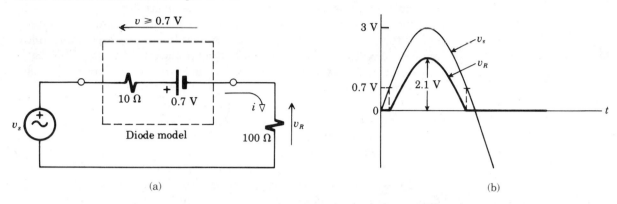

(a)   (b)

Figure 8.2–6b shows the resulting waveform $v_R(t)$ along with $v_s(t)$. Comparing with Fig. 8.1–2b, we see that the real diode reduces the maximum value of $v_R(t)$ and increases its off time. With a larger AC voltage, say $V_m > 10$ V, those differences would be inconsequential.

Exercise 8.2–2   Let the square-law diode in Exercise 8.2–1 be operated in a circuit where $3 \le v \le 5$ V. Find values for V_y and R_f so that the model in Fig. 8.2–5c yields a reasonable approximation of the i-v curve over the operating voltage range. Then use your model to estimate the value of i when $v = 4$ V.

Reverse Breakdown and Zener Diodes

When a large voltage is applied to a pn junction in the reverse direction, the relatively thin depletion layer acquires a high electric field intensity $\mathcal{E}$, as represented by Fig. 8.2–7. Consider a point in the stressed depletion layer near the p boundary of the region. If a thermal ionization is created at that point, then the electron is forced by the field toward the more positive n region. In the relatively short distance between collisions, when the field is high, this electron can gain enough kinetic energy of motion to cause a new

Figure 8.2–7
Avalanche breakdown.

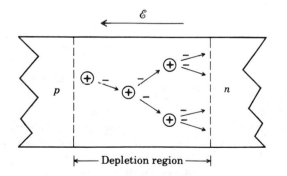

Depletion region

ionization at the next collision. At that point we have the new electron and the original one, both ready to accelerate in the same direction. If, after the next collision, they each produce a new hole-electron pair, then there are now four electrons. At the next collision there could be eight, then sixteen, and so forth in geometric progression. All the holes, of course, travel in the opposite direction of the electrons, adding to the total current. This mechanism, known as **avalanche breakdown,** causes a rapid increase of current when the reverse voltage exceeds a critical value.

An identical result — the rapid increase of current when the reverse voltage exceeds a critical value — occurs in diodes that have narrow depletion layers. As the voltage rises, the field intensity becomes high enough to cause valence electrons to break their bonds, producing large numbers of charge carriers. This mechanism is called **Zener breakdown.**

You should not infer from the term "breakdown" that there is direct damage to the diode. However, the current-voltage product represents power dissipation that does heat the junction, and a diode could be damaged if the temperature remains high enough for a long enough time to cause impurity atoms to diffuse across the junction, thereby degrading it.

Diodes designed expressly to operate in the breakdown region are called **Zener diodes,** irrespective of which mechanism accounts for the breakdown behavior. Such diodes are employed in circuits for the purpose of establishing *reference voltages*. Figure 8.2–8 shows the device symbol along with the linearized *i-v* curve and circuit model.

Figure 8.2–8
Zener diode: (a) symbol; (b) linearized *i-v* curve; (c) circuit model.

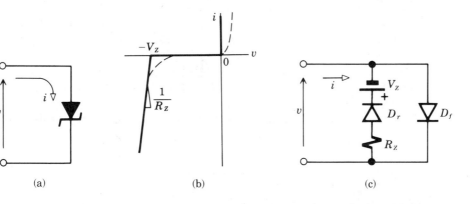

The *i-v* curve indicates that a Zener diode approximates an ideal diode in the forward region. However, when the reverse bias exceeds the **Zener voltage** V_Z, the diode starts to conduct in the reverse direction and acts like a small **reverse resistance** R_Z in series with a battery V_Z. Zener diodes are available with values of V_Z from about 2 V to 200 V with tolerances of ±5% or better.

The circuit model in Fig. 8.2–8c incorporates two ideal diodes, D_f and

D_r, to reflect the forward and reverse characteristics of the Zener diode. The i-v curve thus has *two* breakpoints, one for each ideal diode, and three straight-line segments. With appropriate values of R_Z and V_Z, the left-hand branch of the model could represent any real diode in its breakdown region.

Example 8.2–3
Zener voltage regulator.

Figure 8.2–9a diagrams a simple **voltage regulator** with a Zener diode. Note that the diode is inverted here compared to Fig. 8.2–8, so it will be in reverse breakdown when $V_s - R_s i_{out} > V_Z$. Correspondingly, the forward diode D_f in our model will be off while the reverse diode D_r is on, and we have the equivalent circuit of Fig. 8.2–9b. Straightforward analysis now yields

$$v_{out} = \frac{R_s}{R_s + R_Z}\left(V_Z + \frac{R_Z}{R_s}V_s - R_Z i_{out}\right) \tag{5}$$

Proper operation of this circuit calls for a small reverse resistance R_Z so that

$$R_Z \ll R_s \qquad R_Z\left|\frac{V_s}{R_s} - i_{out}\right| \ll V_Z \tag{6a}$$

in which case

$$v_{out} \approx V_Z \tag{6b}$$

Hence, the Zener diode "regulates" v_{out} by holding it at the fixed Zener voltage V_Z, despite possible variations of V_s or i_{out}.

Figure 8.2–9
Zener voltage regulator:
(a) circuit diagram;
(b) equivalent circuit.

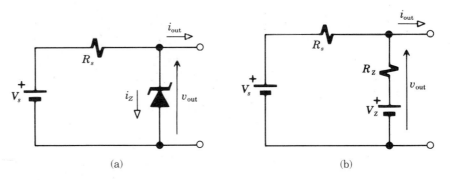

(a) (b)

Exercise 8.2–3

Let the circuit in Fig. 8.2–9 have $V_s = 25$ V, $R_s = 100\ \Omega$, $V_Z = 20$ V, and $R_Z = 4\ \Omega$. Calculate v_{out} for $i_{out} = 0$ and $i_{out} = 50$ mA. Also find the corresponding values of the reverse current i_Z through the Zener diode.

Breakpoint Analysis

When a circuit includes two or more ideal diodes — as in our model for the Zener diode — it has several distinct operating conditions resulting from the OFF and ON states of the diodes. A systematic way of finding those operating conditions is the method of *breakpoint analysis*.

Consider a two-terminal network containing resistors, sources, and N ideal diodes, and driven by a source voltage v. Its i-v curve in general will consist of $N+1$ straight-line segments with N breakpoints. This curve can be constructed by following these steps:

1. Determine the states of all diodes as $v \to +\infty$, and write i in terms of v for his extreme condition. Do the same for $v \to -\infty$.

2. Take one diode to be at its breakpoint, so it has zero voltage drop and zero current, and find the resulting values of i and v at the terminals. Do the same for each of the other diodes.

3. Plot the i-v breakpoints found in step 2, connect them with straight lines, and add the end lines found in step 1.

Having thus constructed the terminal i-v curve, you can also determine the behavior of any other voltage or current of interest within the network.

Figure 8.2–10
Breakpoint analysis:
(a) circuit diagram;
(b) D_1 at breakpoint;
(c) D_2 at breakpoint.

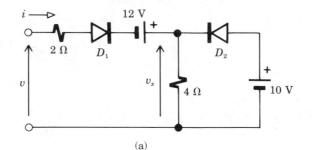

(a)

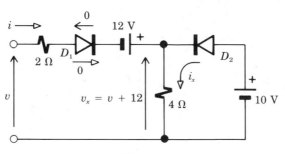

(b)

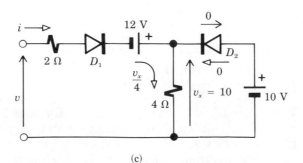

(c)

While doing the second step, you may find in some cases that two or more diodes are at breakpoint conditions simultaneously. When this occurs, the number of breakpoints and line segments of the i-v curve are correspondingly reduced.

Example 8.2–4

To illustrate breakpoint analysis, we'll use it to find the i-v curve of the network in Fig. 8.2–10a.

First, as $v \rightarrow +\infty$, we see that diode D_1 will be forward biased while diode D_2 will be reverse biased by $v_x > 10$ V. Hence, with D_1 on and D_2 off, $v = 2i - 12 + 4i$ so

$$i = \frac{v}{6} + 2 \qquad v \rightarrow +\infty$$

Similarly, by inspection,

$$i = 0 \qquad v \rightarrow -\infty$$

since D_1 will be off and D_2 on.

Next, we set D_1 at its breakpoint by labeling the circuit as shown in Fig. 8.2–10b. Clearly, $i = 0$ and $v_x = v + 12$, but we do not yet know the value of v_x and the state of D_2. If we assume $v_x > 10$ V, then D_2 is off and there is no source for the current $i_x = v_x/4$. We thus conclude that D_2 must be on and, consequently, $v_x = 10$ V. The corresponding i-v breakpoint is at

$$i = 0 \qquad v = -2 \text{ V}$$

which follows from $v = v_x - 12$.

Finally, with D_2 at its breakpoint, Fig. 8.2–10c reveals that $v_x = 10$ V and D_1 must be on to carry $i = v_x/4 = 2.5$ A. We thereby get the second breakpoint at

$$i = 2.5 \text{ A} \qquad v = 3 \text{ V}$$

since $v = 2i - 12 + v_x$.

Figure 8.2–11
i-v curve for Example 8.2–4.

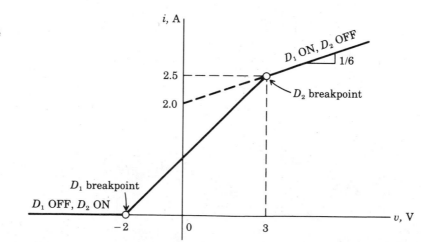

The complete i-v curve based on our results is plotted in Fig. 8.2–11. From the breakpoints and extreme conditions, we now conclude that both D_1 and D_2 will be on over the middle region $-2 < v < 3$.

Exercise 8.2–4   Find the i-v curve for the circuit in Fig. 8.2–10a when the direction of D_2 is reversed.

8.3   SPECIAL-PURPOSE DEVICES †

Although junction diodes behave, in general, like electrical one-way valves, the fabricators have a number of parameters under their control that can alter a diode's characteristics. Variations in such things as impurity concentrations and depletion layer thickness may result in diodes that are less than optimum for some purposes, but serve others very well. In this section we examine some interesting and useful special-purpose diodes and related nonlinear devices (SCRs and other controlled rectifier devices will be described in Section 12.2.)

Junction Capacitance and Varactors

The *reverse*-biased pn junction was seen to have a depletion layer devoid of mobile carriers, but with fixed positive and negative charges (ions) separated by an apparent insulator, as depicted in Fig. 8.3–1a. This fulfills the definition of *capacitance*, and the value of C_R decreases with reverse voltage because more and more impurity atoms are "uncovered" and the width of the depletion layer increases. The ability to change a capacitance with applied voltage is a handy one used to advantage in applications such as the

Figure 8.3–1

Reverse junction capacitance:
(a) uncovered ions;
(b) varactor symbol.

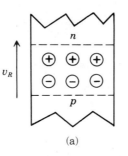

(a)

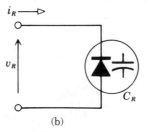

(b)

automatic frequency control (AFC) circuit of an FM radio. Diodes fabricated especially to enhance this property are called **varactors** or **varicaps,** symbolized in Fig. 8.3–1b. The reverse capacitance is given by

$$C_R \approx C_C + \frac{C_0}{\sqrt{1 + 2v_R}}$$

where C_C and C_0 are constants. Reverse capacitance, also known as **transition capacitance,** has nominal values in the range of 1–500 pF.

Another type of capacitance is caused by the high concentration of minority carriers in the depletion region of a *forward*-biased diode. This is called **diffusion capacitance** C_D. Its value is proportional to the DC forward current, and may be much greater than C_R.

Ohmic Contacts and Schottky Diodes

Attaching the leads to a diode requires formation of a *nonrectifying* or *ohmic* junction between the metal and the semiconductor. This is usually accomplished by increasing the impurity concentration to obtain a high-conductivity region adjacent to the metallic contact. Such heavily doped regions would be labeled with the symbols n^+ or p^+.

It is also possible to construct a *rectifying* metal-to-semiconductor junction instead of a *pn* junction. When this is done using *n*-type material, a positive voltage on the metal with respect to the semiconductor causes electrons to be injected into the metal with such high velocity that they are sometimes called *hot carriers*. Since the current flows in the form of majority carriers only, the effective capacitance is very low, as is the forward voltage drop. Such devices, known as **Schottky diodes,** have advantages in high-speed applications.

Figure 8.3–2 shows the structure, symbol, and *i-v* curve of a Schottky

Figure 8.3–2
Schottky diode: (a) construction; (b) symbol; (c) *i-v* curve.

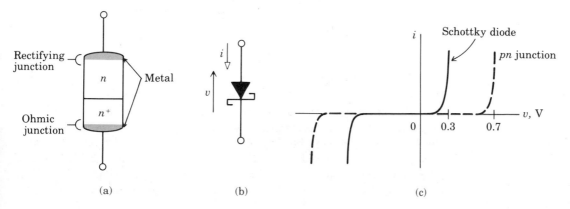

(a) (b) (c)

diode. We see that the metal-to-semiconductor junction results in smaller values of forward threshold and reverse breakdown voltage, when compared to a *pn* junction.

Photo Devices

Light-sensitive and light-emitting devices may be grouped under the general heading of photo devices. Some of the major ones are described here.

It is possible to fabricate a *pn* junction very close to the surface of a crystal. An ohmic contact may then be made by evaporation, for instance, with a surface material so thin that it is transparent to light. When light impinges upon this device, many photons begin to interact with the valence electrons of atoms in the depletion region and generate hole-electron pairs by photoionization. The amount of reverse saturation current, $I_0 + I_p$, now depends on the number of incident photons per second, or light intensity (as well as on the temperature), as illustrated in Fig. 8.3–3. Diodes designed to be used as light sensors are called **photodiodes.** Those intended to produce electric power from light are called **solar cells.**

Figure 8.3–3

Photodiode: (a) symbol; (b) *i-v* curves with load lines for diode operation (*A*), solar cell (*B*), and light sensor (*C*).

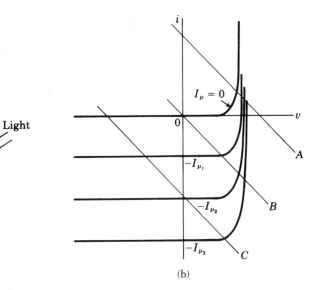

The load line labeled *A* in Fig. 8.3–3b is drawn, as customary, for a fixed resistance and a positive supply voltage. This arrangement produces very little change in voltage or current for varying light levels. The load line labeled *B* is for the same resistance value, but the supply voltage is zero, so the line goes through the origin. Now the illuminated diode generates power $p = vi$ that heats the resistor with what was originally light energy. This is the basic arrangement for a solar-cell application, although in practice many cells would be connected in series in order to raise the available voltage. The third load line, labeled C, is for a negative value of applied voltage. We then have a light sensor that produces both current and voltage changes proportional to the changes of light intensity.

But light sensors do not have to be diodes. **Photocells** are two-terminal photoconductive devices in which a thin layer of a semiconducting material (usually cadmium sulphide) produces many new charge carriers when exposed to light. Thanks to these additional carriers, a photocell's resistance under illumination drops to a small fraction of the "dark" resistance.

Now consider a junction diode operating in the forward direction: Large numbers of electrons are injected into the p region where they recombine with holes; and large numbers of holes are injected into the n region where they recombine with electrons. The recombination process in both cases releases energy, which is not surprising since an external energy source is required to produce hole-electron pairs. Under favorable circumstances, the energy released by a recombination may be in the form of a light photon that has an optically clear path out of the diode. Such diodes act like photocells in reverse, producing light when electrically energized. These are called **light emitting diodes**, or LEDs. LEDs come in a half dozen colors, have forward voltage drops of one to two volts, and — unlike most other light sources — change their intensity very rapidly in response to sudden changes in diode current.

A familiar application is the seven-segment display in Fig. 8.3–4a, where each segment consists of one or several LEDs. Another application

Figure 8.3–4

Light-emitting diodes in: (a) seven-segment display; (b) optoisolator.

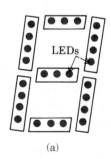

(a)

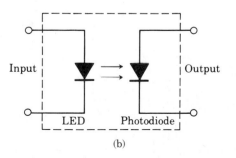

Input   Output

LED   Photodiode

(b)

is the **optoisolator,** Fig. 8.3–4b, which contains an LED whose light output directly illuminates a photodiode inside an opaque package. The result is a four-terminal device in which neither of the two input terminals is connected to an output terminal. This ability to isolate one circuit from another is very useful, but is generally limited to ON/OFF signals. The variations of a continuous signal would be distorted by the nonlinearity of the optoisolator.

Thermistors and Varistors

A **thermistor** is a two-terminal device made of polycrystalline semiconductor material instead of a single crystal. The polycrystalline material produces many new hole-electron pairs when the temperature rises, so the device exhibits a negative temperature coefficient of resistance. Figure 8.3–5 shows the thermistor symbol and a typical curve of resistance versus temperature. A thermistor may be put in one branch of a bridge circuit for the purpose of sensing temperature variations.

Figure 8.3–5
Thermistor symbol and resistance curve.

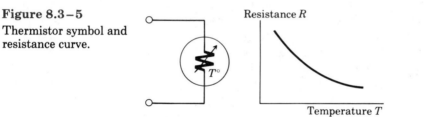

Another useful polycrystalline device is the **varistor** in Fig. 8.3–6. A varistor's resistance is quite large until the applied voltage exceeds a critical value for either polarity. The device then breaks down and conducts heavily, acting something like a Zener diode. Varistors are used to prevent damage to other elements by the sudden high-voltage spikes that sometimes occur on power and communication lines. Most *surge protectors,* for instance, employ varistors to combat voltage spikes on AC power lines.

Figure 8.3–6
Varistor symbol and *i-v* curve.

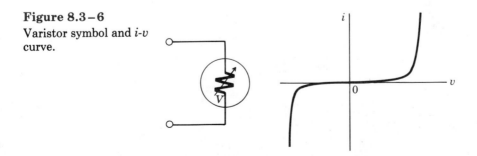

PROBLEMS

8.1–1 A certain semiconductor diode at room temperature has $i = 1$ mA when $v = v_1 > 0.1$ V and $i = 10$ mA when $v = v_2 > v_1$. Calculate the voltage change $v_2 - v_1$.

8.1–2 Do Problem 8.1–1 for $i = 100$ mA at $v = v_2 > v_1$.

8.1–3 Both diodes in Fig. P8.1–3 are at room temperature and have $I_0 = 2\ \mu$A. Find i_1, i_2, and i when $v = 30$ mV.

Figure P8.1–3

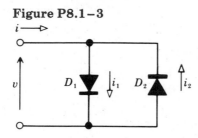

8.1–4 Both diodes in Fig. P8.1–3 are at room temperature and have $I_0 = 2\ \mu$A. Find v, i_2, and i given that $i_1 = 4\ \mu$A.

8.1–5 Both diodes in Fig. P8.1–3 are at room temperature and have $I_0 = 1\ \mu$A. Make a reasonable assumption to find i_2, i_1, and v given that $i = 200\ \mu$A. Check the validity of your assumption.

8.1–6 The op-amp in Fig. P8.1–6 is ideal and the diode is at room temperature.

(a) Show that if $v_{\text{in}} > 0.1$ V, then this circuit acts as an **exponential amplifier** with

$$v_{\text{out}} \approx -RI_0 e^{40v_{\text{in}}}.$$

(b) Let the diode and resistor be interchanged, the diode still pointing to the right. Show that if $v_{\text{in}} \gg RI_0$, then the modified circuit acts as a **logarithmic amplifier** with $v_{\text{out}} \approx -0.025 \ln (v_{\text{in}}/RI_0)$.

Figure P8.1–6

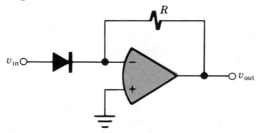

8.1–7 Use the results of Problem 8.1–6 to devise an op-amp system whose output voltage is proportional to the *product* of the positive input voltages v_1 and v_2. Hint: Recall that $e^{(\ln x + \ln y)} = xy$.

8.1–8 Find the resistivity of silicon at room temperature with doping such that $n = 3n_i$. Also calculate and interpret the ratio $n\mu_e/p\mu_p$.

8.1–9 Do Problem 8.1–8 with doping such that $p = 5n_i$.

8.1 – 10   A resistor in an integrated circuit is a bar of doped silicon with area $A = 50\,\mu m^2$ and length $\ell = 100\,\mu m$. Assume that $n \gg n_i$ and find n needed to get $R \approx 4\ k\Omega$. Then check the assumption.

8.1 – 11   Assume that $p \gg n_i$ and find p needed to get $R \approx 1\ k\Omega$ for the resistor in Problem 8.1 – 10. Then check the assumption.

8.1 – 12   If a doped semiconductor has concentration N_d of donor atoms and concentration N_a of acceptor atoms, then charge neutrality requires $n + N_a = p + N_d$. Show that the electron concentration must satisfy the quadratic equation

$$n^2 + (N_a - N_d)n - n_i^2 = 0$$

Then verify that $n \approx N_d$ when $N_a \ll N_d$ and $N_d \gg n_i$.

8.1 – 13   Use the quadratic equation in Problem 8.1 – 12 to find n when $N_a = 8 \times 10^{17}$ and $N_d = 10^{17}$. Then show that $p \approx N_a$.

8.1 – 14   Suppose both diodes in Fig. P8.1 – 3 have $I_0 = 10$ nA at $T = 290$ K. Find i_1, i_2, and i when $T = 340$ K and $v = 0.3$ V.

8.1 – 15   Suppose both diodes in Fig. P8.1 – 3 have $I_0 = 0.1\ \mu A$ at $T = 290$ K. Find i_1, i_2, and i when D_1 operates at 300 K, D_2 operates at 350 K, and $v = 0.2$ V.

8.1 – 16   A diode operating at $T = 290$ K has $i = 20$ mA when $v = 0.5$ V. Find the values of I_0 and i when T increases to 350 K and v is held fixed.

8.1 – 17   A diode operating at $T = 290$ K has $I_0 = 10$ pA and $i = 10$ mA. Find the resulting voltage change Δv when the temperature increases by $\Delta T = 40$ K and i is held fixed.

8.1 – 18   The effect known as **current hogging** occurs when two identical diodes are connected in parallel, both conducting in the forward direction, but one diode is at a higher temperature. To illustrate this effect, find the diode currents i_1 and i_2 when diode D_1 is at $T = 290$ K, diode D_2 is at $T = 310$ K, $v = 0.25$ V, and $i_1 + i_2 = 1$ mA.

8.1 – 19   A diode has $I_0 = 1\ \mu A$ at $T = 290$ K. Obtain an expression for v in terms of T when the temperature varies but the forward current remains constant at $i = 1$ mA.

8.2 – 1   Let $V_s = 10$ V and $R = 5\ \Omega$ in Fig. 8.2 – 1. Solve analytically for v_Q and i_Q with a nonlinear element having $i = 0.02v^2$ for $v \geq 0$.

8.2 – 2   Do Problem 8.2 – 1 with a nonlinear element having $i = 0.1(v + v^2)$.

8.2 – 3   Do Problem 8.2 – 1 with a nonlinear element having $i = 0.4\sqrt{v}$.

8.2 – 4   A certain diode is found to have $i = 0.1$ mA at $v = 0.6$ V and $i = 75$ mA at $v = 1.5$ V.

     (a)   Mode the forward-bias behavior in the form of Fig. 8.2 – 5c.

     (b)   Use your model to estimate the peak value of v_R when the diode is put in a half-wave rectifier like Fig. 8.2 – 6a with $v_s = 2 \cos \omega t$ V. Then calculate the *conduction angle* $\theta = \omega T_c$, where T_c is the amount of time the diode conducts during each cycle.

8.2 – 5   Do Problem 8.2 – 4 for a diode found to have $i = 1$ mA at $v = 0.3$ V and $i = 100$ mA at $v = 1.0$ V.

8.2 – 6   Do Problem 8.2 – 4 for a diode found to have $i = 0.5$ mA at $v = 0.8$ V and $i = 400$ mA at $v = 2.4$ V.

8.2−7   Obtain an expression for i_Z in Fig. 8.2−9b, and use it to derive Eq. (5).

8.2−8   Let the regulator in Fig. 8.2−9 have $R_s = 18\ \Omega$, $R_Z = 2\ \Omega$, and $V_Z = 30$ V, and let V_s vary over 40 to 50 V. Calculate the minimum and maximum values of v_{out} and i_Z when a 60-Ω load resistor is connected across the output.

8.2−9   Suppose the Zener diode in Fig. 8.2−9 is rated for $i_Z \leq 2$ A and has $V_Z = 12$ V and $R_Z = 1\ \Omega$. Determine the minimum allowable value of R_s and the maximum allowable value of i_{out} when V_s varies from 22 to 28 V.

8.2−10 Let the regulator in Fig. 8.2−9 have $R_Z = R_s/9$. Find the Thévenin parameters $v_{out\text{-}oc}$ and R_{eq} when $V_s - R_s i_{out} > V_Z$ and when $V_s - R_s i_{out} < V_Z$.

8.2−11 Suppose that i_{out} in Fig. 8.2−9 varies from 0 to 0.5 A, and the regulator circuit has $V_s = 30$ V, $R_s = 20\ \Omega$, $V_Z = 25$ V, and $R_Z = 5\ \Omega$. Plot v_{out} versus i_{out} by considering the two end points and the Zener's break point.

8.2−12 Plot and label the linearized i-v curve for Fig. P8.2−12, taking the Zener diode to be modeled by Fig. 8.2−8c with $V_Z = 12$ V and $R_Z = 0$.

Figure P8.2−12

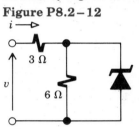

8.2−13 Plot and label the linearized i-v curve for Fig. P8.2−12, taking the Zener diode to be modeled by Fig. 8.2−8c with $V_Z = 18$ V and $R_Z = 6\ \Omega$.

8.2−14 Let the circuit in Fig. P8.2−14 have $R_a = \infty$ and and $R_b = 4\ \Omega$. Plot and label the i-v curve.

Figure P8.2−14

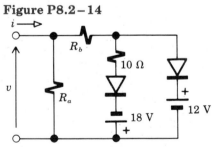

8.2−15 Let the circuit in Fig. P8.2−14 have $R_a = 6\ \Omega$ and $R_b = 0$. Plot and label the i-v curve.

8.2−16 Let the circuit in Fig. P8.2−14 have $R_a = R_b = 6\ \Omega$. Plot and label the i-v curve.

Chapter

9 Transistors and Integrated Circuits

The transistor family embraces a diverse and growing collection of semi-conductor devices with literally thousands of different members. Despite this multiplicity, all transistors are *three-terminal* elements that have the distinctive ability to *control* a voltage or current in accordance with the value of another voltage or current. There are two major types of transistors: **field-effect transistors** (FETs), which are voltage controlled, and **bipolar junction transistors** (BJTs), which are inherently current controlled.

Depending upon the circuit configuration and application, a given FET or BJT may operate in either a linear or nonlinear mode. As the active device in electronic amplifiers and instrumentation systems, the transistor is operated within its linear regime and acts like a *controlled source*. When operated at the extremes of its nonlinear mode, a transistor effectively becomes a *switch* for use in digital computers or other electronic switching systems.

This chapter discusses the internal structure and external characteristics of transistors. We'll start with the class of field-effect devices known as enhancement MOSFETs, after which we'll consider depletion MOSFETs and JFETs. Then we'll examine bipolar junction devices, bringing out the similarities and differences between BJTs and FETs. The closing section describes semiconductor device fabrication and the technology of integrated circuits.

Objectives

After studying this chapter and working the exercises, you should be able to do each of the following:

- Identify the symbol, static characteristics, and operating regions of an enhancement MOSFET (Section 9.1).

- Plot and interpret the transfer curve and waveforms for a simple enhancement MOSFET circuit (Section 9.1).

- Find the operating point of an enhancement MOSFET using load-line analysis or universal curves (Section 9.1).

- Explain how depletion MOSFETs and JFETs differ from enhancement MOSFETs (Section 9.2).

- Identify the symbol, static characteristics, and operating regions of a BJT (Section 9.3).

- Find the operating point of a BJT using an appropriate large-signal model (Section 9.3).

- Describe the processes used to fabricate an integrated circuit (Section 9.4). †

9.1 ENHANCEMENT MOSFETS

The family tree of FETs charted in Fig. 9.1–1 has three main branches and six individual devices, each represented by its own circuit symbol. All of these transistors feature a semiconductor channel, either n-type or p-type, whose conduction is controlled by a field effect. Consequently, all FETs behave in a similar fashion.

Figure 9.1–1
FET family tree.

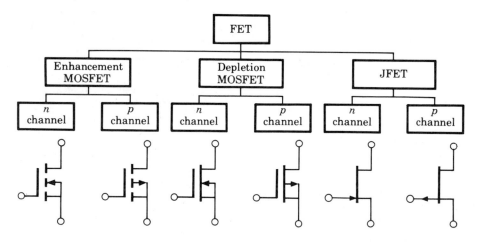

This section introduces field effect in the enhancement MOSFET (pronounced "moss-fet"). We'll show how the internal structure relates to the external characteristics, and how those characteristics lead to amplification and switching action. We'll also present universal curves for enhancement MOSFETs. Our attention will be devoted initially to the more popular n-channel device, after which we'll summarize the differences in the properties of its p-channel sibling.

Structure and Operation

Figure 9.1–2a depicts the cross-sectional structure of an **n-channel enhancement MOSFET.** This device consists of **metallic** and **oxide** layers on a **semiconductor substrate.** The substrate is a piece of silicon with moderate p-type doping. Heavily doped n^+ regions added near each end provide terminal contacts for the **source** (S) and **drain** (D). The **gate** (G) is a metallic film, insulated from the substrate by a thin oxide layer. Hence, a MOSFET is also called an **insulated-gate FET** or IGFET.

Figure 9.1–2
n-channel enhancement MOSFET;
(a) structure;
(b) symbol.

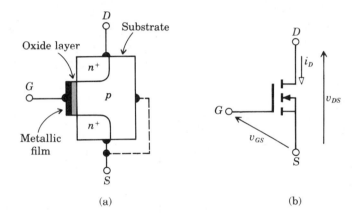

(a)   (b)

The oxide layer has such large resistance — from 10^{12} to 10^{15} Ω — that DC current through the gate is entirely negligible. Furthermore, without voltage applied at the gate, current cannot pass through the substrate because the n^+-p-n^+ path is equivalent to two pn junctions pointing in opposite directions, and one junction or the other will always be reverse biased. An enhancement MOSFET thus acts as a *"normally-off" device* when used for switching purposes.

The MOSFET symbol in Fig. 9.1–2b represents its normally-off state by the broken vertical bar, while the adjacent gap represents the DC open circuit at the gate terminal. The symbol also indicates an electrical connection between the substrate and the source terminal, corresponding to the dashed wire in Fig. 9.1–2a. This substrate-source connection is standard for most discrete MOSFETs, as distinguished from integrated-circuit MOSFETs, and it will be assumed throughout subsequent discussions.

Conduction between the drain and source terminals requires an electric field to create a **channel** in the substrate. We produce this field effect by applying a positive **gate-to-source voltage** v_{GS}. Additionally, in most applications we connect external elements such that the **drain-to-source voltage** v_{DS} will also be positive. The **drain current** i_D then passes through the substrate channel in the direction from drain to source. To explain how the gate voltage controls i_D, we must examine the internal physical picture.

Figure 9.1–3a shows the electric field $\mathscr{E}$ when $v_{GS} > 0$. This transverse field exerts force that pushes holes in the substrate away from the gate and draws mobile electrons toward it. If v_{GS} exceeds the MOSFET's **threshold voltage** V_T, then the field becomes strong enough to establish the situation portrayed by Fig. 9.1–3b. Here, attracted electrons have turned the region along the gate into an ***n*-type channel,** while a **depletion region** isolates the channel from the rest of the substrate.

Figure 9.1–3
(a) Electric field in a MOSFET. (b) Formation of *n*-type channel. (c) Pinched down channel.

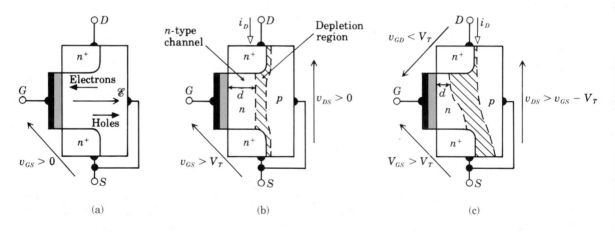

With $v_{GS} > V_T$ and $v_{DS} > 0$, electrons are injected into the channel from the n^+ source region and collected at the n^+ drain region. Since negatively-charged electrons go from source to drain, the equivalent current i_D goes from drain to source as shown. None of the electrons comes from the *p*-type portion of the substrate, which now forms a reverse-biased junction with the *n*-type channel. This substrate-channel junction is represented by an arrowhead on the circuit symbol back in Fig. 9.1–2b.

As the gate voltage increases above V_T, the field increases the channel depth and thereby *enhances* conduction. If v_{GS} is held fixed and the drain voltage is small, then the channel has uniform depth d and it acts like a resistance connected between the drain and source terminals. Accordingly, we say that the MOSFET is operating in the **ohmic state.**

But ohmic operation takes place only with small drain voltage. A larger value of v_{DS} significantly reduces the **gate-to-drain voltage** $v_{GD} = v_{GS} -$

v_{DS}, thus reducing the field strength and channel depth at the drain end of the substrate. Figure 9.1–3c illustrates the resulting "pinched down" condition that occurs with $v_{GD} < V_T$, corresponding to $v_{DS} > v_{GS} - V_T$. The narrowed neck of the channel limits electron flow to the point that i_D becomes essentially constant and independent of v_{DS}. Hence, the MOSFET now operates in a **constant-current state.**

To summarize our description of MOSFET behavior, Fig. 9.1–4 plots i_D versus v_{DS} with v_{GS} held fixed at some value above V_T. This i-v curve rises linearly in the ohmic state, goes through a transitional knee, and flattens off in the constant-current state. Curves plotted for different values of v_{GS} would have the same shape but flatten off at lower or higher currents because the field strength depends upon the gate voltage. However, the field is insufficient to form a channel when $v_{GS} \leq V_T$, so the i-v curve for the normally-off state simply reduces to a horizontal line at $i_D = 0$.

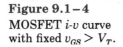
Figure 9.1–4
MOSFET i-v curve
with fixed $v_{GS} > V_T$.

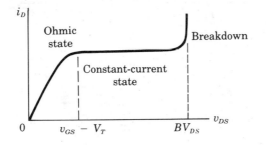

Figure 9.1–4 also shows what happens when v_{DS} exceeds the **drain breakdown voltage** BV_{DS}, usually between 20 and 50 V. The reverse-biased substrate-channel junction then suffers avalanche breakdown, just as in a diode, and the drain current abruptly increases. Continued operation in this condition may generate enough heat to damage the MOSFET.

Another damaging effect (not shown) occurs when v_{GS} exceeds the **gate breakdown voltage,** at about 50 V. Excessive gate voltage causes a sudden and permanent rupture of the oxide layer, completely destroying the device. Even the simple act of handling a MOSFET may result in oxide rupture because static charge accumulates on the insulated gate and quickly builds up to destructive levels. Integrated circuits often include auxiliary diodes for the protection of MOSFET gates.

Static Characteristics

In most applications, MOSFETs have external circuitry arranged such that v_{GS} and v_{DS} may take on any positive values below breakdown. Thus, to analyze a MOSFET circuit, we need to know how i_D changes when either voltage increases or decreases. This information is displayed graphically by the **static characteristics,** a set of plots of i_D versus v_{DS} measured at several different values of the control voltage v_{GS}. The term "static" means that the measurements were made with slowly changing voltages, so internal capacitance has no effect.

The static characteristics of a typical discrete MOSFET are plotted in Fig. 9.1–5. We see that $i_D = 0$ when $v_{GS} \leq 2.5$ V, so this particular MOSFET has $V_T = 2.5$ V. The labeling indicates that the other curves were measured by increasing v_{GS} above V_T in steps of 0.5 V. As expected, these curves rise linearly and go through transitional knees. However, instead of becoming completely horizontal beyond the knees, the curves have a slight upward slope. This slope is produced by a phenomenon known as **channel-length modulation,** but it is usually small enough to be of little consequence.

Figure 9.1–5
Static characteristics of a typical *n*-channel enhancement MOSFET.

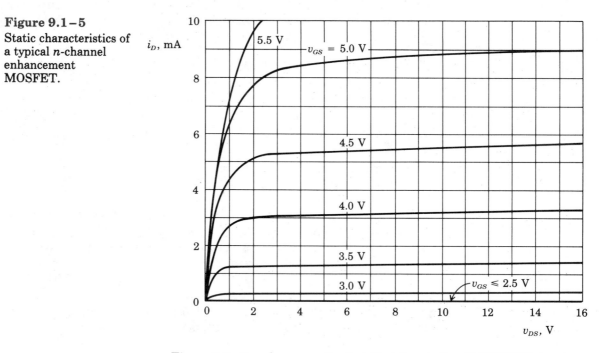

Figure 9.1–5 underscores the fact that the study of MOSFET circuits involves *three* different quantities, i_D, v_{DS}, and v_{GS}. Given values for any two of them, you can determine the value of the third from the characteristic curves. To illustrate, suppose you know that the MOSFET represented by Fig. 9.1–5 happens to operate at $v_{DS} = 9$ V and $i_D = 4$ mA. You would then estimate the gate voltage to be $v_{GS} \approx 4.2$ V, since the point in question falls somewhat less than halfway between the curves for $v_{GS} = 4.0$ and 4.5 V.

Now consider the basic MOSFET circuit in Fig. 9.1–6. The input is a variable gate voltage, and the output load network consists of a supply voltage V_{DD} in series with a resistor R_D. We want to find the dependence of the output current i_D and voltage v_{DS} on the controlling input v_{GS}. Since the

Figure 9.1–6
MOSFET circuit with
variable gate voltage.

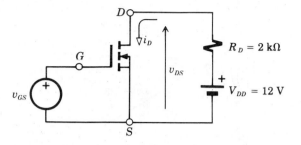

MOSFET has nonlinear drain-to-source characteristics, we must use the graphical load-line approach previously introduced to handle nonlinear diode circuits.

Recall from Section 8.2 that a **load line** represents the i-v relation at the terminals of the load. For the case at hand, the current i_D through the load and the voltage v_{DS} across it are related by the simple load-line equations

$$v_{DS} = V_{DD} - R_D i_D \qquad i_D = \frac{V_{DD} - v_{DS}}{R_D} \tag{1}$$

The load line itself is easily constructed by connecting the open-circuit point ($v_{DS} = V_{DD}$ at $i_D = 0$) to the short-circuit point ($i_D = V_{DD}/R_D$ at $v_{DS} = 0$). Figure 9.1–7a shows this load line drawn on the characteristic curves from Fig. 9.1–5. As before, the intersection of the load line and the device curve defines the **operating point** Q.

But now we have a multitude of device curves, representing different values of the control variable v_{GS}. Hence, the Q point depends upon the specific value of v_{GS}. If $v_{GS} = 4.0$ V, for instance, then we see that the Q point will be at $i_D \approx 3$ mA and $v_{DS} \approx 6$ V. If v_{GS} increases or decreases, then the Q point moves up or down along the load line. A plot of the Q-point values i_D and v_{DS} as a function of v_{GS} thus provides a graphical picture of the input-output relationships. The resulting **transfer curves** are given in Fig. 9.1–7b.

Our transfer curves describe the *complete* MOSFET circuit, including the load elements, and they reveal three more-or-less distinct operating regions:

- The **cutoff region** is characterized by small drain current and large voltage, so

$$i_D \ll V_{DD}/R_D \qquad v_{DS} \approx V_{DD}$$

which corresponds to the MOSFET's normally-off state.

- The **saturation region** is characterized by large drain current and small voltage, so

$$i_D \approx V_{DD}/R_D \qquad v_{DS} \ll V_{DD}$$

which corresponds to the MOSFET's ohmic state.

Figure 9.1–7
(a) MOSFET
characteristics with
load line. (b) Transfer
curves.

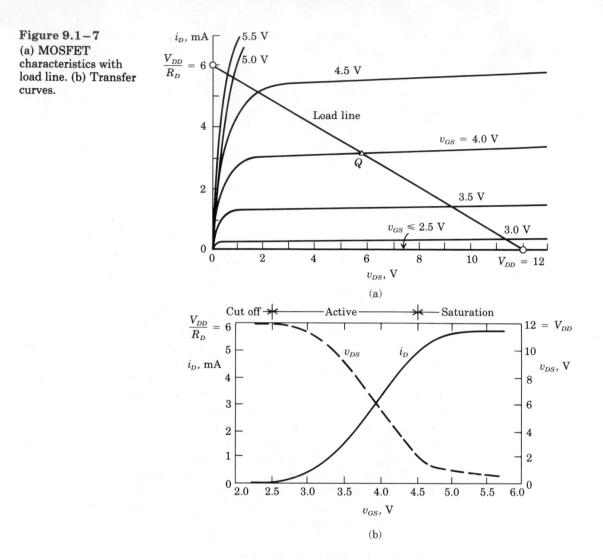

(a)

(b)

■ The **active region** is characterized by intermediate values of drain current and voltage controlled by the gate voltage, which corresponds to the MOSFET's constant-current state.

In view of the correspondence between MOSFET states and the three regions of circuit operation, we frequently use "state" and "region" as interchangeable terms.

Observe from Fig. 9.1–7b that the extreme values of i_D and v_{DS} depend primarily on V_{DD} and R_D. Thus, a MOSFET acts as a *switching device* when v_{GS} drives the circuit from cutoff to saturation, or vice versa. Also observe that the transfer curves are roughly linear near the middle of the active region. Thus, a MOSFET acts as an *amplifying device* when small variations of v_{GS} produce proportional variations of i_D and v_{DS}. Finally, observe that increasing the gate voltage causes the drain voltage to decrease. Thus, both switching and amplifying circuits will have an inherent *voltage inversion* from input to output.

Example 9.1–1
MOSFET switching circuit.

Let the applied gate voltage in Fig. 9.1–6 be the sinusoidal signal $v_{GS}(t) =$ $6 \sin \omega t$ V, whose frequency ω is slow enough to satisfy the static condition. Figure 9.1–8 shows the input waveform along with the resulting output waveforms $i_D(t)$ and $v_{DS}(t)$ obtained from the transfer curves. Severe distortion in the form of top and bottom "clipping" appears in the output waveforms because the circuit is cut off whenever $v_{GS} \leq 2.5$ V and saturated whenever $v_{GS} > 4.5$ V.

Figure 9.1–8
Waveforms for MOSFET switching circuit.

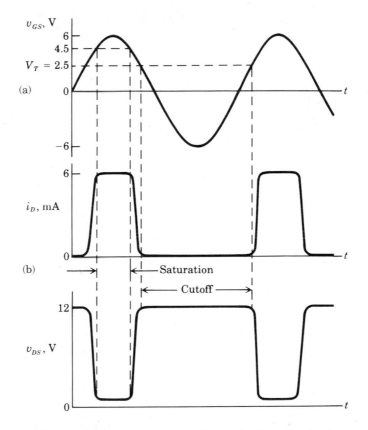

Clearly, what we have here is a switching circuit that goes from OFF to ON when the input changes by about 2.0 V. The MOSFET thereby functions as a switch, turning power off and on to the load resistor R_D, and it does so without drawing power from the controlling input. We'll have more to say about transistor switching circuits and their applications in Chapter 10.

Example 9.1–2
MOSFET amplifier.

For relatively undistorted amplification, our MOSFET circuit must be restricted to signal variations within the active region. Accordingly, suppose we take

$$v_{GS}(t) = 4 + 0.2 \sin \omega t \text{ V}$$

so the gate voltage swings ± 0.2 V relative to a 4-V DC component, as shown in Fig. 9.1–9a.

Figure 9.1–9
Waveforms for
MOSFET amplifier.

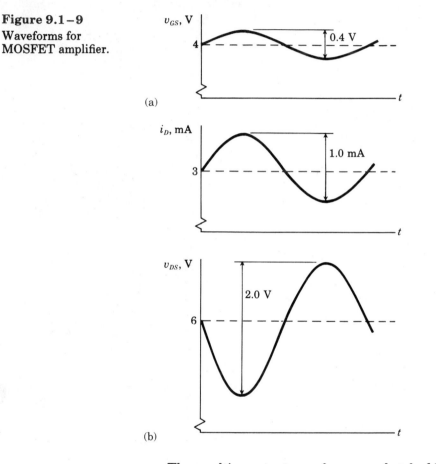

(a)

(b)

The resulting output waveforms are sketched in Fig. 9.1–9b, and they can be expressed as

$$i_D(t) \approx 3 + 0.5 \sin \omega t \text{ mA}$$

$$v_{DS}(t) \approx 6 - 1.0 \sin \omega t \text{ V}$$

We see that the 0.4-V variation of v_{GS} produces a 2.0-V variation of v_{DS} in the opposite direction. Hence, this circuit provides *voltage amplification* by the factor

$$A_v = \Delta v_{DS}/\Delta v_{GS} \approx -2.0/0.4 = -5$$

where the minus sign reflects the voltage inversion. Chapters 11 and 12 pursue transistor amplifier circuits in greater detail.

Exercise 9.1–1   Construct a load line on Fig. 9.1–5 for $V_{DD} = 15$ V and $R_D = 5$ kΩ. Then plot the corresponding voltage transfer curve v_{DS} versus v_{GS} and use it to sketch $v_{DS}(t)$ when $v_{GS}(t) = 3.5 + 0.2 \sin \omega t$ V. Estimate the resulting voltage amplification A_v.

Universal Curves

Drawing upon results from field-effect theory, we now present idealized expressions for the active and ohmic states. These expressions lead to universal curves that approximate the static behavior of any n-channel enhancement MOSFET operated below breakdown.

Field-effect theory says that a MOSFET is in the **active state** when the drain and gate voltages satisfy the conditions

$$v_{DS} > v_{GS} - V_T \qquad v_{GS} > V_T \qquad \textbf{(2)}$$

The drain current then depends on v_{GS} in accordance with the parabolic relation

$$i_D = \frac{I_{DSS}}{V_T^2}(v_{GS} - V_T)^2 = I_{DSS}\left(\frac{v_{GS}}{V_T} - 1\right)^2 \qquad \textbf{(3)}$$

The parameter I_{DSS} signifies the value of i_D when $v_{GS} = 2V_T$, but Eq. (3) may also be written as $i_D = k(v_{GS} - V_T)^2$ where $k = I_{DSS}/V_T^2$. Since v_{GS} controls i_D, independent of v_{DS}, an active MOSFET behaves like a **nonlinear voltage-controlled current source**.

Operation in the **linear ohmic state** takes place at small drain voltages, when

$$|v_{DS}| \leq \tfrac{1}{4}(v_{GS} - V_T) \qquad v_{GS} > V_T \qquad \textbf{(4)}$$

The theory for this case predicts that

$$i_D \approx v_{DS}/R_{DS} \qquad \textbf{(5)}$$

where R_{DS} represents the **equivalent drain-to-source resistance** given by

$$R_{DS} = \frac{V_T^2}{2I_{DSS}(v_{GS} - V_T)} \qquad \textbf{(6)}$$

Hence, the MOSFET acts as a **voltage-controlled resistor**.

The condition on v_{DS} in Eq. (4) includes *negative* drain voltages, and Eq. (5) indicates that i_D will be negative when v_{DS} is negative. Ohmic operation therefore allows *bidirectional* current flow. This bidirectional property is justified by the symmetric physical picture back in Fig. 9.1–3b.

With the help of Eqs. (2)–(6) we can construct the **universal static characteristics** shown in Fig. 9.1–10a. The dashed curve marks the lower

Figure 9.1–10
Universal enhancement MOSFET curves: (a) static characteristics; (b) active transfer characteristic.

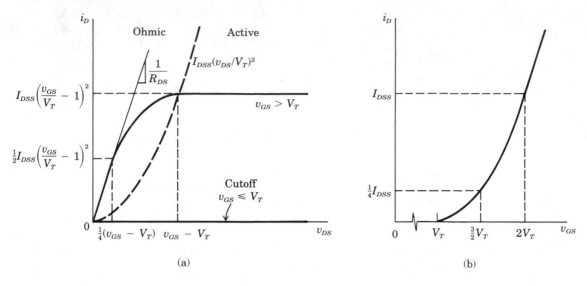

(a)

(b)

edge of the active region, where $v_{DS} = v_{GS} - V_T$ and $i_D = I_{DSS}(v_{DS}/V_T)^2$. We have also included the cutoff region, where $i_D = 0$ for $v_{GS} \leq V_T$.

Another universal curve is the **active transfer characteristic** defined by Eq. (3) and plotted in Fig. 9.1–10b. This curve displays the drain current versus gate voltage for any drain voltage satisfying the active condition $v_{DS} > v_{GS} - V_T$. The active transfer characteristic takes on special importance when we get to amplifier circuit analysis in Chapter 11.

Although our universal curves omit small departures from ideal behavior exhibited by real MOSFETs, they are sufficiently accurate to free us from tedious graphical analysis in most cases of interest. Furthermore, they involve just two MOSFET parameters, I_{DSS} and V_T. Threshold voltages generally fall in the range $V_T \approx 1$–6 V, while I_{DSS} may be less than 100 μA for an integrated-circuit device or greater than 100 mA for a high-power device.

Manufacturers seldom supply characteristic curves with individual transistors. Instead, they provide data from which you can estimate *nominal* values of V_T and I_{DSS} to use with the universal curves. Be aware, however, that variations in the fabrication process may cause the parameters of a particular transistor to differ from the nominal values by as much as a factor of 2.

Example 9.1–3   Consider the MOSFET represented by the characteristic curves back in Fig. 9.1–5, where $V_T = 2.5$ V. When $v_{GS} = 2V_T = 5.0$ V, i_D flattens off around 8.3 mA at the edge of the active region, so we estimate that $I_{DSS} \approx$ 8.3 mA. We can now use the values of V_T and I_{DSS} to predict operating points for the circuit in Fig. 9.1–6, without resorting to graphical methods.

Suppose we want to find v_{GS} such that the MOSFET has $i_D = 4$ mA, given that $V_{DD} = 12$ V and $R_D = 2$ kΩ as before. The corresponding drain voltage will be

$$v_{DS} = V_{DD} - R_D i_D = 4 \text{ V}$$

which is large enough to suggest active-region operation. We therefore start with Eq. (3) and substitute numerical values for I_{DSS}, V_T, and i_D to obtain

$$i_D = 8.3 \text{ mA} \left(\frac{v_{GS}}{2.5} - 1 \right)^2 = 4 \text{ mA}$$

Solving for the required gate voltage then gives

$$v_{GS} = 2.5(1 \pm \sqrt{4/8.3}) \approx 4.24 \text{ V}$$

where we have taken the positive value of the square root because the negative value would yield $v_{GS} < 2.5$ V, corresponding to cutoff rather than active operation. The other active condition in Eq. (2) requires

$$v_{DS} > v_{GS} - V_T \approx 1.7 \text{ V}$$

which is well satisfied by $v_{DS} = 4$ V.

But suppose that the gate voltage increases to 6 V, so i_D increases while v_{DS} decreases and presumably puts the MOSFET in its linear ohmic region. The equivalent resistance is calculated from Eq. (6) to be

$$R_{DS} = \frac{(2.5 \text{ V})^2}{2(8.3 \text{ mA})(6 \text{ V} - 2.5 \text{ V})} = 0.108 \text{ k}\Omega$$

Since this resistance appears in series with R_D, the drain current and voltage are

$$i_D = \frac{V_{DD}}{R_D + R_{DS}} = 5.69 \text{ mA} \qquad v_{DS} = R_{DS} i_D = 0.615 \text{ V}$$

The value of v_{DS} confirms our assumption of linear ohmic operation, which requires $v_{DS} \leq \frac{1}{4} (v_{GS} - V_T) = 0.875$ V.

Example 9.1–4   The MOSFET in Fig. 9.1–11a has been made into a two-terminal device by connecting the gate to the drain. This connection forces $v_{GS} = v_{DS} = v$, and the total current is $i = i_D$ since the gate draws no current.

Ohmic operation is impossible here because $v_{DS} = v_{GS} > v_{GS} - V_T$. Hence, the i-v curve in Fig. 9.1–11b shows cutoff operation for $v \leq V_T$ and active operation for $v > V_T$. In the latter case we have

$$v = V_T(1 + \sqrt{i/I_{DSS}}) \qquad \text{(7)}$$

which follows from Eq. (3).

Figure 9.1–11
(a) MOSFET
connected as a two-
terminal device.
(b) i-v curve.

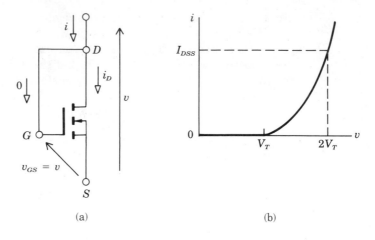

(a)   (b)

Exercise 9.1–2   A certain integrated-circuit MOSFET has $V_T = 2$ V and $I_{DSS} = 400$ μA. Sketch and label i_D versus v_{DS} for $v_{GS} = 3$, 4, and 5 V. Also draw the boundary of the active region.

Exercise 9.1–3   Let the circuit in Fig. 9.1–6 have $V_{DD} = 20$ V and $R_D = 5$ kΩ, and let the MOSFET have $V_T = 4$ V and $I_{DSS} = 8$ mA. Find i_D and v_{DS} for $v_{GS} = 1$, 5, and 9 V.

p-Channel MOSFETs

A **p-channel enhancement MOSFET** differs from an n-channel device in that the doping types are interchanged. The circuit symbol in Fig. 9.1–12a indicates this fact by the direction of the solid arrowhead, which represents the pn junction formed by the p-type channel and the n-type substrate.

Since the doping types have been interchanged, channel conduction now requires *negative* gate-to-source voltage. Alternatively, we can work with more convenient *positive* values by defining the **source-to-gate voltage** $v_{SG} = -v_{GS}$ and the **source-to-drain voltage** $v_{SD} = -v_{DS}$. The reference arrows for these positive quantities are indicated in Fig. 9.1–12a. If v_{SD} is positive, then i_D goes from source to drain, the opposite direction of n-channel drain current.

Figure 9.1–12b shows the p-type channel created in an n-type substrate when v_{SG} exceeds the threshold voltage V_T. The resulting transverse electric field pushes holes toward the gate and draws mobile electrons away from it. The current i_D thus consists of holes injected into the channel from the p^+ source region and collected at the p^+ drain region.

Clearly, a p-channel transistor behaves just like its n-channel counterpart, except for the sign of the mobile charges. Hence, our previous equations remain valid here when we account for sign reversals by writing

Figure 9.1–12

p-channel
enhancement
MOSFET: (a) symbol;
(b) structure.

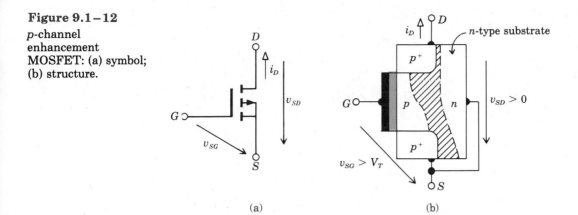

(a)   (b)

v_{SG} and v_{SD} in place of v_{GS} and v_{DS}, respectively. We therefore say that *p*-channel and *n*-channel MOSFETs are **complementary transistors,** having the same general characteristics but opposite current direction and voltage polarities.

However, there is one important practical difference: A *p*-type channel does not conduct as well as an *n*-type channel of the same size because holes are less mobile than electrons. Consequently, *p*-channel devices tend to have smaller values of I_{DSS}. For this reason, *p*-channel MOSFETs are used primarily in circuits that include both *n*-channel and *p*-channel transistors to exploit their complementary properties.

Exercise 9.1–4   Measurements on a certain *p*-channel enhancement MOSFET show that $i_D = 0$ when $v_{GS} \leq -3$ V, and $i_D = 5$ mA when $v_{GS} = v_{DS} = -8$ V. Find the parameter values V_T and I_{DSS}.

9.2   DEPLETION MOSFETS AND JFETS

This section examines the properties of depletion MOSFETs and JFETs. These are *normally-on* transistors in which the field effect reduces conduction by depleting a built-in channel. Nonetheless, the characteristics turn out to be quite similar to those of the normally-off enhancement MOSFET.

Depletion MOSFETS

Figure 9.2–1a depicts an ***n*-channel depletion MOSFET** in cross section. The structure is the same as an enhancement device with the addition of a lightly doped *n*-type channel region. Since the channel has been built in, field effect is not required for conduction between drain and source. The

Figure 9.2–1
n-channel depletion
MOSFET:
(a) structure;
(b) symbol.

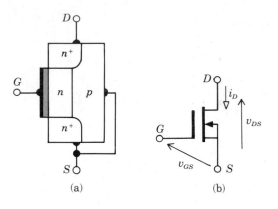

(a)   (b)

circuit symbol in Fig. 9.2–1b represents the normally-on state by an un-broken vertical bar.

If we apply a *positive* gate voltage, then the resulting transverse field draws mobile electrons from the substrate into the channel—thereby *enhancing* conduction. But *negative* gate voltage produces a field in the opposite direction, which pushes holes from the substrate into the channel and draws channel electrons into the substrate. Electron-hole recombinations now create the *two depletion regions* shown in Fig. 9.2–2a. The depletion region along the gate reduces the effective channel depth, so negative gate voltage *depletes* channel conduction and reduces i_D for a given value of v_{DS}.

Figure 9.2–2
Depletion MOSFET channel: (a) depletion regions; (b) pinchoff condition; (c) pinch down condition.

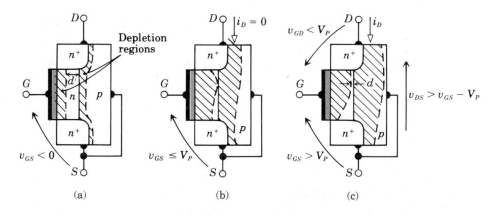

(a)   (b)   (c)

Eventually, when v_{GS} reaches some negative value $-V_P$, we get the condition in Fig. 9.2–2b where the depletion region completely blocks the channel. We call V_P the **pinchoff voltage** because $i_D = 0$ for $v_{GS} \geq -V_P$. (Other texts may take V_P to be a negative quantity and omit the minus sign.) If $v_{GS} > -V_P$ but $v_{GD} < -V_P$, corresponding to $v_{DS} > v_{GS} + V_p$, then the channel becomes partially blocked or pinched down, as illustrated by Fig. 9.2–2c.

Our internal pictures of a depletion MOSFET account for the typical static characteristics plotted in Fig. 9.2–3. (Temporarily ignore the dashed load line.) The pinchoff voltage is seen to be $V_P = 4$ V since $i_D = 0$ for $v_{GS} \leq -4$ V. Hence, this transistor operates in the depletion mode when -4 V $< v_{GS} \leq 0$, while it operates in the enhancement mode when $v_{GS} > 0$. Whether depleted or enhanced, the channel has uniform depth and acts as a resistance at small drain voltages, so i_D initially rises linearly with v_{DS}. Pinch-down then causes i_D to flatten off at larger drain voltages.

Figure 9.2–3

Static characteristics of a typical *n*-channel depletion MOSFET.

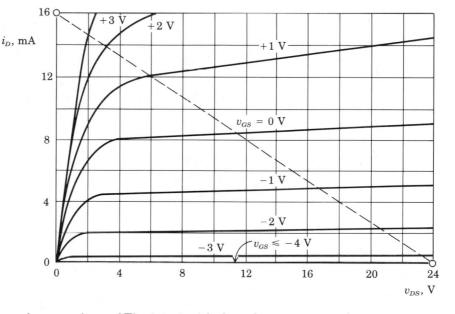

A comparison of Fig. 9.2–3 with the enhancement MOSFET characteristics in Fig. 9.1–5 reveals no difference in the general shape of the curves. The sole practical distinction between depletion and enhancement MOSFETs is the gate voltage range. In particular, the load line in Fig. 9.2–3 brings out the fact that a depletion MOSFET can be in the active region when $v_{GS} = 0$, whereas an enhancement MOSFET must have $v_{GS} > V_T > 0$.

Exercise 9.2–1   Use the load line in Fig. 9.2–3 to sketch the transfer curve v_{DS} versus v_{GS}. Compare this curve with Fig. 9.1–7b.

Junction FETS

JFET (pronounced "jay-fet") stands for **junction field-effect transistor.** The name comes from the fact that a JFET has a *pn* junction between the gate terminal and the channel, rather than the insulating oxide layer

Figure 9.2–4

n-channel JFET: (a) structure; (b) symbol; (c) depletion region.

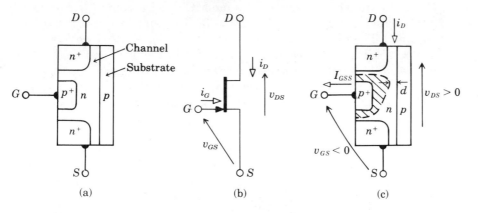

(a)   (b)   (c)

found in MOS devices. The junction is formed in an n-channel JFET by adding a heavily doped p^+ region for the gate contact, as depicted by Fig. 9.2–4a. The p-type substrate, if present, plays no role other than structural.

The JFET symbol in Fig. 9.2–4b represents the gate-to-channel junction by an arrowhead whose direction identifies the channel as n-type. The unbroken vertical bar again denotes a normally-on depletion device. Since the gate is not insulated, the symbol is labeled with a **gate current** i_G to be discussed shortly.

Figure 9.2–4c illustrates what happens inside a JFET when we reverse-bias the junction by applying negative gate voltage. A depletion region with its accompanying electric field penetrates the channel and reduces the effective conducting depth d which, in turn, reduces the drain current. Depending upon the values of v_{DS} and v_{GS}, the channel may become pinched down or completely pinched off. As a result, a JFET behaves much like a depletion MOSFET.

There are, however, several minor differences between JFETs and depletion MOSFETs. First, with $v_{GS} < 0$, the junction carries a **reverse-saturation gate current** $i_G \approx -I_{GSS}$. The value of I_{GSS} is quite small, around 1 nA, and can usually be ignored. Second, any positive gate voltage above about $+0.6$ V would *forward-bias* the junction, resulting in a large forward gate current. Thus, enhancement-mode operation is not possible.

In exchange for these disadvantages, the channel in a JFET has greater conduction than the channel in a MOSFET of the same size, and the characteristic curves are more nearly horizontal in the pinch-down region. Furthermore, JFETs do not necessarily suffer permanent damage from excessive gate voltage, whereas MOSFETs would be destroyed.

We should also mention a special type of junction FET called the **MESFET**. MESFETs are fabricated with a metal-semiconductor junction, like a Schottky diode. This structure reduces junction capacitance and allows operation at higher speeds than can be achieved by other field-effect transistors.

Universal Curves

Since depletion MOSFETs and JFETs are so similar in behavior, they both can be approximated by one set of equations and universal curves. Furthermore, the equations have the same form as those of an enhancement MOSFET with negative pinchoff voltage $-V_P$ in place of positive threshold voltage V_T.

Figure 9.2–5a plots the universal static characteristics for an n-channel depletion MOSFET or JFET. Operation takes place in the active region when

$$v_{DS} > v_{GS} + V_P \qquad v_{GS} > -V_P \tag{1}$$

The drain current is then given by

$$i_D = \frac{I_{DSS}}{V_P^2}(v_{GS} + V_P)^2 = I_{DSS}\left(1 + \frac{v_{GS}}{V_P}\right)^2 \tag{2}$$

where the parameter I_{DSS} now represents the value of i_D when $v_{GS} = 0$. Linear ohmic operation requires

$$|v_{DS}| \le \tfrac{1}{4}(v_{GS} + V_P) \qquad v_{GS} > -V_P \tag{3}$$

and the equivalent drain-to-source resistance is

$$R_{DS} = \frac{V_P^2}{2I_{DSS}(v_{GS} + V_P)} \tag{4}$$

Operation in the cutoff region occurs when $v_{GS} \le -V_P$, so $i_D = 0$.

Figure 9.2–5
Universal depletion MOSFET or JFET curves: (a) static characteristics; (b) active transfer characteristic.

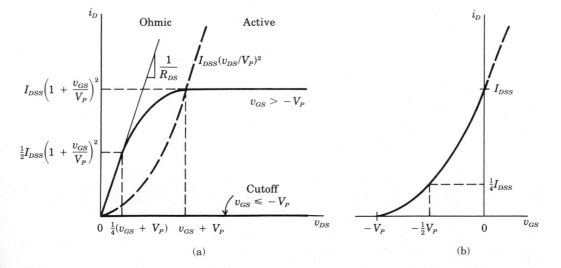

The primary difference between a depletion MOSFET and a JFET is brought out by the active transfer characteristic based on Eq. (2) and plotted in Fig. 9.2–5b. The solid segment of the curve represents the **depletion mode** for either device with negative gate voltage, $-V_P \leq v_{GS} \leq 0$. The dashed segment represents the **enhancement mode** with positive gate voltage, $v_{GS} > 0$, which is not allowed for a JFET due to the resulting gate current. You should also compare Fig. 9.2–5b with the enhancement MOSFET curve back in Fig. 9.1–10b.

Last, we point out that a p-channel depletion MOSFET or JFET behaves like its n-channel counterpart with reversed voltage polarities and current directions. Hence, our universal equations and curves also hold for the p-channel case if we replace v_{GS} and v_{DS} by v_{SG} and v_{SD}, respectively.

Example 9.2–1

Figure 9.2–6 diagrams one way to get negative gate voltage in a JFET circuit with a positive supply voltage. This arrangement makes $v_{GS} + v_K = 0$, where the voltage across R_K is given by $v_K = R_K i_D$ since the gate current is negligible. Thus,

$$v_{GS} = -v_K = -R_K i_D < 0$$

We'll find v_{GS}, i_D, and v_{DS}, given the JFET parameters $V_P = 6$ V and $I_{DSS} = 18$ mA.

Figure 9.2–6
JFET circuit for
Example 9.2–1.

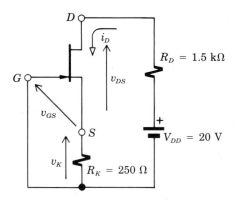

We start with the assumption of active operation, and we draw upon the fact that $R_K i_D = -v_{GS}$. Hence, multiplying both sides of Eq. (2) by R_K eliminates i_D and yields

$$-v_{GS} = \frac{R_K i_{DSS}}{V_P^2} (v_{GS} + V_P)^2$$

After substituting numerical values and rearranging, we get the quadratic equation

$$v_{GS}^2 + 20v_{GS} + 36 = 0$$

whose solution is

$$v_{GS} = \tfrac{1}{2}(-20 \pm \sqrt{20^2 - 4 \times 36}) = -10 \pm 8$$

But we must have $v_{GS} > -V_P = -6$ V for active operation, so we take $v_{GS} = -10 + 8 = -2$ V. Inserting this result back into Eq. (2) gives the corresponding drain current, $i_D = 8$ mA. Finally, we find the drain-to-source voltage from the loop equation

$$v_{DS} = V_{DD} - R_D i_D - R_K i_D = 6 \text{ V}$$

which satisfies the active condition $v_{DS} > v_{GS} + V_P = +4$ V.

Exercise 9.2–2   Let R_K be variable in Fig. 9.2–6, and let $V_P = 6$ V and $I_{DSS} = 16$ mA. Find v_{GS} and the corresponding value of R_K for active operation at $i_D = 2$ mA. Also calculate v_{DS} to confirm the active condition.

9.3 BIPOLAR JUNCTION TRANSISTORS

The family of bipolar junction transistors in Fig. 9.3–1 has just two members, the *npn* BJT and the *pnp* BJT. Both types contain *semiconductor junctions,* and they operate with *bipolar* internal currents consisting of holes and electrons — as distinguished from the unipolar current inside a field-effect transistor.

The structure and characteristics of bipolar junction transistors will be examined here, concentrating primarily on the *npn* case. We'll show that a BJT behaves more linearly than a FET, and we'll take advantage of this linearity to develop useful circuit models for the operating states of a BJT.

Structure and Characteristics

Figure 9.3–1
BJT family tree.

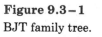
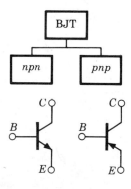

Figure 9.3–2a illustrates the structure of an npn BJT. The **base** (B) is a very thin layer of lightly doped *p*-type material sandwiched between the *n*-type **collector** (C) and the heavily doped n^+-type **emitter** (E). The n^+-*p*-*n* path from emitter to collector goes through two *pn* junctions pointing in opposite directions, similar to the substrate of an enhancement MOSFET before channel formation. Hence, a BJT likewise acts as a *normally-off* device.

Conduction through a BJT requires a forward bias across the emitter junction. This junction is represented on the circuit symbol in Fig. 9.3–2b by the solid arrowhead pointing from base to emitter. Positive **base-to-emitter voltage** v_{BE} produces **base current** i_B, which controls the **collector current** i_C. The base and collector currents sum to form the **emitter current** $i_E = i_C + i_B$. Thus, in contrast to MOSFET operation, we must now deal with at least *two* currents. We must also take account of

Figure 9.3–2

npn BJT: (a) structure; (b) symbol.

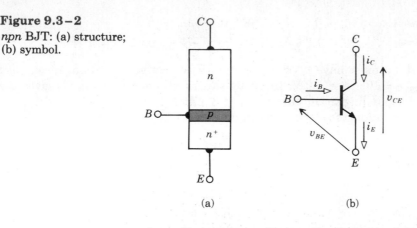

(a)   (b)

the **collector-to-emitter voltage** v_{CE} and the resulting **base-to-collector voltage** $v_{BC} = v_{BE} - v_{CE}$. If v_{CE} exceeds v_{BE}, as it often does, then v_{BC} becomes negative and the collector junction will be reverse biased. Even so, joint emitter-collector action can result in substantial current through the reverse-biased collector junction.

Before examining emitter-collector action, we'll consider the simpler case portrayed by Fig. 9.3–3a where the collector terminal has been open-circuited to block i_C. The remaining base-emitter path therefore behaves like an ordinary semiconductor diode. When an external source supplies $i_B > 0$, internal conduction takes the form of holes and electrons diffusing in opposite directions across the forward-biased emitter junction. The n^+

Figure 9.3–3

Base-emitter behavior with $i_C = 0$: (a) current components; (b) i-v curve.

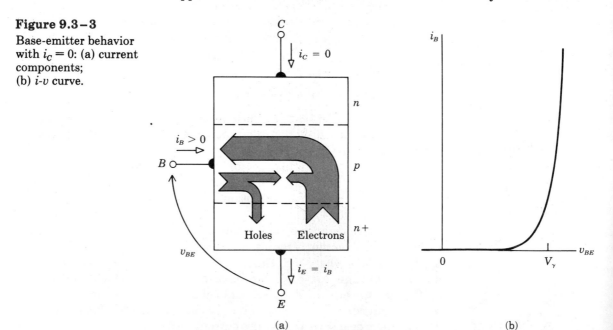

(a)   (b)

doping makes the electron concentration in the emitter much greater than the hole concentration in the lightly-doped base. Accordingly, the electron diffusion current is much larger than the hole diffusion current, and many of the holes recombine with electrons injected into the base.

If the base current is allowed to vary, then the plot of i_B versus v_{BE} has the familiar shape sketched in Fig. 9.3–3b—the i-v curve for a semiconductor diode. This plot emphasizes the fact that, with varying but positive base current, v_{BE} stays nearly constant at the **junction threshold voltage** V_γ. The value of V_γ is about 0.7 V, assuming the usual case of a silicon BJT.

Now let's keep $i_B > 0$, so $v_{BE} \approx V_\gamma$, and let's connect additional circuitry such that $v_{CE} > v_{BE}$. Figure 9.3–4a depicts the internal current components under these conditions. Holes still diffuse from base to emitter, and the reverse bias across the collector junction prevents hole diffusion into the collector. Electrons still diffuse in large numbers from emitter to base, where some of them recombine with holes. But the base region is so thin—typically less than 0.01 mm—that most of the injected electrons approach the collector junction and get swept across by the higher potential. The resulting collector current i_C thus consists of electrons *emitted* at the emitter and *collected* at the collector. This emitter-collector action diverts electron flow from the base terminal, so the internal base current now consists almost entirely of holes.

Figure 9.3–4
Collector-emitter behavior with $i_B > 0$: (a) current components; (b) i-v curve.

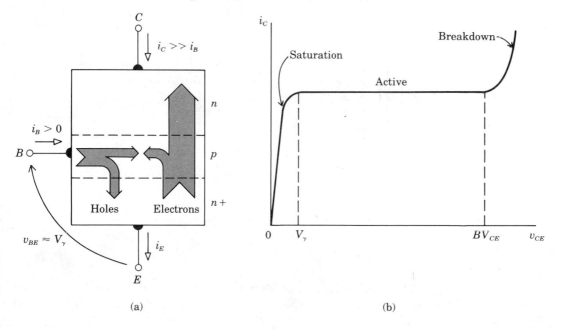

(a)   (b)

In view of the relative electron and hole concentrations, we conclude that i_C will be much greater than i_B. We further conclude that the value of i_C is independent of v_{CE}, as long as the collector has a higher potential than the base. Figure 9.3–4b illustrates how i_C changes with v_{CE}, keeping the base current fixed. This i-v curve rises sharply in the saturation region below $v_{CE} \approx V_\gamma$, and it becomes constant at $i_C \gg i_B$ in the active region. Constant-current behavior ends when v_{CE} reaches the **collector breakdown voltage** BV_{CE} (typically 20–60 V), causing avalanche breakdown of the reverse-biased collector junction.

The static characteristics of a general-purpose *npn* silicon BJT are given in Fig. 9.3–5 for the normal operating range, $0 \le v_{CE} < BV_{CE}$ and $i_C \ge 0$. The bottom curve corresponds to the cutoff region, where $i_C \approx 0$ for $i_B = 0$. The slight upward slope in the active region comes from an effect known as **base-width modulation.** Although these curves resemble FET characteristics in several respects, there are three notable differences. First, the curves have been measured at different values of base current rather than voltage because a BJT acts more like a current-controlled device. Second, the knees are much sharper and occur at smaller voltages, making a very narrow saturation region. Third, the curves have more uniform vertical spacing in the active region.

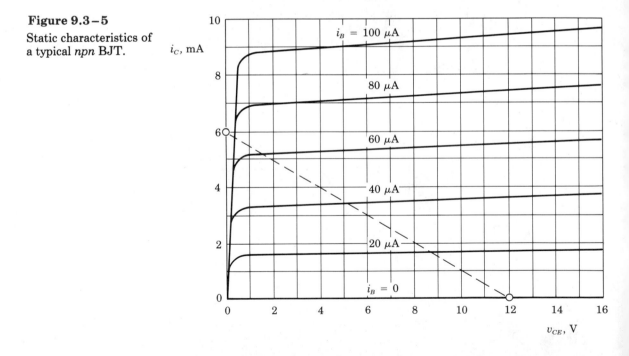

Figure 9.3–5

Static characteristics of a typical *npn* BJT.

The similarities and differences between a BJT and a FET become more evident when we analyze the BJT circuit in Fig. 9.3–6a. This circuit

Figure 9.3–6

(a) Common-emitter BJT circuit. (b) Transfer curves.

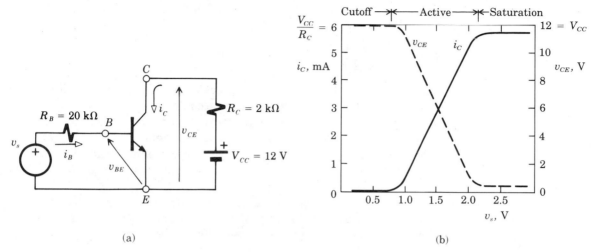

(a)

(b)

is called a **common-emitter configuration** because the emitter serves as the common terminal for the input at the base and the output at the collector. The output load network has the same element values as our previous MOSFET circuit (Fig. 9.1–6) but the labels have been changed to R_C and V_{CC} in accordance with standard convention. The input network consists of a current-limiting resistor R_B in series with the signal source v_s. Input current i_B flows only when the emitter junction is forward biased, so $v_{BE} \approx V_\gamma$. Thus,

$$i_B = \begin{cases} 0 & v_s < V_\gamma \\ (v_s - V_\gamma)/R_B & v_s > V_\gamma \end{cases} \tag{1}$$

which provides the necessary relationship between the input v_s and the control variable i_B.

Figure 9.3–6b shows the transfer curves obtained using Eq. (1) together with the load line superimposed on the BJT characteristics in Fig. 9.3–5. These transfer curves reveal that our BJT circuit goes from cutoff, to active operation, to saturation as the input voltage v_s increases. Therefore, like a FET, a BJT can function as a switching or amplifying device. Indeed, the nearly linear shape of the transfer curves throughout the active region indicates that a BJT amplifier produces less distortion with greater signal swing. Furthermore, the small saturated value of v_{CE} means that a BJT switch acts more nearly like a short circuit when turned on.

But BJTs are not necessarily superior to FETs in all applications. For one reason, they require input current and thus draw power from the input source. For another, they take up much more space on integrated-circuit chips, as we'll see in Section 9.4.

Exercise 9.3 – 1   Calculate i_B in Fig. 9.3–6a when $v_s = 1$ V and 2 V. Then estimate the corresponding values of v_{CE} and i_C from the load line on Fig. 9.3–5, and find the voltage amplification $A_v = \Delta v_{CE}/\Delta v_s$ and the current amplification $A_i = \Delta i_C/\Delta i_B$.

Large-Signal Models

The nearly linear transfer curves suggest the possibility of devising a *linear* circuit model for a BJT — something we could not do for a FET. Actually, we'll develop separate models representing the active, saturated, and cut-off states. These are called **large-signal models** because together they span all of the normal operating range, thereby eliminating the need for universal curves. As a preliminary step, we first consider the current relationships in the active region.

Refering back to Fig. 9.3–4a, our physical picture of an active *npn* BJT indicates that the collector current consists of those emitted electrons that travel completely through the base. But we have ignored the **reverse saturation current** I_{CBO} going from base to collector across the reverse-biased collector junction. This current is quite small at room temperature, around 10 nA in a silicon BJT. However, I_{CBO} increases with temperature per Eq. (6), Section 8.1, and it can be significant at high temperatures.

In general, then, i_C includes I_{CBO} along with a large fraction of i_E. Accordingly, we let α represent the collection fraction and we write

$$i_C = \alpha i_E + I_{CBO}$$

$$i_B = i_E - i_C = (1 - \alpha)i_E - I_{CBO}$$

Eliminating i_E and solving for i_C now yields

$$i_C = \frac{\alpha}{1 - \alpha} i_B + \left(\frac{\alpha}{1 - \alpha} + 1\right) I_{CBO}$$

This result becomes more compact in terms of the **common-emitter current gain**

$$\beta \triangleq \frac{\alpha}{1 - \alpha} \tag{2}$$

Thus,

$$i_C = \beta i_B + I_{CEO} \tag{3}$$

in which

$$I_{CEO} \triangleq (\beta + 1)I_{CBO} \tag{4}$$

We call I_{CEO} the **collector cutoff current** because Eq. (3) also holds for the cutoff region where $i_B = 0$ and $i_C = I_{CEO}$. The meaning of Eqs. (2)–(4) is best illustrated with some representative numbers.

Thanks to the heavy emitter doping and thin base, the collection fraction α typically falls between 0.98 and 0.995. The corresponding values of β calculated via Eq. (2) range from about 50 to 200. (Special supergain construction may even achieve $\beta \approx 1000$.) If a BJT at room temperature has $\beta = 100$, for instance, then Eq. (4) predicts $I_{CEO} \approx 101 \times 10$ nA ≈ 0.001 mA. Such a tiny collector cutoff current goes unnoticed under ordinary conditions, which explains why we previously wrote $i_C \approx 0$ for the cutoff condition. Furthermore, active-region operation usually has $\beta i_B \gg I_{CBO}$, so Eq. (3) simplifies to $i_C \approx \beta i_B$. This expression says that a small base current controls a much larger collector current, thereby justifying the interpretation of β as a current gain.

Based on our conclusions about active operation, we now define the **idealized active state** of an *npn* BJT by the set of conditions

$$v_{BE} = V_\gamma \qquad i_B > 0 \qquad i_C = \beta i_B \qquad v_{CE} > V_\gamma \tag{5}$$

The linear circuit model in Fig. 9.3–7a represents these conditions using a battery for the fixed value of v_{BE} and a current-controlled current source for i_C. The resulting active emitter current is

$$i_E = i_C + i_B = (\beta + 1)i_B \tag{6}$$

so $i_E \approx \beta i_B = i_C$ if $\beta \gg 1$.

Figure 9.3–7
BJT large-signal models: (a) active state; (b) saturated state; (c) cutoff state.

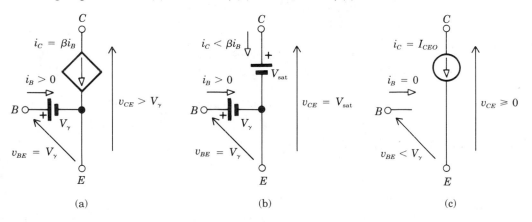

(a) (b) (c)

A saturated BJT has $i_C < \beta i_B$ and $v_{CE} < V_\gamma$. Although the collector voltage may vary somewhat in the saturation region, we usually assume a fixed value $v_{CE} = V_{sat}$ with $V_{sat} \approx 0.2$ V for a silicon BJT. The **idealized saturated state** is thus defined by

$$v_{BE} = V_\gamma \qquad i_B > 0 \qquad i_C < \beta i_B \qquad v_{CE} = V_{sat} \tag{7}$$

The corresponding circuit model in Fig. 9.3–7b contains two batteries, but the collector battery may be replaced by a short circuit when the small value of V_{sat} is inconsequential.

Cutoff occurs when v_{BE} falls below V_γ and the base current becomes negligible. We define the **idealized cutoff state** by the conditions

$$v_{BE} < V_\gamma \qquad i_B = 0 \qquad i_C = I_{CEO} \qquad v_{CE} \geq 0 \tag{8}$$

so the circuit model in Fig. 9.3–7c includes an independent source for the collector cutoff current. However, I_{CEO} may often be ignored at room temperature, in which case the model reduces to an open circuit at all three terminals.

Hereafter, we'll take the idealized large-signal models in Fig. 9.3–7 as reasonable approximations for the behavior of most *npn* BJTs. And, unless otherwise stated, we'll take

$$V_\gamma \approx 0.7 \text{ V} \qquad V_{sat} \approx 0.2 \text{ V} \qquad I_{CEO} \approx 0.001 \text{ mA} \tag{9}$$

which are representative values for a silicon BJT at room temperature.

The only BJT parameter that must be specified is the common-emitter current gain β. In practice, values of β should be viewed as nominal and subject to considerable variation. This variability exists because small changes of the collection fraction α produce large changes of β. For example, if α increases just 0.5% from 0.990 to 0.995, then Eq. (2) shows that the current gain goes from $\beta = 0.990/0.010 = 99$ to $\beta = 0.995/0.005 = 199$ — a 100% increase! Consequently, even BJTs fabricated in the same batch may have quite different values of β. It also turns out that operating a BJT at lower than normal current levels reduces α and results in a substantially reduced current gain.

Example 9.3–1   Suppose we want the circuit in Fig. 9.3–6a to switch from cutoff to saturation. We'll use our models to find the conditions on v_s, given that the BJT has $\beta = 100$.

Figure 9.3–8a shows the circuit diagram with the BJT symbol replaced

Figure 9.3–8
Equivalent circuits for Example 9.3–1: (a) cutoff conditions; (b) saturated conditions.

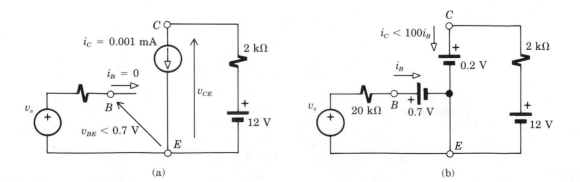

(a)   (b)

by the model for the cutoff state. Since $i_B = 0$, $v_{BE} = v_s$ so we must have $v_s < 0.7$ V for cutoff. Then $i_C = I_{CEO} \approx 0.001$ mA and $v_{CE} = 12$ V $- 2$ k$\Omega \times 0.001$ mA $= 11.998$ V ≈ 12 V. Clearly, we might just as well have omitted I_{CEO}.

With the model for the saturated state inserted in Fig. 9.3–8b, we see that

$$i_B = \frac{v_s - 0.7 \text{ V}}{20 \text{ k}\Omega} \qquad i_C = \frac{12 \text{ V} - 0.2 \text{ V}}{2 \text{ k}\Omega} = 5.9 \text{ mA}$$

Thus, the saturation condition $i_C < 100i_B$ requires

$$v_s > 0.7 \text{ V} + \frac{20 \text{ k}\Omega \times 5.9 \text{ mA}}{100} = 1.88 \text{ V}$$

Had we assumed $V_{\text{sat}} \approx 0$, we would have gotten the slightly higher condition $v_s > 1.9$ V for saturation.

Example 9.3–2 Consider the circuit in Fig. 9.3–9a, where the base current now comes from the supply battery via R_B connected to the collector. Also notice that the load resistor labeled R_E carries $i_E = i_C + i_B$. We'll find the resulting state and operating point when the BJT has $\beta = 80$.

Figure 9.3–9

(a) Circuit diagram for Example 9.3–2. (b) Equivalent circuit with active BJT model.

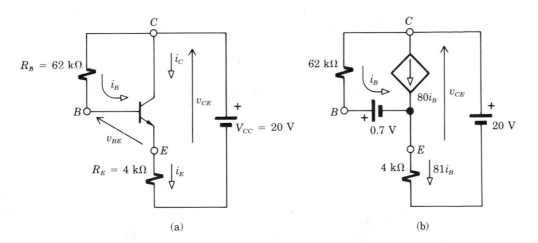

(a)   (b)

A brute-force way of determining the BJT's state would be to try each of the three models in the circuit and see which one gives consistent results. However, we can often take a short cut via some preliminary calculations. For the case at hand, application of KVL yields two helpful relations for v_{BE}, namely,

$$v_{BE} = v_{CE} - R_B i_B = V_{CC} - R_E i_E - R_B i_B$$

If we assume the saturated state, then $v_{CE} = V_{sat}$ and $i_B > 0$, so $v_{BE} = V_{sat} - R_B i_B < 0.2$ V—in violation of the saturation condition $v_{BE} = V_\gamma = 0.7$ V. If we assume the cutoff state, then $i_B = 0$ and $i_E = i_C = I_{CEO}$, so $v_{BE} = V_{CC} - R_E I_{CEO} \approx 20$ V—in violation of the cutoff condition $v_{BE} < V_\gamma$.

Having thereby eliminated saturation and cutoff, we substitute the model for the active state as shown in Fig. 9.3–9b. Now $v_{BE} = 0.7$ V and $i_E = 80i_B + i_B = 81i_B$, so the outer loop equation becomes

$$20 - 62 \times i_B - 0.7 - 4 \times 81 i_B = 0$$

Upon solving for the base current, we obtain $i_B = 0.05$ mA and $i_C = 80i_B = 4$ mA. Hence, $v_{CE} = 20 - 4 \times 81 i_B = 3.8$ V, which satisfies the active condition $v_{CE} > V_\gamma$.

Exercise 9.3–2   Let the *npn* BJT in Fig. 9.3–8a have $\beta = 150$. Find the BJT's state and the values of i_C and v_{CE} when $i_B = 20$ μA and $i_B = 60$ μA. Confirm in each case that your results satisfy the state conditions.

Exercise 9.3–3   Calculate i_E in Fig. 9.3–9a when $R_E = 0$ and $\beta = 50$.

The *pnp* BJT

All of the foregoing discussion is easily modified for a *pnp* BJT. The structure depicted in Fig. 9.3–10a consists of an *n*-type base between a *p*-type collector and a p^+-type emitter. Since the doping types have been interchanged compared to an *npn* transistor, the reference voltage and current arrows are reversed on the symbol in Fig. 9.3–10b. For the same reason, the solid arrowhead representing the emitter junction now points from the p^+-type emitter to the *n*-type base.

If the emitter junction is forward biased by $v_{EB} \approx V_\gamma$, and if $v_{EC} > v_{EB}$, then i_C consists primarily of holes emitted by the emitter and collected at

Figure 9.3–10
pnp BJT: (a) structure; (b) symbol; (c) large-signal active model.

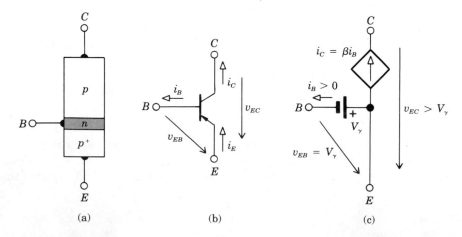

the collector. Hence, *pnp* static characteristics look like the *npn* characteristics back in Fig. 9.3–5, except that the horizontal axis would be labeled v_{EC} instead of v_{CE}. Furthermore, our previous large-signal models also hold here when we reverse all voltage and current arrows, reverse the battery polarities, and replace v_{BE} and v_{CE} with v_{EB} and v_{EC}, respectively. Figure 9.3–10c diagrams the resulting model for a *pnp* BJT in the active state.

Despite the structural similarities, a *pnp* BJT has smaller current gain than a comparable *npn* BJT because holes are less mobile than electrons. Consequently, most applications of *pnp* BJTs involve pairing them with *npn* BJTs to take advantage of complementary operation. Manufacturers use special fabrication techniques to produce **matched pairs** having nearly identical but complementary characteristics.

Example 9.3–3

The circuit in Fig. 9.3–11a has a *pnp* BJT turned upside down to account for the reversed polarities. We are asked to find R_B such that $v_{EC} = 4$ V, given $\beta = 25$.

Figure 9.3–11
(a) Circuit diagram for Example 9.3–3.
(b) Equivalent circuit with active BJT model.

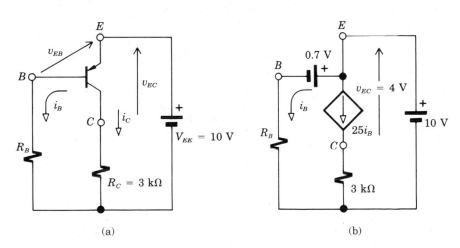

(a)   (b)

The desired value of v_{EC} clearly corresponds to active operation, so we redraw the circuit as shown in Fig. 9.3–11b. Since $v_{CE} = 10 - 3 \times 25 i_B = 4$ V, we must have $i_B = 0.08$ mA. But $i_B = (10 - 0.7)/R_B$, so $R_B = 9.3/0.08 \approx 116$ kΩ.

Exercise 9.3–4

Taking $\beta = 25$, find the condition on R_B in Fig. 9.3–11a such that i_C has the largest possible value.

9.4 DEVICE FABRICATION AND INTEGRATED CIRCUITS †

Integrated circuits and the planar transistor were both invented in 1958, just a decade after the first transistor. Ever since, improvements in planar silicon technology have permitted doubling the number of components per circuit per year, and the cost per component has decreased at nearly the same rate. Today, an advanced, very large-scale integrated circuit might contain over 100,000 components on a chip smaller than a postage stamp and costing only 0.01% of a discrete-device version — which probably could not even be built.

This section surveys some of the major techniques of device fabrication and integrated-circuit construction that have brought about the microelectronics revolution.

Device Fabrication Methods

Although germanium was once used extensively in semiconductor devices, and there are some applications for materials such as alloys of galium, most of the semiconductor industry uses silicon — fortunately a very plentiful material. Silicon is usually obtained by the reduction of common sand with carbon in a high-temperature furnace. The resulting intermediate product, a polycrystalline material, is then cast into ingots, which are purified by a process called **zone refining.** A small zone of each ingot is heated by induction to a temperature just above melting. The melted zone is slowly moved along the ingot and any impurities tend to move along with the melted part, leaving the recrystallized areas more nearly pure silicon but still an intermediate polycrystalline material rather than a single crystal.

The next step is to grow a single crystal. This is commonly done by the **Czochralski method** portrayed in Fig. 9.4–1a. A small seed crystal is very slowly withdrawn from molten silicon while being slowly rotated. As the melt condenses, new crystal growth occurs on the seed along the crystalline axes of the original seed crystal. The result is a single crystal boule of relatively pure silicon, typically 8 cm in diameter and 25 cm long. By purposely adding appropriate donor or acceptor impurities to the melt, the boule is made p- or n-type silicon.

One of the methods once used to fabricate semiconductor devices was based on crystal growth from a doped melt. While the crystal was being slowly pulled, additional impurities were added to the melt, changing it from n-type to p-type, for instance. After a little further growth, the melt could be changed back again, producing a boule with an npn sandwich as in Fig. 9.4–1b. Devices made this way are called **grown-junction** transistors or diodes.

But most devices and ICs are now made from a reasonably homogeneous boule of silicon that has been sliced into wafers before processing. These wafers are smoothed and polished, to be used as the "raw material" for a variety of additional procedures.

One such procedure is to dot the surface of an n-type wafer with tiny pieces of an acceptor material. When the wafer is heated, each dot partially melts into the surface to produce an alloy whose silicon content increases with depth. There is mostly acceptor metal on the surface, and mostly p-type silicon farther inside, with a curved pn junction where the two meet.

Figure 9.4–1
(a) Growing a crystal.
(b) Making a grown-junction device.

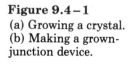

(a)

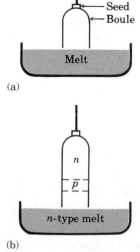

(b)

Figure 9.4–2
Alloy junction transistors: (a) cross section; (b) scribed wafer.

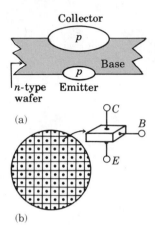

(a)

(b)

This process produces an **alloy junction.** A batch of alloy-junction transistor is obtained by putting an opposing set of donor metal pieces on the other face of the wafer, and heating the wafer just long enough for a thin region of n-type silicon to remain between the two junctions, as illustrated in Fig. 9.4–2a. When scribed into tiny squares, broken at the scribe marks, and with proper leads attached, several hundred transistors are produced from one wafer as shown in Fig. 9.4–2b.

Another method for creating pn junctions is **impurity diffusion.** A wafer is heated in an inert gas atmosphere containing impurity atoms such as boron (for acceptors) or phosphorus (for donors). The impurity atoms diffuse into the silicon, going from a region of high to low concentration just as electrons and holes diffuse across a junction. For example, the npn transistor structure of Fig. 9.3–1a might have been obtained by a two-step process, first diffusing boron into an n-type wafer to get the p-region base and then diffusing phosphorus for the n^+-region emitter.

It is also possible to insert impurities at specific locations inside the crystal via **ion implantation.** This technique, carried out at room temperature, employs a special type of mass spectrometer to bombard the wafer with impurity ions. The energy of the ions determines the depth of penetration.

Finally, a silicon wafer can be used as a substrate upon which additional layers of single-crystal silicon are deposited by **epitaxial growth.** In this process the surface of the wafer is exposed to an appropriate gaseous atmosphere at controlled temperatures. The gas is normally hydrogen and silicon tetrachloride, which yields silicon at low concentrations of the tetrachloride. At high concentrations, the reaction reverses and the treatment etches away rather than builds upon the substrate.

Epitaxial processing has several important advantages. When small quantities of selected impurities are added to the gas, a newly formed layer can be made to have any desired concentration of p or n doping. Since the growth rate is readily predicted, the resulting thickness of the new growth, along with any desired variations of impurity concentration, can be accurately controlled.

Planar Processing

Let's now walk through the steps required to fabricate a batch of diodes on a wafer using the technology known as **planar processing.** We start with a heavily doped n-type substrate. A relatively thick layer, perhaps 25 μm thick, is epitaxially grown on its upper surface. Exposure to steam at high temperature then causes a surface layer of silicon dioxide (SiO_2) to form. Silicon dioxide happens to be one of the best electrical insulators and also provides excellent protection for the top surface.

However, to gain access to the silicon at selected points, we now want to remove some of the oxide. For that purpose we coat the surface with a photosensitive material called **photoresist.** A **mask** is next placed on the

Figure 9.4-3
Planar processing: (a) masking photoresist; (b) after etching; (c) diffusion for
p-type regions; (d) metalization.

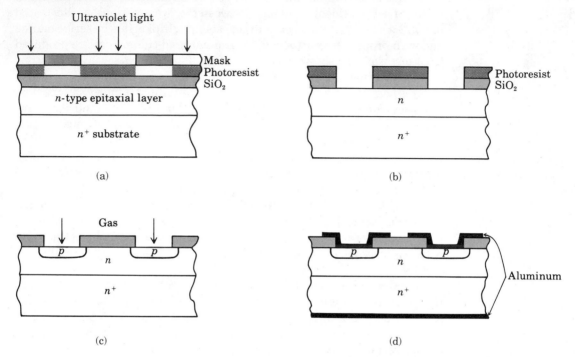

photoresist, resulting in the layered arrangement of Fig. 9.4–3a. In this
hypothetical example, the mask could be a transparent sheet marked with
a regular arrangement of tiny dots at the points where we wish to create
diodes. In practice, the mask would be a photographic negative, produced
from large-scale artwork in a special reducing camera.

We now expose our multilayered sandwich to ultraviolet light, which
polymerizes the photoresist in the areas where the mask is transparent.
The mask is then removed and the photoresist is dissolved, except where it
was exposed. The surface is now ready for an etching solution that removes
the oxide from the diode locations, as in Fig. 9.4–3b. The remaining photoresist may then be washed away chemically.

The next step is diffusion. Since the oxide is highly resistant to penetration by foreign atoms, high-temperature exposure to an acceptor-impu-
rity gas produces *p*-type regions in the epitaxial layer below the oxide-layer
windows. We now have formed *pn* junctions shown in the cross-section
view of Fig. 9.4–3c.

The remaining sequence fabricates the individual terminals for wire
bonding by a procedure called **metallization.** It starts by exposing both
faces of the wafer to evaporating aluminum atoms in vacuum, resulting in a

metal film over the entire surface. Photoresist is applied again, followed with a new mask on the upper surface. This mask is opaque except for an array of transparent circles aligned to cover the positions of the diodes. After exposure and etching, metal coats the entire bottom of the wafer in Fig. 9.4–3d, but it covers only areas slightly larger than the apertures in the oxide layer on the upper side. Note that the circumferential edge of the *pn* junction is still protected from air and moisture by the oxide film.

It is now possible to scribe and break the wafer into hundreds of planar diodes ready for packaging. All of the foregoing processing steps would, generally, be done to many wafers simultaneously, resulting in nearly identical characteristics for an entire batch of, perhaps, ten thousand devices.

Integrated Circuits

Although diodes are relatively simple devices, the very same processes we've just discussed — epitaxial growth, oxidation, impurity diffusion, photo masking, etching, and metallization — are used to make integrated circuits (ICs) of varying degrees of complexity. An IC is a silicon chip containing numerous transistors and, perhaps, resistors and capacitors — all components having been formed by successive processing steps on a wafer about 5 cm in diameter that yields more than 1000 chips of an area of 2 mm². (IC dimensions are usually stated in **mils,** where 1 mil = 0.001 in. = 0.0254 mm.)

Small-scale integration (SSI) would be used for a 20-component op-amp, say, whereas **large-scale integration** (LSI) puts an entire microprocessor with some 10,000 components on a single chip! The benefits derived from integrating many components on an IC chip include low cost, small size, high reliability, and matched characteristics.

To illustrate some of the possibilities, let's consider one *npn* transistor — a small part of an IC on a *p*-type substrate about 0.2 mm thick. The resulting configuration, somewhat idealized, has the cross section shown in Fig. 9.4–4. Fabrication requires six masking steps. All the terminals are on the top, and the substrate is not ordinarily used as an electrode. Instead, it should be connected to the negative terminal of the supply source to prevent conduction through the substrate-to-collector junction.

Figure 9.4–4
Cross section of an IC
npn transistor.

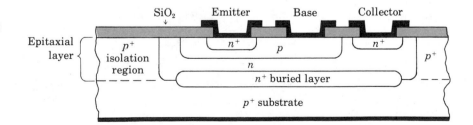

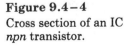

(This junction does, of course, add an effective capacitance to the collector.) The p^+ regions around the outside serve to isolate this transistor from its neighbors. These isolation regions are produced by continuing to diffuse acceptor impurities until they penetrate the entire epitaxial layer and join the substrate. The n^+ **buried layer** serves to reduce the effective series resistance of the collector circuit by increasing the conductivity of much of the collector volume.

The physical size of an integrated bipolar junction transistor would typically be 0.2 mm square. While that may *seem* quite small, the MOSFET device illustrated in Fig. 9.4–5 takes up only 0.04 mm × 0.14 mm of chip area, meaning that about 20 MOSFETs could be put in the place of one BJT. A major space-saving factor here is the *self-isolating* property of MOSFETs. Figure 9.4–5 also shows a low-resistance n-type strip inside the oxide layer that serves as a **buried crossover** for connecting other components.

Figure 9.4–5
Cross section of an IC enhancement MOSFET.

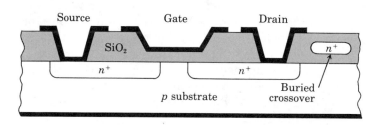

When IC resistors are needed, they may be made along with other devices using the same processing steps. An example of such a resistor is shown in cross section in Fig. 9.4–6a. The value of the resulting resistance will depend on the impurity concentration used, as well as on the shape of its region. To achieve the length necessary for larger resistance values, resistive regions are sometimes folded back and forth several times, as in the top view of Fig. 9.4–6b. The fabrication technique used to make such resistors is relatively inaccurate, but the ratio of values of any two resistances in one IC is likely to match quite well the ratio of resistance values in

Figure 9.4–6
IC resistor: (a) cross section; (b) top view.

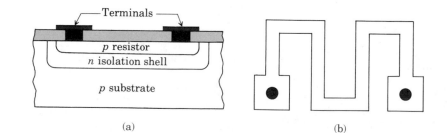

(a)   (b)

another similar circuit. Precision resistances are usually created by evaporating a thin alloy film with a low temperature coefficient onto an oxide layer, and then masking and etching the film to the desired configuration. If necessary, the resistance value can be modified in a controlled manner after processing by nibbling away at it with a laser beam during final testing.

Integrated capacitors with relatively small capacitance values are most easily made by using the capacitance of a reverse biased junction. However, these components suffer from variations in their capacitance value with applied voltage (the varactor effect), and from the presence of diode behavior if the applied voltage forward-biases the junction. Much larger and more stable capacitors are formed of thin metallic films over silicon dioxide as a dielectric, as in Fig. 9.4–7. Such capacitors do require substantial areas on the substrate, and have values in the neighborhood of 500 pF per square millimeter. (One square millimeter is quite a bit of "real estate" to devote to one circuit element.)

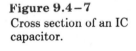

Figure 9.4–7
Cross section of an IC capacitor.

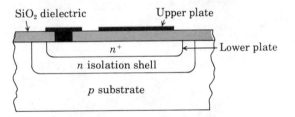

But integrated resistors may require even larger areas than capacitors, so special switching techniques have been developed to make capacitors behave like RC circuits. Other kinds of active circuits may be used to convert IC capacitance into equivalent inductance, since inductors cannot be implemented effectively in IC form.

Medium and large-scale ICs generally include a network of metallic connecting bridges that join their devices. These bridges form an intricate, often artistic pattern as seen in the magnified top view of a typical chip in Fig. 9.4–8. It's rather sad that those views must be concealed by protective packaging.

Of the many IC packaging techniques, perhaps the most popular is the **dual-in-line package** (DIP) illustrated in Fig. 9.4–9. It consists of a rectangular plastic or ceramic case enclosing the IC, with protruding pin terminals that make it look something like a centipede. These pins can be soldered directly to the external circuitry, or inserted into standard sockets that, in turn, are either soldered or wirewrapped for connection to other sockets. This particular unit has 16 pins, and is about 8 mm wide and 18 mm long. Larger DIPS with 24 to 64 pins accommodate the many external connections needed for an LSI chip such as a microprocessor.

Figure 9.4–8
Microphotograph of an
integrated circuit
containing 132
MOSFETs and one
resistor (near upper
right corner) on a
1 mm × 2 mm chip
with 46 bonding pads
for external
connections.

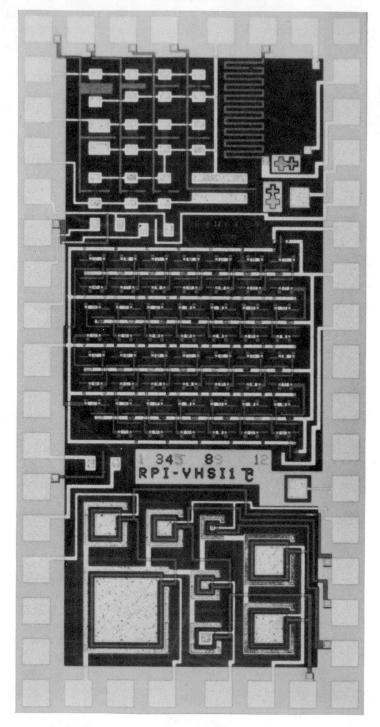

Figure 9.4–9
Dual in-line package
with top removed.
(Courtesy of Texas
Instruments).

PROBLEMS

9.1–1   The FET in Fig. P9.1–1 has the characteristics given in Fig. 9.1–5. Construct a load line to estimate the value of R_D such that $v_{DS} = 5$ V when $v_{GS} = 4$ V and $V_{DD} = 10$ V.

Figure P9.1–1

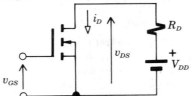

9.1–2   Do Problem 9.1–1 for $i_D = 5$ mA when $v_{GS} = 4.5$ V and $V_{DD} = 5$ V.

9.1–3   Do Problem 9.1–1 for $v_{DS} = 10$ V when $v_{GS} = 3.5$ V and $V_{DD} = 15$ V.

9.1–4   The FET in Fig. P9.1–1 has the characteristics given in Fig. 9.1–5. Use a load line to plot v_{DS} versus t when $V_{DD} = 9$ V, $R_D = 1.5$ kΩ, and v_{GS} has the waveform in Fig. P9.1–4 with $V_0 = 3.5$ V and $V_1 = 1$ V.

Figure P9.1–4

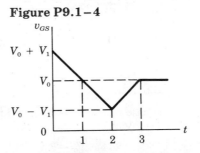

9.1–5   The FET in Fig. P9.1–1 has the characteristics given in Fig. 9.1–5. Use a load line to plot v_{DS} versus t when $V_{DD} = 16$ V, $R_D = 4$ kΩ, and v_{GS} has the waveform in Fig. P9.1–4 with $V_0 = 4$ V and $V_1 = 1.5$ V.

9.1–6   The FET in Fig. P9.1–1 has the characteristics given in Fig. 9.1–5. Use a load line to plot v_{DS} versus t when $V_{DD} = 6$ V, $R_D = 0.75$ kΩ, and v_{GS} has the waveform in Fig. P9.1–4 with $V_0 = 3.5$ V and $V_1 = 2$ V.

9.1–7   Suppose that $v_{GS} = 4 + 0.1 \cos \omega t$ V in Fig. 9.1–6. Use Eq. (3) to obtain an expression for i_D when the FET has $I_{DSS} = 4$ mA and $V_T = 2$ V. Then sketch v_{DS} versus t, and calculate $A_v = \Delta v_{DS}/\Delta v_{GS}$.

9.1–8   Do Problem 9.1–7 when the FET has $I_{DSS} = 27$ mA and $V_T = 3$ V.

9.1–9   Do Problem 9.1–7 when the FET has $I_{DSS} = 10$ mA and $V_T = 2.5$ V.

9.1–10   Let $V_{DD} = 6$ V and $R_D = 2$ kΩ in Fig. P9.1–1, and let the FET have $V_T = 3$ V. Determine the FET's state and evaluate I_{DSS} given that $v_{GS} = 5$ V and $v_{DS} = 3.6$ V.

9.1–11   Do Problem 9.1–10 given that $v_{GS} = 9$ V and $v_{DS} = 1.2$ V.

9.1–12   Find the operating state and the values of i_D and v_{DS} in Fig. P9.1–1 when $V_{DD} = 12$ V, $R_D = 2$ kΩ, and the FET has $V_T = 2$ V, $I_{DSS} = 0.5$ mA, and $v_{GS} = 6$ V.

9.1–13   Do Problem 9.1–12 with $v_{GS} = 12$ V.

9.1–14   Find the operating state and the values of i_D and v_{DS} in Fig. P9.1–1 when $V_{DD} = 9$ V, $R_D = 3$ kΩ, and the FET has $V_T = 1.5$ V, $I_{DSS} = 0.75$ mA, and $v_{GS} = 6$ V.

9.1–15   Let $V_{DD} = 25$ V and $R_D = 4$ kΩ in Fig. P9.1–1, and let the FET have $V_T = 3$ V and $I_{DSS} = 5$ mA. What is the operating state and the value of v_{GS} if $v_{DS} = 20$ V?

9.1–16   Do Problem 9.1–15 with $v_{DS} = 9$ V.

9.1–17   Do Problem 9.1–15 with $v_{DS} = 1$ V.

9.1–18   The circuit in Fig. P9.1–18 uses transistor T_2 as an **active load** to achieve linear amplification with T_1. Let T_1 have $I_{DSS} = 9$ mA and $V_T = 6$ V, and let T_2 have $I_{DSS} = 9$ mA and $V_T = 1.5$ V. Obtain an expression for v_{out} in terms of v_{in} when T_1 is active, and determine the allowable range of v_{in}. Hint: Use Eq. (7) for T_2.

Figure P9.1–18

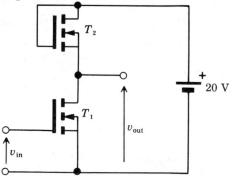

9.1–19   Do Problem 9.1–18 for the case where T_1 has $I_{DSS} = 5$ mA and $V_T = 2$ V, and T_2 has $I_{DSS} = 0.2$ mA and $V_T = 2$ V.

9.1–20   The p-channel FET in Fig. P9.1–20 has $V_T = 3$ V and $I_{DSS} = 0.75$ mA. What value of v_{in} yields $v_{\text{out}} = 15$ V?

9.1–21   Do Problem 9.1–20 for $v_{\text{out}} = 7$ V.

9.1–22   Do Problem 9.1–20 for $v_{\text{out}} = 1$ V.

Problems **321**

Figure P9.1–20

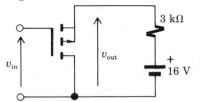

9.2–1 The MOSFET in Fig. P9.2–1 has the characteristics given in Fig. 9.2–3. Construct an appropriate load line to estimate the value of v_{in} that yields $v_K = 12$ V when $R_K = 1.5$ kΩ, $R_D = 0.5$ kΩ, and $V_{DD} = 20$ V.

Figure P9.2–1

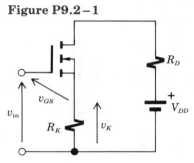

9.2–2 The MOSFET in Fig. P9.2–1 has the characteristics given in Fig. 9.2–3. Construct an appropriate load line to estimate the value of v_{in} that yields $v_K = 3$ V when $R_K = 0.25$ kΩ, $R_D = 0.5$ kΩ, and $V_{DD} = 12$ V.

9.2–3 The MOSFET in Fig. P9.2–1 has the characteristics given in Fig. 9.2–3. Construct an appropriate load line to estimate the value of v_{in} that yields $v_K = 3$ V when $R_K = 0.5$ kΩ, $R_D = 1$ kΩ, and $V_{DD} = 18$ V.

9.2–4 Let the JFET in Fig. 9.2–6 have $V_P = 6$ V and $I_{DSS} = 18$ mA. Assume active operation to find v_{GS} and i_D when $R_K = 2$ kΩ and $R_D = 3$ kΩ. Then determine the condition on V_{DD} for active operation.

9.2–5 Do Problem 9.2–4 with $R_K = 80$ Ω and $R_D = 400$ Ω.

9.2–6 Determine the operating state and the value of i_D in Fig. 9.2–6 when $R_K = 0$, $R_D = 4$ kΩ, and $V_{DD} = 9$ V, given that the JFET has $V_P = 5$ V and $I_{DSS} = 5$ mA.

9.2–7 Do Problem 9.2–6 given that the JFET has $V_P = 8$ V and $I_{DSS} = 4$ mA.

9.2–8 Let the MOSFET in Fig. P9.2–1 have $V_P = 5$ V and $I_{DSS} = 2.5$ mA, and let $R_K = 2.5$ kΩ, $R_D = 1.5$ kΩ, and $V_{DD} = 30$ V. Find i_D and v_{DS} when $v_{\text{in}} = 3$ V.

9.2–9 Do Problem 9.2–8 with $v_{\text{in}} = 10$ V.

9.3–1 The BJT in Fig. P9.3–1 has the characteristics given in Fig. 9.3–5. Plot i_C and v_{CE} versus t when $V_{BB} = 1.5$ V, $R_B = 10$ kΩ, $R_C = 1$ kΩ, $V_{CC} = 9$ V, and $v_s = 0.2 \cos \pi t/2$ V.

9.3–2 Do Problem 9.3–1 with $V_{BB} = 1.0$ V, $R_B = 30$ kΩ, $R_C = 4$ kΩ, $V_{CC} = 16$ V, and $v_s = 0.6 \cos \pi t/2$ V.

9.3–3 Do Problem 9.3–1 with $V_{BB} = 5.5$ V, $R_B = 80$ kΩ, $R_C = 3$ kΩ, $V_{CC} = 15$ V, and $v_s = 3.2 \cos \pi t/2$ V.

Figure P9.3–1

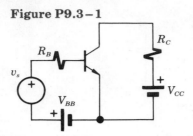

9.3–4   Let $v_s = 0.2 \cos \omega t$ V in Fig. P9.3–1, and let $R_B = 10$ kΩ, $R_C = 5$ kΩ, and $V_{CC} = 30$ V. Use the large-signal active model with $V_\gamma = 0.7$ V to obtain expressions for i_C and v_{CE} and to calculate $A_v = \Delta v_{CE}/\Delta v_s$ when $V_{BB} = 1.2$ V and $\beta = 80$.

9.3–5   Do Problem 9.3–4 with $V_{BB} = 1.0$ V and $\beta = 120$.

9.3–6   Do Problem 9.3–4 with $V_{BB} = 1.5$ V and $\beta = 60$.

9.3–7   Let $V_{CC} = 10$ V and $R_C = 4.9$ kΩ in Fig. P9.3–1. Using large-signal models with $V_\gamma = 0.7$ V and $V_{sat} = 0.2$ V, determine the conditions on v_s to achieve cutoff and saturation of the transistor when $V_{BB} = 1.0$ V, $R_B = 30$ kΩ, and $\beta = 40$.

9.3–8   Do Problem 9.3–7 with $V_{BB} = 0.5$ V, $R_B = 10$ kΩ, and $\beta = 50$.

9.3–9   Do Problem 9.3–7 with $V_{BB} = 2.5$ V, $R_B = 90$ kΩ, and $\beta = 100$.

9.3–10 Let $R_E = 0$ in Fig. P9.3–10. Find i_B, i_C, and v_{CE} when $R_B = 810$ kΩ and $R_C = 12$ kΩ.

Figure P9.3–10

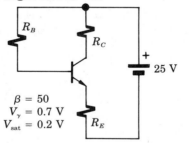

$\beta = 50$
$V_\gamma = 0.7$ V
$V_{sat} = 0.2$ V

9.3–11 Do Problem 9.3–10 with $R_B = 405$ kΩ and $R_C = 10$ kΩ.

9.3–12 Let $R_B = 0$ in Fig. P9.3–10. Find i_B, i_C, and v_{CE} when $R_C = 500$ Ω and $R_E = 9$ kΩ.

9.3–13 Do Problem 9.3–12 with $R_C = 250$ Ω and $R_E = 6$ kΩ.

9.3–14 Find i_B, i_C, and v_{CE} in Fig. P9.3–10 with $R_B = 300$ kΩ, $R_E = 10$ kΩ, and $R_C = 2$ kΩ.

9.3–15 Find i_B, i_C, and v_{CE} in Fig. P9.3–10 with $R_B = 90$ kΩ, $R_E = 3$ kΩ, and $R_C = 1.5$ kΩ.

9.3–16 Find i_B, i_C, and v_{CE} in Fig. P9.3–10 with $R_B = 30$ kΩ and $R_E = R_C = 1$ kΩ.

9.3–17 Let $R_C = 2$ kΩ, $R_E = 3$ kΩ, and $i_E = 4$ mA in Fig. P9.3–10.
(a) Calculate v_{CE} to confirm active operation.
(b) Find the corresponding value of R_B.
(c) What would be the value of i_E if β decreases to 25 or increases to 100?

9.3–18 Do Problem 9.3–17 with $R_C = 4$ kΩ, $R_E = 10$ kΩ, and $i_E = 1.5$ mA.

9.3–19   Do Problem 9.3–17 with $R_C = 1.5$ kΩ, $R_E = 6$ kΩ, and $i_E = 3$ mA.

9.3–20   The circuit in Fig. P9.3–20 is called a **current mirror** because it maintains $I \approx I_{\text{ref}}$ despite possible variations of R_L, as long as T_2 remains active. The transistors are identical and $v_{BE2} = v_{BE1}$, so $i_{B2} = i_{B1}$. Determine values for V_{CC} and R to get $I = 20$ mA with $\beta = 40$ and $R_L \le 600$ Ω. Take V_{CC} to be the smallest multiple of 5 V that keeps T_2 active.

Figure P9.3–20

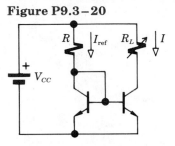

9.3–21   Do Problem 9.3–20 with $I = 100$ mA, $\beta = 50$, and $R_L \le 75$ Ω.

9.3–22   Assuming that both transistors in Fig. P9.3–22 are active and have $V_\gamma = 0.7$ V, find v_1 and v_{in} given that $v_2 = 10$ V.

Figure P9.3–22

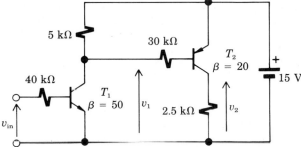

9.3–23   Assuming that both transistors in Fig. P9.3–22 are active and have $V_\gamma = 0.7$ V, find v_2 and v_{in} given that $v_1 = 10$ V.

9.3–24   Assuming that both transistors in Fig. P9.3–22 are active and have $V_\gamma = 0.7$ V, find v_1 and v_2 given that $v_{\text{in}} = 1.5$ V.

Chapter

10 Nonlinear Electronic Circuits

Most electronic circuits belong to one of two main categories, *linear* or *nonlinear*. Within each category, a further distinction can be based upon the *power levels* involved. Chapter 12 will consider the distinctive problems of power electronics in general, while Chapter 11 discusses low-power electronic circuits that function in a linear mode for purposes of amplification. This chapter is devoted to *nonlinear low-power electronic circuits*—circuits that exploit *switching action* to perform a wide variety of practical tasks.

Specifically, we'll introduce here the operating principles of simple diode and transistor circuits designed for electronic switching, waveshaping, triggering, and gating. Both analog and digital gates will be examined to illustrate switching concepts and to pave the way for our study of signal processing and digital systems in Part III. We'll also look at two aspects of transients resulting from energy storage in switching circuits. On the one hand, these transients limit the performance of high-speed gates. On the other hand, they play an essential role in timing circuits and waveform generators.

Objectives

After studying this chapter and working the exercises, you should be able to do each of the following:

- Design diode waveshapers and clamps (Section 10.1).
- Explain the principles of basic digital gates (Sections 10.1 and 10.2).
- Show how diodes or FETs can be used to build an analog gate (Sections 10.1 and 10.2).
- Analyze the OFF and ON states of a transistor switch (Section 10.2).

- Sketch the waveforms associated with a bistable flip-flop (Section 10.2).
- Design a regenerative comparator circuit with specified hysteresis (Section 10.3).
- Identify the differences between monostable and astable circuits (Section 10.3).
- Calculate the time intervals for an RC timing circuit or waveform generator (Section 10.3).

10.1 DIODE SWITCHING CIRCUITS

This section focuses on nonlinear circuits containing diodes as switching devices. Applications include clippers and waveshapers, digital and analog gates, and peak detectors and clamps. Except when otherwise stated, we assume operation at sufficiently large voltages that the real diodes behave like ideal diodes.

Clippers and Waveshapers

Consider the elementary waveshaping circuit in Fig. 10.1–1a. The diode goes on when v_{in} exceeds the battery voltage V_B, so the input current is

Figure 10.1–1

Diode waveshaper:
(a) circuit; (b) voltage
transfer characteristic;
(c) illustrative
waveforms.

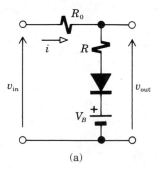

(a)

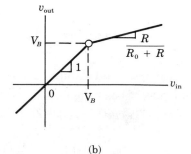

(b)

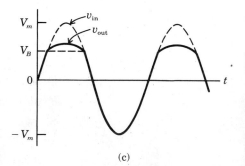

(c)

$i = (v_{in} - V_B)/(R_0 + R)$ and the output voltage becomes

$$v_{out} = v_{in} - R_0 i = \frac{R}{R_0 + R} v_{in} + \frac{R_0}{R_0 + R} V_B \qquad v_{in} > V_B \qquad (1)$$

Otherwise, $i = 0$ and $v_{out} = v_{in}$ since the diode goes off when $v_{in} < V_B$. Figure 10.1–1b displays the resulting voltage transfer characteristic, v_{out} versus v_{in}. The waveshaping action is illustrated in Fig. 10.1–1c taking a sinusoidal input $v_{in} = V_m \sin \omega t$ with $V_m > V_B$. Consequently, the circuit reduces positive output peaks by the factor $R/(R_0 + R)$. If $R = 0$, then we have a **positive peak clipper** forcing $v_{out} = V_B$ whenever $v_{in} > V_B$.

A **two-sided clipper** that clips both positive and negative peaks requires two diode-battery branches, arranged as in Fig. 10.1–2a. The corresponding transfer curve in Fig. 10.1–2b emphasizes that v_{out} is now constrained by

$$v_{out} = \begin{cases} -V_{BN} & v_{in} < -V_{BN} \\ v_{in} & -V_{BN} < v_{in} < V_{BP} \\ V_{BP} & V_{BP} < v_{in} \end{cases} \qquad (2)$$

Thus, a large sinusoidal input will be top-clipped at $+V_{BP}$ and bottom-clipped at $-V_{BN}$, producing the "squarish" output waveform sketched in Fig. 10.1–2c. Symmetrical two-sided clippers with $V_{BN} = V_{BP}$ are called **limiters.** Limiters are incorporated in some waveform generators to get flat-topped outputs and in FM radios to suppress unwanted noise.

Figure 10.1–2
(a) Two-sided clipper.
(b) Transfer curve.
(c) Waveforms.
(d) Implementation with Zener diodes.

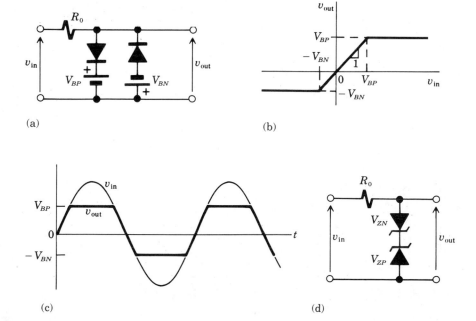

(a)

(b)

(c)

(d)

If the batteries are omitted in Fig. 10.1–2a, then the circuit still clips the output at about ± 0.7 V due to the threshold voltage of real diodes. By the same reasoning, we can draw upon the properties of reverse-biased Zener diodes to eliminate the batteries. In particular, Fig. 10.1–2d diagrams a two-sided clipper with Zener diodes connected "front-to-front" in series. The upper Zener stays off when $-V_{ZN} < v_{in} \leq 0$ and the inverted lower Zener stays off when $0 \leq v_{in} < V_{ZP}$, so $v_{out} = v_{in}$ for $-V_{ZN} < v_{in} < V_{ZP}$. But if $v_{in} > V_{ZP}$, then $v_{out} \approx +V_{ZP}$ because the upper Zener is forward biased and the lower one is in reverse breakdown. Conversely, $v_{out} \approx -V_{ZN}$ for $v_{in} < -V_{ZN}$. Effective clipping action requires a small reverse resistance R_Z compared to R_0.

Diodes are also employed to build **piecewise-linear waveshapers,** which contain resistor-diode-battery branches connected in parallel. For instance, Fig. 10.1–3a is a two-branch waveshaper with battery voltages $V_1 < V_2$. Clearly, diode D_1 conducts when $v_{out} > V_1$, while D_2 conducts when $v_{out} > V_2 > V_1$. Application of the breakpoint analysis method from Section 8.2 then yields the piecewise-linear transfer characteristic plotted in Fig. 10.1–3b for $v_{in} \geq 0$. The line slopes are labeled in terms of their reciprocals a_1 and a_2, given by

$$a_1 = 1 + \frac{R_0}{R_1} \qquad a_2 = 1 + \frac{R_0}{R_1} + \frac{R_0}{R_2}$$

For the general case of three or more branches with $V_1 < V_2 < V_3 \ldots$, the nth diode conducts when $v_{out} > V_n$ and the reciprocal of the slope is

$$a_n = 1 + \frac{R_0}{R_1} + \frac{R_0}{R_2} + \cdots + \frac{R_0}{R_n} \tag{3}$$

which holds over the range $V_n < v_{out} < V_{n+1}$.

Figure 10.1–3

Two-branch waveshaper; (a) circuit; (b) transfer curve.

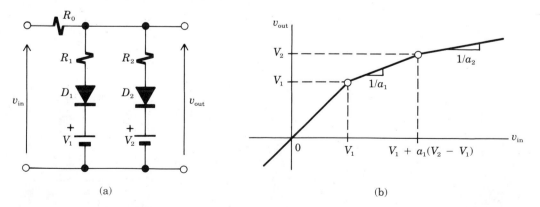

(a)   (b)

When v_{in} takes on negative values, negative output values can likewise be shaped by adding *inverted* resistor-diode-battery branches to Fig. 10.1–3a. Furthermore, when desired, an amplifier inserted at the input or output will multiply all slopes by the voltage gain A_v. Note, however, that $a_1 < a_2 < a_3 < . . .$, so the slopes become progressively *smaller* as $|v_{in}|$ increases. An op-amp circuit for increasing slopes is given in Problem 10.1–8.

Example 10.1–1

Triangular-to-sinusoidal waveshaper.

Some laboratory instruments generate a triangular waveform, which is then converted into a nearly sinusoidal waveform by a piecewise-linear waveshaper. To illustrate this technique, we'll design a diode circuit to produce

$$v_{out} \approx K \sin \frac{\pi v_{in}}{24} \qquad -12 \le v_{in} \le 12 \text{ V}$$

where the proportionality constant K is yet to be determined.

We start with a plot of the desired function $K \sin (\pi v_{in}/24)$ for $0 \le v_{in} \le 12$ V, as shown in Fig. 10.1–4a. Also shown is a *piecewise-linear approximation* consisting of three segments with breakpoints at $v_{in} = 4$ V and 8 V. This approximation can therefore be created by a two-branch

Figure 10.1–4

Triangular-to-sinusoidal waveshaper: (a) piecewise-linear approximation; (b) circuit with Zener diodes; (c) waveforms.

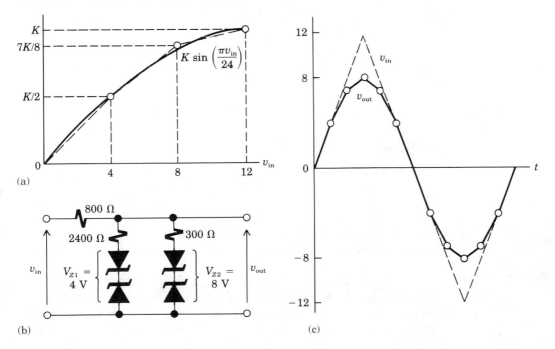

(a)

(b)

(c)

waveshaper. A comparison of Figs. 10.1–3b and 10.1–4a immediately suggests that we should take $K/2 = V_1 = 4$ V and $7K/8 = V_2$, so

$$K = 8 \qquad V_1 = 4 \text{ V} \qquad V_2 = 7 \text{ V}$$

The reciprocal slopes over $4 < v_{\text{in}} < 8$ and $8 < v_{\text{in}} \leq 12$ are now found via

$$a_1 = \frac{8 - 4}{7K/8 - K/2} = \frac{4}{3} = 1 + \frac{1}{3}$$

$$a_2 = \frac{12 - 8}{K - 7K/8} = 4 = 1 + \frac{1}{3} + \frac{8}{3}$$

Hence, from Eq. (3), we want $R_0/R_1 = \frac{1}{3}$ and $R_0/R_2 = \frac{8}{3}$, so

$$R_1 = 3R_0 \qquad R_2 = 3R_0/8$$

which completes the design of the positive waveshaper.

Since the sine function has odd symmetry about the origin, the negative output values are suitably shaped by identical but inverted branches. Alternatively, each pair of identical branches can be combined into one branch containing front-to-front Zener diodes for positive and negative shaping. This implementation is diagrammed in Fig. 10.1–4b taking $R_0 = 800$ Ω.

Finally, Fig. 10.1–4c plots the 12-V triangular wave input and the resulting output waveform. Our sinusoidal approximation differs from a true sinusoid by less than 4% error at any point. Actual laboratory generators produce even better approximations using a three-branch waveshaper.

Exercise 10.1–1   Design a two-branch waveshaper to get $v_{\text{out}} \approx \sqrt{v_{\text{in}}}$ for $0 \leq v_{\text{in}} \leq 9$ V. Take $R_0 = 2$ kΩ and put the breakpoints such that $v_{\text{out}} = \sqrt{v_{\text{in}}}$ at $v_{\text{in}} = 1$ V and 4 V.

Digital and Analog Gates

A **gate** is a nonlinear circuit that receives one or more time-varying input waveforms and delivers a modified waveform at the output. These time-varying waveforms are commonly known as signals, and nature of the gate circuit itself depends upon whether the signals in question are digital or analog. We'll have much more to say about the applications of digital signals in Chapter 14. Our purpose here is to introduce simple digital and analog gates built with diodes.

Figure 10.1–5 depicts a digital diode gate with voltage inputs v_A and v_B. Each input is a **digital signal** whose values are restricted to *two narrow ranges,* a low level and a high level. The low level is usually around zero volts relative to ground, while the high level is several volts above ground. If v_A and v_B are both low, then neither diode conducts and v_{out} is low. But if

Figure 10.1–5
Diode OR gate.

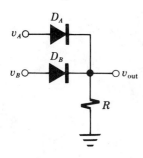

Figure 10.1−6
Diode AND gate.

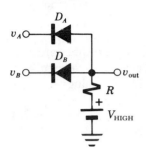

v_A is high, then diode D_A conducts and v_{out} becomes high — regardless of the level of v_B. Furthermore, regardless of v_A, v_{out} goes high whenever v_B goes high. This circuit is called an **OR gate** because v_{out} will be high when v_A is high *or* when v_B is high *or* when both inputs are high. Similarly, an OR gate with three or more inputs would produce a high output when at least one of the inputs goes high.

The gate circuit in Fig. 10.1−6 also has two digital input signals. However, a high output here requires both inputs to be high. Otherwise, a low input pulls v_{out} down to the low level through a conducting diode. This circuit is called an **AND gate** because v_{out} will be high only when v_A *and* v_B are high. Similarly, an AND gate with three or more inputs would produce a high output only when all inputs go high.

To clarify the difference between OR and AND gates, let the high input level be at +5 V and let the signals v_A and v_B have the pulsed waveforms sketched in Fig. 10.1−7a. The resulting output waveforms are drawn in parts *b* and *c* of the figure, assuming gates with ideal diodes. The dashed lines indicate what happens when we account for the threshold voltage of real diodes. This voltage drop reduces the high output of the OR gate by $V_\gamma \approx 0.7$ V, while it raises the low output of the AND gate by the same amount.

Figure 10.1−7
(a) Digital input waveforms. (b) OR-gate output. (c) AND-gate output.

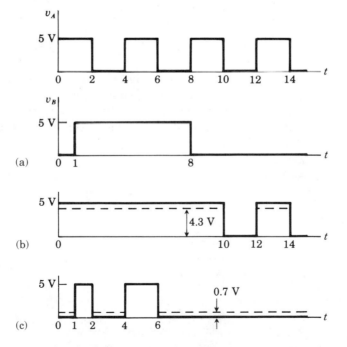

When the output of a real diode gate drives the input of another diode gate, the combination of V_γ drops and loading effects may significantly degrade the final output levels. Consequently, digital circuits containing multiple gates always employ transistors rather than diodes for the switching devices. Digital transistor gates will be discussed in the next section. Meanwhile, we turn to the topic of analog gates.

In contrast to the pulsed shape of digital signals, an **analog signal** typically exhibits more-or-less smooth time variations over some continuous range. The task of an **analog gate** is to extract short segments or **samples** of an analog waveform. Figure 10.1–8 illustrates analog sampling with a mechanical switch, so $v_{out} = 0$ when the switch is OFF and $v_{out} = v_{in}$ when the switch is ON. The resulting sampled output has the form needed to begin other processing operations, including the important case of analog-to-digital conversion.

Figure 10.1–8
Analog gate: (a) switching circuit; (b) waveforms.

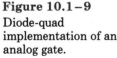

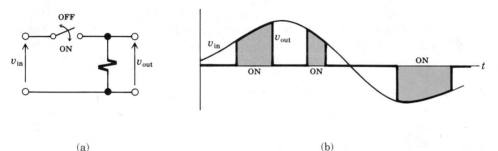

(a)   (b)

Rapid sampling obviously calls for electronic rather than mechanical switches. A popular high-speed electronic sampling circuit is the **diode-quad gate** diagrammed in Fig. 10.1–9. The diodes are fabricated on the same IC chip and have nearly identical characteristics. Switching action is governed by the *control voltage* v_c, which alternates between $+ V_C$ and $- V_C$ at the upper control terminal. Simultaneously, the inverted voltage $- v_c$ alternates between $- V_C$ and $+ V_C$ at the lower control terminal.

Figure 10.1–9
Diode-quad implementation of an analog gate.

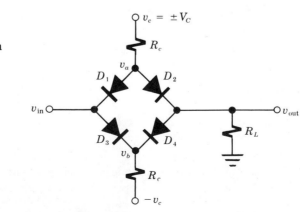

All four diodes should be reverse-biased and turned off when $v_c = -V_C$ (so $-v_c = +V_C$). Hence, the output terminal is electrically isolated and $v_{out} = 0$. On the other hand, all four diodes should be forward-biased and turned on when $v_c = +V_C$. The equal threshold voltage drops across D_1 and D_2 then result in $v_a = v_{in} + V_\gamma$ and $v_{out} = v_a - V_\gamma = v_{in}$, as desired for sampling. However, to ensure proper operation, the input voltage must fall within the range

$$|v_{in}| \leq V_{max} = \frac{R_L}{R_L + R_c} (V_C - V_\gamma) \qquad (4)$$

The upper limit occurs when a large positive input causes D_1 and D_4 to go off, so $v_{out} = +V_{max}$, or when a large negative input causes D_2 and D_3 to go off, so $v_{out} = -V_{max}$.

Exercise 10.1–2 Let the OR gate in Fig. 10.1–5 have $R = 1$ kΩ and let the diodes be modeled by Fig. 8.2–5c with $V_\gamma = 0.7$ V and $R_f = 100$ Ω. Draw the equivalent circuit and find v_{out} when $v_A = 5$ V and $v_B = 0$ V.

Exercise 10.1–3 Suppose a diode-quad gate has $R_c = 2$ kΩ, $R_L = 8$ kΩ, and $V_\gamma = 0.7$ V. Find the diode that has the largest reverse voltage v_R when $v_{in} = +V_{max}$ and $v_c = -V_C$—so all diodes are off and $v_{out} = 0$. Then calculate the maximum allowable values of V_C and V_{max} if the diodes must have $v_R \leq 6$ V to prevent reverse breakdown.

Peak Detectors and Clamps

Several interesting and practical circuits combine diode switching with capacitance energy storage. These circuits draw upon the principle that a capacitor can be charged through a forward-biased diode but it cannot discharge back through the diode in the reverse direction.

The simplest diode-capacitor circuit is the **peak detector** in Fig. 10.1–10a, where v_{out} is taken across C. If the capacitor is initially uncharged and v_{in} has the illustrative waveform sketched in Fig. 10.1–10b, then the diode won't conduct until $v_{in} \geq 0$. The output voltage now follows

Figure 10.1–10
Peak detector: (a) circuit; (b) waveforms.

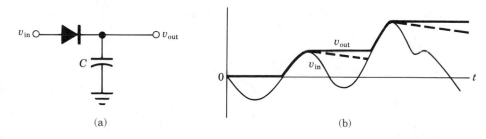

v_{in} up to the first peak, where it remains constant because the capacitor has no discharge path. Actually, as indicated by the dashed curve, the capacitor might discharge slowly through its own leakage resistance or through an external load resistor. Nonetheless, the diode becomes reverse biased and stays off until v_{in} rises above the current value of v_{out}. Hence, the circuit detects and stores the largest previous peak value of the input. A variation of this circuit performs *envelope detection* in AM radios.

Another variation of the peak detector is the **clamp circuit** in Fig. 10.1–11a. Clamps are designed to operate on input waveforms like the one in Fig. 10.1–11b, having positive peaks of height V_p and spacing T. The diode keeps $v_{out} \leq 0$ and allows v_C to rise to the first voltage peak. The capacitor then discharges slowly through the resistor R until the next peak comes along, recharging the capacitor. The values of R and C are chosen to satisfy

$$RC \gg T \tag{5}$$

in which case v_C stays nearly constant at V_p between peaks. Thus, after the initial start-up interval, we have

$$v_{out} = v_{in} - v_C \approx v_{in} - V_p \tag{6}$$

The resulting output waveform in Fig. 10.1–11c follows the same shape as v_{in}, but with the positive peaks clamped at zero volts. This type of clamp is used in television receivers to restore the proper DC level of the video waveform by clamping the peaks of the synchronization pulses.

Figure 10.1–11
Clamp circuit and waveforms.

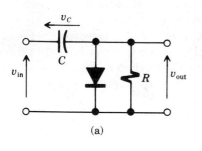

(a)

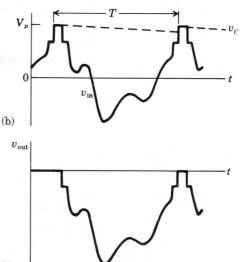

(b)

(c)

Figure 10.1–12
Biased clamp circuit.

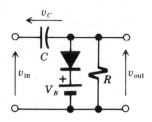

To clamp periodic positive peaks at some fixed voltage $V_B < V_p$, we simply put a battery in series with the diode. The resulting circuit in Fig. 10.1–12 is called a **biased clamp**. If $RC \gg T$, then the voltage across the capacitor builds up to $v_C \approx V_p - V_B$ and

$$v_{out} \approx v_{in} - V_p + V_B \tag{7a}$$

Inverting the diode gives a circuit that takes periodic negative peaks at $v_{in} = -V_n$ and clamps them to V_B, so

$$v_{out} \approx v_{in} + V_n + V_B \tag{7b}$$

In either case, V_B can be made negative by inverting the battery.

Example 10.1–2
AC/DC voltmeter.

Figure 10.1–13 shows how a clamp circuit allows the measurement of AC voltages with a DC voltmeter. Here, the negative sinusoidal input peak at $-V_m$ is clamped to zero by the inverted diode, so $v_D \approx v_{in} + V_m + 0 = V_m \cos \omega t + V_m$. The lowpass filter consisting of R_1 and C_1 rejects the AC component of v_D, leaving the average value $v_{av} = V_m$ applied to the DC meter.

Figure 10.1–13
AC/DC voltmeter
circuit and waveforms.

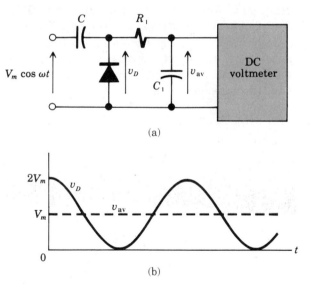

(a)

(b)

Exercise 10.1–4

Let the input to the circuit in Fig. 10.1–11a be

$$v_{in} = \begin{cases} 5 \sin \omega t & 0 \le t < T \\ 10 \sin \omega t & T \le t < 2T \\ 5 \sin \omega t & 2T \le t < 3T \\ 0 & 3T \le t \end{cases}$$

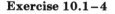

where $\omega = 2\pi/T$ and $T \ll RC$. Sketch the waveforms v_C and v_{out} starting at $t = 0$ with $v_C = 0$.

10.2 TRANSISTOR SWITCHING CIRCUITS

We saw in Chapter 9 that FETs and BJTs have a nonconducting or cutoff state and a strongly conducting or saturated state. The switching circuits considered in this section rely on changes at the input to drive a transistor from one of these states to the other.

We'll study gate circuits built with transistor switches, and we'll show how a bistable flip-flop circuit acts as a digital memory. (More sophisticated digital gates and memory units are covered in Chapters 14 and 15.) We'll also investigate some transients that occur when a transistor switching circuit includes a capacitor or inductor.

BJT Switches and Gates

Figure 10.2–1a depicts an elementary switch using an *npn* bipolar junction transistor. The emitter terminal of the BJT is grounded, and an input drives base current through resistor R_B. By convention, the source of the supply voltage V_{CC} has been omitted from the circuit diagram, but it is actually connected between the upper terminal and the ground point. We'll use information from our large-signal models back in Fig. 9.3–7 to analyze the switching operation.

Figure 10.2–1
BJT switch or inverter:
(a) circuit; (b) voltage
transfer curve.

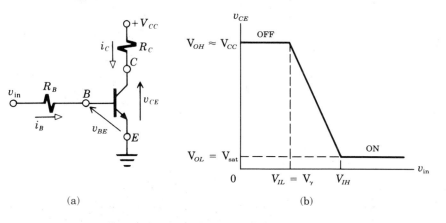

(a)   (b)

Any input voltage less than the base-emitter threshold V_γ leaves the BJT in cutoff. This OFF state is characterized by $v_{BE} < V_\gamma$, $i_B = 0$, and $i_C = I_{CEO} \approx 0$. Hence, $v_{CE} = V_{CC} - R_C i_C \approx V_{CC}$ for $v_{\text{in}} < V_{IL}$ where $V_{IL} \approx V_\gamma$. As v_{in} increases, the transistor passes through its active region and eventually reaches the saturated or ON state, with $v_{BE} = V_\gamma$, $i_B > 0$, $v_{CE} =$

V_{sat}, and $i_C < \beta i_B$. The base and collector currents are then found from the diagram to be

$$i_B = \frac{v_{in} - V_\gamma}{R_B} \qquad i_C = \frac{V_{CC} - V_{sat}}{R_C} \qquad (1)$$

Since saturation requires $i_C < \beta i_B$, the input must satisfy the condition

$$v_{in} > V_{IH} = V_\gamma + \frac{R_B}{\beta R_C}(V_{CC} - V_{sat}) \qquad (2)$$

If v_{in} falls between V_{IL} and V_{IH}, then the BJT is in neither the OFF state nor the ON state.

Some applications employ a BJT switch to control the collector current i_C. Thus, i_C switches from zero when $v_{in} < V_{IL}$ to its maximum possible value when $v_{in} > V_{IH}$. This switched current might, for instance, actuate the relay connected to one hammer of a dot-matrix printer. The BJT thereby functions as a relay driver, whose transient behavior we'll examine subsequently.

Other applications employ a BJT switch to control the collector voltage v_{CE}. The corresponding plot of v_{CE} versus v_{in} in Fig. 10.2–1b emphasizes the resulting voltage inversion effect. If v_{in} happens to be a digital signal, then a low input ($v_{in} < V_{IL}$) produces a high output ($v_{CE} = V_{OH} \approx V_{CC}$) and a high input ($v_{in} > V_{IH}$) produces a low output ($v_{CE} = V_{OL} = V_{sat}$). Accordingly, we say that the BJT switch acts as a **digital inverter.** An inverter is also called a **NOT gate** because its output is not low (i.e., high) when the input is low, and vice versa.

But our elementary inverter circuit has the disadvantage that the value of V_{IH} depends upon the transistor's current gain β—a rather unreliable parameter. The improved circuit in Fig. 10.2–2a overcomes that problem with the help of diodes D_i, D_1, and D_2. This circuit is designed such that a high input turns D_i off, allowing the transistor to be saturated by base current coming from V_{CC} through R_B, D_1, and D_2. Since each

Figure 10.2–2
BJT inverter with diodes: (a) circuit; (b) voltage transfer curve.

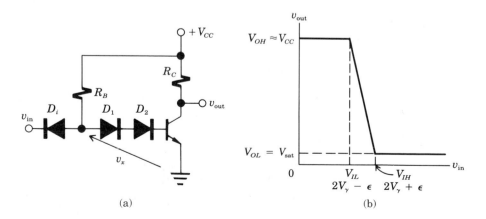

(a)

(b)

conducting diode introduces another threshold voltage drop V_γ, the voltage v_x equals $V_\gamma + V_\gamma + v_{BE} = 3V_\gamma$ in that situation. Hence, to keep D_i off and get $v_{out} = V_{OL}$, we must have $v_{in} > v_x - V_\gamma = 2V_\gamma$. Conversely, D_i goes on when $v_{in} < 2V_\gamma$, which makes $v_x < 3V_\gamma$ and puts the BJT in cutoff, thereby yielding $v_{out} = V_{OH}$.

The inverter's transfer curve in Fig. 10.2–2b depicts the abrupt output transition from V_{OH} for $v_{in} \le V_{IL} = 2V_\gamma - \epsilon$ to V_{OL} for $v_{in} \ge V_{IH} = 2V_\gamma + \epsilon$. The value of ϵ turns out to be about 75 mV, so v_{out} switches completely when v_{in} increases by $V_{IH} - V_{IL} = 2\epsilon \approx 150$ mV, independent of β. However, we have implicitly assumed that β satisfies the saturation condition

$$\beta > i_C/i_B \tag{3}$$

Otherwise, the low output would not go down to V_{sat}.

Now suppose we connect two parallel digital inputs to the BJT inverter, as diagrammed in Fig. 10.2–3. Here, both input diodes D_A and D_B must be off to get $v_x = 3V_\gamma$ and $v_{out} = V_{OL}$. We call this circuit a **NAND gate** — short for NOT-AND gate — because the resulting output is not high only if v_A *and* v_B are high. When an AND gate is needed, it can be implemented by putting a second inverter at the output of the NAND gate to cancel the first inversion. A NOT-OR or **NOR gate** is considered in the following example.

Figure 10.2–3
BJT NAND gate.

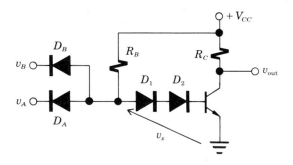

Example 10.2–1

The circuit in Fig. 10.2–4a consists of a NOR gate followed by an inverter to make an OR gate. We'll analyze its operation assuming that $V_\gamma = 0.7$ V and $V_{sat} = 0.2$ V, the nominal values for silicon diodes and transistors.

First, let v_A and v_B be low, so transistor T_1 is cutoff, $v_1 = V_{OH} = 5$ V, and D_i is off. Hence,

$$i_{B2} = (V_{CC} - 3V_\gamma)/R_{B2} = 0.29 \text{ mA}$$

which should saturate T_2 and give the desired low output $v_2 = V_{OL} = 0.2$ V. If T_2 does saturate, then

$$i_{C2} = (V_{CC} - V_{sat})/R_{C2} = 2.4 \text{ mA}$$

Figure 10.2−4
Circuits for Example
10.2−1.

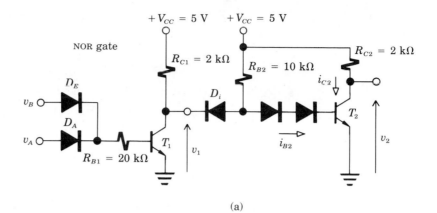

(a)

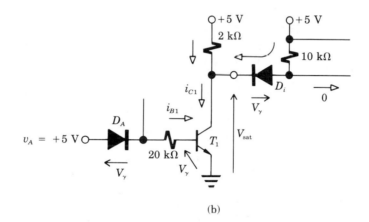

(b)

Consequently, saturation requires T_2 to have gain $\beta_2 > i_{C2}/i_{B2} \approx 8.3$. (Such a modest gain requirement is typical of integrated-circuit BJT gates.)

Next, let v_A be high and assume that T_1 saturates, so D_i goes on and T_2 is cutoff to yield the desired high output $v_2 = V_{OH}$. Figure 10.2−4b shows the relevant portion of the circuit taking $v_A = 5$ V, which results in

$$i_{B1} = (v_A - 2V_\gamma)/R_{B1} = 0.18 \text{ mA}$$

But note that i_{C1} includes the current through D_i as well as through R_{C1}. Thus,

$$i_{C1} = \frac{V_{CC} - V_{\text{sat}}}{R_{C1}} + \frac{V_{CC} - V_\gamma - V_{\text{sat}}}{R_{B2}} = 2.81 \text{ mA}$$

and T_1 must have gain $\beta_1 > i_{C1}/i_{B1} \approx 15.6$. More gain would be needed if the high value of v_A happens to be less than 5 V.

Finally, we observe that the situation in Fig. 10.2−4b also holds for $v_B = 5$ V or for $v_A = v_B = 5$ V. This observation confirms that v_1 will be low when either input is high, as expected of a NOR gate.

Exercise 10.2−1

Let $v_B = 0$ in Fig. 10.2–4a, and let the circuitry connected to T_2 be in the form of Fig. 10.2–1a with $R_B = 20$ kΩ, $R_C = 2$ kΩ, and $V_{CC} = 5$ V.

(a) Obtain the condition on v_A such that R_1 is cutoff. Then find v_1 and the inverter current gain β_2 required for saturation of T_2.

(b) Obtain the condition on v_A such that T_1 is saturated, assuming $\beta_1 = 10$.

(c) Use results from (a) and (b) to sketch the transfer curve v_1 versus v_A.

FET Switches and Gates

Switches built with field-effect transistors differ from BJT switches in two major respects. First, being a voltage-controlled device, the FET does not draw input current. Second, the ON state of a FET corresponds to its linear ohmic region, which allows current in either direction through the channel. This bidirectional feature leads to convenient implementation of analog gates with FET switches. However, for both analog and digital applications, the currents must be restricted to rather small values to keep the FET in its ON state.

As a case in point, consider an n-channel enhancement MOSFET with parameters V_T and I_{DSS}. Operation in the linear ohmic region occurs when the gate-to-source and drain-to-source voltages satisfy

$$v_{GS} > V_T \qquad |v_{DS}| \le \tfrac{1}{4}(v_{GS} - V_T) \tag{4a}$$

The channel then acts like a resistance

$$R_{DS} = V_T^2/2I_{DSS}(v_{GS} - V_T) \tag{4b}$$

so the drain current will be $i_D = v_{DS}/R_{DS}$. The limitation on $|v_{DS}|$ thus permits positive or negative drain current but limits the magnitude $|i_D|$. Similar limitations hold for depletion MOSFETs and JFETs.

Figure 10.2–5 shows how easy it is to perform analog sampling with an enhancement MOSFET controlled by voltage v_c. When v_c puts the MOSFET in its cutoff or OFF state, we have $i_D = 0$ and $v_{\text{out}} = 0$ so $v_{GS} = v_c$. Hence, the control voltage may have any value $v_c < V_T$. The input is sampled when v_c puts the MOSFET into its ON state in the ohmic region. If $R_{DS} \ll R_L$, then

$$v_{\text{out}} = \frac{R_L}{R_L + R_{DS}} v_{\text{in}} \approx v_{\text{in}} \tag{5a}$$

and

$$|v_{DS}| = |v_{\text{in}} - v_{\text{out}}| \approx \frac{R_{DS}}{R_L}|v_{\text{in}}| \ll |v_{\text{in}}| \tag{5b}$$

The control voltage required for sampling can be determined from Eqs. (4a) and (4b) with $v_{GS} = v_c - v_{\text{out}} \approx v_c - v_{\text{in}}$. But note that R_{DS} varies with v_{GS}, which means that the switch causes some *nonlinear distortion* in the output signal.

Figure 10.2–5
Analog gate with
MOSFET switch.

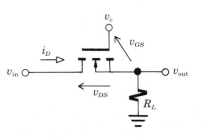

A better analog gate employs *complementary* n-channel and p-channel MOSFETs connected per Fig. 10.2–6a. This unit, symbolized by Fig. 10.2–6b, is called a **CMOS transmission gate** (pronounced "see-moss"). Like a diode-quad circuit, the CMOS transmission gate requires two control voltages, v_c and $-v_c$. With $v_c = +V_C$, the gate has nearly constant resistance and transmits any input voltage in the range $|v_{in}| \leq V_C$. With $v_c = -V_C$, the gate blocks any voltage in the same range.

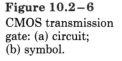
Figure 10.2–6
CMOS transmission
gate: (a) circuit;
(b) symbol.

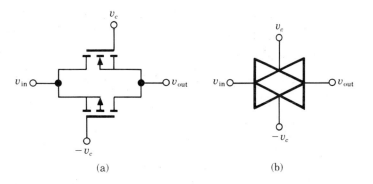

(a)   (b)

Complementary MOSFETs are also used for *digital* CMOS gates, which will be discussed in Section 14.4. Another digital circuit family incorporates n-channel enhancement and depletion MOSFETs, but no resistors! This family's building block is the **NMOS inverter** diagrammed in Fig. 10.2–7a, the gate leads having been drawn at the middle of the gate bars for clarity. Switching is performed by the enhancement MOSFET (T_1). A low input $v_{in} = v_{GS1}$ leaves T_1 off and produces a high output $v_{out} = v_{DS1}$, while a sufficiently high input turns T_1 on and produces a low output. The depletion MOSFET (T_2) serves as an **active load** and limits the current through T_1.

The current-limiting property of T_2 is achieved by connecting its gate and source terminals together, so $v_{GS2} = 0$. The i-v curve for T_2 thus takes the form of Fig. 10.2–7b, where $i_{D2} \leq I_{DSS2}$ (see Fig. 9.2–5a). But the circuit is arranged such that $v_{DS2} = V_{DD} - v_{DS1}$ and $i_{D2} = i_{D1}$. We may therefore superimpose i_{D2} versus $v_{DS1} = V_{DD} - v_{DS2}$ on the characteristic curves of T_1 in Fig. 10.2–7c. This figure brings out the fact that the T_2 curve becomes a *nonlinear load line* for T_1.

Figure 10.2–7

NMOS inverter: (a) circuit; (b) i-v curve for T_2; (c) characteristic curves of T_1; (d) voltage transfer curve.

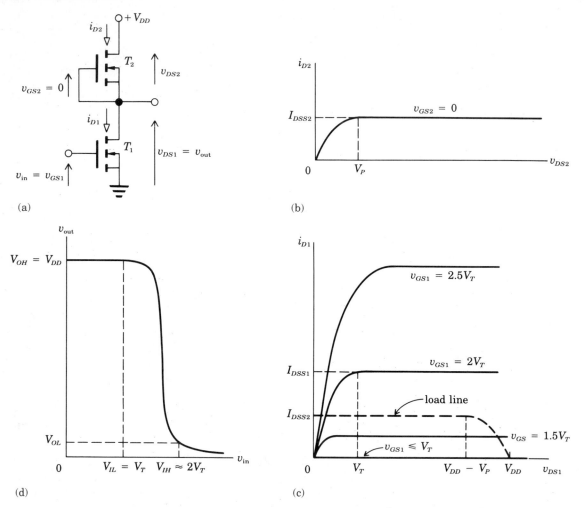

Now, from the intersections of the characteristic curves with the load curve, we construct the transfer curve plotted in Fig. 10.2–7d. The high output of the NMOS inverter is $V_{OH} = V_{DD}$ for $v_{in} < V_{IL} = V_T$. The low output is not uniquely defined because v_{out} continues to decrease as v_{in} increases. However, if I_{DSS1} is greater than I_{DSS2} as assumed in Fig. 10.2–7c, then we might arbitrarily take $V_{IH} \approx 2V_T$ so $V_{OL} \ll V_{DD}$ for $v_{in} > V_{IH}$.

Multi-input NMOS gates are fabricated simply by adding more enhancement MOSFETs in series or parallel. Figure 10.2–8 shows the configurations for a two-input NAND gate and a two-input NOR gate. These NMOS gates take up much less space on an IC chip than would be required

Figure 10.2–8
(a) NMOS NAND gate.
(b) NMOS NOR gate.

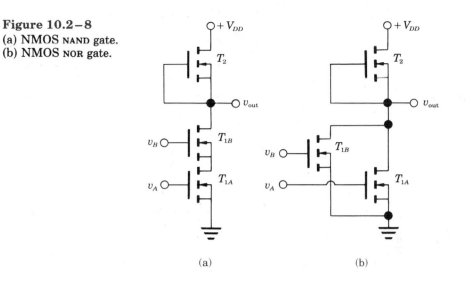

(a)   (b)

for BJT gates because MOSFETs are smaller than BJTs and much smaller than resistors.

Exercise 10.2–2

A typical NMOS inverter might have $I_{DSS1} = 50$ μA, $I_{DSS2} = 20$ μA, $V_T = V_P = 1.6$ V, and $V_{DD} = 5$ V. Use Eq. (4b) to find v_{out} when $v_{in} = 3.2$ V. Also confirm that T_1 is in its linear ohmic region.

Bistable Flip-Flops

Suppose the output of one BJT inverter drives the input of another, as shown in Fig. 10.2–9. When v_{in} is high, T_1 is on and v_1 is low, so T_2 is off and v_2 is high — just like v_{in}. Conversely, T_1 goes off and T_2 goes on so v_2 is low when v_{in} is low. In either case, we could eliminate the external signal v_{in} by making the dashed-line connection from the output of T_2 back to the input of T_1. We then have a *bistable* circuit that stays indefinitely in one of two possible states, since either T_1 will be on and T_2 off or T_1 will be off and T_2 on.

Figure 10.2–9
Bistable BJT circuit.

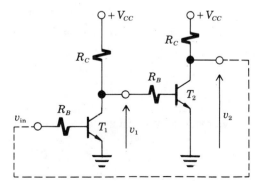

Figure 10.2–10

Bistable flip-flop: (a) BJT circuit; (b) NMOS circuit; (c) illustrative waveforms.

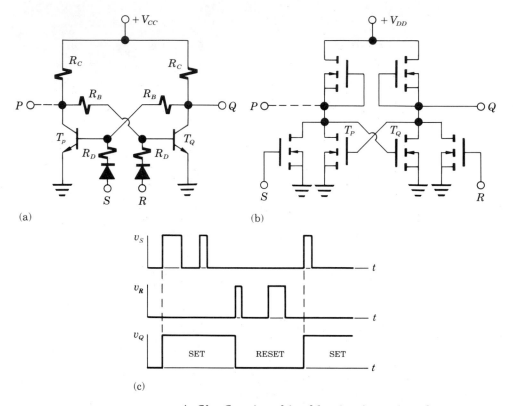

(a)

(b)

(c)

A **flip-flop** is a bistable circuit equipped to accept auxiliary input pulses. The diagrams of a simple BJT flip-flop and an NMOS flip-flop are given in Figs. 10.2–10a and 10.2–10b, where the inverter transistors have been labeled T_Q and T_P. The external output is taken at terminal Q, but an output at P may also be available. We change the state of these flip-flops by applying appropriate trigger pulses as follows:

- A positive pulse v_S at input S establishes the SET state, with T_P on and T_Q off — so v_Q goes high and v_P goes low.

- A positive pulse v_R at input R establishes the RESET state, with T_P off and T_Q on — so v_Q goes low and v_P goes high.

Simultaneous applications of S and R pulses must be avoided because the flip-flop cannot be in both states at the same time.

The waveforms in Fig. 10.2–10c illustrate a sequence of SET/RESET operations for a flip-flop initially in the RESET state. The first S pulse triggers the SET state and v_Q goes high. The flip-flop now remains SET until an R pulse triggers the RESET state and v_Q goes low. Any intervening S pulses therefore have no effect. However, an S pulse after an R pulse once again establishes the SET state.

These waveforms help explain why we say that flip-flops possess **digital memory.** A flip-flop "remembers" that it has been SET or RESET, and it stays that way until triggered into the other state — just as a mechanical switch "remembers" whether it is OFF or ON and stays that way until you turn the switch. The following example describes an important flip-flop application in conjunction with mechanical switches.

Example 10.2−2
Switch debouncing.

Figure 10.2−11a represents a digital circuit with one input generated manually via a mechanical toggle switch. When you turn the switch from one position to the other, say at time t_0, a miniature hammer strikes a tiny anvil and bounces several times before making continuous contact at some instant $t > t_0$. The resulting waveform v_S sketched in Fig. 10.2−11b looks like a digital signal that initially jumps up and down between the high and low levels.

Figure 10.2−11
Switch debouncing: (a) block diagram; (b) waveforms.

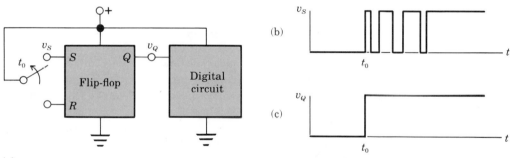

(a)

To eliminate these unwanted signal transitions, a flip-flop has been inserted between the switch and the rest of the circuit. The flip-flop's output signal v_Q shown in Fig. 10.2−11c goes high at t_0 and stays high as desired. Similarly, if the switch is turned to the other position, then v_Q will go low and stay low despite contact bounce. Hence, the flip-flop performs the function of **switch debouncing.**

Exercise 10.2−3

Let the BJT flip-flop circuit in Fig. 10.2−10a have $V_{CC} = 10$ V, $R_C = 5$ kΩ, and $R_B = 26$ kΩ, and let both transistors have $\beta \geq 10$. Confirm that the SET state is stable and find the high value of v_Q by taking $v_S = v_R = 0$ and assuming that T_Q is cut off and T_P is saturated.

Switching Transients [†]

Switching circuits seldom include capacitors or inductors because these energy-storage elements would inhibit rapid changes from one state to another. However, electronic devices always have some energy storage in the form of stray capacitance and stored charge carriers. The resulting unavoidable switching transients put upper limits on the performance of high-speed gates. Significant transient effects also arise in other applications when a transistor switch must drive a capacitive or inductive load.

The analysis of switching transients often becomes very complicated, especially in the case of a FET switch whose large-signal characteristics defy reasonable linearizing approximations. Consequently, we will not attempt a comprehensive treatment here. Instead, we'll look at two simplified but nonetheless informative examples with a BJT switch.

Our first example will illustrate how an inductive load slows down current switching, while the second illustrates how a capacitive load slows down voltage switching. Both examples involve discontinuous jumps that take the operating point of the BJT away from the usual load line. To focus on the important concepts, we'll assume a current source for the input signal and we'll take $V_{\text{sat}} = 0$ so the saturated transistor acts like a short circuit at the output.

Example 10.2–3
BJT relay driver.

Figure 10.2–12a depicts a BJT switch connected to an RL load representing the coil of a relay or a similar electromechanical device. An input current pulse of duration D goes on at time $t = 0$, producing output current i_L which energizes the coil and actuates the device in question. The diode is ordinarily reverse-biased and carries no current except when the coil is de-energized at the end of the pulse. The need for this diode will be seen in due course.

We begin our analysis in the steady-state condition with $i_B = 0$ for $t < 0$, so $i_L = i_C = 0$ and $v_{CE} = V_{CC}$. This initial operating point is indicated on the idealized BJT characteristic curves in Fig. 10.2–12b. The input source now goes on for $0^+ \leq t \leq D$, and the transistor immediately becomes saturated by base current $i_B = I_{\text{in}} > V_{CC}/\beta R$. However, the diode is still reverse biased, and continuity of current through L requires $i_C(0^+) = i_L(0^+) = 0$. Hence, at $t = 0^+$, the operating point must jump to the origin of the characteristic curves — the only point where $i_B = I_{\text{in}}$ and $i_C = 0$.

Since $v_{CE} = 0$ at this point, and since $V_{\text{sat}} \approx 0$, we get the equivalent circuit in Fig. 10.2–12c. It thus follows that

$$i_C(t) = i_L(t) = \frac{V_{CC}}{R}\left(1 - e^{-t/\tau}\right) \qquad 0^+ \leq t \leq D \qquad \textbf{(6)}$$

where $\tau = L/R$. If the pulse duration is long enough, say $D \geq 5\tau$, then the circuit reaches a new steady-state operating point with $i_C \approx V_{CC}/R$ and $v_{CE} \approx 0$ by the time the pulse ends at $t = D$.

Figure 10.2−12
BJT relay driver: (a) circuit; (b) operating points; (c) saturated conditions; (d) cutoff conditions.

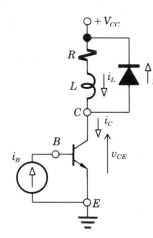

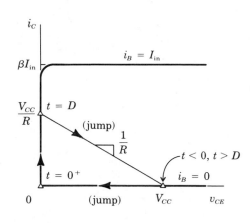

(a)

(b)

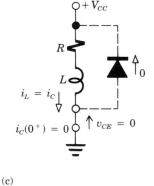

(c)

(d)

Finally, with $i_B = 0$ for $t > D$, the transistor goes off and $i_C = 0$. But i_L cannot change instantaneously, so the diode must go on at this time and we have the new equivalent circuit in Fig. 10.2−12d. Since the diode now forms a closed loop with the coil, the exponential decay of i_L is given by

$$i_L(t) = i_L(D)e^{-(t-D)/\tau} \qquad t > D \tag{7}$$

Meanwhile, the transistor jumps back to its initial operating point.

The various waveforms for the BJT relay driver are plotted in Fig. 10.2−13. Note that i_L would not reach its maximum value if the pulse duration is too short, say $D < \tau$. In that case, the current might not be sufficient to actuate the relay. Also note that i_C is allowed to jump at $t = D$ because the diode goes on and preserves continuity of i_L. In such situations,

Figure 10.2–13
Waveforms for BJT
relay driver.

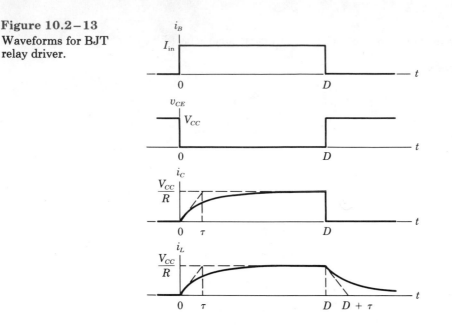

the diode is said to be **free-wheeling.** Had we omitted the free-wheeling diode, the abrupt change of i_L could produce a large voltage spike that might damage the transistor.

Example 10.2–4
BJT switch with
capacitive load.

Many BJT switches must drive capacitive loads like the situation in Fig. 10.2–14a. Here, C represents all stray and load capacitance between the collector and emitter. We'll investigate the transients produced by an input current pulse $i_B = I_{in} > V_{CC}/\beta R$ for $0^+ \le t \le D$. As in the previous example, we start at the steady-state condition with $i_B = i_C = 0$ and $v_{CE} = V_{CC}$ for $t < 0$.

When i_B goes on, continuity of voltage across C prevents immediate transistor saturation—which would require $v_{CE} = V_{sat} \approx 0$. Hence, the transistor initially jumps to the *active* operating point shown in Fig. 10.2–14b at $v_{CE}(0^+) = V_{CC}$ and $i_C(0^+) = \beta I_{in}$. The corresponding equivalent circuit in Fig. 10.2–14c reveals that C discharges via the constant current i_C. Including the effect of V_{CC} and R, we find that v_{CE} decays with time constant $\tau = RC$ and heads towards the negative steady-state value $-(R\beta I_{in} - V_{CC})$.

But active operation ends when $v_{CE} = 0$. We therefore write

$$v_{CE}(t) = -(R\beta I_{in} - V_{CC}) + R\beta I_{in} e^{-t/\tau} \qquad 0^+ \le t \le t_0 \tag{8}$$

where t_0 denotes the instant at which $v_{CE}(t_0) = 0$. If the input pulse duration D exceeds t_0, then the transistor becomes saturated at time t_0^+ and the collector current suddenly drops. The circuit then remains in the saturated steady-state condition with $i_C = V_{CC}/R$ and $v_{CE} = 0$ for $t_0 < t \le D$.

Figure 10.2–14

BJT switch with capacitive load: (a) circuit; (b) operating points;
(c) active conditions; (d) cutoff conditions.

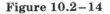

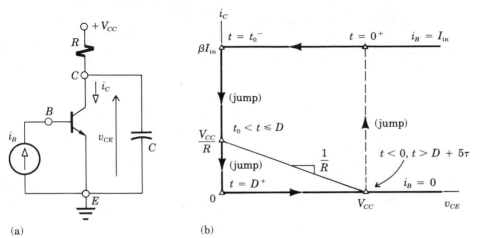

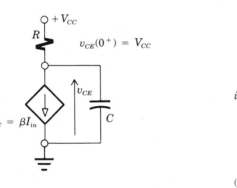

(a)

(b)

(c)

(d)

When i_B goes off at $t = D^+$, the collector current jumps to $i_C = 0$ while continuity keeps the collector voltage at $v_{CE} = 0$. From the new equivalent circuit in Fig. 10.2–14d we easily obtain

$$v_{CE}(t) = V_{CC}[1 - e^{-(t - D)/\tau}] \qquad t > D \qquad (9)$$

so $v_{CE} \approx V_{CC}$ for $t > D + 5\tau$.

The waveforms for the entire sequence are sketched in Fig. 10.2–15. The voltage waveform has important implications when v_{CE} happens to be the output of a BJT inverter with a digital input signal. We see that the output transition from high ($v_{CE} = V_{CC}$) to low ($v_{CE} \approx 0$) does not take place instantaneously at $t = 0$ but is delayed by t_0 seconds. Furthermore, the transition back to the high output starts at $t = D$ but it takes about 5τ seconds to arrive at $v_{CE} \approx V_{CC}$. Consequently, the design of high-speed digital gates must minimize capacitive loading.

Figure 10.2–15
Waveforms for BJT
switch with capacitive
load.

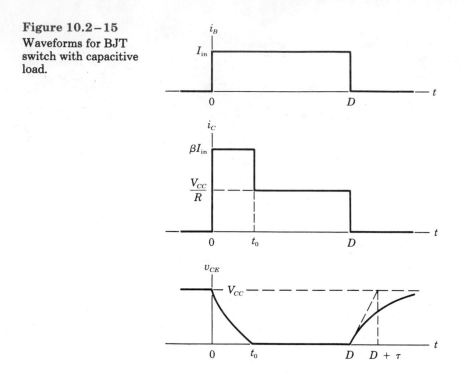

10.3 TRIGGER AND TIMING CIRCUITS

Many electronic systems include triggers and timers working hand in hand. A trigger circuit detects when a certain event has occurred, while a timing circuit issues a command at some specified time interval after being triggered. Electronic systems may also need waveform generators to supply period time markers, like a clock, or to generate a periodic signal with a particular waveshape.

This section begins with simple comparators and triggers. Then we combine triggers and comparators with RC circuits to build various timers, monostable circuits, and astable waveform generators. Consistent with modern practice, our emphasis will be on circuits that incorporate readily available IC units rather than assemblies of discrete components. Fittingly, the section closes with a discussion of the widely-used 555 timer.

Comparators and Triggers

The **ideal comparator** symbolized by Fig. 10.3–1a has two input terminals, neither of which draws input current. This comparator compares the input voltages v_p and v_n to produce the digital output

$$v_{\text{out}} = \begin{cases} V_{OL} & v_p < v_n \\ V_{OH} & v_p > v_n \end{cases} \tag{1}$$

Figure 10.3–1
Ideal comparator:
(a) symbol; (b) voltage
transfer curve.

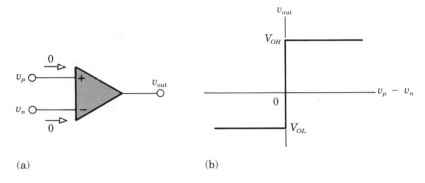

(a)   (b)

The plot of v_{out} versus $v_p - v_n$ in Fig. 10.3–1b emphasizes that the output state depends entirely on the polarity of the voltage difference across the input. Such behavior can be approximated by a high-gain op-amp without negative feedback, as previously seen in Fig. 4.2–2. Better comparator performance is obtained with IC units built explicitly for this purpose, which have faster switching times and outputs designed to drive standard digital gates.

Comparators are often used to detect when a time-varying analog signal goes above or below some fixed reference level. Consider, for example, the circuit in Fig. 10.3–2 where $v_n = v_{in}(t)$. The battery establishes the reference voltage $v_p = V_{ref}$, so v_{out} switches from high to low whenever $v_{in}(t)$ goes above V_{ref}. (A low-to-high transition could be obtained by interchanging the input connections.) The switched output might then trigger the horizontal sweep on an oscilloscope or activate a digital system. However, in such applications, we may need to account for the fact that analog signals sometimes exhibit random fluctuations caused by spurious noise. Thus, when $v_{in}(t) \approx V_{ref}$, the small input fluctuations will produce rapid and unwanted output switching known as "chatter."

Figure 10.3–2
Comparator with
reference voltage:
(a) circuit; (b) transfer
curve.

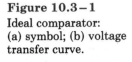

(a)   (b)

Output chatter can be suppressed by the **regenerative comparator** diagrammed in Fig. 10.3–3a. This circuit incorporates positive feedback from the output to the noninverting input such that

$$v_p = V_B + \frac{R_1}{R_1 + R_F} (v_{out} - V_B) = V_{ref} + \alpha v_{out} \tag{2a}$$

Figure 10.3–3
Regenerative comparator: (a) circuit; (b) transfer curve with hysteresis.

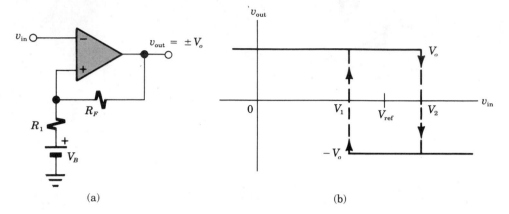

(a)

(b)

where

$$V_{ref} = \frac{R_F}{R_1 + R_F} V_B \qquad \alpha = \frac{R_1}{R_1 + R_F} \tag{2b}$$

The comparator itself has $V_{OH} = +V_o$ and $V_{OL} = -V_o$. Hence, the low output $v_{out} = -V_o$ requires $v_{in} > v_p = V_{ref} - \alpha V_o$, while the high output $v_{out} = +V_o$ requires $v_{in} < v_p = V_{ref} + \alpha V_o$. Positive feedback thereby creates two fixed comparison voltages

$$V_1 \triangleq V_{ref} - \alpha V_o \qquad V_2 \triangleq V_{ref} + \alpha V_o \tag{3}$$

and the input-output relation becomes

$$v_{out} = \begin{cases} -V_o & v_{in} > V_1 \\ +V_o & v_{in} < V_2 \end{cases} \tag{4}$$

We call V_1 and V_2 the **trigger levels.**

Equation (4) leads to the transfer curve for a regenerative comparator plotted in Fig. 10.3–3b. This figure clearly shows that the output may be either high or low when the input falls between the trigger levels V_1 and V_2. Consequently, we have a **hysteresis effect** in the sense that the value of v_{out} now depends upon the past behavior of v_{in}. If v_{in} starts below V_1 and increases past V_2, then v_{out} jumps from high to low when $v_{in} = V_2$. Or if v_{in} starts above V_2 and decreases past V_1, then v_{out} jumps from low to high when $v_{in} = V_1$.

The waveforms in Fig. 10.3–4 illustrate how hysteresis suppresses chatter at the output of a regenerative comparator whose trigger levels form a narrow band around the nominal reference voltage V_{ref}. The hys-

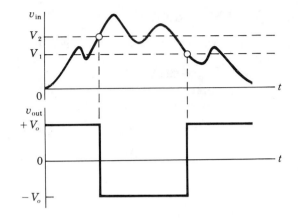

teresis width is $V_2 - V_1 = 2\alpha V_o$, which should exceed the peak-to-peak input fluctuations for a chatter-free output.

Hysteresis effect also occurs in modified flip-flops called **Schmitt triggers,** and most families of digital ICs include at least one such device. By way of example, Fig. 10.3–5 shows the symbol and transfer characteristics of an inverting Schmitt trigger from the CMOS family. When operated with supply voltage V_{DD}, the triggering and output levels are

$$V_1 \approx 0.4 V_{DD} \qquad V_2 \approx 0.6 V_{DD}$$

$$V_{OL} \approx 0 \qquad V_{OH} \approx V_{DD} \tag{5}$$

Hence, this unit simply acts as an ordinary inverter if the input happens to be a digital signal switching between the low and high levels. Although the CMOS trigger draws no steady-state input current, a large resistance should be connected in series with the input to guard against excessive current during switching transients.

Figure 10.3–5
CMOS Schmitt-trigger
inverter: (a) symbol;
(b) transfer curve.

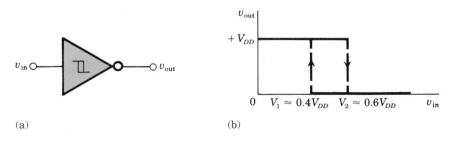

(a)   (b)

Exercise 10.3–1   Given a regenerative comparator like Fig. 10.3–3a with $V_o = 5$ V and $R_1 = 1$ kΩ, find the values of R_F and V_B to obtain chatter-free output switching at $v_{in} \approx 3$ V when $v_{in}(t)$ has noise fluctuations smaller than 0.4 V peak to peak.

Timers and Monostables

The basic purpose of a timing circuit is to generate an output transition at a specified time after a triggering input event. These circuits depend upon some form of hysteresis together with the predictable behavior of RC transients. Recall that if $v(t)$ is a transient voltage heading toward V_{ss} with time constant $\tau = RC$, and if $v(t_0) = V_0$, then

$$v(t) = V_{ss} + (V_0 - V_{ss})e^{-(t - t_0)/\tau} \qquad t > t_0 \qquad \textbf{(6a)}$$

Consequently, the time interval T satisfying a triggering condition $v(t_0 + T) = V_T$ is given by

$$T = \tau \ln \frac{V_0 - V_{ss}}{V_T - V_{ss}} \qquad \textbf{(6b)}$$

We'll have frequent use for these relations.

As an introduction to timing concepts, consider the RC circuit connected to a CMOS Schmitt trigger in Fig. 10.3–6a. Resistor R_p protects the trigger but plays no other role. If the input is a pulse of height V_{DD} and its duration D_{in} is longer than the time constant τ, then v_C has the familiar transient waveform sketched in Fig. 10.3–6b. The corresponding trigger

Figure 10.3–6

(a) Timing circuit. (b) Waveforms.

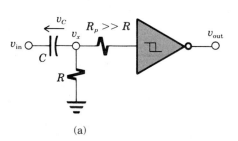

(a)

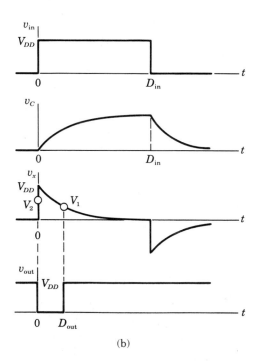

(b)

input waveform is constructed by noting that $v_x = v_{in} - v_C$. Since v_x jumps to $V_{DD} > V_2 = 0.6V_{DD}$ at $t = 0^+$, the output immediately switches from high to low. The output goes high again when v_x decays past $V_1 = 0.4V_{DD}$. Hence, we get an inverted output pulse of shorter duration

$$D_{out} = \tau \ln \frac{V_{DD}}{V_1} \approx \tau \ln 2.5 \approx 0.916 \, RC \qquad (7)$$

which follows from Eq. (6).

The value of D_{out} is easily controlled by the choice of R and C, independent of V_{DD}. When desired, a noninverted pulse can be obtained with the help of an inverter or another Schmitt trigger. The transients at the end of the input pulse have no effect on the output, but the capacitor should discharge completely before the next input pulse arrives. This timing circuit therefore requires a **recovery time** of about 5τ seconds.

To generate a longer output pulse, we need the type of timing circuit called a **monostable** — also known as a **monostable multivibrator** or a **one-shot.** A monostable circuit has two states, one being stable, the other being temporary or unstable. When the unstable state is activated, it lasts for a well-defined time interval until the circuit spontaneously returns to the stable state. We'll clarify monostable operation by analyzing the circuit in Fig. 10.3–7, which uses two CMOS Schmitt triggers and feedback from the output.

The pair of diodes and the first trigger just constitute a NOR gate, so v_{NOR} goes low when v_{in} goes high or when v_{out} goes high. The stable state

Figure 10.3–7
(a) Monostable circuit. (b) Waveforms.

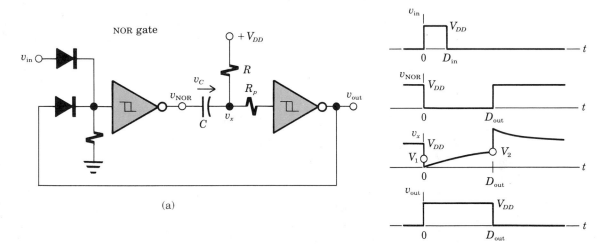

(a)

(b)

with $v_{in} = 0$ results in $v_{NOR} = v_x = V_{DD}$ and $v_{out} = 0$, as demonstrated subsequently. The unstable state is activated by an input pulse, producing $v_{NOR} = 0$, $v_x = 0$, and $v_{out} = V_{DD}$ at $t = 0^+$. The resulting high output now keeps v_{NOR} low after the end of the input pulse. Meanwhile, the capacitor starts to charge towards $+V_{DD}$, so the output goes low when $v_x = V_2$ at $t = D_{out}$ where

$$D_{out} = \tau \ln \frac{V_{DD}}{V_{DD} - V_2} \approx \tau \ln 2.5 \approx 0.916 \, RC \qquad \textbf{(8)}$$

When the output pulse ends, v_{NOR} goes high and continuity of v_C initially forces the jump to $v_x = v_{NOR} + v_C = V_2 + V_{DD}$. The capacitor then discharges and the circuit finally reaches its stable state with $v_{NOR} = v_x = V_{DD}$ and $v_{out} = 0$ after a recovery time of about 5τ seconds.

Equation (8) yields the same numerical value as Eq. (7) because CMOS Schmitt triggers have $V_{DD} - V_2 \approx V_1$. However, the monostable's input pulse may have any duration $D_{in} < D_{out}$, whereas the timer in Fig. 10.3–6 requires $D_{in} > D_{out}$. For both circuits, triggering occurs at the leading or rising edge of the input.

Other monostable circuits employ BJT switches and are available as IC packages containing everything except the timing elements R and C attached by the user. Figure 10.3–8a represents a BJT monostable that triggers on a trailing or falling edge of the input and generates output

Figure 10.3–8
(a) BJT monostable. (b) Input waveform. (c) Output waveform with short pulses. (d) Nonretriggerable output with long pulses. (e) Retriggerable output with long pulses.

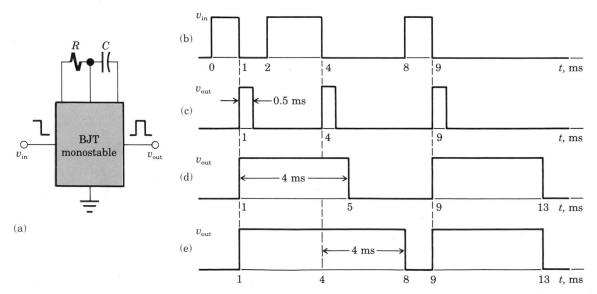

pulses of duration

$$D_{\text{out}} = RC \ln 2 \approx 0.693\, RC \qquad (9)$$

The recovery time is very small compared to D_{out}.

If the input is a train of pulses like Fig. 10.3–8b, and if $D_{\text{out}} = 0.5$ ms, then the output is the train of short pulses shown in Fig. 10.3–8c. But if RC is increased such that $D_{\text{out}} = 4$ ms, then the resulting output depends on whether the monostable is non-retriggerable or retriggerable. A **nonretriggerable** monostable cannot be triggered while in its unstable state, so Fig. 10.3–8d would be the output waveform. Figure 10.3–8e shows the output of a **retriggerable monostable** whose unstable interval starts over again at each input trigger.

Exercise 10.3–2　　　Suppose the positions of R and C are interchanged in Fig. 10.3–6, so $v_x = v_C$.

(a) Sketch the resulting output waveform when $D_{\text{in}} \geq 5\tau$.
(b) What happens when $D_{\text{in}} < 0.9\tau$?

Astable Waveform Generators

As the name implies, an **astable circuit** has *no stable states*. Instead, it continuously switches back and forth between two unstable states — thereby producing a periodic output signal without any input signal. Such circuits are also called **relaxation oscillators,** and they exist in many different and clever versions. Here, we'll examine two illustrative astable waveform generators selected for their practical significance.

Figure 10.3–9
CMOS clock: (a) circuit; (b) waveforms.

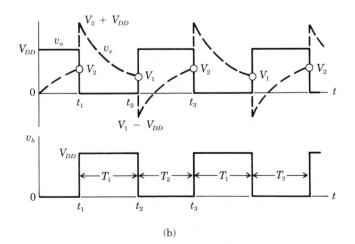

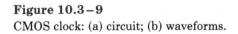
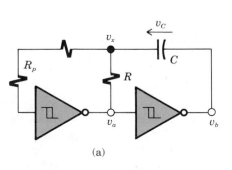

(a)　　　　　　　　　　　　　　　　　　(b)

The astable circuit in Fig. 10.3–9a is known as a **CMOS clock** because it generates a square pulse train like the clock signal needed for many digital systems. Either v_a or v_b may be taken to be the output, since the second Schmitt trigger merely acts as a digital inverter. Astable operation is established by positive feedback through the capacitor, so v_C charges towards $v_a - v_b$ and v_x always heads toward v_a. We'll first confirm this astable property by considering the two possible values of v_a. On the one hand, we must have $v_x < V_2$ when v_a is high; but v_x then rises toward $v_a = V_{DD}$, causing v_a to switch when $v_x = V_2$. On the other hand, we must have $v_x > V_1$ when v_a is low; but v_x then falls toward $v_a = 0$, causing v_a to switch when $v_x = V_1$.

For quantitative analysis of the astable operations, Fig. 10.3–9b shows the waveforms starting at $t = 0$ with $v_a = V_{DD}$, $v_b = 0$, and $v_C = 0$, so $v_x = 0$. Switching first occurs when v_x rises to V_2 at time $t_1 \approx \tau \ln 2.5$, which follows from Fig. 10.3–7b and Eq. (8). Now v_b goes high and continuity of v_C forces $v_x(t_1^+) = V_2 + V_{DD}$, so v_x decays toward $v_a = 0$ until the next switching time $t_2 = t_1 + T_1$ when $v_x(t_2) = V_1$. Hence, the duration of this half cycle is

$$T_1 = \tau \ln \frac{V_{DD} + V_2}{V_1} \approx \tau \ln 4 \qquad (10a)$$

The next half cycle begins at t_2^+, when v_b goes low and v_x jumps down to $V_1 - V_{DD}$ before rising to V_2 at $t_3 = t_2 + T_2$, where

$$T_2 = \tau \ln \frac{2V_{DD} - V_1}{V_{DD} - V_2} \approx \tau \ln 4 \qquad (10b)$$

Clearly, the situation at t_3^+ is the same as at t_1^+, so the T_1 and T_2 half cycles continue to repeat periodically. Therefore, after the start-up interval t_0, we have a digital output pulse train with $T_2 \approx T_1$ and repetition frequency

$$f = \frac{1}{T_1 + T_2} \approx \frac{1}{2\tau \ln 4} \approx \frac{0.361}{RC} \qquad (11)$$

If $R = 1$ kΩ and $C = 1$ nF, for instance, then $f \approx 361$ kHz.

Astable circuits can also be designed for analog applications. In particular, Fig. 10.3–10a diagrams a circuit of the type often used to generate both *square waves* and *triangular waves*. The square wave comes from the regenerative comparator, whose output v_a switches between $+V_o$ and $-V_o$. The op-amp integrator produces the triangular wave v_b by integrating v_a since, from Example 5.1–2, the slope of v_b is given by

$$\frac{dv_b}{dt} = \begin{cases} -V_o/RC & v_a = +V_o \\ +V_o/RC & v_a = -V_o \end{cases}$$

Connecting v_b to the noninverting input of the comparator provides the positive feedback necessary for astable operation.

We begin the analysis by drawing upon Eq. (2) to formulate an expression for v_p. Setting $R_1 = kR_F$, $V_B = v_b$, and $v_{out} = v_a$ yields the result $v_p =$

Figure 10.3–10

Analog waveform generator: (a) circuit; (b) transfer characteristic of comparator; (c) waveforms.

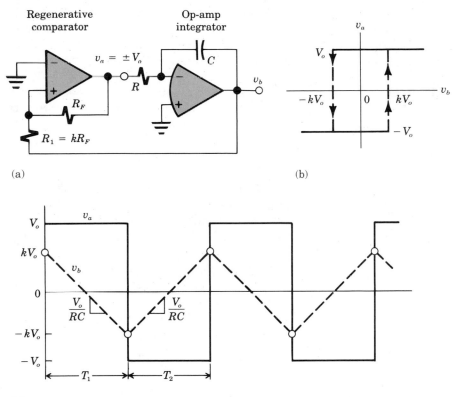

(a)

(b)

(c)

$(v_b + kv_a)/(k + 1)$. Next, noting that the ground at the comparator's inverting input makes $v_n = 0$, we see that the high output $v_a = +V_o$ requires $v_p = (v_b + kV_o)/(1 + k) > 0$ or $v_b > -kV_o$, while the low output $v_a = -V_o$ requires $v_b < +kV_o$. The resulting transfer characteristic plotted in Fig. 10.3–9b exhibits hysteresis effect, as expected, and it indicates that v_b must remain within the range $-kV_o \leq v_b \leq +kV_o$ during the astable cycles.

Now, to skip over the start-up interval, let $v_b = +kV_o$ at $t = 0$ and assume that v_a has just jumped to $+V_o$. Hence, v_b decreases linearly with slope $+V_o/RC$ during the first half cycle, as drawn in Fig. 10.3–10c. This behavior ends at $t = T_1$ when $v_b = -k\,V_o$ and v_a goes low. Then v_b increases linearly with slope $+V_o/RC$ during the second half cycle, until $t = T_1 + T_2$ when $v_b = +kV_o$ and the first half cycle begins again. Since v_b changes by $2kV_o$ over each half cycle, we see that $T_1 = T_2 = 2kRC$ and the frequency of the two waveforms is

$$f = \frac{1}{T_1 + T_2} = \frac{1}{4kRC} = \frac{R_F}{4R_1 RC} \qquad (12)$$

A sinusoidal waveform at the same frequency can be obtained by applying v_b to a waveshaper like the one back in Example 10.1–1.

Simple modifications of the circuit in Fig. 10.3–10a yield asymmetrical waveforms with $T_1 \neq T_2$ (see Problems 10.3–16, 17, and 18). Another asymmetrical waveform generator will be discussed in conjunction with the 555 timer.

Exercise 10.3–3   Note that $v_x(t)$ in Fig. 10.3–9b is heading toward $v_a = V_{DD}$ during the interval $t_2 < t \leq t_3$. Use Eqs. (6a) and (6b) to obtain an expression for $v_x(t)$ in this interval and to derive Eq. (10b).

The 555 Timer [†]

The 555 timer (pronounced "five fifty-five") is a popular IC unit designed to accommodate numerous applications, with timing intervals from about one microsecond to one minute. It is easily connected such that the input signal controls the duration or position of the output pulses, and it has sufficient power-handling capability to drive an electromechanical relay or power transistor. The required supply voltage can be any value in the 5–15 V range. For further flexibility, the 556 timer contains two 555 timers in one package.

Figure 10.3–11 gives a somewhat simplified picture of the 555 timer. A careful inspection from bottom to top reveals the following properties:

- If $v_{\text{trig}} < V_1$, then the output of comparator C1 goes high, so the flip-flop is SET and $v_{\text{out}} \approx V_{CC}$.

- If $v_{\text{thresh}} > V_2$, then the output of comparator C2 goes high, so the flip-flop is RESET and $v_{\text{out}} \approx 0$.

- In absence of an external input applied to the control terminal, $V_2 = 2V_{CC}/3$ and $V_1 = V_{CC}/3$. Otherwise, $V_2 = v_{\text{cont}}$ and $V_1 = \frac{1}{2} v_{\text{cont}}$.

- The diode connected to the output forces $v_{\text{dis}} \leq v_{\text{out}} + V_\gamma \approx v_{\text{out}}$.

Figure 10.3–11
Diagram of 555 timer.

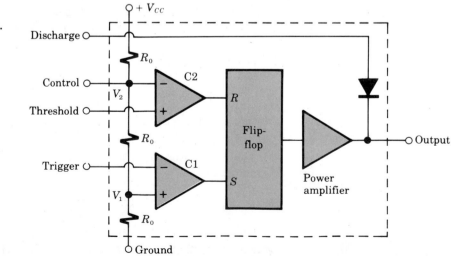

We'll apply these properties to explain the operation of the 555 timer in a monostable and an astable configuration.

The external hook-up and waveforms for monostable operation are given in Fig. 10.3–12, where $v_{dis} = v_{thresh} = v_C$. The control terminal is left unconnected, so $V_2 = 2V_{CC}/3$ and $V_1 = V_{CC}/3$. The RESET state with $v_{out} \approx 0$ is stable here, and the diode keeps $v_C \approx 0$. The unstable SET state is activated at $t = 0$ when v_{in} switches from high to low. This downward step is coupled through a small capacitor C_i, so v_{trig} momentarily drops below V_1 and v_{out} goes high. With the diode now reverse biased, the timing capacitor C charges toward V_{CC} until $v_C = V_2^+$ at $t = D$. Then v_{out} goes low again and v_C rapidly discharges through the diode, returning the circuit to its stable state. The resulting output pulse has duration

$$D = RC \ln 3 \approx 1.10 \, RC \qquad (13)$$

A subsequent rising input edge may occur at any time $t > 0$ and has no effect on the output. However, for quick recovery of the trigger signal, $R_i C_i$ should be small compared to RC.

Figure 10.3–12
555 timer operated as a monostable.

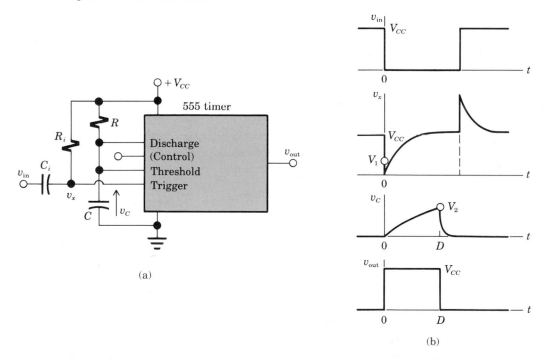

(a)

(b)

Astable operation requires one capacitor and two resistors connected to the 555 timer as shown in Fig. 10.3–13a. During the SET interval, with $v_{out} = V_{CC}$, the capacitor charges through R_A and R_B until $v_C = V_2 =$

Figure 10.3–13

555 timer operated as a waveform generator.

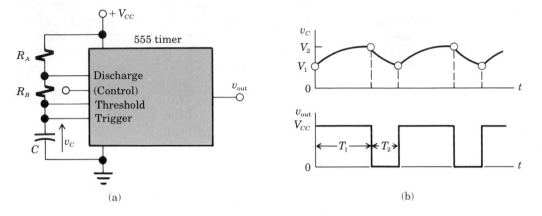

(a)   (b)

$2V_{CC}/3$. During the RESET interval, with $v_{out} = 0$, the capacitor discharges through R_B and the diode until $v_C = V_1 = V_{CC}/2$. The asymmetric waveforms in Fig. 10.3–13b reflect the fact that two different time constants are involved here. Thus,

$$T_1 = (R_A + R_B)C \ln 2 \qquad T_2 = R_B C \ln 2 \qquad \textbf{(14a)}$$

and

$$f = \frac{1}{T_1 + T_2} \approx \frac{1.44}{(R_A + 2R_B)C} \qquad \textbf{(14b)}$$

Practical values of f range from about 0.1 Hz to 100 kHz.

If an external voltage is applied to the control terminal, then the values of V_1 and V_2 are altered but v_C still charges toward V_{CC} and discharges toward zero. Consequently, the frequency depends upon v_{cont} and the circuit becomes a **voltage-to-frequency converter** or **voltage-controlled oscillator**.

PROBLEMS

10.1–1   Suppose the waveform in Fig. P10.1–1 is applied to the clipper in Fig. 10.1–1 with $R_0 = 10$ kΩ. Plot v_{out} versus t for $V_B = 12$ V and $R = 5$ kΩ.

10.1–2   Do Problem 10.1–1 for $V_B = 18$ V and $R = 2$ kΩ.

10.1–3   Do Problem 10.1–1 for $V_B = 6$ V and $R = 10$ kΩ.

10.1–4   Design a three-branch version of the waveshaper in Fig. 10.1–3a to get $v_{out} = K \ln (v_{in} + 1)$ at $v_{in} = 1, 3, 7,$ and 15 V. Take $K = 1/(\ln 2)$ and $R_0 = 1$ kΩ.

10.1–5   Design a three-branch version of the waveshaper in Fig. 10.1–3a to get $v_{out} = K \arctan (0.5v_{in})$ at $v_{in} = 1, 2, 3,$ and 4 V. Take $K = 1/(\arctan 0.5)$ and $R_0 = 1$ kΩ.

Figure P10.1–1

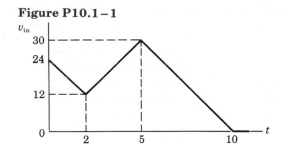

10.1–6 Suppose battery V_1 is inverted in Fig. 10.1–3a. Plot v_{out} versus v_{in} when $R_0 = R_1 = R_2 = 1$ kΩ, $V_1 = 2$ V, and $V_2 = 4$ V.

10.1–7 Suppose diode D_1 is inverted in Fig. 10.1–3a. Plot v_{out} versus v_{in} when $R_0 = R_1 = R_2 = 1$ kΩ, $V_1 = 2$ V, and $V_2 = 4$ V.

10.1–8 Figure P10.1–8 gives the circuit diagram and transfer curve of a waveshaper with *increasing* slopes. Assume an ideal op-amp to show that

$$b_1 = b_0 \left(1 + \frac{R_F}{R_1} \right) \qquad b_2 = b_0 \left(1 + \frac{R_F}{R_1} + \frac{R_F}{R_2} \right)$$

where $b_0 = R_a/(R_a + R_b)$.

Figure P10.1–8

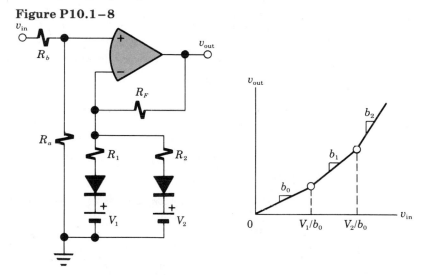

10.1–9 Taking $R_b = R_F = 1$ kΩ, find R_a, R_1, R_2, and V_1, and V_2 so the waveshaper in Problem 10.1–8 produces $v_{out} = 2(v_{in}/3)^2$ at $v_{in} = 0, 3, 6,$ and 9 V.

10.1–10 Taking $R_b = R_F = 1$ kΩ, find R_a, R_1, R_2, V_1, and V_2 so the waveshaper in Problem 10.1–8 produces $v_{out} = 2^{0.5v_{in}} - 1$ at $v_{in} = 0, 2, 4,$ and 6 V.

10.1–11 Let the inputs in Fig. 10.1–5 be $v_A = 10 \sin \omega t$ V and $v_B = 10 \sin 2\omega t$ V, and let $V_y = 0$. Sketch v_A, v_B, and v_{out} versus ωt on the same set of axes.

10.1–12 Let the inputs in Fig. 10.1–6 be $v_A = 10 \sin \omega t$ V and $v_B = 10 \sin 2\omega t$ V, and let $V_y = 0$ and $V_{HIGH} = 5$ V. Sketch v_A, v_B, and v_{out} versus ωt on the same set of axes.

10.1–13   Figure P10.1–13 represents a cascade of digital gates like Figs. 10.1–5 and 10.1–6. The diodes have $R_f = 0$ and $V_\gamma = 0.7$ V, all resistors are equal, and $V_{HIGH} = 5$ V. The input terminals of the first gate have been tied together, so there are four possible input combinations, namely: $v_1 = v_2 = 0$ V; $v_1 = 0$ and $v_2 = 5$ V; $v_1 = 5$ V and $v_2 = 0$; $v_1 = v_2 = 5$ V. Find the value of v_{out} for each input combination when:

(a) Both gates are OR gates;

(b) Gate 1 is an OR gate and gate 2 is an AND gate.

Figure P10.1–13

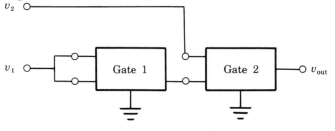

10.1–14   Do Problem 10.1–13 when:

(a) Both gates are AND gates;

(b) Gate 1 is an AND gate and gate 2 is an OR gate.

10.1–15   Derive Eq. (4) for the diode-quad gate in Fig. 10.1–9 by taking diodes D_1 and D_4 to be at their breakpoints ($v = V_\gamma$, $i = 0$) when $v_{out} = v_{in} = +V_{max}$ and $v_c = +V_C$.

10.1–16   Find $R_{in} = v_{in}/i_{in}$ for the diode-quad gate in Fig. 10.1–9 when $v_c = +V_C$ and $|v_{in}| \leq V_{max}$. Hint: Sum the currents entering and leaving the diode array.

10.1–17   Let $v_{in} = V_m \cos \omega t$ be applied to the clamp circuit in Fig. 10.1–11, and let $RC \gg T = 2\pi/\omega$.

(a) Write an expression for the steady-state output voltage when $V_m = 3$ V and $V_m = 5$ V;

(b) Repeat part (a) with the diode inverted.

10.1–18   Do Problem 10.1–17 for the biased clamp circuit in Fig. 10.1–12 with $V_B = 4$ V.

10.1–19   Do Problem 10.1–17 for the biased clamp circuit in Fig. 10.1–12 with $V_B = -4$ V.

10.1–20   The circuit in Fig. 10.1–13a becomes a **voltage doubler** when R_1 is replaced by another diode D_1 conducting from left to right. Find v_{av} and the waveform $v_D - v_{av}$ that appears across D_1.

10.2–1   Suppose the BJT switch in Fig. 10.2–1 has $\beta = 20$ and $V_{CC} = 5$ V. If v_{in} switches between 0 and V_{CC}, and if we want $i_B \leq 0.1$ mA, then what are the conditions on R_B and R_C for proper operation?

10.2–2   Do Problem 10.2–1 with $V_{CC} = 9$ V.

10.2–3   Let the inverter in Fig. P10.2–3 have $R_B = 5.8$ kΩ and $R_{C1} = R_{C2} = 4$ kΩ. Find the minimum values of β_1 and β_2 to ensure that T_1 and T_2 saturate when v_{in} is high.

10.2–4   Let the inverter in Fig. P10.2–3 have $R_B = 4$ kΩ, $R_{C1} = 2.5$ kΩ, and $\beta_1 = \beta_2 = 4$. Show that T_1 is saturated when v_{in} is high, and find the condition on R_{C2} so that T_2 is saturated.

Figure P10.2-3

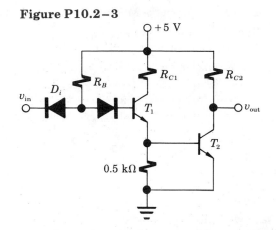

10.2-5 Let the inverter in Fig. P10.2-3 have $R_{C1} = R_{C2} = 2$ kΩ and $\beta_1 = \beta_2 = 10$. Show that T_2 is saturated when v_{in} is high, and find the condition on R_B so that T_1 is saturated.

10.2-6 Figure P10.2-6 represents N gates being driven by the voltage v_0 across T_0—a situation known as **fan-out**. Find the maximum allowed value of N for proper operation when $i_{B0} = 0.5$ mA, T_0 has $\beta_0 = 20$, and the driven gates are like Fig. 10.2-2a with $V_{CC} = 5$ V, $R_B = 10$ kΩ, and $R_C = 2$ kΩ.

Figure P10.2-6

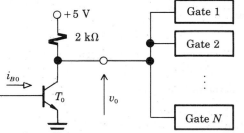

10.2-7 Do Problem 10.2-6 for the case when $i_{B0} = 0$ and the driven gates are inverters like Fig. 10.2-1a with $V_{CC} = 5$ V, $R_B = 10$ kΩ, $R_C = 2$ kΩ, and $\beta = 20$.

10.2-8 Let the gate circuit in Fig. 10.2-4a have $v_B = 0$ and $\beta_1 = \beta_2 = 40$.
  (a) Assume that T_2 is barely saturated, with $v_2 = V_{sat}$ and $i_{C2} = \beta_2 i_{B2}$. Show that D_i must be on and T_1 must be active. Then find the corresponding value of v_A.
  (b) Now assume that T_2 is barely cut off, with $i_{C2} \approx 0$ and $i_{B2} = 0^+$. Determine the states of D_i and T_1, and find the corresponding value of v_A.
  (c) Use the results of (a) and (b) to plot v_2 versus v_A.

10.2-9 Do Problem 10.2-8 with $\beta_1 = \beta_2 = 24$.

10.2-10 Suppose that v_{in} in Fig. 10.2-5 varies over 0–5 V, and $R_L = 10$ kΩ. If the MOSFET has $V_T = 4$ V and $I_{DSS} = 1$ mA, then what is the minimum value of v_c needed to get $R_{DS} \leq 1$ kΩ during sampling?

10.2–11   Do Problem 10.2–10 when the MOSFET has $V_T = 3$ V and $I_{DSS} = 1$ mA.

10.2–12   Find the condition on the ratio I_{DSS2}/I_{DSS1} such that the NMOS inverter in Fig. 10.2–7 has $v_{out} \leq V_{DD}/10$ when $v_{in} = 2V_T$.

10.2–13   If the NMOS inverter in Fig. 10.2–7 has $I_{DSS1} = I_{DSS2}$ and $V_{DD} = 2.5V_T$, then what value of v_{in} yields $v_{out} = V_{DD}/10$?

10.2–14   Let both NMOS gates in Fig. 10.2–8 have $V_{DD} = 5$ V, $V_T = V_P = 1.6$ V, $I_{DSS1} = 50\ \mu$A, and $I_{DSS2} = 20\ \mu$A. Calculate v_{out} for each gate when: (a) $v_A = 3.2$ V and $v_B = 0$; (b) $v_A = v_B = 3.2$ V.

10.2–15   Do Problem 10.2–14 with $I_{DSS1} = 100\ \mu$A and $I_{DSS2} = 10\ \mu$A.

10.2–16   Given that the BJT flip-flop in Fig. 10.2–10a has $V_{CC} = 9$ V, $R_C = 4$ kΩ, and $\beta = 10$, determine the condition on R_B for proper operation, and find the corresponding limit on the high value of v_Q.

10.2–17   Suppose the BJT flip-flop in Fig. 10.2–10a has $V_{CC} = 5$ V, $R_C = 2$ kΩ, and $\beta = 10$. Find the allowable range of R_B for proper operation with the high value of v_Q being at least 4.5 V.

10.2–18   Find $i_L(t)$ for the relay driver in Fig. 10.2–12 given that $V_{CC} = 10$ V, $L = 50$ mH, $R = 25\ \Omega$, $\beta = 100$, and $i_B = 5$ mA for $0^+ \leq t \leq 4$ ms.

10.2–19   Suppose the relay being driven in Fig. 10.2–12 closes when i_L reaches the 400-mA "pull-in" current and opens when i_L falls to the 150-mA "drop-out" current. If $V_{CC} = 15$ V, $L = 150$ mH, $R = 30\ \Omega$, $\beta = 100$, and $i_B = 8$ mA for $0^+ \leq t \leq 30$ ms, then when will the relay close and open?

10.2–20   Find $i_C(t)$ and $v_{CE}(t)$ in Fig. 10.2–14 given that $V_{CC} = 12$ V, $R = 1$ kΩ, $C = 100$ pF, $\beta = 50$, and $i_B = 1$ mA for $0^+ \leq t \leq 1\ \mu$s.

10.2–21   Derive Eq. (8) from Fig. 10.2–14c and Eq. (12), Section 7.3.

10.2–22   Obtain expressions for $i_C(t)$ and $v_{CE}(t)$ in Fig. 10.2–14a when $i_B = I_{in}$ for $0^+ \leq t \leq D$ with $I_{in} < V_{CC}/\beta R$ and $D \gg RC$.

10.3–1   Given that the regenerative comparator in Fig. 10.3–3a has $V_o = 15$ V and $R_F = 10$ kΩ, find R_1 and V_B needed to get trigger levels at $V_1 = 5$ V and $V_2 = 11$ V.

10.3–2   Do Problem 10.3–1 for $V_1 = 4$ V and $V_2 = 6$ V.

10.3–3   The regenerative comparator in Fig. P10.3–3 obtains its reference from the power supply V_o. Use superposition to find v_p with $v_{out} = \pm V_o$ and show that $V_1 = (R_1 - R_3)V_o/R_o$ and $V_2 = (R_1 + R_3)V_o/R_o$ where $R_o = (R_1R_2 + R_1R_3 + R_2R_3)/R_2$.

Figure P10.3–3

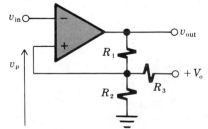

10.3–4   Derive Eq. (6b) from Eq. (6a).

10.3–5   Redraw the waveforms v_C, v_x, and v_{out} in Fig. 10.3–6 for the case where the lower end of R is connected to $+V_{DD}$ rather than to ground. Assume $D_{in} \gg RC$.

10.3–6   Suppose the timing circuit in Fig. 10.3–6a has $V_{DD} = 10$ V, $R = 20$ kΩ, and $C = 0.25\ \mu$F. Sketch v_x and v_{out}, and find the durations of the output pulses produced by the input signal shown in Fig. P10.3–6 with $V_{HIGH} = 10$ V.

Figure P10.3–6

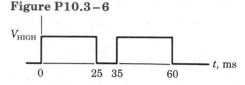

10.3–7   Do Problem 10.3–6 with $C = 0.5\ \mu$F.

10.3–8   Consider a nonretriggerable BJT monostable described by Eq. (9) with $C = 300$ pF. The monostable triggers on falling edges and has negligible recovery time. Given the periodic input signal shown in Fig. P10.3–8, sketch the output waveform and calculate its repetition frequency f when: (a) $R = 4$ kΩ; (b) $R = 15$ kΩ.

Figure P10.3–8

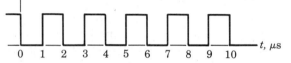

10.3–9   Do Problem 10.3–8 with $C = 600$ pF.

10.3–10 Using BJT monostables like Fig. 10.3–8a with $C = 0.01\ \mu$F, design a system to produce a 10-μs output pulse that begins 40-μs after the trailing edge of an input pulse.

10.3–11 Using BJT monostables like Fig. 10.3–8a with $C = 0.05\ \mu$F, design a system to produce a 100-μs output pulse that begins 20-μs after the trailing edge of an input pulse.

10.3–12 Using a diode OR gate and BJT monostables like Fig. 10.3–8a with $C = 2\ \mu$F, design a system to produce the output waveform in Fig. P10.3–6 after the trailing edge of an input pulse.

10.3–13 Modify Eq. (11) for the case where the Schmitt triggers in Fig. 10.3–9a have $V_1 = V_{DD}/3$ and $V_2 = 2V_{DD}/3$.

10.3–14 Do Problem 10.3–13 with $V_1 = 0.2V_{DD}$ and $V_2 = 0.7V_{DD}$.

10.3–15 Suppose resistor R in Fig. 10.3–9a is replaced by the unit in Fig. P10.3–15 with diode D_A at the top. Revise Eqs. (10) and (11) to reflect this modification, assuming ideal diodes.

Figure P10.3–15

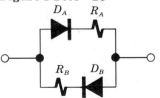

10.3 – 16   Suppose resistor R in Fig. 10.3 – 10a is replaced by the unit in Fig. P10.3 – 15 with diode D_A at the left. Find the new values of dv_b/dt and revise Eq. (12) to reflect this modification, assuming ideal diodes.

10.3 – 17   Let Fig. 10.3 – 10a be modified by connecting a source voltage to the noninverting input of the op-amp, so $v_p = \alpha V_o$ with $|\alpha| < 1$. Find the new values of dv_b/dt and revise Eq. (12) to reflect this modification.

10.3 – 18   Let Fig. 10.3 – 10a be modified by connecting an additional resistor $R_o = R/\alpha$ with $\alpha < 1$ from the inverting input of the op-amp to a supply voltage $+ V_o$. Find the new values of dv_b/dt and revise Eq. (12) to reflect this modification.

10.3 – 19   Derive Eq. (13).

10.3 – 20   What should be the values of R_A and R_B in Fig. 10.3 – 13a to get $T_2 = 0.5T_1$ and $f = 5$ Hz if $C = 10\ \mu F$?

10.3 – 21   What should be the values of R_A and R_B in Fig. 10.3 – 13a to get $T_2 = 0.8T_1$ and $f = 20$ kHz if $C = 0.01\ \mu F$?

10.3 – 22   Suppose Fig. 10.3 – 13a is modified by adding an ideal diode conducting downward in parallel with R_B. Show that it is now possible to get $T_2 = T_1$, and revise Eq. (14b) for this condition.

10.3 – 23   Let an external voltage v_{cont} be applied to the control terminal in Fig. 10.3 – 13a, so $V_2 = v_{cont}$. Since $V_1 = \frac{1}{2}V_2$ under any conditions, the RESET interval duration T_2 is unchanged. However, the SET interval duration becomes

$$T_1 = (R_A + R_B)C \ln \frac{V_{CC} - \frac{1}{2}v_{cont}}{V_{CC} - v_{cont}}$$

Derive this expression.

Chapter

11 Transistor Amplifiers

A complete amplifier usually contains several different transistor circuits called *stages*. There are input stages designed to accept signals coming from various sources, there are intermediate stages designed to provide most of the amplification, and there are output stages designed to drive various loads. Most of these stages fall in the category of *small-signal amplifiers,* meaning that the signal variations are just a fraction of the power-supply voltage. In contrast, the large-signal excursions of a power amplifier may be comparable to the supply voltage.

This chapter is devoted to the study of small-signal amplifiers, including some design techniques along with the operating principles, small-signal models, and analysis methods. Primary emphasis is given to voltage amplification using n-channel FETs or npn BJTs. We'll also examine the use of negative feedback in amplifier circuits, and we'll investigate amplifier frequency response. The closing section introduces differential amplifiers and takes a look at the circuitry inside an operational amplifier.

Objectives

After studying this chapter and working the exercises, you should be able to do each of the following:

- Draw the small-signal model of a FET or BJT and calculate the element values (Sections 11.1 and 11.2).

- Analyze single-stage and cascade amplifiers with common-source FET circuits or common-emitter BJT circuits (Sections 11.1 and 11.2).

- Design a simple common-source or common-emitter amplifier to meet stated specifications (Sections 11.1 and 11.2).

- State the advantages of negative feedback for transistor amplifiers (Section 11.3).

- Evaluate the small-signal properties of a single-stage feedback amplifier or voltage follower (Section 11.3).

- State the major factors that limit the frequency response of an amplifier, and select coupling capacitors to achieve a given lower cutoff frequency (Section 11.4).

- Estimate the upper cutoff frequency of a simple transistor amplifier (Section 11.4).†

- Identify the properties of a differential amplifier (Section 11.5).†

11.1 SMALL-SIGNAL FET AMPLIFIERS

We saw in Chapter 9 that a FET produces reasonably linear amplification under the restriction of small voltage and current variations within the active region. Now we'll examine FET circuits designed explicitly to function as linear small-signal amplifiers. Besides being building blocks in multistage amplifiers, these circuits are important in their own right for applications such as the interface between a signal source and a measuring or display instrument.

Our investigation begins with enhancement MOSFET amplifiers. We'll introduce the principles of small-signal operation, derive a linearized small-signal model, and analyze the basic common-source amplifier circuit. We'll also discuss biasing arrangements and develop a procedure for bias design. Then we'll briefly consider amplifier circuits built with depletion MOSFETs and JFETs.

Small-Signal Operation

As an introduction to small-signal operation, consider the elementary n-channel enhancement MOSFET circuit in Fig. 11.1–1a. The usual load line is established by R_D and a DC voltage supply V_{DD} connected between the upper terminal and the common ground point. The input signal comes from a time-varying voltage source v_s whose waveform might look like Fig. 11.1–1b, containing both positive and negative excursions.

Recall, however, that an active n-channel enhancement MOSFET requires positive gate voltage $v_{GS} > V_T$. Accordingly, an auxiliary battery has been put in series with the input so that

$$v_{GS} = V_{GS} + v_{gs} \tag{1a}$$

The total gate voltage thus consists of a **DC bias** V_{GS} plus a **signal component** v_{gs}. The signal component equals the external source voltage v_s in this case because the MOSFET draws no input current. The DC bias ensures that v_{GS} satisfies the active condition, despite signal variations. The drain

Figure 11.1−1

MOSFET amplifier: (a) circuit; (b) typical input waveform; (c) load line and Q point; (d) signal waveforms.

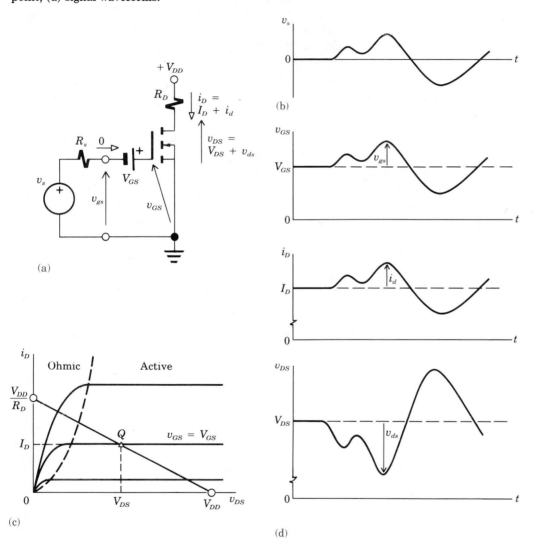

current and voltage are then controlled by v_{GS}, and they have the similar form

$$i_D = I_D + i_d \qquad v_{DS} = V_{DS} + v_{ds} \tag{1b}$$

As in Eq. (1a), we symbolize DC bias components by capital letters with capital subscripts (I_D and V_{DS}), while signal components are symbolized by lower-case letters with lower-case subscripts (i_d and v_{ds}).

The three DC bias components are readily interpreted using the load line drawn on the MOSFET's static characteristics in Fig. 11.1–1c. In absense of signal variations, the DC gate voltage V_{GS} biases the MOSFET at some point Q where $i_D = I_D$ and $v_{DS} = V_{DS}$. The DC bias thereby constitutes a fixed **operating point** in the active region.

When v_{gs} is added to the DC gate voltage, the resulting signal components i_d and v_{ds} appear as variations around the Q point. These variations are illustrated by the waveforms sketched in Fig. 11.1–1d for the case of small input excursions. Under this condition, i_d is essentially proportional to v_{gs}, and v_{ds} is an amplified and inverted version of the input waveform.

Having presented a qualitative picture of our elementary amplifier, we turn to quantitative analysis using the mathematical description of an active MOSFET. Specifically, Eq. (3), Section 9.1, states that if $v_{GS} > V_T$ and $v_{DS} > V_{GS} - V_T$, then

$$i_D = k(v_{GS} - V_T)^2 \qquad k = I_{DSS}/V_T^2 \tag{2}$$

Since Eq. (2) holds with or without signal variations, the DC drain current is related to the gate bias voltage by

$$I_D = k(V_{GS} - V_T)^2 \tag{3}$$

Given V_{GS} and the MOSFET parameters k and V_T, Eq. (3) allows us to calculate I_D without resorting to graphical methods. The corresponding value of V_{DS} is determined via the load-line equation $R_D I_D + V_{DS} = V_{DD}$.

We can also study the signal variations from Eq. (2) by plotting the active transfer curve i_D versus v_{GS} in Fig. 11.1–2. Setting $i_D = I_D + i_d$ and $v_{GS} = V_{GS} + v_{gs}$ then shows how i_d varies with v_{ds}. For small values of $|v_{gs}|$, the curve differs only slightly from a linear approximation over the signal range. Hence, we write

$$i_d = g_m v_{gs} \tag{4}$$

where g_m equals the slope of the tangent line passing through the Q point. Expressing this slope mathematically and using Eqs. (2) and (3) yields

$$g_m \triangleq \left.\frac{di_D}{dv_{GS}}\right|_Q = 2k(V_{GS} - V_T) = 2\sqrt{kI_D} \tag{5}$$

Figure 11.1–2

Active transfer curve with Q point.

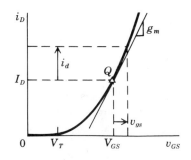

We call g_m the **transconductance** of the MOSFET because it has the same units as conductance. Typical values go from about 0.1 to 10 millisiemens (mS), depending upon the MOSFET parameters and the Q point.

Based on Eq. (4), we now construct the **small-signal model** for a MOSFET diagrammed in Fig. 11.1–3—a linear circuit model depicting the relationships between the signal components. The voltage-controlled current source represents the dependence of i_d on v_{gs}, while the open circuit at the gate terminal reflects the fact that a MOSFET draws no gate current.

Figure 11.1–3

Small-signal model of
n-channel
enhancement
MOSFET.

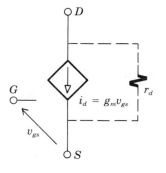

A more refined MOSFET model would also include the **dynamic resistance** r_d, shown by the dashed connection. This resistance accounts for channel-length modulation, which causes i_d to increase slightly with v_{ds}. However, we'll ignore that effect because values of r_d are usually much larger than the parallel external resistors found in most amplifier circuits. Additional modifications to account for the MOSFET's internal capacitance will be discussed in Section 11.4.

As a final point, it should be noted that all of our results here are easily modified for the case of a p-channel enhancement MOSFET. You just reverse the direction of the voltage and current arrows and interchange the voltage subscripts, writing V_{SG} in place of V_{GS}, etc. Figure 11.1–4 shows the corresponding small-signal model.

Figure 11.1–4

Small-signal model of
p-channel
enhancement
MOSFET.

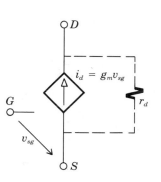

Common-Source Amplifiers

Figure 11.1–5 gives the schematic diagram of an n-channel enhancement MOSFET amplifier. The MOSFET is said to be in the **common-source configuration** because its source terminal is connected to the ground point common to the input and output. Unlike Fig. 11.1–1a, this amplifier does not need an auxiliary biasing battery. Instead, a voltage divider formed by the **bias resistors** R_1 and R_2 provides the DC gate voltage $V_{GS} = R_2 V_{DD}/(R_1 + R_2)$.

Figure 11.1–5
Common-source amplifier with bias resistors and coupling capacitors.

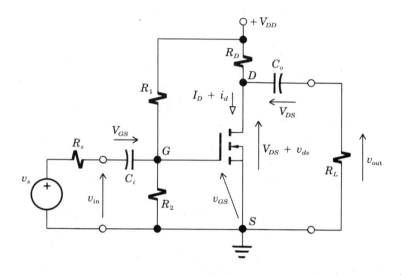

The input signal from v_s is added to v_{GS} through a **coupling capacitor** C_i, which also serves as a DC block. Ideally, C_i is so large that it acts like a short circuit for any time-varying signal and only the DC voltage V_{GS} appears across C_i. Hence, the signal component at the gate is

$$v_{gs} = v_{GS} - V_{GS} = v_{in}$$

Likewise, the output signal is coupled to the load R_L through another large capacitor C_o. This output capacitor blocks the DC component of v_{DS} but passes the time-varying component, so

$$v_{out} = v_{DS} - V_{DS} = v_{ds}$$

Our objective here is to find the small-signal voltage gain v_{out}/v_s.

Amplifier analysis becomes easier with the help of the **small-signal equivalent circuit,** which omits all DC components and focuses attention directly on the signal components. To obtain this circuit from the schematic diagram, you proceed as follows:

- Replace the transistor by its small-signal model.
- Replace the coupling capacitors by short circuits.
- Replace the DC supply voltage by a short circuit.

Short circuits are substituted for the coupling capacitors and supply voltage to represent the fact that zero signal voltage appears across those elements.

Applying these steps to Fig. 11.1–5 yields the small-signal equivalent circuit in Fig. 11.1–6a. Careful inspection now reveals that the signal short from the supply terminal to ground puts R_1 in parallel with R_2 and puts R_D in parallel with R_L. Thus, after redrawing the circuit in the form of Fig. 11.1–6b, we easily identify the equivalent input and output resistances

$$R_i = R_1 \| R_2 \qquad R_o = R_D \tag{6}$$

Norton-to-Thévenin conversion then gives the standard voltage amplifier model in Fig. 11.1–6c, where we've introduced the open-circuit amplification factor

$$\mu \triangleq \frac{v_{\text{out}-\text{oc}}}{v_{\text{in}}} = -g_m R_D \tag{7}$$

The negative value of μ incorporates the inherent voltage inversion.

Figure 11.1–6
Evolution of small-signal equivalent circuit.

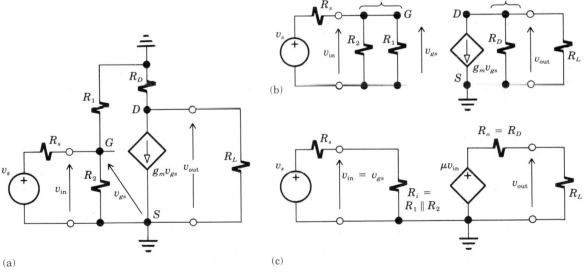

(a)

(b)

(c)

Although the amplifier has finite input resistance, the bias resistors R_1 and R_2 can and should be large enough to minimize loading of the input source. We'll therefore assume that

$$R_1 \| R_2 \gg R_s \qquad v_{\text{in}} \approx v_s$$

However, we cannot ignore output loading, so the **small-signal voltage**

gain is

$$A_v \triangleq \frac{v_{\text{out}}}{v_s} = \frac{v_{\text{in}}}{v_s} \times \frac{v_{\text{out}}}{v_{\text{in}}} \approx \mu \frac{R_L}{R_D + R_L} \tag{8}$$

Alternatively, upon inserting $\mu = -g_m R_D$, we have

$$A_v \approx -\frac{g_m R_D R_L}{R_D + R_L} = -g_m (R_D \| R_L) \tag{9}$$

This expression clearly shows that the voltage gain is proportional to the parallel combination of R_D and R_L. Furthermore, since $g_m = 2\sqrt{kI_D}$, Eq. (9) brings out the fact that A_v depends upon the Q-point current I_D.

When one MOSFET does not provide sufficient gain, you can connect two or more in a **cascade configuration.** As an example, Fig. 11.1–7 gives the schematic diagram and equivalent circuit of a two-stage cascade amplifier. The overall voltage gain is

$$A_v = \frac{v_{\text{in}}}{v_s} \times \frac{v_{12}}{v_{\text{in}}} \times \frac{v_{\text{out}}}{v_{12}} \approx \mu_1 \mu_2 \frac{R_L}{R_{D2} + R_L}$$

Figure 11.1–7
Cascade of two
MOSFET stages:
(a) circuit diagram;
(b) small-signal
equivalent circuit.

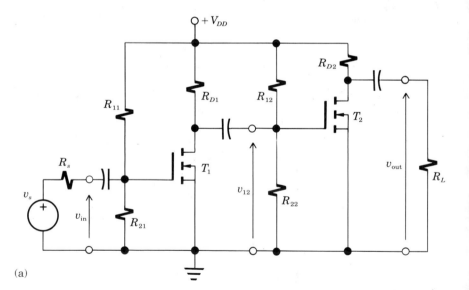

(a)

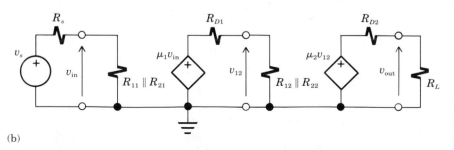

(b)

By extrapolation, a cascade of n identical stages has

$$A_v \approx \mu^n \frac{R_L}{R_D + R_L} \tag{10}$$

provided that the bias resistors are large enough to minimize loading between stages as well as at the input.

In all cases — single-stage or cascade — the gate voltage excursions must satisfy our assumption that $|v_{gs}|$ is small. This condition in turn limits $|v_{out}|_{max}$ to a few volts. Otherwise, the output signal may suffer appreciable nonlinear distortion.

Example 11.1−1

Consider an enhancement MOSFET having the parameter values

$$V_T = 2 \text{ V} \qquad I_{DSS} = 6 \text{ mA} \qquad k = I_{DSS}/V_T^2 = 1.5 \text{ mA/V}^2$$

If it is used in a single-stage amplifier with $I_D = 1.5$ mA and $R_D = R_L = 4$ kΩ, then Eqs. (5) and (9) yield

$$g_m = 2\sqrt{1.5 \text{ mA/V}^2 \times 1.5 \text{ mA}} = 3 \text{ mS}$$

$$A_v \approx -3 \text{ mS} \times (4\|4) \text{ k}\Omega = -6$$

A cascade of two such stages would provide much more gain since $\mu = -3 \text{ mS} \times 4 \text{ k}\Omega = -12$ and Eq. (10) predicts

$$A_v \approx (-12)^2 \frac{4}{4+4} = +72$$

Note that A_v is positive because the two voltage inversions cancel out.

Exercise 11.1−1

Suppose an enhancement MOSFET having $V_T = 4$ V and $I_{DSS} = 10$ mA is operated at $I_D = 2.5$ mA. Calculate A_v for a single-stage amplifier and for a three-stage cascade with $R_D = 4$ kΩ and $R_L = 6$ kΩ.

Exercise 11.1−2

Modify Fig. 11.1−6b to include the MOSFET's dynamic resistance r_d and show that $\mu = -g_m(r_d\|R_D)$.

Biasing and Design Considerations

The biasing circuit back in Fig. 11.1−5 creates a *fixed* DC gate voltage V_{GS} which, in principle, establishes the desired Q point at $I_D = k(V_{GS} - V_T)^2$. In practice, the MOSFET parameters exhibit considerable variation with temperature and from unit to unit. Consequently, the actual values of k and V_T may differ by as much as a factor of 2 from stated nominal values. If V_T happens to be smaller than expected, then I_D will be larger than expected — perhaps so large that the MOSFET goes into its ohmic region, thereby causing serious nonlinear distortion of the amplified signal.

A better arrangement that stabilizes I_D is the **self-biased circuit** in Fig. 11.1−8. Here, we have connected another resistor R_K and a parallel

Figure 11.1–8
Self-biased circuit.

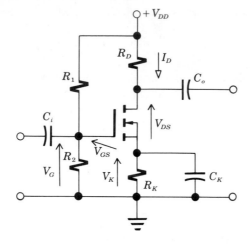

bypass capacitor C_K at the MOSFET's source terminal. Since the bypass capacitor acts as a signal short to ground, R_K does not affect amplification. However, from the DC values labeled on the diagram, the gate bias voltage now becomes

$$V_{GS} = V_G - V_K = \frac{R_2}{R_1 + R_2}\, V_{DD} - R_K I_D \tag{11}$$

A larger than wanted value of I_D therefore causes V_{GS} to decrease which, in turn, decreases I_D. Conversely, a smaller value of I_D causes V_{GS} to increase, which then increases I_D. Either way, the self-biased circuit automatically performs corrective action.

As a graphical demonstration of self bias versus fixed bias, Fig. 11.1–9 plots the active transfer curve for a nominal value of V_T and for another value $V_T' < V_T$. The bias line defined by Eq. (11) intersects the solid curve at the desired operating point Q. If the MOSFET actually has the smaller parameter value V_T', then the operating point shifts only slightly along the

Figure 11.1–9
Q-point shift for self bias and fixed bias.

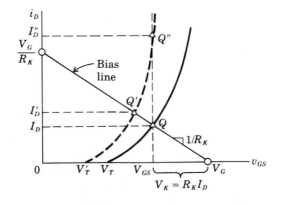

bias line to the intersection with the dashed curve at Q'. The larger the value of R_K, the smaller the change from I_D to I'_D. In contrast, fixed bias holds V_{GS} constant, so the operating point would jump to Q'' on the dashed curve where I''_D is considerably larger than the desired value I_D.

Having seen the importance of stabilized biasing, we turn to the task of bias circuit design. This task involves an iterative "cut-and-try" process to account for the relationship between the bias circuit and the small-signal amplification properties. Thus, in addition to the nominal values of the MOSFET parameters, you need to know the source and load resistances and the desired gain and output swing. If the supply voltage V_{DD} has been specified, and if extreme temperature variations are not likely to occur, then you can use the following simplified design procedure.

Step 1 Determine if a cascade is needed. Since R_D is unknown at the start, initially assume that $R_D \leq R_L$ so $R_D\|R_L \leq R_L/2$. Thus, to get gain A_v with a single-stage amplifier, Eq. (9) requires

$$I_D \geq A_v^2/kR_L^2 \tag{12}$$

A cascade amplifier will be necessary if this current is excessive, so go to step 2b. Otherwise, go to step 2a.

Step 2a Select I_D and R_D for a single-stage amplifier. Choose a value of I_D that satisfies Eq. (12) and evaluate $g_m = 2\sqrt{kI_D}$. Now choose a value of R_D to satisfy Eq. (9) rewritten in the form

$$R_D \approx \frac{|A_v|R_L}{g_m R_L - |A_v|} \tag{13}$$

However, if Eq. (13) requires $R_D > V_{DD}/2I_D$, then it may be desirable to increase I_D and repeat the calculations before going to step 3.

Step 2b Select I_D and R_D for a cascade amplifier. Pick a trial value of I_D, evaluate $g_m = 2\sqrt{kI_D}$, and take

$$R_D \approx V_{DD}/2I_D \tag{14}$$

Then use Eq. (10) to choose I_D and R_D for the desired value of A_v with an appropriate number of identical stages.

Step 3 Select V_{DS} and R_K. Calculate V_{GS} from Eq. (3) rewritten as

$$V_{GS} = V_T + \sqrt{I_D/k} \tag{15}$$

To ensure active-region operation over the desired output signal swing, let

$$V_{DS} \geq V_{GS} - V_T + |v_{\text{out}}|_{\text{max}} \tag{16}$$

To provide reasonable bias stability, let

$$R_K \geq R_D/4 \tag{17}$$

Now choose specific values for V_{DS} and R_K that satisfy Eqs. (16) and (17) together with the DC load-line equation

$$(R_D + R_K)I_D + V_{DS} = V_{DD} \tag{18}$$

A larger supply voltage will be needed if Eqs. (16)–(18) cannot be satisfied with the given value of V_{DD}.

Step 4   Select R_1 and R_2. Since R_1 and R_2 must satisfy the biasing condition in Eq. (11), and since the input resistance is $R_i = R_1 \| R_2$, it follows that

$$R_1 = cR_i \qquad R_2 = \frac{c}{c-1} R_i \tag{19a}$$

where

$$c = \frac{V_{DD}}{V_{GS} + R_K I_D} \tag{19b}$$

Choose values for R_1 and R_2 from Eq. (19) by letting R_i be any convenient value consistent with $R_i \gg R_s$. However, keep the sum $R_1 + R_2$ less than about 10 MΩ so that incidental DC gate current does not appreciably shift the Q point.

Small adjustments should be made throughout the design procedure to conform with readily available components. In this regard, standard values of carbon resistors are listed in a table at the back of the book. Since the actual resistor values may vary by as much as $\pm 10\%$, it does no harm to round-off a calculated value to the closest standard value.

When the value of I_D or R_D has been specified in advance, the appropriate steps of the procedure may be used to determine the other component values. If V_{DD} is not initially specified, then the procedure can still be used by selecting a trial value for V_{DD} and testing it via Eqs. (16)–(18).

Example 11.1–2 A MOSFET amplifier design.	Suppose you want to design a battery-powered amplifier with $V_{DD} = 9$ V that delivers $|v_{out}|_{max} \approx 1$ V across $R_L = 10$ kΩ. The input signal comes from a transducer having $R_s = 5$ kΩ and $|v_s|_{max} \approx 25$ mV, so the amplifier must provide voltage gain $|A_v| \approx (1\ \text{V})/(25\ \text{mV}) = 40$. Available for this purpose are some n-channel enhancement MOSFETs with nominal parameter values $V_T = 2.5$ V and $k = 0.7$ mA/V^2. The data sheet indicates that these MOSFETs are suitable for small-signal operation at $I_D = $ 1-3 mA.

Starting at step 1 of the design procedure, Eq. (12) calls for $I_D \geq$ 25 mA. Since this is clearly excessive, a cascade circuit will be necessary. You therefore go to step 2b and try the minimum current $I_D = 1$ mA, in which case $g_m = 1.67$ mS. Equation (14) then yields $R_D \approx 4.5$ kΩ, but the closest standard resistor values are 4.3 kΩ and 4.7 kΩ. Upon testing these possibilities in Eq. (10), you find that a cascade of two identical stages with $R_D = 4.7$ kΩ has $A_v \approx 42$ and meets your requirements.

Next, going on to step 3, Eqs. (15)–(17) give $V_{GS} = 3.7$ V, $V_{DS} \geq 2.2$ V, and $R_K \geq 1.2$ kΩ, so $(R_D + R_K)I_D + V_{DS} \geq 8.1$ V whereas $V_{DD} = 9$ V. Hence, you select the larger standard value $R_K = 1.5$ kΩ to get better stability, while still allowing $V_{DS} = V_{DD} - (R_D + R_K)I_D = 2.8$ V.

Finally, in step 4, you calculate $R_1 = 1.73\,R_i$ and $R_2 = 2.37\,R_i$ from Eq. (19). Choosing $R_1 = 1.8\,\text{M}\Omega$ corresponds to taking $R_i = R_1/1.73 = 1.04\,\text{m}\Omega \gg R_s$. Then $R_2 = 2.47\,\text{m}\Omega$, which can be rounded to the standard value $2.4\,\text{m}\Omega$.

Exercise 11.1−3   A single-stage amplifier with $R_L = 10\,\text{k}\Omega$ and $V_{DD} = 12\,\text{V}$ is to have $|A_v| \approx 5$, $|v_{\text{out}}|_{\text{max}} \approx 2\,\text{V}$, and $R_i \approx 1\,\text{M}\Omega$. Select standard values for R_D, R_K, R_1, and R_2 using the MOSFET in Example 11.1−2 operated at $I_D = 2\,\text{mA}$. Take R_K to be as large as possible.

Depletion MOSFET and JFET Amplifiers

Most of the foregoing analysis and design relations also apply to small-signal amplifiers built with depletion MOSFETs or JFETs. The only major difference is the DC gate bias voltage.

Consider an n-channel depletion MOSFET or JFET having negative pinchoff voltage $-V_P$. Operation in the active region occurs when $v_{DS} > v_{GS} + V_P$ and $v_{GS} > -V_P$, which produces the drain current

$$i_D = k(v_{GS} + V_P)^2 \quad k = I_{DSS}/V_P^2$$

Figure 11.1−10 shows the corresponding active transfer curve with $i_D = I_D + i_d$ and $v_{GS} = V_{GS} + v_{gs}$. Clearly, we can still write the small-signal approximation $i_d = g_m v_{gs}$, and we can still use the small-signal model from Fig. 11.1−3. Hence, all of our previous amplifier equations are valid here with the transconductance now given by

$$g_m \triangleq \left.\frac{di_D}{dv_{GS}}\right|_Q = 2k(V_{GS} + V_P) = 2\sqrt{kI_D} \qquad \textbf{(20)}$$

However, in the case of a JFET, we have the additional limitations

$$V_{GS} \le 0 \qquad g_m \le 2kV_P$$

Figure 11.1−10
Active transfer curve with Q point for depletion MOSFET or JFET.

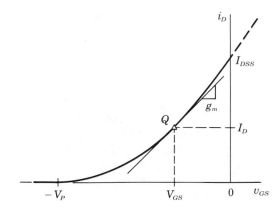

since positive gate voltage would forward-bias the gate-to-channel junction.

A simple method for biasing a JFET is diagrammed in Fig. 11.1–11, where C_K again acts as a signal bypass. The input resistor R_i carries no DC current and merely serves to keep the gate at the ground potential when $v_{in} = 0$, so $V_{GS} = -V_K = -R_K I_D$. For a specified value of I_D, the required DC gate-to-source voltage is

$$V_{GS} = -V_P + \sqrt{I_D/k} \tag{21}$$

which can be obtained by taking

$$R_K = -V_{GS}/I_D \tag{22}$$

Obviously, the same circuit could be used to bias a depletion MOSFET at $V_{GS} < 0$.

Figure 11.1–11
JFET bias circuit.

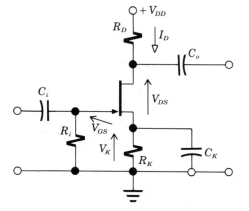

But better bias stability is obtained with the self-bias circuit back in Fig. 11.1–8. Equation (11) confirms that this circuit is also suitable for JFETs or depletion MOSFETs because V_{GS} will be negative when we make R_K large enough that $V_K > V_G$. The design procedure is the same as before except that $-V_P$ replaces V_T in Eqs. (15) and (16), and we must add the stipulation $R_K > -V_{GS}/I_D$ to Eq. (17).

11.2 SMALL-SIGNAL BJT AMPLIFIERS

This section examines amplifiers built with bipolar junction transistors in the common-emitter configuration. We'll develop the small-signal model, analyze the amplification properties, and design a self-bias circuit.

In contrast to voltage-controlled FETs, BJTs are current controlled so they draw input current from the signal source — a property that complicates analysis and design. However, we'll see that BJT amplifiers often provide much more voltage gain than FET amplifiers.

Small-Signal BJT Models

Consider the *npn* BJT in Fig. 11.2–1, which is labeled with four variables: the base current and voltage, and the collector current and voltage. For small-signal amplification, each of these variables consists of a DC bias component plus a signal component. Accordingly, we write

$$i_B = I_B + i_b \qquad v_{BE} = V_{BE} + v_{be}$$
$$i_C = I_C + i_c \qquad v_{CE} = V_{CE} + v_{ce} \tag{1}$$

The task at hand is to develop a model relating the signal components.

Figure 11.2–1
npn BJT with DC and signal components.

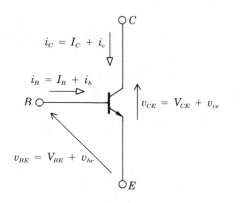

A BJT operating in the active region must be biased such that $v_{CE} > V_{sat}$ and $i_B > 0$, despite signal variations. The total collector current is then

$$i_C = \beta i_B + I_{CEO}$$

where β is the current gain and I_{CEO} is the collector cutoff current. BJT amplifiers usually have $\beta \gg 1$ and $I_C \gg I_{CEO}$. Setting $i_B = I_B + i_b$ and $i_C = I_C + i_c$ immediately yields

$$I_C = \beta I_B + I_{CEO} \approx \beta I_B \tag{2}$$
$$i_c = \beta i_b \tag{3}$$

Hence, both components of i_C are directly proportional to the respective components of i_B. Nonlinearity and the need for linearization enters the picture when the input signal comes from a voltage source. Accordingly, we turn to current-voltage relationship at the base.

With $i_B > 0$, the base-emitter path acts essentially like a forward-biased pn junction, so

$$i_B \approx I_0 e^{\lambda v_{BE}} \qquad \lambda \approx 40 \text{ V}^{-1} \tag{4}$$

as follows from Eqs. (7) and (8), Section 8.1, assuming room temperature. The familiar diode *i-v* curve plotted in Fig. 11.2–2 shows the signal variations i_b and v_{be} relative to the Q-point values I_B and $V_{BE} \approx V_\gamma$. If $|v_{be}|$ is

Figure 11.2–2
Base-emitter i-v curve
with Q point.

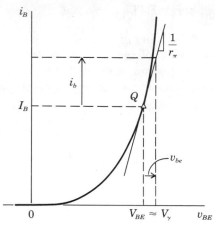

small, then we can use the linear approximation

$$i_b = v_{be}/r_\pi \tag{5}$$

The slope of the tangent line passing through the Q point is $di_B/dv_{BE}|_Q = \lambda I_B$, so

$$r_\pi = \frac{1}{\lambda I_B} \approx \frac{1}{40 I_B} = \frac{\beta}{40 I_C} \tag{6}$$

Physically, we interpret r_π as the **small-signal base resistance.**

Figure 11.2–3 diagrams the small-signal npn BJT model obtained directly from Eqs. (3) and (5). Alternatively, for comparison with our voltage-controlled FET model, we note that $i_c = \beta i_b = \beta(v_{be}/r_\pi)$. Hence, we can write

$$i_c = g_m v_{be}$$

where the equivalent BJT transconductance is defined by

$$g_m \triangleq \beta/r_\pi = 40 I_C \tag{7}$$

The same model could also represent a pnp BJT if all polarity arrows are reversed, as was done back in Fig. 9.3–10.

Figure 11.2–3
Small-signal model for
npn BJT.

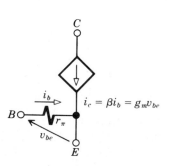

A more elaborate BJT model that includes several second-order effects is shown in Fig. 11.2–4 and labeled with the **common-emitter hybrid parameters.** The parameter h_{ie} corresponds to r_π, while h_{fe} represents the small-signal current gain — which may differ somewhat from the static current gain β. The dynamic collector-to-emitter resistance $1/h_{oe}$ accounts for base-width modulation that causes i_c to increase slightly as v_{ce} increases. The reverse coupling factor h_{re} reflects the fact that v_{be} also increases slightly as v_{ce} increases. But we can usually ignore these small second-order effects and take

$$h_{ie} = r_\pi \qquad h_{re} = 0 \qquad h_{fe} = \beta \qquad 1/h_{oe} = \infty$$

Figure 11.2–4 then reduces to the simplified model in Fig. 11.2–3. The modifications required for high-frequency effects will be discussed in Section 11.4.

Figure 11.2–4
Common-emitter hybrid parameter BJT model.

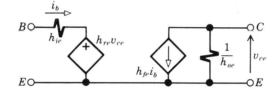

Example 11.2–1

A typical BJT with $\beta = 100$ might be biased at $I_B = 20~\mu A = 0.02$ mA, so $I_C = 2$ mA. Equation (7) then yields

$$g_m = 40~\text{V}^{-1} \times 2~\text{mA} = 80~\text{mS} \qquad r_\pi = \beta/g_m = 1.25~\text{k}\Omega$$

The large value of g_m implies much greater voltage amplification than a FET. However, judging from the small value of r_π, we suspect that the BJT may cause appreciable loading of the input signal source. These observations are developed further in our analysis of common-emitter amplifiers.

Common-Emitter Amplifiers

Figure 11.2–5a diagrams a simple amplifier circuit with an *npn* BJT in the common-emitter configuration. The power supply V_{CC} establishes the DC load line in conjunction with R_C, and it also provides DC base current via R_B to forward-bias the base-emitter junction. The DC Q-point values are thus given by

$$V_{BE} \approx V_\gamma \qquad I_B = (V_{CC} - V_{BE})/R_B$$

$$I_C \approx \beta I_B \qquad V_{CE} = V_{CC} - R_C I_C$$

The large coupling capacitors C_i and C_o serve as DC blocks while passing the time-varying signal components, so

$$v_{be} = v_{BE} - V_{BE} = v_{\text{in}} \qquad v_{\text{out}} = v_{CE} - V_{CE} = v_{ce}$$

We want to find the voltage gain v_{out}/v_s.

Figure 11.2-5
Common-emitter amplifier and small-signal equivalent circuits.

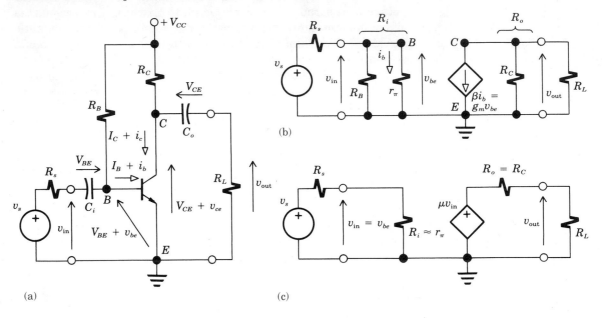

(a)　　　　(b)　　　　(c)

To begin the small-signal analysis, we put our model from Fig. 11.2-3 in place of the transistor, and we substitute short circuits for the coupling capacitors and supply voltage. The resulting small-signal equivalent circuit in Fig. 11.2-5b shows that R_B appears in parallel with r_π at the input. Since most BJT amplifiers have $R_B \gg r_\pi$, the equivalent input resistance becomes

$$R_i = R_B \| r_\pi \approx r_\pi \tag{8}$$

Norton-to-Thévenin conversion now yields the standard voltage amplifier model in Fig. 11.2-5c with $R_o = R_C$ and

$$\mu = -g_m R_C = -\beta R_C / r_\pi \tag{9}$$

which represents the open-circuit voltage amplification factor.

Including loading effects at the input as well as the output, the small-signal voltage gain is found from Fig. 11.2-5c to be

$$A_v = \frac{v_{in}}{v_s} \times \frac{v_{out}}{v_{in}} = \frac{R_i}{R_s + R_i} \mu \frac{R_L}{R_C + R_L} \tag{10}$$

Inserting $R_i \approx r_\pi$ and $\mu = -\beta R_C / r_\pi$ then gives

$$A_v \approx -\frac{r_\pi}{R_s + r_\pi} \frac{\beta}{r_\pi} \frac{R_C R_L}{R_C + R_L} = -\beta \frac{R_C \| R_L}{R_s + r_\pi} \tag{11}$$

Further examination of this expression shows that BJT voltage amplification depends on the allowable amount of source loading.

Suppose, on the one hand, that we can take $r_\pi \ll R_s$ to draw maximum input current. We thereby get the maximum gain, $|A_v|_\mathrm{max} \approx \beta(R_C\|R_L)/R_s$. Furthermore, if $r_\pi \ll R_s$ and $R_B \gg r_\pi$ in Fig. 11.2–5b, then $i_b \approx v_s/R_s$ so the input voltage source acts essentially like a current source driving the base. Since the current relationship $i_c = \beta i_b$ does not involve the small-signal limitation on $|v_{be}|$, linear amplification is possible over a much larger output signal swing. On the other hand, suppose that we must take $r_\pi \gg R_s$ to minimize source loading. Then the gain will be appreciably less than $|A_v|_\mathrm{max}$, and the output will suffer from nonlinear distortion unless $|v_{be}|$ is small.

Finally, consider a cascade built with two or more common-emitter BJT stages. Such amplifiers can be analyzed using the equivalent circuit from Fig. 11.2–5c for each stage. When there are n identical stages, the overall voltage gain becomes

$$A_v \approx \frac{(R_i \mu)^n}{(R_s + R_i)(R_C + R_i)^{n-1}} \frac{R_L}{R_C + R_L} \tag{12}$$

This expression is more complicated than the gain of a FET cascade because we must now include input and interstage loading.

Example 11.2–2

Consider a common-emitter BJT amplifier with $\beta = 100$, $R_s = 0.5$ kΩ, and $R_C\|R_L = 1.5$ kΩ. If the input source tolerates loading by $R_i \approx r_\pi = 0.5$ kΩ, then Eq. (6) shows that we can take $I_C = \beta/40r_\pi = 5$ mA. Correspondingly, from Eq. (11),

$$A_v \approx -100(1.5\ \mathrm{k}\Omega/1\ \mathrm{k}\Omega) = -150$$

However, if we need $r_\pi \geq 2.5$ kΩ to reduce source loading, then we must reduce the Q-point current to $I_C \leq 1$ mA and the gain drops to $|A_v| \leq 50$.

Exercise 11.2–1

Derive an expression for A_v for a cascade of two identical common-emitter stages. Check your result by setting $n = 2$ in Eq. (12). Then show that $A_v \approx \beta^2(R_C\|R_L)/R_s$ when $R_B \gg r_\pi$, $R_s \gg r_\pi$, and $R_C \gg r_\pi$.

Biasing and Design Considerations

The simple BJT biasing arrangement previously given in Fig. 11.2–5 produces a fixed DC base current I_B to establish the DC collector current $I_C = \beta I_B + I_{CEO}$. But β is an unreliable parameter whose actual value may differ appreciably from the stated nominal value, so I_C may turn out to be too small or too large. Additionally, a fixed-bias BJT has the potential threat of **thermal runaway** caused by the temperature dependence of the collector cutoff current I_{CEO}. Although I_{CEO} is usually small enough to be ignored at room temperature, it grows exponentially when the temperature increases. The resulting increased value of I_C produces more power dissipation in the transistor, which raises the temperature further and causes I_C to increase even more until, eventually, the transistor may be destroyed by overheating.

The self-bias circuit in Fig. 11.2–6a includes an emitter resistor R_E that stabilizes I_C and prevents thermal runaway. This biasing method should therefore be employed for most BJT amplifiers. The parallel bypass capacitor C_E acts as a signal short, so the small-signal equivalent circuit will be identical to Fig. 11.2–5b with $R_B = R_1 \| R_2$. The stabilizing properties and resulting value of I_C can be found from the DC equivalent base circuit in Fig. 11.2–6b.

Figure 11.2–6
Self-biased circuit and DC equivalent base circuit.

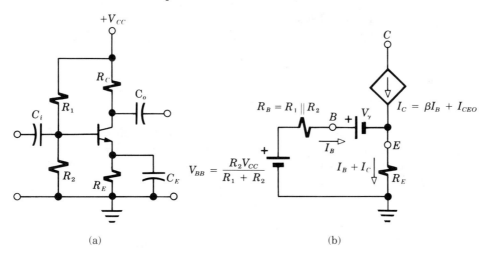

(a)   (b)

Here, we have replaced the BJT with its large-signal model. We have also formed the Thévenin equivalent biasing source by looking back from the base to obtain the open-circuit voltage and equivalent resistance

$$V_{BB} = \frac{R_2}{R_1 + R_2} V_{CC} \qquad R_B = R_1 \| R_2$$

Setting $I_B = (I_C - I_{CEO})/\beta$ and solving for I_C yields

$$I_C = \frac{\beta(V_{BB} - V_\gamma)}{\beta R_E + R_E + R_B} + \frac{R_E + R_B}{\beta R_E + R_E + R_B} I_{CEO} \qquad (13)$$

Effective stabilization requires

$$\beta R_E \gg R_E + R_B$$

so that

$$I_C \approx \frac{V_{BB} - V_\gamma}{R_E} + \frac{R_E + R_B}{\beta R_E} I_{CEO}$$

Thus, when I_{CEO} is negligible, self-biasing results in $I_C \approx (V_{BB} - V_\gamma)/R_E$ which is independent of β as desired. Furthermore, if I_{CEO} starts to increase, then its contribution to I_C is reduced by the factor $(R_E + R_B)/\beta R_E \ll 1$— which stops thermal runaway.

Equation (13) contains some of the information needed to design the biasing circuit for a common-emitter BJT amplifier. However, as was true of FET amplifiers, bias design also involves the relationship between the bias circuit and the small-signal amplification properties. The following procedure facilitates design, given the supply voltage V_{CC}. It is assumed that the BJT to be used has $\beta \gg 1$ and $I_{CEO} \ll I_C$, and that extreme temperature variations are not likely.

Step 1 Determine if a cascade is needed. Since R_C is unknown at the start, initially assume that $R_C \le R_L$ so $R_C \| R_L \le R_L/2$. Thus, to get gain A_v with a single-stage amplifier, Eq. (11) requires

$$I_C \ge \frac{0.05\beta|A_v|}{\beta R_L - 2|A_v|R_s} \tag{14}$$

A cascade amplifier will be necessary if this current is excessive or if $\beta R_L - 2|A_v|R_s < 0$, so go to step 2b. Otherwise, go to step 2a.

Step 2a Select I_C and R_C for a single-stage amplifier. Choose a value of I_C that satisfies Eq. (14) and evaluate $r_\pi = \beta/40I_C$. Now choose a value of R_C to satisfy Eq. (11) rewritten in the form

$$R_C \approx \frac{|A_v|(r_\pi + R_s)R_L}{\beta R_L - |A_v|(r_\pi + R_s)} \tag{15}$$

However, if Eq. (15) requires $R_C > V_{CC}/2I_C$, then it may be desirable to increase I_C and repeat the calculations before going to step 3.

Step 2b Select I_C and R_C for a cascade amplifier. Pick a trial value of I_C, evaluate $r_\pi = \beta/40I_C$, and take

$$R_C \approx V_{CC}/2I_C \tag{16}$$

Then use Eq. (12) with $R_i \approx r_\pi$ and $R_i\mu \approx -\beta R_C$ to choose I_C and R_C for the desired value of A_v with an appropriate number of identical stages.

Step 3 Select V_{CE} and R_E. To ensure active-region operation over the desired output signal swing, let

$$V_{CE} \ge V_\gamma + |v_{\text{out}}|_{\text{max}} \tag{17}$$

To provide reasonable bias stability, let

$$R_E \ge R_C/4 \tag{18}$$

Choose specific values for V_{CE} and R_E that satisfy Eqs. (17) and (18) together with the DC load-line equation

$$(R_C + R_E)I_C + V_{CE} = V_{CC} \tag{19}$$

A larger supply voltage will be needed if Eqs. (17)–(19) cannot be satisfied with the given value of V_{CC}.

Step 4 Select R_1 and R_2. To satisfy the stability requirement $R_B \ll (\beta - 1)R_E$, let

$$R_B \approx \beta R_E/10 \tag{20}$$

Then calculate the base bias voltage

$$V_{BB} = (R_E + R_B/\beta)I_C + V_\gamma \tag{21}$$

Now choose values for R_2 and R_1 such that

$$R_1 = cR_B \qquad R_2 = \frac{c}{c-1}R_B \tag{22a}$$

where

$$c = V_{CC}/V_{BB} \tag{22b}$$

If the value of R_B in Eq. (20) does not satisfy the assumption $R_B \gg r_\pi$, then the resulting gain should be checked using Eq. (11) or (12) with $R_i = R_B \| r_\pi$.

This procedure is easily adapted to situations where I_C or R_C has been specified, or when V_{CC} has not been specified in advance. In any case, calculated resistor values may be rounded-off to standard values.

Example 11.2–3
A BJT amplifier design.

In Example 11.1–2 we found that a MOSFET amplifier powered by a 9-V battery had to have two stages to produce $|A_v| \approx 40$ and $|v_{out}|_{max} \approx 1$ V when $R_s = 5$ kΩ and $R_L = 10$ kΩ. Now we'll rework that design problem using a BJT with $\beta = 120$ intended for small-signal operation at $I_C = 1$–3 mA.

Step 1 of the design procedure calls for $I_C \geq 0.3$ mA, per Eq. (14), so we can get by here with a single-stage amplifier. If we try $I_C = 1$ mA in step 2a, then $r_\pi = 3$ kΩ and Eq. (15) yields $R_C \approx 3.64$ kΩ. The closest standard values are 3.6 kΩ and 3.9 kΩ, both being less than $V_{CC}/2I_C = 4.5$ kΩ. We'll choose $R_C = 3.9$ kΩ to ensure sufficient gain.

Equations (17) and (18) in step 3 require $V_{CE} \geq 1.7$ V and $R_E \geq 0.98$ kΩ, so $(R_c + R_E)I_C + V_{CE} \geq 6.58$ V whereas $V_{CC} = 9$ V. Taking $R_E = 3$ kΩ should therefore provide very good stability while allowing $V_{CE} = V_{CC} - (R_C + R_E)I_D = 2.1$ V.

Proceeding to step 4, we calculate $R_B = 36$ kΩ and $V_{BB} = 4.0$ V from Eqs. (20) and (21). Equation (22) then gives $R_1 = 81$ kΩ and $R_2 = 64.8$ kΩ, which round off to standard values of 82 kΩ and 68 kΩ, respectively. Finally, to check the design, we insert $R_i = R_B \| r_\pi = 2.77$ kΩ and $\mu = -\beta R_C/r_\pi = -156$ back into Eq. (10) which yields $A_v \approx -40$.

Exercise 11.2−2 Use Eq. (13) to find the maximum and minimum values of I_C for the amplifier in Example 11.2−2 when β varies from 90 to 180. Assume that I_{CEO} is negligible.

Exercise 11.2−3 A single-stage amplifier with $R_s = 2\ \text{k}\Omega$, $R_L = 10\ \text{k}\Omega$, and $V_{CC} = 12$ V is to have $|A_v| \approx 100$ and $|v_{out}|_{max} \approx 1.5$ V. Select standard values for R_C, R_E, R_1, and R_2 using a BJT with $\beta = 200$ operated at $I_C = 2.5$ mA. Take R_E to be as large as possible.

11.3 AMPLIFIERS WITH FEEDBACK

The simple FET and BJT amplifiers covered so far have several potential liabilities that can be overcome with the help of negative feedback. Consequently, almost all practical amplifier circuits include some form of negative feedback.

This section starts with the benefits of negative feedback and the price that must be paid for those benefits. Then we'll show how feedback is incorporated in single-stage transistor amplifiers and voltage followers, and how such stages improve the performance of multistage amplifiers.

Negative Feedback

Conceptually, a voltage amplifier with negative feedback can be represented by the generalized block diagram in Fig. 11.3−1. Here, a transistor amplifying unit produces the output signal v_{out}, and a feedback unit connected at the output delivers a proportional signal v_f back to the input. Subtracting v_f from the external input v_{in} then forms the difference signal v_d applied to the amplifying unit. Feedback arranged in this manner creates a self-correcting amplification system that compensates for various shortcomings of the amplifying unit. Specific advantages may include:

- Less sensitivity to transistor parameter variations.
- Improved linearity of the output signal.

Figure 11.3−1
Amplifier with negative feedback.

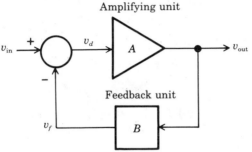

- Reduction of input or output loading effects.
- Better low-frequency and high-frequency response.

The first two of these advantages will be demonstrated here. The others are explored subsequently.

To carry out the analysis of Fig. 11.3–1, we start with the output signal written as

$$v_{\text{out}} = Av_d \tag{1}$$

where A represents the gain of the amplifying unit by itself. Next, we define the feedback factor B such that

$$v_f = Bv_{\text{out}} \tag{2}$$

Thus, the input to the amplifying unit is

$$v_d = v_{\text{in}} - v_f = v_{\text{in}} - Bv_{\text{out}} \tag{3}$$

which produces the output

$$v_{\text{out}} = A(v_{\text{in}} - Bv_{\text{out}})$$

Rewriting this expression as $(1 + AB)v_{\text{out}} = Av_{\text{in}}$ now yields the gain with feedback

$$A_f \triangleq \frac{v_{\text{out}}}{v_{\text{in}}} = \frac{A}{1 + AB} \tag{4}$$

We also note from Eq. (1) that $v_d = v_{\text{out}}/A$ so

$$\frac{v_d}{v_{\text{in}}} = \frac{1}{1 + AB} \tag{5}$$

a relationship that will be helpful subsequently.

The product AB appearing in Eqs. (4) and (5) is known as the **loop gain** because $v_f = B(Av_d) = (AB)v_d$. In order to obtain *negative* feedback, the loop gain must be a *positive* quantity. But with $AB > 0$ in Eq. (4), it follows that $|A_f| < |A|$. Hence, a negative-feedback amplifier always has less gain than the amplifying unit. Electronic engineers willingly sacrifice gain in exchange for the benefits of negative feedback.

Those benefits become most pronounced when the amplifying unit has large gain, large enough that the loop gain satisfies the condition

$$AB \gg 1$$

Then $1 + AB \approx AB$ and Eq. (4) simplifies to

$$A_f \approx 1/B$$

We still get useful amplification in this case, provided that the feedback unit has $|B| < 1$ so $|A_f| \approx 1/|B| > 1$. And since A_f remains essentially constant as long as $AB \gg 1$, feedback greatly reduces the effect of transistor parameter variations in the amplifying unit.

Feedback also reduces the effect of nonlinear distortion in the amplifying unit. However, a general analysis of this property is difficult because the linear relation $v_{out} = Av_d$ no longer holds when the amplifying unit introduces significant nonlinearity. Graphical or numerical methods must therefore be employed to study individual cases, as illustrated in the following example.

Example 11.3–1   Suppose a transistor amplifying unit has the nonlinear input-output relationship

$$v_{out} = 50(v_d + 3v_d^2)$$

The transfer characteristic in Fig. 11.3–2a shows that $v_{out} \approx 50v_d$ for $|v_d| \leq 0.02$ V, which limits the undistorted output to $|v_{out}|_{max} \approx 1$ V. But larger output excursions without excessive distortion are possible if we add a negative feedback loop so that $v_d = v_{in} - Bv_{out}$.

Figure 11.3–2

(a) Nonlinear transfer characteristic. (b) Improved linearity with feedback.

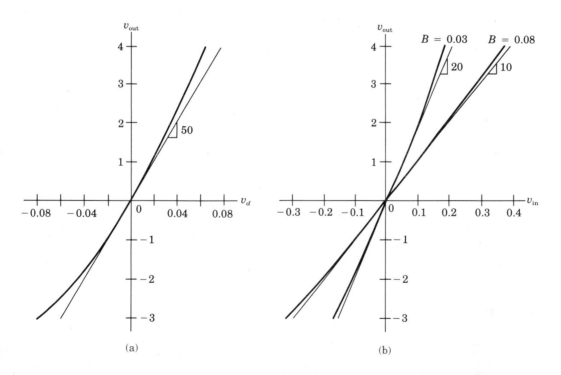

(a)   (b)

We find the resulting characteristic of the feedback amplifier by working backward from specific values of v_d and v_{out} to calculate the corresponding value of $v_{in} = v_d + Bv_{out}$. Figure 11.3–2b gives the curves obtained taking $B = 0.03$ and 0.08. These curves clearly demonstrate that negative feedback reduces nonlinear distortion, for we now have a nearly linear output up to $|v_{out}|_{max} \approx 2$ V with $B = 0.03$ or $|v_{out}|_{max} \approx 3$ V with $B = 0.08$.

Of course, we must pay the price of reduced gain to get the improved linearity. Indeed, since $v_{out}/v_d \approx 50$ for small signals, Eq. (4) predicts that the gain with feedback will be

$$A_f \approx \frac{50}{1 + 50B} = \begin{cases} 20 & B = 0.03 \\ 10 & B = 0.08 \end{cases}$$

The slopes in Fig. 11.3–2b agree with these predicted gains. If necessary, a small-signal preamplifier can be connected at the input of the feedback amplifier to make up for the reduced gain.

Exercise 11.3–1   Find the feedback factor B needed to get $A_f = -20$ in Fig. 11.3–1 when $A = -100$. Then find the resulting range of A_f when transistor parameter variations cause A to fluctuate between -50 and -200.

Feedback Amplifier Circuits

The simplest way to incorporate negative feedback in a transistor amplifier is to use a self-biased circuit with an unbypassed resistor at the emitter terminal of a BJT or the source terminal of a FET. The properties of these single-stage feedback amplifiers will be analyzed here.

Consider first the BJT feedback amplifier diagrammed in Fig. 11.3–3a. The total emitter resistance for DC biasing is $R_{E1} + R_{E2}$, but capacitor C_E bypasses only R_{E2}. The unbypassed resistance R_{E1} establishes a feed-

Figure 11.3–3
BJT feedback amplifier and equivalent circuit.

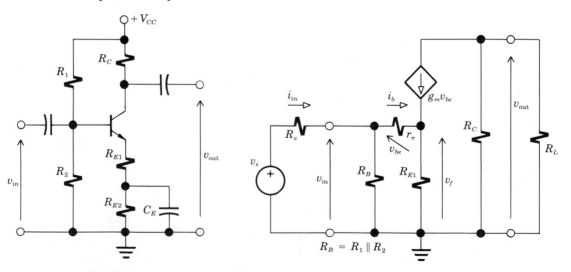

(a)   (b)

back signal voltage v_f proportional to v_{out}. Figure 11.3–3b shows the resulting small-signal equivalent circuit, including the source and load. Since v_f appears across R_{E1}, the BJT is controlled by the difference voltage

$$v_{be} = v_{in} - v_f \qquad (6)$$

and amplification produces the output voltage

$$v_{out} = -g_m(R_C\|R_L)v_{be} \qquad (7)$$

To confirm that v_f is proportional to v_{out}, we write $i_b = v_{be}/r_\pi = g_m v_{be}/\beta$ so

$$v_f = R_{E1}(g_m v_{be} + i_b) = g_m R_{E1}(1 + 1/\beta)v_{be}$$

Substituting $v_{be} = -v_{out}/g_m(R_C\|R_L)$ from Eq. (7) then gives

$$v_f = -\frac{R_{E1}(1 + 1/\beta)}{R_C\|R_L} v_{out} \approx -\frac{R_{E1}}{R_C\|R_L} v_{out} \qquad (8)$$

where the approximation $1 + 1/\beta \approx 1$ corresponds to $\beta \gg 1$.

Comparing Eqs. (6)–(8) with Eqs. (1)–(3), we see that the BJT feedback amplifier can be represented by our block diagram back in Fig. 11.3–1 with

$$v_d = v_{be} \qquad A = -g_m(R_C\|R_L) \qquad B \approx -R_{E1}/(R_C\|R_L)$$

Hence, the loop gain is

$$AB \approx g_m R_{E1}$$

Now the open-circuit voltage amplification factor is easily obtained by letting $R_L \to \infty$, so $v_{out} = v_{out-oc}$ and $A = -g_m R_C$. It then follows from Eq. (4) that

$$\mu = \frac{v_{out-oc}}{v_{in}} \approx \frac{-g_m R_C}{1 + g_m R_{E1}} = \frac{-R_C}{R_{E1} + 1/g_m} \qquad (9a)$$

But BJTs usually have large enough transconductance that $R_{E1} \gg 1/g_m$ and

$$\mu \approx -R_C/R_{E1} \qquad (9b)$$

Since μ depends only on the ratio of two resistors in this case, the gain with feedback is easily adjusted and insensitive to variations of β. Additionally, judging from Example 11.1–1, we conclude that feedback also improves the linearity of the amplified signal.

Returning to Fig. 11.3–3b, the equivalent output resistance is seen to be $R_o = R_C$—the same as without feedback. However, feedback does modify the equivalent input resistance $R_i = v_{in}/i_{in}$. We calculate R_i by noting that $i_{in} = (v_{in}/R_B) + (v_{be}/r_\pi)$, and we now use Eq. (5) to write $v_{be} = v_{in}/(1 + AB)$. Thus,

$$\frac{i_{in}}{v_{in}} \approx \frac{1}{R_B} + \frac{1}{r_\pi(1 + g_m R_{E1})} = \frac{1}{R_B} + \frac{1}{\beta R_{E1} + r_\pi}$$

from which we get

$$R_i \approx R_B \| (\beta R_{E1} + r_\pi) \tag{10}$$

as compared to $R_i = R_B \| r_\pi$ without feedback. Clearly, feedback increases the input resistance and may thereby minimize source loading.

When $R_i \gg R_s$, $\beta \gg 1$, and $g_m R_{E1} \gg 1$, the overall voltage gain of a BJT feedback amplifier becomes

$$A_v = \frac{v_{\text{out}}}{v_s} \approx \mu \frac{R_L}{R_C + R_L} = -\frac{R_C \| R_L}{R_{E1}} \tag{11}$$

In contrast, a conventional common-emitter amplifier with $r_\pi \gg R_s$ would have $A_v \approx -\beta (R_C \| R_L)/r_\pi = -g_m(R_C \| R_L)$. Thus, feedback stabilizes the gain but decreases it by the factor $g_m R_{E1} \gg 1$.

Next, consider the FET feedback amplifier diagrammed in Fig. 11.3–4a. Although an enhancement MOSFET is shown here, the small-signal equivalent circuit in Fig. 11.3–4b could equally well represent a depletion MOSFET or JFET amplifier. Furthermore, Fig. 11.3–4b has the same structure as the BJT circuit in Fig. 11.3–3b when we let $v_{be} = v_{gs}$, $r_\pi \rightarrow \infty$, $R_{E1} = R_{K1}$, and $R_C = R_D$. Making these replacements in Eq. (9a) immediately yields

$$\mu = \frac{-g_m R_D}{1 + g_m R_{K1}} = \frac{-R_D}{R_{K1} + 1/g_m} \tag{12}$$

Figure 11.3–4
FET feedback amplifier and equivalent circuit.

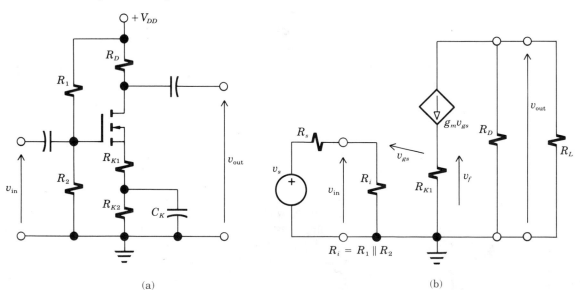

(a)   (b)

The loop gain is $AB = g_m R_{K1}$, but it does not necessarily satisfy the condition $AB \gg 1$ because FETs generally have smaller transconductance than BJTs.

The output resistance in Fig. 11.3–4b is still $R_o = R_D$ and the input resistance remains unchanged at $R_i = R_1 \| R_2$. Since a FET amplifier usually has $R_1 \| R_2 \gg R_s$, the overall voltage gain with feedback will be

$$A_v \approx \mu \frac{R_L}{R_D + R_L} = -\frac{R_D \| R_L}{R_{K1} + 1/g_m} \tag{13}$$

The primary benefits of feedback here are reduced nonlinear distortion and less sensitivity to FET parameter variations.

Example 11.3–2

Suppose you need a BJT feedback amplifier to get $A_v \approx -10$ and $R_i \geq 10$ kΩ when $R_s = 1$ kΩ and $R_L = 5$ kΩ. You have available a BJT with $\beta = 120$ which you decide to operate at $I_C = 1.5$ mA, so $g_m = 60$ ms and $r_\pi = 2$ kΩ.

If you initially try $R_C = R_L$, then you can determine R_{E1} via Eq. (11) written as

$$R_{E1} \approx -(R_C \| R_L)/A_v \approx (2.5 \text{ k}\Omega)/10 = 0.25 \text{ k}\Omega$$

The assumed conditions are satisfied since $\beta \gg 1$, $g_m R_{E1} = 15 \gg 1$, and $R_i \gg R_s$ if $R_i \geq 10$ kΩ. But good bias stability calls for $R_{E1} + R_{E2} \geq R_C/4 = 1.25$ kΩ. Hence, you might include the bypassed emitter resistance $R_{E2} = 1$ kΩ and take $R_B \approx \beta(R_{E1} + R_{E2})/10 = 15$ kΩ. Then, from Eq. (10),

$$R_i \approx (15 \text{ k}\Omega) \| (120 \times 0.25 \text{ k}\Omega) = 15 \| 32 \approx 10.2 \text{ k}\Omega$$

so $R_i \geq 10$ kΩ as desired. To obtain a larger input resistance, you would need to increase R_{E2} and R_B.

Exercise 11.3–2

Obtain expressions for $A = v_{\text{out}}/v_{gs}$ and $B = v_f/v_{\text{out}}$ in Fig. 11.3–4b, and draw a block diagram representing the FET feedback amplifier. Then let $R_L \to \infty$ to derive Eq. (12).

Voltage Follower Circuits

The relatively high output resistance of most transistor amplifiers often causes excessive loading when they must drive low-resistance loads. An effective cure for the problem of output loading is the modified feedback arrangement known as a **voltage follower.** We'll show here that followers have the desirable property of small output resistance, but they do not provide any voltage gain. Consequently, followers are used exclusively for interfacing purposes, either by themselves or at the output of multistage amplifiers.

Figure 11.3–5a gives the schematic diagram of the BJT feedback circuit called an **emitter follower.** Observe that the output is connected to

Figure 11.3–5
Emitter follower and equivalent circuit.

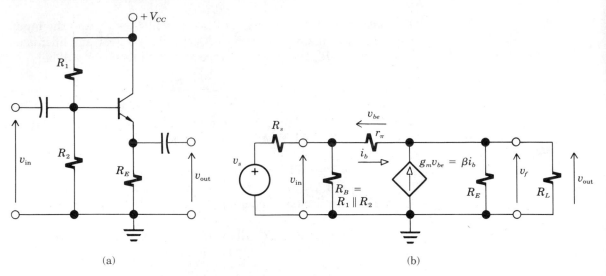

(a)   (b)

the emitter terminal, and that the collector resistor is absent. The small-signal equivalent circuit in Fig. 11.3–5b therefore has the collector terminal at ground, and we say that the BJT is in the **common-collector configuration.**

Putting R_L in parallel with R_E forces $v_f = v_{out}$ and $v_{be} = v_{in} - v_{out}$. This means that the emitter follower has unity feedback factor, $B = 1$. But

$$v_{out} = (R_E\|R_L)(g_m v_{be} + i_b) = g_m(R_E\|R_L)(1 + 1/\beta)v_{be}$$

so the loop gain is

$$AB = A = g_m(R_E\|R_L)(1 + 1/\beta)$$

Setting $R_L = \infty$ and again using Eq. (4), we obtain the open-circuit amplification factor

$$\mu = \frac{g_m R_E(1 + 1/\beta)}{1 + g_m R_E(1 + 1/\beta)} = \frac{1 + 1/\beta}{1 + 1/\beta + 1/g_m R_E} \tag{14}$$

If $g_m R_E \gg 1$, then $\mu \approx 1$ and $v_{out-oc} = \mu v_{in} \approx v_{in}$. Thus, the output voltage "follows" the input, without polarity inversion or voltage gain. By giving up voltage gain, the follower circuit achieves both high input resistance and low output resistance.

The input resistance is found by the same method used for the BJT feedback amplifier, the result being

$$R_i \approx R_B\|[\beta(R_E\|R_L) + r_\pi] \tag{15}$$

Note that R_i depends in part upon the load resistance because R_L is in parallel with R_E. The value of R_i will usually be much greater than r_π, but less than the input resistance of a standard BJT feedback amplifier.

The output resistance can be found via the general definition $R_o = v_{\text{out}-\text{oc}}/i_{\text{out}-\text{sc}}$, expressing $v_{\text{out}-\text{oc}}$ and $i_{\text{out}-\text{sc}}$ in terms of the source voltage v_s. To expedite this task, we redraw the small-signal circuit as shown in Fig. 11.3–6, where everything to the left of the controlled source has been replaced by the Thévenin equivalent voltage and series resistance

$$v'_s = \frac{R_B}{R_B + R_s} v_s \qquad R'_s = R_s \| R_B + r_\pi$$

Routine analysis with $R_L = 0$ and $R_L = \infty$ then yields

$$i_{\text{out}-\text{sc}} = \frac{(\beta + 1)v'_s}{R'_s} \qquad v_{\text{out}-\text{oc}} = \frac{R_E(\beta + 1)v'_s}{(\beta + 1)R_E + R'_s}$$

Thus,

$$\frac{i_{\text{out}-\text{sc}}}{v_{\text{out}-\text{oc}}} = \frac{(\beta + 1)R_E + R'_s}{R_E R'_s} \approx \frac{\beta}{R'_s} + \frac{1}{R_E}$$

and we can write

$$R_o \approx \left(\frac{R'_s}{\beta}\right) \Bigg\| R_E = \left(\frac{R_s \| R_B + r_\pi}{\beta}\right) \Bigg\| R_E \qquad \textbf{(16a)}$$

This rather formidable expression simplifies considerably in the usual case where $(R_s \| R_B + r_\pi)/\beta \ll R_E$, so

$$R_o \approx (R_s \| R_B + r_\pi)/\beta \qquad \textbf{(16b)}$$

Typically, R_o will be less than 100 Ω, in contrast to the several kilohms output resistance when $R_o = R_C$.

Figure 11.3–6
Reduced equivalent circuit of the emitter follower.

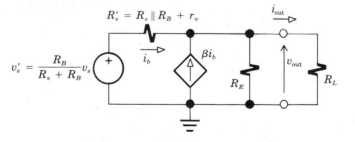

Figure 11.3–7 diagrams the FET version of a follower. This circuit is called a **source follower**—the output being connected to the FET's source terminal—and the FET is in the **common-drain configuration**. The resulting open-circuit amplification factor and the output resistance

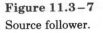

Figure 11.3–7
Source follower.

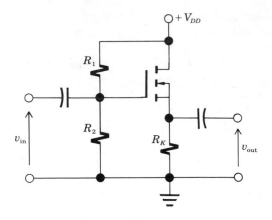

are easily found to be

$$\mu = \frac{g_m R_K}{1 + g_m R_K} = \frac{1}{1 + 1/g_m R_K} \qquad R_o = \left(\frac{1}{g_m}\right)\left\|R_K = \frac{R_K}{1 + g_m R_K}\right. \qquad (17)$$

The input resistance is still $R_i = R_1\|R_2$.

Exercise 11.3–3   Consider a source follower with $g_m = 4.5$ mS, $R_K = 2$ kΩ, and $R_i = 500$ kΩ. Find v_{out}/v_s when $R_L = 1$ kΩ and $R_s = 25$ kΩ. For comparison purposes, also calculate v_{out}/v_s when the follower is omitted.

Multistage Amplifiers

In contrast to a cascade of identical stages, a multistage amplifier contains different types of stages. These stages are arranged in three distinct functional sections, as follows:

- The input stage presents an appropriate input resistance to the external source.

- The middle section consists of one or more stages that provide most of the gain.

- The output stage serves as the interface with the load.

Usually, the input stage must have a high input resistance, whereas the output stage must have a low output resistance.

To analyze a multistage circuit, you can represent each individual stage as a voltage amplifier with input resistance R_i, open-circuit amplification factor μ, and output resistance R_o. As a convenient reference, Table 11.3–1 lists the relevant equations for the FET and BJT stages discussed so far. However, care must be exercised in the case of an emitter follower, since its input resistance depends on the attached load and its output resistance depends upon the equivalent source resistance of the previous

Table 11.3–1

Amplifier stage equations.

Configuration	R_i	μ	R_o	
FET common source (Figs. 11.1–5, 11.1–8)	$R_1 \| R_2$	$-g_m R_D$	R_D	$g_m = 2\sqrt{kI_D}$
FET feedback circuit (Fig. 11.3–5a)	$R_1 \| R_2$	$\dfrac{-R_D}{R_{K1} + 1/g_m}$	R_D	
FET source follower (Fig. 11.3–7)	$R_1 \| R_2$	$\dfrac{1}{1 + 1/g_m R_K}$	$\dfrac{1}{g_m} \bigg\| R_K$	
BJT common emitter (Figs. 11.2–5a, 11.2–6a)	$R_B \| r_\pi$	$-g_m R_C$	R_C	$g_m = \dfrac{\beta}{r_\pi} \approx 40 I_C$
BJT feedback circuit (Fig. 11.3–4a)	$R_B \| (\beta R_{E1} + r_\pi)$	$\dfrac{-R_C}{R_{E1} + 1/g_m}$	R_C	
BJT emitter follower (Fig. 11.3–6a)	$R_B \| [\beta(R_E \| R_L) + r_\pi]$	$\dfrac{1 + 1/\beta}{1 + 1/\beta + 1/g_m R_E}$	$\left(\dfrac{R_s \| R_B + r_\pi}{\beta}\right) \bigg\| R_E$	

stage. After calculating R_i, μ, and R_o for each stage, a simple chain equation yields the overall amplification.

Example 11.3–3

A three-stage amplifier.

Figure 11.3–8 shows the schematic diagram of a 3-stage amplifier, all resistance values being given in kilohms. The amplifier is designed to deliver $|v_{\text{out}}|_{\max} \approx 3$ V across $R_L = 0.6$ k$\Omega = 600$ Ω, with gain adjustable up to $A_v \approx 200$. Gain control is implemented by an input potentiometer whose total resistance is very large compared to the 10-kΩ resistance of the source.

A FET feedback circuit was chosen for the first stage to get high input resistance and modest but stable voltage gain. The FET has $k = 1$ mA/V^2 and is biased at $I_D = 1$ mA, so $g_m = 2$ mS. Thus, with $R_D = 10$ kΩ and $R_{K1} = 1.5$ kΩ, the amplification factor is $\mu_1 = -10/(1.5 + 1/2) = -5$. The input and output resistances are $R_{i1} = 4300\|5100 = 2300$ kΩ and $R_{o1} = 10$ kΩ.

The rest of the gain is provided by the second stage, a BJT common-emitter circuit having $I_C = 2$ mA, $\beta = 200$, $g_m = 80$ mS, $r_\pi = 2.5$ kΩ, and $R_B = 51\|120 = 35.8$ kΩ. Hence, $R_{i2} = 35.8\|2.5 = 2.34$ kΩ, $\mu_2 = -80 \times 3 = -240$, and $R_{o2} = 3$ kΩ. Although this stage does not include feedback, nonlinear distortion will be minimal because r_π is considerably smaller than R_{o1}. The first stage therefore acts like a current source driving the base of the second stage.

The emitter follower in the third stage has the low output resistance needed to interface with the load. The BJT operates at $I_C = 7.2$ mA and is biased by DC base current coming through the collector resistor in the second stage — a direct-coupled arrangement that eliminates the coupling

Figure 11.3–8
Three-stage amplifier for Example 11.3–3 (all resistances in kilohms).

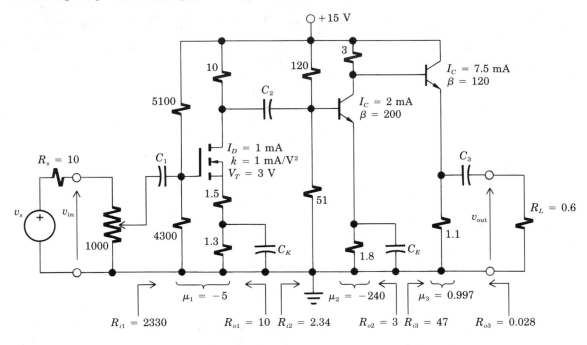

capacitor and bias resistors. Consequently, $R_B = \infty$ in the small-signal equivalent circuit. Since $\beta = 120$, $g_m = 300$ mS, and $r_\pi = 0.4$ kΩ, and since R_{o2} is the equivalent source resistance, the stage equations yield

$$R_{i3} = \infty \| [120(1.1\|0.6) + 0.4] = 47 \text{ k}\Omega$$

$$\mu_3 = \frac{1 + 1/120}{1 + 1/120 + 1/(300 \times 1.1)} = 0.997$$

$$R_{o3} = \left(\frac{3\|\infty + 0.4}{120}\right) \Big\| 1.1 = 0.028 \text{ k}\Omega = 28 \text{ }\Omega$$

These values are indicated at the bottom of Fig. 11.3–8 along with those for the other stages.

Now we can calculate the maximum overall voltage gain, obtained with the potentiometer wiper at the upper end. Specifically,

$$A_v = \frac{R_{i1}\|1000}{R_s + R_{i1}\|1000} \mu_1 \frac{R_{i2}}{R_{o1} + R_{i2}} \mu_2 \frac{R_{i3}}{R_{o2} + R_{i3}} \mu_3 \frac{R_L}{R_{o3} + R_L}$$

$$= \frac{700}{710}(-5)\frac{2.34}{12.34}(-240)\frac{47}{50}0.997\frac{0.6}{0.628} = 201$$

If the potentiometer is linear and set at the middle, then half of its resistance will be in series with R_s while the other half remains in parallel with the FET bias resistors. Hence, $v_{in}/v_s = 412/(412 + 510) = 0.447$ and $A_v = 91$.

Exercise 11.3–4 Find the gain of the amplifier in Fig. 11.3–8 when the potentiometer is removed and the first stage is a BJT emitter follower with $I_C = 1$ mA, $\beta = 200$, $R_E = 7.5$ kΩ, and $R_B = 150$ kΩ

11.4 AMPLIFIER FREQUENCY RESPONSE

This section investigates the factors that cause the gain of an amplifier to vary with signal frequency. Following an overview of frequency-response characteristics, we'll develop and analyze circuit models for amplifier behavior at low frequencies and at high frequencies.

Frequency-Response Characteristics

Our previous calculations of amplifier voltage gain implicitly assumed operation of "moderate" signal frequencies — neither too low nor too high. However, the gain actually depends upon frequency, as represented by the AC amplitude ratio

$$|V_{out}/V_s| = |A_v||H(\omega)| \tag{1}$$

Here, $|A_v|$ is the maximum gain and $|H(\omega)|$ is the frequency-response function normalized such that $|H(\omega)|_{max} = 1$. Although we work symbolically with angular frequency ω for convenience, we often express values in terms of the cyclical frequency $f = \omega/2\pi$ measured in hertz.

Usually, two major factors contribute to amplifier frequency response. At low signal frequencies, the coupling and bypass capacitors have significant impedance so they tend to block signal current as well as DC current. Consequently, in the case of a capacitor-coupled amplifier, $|H(\omega)| \to 0$ as $\omega \to 0$. At high frequencies, various shunt capacitances come into play and tend to short out the signal because they appear in parallel at the input or output side. The internal capacitances associated with transistors have this shunting effect. Consequently, for any transistor amplifier, we have $|H(\omega)| \to 0$ as $\omega \to \infty$.

Taking both factors into account, a typical plot of amplifier gain versus frequency looks like Fig. 11.4–1a. This plot exhibits a **midband region** with essentially maximum gain. The gain falls off in the **low-frequency region** below ω_ℓ and in the **high-frequency region** above ω_u. The lower and upper cutoff frequencies are defined by the standard convention

$$|H(\omega)| = 1/\sqrt{2} = 0.707 \qquad \omega = \omega_\ell, \omega_u \tag{2a}$$

or, converting to decibels,

$$20 \log |H(\omega)| \approx -3 \text{ dB} \qquad \omega = \omega_\ell, \omega_u \tag{2b}$$

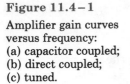

Figure 11.4–1

Amplifier gain curves
versus frequency:
(a) capacitor coupled;
(b) direct coupled;
(c) tuned.

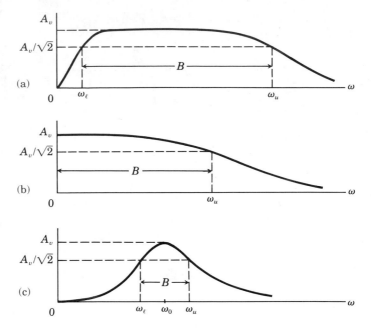

The region between the two cutoff points constitutes the amplifier's bandwidth, defined as $B = \omega_u - \omega_\ell$.

If the signals to be amplified include significant DC components or components at very low frequencies, then the amplifier should be **direct coupled** to obtain the lowpass response illustrated in Fig. 11.4–1b. Operational amplifiers have this type of gain curve. Direct coupling is usually achieved via differential amplification, as will be discussed in Section 11.5.

Another type of frequency response is exhibited by a **tuned amplifier,** whose gain curve is sketched in Fig. 11.4–1c. Tuned amplifiers incorporate bandpass circuits to amplify a narrow frequency band centered at ω_0. Radio and TV receivers include several tuned amplifiers, some with variable center frequencies for channel selection. The popular graphic equalizers found in audio systems contain a bank of tuned amplifiers arranged so that you can independently adjust the gain for each frequency band.

But most applications call for a characteristic like Fig. 11.4–1a with $\omega_u \gg \omega_\ell$. By way of example, a high-fidelity audio amplifier might have $f_\ell = \omega_\ell/2\pi \approx 20$ Hz and $f_u = \omega_u/2\pi \approx 20$ kHz, while the video amplifier in a TV set accommodates the range from about 30 Hz to 4.2 MHz. In such cases, the large separation between the low-frequency and high-frequency regions allows us to analyze the frequency response in three separate steps:

1. Find the midband voltage gain A_v by ignoring all capacitances.

2. Find the lower cutoff ω_ℓ from a low-frequency model that includes coupling and bypass capacitors but ignores shunt capacitances.

3. Find the upper cutoff ω_u from a high-frequency model that includes shunt capacitances but ignores coupling and bypass capacitors.

Since we have already studied midband behavior in Sections 11.1–11.3, the remainder of this section focuses on the low-frequency and high-frequency analysis.

Low-Frequency Response

Coupling and bypass capacitors limit the low-frequency response of an amplifier. We'll analyze their effect here and develop design guidelines for choosing capacitor values to achieve a specified lower cutoff frequency.

Coupling capacitors generally appear in a configuration equivalent to Fig. 11.4–2. This circuit is simply an *RC highpass filter* with an additional resistance R'. The amplitude ratio is easily found to be

$$\frac{|V_2|}{|V_1|} = \frac{R}{R' + R} \frac{\omega}{\sqrt{\omega^2 + 1/[(R' + R)C]^2}}$$

Hence, we write the normalized response as

$$|H(\omega)| = \frac{\omega}{\sqrt{\omega^2 + \omega_{co}^2}} = \frac{1}{\sqrt{1 + (\omega_{co}/\omega)^2}} \tag{3a}$$

where

$$\omega_{co} = 1/(R' + R)C \tag{3b}$$

which is the cutoff frequency of the highpass filter.

Figure 11.4–2
Highpass filter.

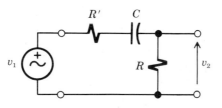

Now consider a single-stage voltage amplifier with coupling capacitors at the input and output. To determine the low-frequency response, we'll use the circuit model in Fig. 11.4–3a. The amplifier has been represented by its input resistance R_i, amplification factor μ, and output resistance R_o, so our model holds for several different FET or BJT amplifiers. Comparing Fig. 11.4–3a to Fig. 11.4–2, we see that the input and output sides each act like separate highpass filters described by Eq. (3). Hence, the normalized overall low-frequency amplitude response becomes the product

$$|H(\omega)| = \frac{1}{\sqrt{1 + (\omega_1/\omega)^2} \sqrt{1 + (\omega_2/\omega)^2}} \tag{4a}$$

Figure 11.4–3
Capacitor-coupled
amplifier:
(a) equivalent circuit;
(b) Bode plot.

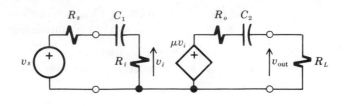

(a)

(b)

with

$$\omega_1 = \frac{1}{(R_s + R_i)C_1} \qquad \omega_2 = \frac{1}{(R_o + R_L)C_2} \qquad \textbf{(4b)}$$

We call ω_1 and ω_2 the **break frequencies** of $|H(\omega)|$. Note that if negative feedback increases R_i, then ω_1 is decreased and the low-frequency response may be improved.

The Bode plot of Eq. (4a) is given in Fig. 11.4–3b for the case of $\omega_2 > 4\omega_1$. The curve passes through -3 dB at $\omega \approx \omega_2$, and the amplifier's lower cutoff frequency is $\omega_\ell \approx \omega_2$. The other break point at ω_1 occurs sufficiently below ω_2 to have no appreciable effect on the location of ω_ℓ. Generalizing from this observation, it follows that if the break frequencies differ by more than a factor of 4 (two octaves on the Bode plot), then

$$\omega_\ell \approx \begin{cases} \omega_2 & \omega_2 > 4\omega_1 \\ \omega_1 & \omega_1 > 4\omega_2 \end{cases} \qquad \textbf{(5)}$$

Accordingly, we say that the *larger* break frequency dominates the low-frequency response.

Equation (5) has practical significance for two reasons. First, certain undesirable side effects may occur when the break frequencies are too close together. Second, the size and cost of coupling capacitors will be minimized by associating ω_ℓ with the side of the amplifier that has the smaller series resistance. For instance, to obtain a particular value of ω_ℓ when $R_o + R_L < R_s + R_i$, you could let $\omega_2 = \omega_\ell$ and $\omega_1 < \omega_\ell/4$ and take

$$C_2 = 1/\omega_\ell(R_o + R_L) \qquad C_1 > 4/\omega_\ell(R_s + R_i)$$

These expressions bring out the fact that large capacitances will be needed when you want a low ω_ℓ or when the amplifier has low input resistance. Fortunately, thanks to the fixed DC bias voltage across coupling capacitors, compact and inexpensive tantalum capacitors can be used in this application.

If the amplifier circuit includes a bypass capacitor, then it no longer provides perfect signal bypassing at very low frequencies and the resulting negative feedback reduces the amplification. Thus, the bypass capacitor introduces another break frequency

$$\omega_{\text{by}} = 1/R_{\text{eq}}C_{\text{by}} \tag{6}$$

Here, C_{by} is the capacitor in question (C_K or C_E) and R_{eq} is the equivalent resistance across the terminals of C_{by}.

Since a bypass capacitor appears in the same position as the load in a follower circuit, we can calculate R_{eq} using Eqs. (16) and (17), Section 11.3. Accordingly, for a FET amplifier with bypassed resistor R_{K2} and, possibly, an unbypassed resistor R_{K1}, we have

$$R_{\text{eq}} = \left(\frac{1}{g_m} + R_{K1}\right)\bigg\| R_{K2} \tag{7a}$$

Similarly, for a BJT amplifier,

$$R_{\text{eq}} = \left(\frac{R_s\|R_B + r_\pi}{\beta} + R_{E1}\right)\bigg\| R_{E2} \tag{7b}$$

When a BJT amplifier includes an input coupling capacitor, the value of R_{eq} may be somewhat larger than predicted by Eq. (7b) — an effect we can safely ignore.

Usually, R_{eq} is quite small and we need a correspondingly large bypass capacitance. But the DC voltage drop across C_{by} seldom exceeds a few volts, which permits economical implementation using a miniature electrolytic capacitor with as much as 10,000 μF. Hence, a conservative strategy that avoids interaction with other break frequencies is to make the value of C_{by} large enough that ω_{by} falls more than a decade below ω_ℓ.

Our design strategies readily extend to cascade or multistage amplifiers with several coupling and bypass capacitors, as follows. Let R_n be the resistance in series with coupling capacitor C_n, and let R_m be the smallest series resistance. To achieve a cutoff frequency no higher than ω_ℓ, take

$$C_m > 1/\omega_\ell R_m \tag{8a}$$

and choose all other coupling capacitors such that

$$C_n > 4/\omega_\ell R_n \qquad n \neq m \tag{8b}$$

Then choose each bypass capacitor such that

$$C_{by} > 10/\omega_\ell R_{eq} \qquad\qquad (8c)$$

where R_{eq} is calculated from Eq. (7a) or (7b). Standard capacitor values are listed for reference purposes at the back of the book.

Example 11.4–1

Suppose the multistage amplifier diagrammed back in Fig. 11.3–8 is to have $f_\ell \leq 50$ Hz, so $\omega_\ell \leq 100\pi = 314$ rad/s. We see from the circuit diagram that coupling capacitors C_3 and C_2 are in series with

$$R_3 = R_{o3} + R_L = 0.628 \text{ k}\Omega \qquad R_2 = R_{o1} + R_{i2} = 12.34 \text{ k}\Omega$$

The resistance in series with C_1 depends upon the potentiometer setting, but its smallest value is

$$R_{1-min} = R_{i1} = 2330 \text{ k}\Omega$$

Accordingly, from Eqs. (8a) and (8b), we should take

$$C_3 > \frac{1}{\omega_\ell R_3} \approx 5 \text{ } \mu F \qquad C_2 > \frac{4}{\omega_\ell R_2} \approx 1 \text{ } \mu F \qquad C_1 > \frac{4}{\omega_\ell R_1} \approx 0.005 \text{ } \mu F$$

Appropriate tantalum capacitors would be $C_3 = 6.8 \text{ } \mu F$, $C_2 = 1.5 \text{ } \mu F$, and $C_1 = 0.01 \text{ } \mu F$, which result in the dominant break frequency $f_\ell \approx \omega_3/2\pi \approx 37$ Hz and lower break frequencies at $f_2 \approx 13$ Hz and $f_1 < 10$ Hz.

The FET bypass capacitor has $R_{eq} = (0.5 + 1.5)\|1.3 = 0.79$ kΩ, so Eq. (8c) requires $C_K > 40 \text{ } \mu F$. We similarly find for the BJT that $R_{eq} = 0.05$ kΩ and $C_E > 640 \text{ } \mu F$. Choosing miniature electrolytic capacitors of 47 μF and 680 μF puts both bypass break frequencies below 5 Hz.

Exercise 11.4–1

Suppose that $\omega_2 = \omega_1$ in Eq. (4a). Show that $\omega_\ell = 1.55\omega_1$.

Exercise 11.4–2

Select standard values for the coupling and bypass capacitors such that the BJT feedback amplifier in Example 11.3–2 has $f_\ell \leq 40$ Hz. Then calculate the resulting break frequencies in hertz.

High-Frequency Response †

The high-frequency response of an amplifier is limited by shunt capacitance that appears in circuits equivalent to Fig. 11.4–4. Here, C might represent stray capacitance or the internal capacitance of a transistor. Or

Figure 11.4–4
Lowpass filter.

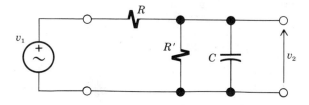

C might be a capacitor intentionally added to the circuit for purposes of high-frequency tone control. In any case, we have a modified *lowpass filter* described by the amplitude ratio

$$\frac{|V_2|}{|V_1|} = \frac{R'}{R' + R} \frac{1}{\sqrt{1 + \omega^2[(R\|R')C]^2}}$$

Hence, the normalized response is

$$|H(\omega)| = \frac{1}{\sqrt{1 + (\omega/\omega_{co})^2}} \qquad \textbf{(9a)}$$

with cutoff frequency

$$\omega_{co} = 1/(R\|R')C \qquad \textbf{(9b)}$$

Note that ω_{co} involves the parallel resistance $R\|R'$ rather than series resistance as in Eq. (3b).

Now suppose that a single-stage amplifier has shunt capacitance on both the input and output sides. By extension of Eq. (9a), the overall normalized high-frequency amplitude response will take the form

$$|H(\omega)| = \frac{1}{\sqrt{1 + (\omega/\omega_1)^2} \sqrt{1 + (\omega/\omega_2)^2}} \qquad \textbf{(10)}$$

in which ω_1 and ω_2 are the two upper break frequencies. If these break frequencies differ by more than a factor of 4, then the *smaller* one dominates the high-frequency response and the upper cutoff frequency is

$$\omega_u \approx \begin{cases} \omega_1 & \omega_1 < \omega_2/4 \\ \omega_2 & \omega_2 < \omega_1/4 \end{cases} \qquad \textbf{(11)}$$

The Bode plot of $|H(\omega)|$ in Fig. 11.4-5 illustrates this point for the case of $\omega_1 < \omega_2/4$.

Figure 11.4-5
Bode plot with two upper break frequencies.

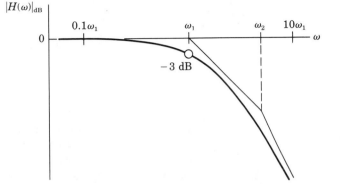

In absence of other types of shunt capacitance, the upper break frequencies of an amplifier are determined by internal transistor capaci-

Figure 11.4−6
High-frequency
transistor models:
(a) BJT; (b) FET.

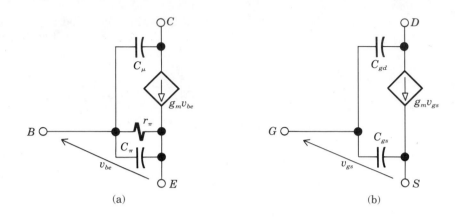

(a)   (b)

tances associated with reverse-biased *pn* junctions. The high-frequency
small-signal BJT and FET models diagrammed in Fig. 11.4−6 show these
junction capacitances connected from the control terminal to the other two
terminals. These models have been simplified by omitting second-order
effects that would greatly complicate pencil-and-paper calculations. More
elaborate models and computer-aided analysis become necessary at very
high frequencies.

Typical values for the base-to-emitter and base-to-collector capacitances in the BJT model are

$$C_\pi = 10\text{--}500 \text{ pF} \qquad C_\mu = 1\text{--}20 \text{ pF}$$

where $1 \text{ pF} = 10^{-12} \text{ F}$. Such small capacitances present very large impedance at low and midband frequencies, but the signal current through C_π and
C_μ increases with frequency. Eventually, the current gain of a BJT drops to
unity at the **transition frequency** given by

$$f_T = g_m/2\pi(C_\pi + C_\mu) \qquad \textbf{(12)}$$

Transition frequencies range from about 50 MHz to 1 GHz. For the FET
model, typical capacitance values are

$$C_{gs} = 2\text{--}10 \text{ pF} \qquad C_{gd} = 0.2\text{--}2 \text{ pF}$$

and the **maximum operating frequency** is

$$f_{\max} = g_m/2\pi(C_{gs} + C_{gd}) \qquad \textbf{(13)}$$

At $f > f_{\max}$, a FET no longer provides useful amplification.

Two additional comments should be made about Fig. 11.4−6. First,
these simplified models only hold at frequencies below about $f_T/5$ for a
BJT or $f_{\max}/5$ for a FET. Second, from a comparison of the two networks, it
follows that results derived using the BJT model can be extrapolated to
FETs by letting $C_\pi = C_{gs}$, $C_\mu = C_{gd}$, and $r_\pi \to \infty$. This observation allows us
to concentrate on BJT amplifiers.

Figure 11.4−7
High-frequency
equivalent circuits of a
common-emitter
amplifier.

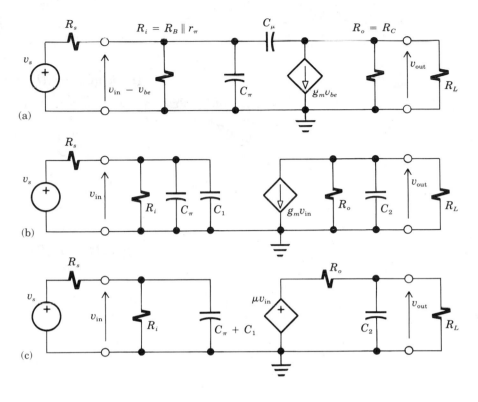

Figure 11.4−7a shows the high-frequency equivalent circuit of a single-stage common-emitter BJT amplifier. Although C_μ does not appear to be in a shunting position, the **Miller effect** previously discussed in Example 5.3–3 causes C_μ to act like two larger shunt capacitors (see Fig. 5.3–5). We therefore redraw the circuit in the form of Fig. 11.4–7b, putting $C_1 = (1 + A)C_\mu$ on the input side and $C_2 = (1 + 1/A)C_\mu$ on the output side. The values of C_1 and C_2 depend upon $A = -v_{out}/v_{in}$, for which we'll use the midband approximation $A \approx g_m(R_o\|R_L)$ —a conservative approximation since A actually decreases at higher frequencies. Hence,

$$C_1 \approx [1 + g_m(R_o\|R_L)]C_\mu \qquad C_2 \approx [1 + 1/g_m(R_o\|R_L)]C_\mu \qquad (14)$$

Combining the parallel capacitors on the input side and performing a Norton-to-Thévenin conversion on the output side yields our final diagram in Fig. 11.4–7c. The two break frequencies are then given by

$$\omega_1 = \frac{1}{(R_s\|R_i)(C_\pi + C_1)} \qquad \omega_2 = \frac{1}{(R_o\|R_L)C_2} \qquad (15)$$

For a common-source FET amplifier, we calculate ω_1 and ω_2 by setting $C_\pi = C_{gs}$ and $C_\mu = C_{gd}$ in Eqs. (14) and (15).

Most single-stage BJT amplifiers have $g_m(R_o\|R_L) \gg 1$ and $\omega_1 \ll \omega_2$. The upper cutoff frequency thus becomes

$$\omega_u = \omega_1 \approx \frac{1}{(R_s\|R_i)[C_\pi + g_m(R_o\|R_L)C_\mu]} \tag{16}$$

An examination of this expression reveals that the designer has little control over ω_u, save through the choice of the transistor and the values of g_m and R_C. But recall that the overall midband voltage gain is $|A_v| \approx R_i g_m(R_C\|R_L)/(R_i + R_s)$, so reducing g_m or R_C to increase ω_u carries the penalty of decreased gain. We emphasize this interrelationship between gain and bandwidth by forming the **gain-bandwidth product**

$$|A_v|\omega_u \approx \frac{g_m(R_o\|R_L)}{R_s[C_\pi + g_m(R_o\|R_L)C_\mu]} < \frac{1}{R_s C_\mu} \tag{17}$$

where the upper bound corresponds to $g_m(R_o\|R_L)C_\mu \gg C_\pi$. By way of example, suppose you want $|A_v| = 40$ and $f_u = 250$ kHz, given that $R_s = 2$ kΩ. Since $|A_v|\omega_u = 40 \times 2\pi \times 500$ kHz $= 2\pi \times 10^7$, the transistor must have $C_\mu < 1/(2000 \times 2\pi \times 10^7) \approx 8$ pF.

A further implication of the gain-bandwidth product is that negative feedback increases bandwidth because it reduces gain and Miller-effect multiplication of capacitance. Consequently, when an amplifier contains high-gain stages along with feedback or follower stages, the high-frequency response is usually dominated by the stage having the largest voltage gain.

Example 11.4−2

Again consider the multistage amplifier back in Fig. 11.3−8. The common-emitter stage provides most of the gain, and we'll evaluate its break frequencies taking $C_\pi = 100$ pF and $C_\mu = 4$ pF.

From the values given on the diagram, we find for the second stage that $R_s\|R_i = R_{o1}\|R_{i2} \approx 1.0$ kΩ, $R_o\|R_L = R_{o2}\|R_{i3} \approx 2.8$ kΩ, and $g_m(R_o\|R_L) \approx 224$. Substituting into Eqs. (14) and (15) yields $C_1 \approx 225C_\mu = 900$ pF, $C_2 \approx C_\mu = 4$ pF, $\omega_1 \approx 5.3 \times 10^5$, and $\omega_2 \approx 8.9 \times 10^7$. Since $\omega_2 \gg \omega_1$, we conclude that $f_u = \omega_1/2\pi \approx 84$ kHz. This upper cutoff is well below the BJT's transition frequency $f_T \approx 122$ MHz, as calculated from Eq. (12).

Exercise 11.4−3

Suppose a FET amplifier has $R_i \gg R_s = 1$ kΩ, $R_o\|R_L = 3$ kΩ, $g_m = 3$ mS, $C_{gs} = 20$ pF, and $C_{gd} = 2$ pF. Find f_u and compare it with f_{max}. Also evaluate the gain-bandwidth product and compare it with the upper bound in Eq. (17).

11.5   DIFFERENTIAL AMPLIFIERS AND OP-AMPS †

This section introduces the principles of differential amplification and operational amplifiers. We'll show how differential amplification makes it possible to build a direct-coupled circuit that amplifies the difference between two input voltages. Then we'll show how those properties are exploited in the design of operational amplifiers, which always include at least one stage of differential amplification.

Differential Amplification

A **differential amplifier** accepts two input voltages, denoted v_p and v_n. These inputs can be expressed in general as

$$v_p = \tfrac{1}{2}v_d + v_{\text{cm}} \qquad v_n = -\tfrac{1}{2}v_d + v_{\text{cm}} \tag{1}$$

where

$$v_d \triangleq (v_p - v_n) \qquad v_{\text{cm}} \triangleq \tfrac{1}{2}(v_p + v_n) \tag{2}$$

We call v_d the **difference voltage** and v_{cm} the **common-mode voltage**. Frequently, the common-mode voltage corresponds to unwanted noise or AC hum. Thus, the purpose of differential amplification is to amplify v_d and reject v_{cm}.

Differential amplification involves two amplifying transistors, either BJTs or FETs. Figure 11.5–1 diagrams a simple BJT differential amplifier. The inputs are applied to the base terminals, and the load is connected between the collector terminals. The emitter terminals join together at node X, where a DC current source goes to the negative power supply. Two power supplies must be used because there are no DC blocking capacitors.

Figure 11.5–1
BJT differential
amplifier.

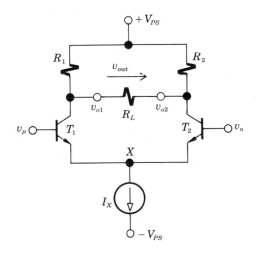

Ideally, a differential amplifier circuit has perfect symmetry, which requires $R_2 = R_1$ and transistors with identical characteristics. The output under these ideal circumstances would be

$$v_{\text{out}} = A_d(v_p - v_n) = A_d v_d$$

where A_d is the **differential gain.** But real circuits invariably fall short of perfection and the output actually becomes

$$v_{out} = A_d v_d + A_{cm} v_{cm} + V_{out} \tag{3}$$

Here, A_{cm} stands for the **common-mode gain** and the DC component V_{out} is the **offset voltage** at the output.

A well-designed differential amplifier should have very small offset and $|A_{cm}| \ll |A_d|$, so that $v_{out} \approx A_d v_d$. Accordingly, we'll analyze our BJT circuit from three different viewpoints to find the DC conditions, the small-signal differential gain, and the common-mode gain. We'll initially work with an unloaded output, for simplicity. We'll also assume identical transistors and let all departures from symmetry be represented by taking $R_2 \neq R_1$, a fair assumption in the usual case of integrated-circuit fabrication.

Consider first the DC circuit in Fig. 11.5–2 where the input terminals have been grounded—equivalent to setting $v_p = v_n = 0$. The ground connection carries DC current from the negative supply into the base terminals, thereby putting the BJTs into the active state. (When signal sources are applied at the inputs, they must pass the DC base currents.) The lower portion of the circuit is symmetric, so both BJTs have the same value of I_B and $I_C = \beta I_B$. Then, since both emitters are at $-V_\gamma$ relative to ground, the collector-to-emitter voltages are

$$V_{CE1} = V_{PS} - R_1 I_C + V_\gamma \qquad V_{CE2} = V_{PS} - R_2 I_C + V_\gamma \tag{4a}$$

We find I_C by summing currents at node X, which yields $I_X = 2I_C + 2I_B =$

Figure 11.5–2
DC conditions in a differential amplifier.

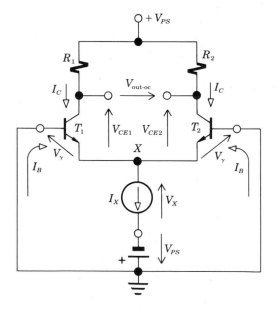

$2(I_C + I_C/\beta)$ so

$$I_C = I_X/2(1 + 1/\beta) \approx \tfrac{1}{2}I_X \tag{4b}$$

Hence, the source current I_X establishes the DC bias values.

There are two other DC values to be found from Fig. 11.5–2. The open-circuit output voltage is

$$V_{out-oc} = V_{CE2} - V_{CE1} = (R_2 - R_1)I_C$$

Minimizing the DC offset therefore requires closely matched resistors R_1 and R_2 as well as nearly identical transistors. Next, applying KVL around either lower loop, we get the DC voltage across the current source

$$V_X = V_{PS} - V_\gamma$$

Since V_X will be a positive quantity, the current source actually absorbs power and does not require an independent source of energy.

To pursue this point, recall that a transistor acts essentially like a constant-current source when it is biased in the active region without an input signal. Drawing upon this property, Fig. 11.5–3 gives possible implementations of the DC current source with a BJT or a JFET (which could also be a depletion MOSFET). The accompanying plot of i_X versus v_X illustrates the constant-current approximation. Notice, however, that if v_X includes a small-signal variation v_x relative to the DC value V_X, then i_X will have a small-signal variation i_x relative to I_X. These variations are related by the **dynamic resistance** $r_x = v_x/i_x$, whose reciprocal equals the slope of the i_x-v_x curve. Typical values of r_x for transistor current sources range from about 20 kΩ up to several megohms.

Figure 11.5–3
DC current source:
(a) JFET circuit;
(b) BJT circuit; (c) *i-v*
curve with dynamic
resistance.

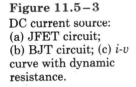
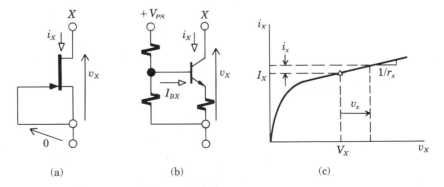

(a)   (b)   (c)

Having analyzed the DC conditions, we begin the small-signal analysis based on the equivalent circuit in Fig. 11.5–4a. This small-signal circuit is obtained from Fig. 11.5–1 in the usual fashion, with the additional provision that the dynamic resistance r_x replaces the DC current source. The open-circuit output voltage is easily found by noting that $v_{o1} = R_1(-g_m v_{be1})$

Figure 11.5–4

(a) Small-signal equivalent circuit of a differential amplifier. (b) Differential
input voltage. (c) Common-mode input voltage.

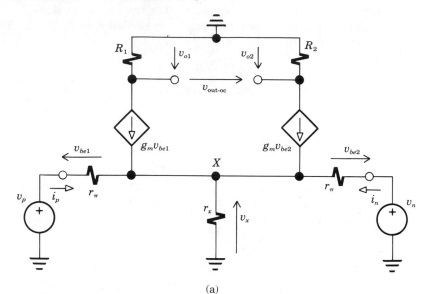

(a)

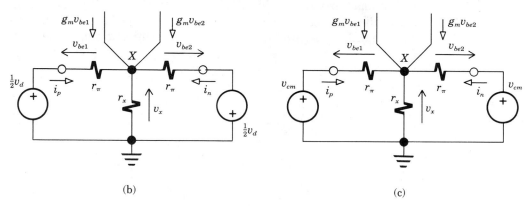

(b)   (c)

and $v_{o2} = R_2(-g_m v_{be2})$, so

$$v_{\text{out-oc}} = v_{o2} - v_{o1} = g_m(R_1 v_{be1} - R_2 v_{be2}) \tag{5}$$

Now we must obtain v_{be1} and v_{be2} in terms of v_p and v_n or, better yet, in terms
of v_d and v_{cm}.

On the one hand, suppose that $v_{cm} = 0$. Then Eq. (1) yields $v_p = \frac{1}{2}v_d$ and
$v_n = -\frac{1}{2}v_d$, and the input portion of the circuit can be redrawn as shown in
Fig. 11.5–4b. Since the two input loops contain equal resistances and
opposite source voltages, it follows that $i_n = -i_p$ and $v_x = 0$. Hence, $v_{be1} =$

$\frac{1}{2}v_d$ and $v_{be2} = \frac{1}{2}v_d$, and Eq. (5) becomes $v_{out-oc} = \frac{1}{2}g_m (R_1 + R_2)v_d$. We therefore define

$$\mu_d \triangleq v_{out-oc}/v_d = \tfrac{1}{2}g_m(R_1 + R_2) \qquad (6)$$

which represents the open-circuit differential amplification factor.

On the other hand, setting $v_d = 0$ in Eq. (1) yields $v_p = v_n = v_{cm}$. The corresponding input circuit in Fig. 11.5–4c now has $i_n = i_p$ and $v_{be2} = v_{be1}$. Thus, using KCL at node X,

$$v_x = r_x(i_p + g_m v_{be1} + g_m v_{be2} + i_n) = 2(g_m v_{be1} + v_{be1}/r_\pi)$$

After inserting this expression into the loop equation $v_{cm} = v_{be1} + v_x$ we get

$$v_{be2} = v_{be1} = \frac{v_{cm}}{2r_x(g_m + 1/r_\pi) + 1} \approx \frac{v_{cm}}{2g_m r_x}$$

where we have reasonably assumed that $g_m \gg 1/r_\pi$ and $g_m r_x \gg 1$. The resulting open-circuit common-mode amplification factor is then found via Eq. (5) to be

$$\mu_{cm} \triangleq v_{out-oc}/v_{cm} \approx (R_1 - R_2)/2r_x \qquad (7)$$

A comparison of Eqs. (6) and (7) reveals that $|\mu_{cm}| \ll |\mu_d|$ when $g_m r_x \gg 1$.

The perfect symmetry of an ideal differential amplifier results in $\mu_{cm} = 0$, as seen from Eq. (7) with $R_2 = R_1$. The quality of a real circuit having $\mu_{cm} \neq 0$ is measured in terms of its **common-mode rejection ratio** expressed in decibels and defined by

$$\text{CMRR}_{dB} \triangleq 10 \log (\mu_d/\mu_{cm})^2 \qquad (8)$$

Values of the common-mode rejection ratio usually fall between 80 and 100 dB.

Finally, we return to Fig. 11.5–4a and take $R_2 = R_1 = R$, so $\mu_{cm} = 0$ and

$$\mu_d = g_m R \qquad (9a)$$

The equivalent resistance seen looking back into the output terminals is

$$R_o = 2R \qquad (9b)$$

Hence, with a load resistance R_L connected to the output terminals, the differential voltage gain becomes

$$A_d = \frac{v_{out}}{v_d} = \mu_d \frac{R_L}{R_o + R_L} = \frac{g_m R R_L}{2R + R_L} \qquad (9c)$$

All of these results also hold for a differential amplifier built with FETs, in which case $r_\pi \to \infty$.

Exercise 11.5–1   Consider a BJT differential amplifier with $I_C = 0.5$ mA, $g_m = 20$ mS, $R_1 = 4.9$ kΩ, $R_2 = 5.1$ kΩ, and $r_x = 50$ kΩ.

(a) Evaluate μ_d, μ_{cm}, and CMRR_{dB}.

(b) Find the resulting output voltage, including the DC offset, when $v_p = \cos \omega_{cm}t + 0.01 \cos \omega_d t$, $v_n = \cos \omega_{cm}t - 0.01 \cos \omega_d t$, and $R_L = \infty$.

Integrated-Circuit Op-Amps

Integrated-circuit amplifiers should be direct-coupled internally because coupling capacitors require so much space on an IC chip. For this reason, an IC amplifier usually features a differential amplifier circuit at the input. The most important type of linear IC amplifier is, of course, the popular and versatile op-amp.

Figure 11.5–5 depicts in block-diagram form the general structure of a complete IC op-amp. The differential input stage provides common-mode rejection and some of the gain. Additional gain comes from the high-gain second stage and sometimes from the level-shifting third stage. However, the main purpose of the level shifter is to establish the correct DC value. The last stage drives the load and has the low output resistance needed to minimize loading. Accordingly, the output stage might be a follower circuit. Or it might be a push-pull power amplifier, which we'll discuss in the next chapter.

Figure 11.5–5
Block diagram of an operational amplifier.

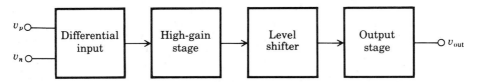

As implied by our diagram, an op-amp has a **single-ended output** — meaning that the load is connected from the single output terminal to ground. But standard differential amplifiers have two output terminals, neither of which can be grounded. (Indeed, there are no ground points within an IC op-amp). The circuitry must therefore include conversion from the differential mode to the single-ended mode.

One technique for mode conversion involves the modified differential amplifier shown in Fig. 11.5–6. Here, the first collector resistor has been omitted and the single-ended output is taken at the second collector. The lack of symmetry results in significant common-mode amplification, namely,

$$\mu_{cm} \approx -R/2r_x \tag{10}$$

Nonetheless, this circuit could be used as the second stage since the differential first stage largely eliminates common-mode voltage. The relevant

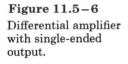

Figure 11.5–6
Differential amplifier
with single-ended
output.

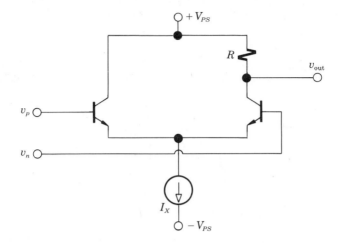

small-signal properties of the second stage then become

$$R_i = 2r_\pi \qquad \mu_d = \tfrac{1}{2}g_m R \qquad R_o = R \qquad (11)$$

The input resistance R_i follows from Fig. 11.5–4a, where $(v_p - v_n)/i_p = 2r_\pi$.

High gain is often achieved by replacing the collector resistor with another current source, so the small-signal equivalent value of R equals the very large dynamic resistance. Transistor current sources functioning in that capacity are called **active loads.**

Taking account of amplifying transistors, biasing current sources, and active loads, an IC op-amp may contain 20 or more transistors. The transistors may be BJTs, MOSFETs, JFETs, or a mixture of bipolar and field-effect devices. The number of resistors is intentionally kept as small as possible because resistors consume more chip space than transistors.

Example 11.5–1

A simplified but illustrative IC op-amp circuit is diagrammed in Fig. 11.5–7, with all resistance values in kilohms. DC voltages at selected points are shown in parentheses, and the indicated DC currents are given in milliamps.

The first stage is a standard MOSFET differential amplifier. Using MOSFETs rather than BJTs eliminates the need for DC bias currents through the input terminals and creates a very large input resistance, theoretically infinite. The output resistance is $R_o = 2R_D = 50$ kΩ and the MOSFETs are biased such that $g_m = 0.8$ mS, so $\mu = \mu_d = g_m R_D = 20$. However, in view of the interchanged connections to the next stage, we consider this stage to have $\mu = -20$.

The second stage is a BJT differential amplifier with a single-ended output and an active load for high gain. The small DC base currents come from the positive supply via the first stage. The BJTs have $g_m = 16$ mS,

Figure 11.5-7
Simplified IC op-amp circuit.

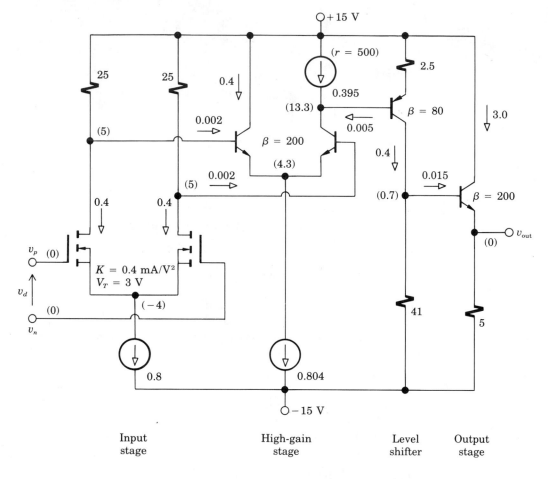

Input stage   High-gain stage   Level shifter   Output stage

and the dynamic resistance of the active load is $r = 500$ kΩ. Thus, from Eq. (11), $R_i = 2r_\pi = 25$ kΩ, $R_o = r$, and $\mu = \frac{1}{2}g_m r = 4000$.

The third stage is a level-shifting amplifier employing an inverted *pnp* BJT. This creative arrangement resets the DC level from 13.3 V at the output of the second stage down to 0.7 V needed at the input of the last stage. But the inverted configuration does not affect the small-signal properties, which are the same as those of an *npn* BJT feedback amplifier with $R_B = \infty$ (by virtue of direct coupling), $R_E = 2.5$ kΩ, and $R_o = R_C = 41$ kΩ. Hence, referring to Eqs. (9) and (10), Section 11.3, we find that $\mu = -16$ and $R_i = 205$ kΩ.

By the way, notice that the *pnp* BJT has a smaller value of β than the *npn* devices. This apparent discrepancy occurs when *pnp* and *npn* transistors are allocated equal amounts of chip space because the emitted holes in

Table 11.5–1
Data for Figure 11.5–7.

	Input stage	High-gain stage	Level shifter	Output stage
I_D, I_C (mA)	0.4	0.4	0.4	3
V_{DS}, V_{CE} (V)	9	10.7, 9	13.3	15
g_m (mS)	0.8	16	16	120
r_π (kΩ)	∞	12.5	5	1.7
R_i (kΩ)	∞	25	205	1000*
μ	-20	4000	-16	0.998
R_o (kΩ)	50	500	41	0.2
Gain	-6.67	1160	-15.4*	0.998*

* Calculated with the output unloaded.

the *pnp* transistors are less mobile than the emitted electrons in the *npn* transistors.

Finally, the last stage is a direct-coupled emitter follower similar to the one back in Example 11.3–3. With no load connected, this stage has $R_i = 1000$ kΩ, $\mu = 0.998$, and $R_o = 0.2$ kΩ.

Table 11.5–1 summarizes the DC and small-signal properties of the four stages. The table also lists the gain of each stage calculated as $\mu R_L/(R_o + R_L)$, where R_L stands for the loading input resistance of the next stage. Taking the product of the stage gains then yields the overall voltage amplification $v_{\text{out-oc}}/v_d \approx 120{,}000$.

Exercise 11.5–2   Suppose the active load for the second stage in Fig. 11.5–7 is replaced by a resistor R. Find the value of R that keeps the DC values unchanged. Then calculate the resulting gain of the second stage, including third-stage loading.

PROBLEMS

11.1–1   Substitute $v_{GS} = V_{GS} + v_{gs}$ into Eq. (2) to show that $i_D \approx I_D + g_m v_{gs}$ when $|v_{gs}| \ll g_m/k$.

11.1–2   Suppose the MOSFET in Fig. 11.1–5 operates at $V_{DS} = \frac{1}{2}V_{DD}$. Show that $\mu = -V_{DD}\sqrt{k/I_D}$.

11.1–3   Let the element values in Fig. 11.1–5 be $R_s = 10$ kΩ, $V_{DD} = 12$ V, $R_1 = 3$ MΩ, $R_2 = 1$ MΩ, $R_D = 4$ kΩ, and $R_L = 6$ kΩ. Find I_D, V_{DS}, and A_v when the MOSFET has $I_{DSS} = 6$ mA and $V_T = 2$ V.

11.1–4   Do Problem 11.1–3 with $I_{DSS} = 2$ mA and $V_T = 1.5$ V.

11.1–5   Suppose the amplifier in Fig. 11.1–7a consists of identical stages with $R_1 = 3.3$ MΩ, $R_2 = 1.2$ MΩ, $R_D = 4$ kΩ, and $V_{DD} = 9$ V. Find I_D, V_{DS}, and A_v when $R_s = 5$ kΩ, $R_L = 1$ kΩ, and the MOSFET has $I_{DSS} = 4$ mA and $V_T = 1.6$ V.

11.1–6   Determine the values of R_K and V_G in Fig. 11.1–8 so that I_D stays in the range 1.75–2.25 mA when the MOSFET has $I_{DSS} = 20$ mA and V_T varies between 2.2 and 3.8 V. Hint: Note in Fig. 11.1–9 that $R_K = -(\Delta V_{GS}/\Delta I_D)$.

11.1−7   Do Problem 11.1−6 including variation of I_{DSS} between 15 and 25 mA.

11.1−8   Derive Eq. (19) by inserting $R_i = R_1 \| R_2$ into Eq. (11).

11.1−9   Design a self-biased amplifier using $V_{DD} = 18$ V and one or more n-channel enhancement MOSFETs with $k \approx 1.6$ mA/V², $V_T \approx 2$ V, and $I_D \le 2.5$ mA. The amplifier is to have $R_i \approx 500$ kΩ, $|v_{\text{out}}|_{\text{max}} = 1.8$ V, and $|A_v| \approx 8$ when $R_L = 5$ kΩ and $R_s \ll R_i$. Select standard resistor values, and make R_K as large as possible.

11.1−10   Do Problem 11.1−9 for $|A_v| \approx 160$.

11.1−11   Design a self-biased amplifier with one or more n-channel enhancement MOS-FETs operated at $I_D \approx 1.2$ mA and having $k \approx 2.7$ mA/V² and $V_T \approx 2.5$ V. The amplifier is to have $R_i \approx 200$ kΩ, $|v_{\text{out}}|_{\text{max}} = 2$ V, and $|A_v| \approx 15$ when $R_L = 10$ kΩ and $R_s \ll R_i$. Select standard component values, and make R_K as large as possible.

11.1−12   Given that the JFET in Fig. 11.1−11 has $k = 4.5$ mA/V² and $V_P = 3$ V, find standard values for R_K, R_D, and V_{DD} to get $I_D \approx 2$ mA, $A_v \approx -20$, and $|v_{\text{out}}|_{\text{max}} = 1.5$ V when $R_L = 10$ kΩ and $R_s \ll R_i$.

11.1−13   Design a self-biased amplifier using $V_{DD} = 15$ V and one or more n-channel JFETs with $k \approx 3$ mA/V², $V_P \approx 2.5$ V, and $I_D \le 3$ mA. The amplifier is to have $R_i \approx 100$ kΩ, $|v_{\text{out}}|_{\text{max}} = 2$ V, and $|A_v| \approx 12$ when $R_L = 20$ kΩ and $R_s \ll R_i$. Select standard resistor values, and make R_K as large as possible.

11.1−14   Do Problem 11.1−13 for $|A_v| \approx 160$.

11.1−15   Redraw Fig. 11.1−9 for the case of a JFET whose transfer curve is given in Fig. 11.1−10. Extend the bias line to get Q and Q' at $V_{GS} < 0$. Then draw another bias line through Q and the origin, corresponding to the circuit in Fig. 11.1−11. Which bias circuit has less sensitivity to variations of the JFET parameters?

11.2−1   Figure P11.2−1 is another small-signal BJT model, equivalent to Fig. 11.2−3. Find the parameters α and r_e in terms of β and I_C.

11.2−2   Figure P11.2−2 shows two BJTs connected as a **Darlington pair,** which sometimes replaces a single BJT in high-gain amplifiers. Assume that $I_B = 10$ μA and $\beta_1 = \beta_2 = 50$.

(a) Find I_C and evaluate $r_{\pi 1}$ and $r_{\pi 2}$.

(b) Use the small-signal model for each BJT to calculate the equivalent values of β, r_π, and g_m for the Darlington pair.

Figure P11.2−1

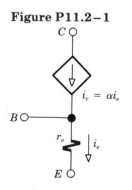

Figure P11.2−2

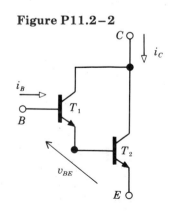

11.2−3 Do Problem 11.2−2 with $I_B = 0.5$ μA, $\beta_1 = 30$, and $\beta_2 = 100$.

11.2−4 Let the element values in Fig. 11.2−5a be $R_s = 4$ kΩ, $R_C = 3$ kΩ, $R_L = 2$ kΩ, and $V_{CC} = 12$ V. Find I_C, R_i, and A_v when $R_B = 332$ kΩ and the BJT has $\beta = 100$. Also determine the limitation on v_s such that $v_{CE} \geq 1$ V.

11.2−5 Do Problem 11.2−4 with $R_B = 664$ kΩ and $\beta = 160$.

11.2−6 A cascade amplifier consists of identical stages like Fig. 11.2−5a with $V_{CC} = 6$ V, $R_C = 2$ kΩ, $R_B = 1060$ kΩ, and $\beta = 200$. The source has $R_s = 5$ kΩ and the load is $R_L = 1$ kΩ. Find I_C, V_{CE}, μ, and R_i. Then calculate A_v when there are two stages and three stages.

11.2−7 Do Problem 11.2−6 with $R_B = 530$ kΩ and $\beta = 150$.

11.2−8 The circuit in Fig. 11.2−5a can be stabilized by connecting the upper end of R_B to the collector rather than directly to the power supply. Perform a DC analysis on this configuration with $I_B = (I_C - I_{CEO})/\beta$ to obtain an expression similar to Eq. (13).

11.2−9 Derive Eq. (14) by inserting $R_C \| R_L \leq R_L/2$ into Eq. (14) and using Eq. (6) to eliminate r_π.

11.2−10 Design a single-stage self-biased amplifier using $V_{CC} = 15$ V and a BJT with $\beta \approx 120$ and $I_C = 1.5-3.0$ mA. The amplifier is to have $|v_{out}|_{max} = 3$ V and $A_v \approx -80$ when $R_s = 1$ kΩ and $R_L = 5$ kΩ. Select standard resistor values, and make R_E as large as possible.

11.2−11 Do Problem 11.2−10 with $V_{CC} = 24$ V and $A_v \approx -150$.

11.2−12 Design a single-stage self-biased amplifier using a BJT with $\beta \approx 160$ operated at $I_C = 1$ mA. The amplifier is to have $|v_{out}|_{max} = 1.5$ V and $A_v \approx -60$ when $R_s = 2$ kΩ and $R_L = 5$ kΩ. Select standard resistor values, and make R_E as large as possible with the smallest standard supply voltage.

11.2−13 Do Problem 11.2−12 with $I_C = 2$ mA and $A_v \approx -100$.

11.3−1 Let the BJT feedback amplifier in Fig. 11.3−3a have $R_1 \| R_2 = 30$ kΩ, $R_C = 2$ kΩ, $R_{E1} = 100$ Ω, and $\beta = 80$. Calculate R_i and μ when $I_C = 1$ mA and when $I_C = 2$ mA.

11.3−2 Do Problem 11.3−1 with $R_{E1} = 80$ Ω and $\beta = 120$.

11.3−3 Select standard values for R_{E1}, R_{E2}, and R_C so that the BJT feedback amplifier in Fig. 11.3−3a has $R_i \approx 5$ kΩ, $\mu \approx -25$, and $R_{E1} + R_{E2} \approx R_C/3$ when $I_C = 1.5$ mA, $\beta = 60$, and $R_B \gg R_i$.

11.3−4 Suppose the BJT feedback amplifier in Fig. 11.3−3a has $R_{E2} = 0$, eliminating the need for C_E. Select standard values for R_{E1} and R_C to get $R_i \approx 10$ kΩ and $\mu \approx -5$ when $R_B = 10R_{E1}$, $\beta = 100$, and $g_m R_{E1} \gg 1$. Then determine the condition on I_C such that $g_m R_{E1} \geq 10$.

11.3−5 Consider the FET feedback amplifier in Fig. 11.3−4a with $R_{K2} = 0$ and $V_{DD} = 18$ V. Given that $g_m \approx 5$ mS and $I_D = 2$ mA, select standard values for R_{K1} and R_D to get $\mu \approx -5$ and $V_{DS} \approx 4$ V. Then find the range of $|\mu|$ when g_m varies from its nominal value by $\pm 20\%$.

11.3−6 Do Problem 11.3−5 with $g_m \approx 4$ mS and $I_D = 1.5$ mA.

11.3−7 Suppose the FET's dynamic resistance r_d is added in parallel with the controlled source in Fig. 11.3−4b. Use a Norton-to-Thévenin conversion to obtain an expression for v_f/v_{in} when $R_L = \infty$. Then show that

$$\mu = \frac{-R_D}{R_{K1} + (r_d + R_D + R_{K1})/g_m r_d}$$

11.3–8   Suppose the emitter follower in Fig. 11.3–5 has $R_s = 2$ kΩ, $R_B = 18$ kΩ, $R_E = 600$ Ω, and $R_L = 200$ Ω. Evaluate R_i, μ, R_o, and A_v when $I_C = 1$ mA and $\beta = 120$. Then use your results to find the current amplification i_{out}/i_{in}.

11.3–9   Do Problem 11.3–8 with $I_C = 5$ mA and $\beta = 80$.

11.3–10   Derive from Fig. 11.3–6 the expressions for i_{out-sc} and v_{out-oc}.

11.3–11   If the source follower in Fig. 11.3–7 has $R_i \gg R_s$, $R_K = 2$ kΩ, and $g_m = 4.5$ mS, then what's the condition on R_L such that $A_v \geq 0.7$?

11.3–12   If the source follower in Fig. 11.3–7 has $R_i \gg R_s$, $g_m = 5$ mS, and $R_L = 400$ Ω, then what's the condition on R_K such that $A_v \geq 0.6$?

11.3–13   Draw the small-signal model of the source follower in Fig. 11.3–7 and derive the expression for R_o given in Eq. (17).

11.3–14   Both FETs in Fig. P11.3–14 have $k = 4$ mA/V^2 and $V_T = 3$ V. Find R_1 and R_K so that $I_{D1} = 1.5$ mA and $I_{D2} = 9$ mA when $R_{K1} = 0.4$ kΩ. Then calculate the overall gain A_v.

Figure P11.3–14

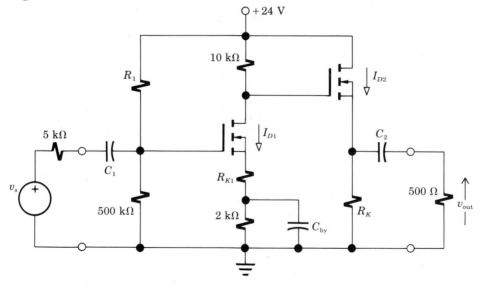

11.3–15   Do Problem 11.3–14 for $I_{D1} = 1$ mA and $I_{D2} = 12$ mA when $R_{K1} = 0.2$ kΩ.

11.3–16   Both BJTs in Fig. P11.3–16 have $\beta = 100$. Find R_{E2} and R_B so that $I_{C1} = 1$ mA and $I_{C2} = 12$ mA when $R_{E1} = 0.15$ kΩ and $R_E = 0.75$ kΩ. Then calculate the overall gain A_v.

11.3–17   Do Problem 11.3–16 for $I_{C1} = 1.2$ mA and $I_{C2} = 6$ mA when $R_{E1} = 0.1$ kΩ and $R_E = 1.0$ kΩ.

Figure P11.3–16

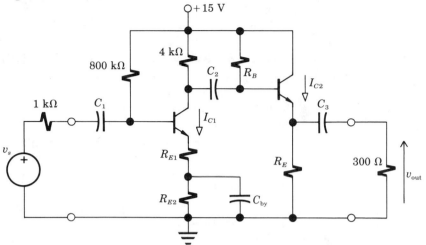

11.4–1 Consider cascade of n identical stages whose normalized low-frequency response is $|H(\omega)| \approx [1/\sqrt{1 + (\omega_1/\omega)^2}]^n$, where ω_1 is the dominant higher break frequency of each stage. Show that $\omega_\ell/\omega_1 = 1/\sqrt{2^{1/n} - 1}$ and evaluate ω_ℓ/ω_1 for $n = 4$ and 8.

11.4–2 Let the FET amplifier in Fig. P11.3–14 have $R_1 = 1.2$ MΩ, $R_{K1} = 0.4$ kΩ, $R_K = 0.5$ kΩ, $g_{m1} = 5$ mS, and $g_{m2} = 10$ mS. Select standard values for C_1, C_2, and C_{by} to get $f_\ell \le 20$ Hz.

11.4–3 A BJT cascade amplifier consists of two identical stages like Fig. 11.1–6a with $R_1 \| R_2 = 16$ kΩ, $R_C = 3$ kΩ, $R_E = 1$ kΩ, $g_m = 50$ mS, and $\beta = 200$. An interstage coupling capacitor C_{12} replaces both C_{o1} and C_{i2}. Select standard values for C_{i1}, C_{12}, C_{o2}, and C_K to get $f_\ell \le 100$ Hz, given that $R_s = 0.2$ kΩ and $R_L = 1$ kΩ.

11.4–4 Let the BJT amplifier in Fig. P11.3–16 have $R_{E1} = 0.1$ kΩ, $R_{E2} = 4$ kΩ, $R_B = 80$ kΩ, $R_E = 1$ kΩ, $g_{m1} = 50$ mS, $g_{m2} = 250$ mS, and $\beta_1 = \beta_2 = 100$. Select standard values for C_1, C_2, C_3, and C_{by} to get $f_\ell \le 40$ Hz.

11.4–5 Suppose the FET amplifier in Fig. P11.3–14 has an adjustable capacitor for C_1 in order to vary f_ℓ from about 40 Hz to 200 Hz. Determine the minimum and maximum values of C_1 given that $R_1 = 1.5$ MΩ, $R_{K1} = 0.2$ kΩ, $g_{m1} = 4$ mS, $C_{by} = 100$ μF, $R_K = 0.75$ kΩ, $g_{m2} = 10$ mS, and $C_2 = 6.8$ μF.

11.4–6 Suppose the FET amplifier in Fig. 11.3–4 has adjustable gain implemented using a linear potentiometer for $R_{K1} + R_{K2}$, with C_K connected to the wiper. Estimate the value of f_ℓ when the wiper is at the bottom, middle, and top position, given that $R_s = 25$ kΩ, $C_i = 0.1$ μF, $R_1 \| R_2 = 500$ kΩ, $R_{K1} + R_{K2} = 4$ kΩ, $C_K = 4.7$ μF, $R_D = R_L = 5$ kΩ, $C_o = 0.47$ μF, and $g_m = 5$ mS. Hint: see Exercise 11.4–1.

11.4–7 A BJT having $C_\pi = 50$ pF, $C_\mu = 5$ pF, and $\beta = 100$ is operated at $I_C = 5$ mA in a common-emitter amplifier with $R_s = 500$ Ω, $R_B = 60$ kΩ, and $R_C = 2$ kΩ. Evaluate f_T. Then find ω_1, ω_2, f_u, and the gain-bandwidth product when $R_L = 2$ kΩ and when $R_L = 0.5$ kΩ.

11.4–8 A BJT having $C_\pi = 300$ pF, $C_\mu = 10$ pF, and $\beta = 120$ is operated at $I_C = 1$ mA in a common-emitter amplifier with $R_B = 100$ kΩ, and $R_C = R_L = 10$ kΩ. Evaluate f_T.

Then find ω_1, ω_2, f_u, and the gain-bandwidth product when $R_s = 3$ kΩ and when $R_s = 5$ Ω.

11.4–9   A FET having $C_{gs} = 6$ pF, $C_{gd} = 1.5$ pF, and $k = 4.5$ mA/V^2 is operated at $I_D = 2$ mA in a common-source amplifier with $R_i = 1$ MΩ, and $R_D = R_L = 4$ kΩ. Evaluate $f_{\max}$. Then find the condition on R_s for $f_u \approx 10$ MHz and for $f_u \approx 50$ MHz.

11.4–10   A BJT having $C_\pi = 300$ pF, $C_\mu = 10$ pF, and $\beta = 120$ is operated at $I_C = 2.5$ mA in a common-emitter amplifier with $R_s = 600$ Ω, $R_B = 100$ kΩ, and $R_C = 2$ kΩ. The amplifier drives a capacitive load consisting of C_L in parallel with $R_L = 2$ kΩ. Find ω_1, ω_2, and f_u when $C_L = 30$ pF and when $C_L = 3000$ pF.

11.4–11   An external capacitor is added to the FET amplifier in Exercise 11.4–3 for the purpose of reducing f_u to 1 MHz. The capacitor can be connected from gate to source, from drain to source, or from gate to drain. Determine the value of C needed for each location.

11.5–1   Suppose that $R_2 = R_1 = R$ in Fig. 11.5–2 but the transistors have $\beta_2 \neq \beta_1$. Then $I_{C2} \neq I_{C1}$ even though symmetry requires $I_{B2} = I_{B1} = I_B$. Find $V_{\text{out}-oc}$ in terms of R, I_X, β_1, and β_2 for this case.

11.5–2   Let the circuit in Fig. 11.5–4 have $g_m = 10$ mS, $r_x = 40$ kΩ, and $R_2 = R_1 - r$ where $r \ll R_1$. What's the condition on r/R_1 so that $\text{CMRR}_{\text{dB}} \geq 80$ dB?

11.5–3   A BJT differential amplifier is fabricated with $\beta = 100$, $R_1 \approx 5$ kΩ, and $R_2 = R_1(1 \pm \epsilon)$ where $\epsilon \leq 0.2$. Determine the conditions on I_X and r_x to get $\mu_d \approx 200$ and $\text{CMRR}_{\text{dB}} \geq 100$ dB.

11.5–4   Referring to Fig. 11.5–4c, show that the common-mode input resistance is $R_{\text{icm}} = v_{\text{cm}}/i_p = r_\pi + 2(\beta + 1)r_x$.

11.5–5   Suppose the BJTs in a differential amplifier have the same value of r_π but unequal transconductances, g_{m1} and g_{m2}. Taking $g_{m1} + g_{m2} \gg 2/r_\pi$ and $r_x(g_{m1} + g_{m2}) \gg 1$, show from Fig. 11.5–4c that $v_{be1} \approx v_{\text{cm}}/r_x(g_{m1} + g_{m2})$. Then obtain an approximate expression for μ_{cm} when $R_2 = R_1 = R$.

11.5–6   Consider the circuit in Fig. 11.5–4a with the right-hand input grounded (so $v_n = 0$) and the left-hand input connected to a voltage source v_s with resistance R_s (so $v_p = v_s - R_s i_p$). Taking $\beta = g_m r_\pi \gg 1$ and $r_x g_m \gg 1$, show that $v_x \approx v_{be1}$. Then obtain an approximate expression for $v_{\text{out}-oc}/v_s$ when $R_2 = R_1 = R$.

11.5–7   Find $v_{\text{out}-oc}/v_d$ when the op-amp in Fig. 11.5–7 drives $R_L = 1$ kΩ.

11.5–8   It is proposed to modify the input stage in Fig. 11.5–7 by replacing the two 25-kΩ resistors with 0.402-mA current sources having dynamic resistance $r = 500$ kΩ. Explain from gain considerations why this proposal is not very practical.

11.5–9   Suppose the level shifter in Fig. 11.5–7 is modified by replacing the 41-kΩ resistor with a 0.385-mA current source having dynamic resistance $r = 500$ kΩ. Find the values in Table 11.5–1 that are changed, and calculate $v_{\text{out}-oc}/v_d$.

11.5–10   Suppose the level shifter in Fig. 11.5–7 is modified by replacing the 2.5-kΩ resistor with a 1-kΩ resistor, keeping $I_C = 0.4$ mA. Find the values in Table 11.5–1 that are changed, and calculate $v_{\text{out}-oc}/v_d$.

12 Power Electronics

Concluding our study of electronic circuits, we turn to the special considerations that arise when semiconductor devices handle large amounts of power. We'll start with the linear power amplifiers needed to drive loudspeakers and various other large-signal output devices. Then we'll discuss nonlinear circuits that perform AC-to-DC power conversion. Maximum electrical and thermal ratings of semiconductor devices become an important concern throughout all of power electronics. The last section is therefore devoted to thermal limitations and derating.

Objectives

After studying this chapter and working the exercises, you should be able to do each of the following:

- Identify the significance of maximum symmetric signal swing and transistor ratings for large-signal amplification (Section 12.1).

- State the major differences between class-A transformer-coupled amplifiers and class-B push-pull amplifiers (Section 12.1).

- Design a simple amplifier to obtain a specified output power (Section 12.1).

- Calculate the DC and ripple component at the output of a rectifier circuit with capacitive or inductive smoothing, and determine the required diode ratings (Section 12.2).

- Analyze the performance of a series voltage regulator (Section 12.2).

- Describe how a controlled rectifier can be used to control the power delivered to a load (Section 12.2).†

- Calculate the maximum allowable power dissipation of a diode or transistor, given its derating curve (Section 12.3).†

- Explain why and how heat sinks are used with power semiconductor devices (Section 12.3).†

12.1 POWER AMPLIFIERS

This section concerns the concepts and problems associated with amplifiers capable of delivering more than about one-half watt of output signal power. Following an introductory treatment of large-signal amplification, we discuss two major types of transistor power amplifiers: transformer-coupled circuits, and push-pull circuits.

Large-Signal Amplification

A power amplifier is a large-signal amplifier whose performance hinges upon three critical aspects:

- Linear behavior over large voltage and current swings;
- Efficient conversion of input supply power to output signal power; and
- Limitations imposed by the electrical and thermal ratings of transistors.

We begin our study of power amplifiers by discussing these distinctive aspects of large-signal amplification.

Clearly, the amplifying transistor must be selected with care to minimize nonlinear distortion of large signals. Bipolar junction transistors intended for such applications exhibit suitable linearity when operated as emitter followers or when driven by current signals. Power BJTs usually have less current gain and larger saturation voltages than small-signal BJTs, but they accommodate much greater voltages and currents. Two special types of enhancement MOSFETs have also been developed for power circuits, the vertical-groove VMOS transistor and the double-diffused DMOS transistor. Unlike ordinary FETs, these power MOSFETs behave quite linearly over most of the active region. However, in the interest of simplicity, we'll confine our attention to amplifiers built with power BJTs.

As an introduction to large-signal amplification, consider the common-emitter circuit diagrammed in Fig. 12.1–1a. The large input signal v_{in} comes from an appropriate driver stage, and the resistor R_3 has been included to get an input current proportional to v_{in}. We want to find the conditions required for maximum signal swing at the output. Additionally, we want to find the power delivered to the load and the power drawn from the supply. To establish a convenient basis for power calculations, we'll work with the AC input signal $v_{in} = V_{in} \sin \omega t$.

Figure 12.1–1

(a) Large-signal common-emitter amplifier. (b) Equivalent AC circuit.

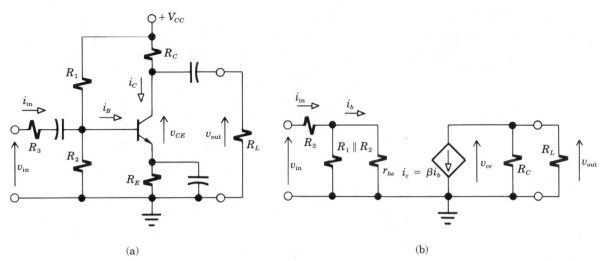

(a)   (b)

Since the coupling and bypass capacitors ideally act as short circuits for AC currents, the AC equivalent circuit takes the form of Fig. 12.1–1b. This circuit differs from a small-signal model in that the signals are large and r_{be} represents the *nonlinear* base-emitter characteristic. Although r_{be} varies with i_b, its average value equals r_π and will be quite small for a power BJT. We therefore assume that $r_{be} \ll R_1 \| R_2$ and $R_3 \gg r_{be}$, so $v_{in}/i_{in} \approx R_3$ and

$$i_b \approx i_{in} \approx v_{in}/R_3$$

Then, with $v_{in} = V_{in} \sin \omega t$,

$$i_c = \beta i_b = I_m \sin \omega t$$

where

$$I_m \approx \beta V_{in}/R_3 \qquad (1)$$

The **AC equivalent load resistance** is

$$R_{AC} \triangleq R_C \| R_L \qquad (2)$$

and the resulting signal voltage is

$$v_{out} = v_{ce} = -V_m \sin \omega t \qquad (3a)$$

with

$$V_m = R_{AC} I_m \qquad (3b)$$

This equation relates the AC amplitudes at the output.

Returning to Fig. 12.1–1a, the biasing circuitry supplies DC base current I_B and sets the Q point at $I_C = \beta I_B \gg I_B$. The DC components are related by

$$V_{CE} = V_{CC} - R_C I_C - R_E(I_C + I_B) \approx V_{CC} - R_{DC} I_C$$

where we've introduced the **DC load-line resistance**

$$R_{DC} \triangleq R_C + R_E \tag{4}$$

Adding the DC components to the signal components yields the total collector current and voltage

$$i_c = I_C + I_m \sin \omega t \qquad v_{CE} = V_{CE} - V_m \sin \omega t \tag{5}$$

Our next task is to determine the bounds on the signal amplitudes I_m and V_m.

Figure 12.1–2a displays the DC and AC relationships together in graphical form. Here, we have constructed the usual DC load line with slope $1/R_{DC}$. But the signal variations fall along a different line defined by

$$(i_c - I_C)/(v_{CE} - V_{CE}) = i_c/v_{ce} = -1/R_{AC}$$

This **AC load line** passes through the Q point with slope $1/R_{AC}$, a steeper slope than the DC load line since $R_{AC} < R_{DC}$. Accordingly, the horizontal intercept is $V_0 < V_{CC}$ and the vertical intercept is $I_0 > V_{CC}/R_{DC}$. As indicated on the figure, we want those intercepts to be at $V_0 = 2V_{CE}$ and $I_0 = 2I_C$ so the Q point will be *centered* on the AC load line. Simple geome-

Figure 12.1–2

(a) DC and AC load lines. (b) Collector-current swing.

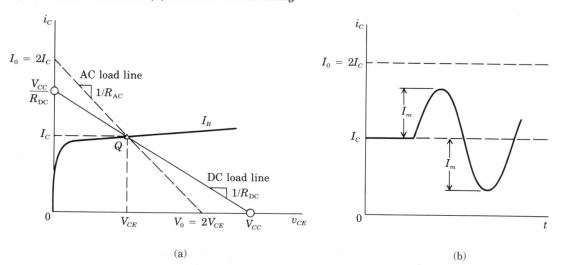

(a)   (b)

try then gives the corresponding values of V_{CE} and I_C as

$$I_C = \frac{V_{CC}}{R_{AC} + R_{DC}} \qquad V_{CE} = R_{AC}I_C = \frac{R_{AC}}{R_{AC} + R_{DC}} V_{CC} \qquad \textbf{(6)}$$

These Q-point values are established by $I_B = I_C/\beta$ with I_C calculated from Eq. (6).

Centering the Q point on the AC load line is important for large-signal amplification because it allows **maximum symmetric swing.** To clarify this concept, consider the waveform i_C sketched in Fig. 12.1–2b. The positive signal swing must satisfy $I_C + I_m \leq I_0$, while the negative swing must satisfy $I_C - I_m \geq 0$. Likewise, the signal swings of v_{CE} must satisfy $V_{CE} + V_m \leq V_0$ and $V_{CE} - V_m \geq 0$. Hence, if $I_0 = 2I_C$ and $V_0 = 2V_{CE}$, then we can get maximum symmetric signal swings with amplitudes bounded by

$$I_m \leq I_C \qquad V_m \leq V_{CE} \qquad \textbf{(7)}$$

In practice, the maximum amplitudes should be somewhat smaller to avoid the nonlinearities near saturation and cutoff.

Now, proceeding to the power calculations, we note that the voltage across the load R_L is $v_{out} = -V_m \sin \omega t$. Therefore, the average signal power delivered to the load will be

$$P_L = V_m^2/2R_L \qquad \textbf{(8a)}$$

With a centered Q point and $V_m = V_{CE}$, the **maximum possible load power** is

$$P_{LM} = V_{CE}^2/2R_L \qquad \textbf{(8b)}$$

This AC power actually comes from the power supply, which also provides DC power dissipated in the circuit. We'll ignore the small DC base current, and we'll find the average supplied power by first writing the instantaneous power

$$p_{CC}(t) = V_{CC}\left(i_C + \frac{v_{out}}{R_L}\right) = V_{CC}I_C + V_{CC}\left(I_m - \frac{V_m}{R_L}\right)\sin \omega t$$

The entire second term averages to zero, so the average value of $p_{CC}(t)$ simply becomes

$$P_{CC} = V_{CC}I_C \qquad \textbf{(9)}$$

The amplifier's **power conversion efficiency** is then defined by the ratio

$$\text{Eff} \triangleq P_L/P_{CC} \qquad \textbf{(10)}$$

As illustrated shortly, the basic common-emitter circuit has very low efficiency — generally less than 8%.

Our final consideration here concerns the **transistor ratings.** Every BJT is rated for maximum allowable values of v_{CE} and i_C. Furthermore, there is a maximum allowable value of the average **transistor power dissipation** $P_D = (v_{CE}i_C)_{av}$. Taken together, $i_{C_{max}}$, $v_{CE_{max}}$, and $P_{D_{max}}$ define the safe operating region shown in Fig. 12.1–3. Observe that the dissipation rating plots as the hyperbolic curve $v_{CE}i_C = P_{D_{max}}$. Not shown here is the fact that $P_{D_{max}}$ depends upon temperature, and the rated value must be reduced when the transistor operates above normal room temperature. The use of heat sinks to keep temperature down will be covered in Section 12.3.

Figure 12.1–3

Safe operating region of a BJT.

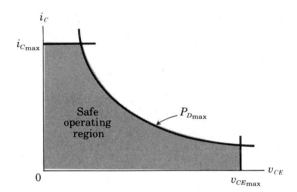

Comparing Fig. 12.1–3 with the AC load line in Fig. 12.1–2a, we see that the common-emitter circuit must have $I_0 \leq i_{C_{max}}$ and $V_0 \leq v_{CE_{max}}$ in order to stay within the safe operating region. As for the transistor power dissipation, its instantaneous value is found from Eq. (5) to be

$$p_D(t) = v_{CE}i_C$$
$$= V_{CE}I_C + (V_{CE}I_m - V_mI_C)\sin \omega t - V_mI_m \sin^2 \omega t$$

Averaging this expression yields

$$P_D = V_{CE}I_C - \tfrac{1}{2}V_mI_m \tag{11a}$$

so maximum dissipation occurs in the no-signal or *idle* condition when $V_m = I_m = 0$ and

$$P_{DM} = V_{CE}I_C \tag{11b}$$

Thus, we have the additional requirement $P_{DM} \leq P_{D_{max}}$ for safe operation. When this requirement is satisfied, the entire AC load line falls below the $P_{D_{max}}$ hyperbola.

Example 12.1–1

Suppose the element values of a large-signal common-emitter amplifier are

$$V_{CC} = 50 \text{ V} \qquad R_C = 150 \ \Omega \qquad R_E = 40 \ \Omega \qquad R_L = 100 \ \Omega$$

Since $R_{DC} = 190 \ \Omega$ and $R_{AC} = 60 \ \Omega$, maximum symmetric swing is obtained with a centered Q point per Eq. (6) at $I_C = 0.2$ A and $V_{CE} = 12$ V.

Hence, the transistor must be rated for $i_{C_{max}} \geq 2I_C = 0.4$ A, $v_{CE_{max}} \geq 2V_{CE} = 24$ V, and $P_{D_{max}} \geq P_{DM} = V_{CE}I_C = 2.4$ W. Equations (8b) and (9) then give $P_{LM} = 0.72$ W and $P_{CC} = 10$ W, so Eff $\leq P_{LM}/P_{CC} = 7.2\%$.

The major factor contributing to this poor efficiency is power lost as ohmic heating in R_C and R_E. For instance, the load receives $P_L = P_{LM} = 0.72$ W under the maximum signal conditions $V_m = V_{CE}$ and $I_m = I_C$, and Eq. (11a) indicates that the transistor dissipates $P_D = \frac{1}{2}V_{CE}I_C = 1.2$ W. It therefore follows that R_C and R_E dissipate $P_{CC} - (P_D + P_L) = 8.08$ W — more than 80% of the supplied power!

Exercise 12.1–1   Let the Q point of the amplifier in Example 12.1–1 be at the middle of the DC load line rather than the AC load line, so that $V_{CE} = V_{CC}/2$ and $I_C = V_{CC}/2R_{DC}$. Redraw Fig. 12.1–2 for this case to find the resulting values of V_0 and I_0 and to show that a symmetric output swing must have $V_m \leq 7.9$ V. Then calculate the maximum load power and conversion efficiency.

Transformer-Coupled Amplifiers

One way of improving the efficiency of a BJT power amplifier is to replace R_C with a transformer coupled to the load. Figure 12.1–4a diagrams such a transformer-coupled circuit. The transformer is assumed to be nearly ideal, and it has been labeled for step-down voltage transformation because

Figure 12.1–4
Transformer-coupled power amplifier: (a) circuit diagram; (b) DC and AC load lines.

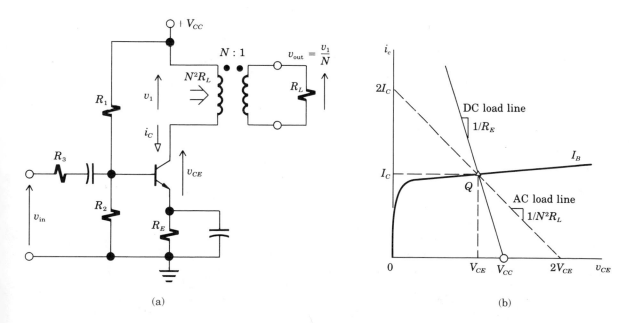

(a)   (b)

most power loads have relatively small values of R_L. We still include a bypassed emitter resistor for stabilization of the operating point.

The DC load-line resistance equals R_E, since the primary of the transformer acts like a short circuit at DC. But at signal frequencies, the transformer reflects into the primary the equivalent AC load resistance N^2R_L. The corresponding load lines are drawn in Fig. 12.1–4b for the normal case of $R_E < N^2R_L$. Consequently, the horizontal intercept of the AC load line falls *above* the supply voltage V_{CC}. The Q point will be at the center of the AC load line, as shown, when

$$I_C = \frac{V_{CC}}{N^2R_L + R_E} \qquad V_{CE} = N^2R_L I_C = \frac{N^2R_L}{N^2R_L + R_E} V_{CC} \qquad (12)$$

per Eq. (6) with $R_{AC} = N^2R_L$ and $R_{DC} = R_E$. If $R_E \ll N^2R_L$, then $V_{CE} \approx V_{CC}$ and $I_C \approx V_{CC}/N^2R_L$.

An AC input signal produces $i_C = I_C + I_m \sin \omega t$ with I_m given by Eq. (1). The resulting AC voltage across the primary is $v_1 = (N^2R_L)I_m \sin \omega t$, and step-down transformation yields

$$v_{\text{out}} = v_1/N = (V_m/N) \sin \omega t \qquad (13a)$$

where

$$V_m = (N^2R_L)I_m \leq (N^2R_L)I_C = V_{CE} \qquad (13b)$$

The load thus receives the AC signal power $P_L = (V_m/N)^2/2R_L = V_m^2/2N^2R_L$. Setting $V_m = V_{CE}$ and using Eq. (12), we obtain the maximum load power

$$P_{LM} = \frac{V_{CE}^2}{2N^2R_L} = \frac{N^2R_L V_{CC}^2}{2(N^2R_L + R_E)^2} \qquad (14)$$

Note that $P_{LM} \approx V_{CC}^2/2N^2R_L$ if $R_E \ll N^2R_L$.

Next, we calculate the supplied power P_{CC} by averaging $p_{CC}(t) = V_{CC}i_C = V_{CC}I_C + V_{CC}I_m \sin \omega t$, from which

$$P_{CC} = V_{CC}I_C = V_{CC}^2/(N^2R_L + R_E)$$

Hence, the power conversion efficiency of the transformer-coupled amplifier is

$$\text{Eff} = \frac{P_L}{P_{CC}} \leq \frac{P_{LM}}{P_{CC}} = \frac{N^2R_L}{2(N^2R_L + R_E)} \leq \frac{1}{2} \qquad (15)$$

In theory, then, the efficiency approaches 50% when $V_m = V_{CE}$ and $R_E \ll N^2R_L$.

Maximum transistor power dissipation occurs in the idle condition and is the same as that of our common-emitter amplifier, namely

$$P_{DM} = V_{CE}I_C = V_{CE}^2/N^2R_L \qquad (16a)$$

Thus, from Eq. (14),

$$P_{DM}/P_{LM} = 2 \qquad (16b)$$

We call this informative ratio the amplifier's **figure of merit.** It tells us that the transistor in a transformer-coupled circuit must be rated to dissipate at least twice the maximum load power.

A related concern is stability of the Q point. The entire AC load line should, of course, be inside the safe operating region. But an unforseen increase of the DC current I_C could drive the Q point up the DC load line such that $V_{CE}I_C > P_{D_{max}}$, and the transistor might be damaged by excessive heat. The self-biased arrangement in Fig. 12.1–4a minimizes this threat, provided that $R_E \gg (R_1 \| R_2)/\beta$.

Finally, it should be mentioned that a *real* transformer has inductive and capacitive reactances that can adversely affect the frequency response of the amplifier. For example, Fig. 12.1–5 shows a frequency-response curve with a resonant peak caused by the transformer's reactances. Even in absence of resonance, the inevitable low-frequency roll-off may be unacceptable for some applications.

Figure 12.1–5

Typical frequency response with transformer coupling.

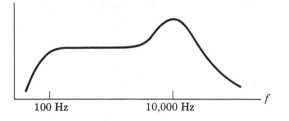

100 Hz   10,000 Hz f

Example 12.1–2

Design of a power amplifier

Consider the following design problem: You want to build a transformer-coupled audio amplifier that delivers up to 5 watts of AC signal power to an 8-Ω loudspeaker, given $|v_{in}| \leq 2$ V. You have available a power BJT with $\beta = 60$ and rated for $i_{C_{max}} = 1$ A, $v_{CE_{max}} = 100$ V, and $P_{D_{max}} = 15$ W. The dissipation rating allows a margin of safety, since $P_{DM} = 2P_{LM} = 10$ W. You need to select appropriate values of I_C, V_{CE}, N, V_{CC}, and the various resistors.

Equation (14) provides a convenient starting point for the design calculations when rewritten as

$$V_{CE} = \sqrt{2N^2 R_L P_{LM}} = \sqrt{80}\, N$$

from which

$$I_C = V_{CE}/N^2 R_L = \sqrt{80}/8N$$

But the conditions $2I_C \leq i_{C_{max}}$ and $2V_{CE} \leq v_{CE_{max}}$ require $2\sqrt{80}/8N \leq 1$ A and $2\sqrt{80}\, N \leq 100$ V, respectively. Combining these conditions yields the allowed range for the transformer's turns ratio as

$$2\sqrt{80}/8 = 2.24 \leq N \leq 100/2\sqrt{80} = 5.59$$

If you select $N = 5$, for instance, then $I_C = 0.224$ A, $V_{CE} = 44.7$ V, and $N^2 R_L = 200$ Ω.

Next, you might take $R_E = N^2 R_L/10 = 20\ \Omega$ so Eff $\leq 200/440 \approx 45\%$, from Eq. (15). The necessary supply voltage is then found via Eq. (12) to be

$$V_{CC} = (N^2 R_L + R_E)I_C \approx 50\ \text{V}$$

Now the bias resistors can be determined using Eqs. (20)–(22), Section 11.2, which yield $R_1 \approx 1100\ \Omega$ and $R_2 \approx 130\ \Omega$ with $R_B = R_1 \| R_2 \approx 120\ \Omega$.

Finally, drawing upon Eq. (1), you set $I_m = I_C$ and $V_s = 2$ V to obtain $R_3 = \beta V_s/I_C \approx 540\ \Omega$. Since the BJT has $r_\pi = \beta/40I_C \approx 7\ \Omega$, you conclude that $r_{be} \ll R_1 \| R_2$ and $R_3 \gg r_{be}$, as required for the current input.

Exercise 12.1–2 Suppose you want V_{CE} to be as small as possible for the amplifier in Example 12.1–2. Find the values of N, I_C, V_{CC}, and R_3, taking $R_E = N^2 R_L/10$.

Push-Pull Amplifiers

As we have seen, a transformer-coupled amplifier must be biased at $I_C \geq I_m$. Consequently, the circuit requires constant supply power, regardless of the inut signal, and the transistor dissipates maximum power when idling. That situation is very wasteful in view of the fact that typical nonsinusoidal signals have frequent idle or low-level intervals. A push-pull amplifier achieves much better overall performance because it draws supply power proportional to the signal and it dissipates no power when idling.

To introduce the underlying strategy of push-pull amplification, consider the simple emitter follower in Fig. 12.1–6a. Here, the load connects directly at the emitter, and a large signal voltage v_{in} is applied to the base through a biasing battery with $V_{BB} = V_\gamma$. If v_{in} is positive, then the BJT

Figure 12.1–6
Class-B emitter follower: (a) circuit; (b) waveforms.

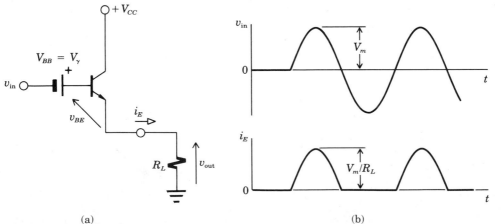

(a)   (b)

becomes active and $v_{BE} = V_\gamma$ so $v_{out} = v_{in} + V_{BB} - v_{BE} = v_{in}$. The resulting emitter current is $i_E = v_{out}/R_L = v_{in}/R_L$. However, if v_{in} is negative, then the BJT goes off so $i_E = 0$ and $v_{out} = 0$. The waveforms v_{in} and i_E sketched in Fig. 12.1–6b show that the transistor conducts during just one-half of each sinusoidal cycle when $v_{in} = V_m \sin \omega t$. Circuits having this characteristic are called **class-B amplifiers,** as distinguished from **class-A amplifiers** in which the transistors conduct over the entire sinusoidal cycle.

A push-pull amplifier consists of *two* class-B circuits arranged such that one conducts during positive signal excursions and the other conducts during negative excursions. Figure 12.1–7a diagrams the arrangement known as a **complementary-symmetry push-pull amplifier.** The upper portion is an *npn* emitter follower identical to Fig. 12.1–6a. The lower portion is a *pnp* emitter follower having complementary characteristics, so the *pnp* transistor acts like the *npn* transistor with all voltages and currents reversed. This configuration permits *direct coupling* at the input and output, but it requires two power supplies as well as a complementary pair of transistors.

Figure 12.1–7

Complementary-symmetry push-pull amplifier: (a) circuit; (b) waveforms.

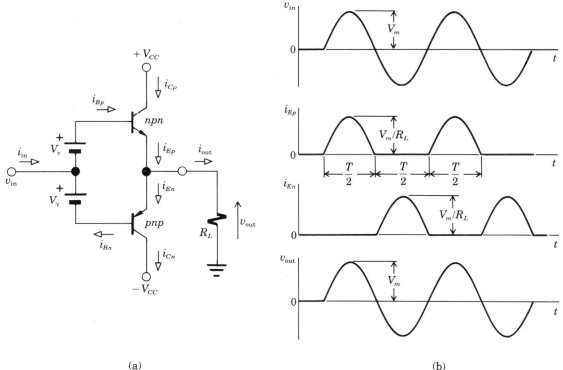

(a)   (b)

The name "push-pull" stems from the property that $i_{out} = i_{Ep} - i_{En}$, where $i_{En} = 0$ when $v_{in} > 0$ and $i_{Ep} = 0$ when $v_{in} < 0$. Hence, the upper transistor "pushes" positive output current while the lower transistor "pulls" negative output current. Figure 12.1–7b depicts the push-pull waveforms with $v_{in} = V_m \sin \omega t$, in which case

$$v_{out} = R_L(i_{Ep} - i_{En}) = V_m \sin \omega t \tag{17}$$

Although the circuit has no voltage gain, it does have current gain since $i_{Ep} = (\beta + 1)i_{Bp}$, $i_{En} = (\beta + 1)i_{Bn}$, and $i_{out} = i_{Ep} - i_{En} = (\beta + 1)(i_{Bp} - i_{Bn}) = (\beta + 1)i_{in}$. Consequently, we can write $v_{out} = R_L i_{out} = R_L(\beta + 1)i_{in}$. But $v_{out} = v_{in}$, so the equivalent input resistance is

$$R_i = v_{in}/i_{in} = (\beta + 1)R_L \tag{18}$$

The large value of R_i justifies ignoring the nonlinear base-emitter resistance r_{be}.

The waveform fidelity of our push-pull amplifier depends upon proper base biasing voltages. Without biasing, neither transistor would conduct when $|v_{in}| < V_\gamma$ and the output signal would exhibit **cross-over distortion** illustrated in Fig. 12.1–8. Additionally, the input amplitude must satisfy

$$V_m \le V_{CC} - V_{sat} \tag{19}$$

This limit keeps the BJTs out of saturation, and the voltage following behavior of the emitter followers thereby ensures good linearity. Thus, with $v_{out} = v_{in} = V_m \sin \omega t$, the average power delivered to the load is

$$P_L = V_m^2/2R_L \le V_{CC}^2/2R_L \tag{20}$$

The upper bound corresponds to $V_{CC} \gg V_{sat}$ and $V_m \approx V_{CC}$.

Figure 12.1–8

Cross-over distortion in push-pull amplification.

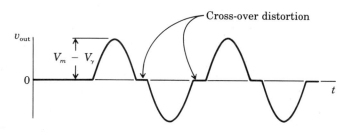

Now, to determine the supplied power, we account for both supplies by writing $p_{CC}(t) = V_{CC}i_{Cp} + V_{CC}i_{Cn}$, where $i_{Cp} \approx i_{Ep}$ and $i_{Cn} \approx i_{En}$ in the usual case of $\beta \gg 1$. Since the waveforms i_{Ep} and i_{En} in Fig. 12.1–7b consist of a sinusoidal half cycle followed by a zero half cycle, their average values are found by performing the integration

$$(i_E)_{av} = \frac{1}{T} \int_0^T i_E \, dt = \frac{1}{T} \int_0^{T/2} \frac{V_m}{R_L} \sin \omega t \, dt$$

But $T = 2\pi/\omega$, so

$$(i_E)_{av} = \frac{\omega V_m}{2\pi R_L} \int_0^{\pi/\omega} \sin \omega t \, dt = \frac{V_m}{\pi R_L} \qquad (21)$$

Therefore, the average supplied power is

$$P_{CC} \approx V_{CC}(i_{Ep})_{av} + V_{CC}(i_{En})_{av} = 2V_{CC}V_m/\pi R_L \qquad (22)$$

This result confirms that P_{CC} varies in proportion to the signal amplitude V_m and, accordingly, the circuit draws no power in the idle condition. Taking the ratio of Eqs. (20) and (22) then yields the conversion efficiency

$$\text{Eff} = \frac{P_L}{P_{CC}} = \frac{\pi V_m}{4V_{CC}} \leq \frac{\pi}{4} \qquad (23)$$

The theoretical maximum efficiency is about 79% when $V_m \approx V_{CC}$.

Besides increasing conversion efficiency, the push-pull arrangement reduces transistor power dissipation. Since each transistor dissipates average power P_D, and since $P_{CC} - P_L = 2P_D$, it follows that

$$P_D = \frac{1}{2}(P_{CC} - P_L) = \frac{V_{CC}}{\pi R_L} V_m - \frac{1}{4R_L} V_m^2 \qquad (24)$$

Figure 12.1–9 plots P_{CC}, P_L, and P_D versus signal amplitude V_m, showing that maximum transistor dissipation occurs at $V_m = 2V_{CC}/\pi$ where

$$P_{DM} = V_{CC}^2/\pi^2 R_L \qquad (25a)$$

The figure of merit is thus

$$P_{DM}/P_{LM} = 2/\pi^2 \approx 0.2 \qquad (25b)$$

so the transistors need only be rated for about 20% of the maximum output

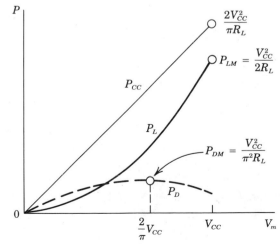

Figure 12.1–9
Plots of supply power, load power, and transistor dissipation versus signal amplitude.

power. The required current and voltage ratings are

$$i_{C_{max}} \geq V_{CC}/R_L \qquad v_{CE_{max}} \geq 2V_{CC} \tag{26}$$

which follow from Fig. 12.1–7a with one transistor off and the other nearly saturated.

Last, we should say a few words about practical circuits. There are several different implementations of the push-pull principle, and none of them requires biasing batteries. One popular way to obtain the biasing voltages involves two diodes connected as shown in Fig. 12.1–10a, where resistors R_D carry DC current that forward biases the diodes. The negative supply voltage can also be eliminated if coupling capacitors are inserted at the input and output. Figure 12.1–10b diagrams such a capacitor-coupled circuit with the single supply voltage labeled $2V_{CC}$ to agree with our previous results.

Figure 12.1–10
Push-pull circuits with:
(a) biasing diodes;
(b) coupling capacitors.

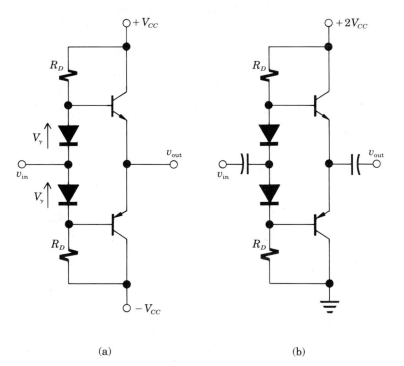

(a)   (b)

Both circuits in Fig. 12.1–10 have $v_{out} = v_{in}$, so the driver stage usually must be a large-signal amplifier — perhaps a common-emitter amplifier or an op-amp. (Incidentally, many op-amps have push-pull output stages similar to Fig. 12.1–10a.) Sometimes, the load is connected via a step-up transformer to get $v_{out} = Nv_{in}$. The reflected load resistance R_L/N^2 then replaces R_L in our previous equations, and Eq. (20) becomes

$$P_L = N^2 V_m^2/2R_L \leq N^2 V_{CC}^2/2R_L \tag{27}$$

Thus, the transformer makes it possible to deliver more output **power** when V_{CC} is small or R_L is large.

Exercise 12.1–3   Let $R_L = 8\Omega$ and $P_{LM} = 25$ W. Find V_{CC}, P_{CC}, and the transistor ratings for the push-pull amplifier in Fig. 12.1–7a. Then, as a comparison, find the same quantities for a class-A transformer-coupled amplifier with $N = 1$ and $R_E \ll R_L$.

12.2 RECTIFIERS AND POWER SUPPLIES

This section concerns AC-to-DC conversion for the purpose of supplying power to electronic apparatus, battery chargers, and other types of electrical loads. Our topics include diode rectifiers, smoothing capacitors, voltage regulators, and DC motor supplies. We'll also discuss power control using controlled rectifiers.

Rectification and Voltage Smoothing

Given an AC source, we can produce a nearly constant supply voltage with the help of a rectifier circuit and a capacitor. The rectifier converts alternating voltage into a pulsating but nonalternating waveform, and the capacitor smooths out the pulsations. These operations are analyzed here, starting with the properties of half-wave and full-wave rectifiers.

Figure 12.2–1a diagrams the circuit of a **half-wave rectifier** with a

Figure 12.2–1
Half-wave rectifier circuit and waveforms.

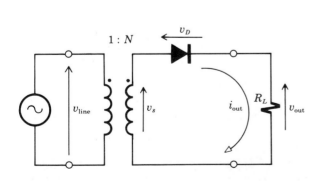

(a)

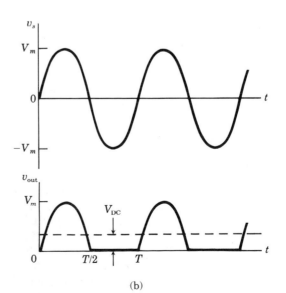

(b)

transformer at the input and a load represented by resistance R_L at the output. The transformer isolates the rectifier from the incoming AC line as well as providing the desired voltage amplitude across its secondary. We'll let

$$v_{\text{line}} = \sqrt{2}V_{\text{rms}} \sin \omega t \qquad \omega = 2\pi f = 2\pi/T \qquad \textbf{(1a)}$$

so transformation produces

$$v_s = V_m \sin \omega t \qquad V_m = N\sqrt{2}V_{\text{rms}} \qquad \textbf{(1b)}$$

A step-down transformer would, of course, have $N < 1$. In any case, we'll assume that V_m is large compared to the diode's forward voltage drop $v_D \approx V_\gamma$. Thus, $v_{\text{out}} = v_s - v_D \approx v_m \sin \omega t$ during the positive half cycles of v_s, while $v_{\text{out}} = 0$ during the negative half cycles.

The half-rectified voltage waveform is plotted in Fig. 12.2–1b. The current waveform is $i_{\text{out}} = v_{\text{out}}/R_L$ and has the same shape. Since v_{out} and i_{out} never go negative, they have nonzero DC components given by

$$V_{\text{DC}} = (v_{\text{out}})_{\text{av}} = \frac{V_m}{\pi} \qquad I_{\text{DC}} = (i_{\text{out}})_{\text{av}} = \frac{V_m}{\pi R_L} \qquad \textbf{(2)}$$

These average values follow from the same integration method used for Eq. (21), Section 12.1.

Two implementations of a **full-wave rectifier** are diagrammed in Fig. 12.2–2 along with the relevant waveforms. The circuit in Fig. 12.2–2a employs a pair of diodes and a transformer with a *center-tapped secondary* and total turns ratio $2N$, so v_s appears across each half of the secondary. Diode D_1 conducts when $v_s > 0$, and D_2 conducts when $v_s < 0$. Hence, i_1 and i_2 are half-rectified currents, with i_2 shifted in time by $T/2$ compared to i_1. The resulting full-rectified current is $i_{\text{out}} = i_1 + i_2 = (V_m/R_L)|\sin \omega t|$, and $v_{\text{out}} = V_m|\sin \omega t|$ as shown. Clearly,

$$V_{\text{DC}} = (v_{\text{out}})_{\text{av}} = \frac{2V_m}{\pi} \qquad I_{\text{DC}} = (i_{\text{out}})_{\text{av}} = \frac{2V_m}{\pi R_L} \qquad \textbf{(3)}$$

since the area under a full-rectified wave is twice that of a half-rectified wave.

The circuit in Fig. 12.2–2b produces the same full-rectified waveforms via four diodes arranged in a *bridge*. Diodes D_1 and D_3 conduct to carry i_1 when $v_s > 0$, while D_2 and D_4 conduct to carry i_2 when $v_s < 0$. Although the bridge configuration involves four diodes, it offers the advantage of a smaller and less costly transformer. Other advantages will emerge subsequently.

Any one of these simple rectifier circuits could be used for battery charging, electroplating, or related applications that require unidirectional but not necessarily constant current. However, DC power supplies require nearly constant voltage, which can be achieved by putting a capacitor in parallel with the load.

Figure 12.2–2

Full-wave rectification: (a) center-tapped circuit; (b) bridge circuit; (c) waveforms.

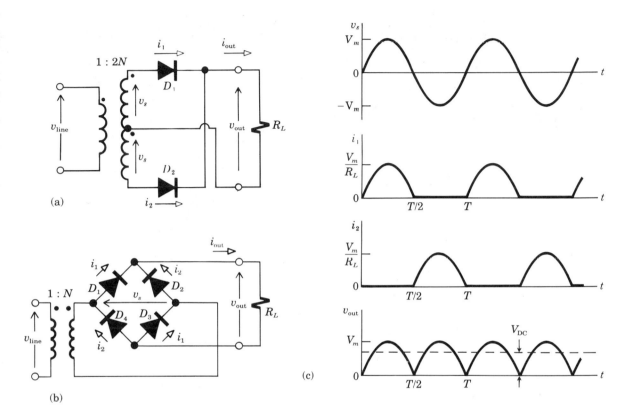

Figure 12.2–3a is a half-wave rectifier with **smoothing capacitor** C. If $R_L C \gg T$, then the circuit essentially becomes a *peak detector* (like Fig. 10.1–10). The steady-state voltage waveforms are plotted in Fig. 12.2–3b, taking $v_s = V_m \cos \omega t$ to get $v_s = V_m$ at $t = 0$ for convenience. The capacitor discharges slightly during the time interval $0 < t < T - \Delta t$, and v_{out} decays from V_m to $V_m - V_r$. The diode will be off throughout this interval because $v_D = v_{\text{out}} - v_s < 0$, so i_{out} comes entirely from the discharging capacitor. The diode goes on when $v_{\text{out}} = v_s$ at $t = T - \Delta t$, and i_D recharges the capacitor until $v_{\text{out}} = V_m$ at $t = T$. These discharge/recharge operations continue periodically for $t > T$.

The output voltage change V_r is called the **peak-to-peak ripple**. A good power supply should have $V_r \ll V_m$, in which case

$$V_{\text{DC}} = (v_{\text{out}})_{\text{av}} \approx V_m - \tfrac{1}{2} V_r \tag{4}$$

To find the value of V_r, we invoke the condition $T/R_L C \ll 1$ together with the fact that the recharging interval Δt will be small compared to T. Thus,

Figure 12.2–3
Half-wave rectifier with smoothing capacitor: (a) circuit; (b) voltage waveforms;
(c) diode current.

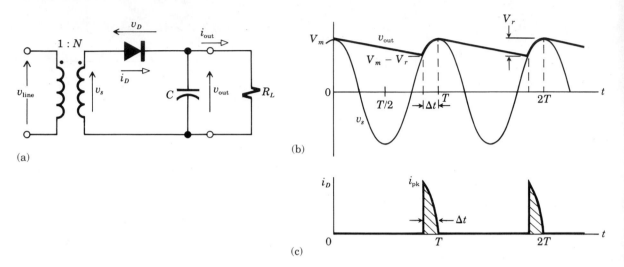

(a)

(b)

(c)

at $t = T - \Delta t$,

$$v_{\text{out}} = V_m - V_r = V_m e^{-(T-\Delta t)/R_L C} \approx V_m e^{-T/R_L C} \approx V_m\left(1 - \frac{T}{R_L C}\right)$$

Solving for V_r yields

$$V_r \approx \frac{V_m}{fR_L C} \approx \frac{I_{\text{DC}}}{fC} \tag{5}$$

where we have inserted $f = 1/T$ and $I_{\text{DC}} = V_{\text{DC}}/R_L \approx V_m/R_L$.

Equation (5) tells us the required value of C, given the maximum allowable ripple voltage and the maximum current drawn by the load. In addition, the combination of Eqs. (4) and (5) leads to

$$V_{\text{DC}} \approx V_m - R_o I_{\text{DC}} \tag{6a}$$

with

$$R_o = \frac{1}{2fC} \tag{6b}$$

Thus, we interpret R_o as the equivalent DC output resistance of the rectifier. Effective voltage smoothing and low output resistance almost always call for the use of large electrolytic capacitors.

Now consider the diode, which will be reverse biased by $v_D = v_s - v_{\text{out}} < 0$ whenever $v_{\text{out}} > v_s$. The waveforms in Fig. 12.2–3b indicate that $(v_{\text{out}} - v_s)_{\text{max}} \approx 2V_m$ at $t \approx T/2$. Consequently, the diode must be rated to withstand the **peak inverse voltage**

$$\text{PIV} \triangleq (-v_D)_{\text{max}} \approx 2V_m \tag{7}$$

Furthermore, the diode must carry the forward current i_D that recharges the capacitor when $v_{\text{out}} = v_s$. This current consists of roughly triangular pulses with height i_{pk} and duration Δt, as sketched in Fig. 12.2–3c.

We evaluate Δt by noting in Fig. 12.2–3b that $V_m - V_r = V_m \cos \omega t$ at $t = T - \Delta t$. Since $\cos \omega(T - \Delta t) = \cos (2\pi - \omega \, \Delta t) = \cos \omega \, \Delta t$, we'll introduce the **conduction angle**

$$\phi \triangleq \omega \, \Delta t = \arccos\left(1 - \frac{V_r}{V_m}\right) \tag{8}$$

Each triangular current pulse transfers charge $\Delta q \approx \tfrac{1}{2} i_{\text{pk}} \, \Delta t$ and increases the voltage across the capacitor by $\Delta v = \Delta q/C = V_r$, so

$$i_{\text{pk}} \approx 2 \, \Delta q/\Delta t = 2CV_r/\Delta t = 4\pi fCV_r/\phi$$

After converting 4π radians to $720°$ and substituting $fCV_r \approx V_m/R_L \approx I_{\text{DC}}$ per Eq. (5), we finally get the diode's **peak forward current**

$$i_{\text{pk}} \approx \frac{720°}{\phi} \frac{V_m}{R_L} \approx \frac{720°}{\phi} I_{\text{DC}} \tag{9}$$

If $V_r \ll V_m$, as desired, then $\phi \ll 720°$ and the diode must be rated for $i_{\text{pk}} \gg I_{\text{DC}}$.

Ripple voltage and peak diode current can be reduced if we smooth the output of a *full-wave* rectifier, rather than a half-wave rectifier. The resulting voltage waveform looks like Fig. 12.2–4, and Eqs. (4) and (8) still hold. But the capacitor now gets recharged *twice* every T seconds. Accordingly, V_r, R_o, and i_{pk} are cut in half to

$$V_r \approx \frac{I_{\text{DC}}}{2fC} \qquad R_o \approx \frac{1}{4fC} \qquad i_{\text{pk}} \approx \frac{360°}{\phi} I_{\text{DC}} \tag{10}$$

The peak inverse voltage across each diode depends upon the circuit con-

Figure 12.2–4
Voltage waveform for a full-wave rectifier with smoothing capacitor.

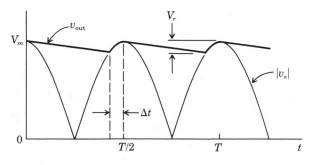

figuration and is given by

$$\text{PIV} \approx \begin{cases} 2V_m & \text{Center-tapped circuit} \\ V_m & \text{Bridge circuit} \end{cases} \qquad \textbf{(11)}$$

Thus, the full-wave bridge circuit allows us to use diodes rated for the smallest possible peak inverse voltage and peak forward current.

Example 12.2–1

A DC power supply is needed to deliver $i_{\text{out}} \leq 0.6$ A at $v_{\text{out}} = 50 \pm 2$ V, given a standard AC input with $V_{\text{rms}} = 120$ V and $f = 60$ Hz. The required capacitor and diode ratings are found by taking $I_{\text{DC}} = 0.6$ A, $V_{\text{DC}} = 50$ V, and $V_r = 4$ V. If a full-wave bridge circuit is used, then

$$C \approx \frac{I_{\text{DC}}}{2fV_r} \approx 1250 \ \mu\text{F} \qquad \text{PIV} \approx V_m \approx V_{\text{DC}} + \tfrac{1}{2}V_r = 52 \text{ V}$$

$$\phi = \arccos\left(1 - \frac{V_r}{V_m}\right) \approx 23° \qquad i_{\text{pk}} \approx \frac{360°}{\phi} I_{\text{DC}} \approx 9.5 \text{ A}$$

The transformer must have

$$N = V_m/\sqrt{2}V_{\text{rms}} \approx 1/3.3$$

which calls for step-down transformation.

Exercise 12.2–1

Suppose the power supply in Example 12.2–1 is built using a full-wave bridge circuit with $V_m = 52$ V and $C = 1250 \ \mu\text{F}$. Find the values of V_r, V_{DC}, and i_{pk} when $R_L = 50 \ \Omega$.

Exercise 12.2–2

Trace the path of i_2 in Figs. 12.2–2a and 12.2–2b when $v_s = -V_m$ and a smoothing capacitor keeps $v_{\text{out}} \approx V_m$. Then justify Eq. (11) by finding the voltages across the nonconducting diodes.

Voltage Regulators

Most DC power supplies include a voltage regulator to hold the output voltage nearly constant, despite fluctuations of the output current or the input AC line voltage. We'll illustrate these features by examining the operating principles of a simple **series regulator.**

Figure 12.2–5a shows a regulator circuit with a BJT for voltage control and a Zener diode for voltage reference. The Zener diode has been labeled in terms of its *reverse* current i_Z and voltage v_Z, related by the *i-v* curve plotted in Fig. 12.2–5b (an inverted version of Fig. 8.2–8b). Resistor R_B carries i_Z plus the BJT base current. Under normal conditions, the unregulated voltage v_{in} exceeds the regulated voltage v_{out} by an amount sufficient to keep the BJT active and to reverse bias the Zener diode. Then $v_{BE} \approx V_\gamma$, $v_Z \approx V_Z$, and

$$v_{\text{out}} = v_Z - v_{BE} \approx V_Z - V_\gamma$$

Figure 12.2–5
Series voltage regulator: (a) schematic diagram; (b) Zener diode reverse $i\text{-}v$ curve; (c) large-signal equivalent circuit.

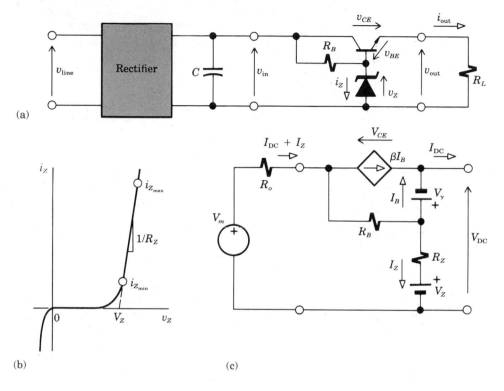

(a)

(b)

(c)

Hence, v_{out} remains essentially constant, independent of any variations of v_{in} or i_{out}. The BJT regulates v_{out} by absorbing the voltage difference $v_{CE} = v_{\text{in}} - v_{\text{out}}$. A power BJT is usually needed to handle $i_C \approx i_{\text{out}}$ and the resulting power dissipation $p_D = v_{CE}i_C \approx (v_{\text{in}} - v_{\text{out}})i_{\text{out}}$.

For a quantitative study of series regulation, we draw upon the large-signal DC equivalent circuit in Fig. 12.2–5c. Here, we have represented the rectifier by its open-circuit DC voltage V_m and its equivalent DC output resistance R_o. We assume that the Zener diode operates on the linear portion of Fig. 12.2–5b and has

$$R_Z \ll R_o + R_B$$

Solving for the regulated DC output voltage then yields

$$V_{\text{DC}} = V_m' - R_o' I_{\text{DC}} \tag{12a}$$

with

$$V_m' \approx V_Z - V_\gamma + \frac{R_Z}{R_o + R_B} V_m \qquad R_o' \approx \frac{R_Z}{R_o + R_B}\left(R_o + \frac{R_B}{\beta + 1}\right) \tag{12b}$$

The expression for V'_m reveals that any variations of V_m are reduced by the factor $R_Z/(R_B + R_o) \ll 1$. Furthermore, the regulator's output resistance is $R'_o \ll R_o + R_B/(\beta + 1)$, indicating that loading effects are substantially reduced. Since R_o is inversely proportional to the smoothing capacitance C, a regulated power supply may get by with a smaller capacitor.

However, the foregoing results hold only when the DC Zener current stays in the range

$$i_{Z_{\min}} \le I_Z \le i_{Z_{\max}} \tag{13a}$$

From further analysis of Fig. 12.2–5c we get

$$I_Z \approx \frac{V_m - V_Z}{R_o + R_B} - \frac{R'_o}{R_Z} I_{\mathrm{DC}} \tag{13b}$$

so I_Z increases with V_m and decreases with I_{DC}. Equation (13a) therefore imposes limits on the allowable ranges of V_m and I_{DC}.

Complete voltage regulators are available as IC packages, including protection circuitry for thermal shutdown if the power transistor gets too hot. Some models have fixed output voltages at a standard value such as 5, 6, 8, 12, 15, 18, or 24 V. Others have provision for variable output voltage, which can be adjusted by the user over a wide range such as 1.2–37 V.

Regulators designed to deliver large amounts of output power often employ a sophisticated high-frequency switching technique to minimize internal power dissipation. These **switching regulators** achieve power conversion efficiencies up to 90%, and they avoid the need for bulky AC line transformers. Switching regulators also work as DC-to-DC converters, the adjustable output voltage being either lower or higher than the input.

Exercise 12.2–3   Suppose the regulator in Fig. 12.2–5 has $R_o = 5\ \Omega$, $R_B = 75\ \Omega$, $R_Z = 2\ \Omega$, $V_Z = 15$ V, $V_\gamma = 0.7$ V, and $\beta = 24$. Find the minimum and maximum values of V_{DC}, I_Z, and V_{CE} when V_m varies over 25–30 V and I_{DC} varies over 0–1 A.

DC Motor Supplies

DC motors generally differ from electronic loads in three respects: They have appreciable inductance, they draw more power, and they need *constant current* rather than constant voltage. Often, the power supply for a DC motor relies on the inherent inductance to smooth the current, thereby eliminating the smoothing capacitor.

Figure 12.2–6a diagrams a half-wave rectifier modified to supply a DC motor winding with inductance L and equivalent series resistance R_L. The rectifying diode D_r conducts when $v_s > 0$ and then goes off. At this point, the **free-wheeling diode** D_{fw} goes on to provide continuity of current through the inductance. Thus, i_{out} circulates around the closed path formed

Figure 12.2-6

Half-wave rectifier with inductive load and free-wheeling diode: (a) circuit; (b) waveforms.

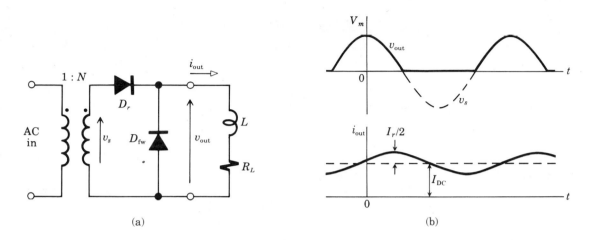

(a)   (b)

by L, R_L, and D_{fw} when $v_s < 0$. This operating cycle repeats periodically, with D_r and D_{fw} switching on and off each time v_s crosses zero.

The waveforms in Fig. 12.2–6b show that v_{out} has a half-rectified shape with $V_{DC} = (v_{out})_{av} = V_m/\pi$. Since there can be no DC voltage component across L in the steady state, V_{DC} appears entirely across R_L and

$$I_{DC} = (i_{out})_{av} = V_{DC}/R_L = V_m/\pi R_L \qquad (14)$$

To estimate the **peak-to-peak ripple current** I_r, we observe that v_{out} has a peak-to-peak swing of V_m volts with period $T = 2\pi/\omega$. Accordingly, we form the approximation

$$v_{out} \approx V_{DC} + \tfrac{1}{2}V_m \cos \omega t$$

Then, by superposition and standard AC circuit analysis,

$$i_{out} \approx I_{DC} + \tfrac{1}{2}I_r \cos (\omega t + \theta)$$

where

$$I_r = \frac{V_m}{\sqrt{R_L^2 + (\omega L)^2}} = \frac{\pi}{\sqrt{1 + (\omega L/R_L)^2}} I_{DC} \qquad (15)$$

Hence, I_r will be quite small compared to I_{DC} if $\omega L/R_L \gg \pi$.

Our approximate expression for v_{out} is justified mathematically from the **Fourier-series expansion** of a half-rectified waveform, as will be discussed in Section 13.1. The complete expansion includes additional AC components at higher frequencies with progressively decreasing amplitudes. However, these components contribute little to i_{out} because the impedance of the inductor increases with frequency.

Example 12.2–2

A certain DC motor has $L = 0.2$ H and $R_L = 4$ Ω. We want to supply it with $I_{DC} \approx 25$ A, given a 60-Hz AC source at $V_{rms} = 220$ V. Equations (14) and (15) yield

$$V_m = \pi R_L I_{DC} \approx 314 \text{ V} \qquad I_r = 0.166 I_{DC} \approx 4 \text{ A}$$

Both diodes must therefore be rated to handle $i_{pk} = I_{DC} + \frac{1}{2} I_r \approx 27$ A. Both diodes must also be rated for PIV $= V_m \approx 314$ V, which is easily determined from Fig. 12.2–6.

Finally, we notice that we can get by here without a transformer since $\sqrt{2} V_{rms} = 311$ V $\approx V_m$. DC motor windings are often designed to operate at $V_{DC} \approx 100$ V, so $V_m = \pi V_{DC} \approx \sqrt{2} \times 220$ V and an input transformer will not needed if 220-V(rms) AC is available.

Exercise 12.2–4

Suppose a full-wave rectifier is connected to the load in Fig. 12.2–6a, so continuity of current does not require a free-wheeling diode. The full-rectified voltage waveform can be approximated by

$$v_{out} \approx \frac{2V_m}{\pi} + \frac{4V_m}{3\pi} \cos 2\omega t$$

Derive an expression for the ratio I_r / I_{DC}.

Controlled Rectifiers[†]

As we have seen, a DC motor draws average current proportional to the average voltage across its terminals. The average power can therefore be controlled by altering the shape of the voltage waveform. Here we'll describe how controlled rectifiers modify waveforms, and how they are used for efficient control of power to AC loads as well as to DC motors.

Figure 12.2–7 shows the structure, symbol, and i-v characteristics of a **silicon controlled rectifier** (SCR). This switching device is a four-layer *pnpn* diode with three external terminals, the **anode** (A), the **cathode** (K), and the controlling **gate** (G). The i-v characteristics indicate that an SCR operates in one of two stable states. A small current pulse injected into the

Figure 12.2–7
Silicon-controlled rectifier: (a) structure; (b) symbol; (c) i-v curves for ON and OFF states.

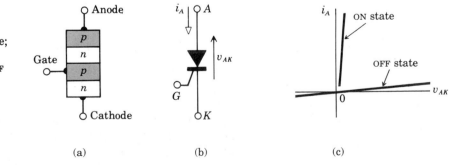

(a)   (b)   (c)

gate activates the forward-conducting ON state, which allows $i_A > 0$ as long as $v_{AK} > 0$. But when v_{AK} goes negative, the SCR switches back to its nonconducting OFF state and $i_A \approx 0$ despite subsequent variations of v_{AK}.

For qualitative explanation of this bistable behavior, we visualize an SCR as consisting of a *pnp* transistor and an *npn* transistor arranged per Fig. 12.2–8. If both i_G and i_{Bp} are initially zero, then $i_{Bn} = i_{Cp} = 0$ so $i_{Cn} = 0$ and $i_A = i_{Bn} + i_{Cn} = 0$. These self-sustaining conditions form the OFF state, independent of v_{AK}. Now let T_p be turned on by a short pulse $i_G > 0$, so $i_{Bn} = i_{Cp} > 0$ and T_n starts to conduct; the collector current i_{Cn} then feeds the base of T_p, keeping it on even after the gate pulse ends. These self-sustaining conditions form the ON state with $i_A = i_{Bn} + i_{Cn} > 0$, provided that $v_{AK} > 0$. The gate voltage needed to trigger conduction is

$$v_{GK} \geq V_{\text{trig}}$$

where V_{trig} ranges from about 1 V to 10 V. Setting $v_{AK} < 0$ restores the OFF state because i_A cannot be negative.

Figure 12.2–8
Model of an SCR using two transistors.

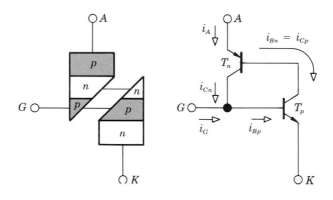

We put SCR switching action to work in the motor-control circuit of Fig. 12.2–9a. The SCR functions as the rectifying diode, and the trigger unit generates a gate pulse shortly after each upward zero-crossing of the AC voltage v_s. The waveforms in Fig. 12.2–9b illustrate steady-state operation with $v_s = V_m \sin \omega t$ and gate pulses at $t = \tau$, $T + \tau$, etc. The first gate pulse triggers the SCR into its ON state, which persists until v_s goes negative at $t = T/2$ so the free-wheeling diode forces $v_{\text{out}} = 0$ and $v_{AK} < 0$. The OFF state then persists until the next gate pulse, when the operating cycle starts over again.

The resulting output voltage waveform looks like a half-rectified sinusoid missing a section of duration τ at the start of each period. Accordingly, we define the **trigger angle** (or **firing angle**)

$$\alpha \triangleq \omega\tau$$

Figure 12.2–9
DC motor control with an SCR: (a) circuit; (b) waveforms.

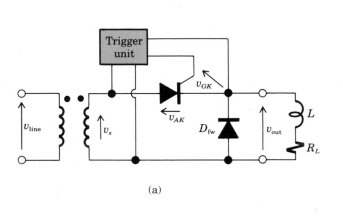

(a)

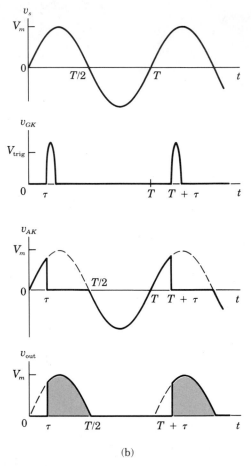

(b)

The average output voltage depends on α, as given by

$$(v_{\text{out}})_{\text{av}} = \frac{1}{T} \int_0^T v \, dt = \frac{1}{T} \int_\tau^{T/2} V_m \sin \omega t \, dt$$

$$= (1 + \cos \alpha) \frac{V_m}{2\pi} \qquad \textbf{(16)}$$

Since this DC voltage appears entirely across R_L, and since most of the time-varying voltage appears across L if $\omega L \gg R_L$, the average output power becomes $P_{\text{out}} \approx (v_{\text{out}})_{\text{av}}^2 / R_L$. Thus, we achieve control of P_{out} by adjusting the trigger angle α.

Silicon controlled rectifiers belong to the family of power switching devices known generically as **thyristors.** Another family member is the

Figure 12.2–10
Incandescent light dimmer with a triac: (a) circuit; (b) waveforms.

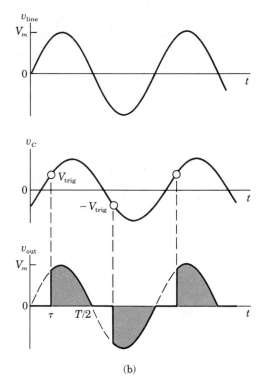

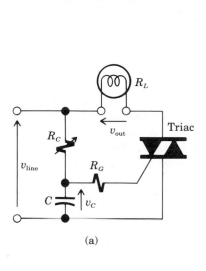

(a)

(b)

triac, which acts like two SCRs connected in parallel but pointing in opposite directions. A positive gate pulse triggers "forward" conduction, while a negative gate pulse triggers "reverse" conduction. This bidirectional capability of a triac lends itself to power control for AC loads such as incandescent lights and small motors.

By way of illustration, Fig. 12.2–10 gives the circuit diagram and waveforms of a simple incandescent light dimmer. The RC network produces voltage v_C that lags v_s and triggers the triac when $v_C \geq V_{\text{trig}}$ or $v_C \leq -V_{\text{trig}}$. Resistor R_G is included to keep the gate current small. The output voltage obviously has no DC component, but its *mean-square value* is

$$(v_{\text{out}}^2)_{\text{av}} = \frac{1}{T} \int_0^T v_{\text{out}}^2 \, dt = \frac{2}{T} \int_\tau^{T/2} V_m^2 \sin^2 \omega t \, dt$$

$$= \left(1 + \frac{\sin 2\alpha}{2\pi} - \frac{\alpha}{180°}\right) \frac{V_m^2}{2} \qquad \textbf{(17)}$$

where the trigger angle α is expressed in degrees. The average output power is $P_{\text{out}} = (v_{\text{out}}^2)_{\text{av}}/R_L$, which can be controlled via the adjustable resistor R_C.

12.3 THERMAL LIMITATIONS[†]

To complete our coverage of power electronics, this section presents a brief discussion of thermal limitations on semiconductor devices. We'll first explain why the power dissipation rating must be reduced when a device operates at high temperature. Then we'll show how heat sinks keep temperatures down to allow more power dissipation.

Thermal Ratings and Derating

Power transistors and diodes frequently dissipate large amounts of electrical power. This power is converted into heat that raises the temperature of internal semiconductor junctions. If a junction gets too hot, then there are adverse consequences such as thermally induced mechanical stress and a migration of impurity atoms across the junction. Eventually, overheating destroys the device.

Safe operation of a semiconductor-junction device entails the observance of two thermal limitations:

- The **power dissipation** P_D must not exceed the rated value P_{DO}.

- The **junction temperature** T_J must not exceed the rated value $T_{J_{max}}$.

These ratings are specified by the manufacturers. Maximum allowable junction temperatures fall in the range 125–200°C for a silicon diode or transistor.

Figure 12.3–1a depicts a junction device dissipating power P_D. The **case temperature** T_C is presumably less than the junction temperature, so heat flows away from the junction and passes out through the case. At

Figure 12.3–1
(a) Heat flow from a junction. (b) Circuit analogy.

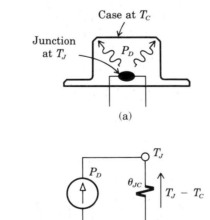

(a)

(b)

thermal equilibrium, the junction temperature becomes

$$T_J = T_C + \theta_{JC} P_D \qquad (1)$$

The proportionality constant θ_{JC} represents the **thermal resistance** from junction to case, expressed in degrees per watt. A helpful interpretation of Eq. (1) is the circuit analogy shown in Fig. 12.3–1b, where we visualize a source current P_D through resistance θ_{JC} that produces the potential difference $T_J - T_C = \theta_{JC} P_D$. Clearly, θ_{JC} should be small to prevent an excessive temperature rise at the junction. Power devices often have one side of the junction bonded directly to the case, thereby providing the good thermal conduction needed to get θ_{JC} down around 0.5–5°C/W.

The manufacturer of a device rates the power dissipation at a specified case temperature T_{CO}. If T_C does not exceed T_{CO} then T_J remains below $T_{J_{\max}}$ for any $P_D \leq P_{DO}$. Accordingly, we write

$$P_{D_{\max}} = P_{DO} \qquad T_C \leq T_{CO} \qquad (2)$$

where $P_{D_{\max}}$ stands for the maximum allowable dissipation. Operation at higher case temperatures is permitted, but the power dissipation must be **derated** (reduced) to keep $T_J \leq T_{J_{\max}}$. Specifically, setting $P_D = P_{D_{\max}}$ and $T_J = T_{J_{\max}}$ in Eq. (1), we obtain

$$P_{D_{\max}} = (T_{J_{\max}} - T_C)/\theta_{JC} \qquad T_{CO} < T_C < T_{J_{\max}} \qquad (3)$$

Equations (2) and (3) define the **power derating curve** plotted versus T_C in Fig. 12.3–2. Any point along or under this curve satisfies the thermal conditions for safe operation. The slope above T_{CO} equals $1/\theta_{JC}$, so

$$\theta_{JC} = (T_{J_{\max}} - T_{CO})/P_{DO} \qquad (4)$$

Given the rated values P_{DO}, $T_{J_{\max}}$, and T_{CO}, you can calculate θ_{JC} directly from Eq. (4).

Figure 12.3–2
Power derating curve.

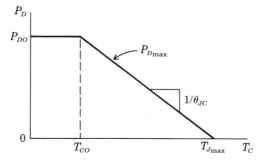

However, the derating curve fails to tell the whole story because T_C increases as heat passes through the case into the surrounding region at **ambient temperature** T_A. Consequently, similar to Eq. (1), we have

$$T_C = T_A + \theta_{CA} P_D \qquad (5)$$

where θ_{CA} stands for the case-to-ambience thermal resistance. If the case simply transfers heat directly into the air, then θ_{CA} typically will be 30–80°C/W. The ambient air temperature is usually constant at 25–30°C, in absence of adjacent heat sources, fan-cooling, etc.

Rewriting Eq. (5) as $P_D = (T_C - T_A)/\theta_{CA}$ gives the dashed line superimposed on the derating curve in Fig. 12.3–3. This dashed line intersects the derating curve at

$$P_Q = \frac{T_{J_{max}} - T_A}{\theta_{JC} + \theta_{CA}} \qquad (6)$$

Since T_C increases with P_D along the dashed line, our diagram reveals that we now must have $P_D \leq P_Q$ for safe operation. The large free-air value of θ_{CA} therefore results in considerable derating, as illustrated by the following numerical example.

Figure 12.3–3

Derating curve with case temperature line.

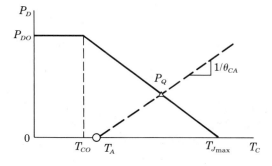

Example 12.3–1

Consider a power transistor rated for $P_{DO} = 100$ W at $T_{CO} = 50$°C and having $T_{J_{max}} = 200$°C. The corresponding junction-to-case thermal resistance is found from Eq. (4) to be

$$\theta_{JC} = (200 - 50)°C/(100 \text{ W}) = 1.5°C/W$$

If $\theta_{CA} = 30$°C/W and $T_A = 25$°C, then Eq. (6) gives the derated power

$$P_D \leq P_Q = \frac{(200 - 25)°C}{(1.5 + 30)°C/W} = 5.6 \text{ W}$$

Thus, when operated without cooling provisions, this 100-watt transistor can safely dissipate less than 6% of its rated power!

Exercise 12.3–1

Assuming that $\theta_{CA} = 30$°C/W, what ambient temperature would allow safe dissipation of 8 W by the transistor in Example 12.3–1? What would be the resulting case temperature?

Heat Sinks

A heat sink improves heat transfer from the case of a power device to the ambient air. An ideal or **infinite heat sink** would have $\theta_{CA} = 0$, so $T_C = T_A$ and the maximum allowable power dissipation would equal the rated power if $T_A \leq T_{CO}$. The nonzero but small thermal resistance of a real heat sink allows $P_{D_{max}}$ to be a significant fraction of P_{DO}.

Most heat sinks are made of extruded aluminum with a series of fins to promote convective cooling. For small power devices, the heat sink is simply clipped to the top of the case. For larger devices, the case is bolted to the sink structure as illustrated in Fig. 12.3–4a. Sometimes, a mica washer must be inserted between case and sink to provide electrical insulation while maintaining good heat conduction. In general,

$$\theta_{CA} = \theta_{CS} + \theta_{SA} \tag{7}$$

where θ_{SA} is the thermal resistance of the heat sink itself and θ_{CS} is the thermal resistance associated with the case-to-sink connection. Combining Eqs. (1), (5), and (7) now gives the junction temperature in the form

$$T_J = T_A + (\theta_{JC} + \theta_{CS} + \theta_{SA})P_D \tag{8}$$

Figure 12.3–4b shows the corresponding circuit analogy with T_A held constant.

Figure 12.3–4
Heat sink: (a) typical structure; (b) circuit analogy.

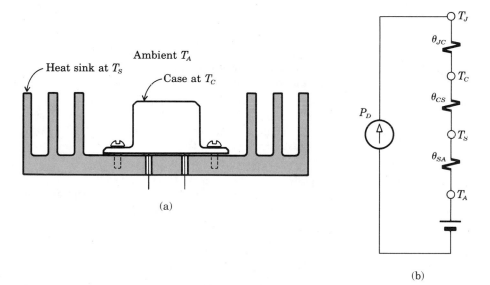

Ambient T_A

Heat sink at T_S

Case at T_C

(a)

T_J

θ_{JC}

T_C

θ_{CS}

P_D

T_S

θ_{SA}

T_A

(b)

Finally, we use Eq. (8) to determine the heat sink required for safe operation at a specified power dissipation $P_D \leq P_{DO}$. Setting $T_J \leq T_{J_{max}}$ yields the design equation

$$\theta_{CS} + \theta_{SA} \leq \frac{T_{J_{max}} - T_A}{P_D} - \theta_{JC} \tag{9}$$

Typical values of θ_{CS} are 0.5 to 2°C/W, where the lower values require coating the surfaces with silicon grease to increase heat conduction. The thermal resistance of a heat sink depends upon its size, and values of θ_{SA} range from about $10°C/W$ down to $1°C/W$. Even smaller values can be achieved with the help of fan cooling. Heat-sink characteristics are often presented graphically, rather than as a single thermal-resistance number, because convective cooling is a nonlinear process.

Example 12.3–2

Suppose we want to get $P_{D_{max}} = 50$ W at $T_A = 30°C$ using the 100-watt transistor from Example 12.3–1. Equation (9) then calls for an efficient heat-sink arrangement with

$$\theta_{CS} + \theta_{SA} \le \frac{200 - 30}{50} - 1.5 = 1.9°C/W$$

The case will still be quite hot, too hot to touch, since $T_C = T_A + (\theta_{CS} + \theta_{SA})P_D \le 125°C$.

Exercise 12.3–2

Redraw Fig. 12.3–3 with $\theta_{CA} = 0$, corresponding to an infinite heat sink. Then find P_Q in terms of P_{DO}, T_{CO}, $T_{J_{max}}$, and T_A.

PROBLEMS

12.1–1 Let the common-emitter amplifier in Fig. 12.1–1a have $V_{CC} = 28$ V, $R_C = 60$ Ω, $R_E = 0$, and $R_L = 40$ Ω. Find the transistor ratings required for maximum symmetric swing. Then calculate the corresponding value of P_{LM} and the upper bound on Eff.

12.1–2 Given that the common-emitter amplifier in Fig. 12.1–1a has $V_{CC} = 32$ V and $R_C = R_L = 100$ Ω, find R_E for maximum symmetric swing with $I_C = 0.2$ A. Also determine the corresponding transistor ratings, the value of P_{LM}, and the upper bound on Eff.

12.1–3 If the common-emitter amplifier in Fig. 12.1–1a has $R_C = 8$ Ω and $R_E = 4$ Ω, then what value of V_{CC} is needed to get $P_{LM} = 1.5$ W when $R_L = 12$ Ω? What are the corresponding transistor ratings and the upper bound on Eff?

12.1–4 Derive from Fig. 12.1–2a the expression for I_C given in Eq. (6).

12.1–5 Show that the efficiency of a common-emitter amplifier is upper bounded by Eff $\le R_{AC}^2/2R_L(R_{AC} + R_{DC})$.

12.1–6 The text states that the entire AC load line falls below the $P_{D_{max}}$ hyperbola when the Q point is centered and $P_{DM} \le P_{Dmax}$. To justify this assertion, let $v_{CE} = V_{CE}$ be an arbitrary point on a load line with end points at $i_C = I_0$ and $v_{CE} = V_0 = R_{AC}I_0$. Obtain an expression for $P_{DM} = V_{CE}I_C$ in terms of I_0, V_{CE}, and R_{AC}. Then show that P_{DM} is maximum when $V_{CE} = V_0/2$ and $I_C = I_0/2$, corresponding to a centered Q point.

12.1–7 Consider a BJT with specified ratings $i_{C_{max}}$, $v_{CE_{max}}$, and $P_{D_{max}}$. By taking account of the end points of the AC load line, show that the allowable range for R_{AC} with a centered Q point is given by

$$(\tfrac{1}{2}v_{CE_{max}})^2/P_{D_{max}} \le R_{AC} \le P_{D_{max}}/(\tfrac{1}{2}i_{C_{max}})^2$$

12.1-8   Given that the transformer-coupled amplifier in Fig. 12.1-4a has $V_{CC} = 20$ V, $N = 3$, and $R_L = 4\ \Omega$, find R_E for maximum symmetric swing with $I_C = 0.5$ A. Then determine the corresponding transistor ratings, the value of P_{LM}, and the upper bound on Eff.

12.1-9   If the transformer-coupled amplifier in Fig. 12.1-4a has $R_E = 10\ \Omega$ and $N = 2.5$, then what value of V_{CC} is needed to get $P_{LM} = 25$ W when $R_L = 8\ \Omega$? What are the corresponding transistor ratings and the upper bound on Eff?

12.1-10   A transformer-coupled amplifier with $|v_{in}| \le 4$ V is to be designed to deliver as much power as possible to a 5-Ω load using a power BJT with $\beta = 75$, $i_{C_{max}} = 2$ A, $v_{CE_{max}} = 100$ V, and $P_{D_{max}} = 20$ W.

(a)   Find N for maximum symmetric swing with $I_C = 0.5$ A, and check the load-line end points against the transistor's ratings.

(b)   Determine V_{CC} and select standard values for R_E, R_1, R_2, and R_3, taking $R_E \approx N^2 R_L/10$.

12.1-11   Do problem 12.1-10 with I_C as large as possible in part (a).

12.1-12   Do Problem 12.1-10 with I_C as small as possible in part (a).

12.1-13   Let the push-pull amplifier in Fig. 12.1-7a have $V_{CC} = 12$ V and $R_L = 5\ \Omega$. Find the maximum possible AC load power $P_{L_{max}}$ when the transistors have $V_{sat} = 2$ V. Then make a table comparing the values of P_{CC}, P_D, and Eff for $P_L = P_{L_{max}}$ and $P_L = \tfrac{1}{2}P_{L_{max}}$.

12.1-14   Do problem 12.1-13 with $V_{CC} = 18$ V.

12.1-15   Suppose the transistors in Fig. 12.1-7a have $v_{CE_{max}} = 60$ V, $i_{C_{max}} = 3$ A, $P_{D_{max}} = 10$ W, $V_{sat} \approx 0$, and $\beta = 50$. Find the value of V_{CC} needed to deliver the largest possible AC signal power when $R_L = 12\ \Omega$. Then calculate the corresponding values of P_{LM}, P_{DM}, P_{CC}, and $|i_{in}|_{max}$.

12.1-16   Do Problem 12.1-15 with $R_L = 6\ \Omega$.

12.1-17   Derive Eq. (25a) by calculating dP_D/dV_m from Eq. (24).

12.1-18   Suppose the input voltage in Fig. 12.1-7 is a square wave, so $v_{in} = +V_m$ throughout one half cycle and $v_{in} = -V_m$ throughout the other half cycle. Obtain modified versions of Eqs. (20)-(23) for this case.

12.1-19   The largest positive output voltage in Fig. 12.1-10a occurs when the upper diode is at its break point and carries negligible current. Using this condition, show that

$$v_{out-max} = \frac{V_{CC} - V_\gamma}{1 + R_D/(\beta + 1)R_L}$$

12.2-1   Suppose both full-wave rectifiers in Fig. 12.2-2 have line voltage $V_{rms} = 120$ V and $N_1 = 100$ turns on the transformer's primary. Further suppose that the diodes have $V_\gamma \approx 1$ V and forward resistance $R_f = 2\ \Omega$. Find the approximate total number of secondary turns N_2 needed to get $I_{DC} = 3$ A with $R_L = 7\ \Omega$.

12.2-2   Do Problem 12.2-1 for $I_{DC} = 0.5$ A with $R_L = 14\ \Omega$.

12.2-3   Use a sketch of v_{out}^2 for the half-wave rectifier in Fig. 12.2-1 to find $(v_{out}^2)_{av}$. Then evaluate the power ratio P_{DC}/P where $P_{DC} = V_{DC}I_{DC}$ and $P = (v_{out}^2)_{av} \times R_L$. Repeat these calculations for the full-wave rectifier in Fig. 12.2–2.

12.2–4   Measurements on the circuit in Fig. 12.2–3a show that $V_{DC} = 100$ V when $I_{DC} = 0$ and $V_{DC} = 90$ V when $I_{DC} = 0.4$ A. What are the values of V_r and i_{pk} when $I_{DC} = 0.2$ A?

12.2–5   Suppose the power supply in Example 12.2–1 has been built with a center-tapped rectifier but one of the diodes becomes disconnected. Find the new values of V_{DC} and i_{pk} when $i_{out} = 0.3$ A.

12.2–6   Given a full-wave bridge rectifier with $V_{rms} = 120$ V and $f = 60$ Hz at the input, determine the values of N, C, and the diode ratings for a power supply that keeps V_{DC} in the range 295–305 V while delivering $I_{DC} \le 1.2$ A.

12.2–7   Given a full-wave center-tapped rectifier with $V_{rms} = 120$ V and $f = 60$ Hz at the input, determine the values of N, C, and the diode ratings for a power supply that keeps V_{DC} in the range 19.8–20.2 V while delivering $I_{DC} \le 240$ mA.

12.2–8   A DC power supply is needed to provide $I_{DC} = 0.5$ A at $V_{DC} = 40$ V. The available diodes are rated for PIV ≤ 60 V and $i_{pk} \le 6$ A. Determine the circuit configuration, the value of V_m with the smallest allowable conduction angle, and the corresponding value of C for $f = 60$ Hz.

12.2–9   The voltage regulator in Fig. 12.2–5a stops regulating when i_{out} is too large and puts the Zener diode out of reverse breakdown, so $i_Z = 0$ and $v_Z < V_Z$. Obtain the resulting expressions for V_{DC}, V_m', and R_o' by removing the R_Z-V_Z branch from Fig. 12.2–5c.

12.2–10   Let the regulator in Fig. 12.2–5 have $R_B = 50\ \Omega$, $R_Z = 2\ \Omega$, and $\beta = 49$. Determine the condition on R_o such that $R_o' \le 0.5\ \Omega$. Then find the corresponding condition on the smoothing capacitor assuming a full-wave rectifier with $f = 60$ Hz. Compare your result with the condition on C needed to get $R_o \le 0.5\ \Omega$ with an unregulated full-wave rectifier.

12.2–11   A regulated power supply is needed to keep V_{DC} in the range 23.8–24.2 V when $I_{DC} \le 0.4$ A and $V_m = 45$ V. Find the values of R_B and R_o, given that $V_Z = 24$ V, $R_Z = 20\ \Omega$, $\beta = 50$, and $V_y = 0.7$ V. If the rectifier is a full-wave circuit with $f = 60$, then what should be the value of C?

12.2–12   Do problem 12.2–11 with $I_{DC} \le 0.2$ A and $V_m = 36$ V.

12.2–13   Working directly from Fig. 12.2–5c, derive an expression for I_Z in terms of V_m, R_o, R_B, β, I_{DC}, R_Z, and V_Z. Then obtain the condition on I_{DC} so that $I_Z \ge i_{Z_{min}}$.

12.2–14   The average power dissipated by the BJT in Fig. 12.2–5 is $P_D = V_{CE}(\beta I_B)$. Assume that $\beta \gg 1$ and $I_Z \ll I_{DC}$ to obtain an expression for P_D in terms of V_m, V_m', R_o, R_o', and I_{DC}. Then show that $P_{D_{max}} \approx (V_m - V_m')^2/4(R_o - R_o')$, which occurs when $I_{DC} \approx (V_m - V_m')/2(R_o - R_o')$.

12.2–15   Suppose the motor in Fig. 12.2–6a has $L = 60$ mH and $R_L = 5\ \Omega$. Find the value of N and the diode ratings needed to deliver 2 kW of DC power to R_L, given a 60-Hz AC input at 120 V (rms).

12.2–16   Estimate the values of R_L and L in Fig. 12.2–6, given that i_{out} varies between 9 A and 15 A when $N = 2$ and the AC input has $V_{rms} = 120$ V and $f = 50$ Hz.

12.2−17   A full-wave center-tapped rectifier with $V_m = \sqrt{2} \times 220$ V and $f = 60$ Hz supplies DC current to the field winding of a motor having $L = 80$ mH and $R_L = 20$ Ω. Use the results of Exercise 12.2−4 to find I_{DC} and I_r. Then determine the required diode ratings.

12.2−18   Suppose the SCR system in Fig. 12.2−9 has a varying load R_L, but the trigger time is adjusted to hold P_{out} constant at 500 W. Assuming $f = 60$ Hz and $\omega L \gg R_L$, find the minimum required value of V_m and the corresponding maximum value of τ for R_L varying over 2–5 Ω.

12.2−19   Do Problem 12.2−18 for R_L varying over 3–25 Ω.

12.2−20   Let diodes D_1 and D_2 in Fig. 12.2−2b be replaced by SCRs S_1 and S_2. Sketch the waveforms i_1, i_2, and v_{out} for the case where S_1 is triggered at $t = \tau,\ T + \tau,\ \ldots$, and S_2 is triggered at $t = 0.5T + \tau,\ 1.5T + \tau,\ \ldots$. Then modify Eq. (16) for this full-wave SCR circuit.

12.2−21   Given that the triac circuit in Fig. 12.2−10 delivers $P_{out} = 100$ W when $\alpha = 0°$, find P_{out} when $\alpha = 45°$, $90°$, and $135°$. Then sketch P_{out} versus α.

12.2−22   Let the triac circuit in Fig. 12.2−10 have $f = 60$ Hz, $V_m = \sqrt{2} \times 120$ V, $R_C = 30$ kΩ, and $C = 0.1$ μF. If $V_{trig} = 10$ V, then what is the value of α?

12.3−1   Use Eqs. (3) and (5) to derive Eq. (6) for P_Q in Fig. 12.3−3.

12.3−2   Consider a power transistor rated for $T_{J_{max}} = 180°$C and $P_{DO} = 40$ W at $T_{CO} = 100°$C. Find the value of θ_{CA} that yields $P_Q = 15$ W when $T_A = 30°$C, and calculate the corresponding value of T_C. Then determine T_J when $P_D = 10$ W.

12.3−3   A power transistor with $P_{DO} = 50$ W, $T_{J_{max}} = 200°$C and $\theta_{JC} = 1°$C/W is operated at $T_A = 30°$C. Find $P_{D_{max}}$ and the corresponding values of T_C and T_J when $\theta_{CA} = 16°$C/W and when $\theta_{CA} = 2°$C/W.

12.3−4   A power transistor with $P_{DO} = 40$ W, $T_{J_{max}} = 200°$C and $\theta_{JC} = 0.5°$C/W is operated at $T_A = 25°$C. Find $P_{D_{max}}$ and the corresponding values of T_C and T_J when $\theta_{CA} = 12°$C/W and when $\theta_{CA} = 3°$C/W.

12.3−5   A transformer coupled power amplifier is to be designed for $P_L \leq 5$ W using a BJT rated for $P_{DO} = 25$ W at $T_{CO} = 100°$C and having $\theta_{CA} = 40°$C/W and $T_{J_{max}} = 175°$C. Show that a heat sink will be required for operation at $T_A = 25°$C. Then find the maximum allowable value of θ_{CA} and the corresponding value of T_C when $\theta_{CS} = 2°$C/W.

12.3−6   A push-pull power amplifier is to be designed for $P_L \leq 75$ W using a pair of BJTs rated for $P_{DO} = 25$ W at $T_{CO} = 100°$C and having $\theta_{CA} = 30°$C/W and $T_{J_{max}} = 180°$C. Show that a heat sink will be required for operation at $T_A = 30°$C. Then find the maximum allowable value of θ_{CA} and the corresponding value of T_C when $\theta_{CS} = 1°$C/W.

Part III Information Systems

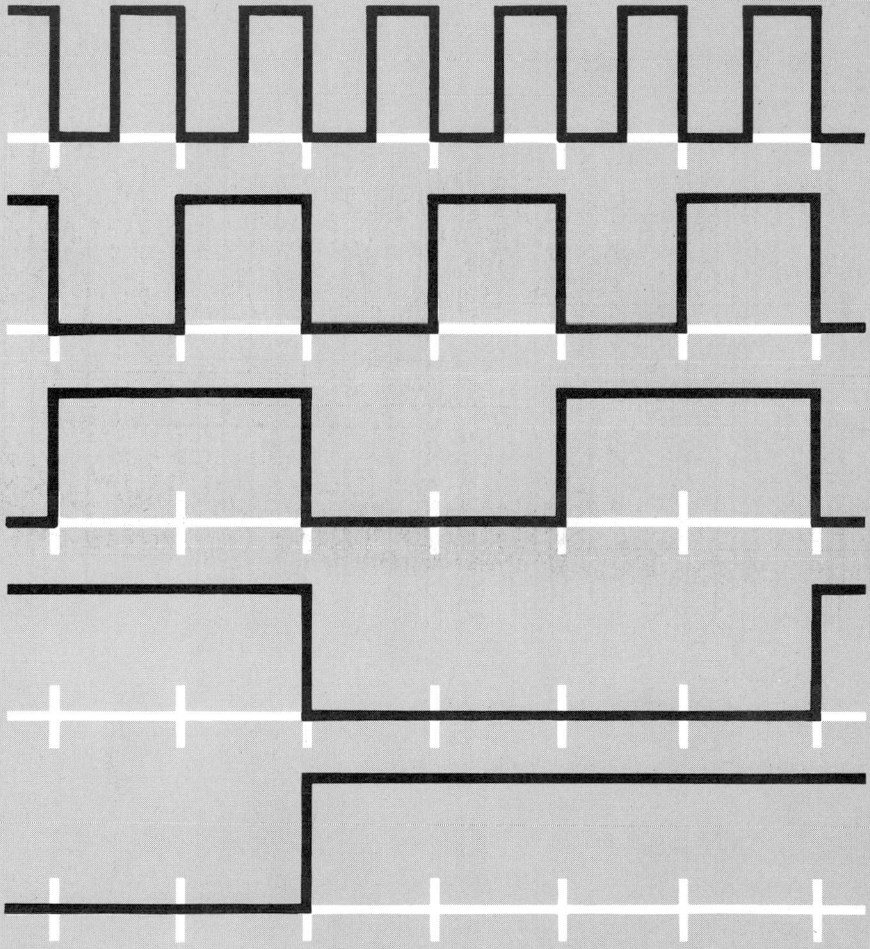

Chapter

13 Signal Processing and Communication Systems

An electrical *signal* is a voltage or current waveform whose time variations correspond to some desired *information*. The information in question, usually from a nonelectrical source, has been converted into electrical form to take advantage of the relative ease and flexibility of processing and transmitting electrical quantities. This chapter deals with systems designed to handle information-bearing signals for purposes of measurement and processing, as in an instrumentation system, or for long distance transmission, as in a communication system. Regardless of their particular details, all of these systems involve certain basic concepts and share certain problems.

Foremost among signal concepts is *spectral analysis:* the representation of signals in terms of their frequency components. That concept serves as a unifying thread throughout this chapter. Specific topics include the problems of distortion, interference, and noise, as well as processing techniques such as equalization, filtering, sampling, modulation, and multiplexing.

We'll concentrate on analog signals, and some of the treatment will be descriptive rather than analytical to keep the mathematics at an appropriate level. Even so, the ideas developed here have value in their own right, and they also establish a background for the discussion of digital instrumentation in Chapter 15.

Objectives

After studying this chapter and working the exercises, you should be able to do each of the following:

- Construct the line spectrum of a periodic signal from its Fourier-series expansion (Section 13.1).

- Interpret the spectrum of a signal, and state the conditions for distortionless transmission (Section 13.1).

- Sketch typical spectra at various points in a system employing product modulation and filtering (Section 13.2).

- State the conditions under which a signal can be sampled and then reconstructed from a pulse-modulated waveform (Section 13.2).

- Identify the major causes of interference and noise, and describe the techniques for minimizing their effects (Section 13.3).†

- Calculate the signal power at the output of a cable or radio transmission system (Section 13.4).†

- Distinguish between the properties of amplitude and frequency modulation, and draw the block diagram of a superheterodyne radio receiver (Section 13.4).†

13.1　SIGNALS AND SPECTRAL ANALYSIS

Our study of signal processing starts with an overview intended to put the various topics of this chapter in perspective. Then we introduce the Fourier series as a way of analyzing periodic signals, and we go on to develop the spectral approach for signal analysis and processing.

Signal-Processing Systems

Figure 13.1–1a depicts an illustrative system for measuring and recording the temperature of an industrial oven. The thermistor inside the oven is a sensing device whose electrical resistance R depends upon the tempera-

Figure 13.1–1
Temperature recording systems.

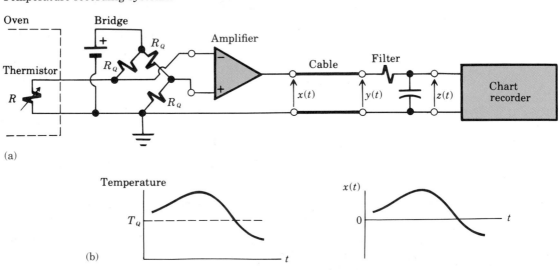

(a)

(b)

ture. This thermistor forms part of a simple bridge circuit designed to be balanced when $R = R_Q$ at the normal operating temperature T_Q. Temperature changes above or below T_Q causes the bridge to become unbalanced and produce the amplified voltage $x(t)$, as sketched in Fig. 13.1–1b. We call $x(t)$ an **analog signal** because the voltage variations are similar or analogous to the temperature variations. A cable connects the amplifier to the recording location where a lowpass filter removes any AC "hum" that has been picked up. The filtered output is finally applied to a chart recorder.

Generalizing from our example, Fig. 13.1–2 diagrams the functional blocks found in most signal-processing systems. The information to be processed resides in a physical quantity—temperature, acoustical pressure, light intensity, etc.—at a source some distance away from the intended destination. An input **transducer** converts the information to an electrical signal, which is then prepared for transmission by the input **processor.** Another processor and transducer at the destination produce the desired output information—a chart record, sound wave, video image, or what have you.

Figure 13.1–2

Functional elements of a signal-processing system.

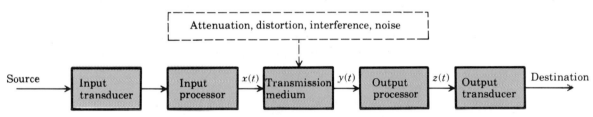

The **transmission medium** serves as the electrical connection spanning the gap from source to destination, be it across the room or across the continent. A pair of wires or a cable usually suffices for the short distances involved in most instrumentation and data-acquisition systems. Transmission over longer distances requires a full-blown communication system, including a transmitter at the source and a receiver at the destination.

Standing in the way of perfect transmission are several problems resulting from attenuation, distortion, interference, and noise. For simplicity, the effects of these phenomena have been concentrated at the center of Fig. 13.1–2, since the transmission medium is often the most vulnerable part of a system, especially long-haul communication systems. But problems may enter the picture anywhere between source and destination, and they can be expected even when distance is not a significant factor.

In general, then, our information-bearing signal $x(t)$ arrives at the destination in the form

$$y(t) = k\,\tilde{x}(t - t_d) + n(t)$$

where k represents the attenuation, $\tilde{x}(t - t_d)$ is a distorted and time-delayed version of $x(t)$, and $n(t)$ represents added interference and noise. **Attenuation,** caused by losses within the system, reduces the size or "strength" of the signal, whereas **distortion** is any alteration of the waveshape itself due to energy storage and/or nonlinearities. **Interference** is contamination by extraneous signals, while **noise** comes from natural sources usually internal to the system. Not all of these effects would be serious in every system, but any one of them could pose a major challenge to the design engineer.

To handle these problems, successful information recovery almost always calls for **signal processing** at the input and output. Common processing operations include:

- **Amplification** to compensate for attenuation.
- **Filtering** to reduce interference and noise, or to extract selected aspects of information.
- **Equalization** to correct certain types of distortion.
- **Frequency translation** or **sampling** to obtain a signal that better suits the characteristics of the system.
- **Multiplexing** to accommodate two or more signals in one system.

Additionally, there is a host of more specialized techniques that enhance the quality of information recovery and presentation—linearizing, averaging, compressing, peak detecting, thresholding, counting, and timing, to name a few.

The rest of this chapter deals with basic methods for analyzing and processing analog signals. Subsequently, in Chapter 15, we'll see how the versatile and inexpensive microprocessor makes it possible to carry out many processing operations in digital form.

Periodic Signals and Fourier Series

The study of analog systems involves predicting how circuits respond to a time-varying voltage or current waveform $x(t)$—a difficult task in general. However, if we can express $x(t)$ as a *sum of sinusoids,* then we can invoke superposition and use the frequency response of the circuit to expedite calculations. Expressing a signal in terms of sinusoidal components is known as **spectral analysis,** which we begin here by considering the Fourier-series expansion of periodic signals.

Figure 13.1–3 shows typical waveforms that might be encountered in a signal-processing system. Although clearly *not* sinusoidal, each signal is **periodic** in that it repeats itself every T seconds. The **Fourier-series theorem** states that almost any periodic signal can be decomposed into an

Figure 13.1–3
Periodic waveforms:
(a) rectangular pulse train;
(b) triangular wave;
(c) sawtooth wave;
(d) square wave;
(e) half-rectified sine wave.

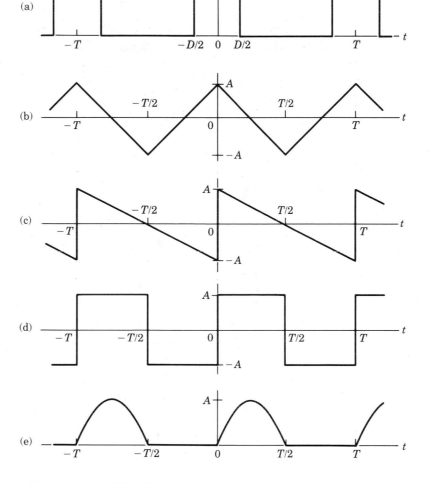

infinite series of the form

$$x(t) = a_0 + (a_1 \cos \omega_1 t + b_1 \sin \omega_1 t)$$
$$+ (a_2 \cos 2\omega_1 t + b_2 \sin 2\omega_1 t)$$
$$+ \cdots$$

$$= a_0 + \sum_{n=1}^{\infty} (a_n \cos n\omega_1 t + b_n \sin n\omega_1 t) \qquad (1)$$

The constants $a_0, a_1, b_1, \ldots$, are the series **coefficients** and ω_1 is called the **fundamental frequency** (in rad/s), related to the period by

$$\omega_1 = 2\pi/T \qquad (2)$$

(Many texts symbolize the fundamental frequency by ω_0, which we avoid to prevent confusion with the resonant frequency of a circuit.)

According to Eq. (1), a periodic signal contains a DC (zero-frequency) component a_0, a sinusoidal component $a_1 \cos \omega_1 t + b_1 \sin \omega_1 t$ at the fundamental frequency, and similar components at $2\omega_1$, $3\omega_1$, ... These integer multiples of ω_1 are known as **harmonics,** $2\omega_1$ being the second harmonic, $3\omega_1$ the third harmonic, and so forth. The DC component and/or some of the harmonics may be absent from the expansion of a particular waveform, but there will never be components other than those listed in Eq. (1).

Calculating the series coefficients requires integrating the waveform over one period. Specifically, the DC component is given by

$$a_0 = \frac{1}{T} \int_0^T x(t) \, dt \tag{3}$$

which is seen to be the *average value* of $x(t)$. The remaining coefficients are computed from the integrals

$$a_n = \frac{2}{T} \int_0^T x(t) \cos n\omega_1 t \, dt \qquad b_n = \frac{2}{T} \int_0^T x(t) \sin n\omega_1 t \, dt \tag{4}$$

for $n = 1, 2, \ldots$

The formulas in Eqs. (3) and (4) are presented primarily for the sake of completeness, since our concern here is not integration techniques but rather the use and interpretation of the Fourier series. To that end, Table 13.1–1 gives coefficient expressions for all the waveforms in Fig. 13.1–3. (Any coefficient not listed equals zero.) More extensive tables are found in mathematics handbooks. Additional entries can also be generated by performing simple operations on waveforms with known coefficients.

Table 13.1–1
Fourier-series
coefficients.

Waveform	Symmetry	a_0	a_n or b_n	
(a) Rectangular pulse train	Even	$\dfrac{DA}{T}$	$a_n = \dfrac{2A}{\pi n} \sin \dfrac{\pi Dn}{T}$	$n = 1, 2, 3, \ldots$
(b) Triangular wave	Even and half-wave	0	$a_n = \dfrac{8A}{\pi^2 n^2}$	$n = 1, 3, 5, \ldots$
(c) Sawtooth wave	Odd	0	$b_n = \dfrac{2A}{\pi n}$	$n = 1, 2, 3, \ldots$
(d) Square wave	Odd and half-wave	0	$b_n = \dfrac{4A}{\pi n}$	$n = 1, 3, 5, \ldots$
(e) Half-rectified sine wave	None	$\dfrac{A}{\pi}$	$b_1 = \dfrac{A}{2}$	
			$a_n = -\dfrac{2A}{\pi(n^2 - 1)}$	$n = 2, 4, 6, \ldots$

As the table implies, various types of *waveform symmetry* cause some of the coefficients to vanish. In particular, the pulse train and triangular wave in Fig. 13.1–3 have symmetry about the vertical axis and would look the same when reversed in time. This **even symmetry** is expressed by

$$x(-t) = x(t) \tag{5a}$$

which results in

$$b_n = 0 \qquad n = 1, 2, 3, \ldots \tag{5b}$$

so the series consists of the even-symmetry components a_0 and $a_n \cos n\omega_1 t$. Conversely, the sawtooth and square wave have symmetry about the origin and would look inverted when reversed in time. This **odd symmetry** is expressed by

$$x(-t) = -x(t) \tag{6a}$$

which results in

$$a_n = 0 \qquad n = 0, 1, 2, \ldots \tag{6b}$$

so the series consists of the odd-symmetry components $b_n \sin n\omega_1 t$. The triangular and square waves also have the property that the second half of each period looks like the first half turned upside down. This **half-wave symmetry** is expressed by

$$x\left(t \pm \frac{T}{2}\right) = -x(t) \tag{7a}$$

in which case

$$a_n = b_n = 0 \qquad n = 2, 4, 6, \ldots \tag{7b}$$

so the series contains only the odd-harmonic components.

You should review Table 13.1–1 in light of these symmetry effects. And while doing that, also observe that the magnitudes of a_n or b_n generally decrease as n increases; in other words, the higher harmonic components tend to be smaller than the lower ones. Furthermore, the coefficients of a relatively "smooth" signal such as the triangular wave decrease more rapidly than those of a "jumpy" or discontinuous one such as the sawtooth wave. We conclude, then, that discontinuous signals have more significant high-frequency content than continuous signals.

Example 13.1–1   The rectangular pulse train in Fig. 13.1–3a consists of pulses of height A and duration D. The ratio D/T is known as the **duty cycle** because the pulse is "on" for D seconds of each period and "off" the remaining time. Such pulse trains are used for timing purposes and to represent digital information.

If a particular pulse train has $A = 3$ and $D = T/3$, then Table 13.1–1 gives

$$a_0 = 1 \qquad a_n = \frac{6}{\pi n} \sin \frac{\pi n}{3} \qquad b_n = 0$$

The Fourier-series expansion thus becomes

$$x(t) = 1 + 1.65 \cos \omega_1 t + 0.83 \cos 2\omega_1 t$$

$$-0.41 \cos 4\omega_1 t - 0.33 \cos 5\omega_1 t + \cdots$$

The terms at $3\omega_1, 6\omega_1, \ldots$, are missing because $a_n = 0$ for $n = 3, 6, \ldots$

We'll interpret this result — and gain confidence in the Fourier-series theorem — by a partial reconstruction of the signal from its expansion. Figure 13.1–4a shows the DC plus fundamental component $a_0 + a_1 \cos \omega_1 t$, along with the second-harmonic term $a_2 \cos 2\omega_1 t$ whose period is $T/2$. Adding these three terms together gives the waveform in Fig. 13.1–4b, a smoothed approximation to the actual pulse train. The approximation is greatly improved when we include all terms through the fifth harmonic, as shown in Fig. 13.1–4c. On the basis of this figure, we infer that all higher harmonics serve primarily to square-up the corners of the pulses.

Figure 13.1–4

Reconstruction of a pulse train: (a) DC component plus first harmonic; (b) sum of first three components; (c) sum of first five components.

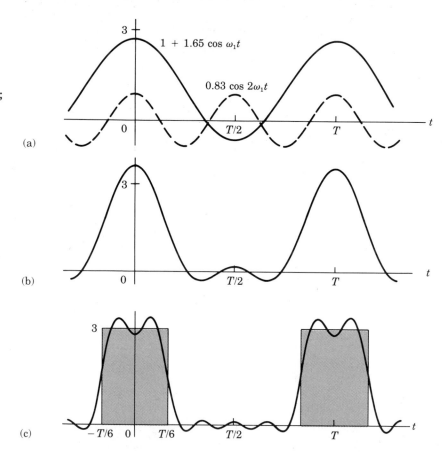

Exercise 13.1–1

Let $v(t) = x(t) - 1$ where $x(t)$ is a rectangular pulse train with $A = 2$ and $D = T/2$. Confirm from a sketch that $v(t)$ is an even-symmetry square wave. Then use the expansion of $x(t)$ to show that $v(t)$ has

$$a_n = \begin{cases} 4/\pi n & n = 1, 5, 9, \ldots \\ -4/\pi n & n = 3, 7, 11, \ldots \end{cases}$$

while all other coefficients equal zero.

Spectral Analysis and Signal Bandwidth

Spectral analysis draws upon the fact that a sinusoidal waveform is completely characterized by three quantities: *amplitude, phase,* and *frequency.* When a signal consists entirely of sinusoids, we can convey all the information about it by plotting *amplitude and phase versus frequency.* This frequency-domain picture helps us identify the signal's bandwidth and other significant properties. We usually let the independent variable be *cyclical* frequency f, rather than angular frequency ω, to incorporate the standard practice of measuring frequency in hertz.

As a specific example, Fig. 13.1–5 shows the frequency-domain picture of the signal

$$x(t) = 3 - 5 \cos (2\pi 20t - 120°) + 2 \sin 2\pi 30t$$

$$= 3 \cos (2\pi 0t + 0°) + 5 \cos (2\pi 20t + 60°) + 2 \cos (2\pi 30t - 90°)$$

This signal consists of sinusoids at 20 and 30 Hz, plus a constant term equivalent to a zero-frequency sinusoid. Hence, the amplitude plot has lines of height 3, 5, and 2 at $f = 0, 20$, and 30, respectively. The phase plot is similarly constructed. Note that amplitude is always a *positive* quantity and that phase angles are referenced to the *cosine* function, in agreement with our phasor convention. The two plots together constitute the **line spectrum** of $x(t)$.

Figure 13.1–5
Amplitude and phase
line spectrum.

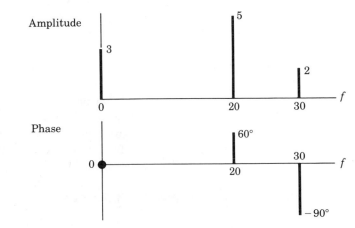

Now let $x(t)$ be any *periodic* signal whose Fourier-series coefficients are known. To construct the spectrum, our cosine-plus-sine series must first be converted into the amplitude-and-phase version

$$x(t) = a_0 + A_1 \cos(\omega_1 t + \phi_1) + A_2 \cos(2\omega_1 t + \phi_2) + \cdots$$

$$= a_0 + \sum_{n=1}^{\infty} A_n \cos(n\omega_1 t + \phi_n) \qquad \text{(8a)}$$

where

$$A_n = \sqrt{a_n^2 + b_n^2} \qquad \phi_n = -\arctan \frac{b_n}{a_n} \qquad \text{(8b)}$$

Equation (8a) follows from Eq. (1) by writing

$$a_n \cos n\omega_1 t + b_n \sin n\omega_1 t = a_n \cos n\omega_1 t + b_n \cos(n\omega_1 t - 90°)$$
$$= A_n \cos(n\omega_1 t + \phi_n)$$

The phasor diagram in Fig. 13.1–6 then yields Eq. (8b).

Figure 13.1–6
Phasor diagram of
Fourier coefficients.

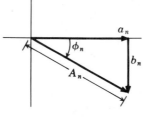

We used the angular frequency ω_1 for compactness in the foregoing expressions. The corresponding fundamental cyclical frequency is, of course, related to the period by

$$f_1 = \frac{\omega_1}{2\pi} = \frac{1}{T}$$

Accordingly, Eq. (8) tells us that the spectrum of a periodic signal contains lines at $f = 0, f_1, 2f_1$, and all higher harmonics of f_1—although some harmonics will be missing whenever $A_n = 0$. The zero-frequency or DC component represents the average value a_0, the components at the first few harmonics represent relatively slow time variations, and the higher harmonics represent more rapid time variations.

By way of illustration, the results of Example 13.1–1 lead to the spectrum in Fig. 13.1–7 for a rectangular pulse train when $T = 1$ ms $= 10^{-3}$ s, so $f_1 = 10^3$ Hz $= 1$ kHz. We see from the spectrum that most of the time variation comes from the large-amplitude components below 6 kHz, as previously observed in Fig. 13.1–4. The higher frequency components have much smaller amplitudes, which account for the stepwise jumps in the pulse train.

Figure 13.1−7
Line spectrum of a
rectangular pulse train.

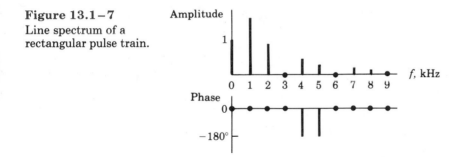

Similar conclusions about the time behavior of *nonperiodic signals* can be inferred from their spectra, which generally take the form of *smooth curves*. Spectral analysis of nonperiodic signals involves *Fourier transform theory* and goes beyond our scope. However, the essential concept is brought out by considering the single rectangular pulse in Fig. 13.1−8a. Intuitively, we could obtain this nonrepeating waveform from a rectangular pulse train (Fig. 13.1−3a) by letting $T \to \infty$ so that all pulses vanish except the one centered at $t = 0$. Since $f_1 = 1/T \to 0$ when $T \to \infty$, the amplitude lines must merge to form a continuous plot like Fig. 13.1−8b.

Figure 13.1−8
(a) Single rectangular
pulse. (b) Continuous
amplitude spectrum.

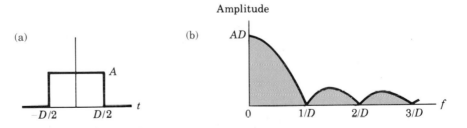

Complicated nonperiodic signals defy theoretical treatment, but some of them have been studied experimentally using spectrum analyzers. For instance, Fig. 13.1−9 shows the amplitude spectrum obtained by measurements on a typical voice signal. Smooth curves such as in Figs. 13.1−8b and 13.1−9 mean that the signal energy is spread over a continuous frequency range, rather than being concentrated in discrete sinusoidal components. The amplitude value at $f = 0$ equals the net area of the nonperiodic signal.

Despite other differences, all of the signals examined so far have spectral peaks at or near $f = 0$ and their amplitude spectra become progressively smaller as frequency increases. These waveforms are known as **low-pass signals,** loosely defined by the following property:

Figure 13.1−9
Amplitude spectrum of
a typical voice signal.

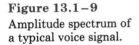

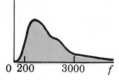

There exists some point W, called the **signal bandwidth,** such that all significant frequency content falls within the range $0 \le f < W$.

Although the value of W for a particular signal is somewhat arbitrary, the concept of signal bandwidth still plays a vital role in signal processing and communication. Table 13.1–2 lists the nominal bandwidths of a few selected signals.

Table 13.1–2
Lowpass signal bandwidths.

Signal type	Bandwidth
Telephone-quality voice	3 kHz
Moderate-quality audio	5 kHz
High-fidelity audio	20 kHz
Television video	4 MHz

Another case of practical concern is the bandwidth of an arbitrary pulse having duration D, but not necessarily rectangular in shape. As a rough but handy rule of thumb, we usually take

$$W \approx 1/D \tag{9}$$

This reciprocal relationship conveys the key point that long pulses have small bandwidths while short pulses have large bandwidths. If $D = 5 \ \mu s = 5 \times 10^{-6}$ s, for example, then $W \approx 0.2 \times 10^{+6}$ Hz $= 200$ kHz. Equation (9) agrees qualitatively with the rectangular pulse spectrum in Fig. 13.1–8b, but it ignores the components above $f = 1/D$. To preserve the square corners of the rectangular pulse shape, we would have to take $W \gg 1/D$.

Exercise 13.1–2

Let the triangular and sawtooth waves in Fig. 13.1–3 both have $A = \pi$ and $T = 0.2$ ms. Sketch and label the two amplitude spectra for $0 \leq f \leq 30$ kHz. Then determine the value of W in each case such that $A_n < A_1/5$ for all $nf_1 > W$.

Filtering, Distortion, and Equalization

We found in Chapter 7 that a two-port network could act as a *filter*, emphasizing or suppressing a particular frequency band. When a signal is applied at the input of such a network, the resulting output waveform may differ appreciably from the input waveform. Any unwanted waveform alteration produced by a frequency-selective network is called **linear distortion** — as distinguished from the distortion produced by *nonlinear* elements. Here we'll investigate filtering and linear distortion from the vantage point of spectral analysis.

Figure 13.1–10 represents an arbitrary linear network with *input signal* $x(t)$ and *output signal* $y(t)$. We want to continue working with cyclical frequency, so the network's AC transfer function $H(j\omega)$ is expressed here in terms of the *amplitude ratio* and *phase shift* written as

$$|H(f)| \triangleq |H(j\omega)| \qquad \theta(f) \triangleq \sphericalangle H(j\omega)$$

where $\omega = 2\pi f$. Thus, when $x(t)$ contains a sinusoidal component of ampli-

Figure 13.1–10
Linear network for
signal transmission.

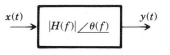

tude A_1 and phase ϕ_1 at frequency $f = f_1$, the corresponding output component has amplitude $|H(f_1)|A_1$ and phase $\phi_1 + \theta(f_1)$.

Now suppose that the input consists of several sinusoids, say

$$x(t) = \sum_n A_n \cos (2\pi f_n t + \phi_n) \qquad \text{(10a)}$$

By superposition, the steady-state response at the output will be

$$y(t) = \sum_n |H(f_n)|A_n \cos [2\pi f_n t + \phi_n + \theta(f_n)] \qquad \text{(10b)}$$

We are not limited here to periodic signals, but Eq. (10b) does include the case of **periodic steady-state response** simply by letting $f_n = nf_1$ and taking $n = 0, 1, 2, \ldots$

Whether periodic or nonperiodic, the output waveform signal is considered to be *undistorted* when

$$y(t) = Kx(t - t_d) \qquad \text{(11)}$$

This relationship says that $y(t)$ has the same shape as $x(t)$ *scaled* by the factor K and *delayed* in time by t_d. It follows from Eq. (11) that **distortionless transmission** through a network requires

$$|H(f)| = K \qquad \theta(f) = -360° \times t_d f \qquad \text{(12)}$$

which must hold for all frequencies in $x(t)$. Thus, a distortionless network has a *constant amplitude ratio* and a *negative linear phase shift* over the frequency range in question.

The bandwidths of lowpass and bandpass filters were defined in Chapter 7 so that Eq. (12) is reasonably well satisfied across the passband. Accordingly, when a *lowpass signal* having bandwidth W is applied to a *lowpass filter,* we get an essentially distortionless output provided that

$$B \geq W$$

where B is the filter's bandwidth measured in hertz. Figure 13.1–11 illustrates this condition — a condition of great practical significance because many information-bearing waveforms are lowpass signals and transmission cables often act like lowpass filters. Furthermore, when a lowpass signal contains unwanted components at $f > W$, they can be removed by lowpass filtering without distorting the filtered waveform.

If $|H(f)|$ does not satisfy Eq. (12), then the output suffers from **amplitude** or **frequency distortion** in that the amplitudes of different frequency components are selectively increased or decreased. If $\theta(f)$ does not satisfy Eq. (12), then the output suffers from **phase** or **delay distortion** in

Figure 13.1–11
Frequency-domain
interpretation of
transmission.

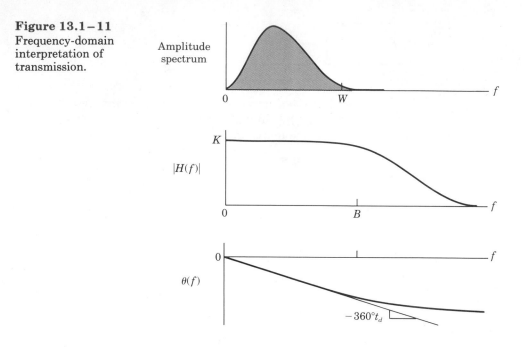

that different frequency components are delayed by different amounts of time. Both types of linear distortion usually occur together.

When linear distortion occurs in signal transmission, it can often be corrected or at least minimized using a network known as an **equalizer.** The equalization strategy is diagrammed in Fig. 13.1–12, where an equalizer has been connected at the output of the transmission medium. The resulting amplitude ratio of the entire system equals the product of the individual amplitude ratios, while the total phase shift equals the sum of the individual phase shifts. Hence, given $|H(f)|$ and $\theta(f)$ for the transmission medium alone, we design the equalizer such that

$$|H(f)||H_{eq}(f)| = K \qquad \theta(f) + \theta_{eq}(f) = -360° \times t_d f \qquad \textbf{(13)}$$

The equalized output signal is then $z(t) = Kx(t - t_d)$, regardless of the distortion in $y(t)$.

Figure 13.1–12
Transmission system
with equalizer.

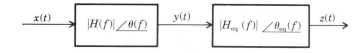

Equalizers also have applications whenever energy storage in a transducer or some other part of a signal-processing system causes linear distortion. For instance, audio equalizers in high-fidelity systems allow you to adjust the amplitude ratio over several frequency bands to correct electrical and acoustical frequency distortion. By the way, phase equalization is not critical in audio systems because the human ear has relatively little sensitivity to delay distortion.

Example 13.1−2

To illustrate linear distortion and demonstrate the use of Eq. (10), let $x(t)$ be the rectangular pulse train from Example 13.1−1 with $f_1 = 1/T = 1$ kHz. We'll find the periodic steady-state response $y(t)$ at the output of a bandpass filter having quality factor $Q = 1$ and tuned to the fundamental signal frequency, so the resonant frequency is $f_0 = \omega_0/2\pi = 1$ kHz.

The amplitude ratio and phase shift at any signal frequency $f_n = nf_1$ are given by

$$|H(f_n)| = \frac{1}{\sqrt{1 + \left(n - \dfrac{1}{n}\right)^2}} \qquad \theta(f_n) = -\arctan\left(n - \frac{1}{n}\right)$$

which follows from Eq. (11b), Section 7.2, with $Q = 1$ and $\omega/\omega_0 = 2\pi nf_1/2\pi f_0 = n$. Figure 13.1−13a shows the amplitude ratio curve $|H(f)|$ along with the input amplitude line spectrum. We see that the filter rejects the DC component of $x(t)$, and it greatly suppresses all components above the second harmonic. Hence, $y(t)$ will have zero average value, and it will lack the higher harmonics needed for square corners.

Figure 13.1−13
Illustration of linear distortion: (a) input spectrum and amplitude ratio curve; (b) input and output waveforms.

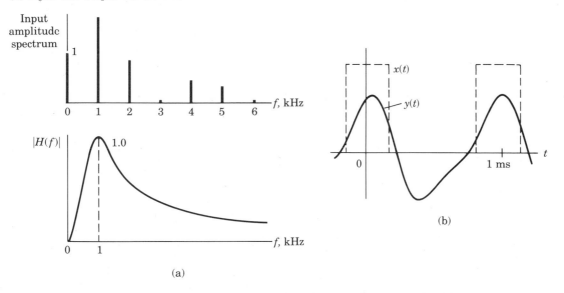

Substituting numerical values for $|H(f_n)|, A_n, \phi_n, \theta(f_n)$ into Eq. (10b) yields

$$y(t) \approx 1.65 \cos 2\pi f_1 t + 0.46 \cos (2\pi 2f_1 t - 56°)$$

where we have omitted the small components above the second harmonic. The resulting distorted waveform is plotted in Fig. 13.1–13b. Equalization could reduce the distortion by restoring some of the higher harmonics. However, an equalizer cannot put back the DC component that was completely eliminated by the bandpass filter.

Exercise 13.1–3   Consider a network with $|H(f)| = 1$ and *constant* phase shift $\theta(f) = -90°$. Find the resulting output $y(t)$ when $x(t) = 3 \cos 2\pi t + \cos (2\pi 3t + 180°)$. Then sketch $x(t)$ and $y(t)$ to demonstrate the effect of delay distortion.

Exercise 13.1–4   The frequency response of a certain transmission system is given by Eq. (5b), Section 7.1, with $f_{co} = \omega_{co}/2\pi = 5$ kHz. Find and sketch the equalizer characteristics needed to satisfy Eq. (13) over $0 \le f \le 10$ kHz with $K = 1$ and $t_d = 0$.

13.2 MODULATION, SAMPLING, AND MULTIPLEXING

Modulation is a processing operation that impresses the information from a signal $x(t)$ on a *carrier* waveform whose characteristics suit the particular application. The carrier may be a sinusoid, in which case we get the phenomenon of *frequency translation*. Or the carrier may be a pulse train, which requires that the signal be *sampled* as part of the modulation process. Frequency translation and sampling have many important uses, and both lend themselves to *multiplexing* that permits one transmission system to handle two or more information-bearing signals simultaneously.

Frequency Translation and Product Modulation

The **product modulator** diagrammed in Fig. 13.2–1a multiplies the signal $x(t)$ and a sinusoidal carrier wave at frequency f_c to produce

$$x_c(t) = x(t) \cos 2\pi f_c t \tag{1}$$

Figure 13.2–1
(a) Product modulator.
(b) Waveforms.

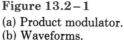

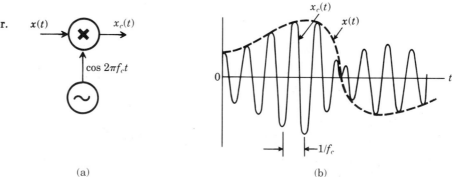

(a)   (b)

Figure 13.2–1b illustrates the relationship between $x_c(t)$ and $x(t)$, taking $x(t)$ to be a lowpass signal with $W \ll f_c$. The modulated wave $x_c(t)$ now has a **bandpass** spectrum resulting from frequency translation.

Frequency translation takes place whenever sinusoids are multiplied. Specifically, the trigonometric identity for the product of cosines gives

$$\cos 2\pi f_1 t \times \cos 2\pi f_2 t = \tfrac{1}{2} \cos 2\pi (f_2 - f_1)t + \tfrac{1}{2} \cos 2\pi (f_2 + f_1)t \qquad \textbf{(2)}$$

so the product consists of the *sum and difference frequencies* $f_2 + f_1$ and $|f_2 - f_1|$. The difference frequency always has the *positive* value $|f_2 - f_1|$ because $\cos 2\pi(f_2 - f_1)t = \cos 2\pi(f_1 - f_2)t$ when $f_1 > f_2$. Thus, the sum and difference frequencies are compactly expressed by writing $|f_2 \pm f_1|$.

To bring out the frequency translation implied in Eq. (1), suppose that $x(t)$ contains a sinusoidal component $A_m \cos 2\pi f_m t$. Multiplication by a carrier wave with $f_c \gg f_m$ then yields

$$A_m \cos 2\pi f_m t \times \cos 2\pi f_c t = \frac{A_m}{2} \cos 2\pi (f_c - f_m)t$$

$$+ \frac{A_m}{2} \cos 2\pi (f_c + f_m)t$$

The various waveforms and line spectra are plotted in Fig. 13.2–2. Note that the product contains neither f_m nor f_c, but consists instead of a *pair* of lines offset from f_c by $\pm f_m$. The low frequency f_m has, as a result, been translated to the higher frequencies $f_c \pm f_m$.

Figure 13.2–2

Frequency translation waveforms and line spectra.

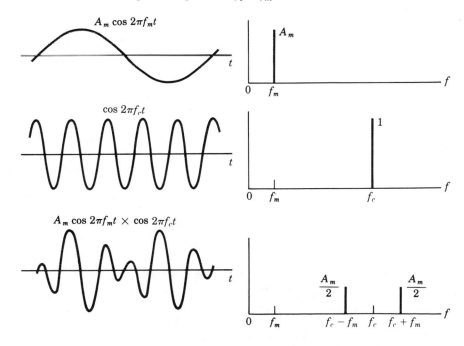

Figure 13.2–3
Amplitude spectra in
double-sideband
modulation: (a) lowpass
modulating signal;
(b) bandpass modulated
signal.

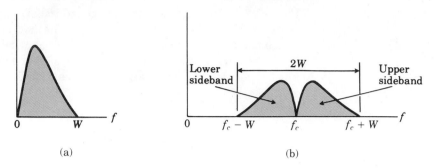

(a)   (b)

Now let $x(t)$ be an arbitrary lowpass signal with the typical amplitude
spectrum of Fig. 13.2–3a. The amplitude spectrum of the modulated wave
$x_c(t)$ will now contain **two sidebands** each of width W on either side of f_c,
as shown in Fig. 13.2–3b. We therefore have a signal that could be transmitted over a *bandpass* system with a minimum bandwidth of

$$B = 2W \tag{3}$$

which is precisely *twice* the bandwidth of the modulating signal. For obvious reasons, this process bears the name **double-sideband modulation
(DSB).** If bandwidth must be conserved, either the lower or upper sideband may be removed by filtering to get **single-sideband modulation
(SSB)** with $B = W$.

The frequency-translation aspect of product modulation, together
with the relative lack of restrictions on f_c, allows an engineer to minimize
distortion and other problems by putting the carrier frequency at a point
where the system has favorable characteristics. This is especially helpful
when $x(t)$ contains important DC and low-frequency components that
would be lost in a system with transformer coupling or coupling capacitors.
The necessary product operation can be implemented electronically in a
number of ways, and certain transducers are easily modified for frequency
translation. As a case in point, the bridge circuit back in Fig. 13.1–1a with a
carrier source instead of the supply battery would produce a DSB signal
that carries the temperature information $x(t)$.

We recover $x(t)$ from $x_c(t)$ using the **product demodulator** of
Fig.13.2–4a, which has a **local oscillator** synchronized in frequency and
phase with the carrier wave. Multiplication then produces both upward
and downward translation, so the input to the low pass filter will be

$$x_c(t) \cos 2\pi f_c t = x(t) \cos^2 2\pi f_c t$$

$$= \tfrac{1}{2} x(t) + \tfrac{1}{2} x(t) \cos 2\pi(2f_c)t$$

as follows from Eq. (2) with $f_1 = f_2 = f_c$. The first term is proportional to
$x(t)$, while the second looks like DSB at carrier frequency $2f_c$. Therefore, if
the filter rejects the high-frequency components and passes $f \leq W$, then
the filtered output has the desired form $z(t) = Kx(t)$.

Figure 13.2–4
(a) Product
demodulation system.
(b) Spectrum before
lowpass filtering.

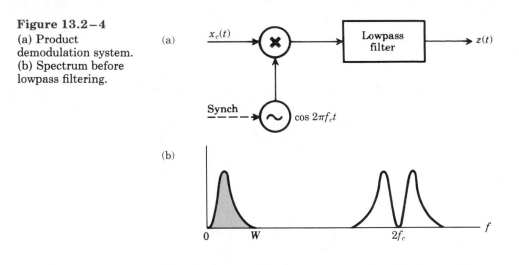

Exercise 13.2–1   Sketch the amplitude spectrum of $x_c(t)$ when $x(t) = 12 \cos 2\pi100t + 8 \cos 2\pi150t$ and $f_c = 600$ Hz. Now list all the frequencies in the product $x_c(t) \cos 2\pi500t$.

Sampling and Pulse Modulation

When engineers plot experimental data, they frequently draw smooth curves through sample data points as a way of interpolating values between them. This familiar process is quite accurate, provided the sample points are "close enough." Rather astonishingly, the same property holds for electrical signals. It is possible to sample an electrical signal, transmit just the *sample values,* and use them to interpolate or reconstruct the *entire* waveform at the destination. Sampling also makes it possible to convert an analog signal to *digital* form, permitting the use of digital processing methods.

Figure 13.2–5a shows a simple switching sampler and waveforms. The switch alternates between the two contacts at the **sampling frequency** $f_s = 1/T_s$ Hz. It dwells at the upper contact for a short interval $D \ll T_s$ and extracts a sample piece of the input signal $x(t)$ every T_s seconds. Since the lower contact is grounded, the output sampled waveform $x_s(t)$ looks like a train of pulses whose tops carry the sample values of $x(t)$. We can analyze this process and prove that $x(t)$ can be recovered from $x_s(t)$ by modeling sampling as *multiplication* in the form

$$x_s(t) = x(t)s(t) \tag{4}$$

where the *switching function* $s(t)$ is nothing more than a unit-height rectangular pulse train, as in Fig. 13.2–5b. Fourier expansion of the periodic switching function yields

$$s(t) = a_0 + a_1 \cos 2\pi f_s t + a_2 \cos 2\pi(2f_s)t + \cdots$$

Figure 13.2–5
(a) Switching sampler. (b) Model using switching function $s(t)$.

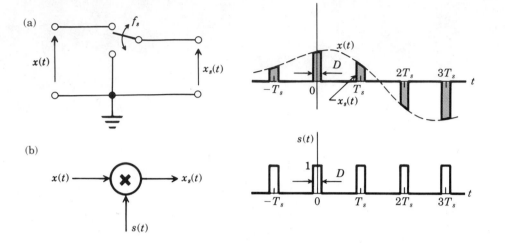

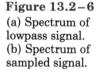

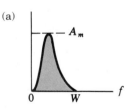

with $a_0 = D/T_s$ and $a_n = (2/\pi n) \sin (\pi Dn/T_s)$ for $n = 1, 2, \ldots$ (see Table 13.1–1). Upon inserting this expansion, Eq. (4) becomes

$$x_s(t) = a_0 x(t) + a_1 x(t) \cos 2\pi f_s t$$
$$+ \, a_2 x(t) \cos 2\pi(2f_s)t + \cdots \qquad \textbf{(5)}$$

an awesome-looking result but easily interpreted if we go to the frequency domain.

Suppose that $x(t)$ has the lowpass amplitude spectrum of Fig. 13.2–6a. The corresponding spectrum of the sampled signal $x_s(t)$ sketched in Fig.

Figure 13.2–6
(a) Spectrum of lowpass signal.
(b) Spectrum of sampled signal.

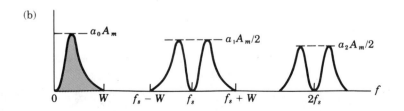

13.2 – 6b is based on a term-by-term examination of Eq. (5). The first term has the same spectrum as $x(t)$ scaled by the factor a_0. The second term is identical to product modulation with a scale factor a_1 and carrier frequency f_s, so it has a double-sideband spectrum over the range $f_s - W \leq f \leq f_s + W$. The third and all other terms have the same DSB interpretation with progressively higher carrier frequencies, $2f_s$, $3f_s$, etc.

Now observe that none of the translated components falls into the signal range $0 \leq f \leq W$, provided that the sampling frequency satisfies the condition

$$f_s \geq 2W \tag{6}$$

Therefore, if we apply the sampled signal $x_s(t)$ to a lowpass filter which removes all components at $f \geq f_s - W$, then the resulting output signal will have exactly the same shape as $a_0 x(t)$. The sample duration D has no effect other than determining the value of the scale factor a_0. We summarize these observations in the following **uniform sampling theorem:**

> A signal having no frequency components at $f \geq W$ is completely described by uniformly spaced sample values taken at the rate $f_s \geq 2W$. The entire signal waveform can be reconstructed from the sampled signal by a lowpass filter that rejects $f \geq f_s - W$.

The minimum sampling frequency $f_s = 2W$ is called the **Nyquist rate.**

When Eq. (6) does not hold, the resulting spectral overlap creates unwanted spurious components in the filtered output. In particular, any component of $x(t)$ originally at $f' > f_s/2$ appears in the output at the lower frequency $|f_s - f'| < W$. This phenomenon is known as **aliasing.** To prevent aliasing, the signal $x(t)$ should be processed by a lowpass filter with bandwidth $B_p \leq f_s/2$ *before* sampling.

Figure 13.2 – 7a shows the elements of a typical **pulse modulation system.** The pulse generator produces a pulse train with the sample values carried by the pulse *amplitude* (**PAM**), *duration* (**PDM**), or relative *position* (**PPM**), as illustrated in Fig. 13.2 – 7b. At the destination, the modulated pulses are converted back to sample values for reconstruction by lowpass filtering.

Pulse-duration and pulse-position modulation have the advantage of being immune to *nonlinear distortion,* since the pulse is either ON or OFF. (This property is also exploited in conjunction with the optoisolator mentioned in Section 8.3.) In exchange, however, the transmission bandwidth must be

$$B \geq 1/D \gg W \tag{7}$$

which is needed to accommodate pulses with duration $D \ll T_s \leq 1/2W$. The pulse-modulated wave may be frequency translated for transmission over a bandpass system, further doubling the bandwidth requirement.

Figure 13.2–7
(a) Pulse modulation system. (b) Waveforms.

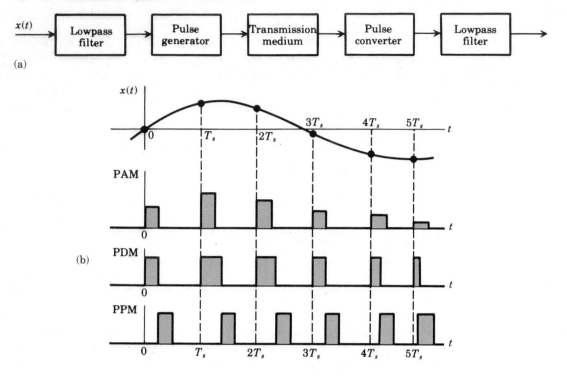

Exercise 13.2–2   Make a careful plot of the signal $x(t) = 3 \cos 2\pi10t - \cos 2\pi30t$, which
approximates a square wave with $W = 30$ Hz. Mark the sample points at
$t = 0, \frac{1}{60}, \frac{2}{60}, \ldots , \frac{6}{60}$, corresponding to $T_s = 1/2W$, and convince yourself
that $x(t)$ could be recovered from these samples. Then mark the sample
points at $t = 0, \frac{1}{40}, \frac{2}{40}, \ldots$, corresponding to $T_s > 1/2W$ and $f_s < 2W$; a
smooth curve drawn through these points will demonstrate aliasing.

Multiplexing Systems

The term *multiplexing* refers to sending two or more signals together on
one transmission facility. **Frequency-division multiplexing (FDM)**
accomplishes this task by translating the various signals to nonoverlap-
ping frequency bands. Bandpass filtering at the destination separates (or
demultiplexes) the signals for individual recovery.

   A simple FDM system of the type used in telephone communications is
diagrammed in Fig. 13.2–8. Each input is first lowpass filtered (LPF),
removing all frequency components above about 3 kHz, and then modulated onto individual **subcarriers** with 4-kHz spacing. The modulation is

Figure 13.2–8
(a) FDM system with three signals.
(b) Spectrum of multiplexed signal with pilot.

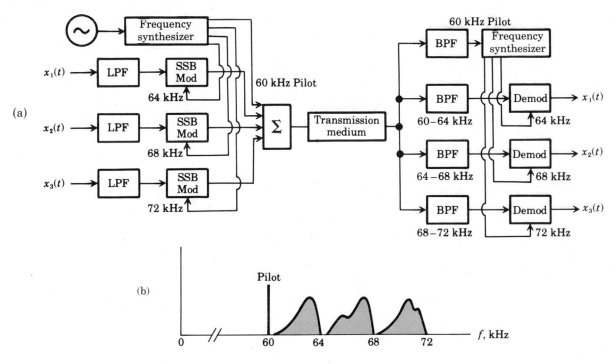

single-sideband (SSB), and all subcarriers are synthesized from a master oscillator. Summing the SSB signals, plus a 60-kHz **pilot carrier,** forms the multiplexed signal, with the typical spectrum as shown. A bank of bandpass filters (BPFs) at the destination isolates each SSB signal for product demodulation. Synchronization is provided by deriving the local-oscillator waveforms from the pilot carrier. As many as 3600 telephone signals have been multiplexed in this manner. (The FDM system for FM stereo broadcasting will be outlined in Section 13.4.)

Time-division multiplexing (TDM) takes advantage of the fact that a sampled signal is OFF most of the time. The intervals between samples are therefore available for the insertion of samples from other signals.

A rudimentary TDM system has the structure diagrammed in Fig. 13.2–9a, where the three input signals are assumed to have equal bandwidths W. An electronic switch or **commutator** sequentially extracts one sample from each input every T_s seconds, producing a multiplexed waveform with interleaved samples, as in Fig. 13.2–9b. A similar switch at the destination separates and distributes the samples to a bank of lowpass

Figure 13.2–9
(a) TDM system with three signals. (b) Multiplexed waveform.

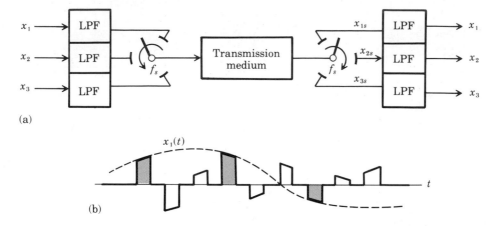

(a)

(b)

filters for individual signal reconstruction. More sophisticated systems convert the sample values to pulse modulation before multiplexing and might include carrier modulation after multiplexing.

TDM has the advantage of simpler implementation than FDM, especially in view of the wide availability of integrated switching circuits. But, of course, the two commutating switches must be synchronized precisely for successful operation.

Exercise 13.2–3   A data telemetry system is to be designed to handle three different signals with bandwidths $W_1 = 1$ kHz, $W_2 = 2$ kHz, and $W_3 = 3$ kHz. Find the transmission bandwidth required using

(a) FDM with DSB subcarrier modulation.
(b) TDM with pulse duration $D = T_s/6$.

13.3 INTERFERENCE AND NOISE[†]

An information-bearing signal often becomes contaminated by externally generated interference or by internally generated noise. This section describes some of the major causes of interference and noise and some methods for dealing with their effects.

Interference, Shielding, and Grounding

Figure 13.3–1 depicts a simple instrumentation system intended to amplify and display the signal $x(t)$ generated by a transducer. The system's electrical environment includes several sources of potential interference

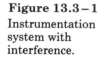

Figure 13.3–1
Instrumentation
system with
interference.

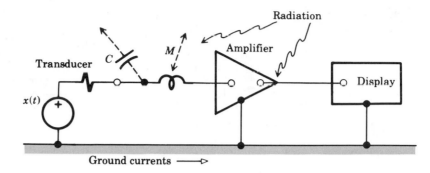

such as AC power lines, rotating machinery, thunder storms, radio trans-
mitters, and other electrical and electronic equipment. Interfering signals
from these sources enter the system primarily through four mechanisms:

- **Capacitive coupling,** via stray capacitance C between the system
  and an external voltage.

- **Magnetic coupling,** via mutual inductance M between the system
  and an external current.

- **Radiative coupling,** from electromagnetic radiation impinging on
  the system.

- **Ground-loop coupling,** from currents flowing between different
  ground points.

Depending on the source and coupling mechanism, interference may take
many forms: AC hum at 60 Hz or its second or third harmonic; higher
frequency "whistles," pulses, and other intelligent-appearing signals; or
erratic waveforms commonly called "static." You may hear some of these
effects if you put an AM/FM radio near a fluorescent light, an automobile
ignition system, or even a seemingly innocent calculator.

Obviously, if possible, we should turn off any offending sources in the
vicinity of the system. Then, to minimize coupling from the inevitable
remaining sources, all exposed elements should be enclosed within conducting **shields,** as illustrated by Fig. 13.3–2. (This drawing assumes that
the amplifier, its power supply, and all other instruments are shielded by
their own metal cases.) When held at a common potential, these shields
greatly reduce most types of interference. However, low-frequency magnetic coupling can induce undesired current flow through the shields
themselves, so we have to interrupt the shield connection, say at the amplifier's output, to avoid a closed-loop current path. Extreme instances of
interference by magnetic coupling may necessitate a layer of special magnetic shielding material in addition to the conducting material.

The grounding terminals (labeled G), the equipment cases, and the
shields have all been tied together at a *single ground point* in Fig. 13.3–2 to
prevent interference caused by ground-loop currents. In contrast, the three

Figure 13.3–2

Shielding and
grounding arrangement.

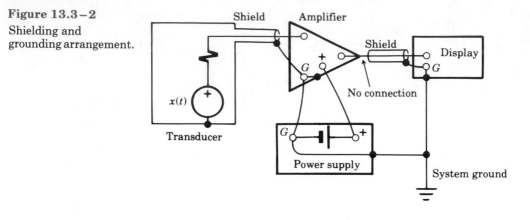

ground points in Fig. 13.3–1 would be at different potentials when the
ground currents (usually at 60 Hz) flow through the small resistances
between them, producing an apparent interference voltage source in series
with $x(t)$. Of course, any instrument connected to an AC power line should
be grounded for safety reasons.

The value of shielding for the reduction of interference via capacitive
coupling is brought out by the equivalent circuits in Fig. 13.3–3. Part a
represents an unshielded transducer with internal resistance R_t connected
to a load R_L and also coupled by stray capacitance C to an interference
voltage v_{int}. Current from v_{int} passes through C and divides between R_t and
R_L, so v_L includes a component proportional to v_{int} along with the desired
signal component proportional to x. Shielding the transducer yields the
equivalent circuit in part b, where C_s represents the capacitance between
the shield and the transducer while R_s represents the shield's resistance
path to ground. Since R_s is very small, virtually all of the current from v_{int} is
now bypassed to ground and has little effect on v_L.

Figure 13.3–3

Equivalent circuits for
capacitive-coupled
interference:
(a) without shielding;
(b) with shielding.

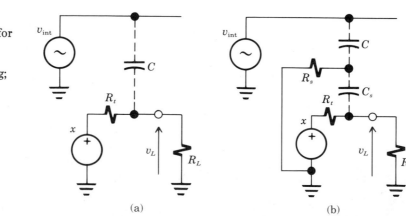

Figure 13.3–4

(a) System with local ground at transducer. (b) Model of transducer with common-mode voltage.

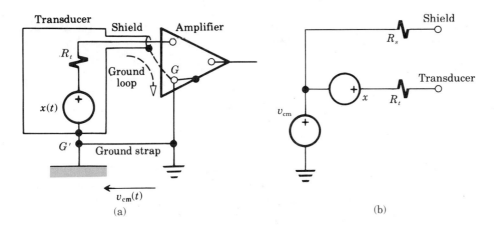

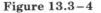
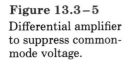

When the transducer has a local ground that cannot be disconnected, we must resort to an arrangement like Fig. 13.3–4a, where a separate **ground strap** connects the local ground G' to the system ground point. (Braided-wire straps are preferred here for their small inductance.) We also disconnect the shield from the amplifier to prevent ground-loop current through the shield. However, since the ground strap has nonzero resistance, any stray current between G' and ground will produce an interference voltage labeled v_{cm}. Figure 13.3–4b diagrams the resulting equivalent circuit for the transducer, omitting stray capacitance. We call v_{cm} a **common-mode voltage** because it appears at both the transducer and shield terminals.

Although v_{cm} is usually quite small, it may still be a problem when the desired signal voltage is equally small. For such cases, we can eliminate the common-mode voltage using the **differential amplifier** in Fig. 13.3–5.

Figure 13.3–5
Differential amplifier to suppress common-mode voltage.

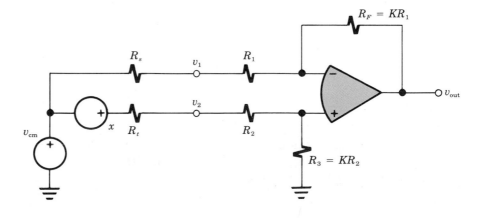

Routine analysis based on the virtual-short model of the op-amp shows that

$$v_{\text{out}} = K(v_2 - v_1) \tag{1}$$

so this circuit amplifies the *difference voltage* $v_2 - v_1$. Under the reasonable assumptions that $R_s \ll R_1$ and $R_t \ll R_2$, we have $v_1 \approx v_{\text{cm}}$ and $v_2 \approx x + v_{\text{cm}}$. Hence,

$$v_{\text{out}} = K[(x + v_{\text{cm}}) - v_{\text{cm}}] = Kx \tag{2}$$

and v_{cm} has been canceled out as desired. This op-amp circuit is also known as an **instrumentation** or **transducer amplifier.**

Finally, it should be obvious that any interference at frequencies outside the signal band can be rejected by appropriate filtering. Less obvious is the last-resort technique of using a **notch filter** to combat interference *within* the signal band. To illustrate, suppose that proper shielding and grounding still leaves bothersome interference at a single frequency f_0, as portrayed by the composite amplitude spectrum in Fig. 13.3–6a. A notch filter having the amplitude ratio shown in Fig. 13.3–6b removes the interference but it also introduces some signal distortion — which may be the only reasonable alternative under the circumstances. Such filtering should precede amplification to prevent possible saturation of the amplifier by the interference.

Figure 13.3–6
(a) Spectrum with single-frequency interference.
(b) Amplitude ratio of notch filter.

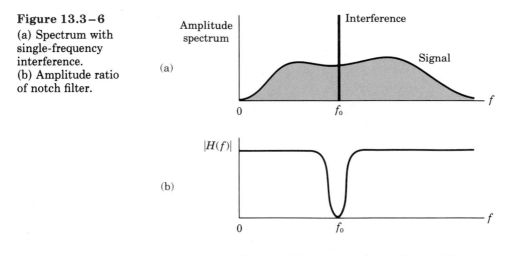

Exercise 13.3–1

Let the transducer in Fig. 13.3–3 have $R_t = 1 \text{ k}\Omega$ and a constant signal voltage $x = 5 \text{ mV}$, and let $v_{\text{int}} = 170 \cos 2\pi 60t$ V and $C = 100 \text{ pF}$.

(a) Use superposition to estimate $v_L(t)$ without shielding when $R_L = \infty$.
(b) Repeat this calculation with a shield having $R_s = 1 \, \Omega$ and $C_s = 10 \text{ pF}$, assuming $C = 100 \text{ pF}$.

Electrical Noise

When a system requires a great deal of amplification for the processing of a very weak signal, the resulting output often includes internally generated noise along with the amplified signal. Every electrical system produces noise, although usually so small that it goes unnoticed. To explain the omnipresent nature of noise, we return to our picture of a metallic lattice with electrons dancing about in random thermal motion (Fig. 2.3–4). Each vibrating electron constitutes a tiny current, and the net effect of billions of random electron currents in a resistive material produces the phenomenon called **thermal noise.** Since we cannot build an electrical system without electrons and resistance, thermal noise is inevitable — like death and taxes.

Figure 13.3–7a illustrates a typical thermal noise waveform $n(t)$. In view of the unpredictable behavior, we must resort to *average* properties for its quantitative description. The average value of $n(t)$ equals zero, so a more useful quantity is the **root-mean-square value** n_{rms} that we obtain by averaging $n^2(t)$ over a long time interval and taking the square root of the result. This quantity has the same interpretation relative to average power as the rms value of a sinusoidal waveform. Specifically, if $n(t)$ appears as a voltage across some resistance R, then the **average noise power** will be

$$N = \frac{n_{\mathrm{rms}}^2}{R} \tag{3}$$

Similarly, for a noise current, we would write $N = R n_{\mathrm{rms}}^2$.

Figure 13.3–7

Thermal noise:
(a) typical waveform;
(b) power spectrum.

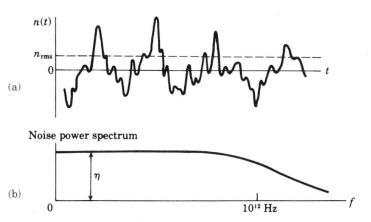

Unlike a sinusoid, the spectrum of thermal noise power is *uniformly spread* over frequency up to the infrared region around 10^{12} Hz, as plotted in Fig. 13.3–7b. This spectral distribution reflects the fact that there will be equal numbers of electrons vibrating at every frequency, on the average, and thus $n(t)$ contains all electrical frequencies in equal proportion. Thermal noise is also called **white noise,** by analogy to white light, which contains all visible frequencies in equal proportion.

The constant η in Fig. 13.3–7b stands for the **noise power spectral density,** expressed in terms of power per unit frequency (W/Hz). We interpret this concept with the help of Fig. 13.3–8a, where resistance R—a thermal noise source—is connected to an amplifier with a matched input resistance. The amplifier is presumed to be noiseless, with power gain G and bandwidth B. Under these conditions, the output noise power is

$$N_{\text{out}} = G\eta B$$

Thus, the amplifier accepts power ηB falling within the passband and amplifies it by the factor G. We then write the source noise power as

$$N = \eta B \qquad\qquad (4)$$

which still includes B because we always "see" thermal noise through the limiting bandwidth of one instrument or another. Equation (4) states that N equals the *area* under the power spectrum curve between the instrument's lower cutoff f_ℓ and upper cutoff $f_u = f_\ell + B$, depicted by Fig. 13.3–8b.

Figure 13.3–8
(a) Thermal noise at amplifier input.
(b) Power spectrum.

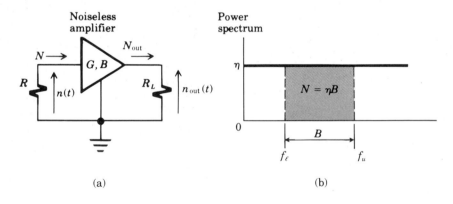

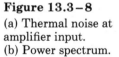

(a)   (b)

As to the value of the density constant η, statistical theory shows that

$$\eta = kT \qquad\qquad (5)$$

where k is Boltzmann's constant and T is the source *temperature* in degrees kelvin. A hot resistance is therefore noisier than a cool one, which agrees with our notion of thermally agitated electrons. Note that η and N do not depend on the resistance R, although the rms noise voltage does. Specifically, combining Eqs. (3)–(5) yields

$$n_{\text{rms}} = \sqrt{RkTB} \qquad\qquad (6)$$

Equation (6) gives the rms noise voltage for a thermal source connected to a matched resistance. The open-circuit voltage would be twice this value.

Prior to amplification, thermal noise is exceedingly minute. For instance, take the case of *room temperature* $T_0 \approx 290$ K (17°C) at which

$$\eta_0 = kT_0 \approx 4 \times 10^{-21} \text{ W/Hz} \tag{7}$$

Thus, the noise power in bandwidth $B = 1$ MHz is only $N = \eta_0 B = 4 \times 10^{-15}$ W, and Eq. (6) predicts $n_{\text{rms}} = 2 \ \mu\text{V}$ if $R = 1$ kΩ. But an amplifier would amplify this noise and contribute additional noise of its own.

Amplifier noise comes from both thermal sources (resistances) and **nonthermal** sources (semiconductor devices). Although nonthermal noise is not related to physical temperature and does not necessarily have a uniform spectrum, it is still convenient to speak of an amplifier's **noise temperature** T_a as a measure of "noisyness" referred to the input. We thus write output noise power caused only by the amplifier in the form

$$N_a = GkT_a B$$

This expression includes the fact that N_a depends on the gain and bandwidth. For the more general situation of Fig. 13.3–9 with input noise $N = \eta B = kTB$ from a source at temperature T, N_a *adds* to the amplified source noise and the output power becomes

$$N_{\text{out}} = GN + N_a = Gk(T + T_a)B \tag{8}$$

Garden-variety amplifiers have $T_a \gg T_0$, meaning that they are very noisy, not physically hot. Therefore, if $T = T_0$, then $N_{\text{out}} \approx N_a$, and the amplifier noise dominates the source noise — a common occurrence.

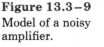
Figure 13.3–9
Model of a noisy amplifier.

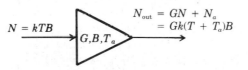

Figure 13.3–10 shows the variation of noise temperature with frequency for a nonthermal source. The pronounced low-frequency rise includes several phenomena lumped together under the heading **one-over-f** **$(1/f)$** **noise.** Transistors generate $1/f$ noise, and so do certain transducers, especially photodiodes and other optical sensors. In addition, slow and

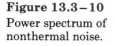
Figure 13.3–10
Power spectrum of nonthermal noise.

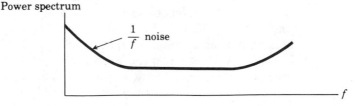

random *equipment drifts* have an effect that can be modeled as $1/f$ noise. Whenever the noise temperature of a source or amplifier varies appreciably with frequency, the value to be used in Eqs. (5)–(8) is the *average* temperature over the frequency range of interest.

Exercise 13.3–2   A certain amplifier has $G = 10^5$ (50 dB), $B = 8$ MHz, and $T_a = 24T_0$. Find the output noise power and rms voltage across a 50-Ω load resistor when $T = T_0$.

Signals in Noise

Now consider the case where a weak information signal is to be amplified by a noisy amplifier. Let P_{in} be the average power of the input signal, so the amplified signal power is

$$P_{out} = GP_{in}$$

But the output also includes source and amplifier noise as given by Eq. (8). We therefore speak of the output **signal-to-noise ratio (SNR)** given by

$$\frac{P_{out}}{N_{out}} \triangleq \frac{GP_{in}}{Gk(T + T_a)B} = \frac{P_{in}}{k(T + T_a)B} \tag{9}$$

Note that the amplifier's power gain cancels out here because both signal and noise power are amplified by G.

The signal-to-noise ratio — usually expressed in decibels — serves as an important system performance measure, for it tells us the signal strength relative to the noise. A large ratio means that the signal may be strong enough to mask the noise and render it inconsequential (or vice versa). For example, intelligible voice communication is possible with $P_{out}/N_{out} \geq 20$ dB, but a reasonably noise-free television image must have $P_{out}/N_{out} \geq 50$ dB. Lower values than these would cause noticeable "static" in the voice signal or a "snowy" TV picture.

Examining Eq. (9) reveals that good performance requires a large value for P_{in} and/or small values for $T + T_a$ and B. Keep in mind, however, that the amplifier's bandwidth B should not be less than the signal bandwidth W. Therefore, we expect noise to be more troublesome when we're dealing with large-bandwidth signals.

When the noise temperature varies with frequency, we can improve system performance by using *frequency translation* to put the signal in a less noisy frequency band. This strategy has particular merit for the case of a lowpass signal whose spectrum falls in the range of an amplifier's $1/f$ noise. Figure 13.3–11a diagrams the implementation with product modulation and demodulation, and Fig. 13.3–11b illustrates the noise reduction in terms of areas under the noise power curve. (It turns out that synchronized product demodulation doubles the final signal-to-noise ratio, which exactly compensates for the fact that product modulation requires band-

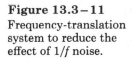

Figure 13.3–11
Frequency-translation
system to reduce the
effect of $1/f$ noise.

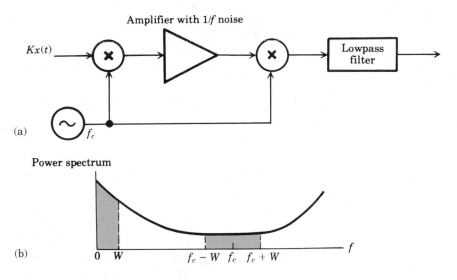

(a)

(b)

width $B = 2W$.) Normally, the two multipliers in Fig. 13.3–11a are implemented using a pair of synchronized switches called **choppers,** which act like the switching sampler back in Fig. 13.2–5.

Another way of improving signal-to-noise ratios when the noise density is higher in one region of the spectrum is the **preemphasis-deemphasis** technique. If we selectively emphasize (amplify) the portion of the signal spectrum that falls in the high noise region before contamination, then we can deemphasize the signal-plus-noise after contamination, thereby restoring the signal spectrum and reducing the noise. In stylus disk recording, for instance, the higher signal frequencies are emphasized so that a lowpass deemphasis filter will suppress the high-frequency surface noise (or hiss) during playback.

Sometimes the signal in question is a *constant* whose value we seek, as in a simple measurement system. Since a constant corresponds to a DC component (or average value) and since noise usually has zero average value, the measurement accuracy will be enhanced by a lowpass filter with the smallest attainable bandwidth B. (Lowpass filtering, then, partially carries out the operation of *averaging*.) But some noise will get through the

Figure 13.3–12
Constant signal with
noise fluctuations.

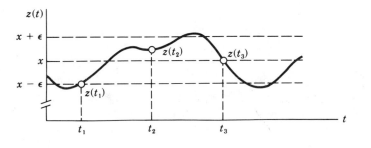

filter and cause the processed signal $z(t)$ to fluctuate about the true value x, illustrated by Fig. 13.3–12.

Any one sample of $z(t)$, say $z_1 = z(t_1)$, normally falls someplace between $x - \epsilon$ and $x + \epsilon$. The quantity ϵ represents the *rms error* and is given by

$$\epsilon = \frac{x}{\sqrt{P_{out}/N_{out}}} \tag{10}$$

This expression follows from the fact that P_{out} is proportional to x^2 and N_{out} is proportional to the square of the rms noise. Suppose, however, that we take M different samples of $z(t)$ and form the arithmetic average

$$z_{av} = \frac{1}{M}(z_1 + z_2 + \cdots + z_M)$$

If the samples are spaced in time by at least $1/B$ seconds, then the noise-induced errors tend to cancel out and the rms error of z_{av} becomes

$$\epsilon_M = \epsilon/\sqrt{M} \tag{11}$$

This averaging method is equivalent to reducing the bandwidth to B/M.

Averaging techniques are also helpful when the signal in question is a sinusoid whose amplitude is to be measured. We then use a narrow *bandpass* filter or a special processor called a **lock-in amplifier.** Other, more sophisticated methods have been devised for extracting information from signals deeply buried in noise, making it possible in essence to "hear a hummingbird in a hurricane." Most of these special methods involve digital processing.

Example 13.3–1

A low-noise transducer at a remote location is connected to the instrumentation system by a cable, delivering the minute signal power $P_{in} = 100$ pW for amplification and processing. The signal has a bandwidth of 5 kHz, and we seek the condition on the amplifier noise temperature T_a such that $P_{out}/N_{out} \geq 50$ dB.

Presumably, the cable generates thermal noise at room temperature, so $T = T_0$ and we can write $k(T + T_a) = kT_0(1 + T_a/T_0)$. After substituting numerical values, Eq. (9) becomes

$$\frac{P_{out}}{N_{out}} = \frac{100 \times 10^{-12}}{4 \times 10^{-21}(1 + T_a/T_0) \times 5 \times 10^3} \geq 10^5$$

which requires that $T_a \leq 49T_0$, a condition not too hard to satisfy with a well-designed amplifier.

Exercise 13.3–3

Suppose the signal in Example 13.3–1 has a bandwidth of 500 kHz. Find P_{out}/N_{out} assuming a noiseless amplifier. Then show that it would be impossible to get $P_{out}/N_{out} \geq 10^5$.

13.4 COMMUNICATION SYSTEMS[†]

Having addressed general aspects of signal analysis and processing, we turn our attention specifically to communication systems. We'll start with the two types of media for long-distance communication, transmission lines and radio transmission. Then we'll examine important features of radio communication, including amplitude modulation (AM), frequency modulation (FM), and some of the associated hardware.

Transmission Lines

Figure 13.4–1 represents a transmission line of length ℓ connecting a signal source to a distant load. The line might take the form of a pair of wires, a coaxial cable, or a hollow waveguide. A crucial factor here is the **characteristic impedance** Z_0, which relates the voltage and current of the wave traveling along the line. For distortionless transmission, Z_0 must be constant and resistive over the frequency range of the signal. Furthermore, the source and load resistances should be *matched* so that $R_s = Z_0$ and $R_L = Z_0$ as indicated. Otherwise, a mismatch at the load ($R_L \neq Z_0$) reflects some of the signal energy back toward the source where any mismatch ($R_s \neq Z_0$) reflects energy in the forward direction again. Impedance matching eliminates these unwanted reflections.

Figure 13.4–1
Transmission line with matched impedances.

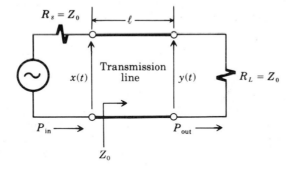

Now assume matched and distortionless conditions with signal voltage $x(t)$ across the input of the line, so the resulting output voltage is $y(t) = Kx(t - t_d)$. The attenuation factor K will always be less than unity because ohmic heating in the line dissipates part of the input signal energy. To emphasize this effect we work with the average signal powers, P_{in} and P_{out}, and we define the **transmission loss** L from the power ratio

$$P_{out}/P_{in} = 1/L \qquad (1)$$

where $L = 1/K^2 > 1$. Hence, $P_{out} = P_{in}/L < P_{in}$.

Regardless of the type of line, the transmission loss increases *exponentially* with distance ℓ and can be written in the form

$$L = 10^{(\alpha\ell/10)} \tag{2a}$$

The parameter α is the line's **attenuation coefficient** in decibels per unit length. Expressing L in dB, similar to power gain, we get

$$L_{dB} = 10 \log \frac{P_{in}}{P_{out}} = \alpha\ell \tag{2b}$$

This equation provides a very simple way of calculating attenuation and explains, in part, why communication engineers usually work with decibels. Typical values of α range from 0.05 to 100 dB/km, depending on the type of cable and the frequency.

But dB values tend to obscure how rapidly output power falls off as distance increases. We underscore this effect by restating Eq. (2b) in the form

$$\frac{P_{out}}{P_{in}} = 10^{-(L_{dB}/10)} = 10^{-(\alpha\ell/10)}$$

which is plotted versus ℓ in Fig. 13.4−2.

Figure 13.4−2
Power ratio versus distance.

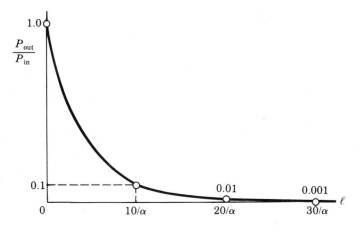

Transmission loss can be overcome with the help of one or more amplifiers connected in cascade with the line. Figure 13.4−3 diagrams a system

Figure 13.4−3
Transmission system with preamplifier and repeater.

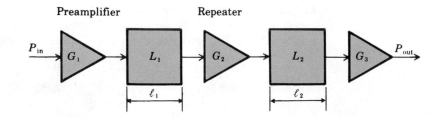

with a transmitting amplifier or **preamplifier** at the source, a receiving amplifier at the destination, and an additional amplifier called a **repeater** inserted at some intermediate point. All amplifiers, of course, must be impedance matched for maximum power transfer.

We find the final output power by a simple chain calculation. If the preamplifier's power gain is G_1, then its output power will be $G_1 P_{in}$, which results in $G_1 P_{in}/L_1$ at the input to the repeater, and so forth. thus,

$$\frac{P_{out}}{P_{in}} = \frac{G_1 G_2 G_3}{L_1 L_2} \tag{3}$$

which shows that we can compensate for the line loss and get $P_{out} \geq P_{in}$ if $G_1 G_2 G_3 \geq L_1 L_2$.

A low-loss system might need only one amplifier, usually at the destination. However, noise considerations often call for a preamplifier to boost the signal level before the noise becomes significant. Long-distance transmission generally requires several repeaters, as in the case of transcontinental telephone links that have literally hundreds of repeaters between source and destination. The following example illustrates some of the practical considerations that arise in the design of cascade transmission systems.

Example 13.4–1

Suppose we want to get $P_{out} = 50$ mW at distance $\ell = 20$ km from a source producing $P_{in} = 2$ mW. A transmission line with appropriate characteristic impedance for the signal frequency is found to have $\alpha = 2.3$ dB/km so, from Eq. (2),

$$\alpha\ell = 2.3 \times 20 = 46 \text{ dB} \qquad L = 10^{(46/10)} \approx 40,000$$

We therefore need a total gain of $G = L(P_{out}/P_{in}) = 10^6$ (60 dB). The amplifiers available for this purpose have adjustable power gain but are subject to two limitations: the input signal power must be at least 1 μW (to overcome internal noise), and the output signal power must not be greater than 1 W (to avoid nonlinear distortion).

We cannot put all the amplification at the destination, because the signal power at the output of the line would be $P_{in}/L = 0.05$ μW, which falls below the amplifier noise level. Nor can we put all the amplification at the source, because the amplified source power $GP_{in} = 2$ kW would exceed the amplifier power rating. However, we could use a preamplifier with $G_1 = 500$, for instance, to get $G_1 P_{in} = 1$ W at the input of the line and $G_1 P_{in}/L = 25$ μW at the output. The output amplifier should then have $G_2 = P_{out}/(25 \mu\text{W}) = 2000$, and a repeater is not needed.

Exercise 13.4–1

Let the system in Fig. 13.4–3 have $P_{in} = 5$ mW, $\ell_1 = 18$ km, $\ell_2 = 20$ km, and $\alpha_1 = \alpha_2 = 1.5$ dB/km. Find $G = G_1 G_2 G_3$ so that $P_{out} = 200$ mW. Then determine the individual gains if the output of each amplifier is 200 mW.

Radio Transmission

Radio transmission involves antennas at source and destination, as sketched in Fig. 13.4–4, and it requires the signal to be modulated on a high-frequency carrier, usually a sinusoid. When driven by an appropriate carrier, the transmitting antenna launches an electromagnetic wave that propagates through space without the help of a transmission line. A small portion of the radiated power can then be collected at the receiving antenna.

Figure 13.4–4
Line-of-sight radio transmission.

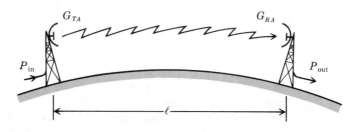

The transmission loss for a line-of-sight radio path of length ℓ is given by

$$L = \frac{P_{\text{in}}}{P_{\text{out}}} = \frac{1}{G_{TA}G_{RA}}\left(\frac{4\pi\ell}{\lambda}\right)^2 \tag{4}$$

where G_{TA} and G_{RA} are the power gains of the transmitting and receiving antennas. The parameter λ is the **wavelength** of the radio wave, related to the carrier frequency f_c by

$$f_c\lambda = c \approx 3 \times 10^5 \text{ km/s} \tag{5}$$

with c being the velocity of light. We see that radio transmission loss differs from that of a transmission line in two respects: It increases as ℓ^2 instead of exponentially, and it may be compensated, in part, by the antenna gains.

Antenna gain depends on both shape and size. **Dipole antennas,** commonly used at lower radio frequencies, consist of a rod or wire of length $\lambda/10$ to $\lambda/2$ and have $G_A = 1.5$–1.64 (1.8–2.1 dB). **Horn antennas** and **parabolic dishes** provide much more gain at higher frequencies, since they have

$$G_A = \frac{4\pi A_{\text{ap}}}{\lambda^2} \tag{6}$$

where A_{ap} is the aperture area of the antenna. Since antenna gain comes from focusing and collecting part of the signal power, a radio system always has $L > 1$ and $P_{\text{out}} < P_{\text{in}}$. Power amplifiers are then required to overcome the loss, just as in a transmission-line system.

Equations (4)–(6) also hold for *optical* radiation where the electromagnetic wave takes the form of a coherent light beam. Figure 13.4–5 shows

Figure 13.4-5
The electromagnetic spectrum. (From *Communication Systems*, by A. B. Carlson. Copyright © by McGraw-Hill. Used with permission of McGraw-Hill Book Co.)

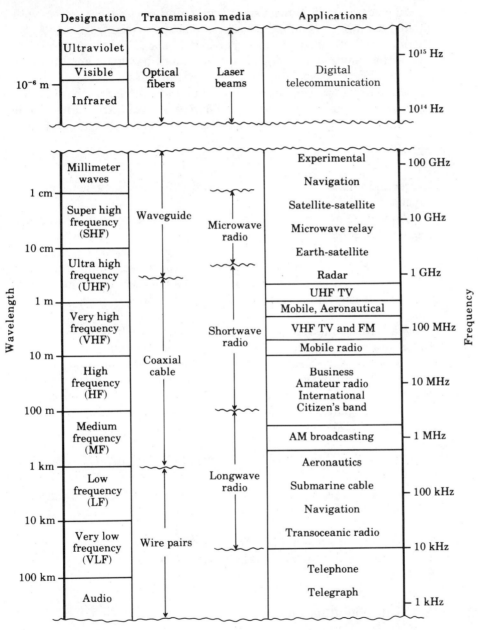

the radio and optical portions of the electromagnetic spectrum used for signal transmission. The figure includes the special designations of various frequency bands, types of transmission media, and representative applications.

In all cases, radio transmission is inherently a *bandpass* process with a limited bandwidth B nominally centered at the carrier frequency f_c. Furthermore, practical considerations result in a significant connection between the values of B and f_c. Radio antennas, bandpass amplifiers, and other hardware associated with bandpass transmission require very careful design to avoid serious distortion if the bandwidth B must be either very large or very small compared to f_c. The **fractional bandwidth** B/f_c becomes a key design factor and, as a rule of thumb, it should fall within

$$\frac{1}{100} \le \frac{B}{f_c} \le \frac{1}{10} \tag{7}$$

Larger and smaller values *can* be achieved, to be sure, but generally at great expense.

An immediate implication of Eq. (7) is that *large signal bandwidths require high carrier frequencies* to satisfy $f_c \ge 10B$. This is one reason why TV signals are transmitted at $f_c \approx 100$ MHz, while AM radio signals get by at $f_c \approx 1$ MHz. Similarly, much of the current interest in optical communication systems stems from the tremendous bandwidth potential ($\approx 10^{12}$ Hz) and the correspondingly high information rate.

Example 13.4−2

A signal with a bandwidth of 100 MHz is to be transmitted 40 km by radio. (Incidentally, 40 km is about the maximum distance for line-of-sight transmission over smooth terrain without tall antenna towers.) Taking $B/f_c = 1/30$, for instance, gives a carrier frequency in the microwave band at $f_c = 30 \times 100$ MHz $= 3$ GHz, so $\lambda = 10^{-4}$ km $= 10$ cm. If we use a circular-aperature parabolic dish with 50-cm radius at each end, then $G_{TA} = G_{RA} = 4\pi(\pi50^2)/10^2 \approx 1000$ and Eq. (4) yields

$$L = \frac{1}{10^6}\left(\frac{4\pi40}{10^{-4}}\right)^2 = 2.5 \times 10^7 = 74 \text{ dB}$$

For comparison purposes, a coaxial cable with 100-MHz bandwidth typically has $\alpha \approx 3$ dB/km and its transmission loss would be $\alpha\ell \approx 120$ dB.

Exercise 13.4−2

Suppose a radio system and a transmission-line system both have $L = 60$ dB when $\ell = 12$ km. Find the loss of each when the distance is increased to 24 km.

Amplitude Modulation

There are several ways to modulate a radio-frequency carrier with an information-bearing lowpass signal. Single-sideband modulation (SSB) maximizes the number of channels in a given frequency band and is widely used for amateur and business communication. However, the synchronized local oscillator needed to demodulate SSB makes the receiver rather expensive. Since broadcast radio involves *many* receivers, the modulation must be designed to simplify receiver hardware. Amplitude modulation (AM) is one such method.

As its name implies, AM carries the modulating signal $x(t)$ in a *time-varying amplitude* of the form

$$A(t) = A_c[1 + m_A x(t)] \tag{8}$$

The constant A_c stands for the unmodulated carrier amplitude, and m_A is the **modulation index.** The resulting modulated wave is

$$
\begin{aligned}
x_c(t) &= A(t) \cos 2\pi f_c t \\
&= A_c \cos 2\pi f_c t + A_c m_A x(t) \cos 2\pi f_c t
\end{aligned}
\tag{9}
$$

which consists of an unmodulated carrier added to double-sideband product modulation. Thus, the AM spectrum has a pair of sidebands (like Fig. 13.2–3) plus a carrier line inserted at $f = f_c$. If $x(t)$ is a lowpass signal with bandwidth W, then the required AM transmission bandwidth is $B = 2W$.

Figure 13.4–6 shows a typical modulating signal and the corresponding AM wave. The dashed line connecting the carrier peaks represents $|A(t)|$, called the **envelope** of $x_c(t)$. We assume here that $1 + m_A x(t)$ never goes negative, so $|A(t)| = A(t) = A_c[1 + m_A x(t)]$. Since the envelope exactly follows the variations of $x(t)$ under this condition, we can demodulate AM by the process known as **envelope detection.**

Figure 13.4–6
AM waveforms:
(a) modulating signal;
(b) modulated wave.

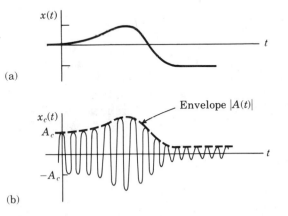

The circuit for a simple envelope detector is diagrammed in Fig. 13.4–7 along with illustrative waveforms. The diode clamps the voltage across R_1 to the positive peaks of $x_c(t)$ and C_1 holds the voltage between peaks, so $x_A(t)$ approximates the envelope $|A(t)|$. Capacitor C_2 then acts as a DC block, removing the unmodulated component A_c to get the desired output

$$x_{\text{out}}(t) = x_A(t) - A_c \approx A_c m_A x(t)$$

Figure 13.4–7
(a) Envelope detector. (b) Waveforms.

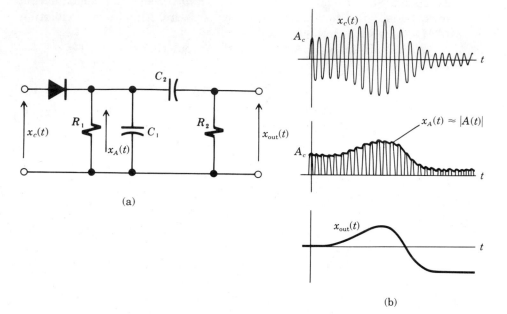

(a)

(b)

Since the DC block would also remove some of the low signal frequencies, envelope detection is not suitable for signals having significant low-frequency content.

Note that the demodulated output is proportional to A_c, which depends upon the received signal power. Additional circuitry (not shown) extracts the value of A_c from $x_A(t)$. This voltage is then fed back to earlier stages in an AM radio for **automatic volume control (AVC)** when the received signal undergoes slow fading.

Amplitude modulation and envelope detection are also used for video transmission in television broadcasting. However, since the video signal has a very large bandwidth ($W \approx 4$ MHz), the bandwidth-doubling effect

Figure 13.4–8
Broadcast spectrum of a TV signal.

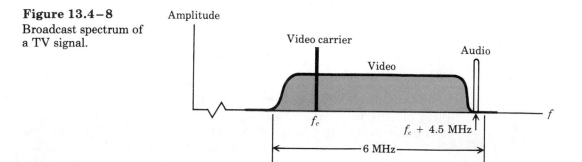

of conventional AM would require excessive transmission bandwidth. TV channel bandwidth is therefore reduced to $B = 6$ MHz by filtering out most of the lower sideband at the transmitter, a modified version of AM known as **vestigial-sideband modulation (VSB)**. Figure 13.4–8 illustrates the broadcast spectrum of a TV signal, including the frequency-modulated audio signal on a separate carrier 4.5 MHz above the video carrier.

Exercise 13.4–3   (a) Sketch one full period of an AM wave with $x(t) = \cos 2\pi f_m t$, $f_m \ll f_c$, and $m_A = 1$. Draw the envelope by connecting the positive peaks of $x_c(t)$. (b) Repeat (a) with $m_A = 2$, in which case the carrier is *over-modulated* and the envelope does not have the same shape as $x(t)$.

Superheterodyne Receivers

A radio receiver must perform three general functions: tuning (station selection), demodulation, and amplification. But most receivers — including commercial AM, FM, and TV sets — are somewhat more complicated, largely due to fractional-bandwidth design considerations. To illustrate these concepts, we'll describe the common **superheterodyne** (or "superhet") AM radio diagrammed in Fig. 13.4–9.

Figure 13.4–9
Superheterodyne AM radio.

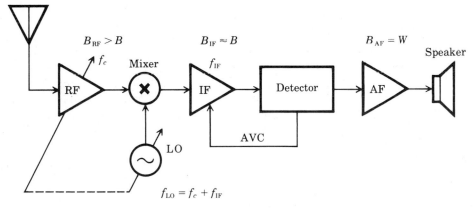

AM broadcasting stations are assigned carrier frequencies from 540 kHz to 1600 kHz with 10-kHz spacing, which allows for $B \approx 10$ kHz and $W = B/2 \approx 5$ kHz. Since a radio antenna picks up many of these signals, the first function of the receiver is to select the desired station. This is partly accomplished by a tunable bandpass filter called the **radio-frequency (RF) stage.** When you tune in the station at $f_c = 800$ kHz, for example, the RF stage may actually pass 780–820 kHz if its bandwidth is

$B_{RF} \approx f_c/20 = 40$ kHz. Considerable amplification and filtering, then, is necessary before the desired signal can be demodulated.

But high-gain tunable amplifiers with small fractional bandwidths are difficult to build. Instead, the superhet has a fixed-tuned **intermediate-frequency (IF) amplifier,** usually with center frequency $f_{IF} = 455$ kHz and bandwidth $B_{IF} = 10$ kHz, so $B_{IF}/f_{IF} \approx 1/50$. A frequency translator consisting of a **mixer** (or multiplier) and **local oscillator** translates the desired f_c down to f_{IF}. Thus, f_{LO} is adjusted simultaneously with the RF stage to produce the difference frequency

$$|f_{LO} - f_c| = f_{IF} \qquad (10)$$

Note that the local oscillator is for frequency translation, not product demodulation, and requires no synchronizing.

Since B_{RF} exceeds 10 kHz, the mixer's output will include components from *adjacent channels* along with the selected channel. The IF amplifier rejects the adjacent-channel signals and amplifies the desired signal. This signal is then applied to the envelope detector and, finally, the demodulated signal is further amplified by an **audio-frequency (AF) amplifier,** which drives the loudspeaker.

Frequency Modulation

A carrier with frequency modulation (FM) has constant amplitude but a time-varying *instantaneous frequency*

$$f(t) = f_c + m_f x(t) \qquad (11)$$

where m_f represents the **frequency modulation index.** Thus, as illustrated by Fig. 13.4–10, the oscillation rate of $x_c(t)$ increases and decreases relative to f_c in proportion to the modulating signal. Such a waveform can be generated by applying $x(t)$ to a **voltage-controlled oscillator (VCO).**

Thanks to its constant-amplitude property, FM offers two important advantages over AM. First, an FM signal is quite *immune to nonlinear*

Figure 13.4–10
FM waveforms.

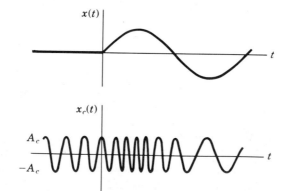

distortion—which often occurs in high-frequency amplifiers. Second, FM can *reduce interference and noise* relative to the signal without increasing the transmitted power. However, to achieve this reduction effect, the transmission bandwidth must be much larger than AM.

Because FM involves more than just direct frequency translation, spectral analysis and bandwidth calculations are difficult in general. Results obtained from a few known cases indicate that the bandwidth of an FM signal depends on the **frequency deviation,** defined by

$$\Delta f \triangleq m_f |x(t)|_{\max}$$

If $x(t)$ has lowpass bandwidth W, and if $\Delta f \geq 2W$, then the FM transmission bandwidth is estimated from

$$B \approx 2\Delta f + 4W \tag{12}$$

We say that FM with $\Delta f \geq 2W$ is *wideband* modulation since it requires $B \geq 8W$. For example, commercial FM broadcasting uses $W = 15$ kHz and $\Delta f = 5W = 75$ kHz, so $B \approx 2 \times 75 + 4 \times 15 = 210$ kHz.

To explain how wideband FM reduces unwanted contaminations, we must first look at the demodulation process—usually accomplished via *FM-to-AM conversion.* Suppose that $x_c(t)$ is applied to a tuned circuit resonant at $f_0 > f_c$. Figure 13.4–11a shows the circuit's amplitude ratio $|H(f)|$ and the straight line slope approximation in the vicinity of f_c. As the instantaneous frequency swings above and below f_c, the tuned circuit con-

Figure 13.4–11
FM-to-AM conversion: (a) amplitude ratio of tuned circuit; (b) waveforms.

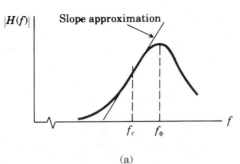

(a)

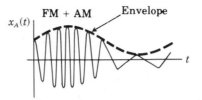

(b)

verts frequency variations to amplitude variations. The resulting wave-form $x_A(t)$ illustrated in Fig. 13.4–11b has amplitude modulation along with the frequency modulation, both proportional to $f(t) = f_c + m_f x(t)$. Hence, we can use an envelope detector to extract the amplitude modulation and recover $x(t)$.

A complete FM demodulator also includes a *limiter* that removes any spurious amplitude variations from $x_c(t)$ before FM-to-AM conversion. But interference and noise affects instantaneous frequency as well as amplitude, so the demodulated output has the form

$$x_{\text{out}}(t) = m_f x(t) + n(t)$$

where $n(t)$ represents the unwanted contaminations. Note that the strength of the signal component depends on the frequency deviation $\Delta f = m_f |x(t)|_{\text{max}}$. When the frequency deviation is large enough, $m_f x(t)$ dominates over $n(t)$ and thereby reduces the contaminations relative to the signal.

Increasing Δf to reduce noise does not increase the transmitted power, but it does increase the transmission bandwidth B. FM therefore exhibits the property known as **wideband noise reduction.** When $\Delta f = 5W$, as in FM broadcasting, the noise-reduction factor turns out to be about 1200 compared to AM — meaning that a 1200-watt AM system could be replaced by a 1-watt FM system with $\Delta f = 5W$ if the bandwidth is increased to $B \approx 14W$. Unfortunately, stereophonic demodulation yields significantly less noise reduction, which explains why many stereo FM receivers allow you to switch to monophonic demodulation of weak signals.

Finally, we describe the multiplexing system used for FM stereophonic broadcasting, starting with the transmitter diagrammed in Fig. 13.4–12a. The left- and right-channel signals, having $W \approx 15$ kHz, are first **matrixed** to form

$$x_1(t) = x_L(t) + x_R(t) \qquad x_2(t) = x_L(t) - x_R(t)$$

This is done so that the sum signal $x_1(t)$ heard on a monophonic receiver will not have any sound "gaps." The difference signal $x_2(t)$ is DSB modulated onto a 38-kHz subcarrier obtained by doubling the frequency from a 19-kHz oscillator. DSB has been employed here to ensure good fidelity at the lower audio frequencies. Combining $x_1(t)$, $x_2(t) \cos 2\pi f_{\text{sc}} t$, and the 19-kHz pilot yields the so-called *baseband signal* of Fig. 13.4–12b, which is applied to the FM modulator for radio transmission with $f_c = 88–108$ MHz.

The stereo FM receiver in Fig. 13.4–12c recovers the baseband signal by FM demodulation. The pilot synchronizes product demodulation of the DSB component, and another matrix reproduces $x_L(t)$ and $x_R(t)$ from $x_1(t)$ and $x_2(t)$. The portion of the receiver prior to the FM demodulator (not shown) is usually a superheterodyne with $f_{\text{IF}} = 10.7$ MHz.

Figure 13.4–12
FM stereo multiplexing: (a) transmitter; (b) multiplexed baseband spectrum; (c) receiver.

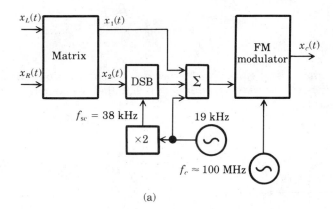

(a)

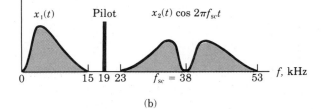

(b)

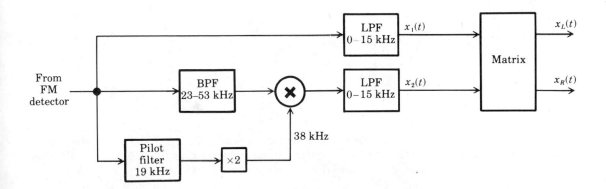

(c)

Exercise 13.4–4   The TV audio signal of Fig. 13.4–8 is frequency-modulated with $\Delta f = 25$ kHz and was $W \approx 10$ kHz. What percentage of the channel bandwidth is occupied by the audio signal?

PROBLEMS

13.1–1   The waveform in Fig. P13.1–1 can be viewed as the combination of a square wave with another periodic wave. Use this approach to find the nonzero Fourier series coefficients.

Figure P13.1–1

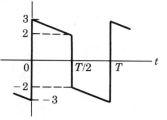

13.1–2   Do Problem 13.1–1 for the waveform in Fig. P13.1–2.

Figure P13.1–2

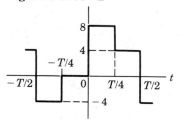

13.1–3   Do Problem 13.1–1 for the waveform in Fig. P13.1–3.

Figure P13.1–3

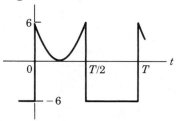

13.1–4   Let the waveform in Fig. 13.1–3b have $A = \pi^2$ and $T = 2.5$ μs. By tabulating values of A_n, determine the bandwidth W such that $A_n < (A_n)_{\max}/10$ for $nf_1 > W$. Then determine W with the criterion changed to $A_n < (A_n)_{\max}/20$ for $nf_1 > W$.

13.1–5   Do Problem 13.1–4 for the waveform in Fig. 13.1–3e with $A = \pi$ and $T = 800$ μs.

13.1–6   Do Problem 13.1–4 for the waveform in Fig. 13.1–3d with $A = \pi$ and $T = 10$ ms.

13.1–7   Do Problem 13.1–4 for the waveform in Fig. 13.1–3a with $A = \pi$, $D = 0.25$ μs, and $T = 0.5$ μs.

13.1–8   The frequency response of a highpass transmission system is given by Eq. (7b), Section 7.1, with $f_{co} = \omega_{co}/2\pi = 100$ Hz. Obtain an approximate expression for the periodic steady state response $y(t)$ when $x(t)$ is a triangular wave with $A = \pi^2/8$ and $T = 25$ ms.

13.1–9   The frequency response of a lowpass transmission system is given by Eq. (5b), Section 7.1, with $f_{co} = \omega_{co}/2\pi = 2$ kHz. Obtain an approximate expression for the periodic steady state response $y(t)$ when $x(t)$ is a rectangular pulse train with $A = \pi/2$, $D = 0.25$ ms, and $T = 0.5$ ms.

13.1–10   The frequency response of a bandpass transmission system is given by Eq. (11b), Section 7.1, with $Q = 2$ and $f_0 = \omega_0/2\pi = 50$ kHz. Obtain an approximate expression for the periodic steady state response $y(t)$ when $x(t)$ is a square wave with $A = \pi/4$ and $T = 25$ μs.

13.1–11   Do Problem 13.1–10 when $x(t)$ is a half-rectified sine wave with $A = \pi$ and $T = 40$ μs.

13.2–1   Suppose the input to the product modulator in Fig. 13.2–1a is $x(t) = 4 \cos 2\pi 3f_m t + 2 \cos 2\pi 7f_m t$ with $f_m = 1$ kHz. Draw and label the amplitude line spectrum of $x_c(t)$ when $f_c = 6$ kHz.

13.2–2   Do Problem 13.2–1 with $f_c = 3$ kHz.

13.2–3   In Fig. P13.2–3 let $x(t) = 3 \cos 2\pi 100t + \cos 2\pi 300t$. Find $x_a(t)$ and $x_b(t)$ when $f_a = 800$, $f_b = 1200$, filter A rejects all $f < 800$, and filter B rejects all $f > 1200$.

Figure P13.2–3

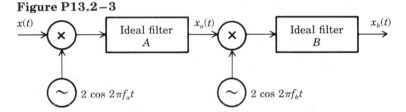

13.2–4   Do Problem 13.2–3 for the case when $f_a = f_b = 600$, filter A rejects all $f > 600$, and filter B rejects all $f < 600$.

13.2–5   A **speech scrambler** converts a voice signal $x(t)$ into an unintelligible signal to ensure privacy of telephone conversations. The system in Fig. P13.2–3 acts as a scrambler when $f_a > W$, $f_b = f_a + W$, filter A rejects all $f < f_a$, and filter B rejects all $f > W$.
   (a)   Sketch the amplitude spectrum at the input and output of each filter, taking the spectrum of $x(t)$ shown in Fig. 13.2–3.
   (b)   Explain why $x_b(t)$ is unintelligible. How can $x(t)$ be recovered from $x_b(t)$?

13.2–6   Suppose the oscillator in Fig. 13.2–4a actually generates $\cos[2\pi(f_c + \Delta f)t + \Delta\phi]$ where Δf and $\Delta\phi$ are synchronization errors. Find $z(t)$ produced by the DSB input $x_c(t) = 4 \cos 2\pi f_m t \cos 2\pi f_c t$. Then let $f_m = 1000$ and investigate what happens to $z(t)$ when: (a) $\Delta f = 200$ and $\Delta\phi = 0$; (b) $\Delta f = 0$ and $\Delta\phi = 90°$.

13.2–7   Do Problem 13.2–6 with the upper-sideband SSB input $x_c(t) = 4 \cos 2\pi(f_c + f_m)t$.

13.2–8   Do Problem 13.2–6 with the lower-sideband SSB input $x_c(t) = 4 \cos 2\pi(f_c - f_m)t$.

13.2–9   Suppose the signal to be sampled in Fig. 13.2–5 is $x(t) = 36 \cos 2\pi 20t + 24 \cos 2\pi 60t$. Draw and label the amplitude line spectrum of $x_s(t)$ for $0 \le f \le 2f_s$

when $f_s = 100$ and $D = T_s/2$. Then find the signal $y(t)$ that would be reconstructed by an ideal lowpass filter that rejects all $f > f_s/2$.

13.2–10   Do Problem 13.2–9 with $f_s = 90$ and $D = T_s/3$.

13.2–11   Let the input to a switching sampler have the continuous amplitude spectrum in Fig. P13.2–11. Draw and label the resulting spectrum of $x_s(t)$ for $0 \leq f \leq 100$ when $f_s = 70$ and $D = T_s/4$. Then explain how $x(t)$ can be reconstructed from $x_s(t)$.

Figure P13.2–11

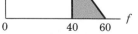

13.2–12   Do Problem 13.2–11 with $f_s = 60$ and $D = T_s/5$.

13.2–13   Let D be the pulse duration of the PPM waveform in Fig. 13.2–7, and let the maximum position shift be $\pm\Delta$. Calculate the maximum permitted value of D and determine the minimum required transmission bandwidth when $\Delta = 1.5D$ and $f_s = 10$ kHz.

13.2–14   Do Problem 13.2–13 when $\Delta = 2D$ and $f_s = 20$ kHz.

13.2–15   Suppose the transmission bandwidth of the TDM system in Fig. 13.2–9 is $B = 250$ kHz. What's the maximum number of voice signals with $W = 4$ kHz that can be multiplexed?

13.2–16   Do Problem 13.2–15 with the additional constraint that the TDM waveform must be OFF at least 50% of the time.

13.3–1   An information signal with significant frequency content for $f \leq 30$ Hz has been contaminated by AC "hum" at 120 Hz. The contaminated signal is applied to a lowpass filter to reduce the hum amplitude by a factor of 0.25. Find the required cutoff frequency f_{co} for: (a) a simple RC filter; (b) the more sophisticated Butterworth filter described in Problem 7.1–6. Which filter would you use?

13.3–2   Do Problem 13.3–1 for the case where the hum amplitude is to be reduced by a factor of 0.1.

13.3–3   A **notch filter** centered at $f_0 = \omega_0/2\pi$ can be built using a resonant circuit arranged such that

$$H(j\omega) = \frac{jQ(\omega/\omega_0 - \omega_0/\omega)}{1 + jQ(\omega/\omega_0 - \omega_0/\omega)}$$

so $|H(f)| = 1$ at $f = 0$ and $f = \infty$. If $Q \gg 1$, then $|H(f)| = 0.707$ at $f_\ell \approx f_0(1 - 1/2Q)$ and at $f_u \approx f_0(1 + 1/2Q)$.
(a) Show that a series RLC circuit acts as a notch filter with $\omega_0 = 1/\sqrt{LC}$ and $Q = (1/R)\sqrt{L/C}$ when the output voltage is taken across L and C.
(b) Given that $R = 50$ Ω, find the values of L and C needed to get $f_\ell = 980$ Hz and $f_u = 1020$ Hz for the purpose of rejecting 1-kHz interference.

13.3–4   Consider the notch filter described in Problem 13.3–3.
(a) Show that the circuit in Fig. 7.1–11a acts as a notch filter with $\omega_0 = 1/\sqrt{LC}$ and $Q = R\sqrt{C/L}$ when the output voltage is taken across R.

(b) Given that $R = 1 \ k\Omega$, find the values of L and C needed to $f_\ell = 58$ Hz and $f_u = 62$ Hz for the purpose of rejecting 60-Hz interference.

13.3–5   A noisy amplifier is found to have $N_{out} = 580 \ \mu W$ when $T = T_0$, but N_{out} drops to $475 \ \mu W$ when the source is immersed in liquid nitrogen at $T = 80$ K. Use these results to calculate the amplifier's noise temperature T_a.

13.3–6   A method for measuring amplifier noise temperature is as follows: Connect a thermal source at temperature T_0 and observe N_{out}; then increase the source temperature to T_R such that N_{out} doubles. Obtain an expression for T_a in terms of T_R and T_0.

13.3–7   Two noisy amplifiers having the same bandwidth are connected in cascade to get the overall gain $G = G_1 G_2$. The individual noise temperatures are T_{a1} and T_{a2}. Find the total output noise power when the input noise to the first amplifier is $N = kTB$. Then compare your result with Eq. (8) to show that the effective noise temperature of the cascade is $T_a = T_{a1} + T_{a2}/G_1$.

13.3–8   A signal with $P = 1 \ \mu W$ and $B = 250$ kHz has been contaminated by white noise at noise temperature $T = 2T_0$. The signal plus noise is applied to an amplifier. Determine the condition on T_a needed to get $P_{out}/N_{out} \ge 80$ dB.

13.3–9   Do Problem 13.3–8 with $P = 0.2 \ \mu W$, $B = 20$ kHz, and $T = 5T_0$.

13.3–10   The **noise figure** of an amplifier is $F = 1 + T_a/T_0$. Write the output signal-to-noise ratio in terms of F, P_{in}, the input noise power N, and the input noise temperature T. Then simplify your result for the case of $T = T_0$.

13.3–11   A system for measuring the constant signal value x has $P_{out}/N_{out} = 40$ dB and $B = 6$ Hz. How long must the output be observed and averaged to get an accuracy of $\pm 0.2\%$?

13.3–12   A system for measuring the constant signal value x has $P_{out}/N_{out} = 30$ dB and $B = 10$ Hz. How long must the output be observed and averaged to get an accuracy of $\pm 0.5\%$?

13.4–1   A transmission system consisting of a preamplifier, cable, and output amplifier is intended to produce $P_{out} = 0.4$ W. The cable has $\alpha = 0.6$ dB/km and a 1-W power limitation. Taking G_1 to be as large as possible, find the required values of G_1 and G_2 in dB when $\ell = 50$ km and $P_{in} = 2$ mW.

13.4–2   Do Problem 13.4–1 with $\ell = 55$ km and $P_{in} = 5$ mW.

13.4–3   Let the transmission system in Fig. 13.4–3 have $G_1 = 23$ dB, $\alpha_1 = \alpha_2 = 2.5$ dB/km, and $\ell_1 + \ell_2 = 30$ km. Given that $P_{in} = 10$ mW, determine the values of ℓ_1, G_2, and G_3 such that $P_{out} = 100$ mW and the signal power equals $20 \ \mu W$ at the input to G_2 and G_3. Express G_2 and G_3 in dB.

13.4–4   Do Problem 13.4–3 with $P_{in} = 1$ mW and $P_{out} = 50$ mW.

13.4–5   Figure P13.4–5 represents a long-haul **repeater system** consisting of M identical cable sections and M identical amplifiers, so $L_1 = L_2 = \cdots = L_M$ and $G_1 = G_2 = \cdots = G_M$. The total distance is 400 km, and the signal power at the input to each

Figure P13.4–5

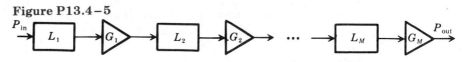

amplifier must be at least 50 μW. What is the minimum number of amplifiers and the corresponding amplifier gain in dB needed to get $P_{out} = 200$ mW when $P_{in} = 200$ mW and $\alpha = 1.8$ dB/km?

13.4–6   Do Problem 13.4–5 with $P_{out} = 1$ W, $P_{in} = 50$ mW, and $\alpha = 0.6$ dB/km.

13.4–7   A line-of-sight radio system with dipole antennas has $P_{in} = 10$ W and $\ell = 20$ km. Calculate P_{out} when $f_c = 100$ MHz and $G_{TA} = G_{RA} = 1.5$ dB.

13.4–8   Do Problem 13.4–7 with $f_c = 500$ MHz and $G_{TA} = G_{RA} = 2.0$ dB.

13.4–9   A satellite radio transmitter has $P_{in} = 5$ W and $G_{TA} = 30$ dB. The receiving antenna is a circular aperture with radius r at the ground station 40,000 km away. Find r in meters needed to $P_{out} = 50$ pW.

13.4–10   A microwave relay system consists of identical horn antennas mounted on towers spaced by 30 km. If $f_c = 6$ GHz and each relay hop has $L = 60$ dB, then what is the antenna aperture area A_{ap} in square meters?

13.4–11   A radar system uses pulses of duration D to modulate the amplitude of a radio carrier wave. The system can distinguish between targets spaced by distance $d \geq cD$. In view of Eq. (7), what's the minimum practical value of f_c so that $d_{min} = 30$ meters?

13.4–12   A TDM signal like Fig. 13.2–9b is formed by sampling M voice signals at $f_s = 8$ kHz. The TDM signal then modulates the amplitude of a 4-MHz carrier for radio transmission. Determine the upper limit on M that satisfies Eq. (7).

13.4–13   Figure P13.4–13 shows a way to generate AM using a nonlinear device that produces $z = y + ay^2$. Taking $x(t) = \cos 2\pi f_m t$, show that appropriate bandpass filtering yields an AM wave in the form of Eq. (9).

Figure P13.4–13

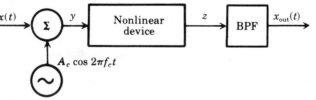

13.4–14   Suppose the nonlinear device in Fig. P13.4–13 produces $z = ay^2$. Taking $x(t) = A_m \cos 2\pi f_m t$, determine how the system must be augmented to get an AM wave in the form of Eq. (9).

13.4–15   Show that *two* values of f_c satisfy Eq. (10) for any given values of f_{LO} and f_{IF}. The higher f_c is called the **image frequency,** and it must not be allowed to get to the mixer in a superheterodyne receiver. What's the corresponding condition on B_{RF}?

13.4–16   A TDM signal like Fig. 13.2–9b is formed by sampling 5 voice signals at $f_s = 8$ kHz. The TDM signal is then transmitted via FM on a radio channel with 400 kHz bandwidth. Estimate the maximum allowable frequency deviation Δf.

13.4–17   Suppose that a video signal having $W = 5$ MHz is transmitted via FM with $\Delta f = 20$ MHz. Determine the bounds on the carrier frequency consistent with fractional-bandwidth considerations.

Chapter

14 Digital Logic

One of the fastest growing and most exciting areas of electrical engineering is *digital information systems*. These systems differ from analog systems in that they operate with discrete or *digitized* variables rather than continuously varying signals. Thus, a digital signal takes on only a few specified values, and we say that a digital system has a *finite number of distinct states*.

By restricting an electronic system to distinct states, we gain several potential advantages over analog systems. These advantages include:

- More economical, flexible, and compact system implementation.
- Improved reliability in the face of contaminating signals or hardware imperfections.
- The ability to make logical decisions, carry out numerical computations, and store the results.

Each new day seems to bring another technological advance in digital electronics that lowers cost, reduces size, improves performance, or expands the scope of applications. Those applications range from pocket calculators and supercomputers to household appliances, scientific instruments, and automated manufacturing.

Our study of digital information systems begins here with the underlying *digital logic*. We'll introduce the basic concepts of logic variables and operations, and we'll develop the tools needed to analyze and design combinational logic networks. The closing section discusses the properties of logic gates implemented as integrated circuits.

Objectives

After studying this chapter and working the exercises, you should be able to do each of the following:

- Define a logic variable and identify the truth table and gate symbol for the AND, OR, and NOT operations (Section 14.1).

- Perform conversions between the decimal, binary, octal, and hexadecimal number systems, and express numbers in BCD code (Section 14.1).

- Use Boolean algebra to analyze a combinational logic network or manipulate a logic expression (Section 14.2).

- Construct the truth table for a logic function, derive and simplify the SOP or POS expression, and draw the corresponding logic network with NAND or NOR gates (Section 14.2).

- Minimize a function of three or four variables using a Karnaugh map (Section 14.2).†

- Identify the functions performed by exclusive-OR gates, decoders, demultiplexers, and multiplexers (Section 14.3).

- Program a MUX or a ROM to generate a function of three variables (Section 14.3).

- Describe the properties of IC logic gates, and compare the performance of TTL, ECL, and CMOS logic families (Section 14.4).†

14.1 DIGITAL LOGIC CONCEPTS

Most digital systems contain thousands of electronic devices that switch between two states. Associated with each switching device is a quantity called a **logic variable,** having just two values. By combining and operating on many logic variables, a digital system can accomplish a wide variety of information-processing tasks.

This section introduces logic variables and operations in the context of simple switching circuits. We'll also develop the relationship between logic variables, binary digits, and decimal numbers.

Logic Variables and Gates

Figure 14.1–1a shows an elementary two-state circuit in which a light is either ON or OFF, depending on whether a switch is UP or DOWN. We'll represent the states of the switch and light by *logic variables A* and *F*, where

Figure 14.1–1

Simple binary systems:
(a) $F = A$; (b) $F = \overline{A}$.

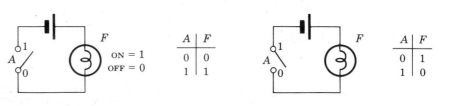

(a)   (b)

the output state F is a function of the input state A. Instead of words like ON/OFF or UP/DOWN, we'll identify each of the two possible input and output states with the **binary digits** 1 and 0. In particular, let $F = 1$ mean that the light is ON, while $F = 0$ means that it's OFF. Similarly, the switch has been labeled $A = 1$ for the UP position and $A = 0$ for DOWN.

The input-output relationship in Fig. 14.1–1a is displayed by a table listing the two values of A and the resulting values of F. It is both obvious and trivial here that $F = A$, since the light is OFF ($F = 0$) when the switch is DOWN ($A = 0$) and the light is ON ($F = 1$) when the switch is UP ($A = 1$). The opposite relationship holds in Fig. 14.1–1b, where the switch has been inverted and must be in the DOWN position to put the light ON. We than write $F = \bar{A}$ to symbolize that $F = 1$ when $A = 0$ and vice versa.

Now consider a light controlled by three switches, A, B, and C, arranged as in Fig. 14.1–2a. Each switch has two states, so there are $2 \times 2 \times 2 = 2^3 = 8$ different input combinations. But the light goes ON only when switch A is DOWN ($A = 0$) and B is UP ($B = 1$) and/or C is UP ($C = 1$). Otherwise, the light is OFF. Hence, the table in Fig. 14.1–2b has $F = 1$ for just three of the eight input combinations. Such a list of every possible input-output condition is known as a **truth table**.

Figure 14.1–2
(a) Logic system with three inputs. (b) Truth table.

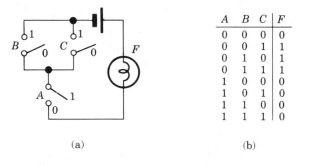

A	B	C	F
0	0	0	0
0	0	1	1
0	1	0	1
0	1	1	1
1	0	0	0
1	0	1	0
1	1	0	0
1	1	1	0

(a)   (b)

The circuit at hand is capable of making a *logical decision* if we attach appropriate interpretations to the switch positions. When used in an automobile, for example, this circuit could activate a buzzer ($F = 1$) alerting the driver that the headlights have been left on ($B = 1$) or a door is ajar ($C = 1$) when the ignition key is turned off ($A = 0$).

In general, a logic network operates on input logic variables to produce some desired logical output. No matter how complicated the input-output relationship is, it can always be broken down in terms of *three basic logic operations:* AND, OR, and NOT. Simple diode and transistor circuits that perform these operations were discussed in Chapter 10, where we worked with digital signals having two possible levels. Here, we'll define logic gates in terms of operations on logic variables.

Figure 14.1–3a gives the schematic symbol for an **AND gate** with two inputs, A and B. By definition, the output F equals 1 only when $A = 1$ *and*

Figure 14.1–3

AND gate with two
inputs: (a) symbol;
(b) truth table;
(c) equivalent switching
circuit.

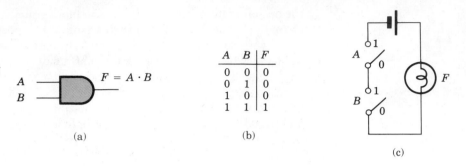

A	B	F
0	0	0
0	1	0
1	0	0
1	1	1

(a)   (b)   (c)

$B = 1$. We denote this input-output relationship as

$$F = A \cdot B \tag{1a}$$

where $A \cdot B$ stands for "A AND B." The truth table in Fig. 14.1–3b lists the
four possible input combinations and the resulting output of the AND gate.
Note that the AND operation acts like ordinary multiplication of binary-
digit logic values, since

$$0 \cdot 0 = 0 \qquad 0 \cdot 1 = 0 \qquad 1 \cdot 0 = 0 \qquad 1 \cdot 1 = 1 \tag{1b}$$

It then follows that a two-input AND gate could be built by connecting two
switches in *series,* as diagrammed in Fig. 14.1–3c.

An AND gate with $n > 2$ inputs is equivalent to n series-connected
switches, so the operation is commutative and associative. Thus, for instance, the output of a three-input AND gate can be written as

$$F = (A \cdot B) \cdot C = A \cdot (B \cdot C) = A \cdot B \cdot C$$

Since $F = 1$ only when *all* inputs equal 1, we say that the logical AND is an
"all-or-nothing" operation.

Our second basic operation is performed by the two-input **OR gate**
symbolized in Fig. 14.1–4a. This operation produces $F = 1$ when $A = 1$ *or*
$B = 1$. We write the logical OR as

$$F = A + B \tag{2a}$$

Figure 14.1–4

OR gate with two
inputs: (a) symbol;
(b) truth table;
(c) equivalent switching
circuit.

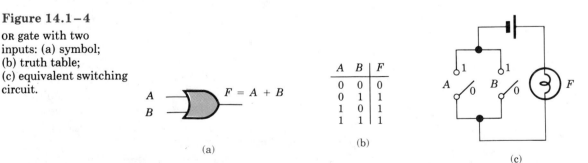

A	B	F
0	0	0
0	1	1
1	0	1
1	1	1

(a)   (b)   (c)

where $A + B$ stands for "A OR B." The truth table in Fig. 14.1–4b shows that

$$0 + 0 = 0 \qquad 0 + 1 = 1 \qquad 1 + 0 = 1 \qquad 1 + 1 = 1 \qquad \textbf{(2b)}$$

so the OR operation acts like ordinary addition except for $1 + 1 = 1$. The logic equation $1 + 1 = 1$ becomes more understandable from the OR-gate circuit in Fig. 14.1–4c, which has two switches in *parallel*. If we start with one switch closed, corresponding to $0 + 1 = 1$ or $1 + 0 = 1$, then the output undergoes no change when we close the other switch, thereby resulting in $1 + 1 = 1$.

An OR gate with $n > 2$ inputs is equivalent to n parallel-connected switches. Again we have a commutative and associative operation, so the output of a three-input OR gate can be written as

$$F = (A + B) + C = A + (B + C) = A + B + C$$

Since $F = 1$ when *any* input equals 1, we say that the logical OR is an "*any-or-all*" operation.

The logical NOT operation is performed by the **NOT gate** or **inverter** symbolized in Fig. 14.1–5a. An inverter has just one input, and its output always takes on the opposite state of the input. We write this logical inversion in the form

$$F = \overline{A} \qquad \textbf{(3a)}$$

where $\overline{A}$ stands for "A NOT"—also called the **negation** or **complement** of A. The truth table in Fig. 14.1–5b emphasizes that the NOT operation changes 0 to 1 and 1 to 0, so

$$\overline{0} = 1 \qquad \overline{1} = 0 \qquad \textbf{(3b)}$$

The mechanical version of a NOT gate would be a "normally-closed" switch like the one in Fig. 14.1–1b.

Figure 14.1–5

(a) NOT gate or inverter. (b) Truth table. (c) Cascade connection.

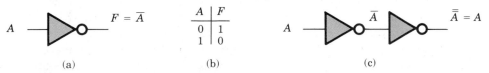

A	F
0	1
1	0

(a)   (b)   (c)

If two NOT gates are connected in the cascade arrangement of Fig. 14.1–5c, then the final output is

$$\overline{\overline{A}} = \overline{(\overline{A})} = A \qquad \textbf{(4)}$$

This expression merely says that the second negation cancels the first.

Example 14.1–1

Let's return to the switching circuit in Fig. 14.1–2a and describe it using the three basic logic operations. To get $F = 1$ we must have $A = 0 = \overline{1}$ *and* $B = 1$ *or* $C = 1$. Hence, we can now write the logic function

$$F = \overline{A} \cdot (B + C)$$

Figure 14.1–6 gives the corresponding gate diagram.

Figure 14.1–6

Gate diagram for Example 14.1–1.

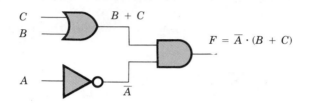

To check this logic function, take the fourth line from the truth table in Fig. 14.1–2b, where $A = 0$, $B = 1$, and $C = 1$. Inserting these values into the expression for F yields

$$F = \overline{0} \cdot (1 + 1) = 1 \cdot (1 + 1) = 1 \cdot 1 = 1$$

which is the correct value. You may find it informative to repeat this check for other lines of the truth table.

Exercise 14.1–1

A logic network with inputs A, B, and C sounds an alarm ($F = 1$) whenever the temperature in an industrial oven exceeds an upper limit ($A = 1$), or when the temperature exceeds a lower limit ($B = 1$) and the conveyor is off ($C = 0$). Construct the truth table using the three-variable input pattern from Fig. 14.1–2. Then devise the logic function and draw the gate diagram.

Binary and Decimal Numbers

Besides logical decisions, logic networks are capable of performing arithmetic operations. Such arithmetic networks will be described in Section 15.3. Here, we explain the underlying concept of treating logic variables as binary digits — called **bits,** for short — and grouping bits together to form **binary numbers.** By drawing upon the positional notation of number systems, we can convert back and forth between the binary numbers inside an arithmetic circuit and the decimal numbers of the outside world.

We construct a number in the decimal system using the ten digits 0, 1, . . . , 9, and the position of each digit signifies a power of 10. To illustrate, 105.8 actually means

$$105.8_{10} = 1 \times 10^2 + 0 \times 10^1 + 5 \times 10^0 + 8 \times 10^{-1}$$

The subscript 10 has been tagged onto 105.8 to emphasize that we are talking about a **base-10** number. Binary numbers have the same structure except that we use only two digits, 0 and 1, and digit position signifies a power of 2. Thus, for instance,

$$110.1_2 = 1 \times 2^2 + 1 \times 2^1 + 0 \times 2^0 + 1 \times 2^{-1}$$

Here, the dot must be called the **binary point** (rather than decimal point) because this is a **base-2** number. It then follows that

$$110.1_2 = (4 + 2 + 0 + 0.5)_{10} = 6.5_{10}$$

Note that binary-to-decimal conversion involves nothing more than summing the base-10 expansion of the binary number.

Now consider an ordered sequence of logic variables such as $[B_2 B_1 B_0 B_{-1}]$, which constitutes a **binary word.** To interpret the word as a binary number, we let the subscript i denote the **weight** 2^i assigned to each bit or, equivalently, the bit position relative to the binary point. The rule for converting our binary number $[B_2 B_1 B_0 B_{-1}]$ to the corresponding decimal quantity x_{10} then becomes

$$x_{10} = B_2 \times 2^2 + B_1 \times 2^1 + B_0 \times 2^0 + B_{-1} \times 2^{-1} \qquad \textbf{(5)}$$

This expression easily generalizes for a binary word with an arbitrary number of bits before or after the binary point.

Decimal-to-binary conversion is somewhat more difficult, especially for noninteger numbers. Small decimal integers can be converted by referring to a table of decimal and binary integers. Large decimal integers are converted by the following procedure known as **double-dabble:**

> Divide the decimal integer by 2, write down the remainder (0 or 1), and take the quotient as a new integer. Repeat this process until the quotient is reduced to zero. The remainders then constitute the binary word written in reverse order.

For example, this procedure is applied to 105_{10} in Fig. 14.1 – 7. Reading the remainders from bottom to top gives $105_{10} = 1101001_2$.

Figure 14.1 – 7

		Remainder
2	105	1
	52	0
	26	0
	13	1
	6	0
	3	1
	1	1
	0	

Converting decimal fractions is complicated by the fact that an exact result requires an infinite string of bits, unless the fraction happens to equal a sum of powers of ½. In any case, fractional conversion using **reverse double-dabble** goes as follows:

> Multiply the decimal fraction by 2, write down the integer carry (0 or 1), and subtract the carry from the product to get a new fraction. Repeat this process for each new fraction. The carries then constitute the binary word written in forward order.

Figure 14.1–8 illustrates reverse double-dabble starting with 0.8_{10}. Upon noting the repeating pattern of the carries, we conclude that $0.8_{10} = 0.11001100 \ldots _2$. Truncating this result after six bits, say, yields the finite-length approximation

$$0.8_{10} \approx 0.110011_2 = 0.796875_{10}$$

which has an error of about 0.4%. Such **truncations errors** sometimes show up in the output of digital-computer calculations, although they are usually much smaller.

Figure 14.1–8

$$
\begin{array}{lll}
 & & \textit{Carry} \\
2 \times 0.8 = 1.6 & & 1 \\
\times 0.6 = 1.2 & & 1 \\
\times 0.2 = 0.4 & & 0 \\
\times 0.4 = 0.8 & & 0 \\
\times 0.8 = 1.6 & & 1 \\
\times 0.6 = 1.2 & & 1 \\
\vdots \quad \vdots & & \vdots
\end{array}
$$

Converting a number with integer and fractional parts requires both direct and reverse double dabble. For instance.

$$
\begin{aligned}
105.8_{10} &= 105_{10} + 0.8_{10} \\
&= 1101001_2 + 0.11001100 \ldots _2 \approx 1101001.110011_2
\end{aligned}
$$

Since fractional parts often lead to unwieldy binary numbers, we'll limit our further discussion primarily to integers.

Table 14.1–1 lists the binary words for 0 through 15_{10}, along with two other number systems. Note that 2-bit words are needed to represent the decimal integers 2 and 3, while 3-bit words are needed for 4 through 7. In general, an n-bit binary integer can represent any decimal integer up to

$$x_{\max} = 2^n - 1 \tag{6}$$

Consequently, on the bottom line of the table we find $1111_2 = 15_{10} = 2^4 - 1$. Leading zeros are added when a long word represents a small decimal number, as in $0011_2 = 3_{10}$.

The table also brings out the natural pattern of binary integers, as follows: The **least significant bit** (LSB) alternates between 0 and 1 down its column, the next bit alternates in pairs, the next in groups of four, and so on. Thus, the **most significant bit** (MSB) of an n-bit word alternates

Table 14.1–1
Number systems.

Decimal	Binary	Hexadecimal	Octal
0	0	0	0
1	1	1	1
2	10	2	2
3	11	3	3
4	100	4	4
5	101	5	5
6	110	6	6
7	111	7	7
8	1000	8	10
9	1001	9	11
10	1010	A	12
11	1011	B	13
12	1100	C	14
13	1101	D	15
14	1110	E	16
15	1111	F	17

in groups of $2^n - 1$. This same pattern is adopted for the input variables in truth tables, such as in Figs. 14.1–2b and 14.1–3b.

Digital computers work with binary words 32 to 64 bits long, usually subdivided into 8-bit units called **bytes.** Intermediate printouts for diagnostic purposes may then be presented in **hexadecimal numbers** (base-16), without going through binary-to-decimal conversion. To express a binary word in hexadecimal notation, you start at the binary point and separate the word into 4-bit groups, adding leading or trailing zeros as needed. Then you replace each 4-bit group with the appropriate hexadecimal digit from Table 14.1–1 where, by convention, the letters A through F represent 10_{10} through 15_{10}. This simple process is illustrated by

$$1011101.1_2 = 0101\ 1101\ .\ 1000 = 5D.8_{16}$$

The corresponding decimal number can then be obtained via the base-16 expansion $5D.8_{16} = 5 \times 16^1 + 13 \times 16^0 + 8 \times 16^{-1} = 93.5_{10}$.

Microprocessor systems and personal computers often employ **octal numbers** (base-8). For binary-to-octal conversion, you subdivide the binary word into 3-bit groups and replace each group with one of the octal digits, 0 through 7. Thus,

$$1011101.1_2 = 001\ 011\ 101\ .\ 100 = 135.4_8$$

and $135.4_8 = 1 \times 8^2 + 3 \times 8^1 + 5 \times 8^0 + 4 \times 8^{-1} = 93.5_{10}$.

Exercise 14.1–2　　Verify each of the following number conversions:

(a) $10101011_2 = 171_{10} = AB_{16} = 253_8$
(b) $0.1011_2 = 0.6875_{10}$
(c) $200_{10} = 11001000_2$
(d) $0.38_{10} \approx 0.01100001_2$

BCD Codes

Standard decimal-to-binary conversion is not suitable for calculators and digital instruments because it requires the entire decimal number to be entered before the conversion process begins. **Binary-coded decimal (BCD)** gets around this problem by encoding single decimal digits one at a time, using a 4-bit or 5-bit code word for each digit. Table 14.1–2 lists a few of these BCD codes.

Table 14.1–2
BCD Codes.

Decimal Digit	4-Bit Codes			5-Bit Codes	
	8421	**5311**	**Excess-3**	**8421P**	**2-of-5**
0	0000	0000	0011	00000	00011
1	0001	0001	0100	00011	00101
2	0010	0011	0101	00101	00110
3	0011	0100	0110	00110	01001
4	0100	0101	0111	01001	01010
5	0101	1000	1000	01010	01100
6	0110	1001	1001	01100	10001
7	0111	1011	1010	01111	10010
8	1000	1100	1011	10001	10100
9	1001	1101	1100	10010	11000

The widely-used 8421 code corresponds to "natural" decimal-to-binary conversion since the 4 bits in the code word are weighted by the factors $8 = 2^3$, $4 = 2^2$, $2 = 2^1$, and $1 = 2^0$, reading from left to right. An m-digit decimal number is then encoded as a sequence of m code words, with one 4-bit word for each decimal digit. Thus, for instance,

$$105_{10} \longrightarrow 0001\ 0000\ 0101$$

Note that this 8421 BCD sequence has a total of 12 bits, as contrasted with the direct conversion $105_{10} = 1101001_2$, which requires only 7 bits. The extra bits in BCD is the price paid for digit-by-digit coding.

The 5311 code is another **weighted code,** but the weighting factors differ from conventional binary numbers. The Excess-3 code is an **unweighted code** formed by adding 3_{10} to the decimal digit and converting the resultant to the corresponding 4-bit binary word.

The 5-bit codes in the table are examples of **error-detecting codes** that guard against errors caused by noise or hardware "glitches." The 8421P code is the same as the 8421 code with a **parity bit** P added at the end such that all code words contain an even number of 1's. Accordingly, we say that the code has **even parity.** If an error occurs in one of these words, changing a 0 to a 1 or a 1 to a 0, then the parity becomes odd. Hence, a parity-checking network can detect the presence of a single error — but not its location. Double errors preserve parity and go undetected. The 2-of-5 code is an unweighted code with even parity and exactly two 1's in every word. This feature allows the detection of many double errors as well as single errors.

Exercise 14.1−3 Find the largest decimal integer that can be represented by 20 binary digits using: (a) direct decimal to binary conversion; (b) the 8421 BDC code; (c) the 2-of-5 BDC code.

14.2 COMBINATIONAL LOGIC

An interconnected group of gates performs *combinational logic* in the sense that the combination of input values at any particular instant completely determines the output values. This section is devoted to the study of combinational logic. We'll develop tools for manipulating logic expressions and for designing combinational gate networks. NAND and NOR logic are also introduced.

Boolean Algebra

Mathematical analysis of combinational logic involves **Boolean algebra,** a special algebra with two distinctive features:

- All variables are logic variables restricted to the two values 0 and 1.
- All operations are defined in terms of the logical AND, OR, and NOT.

Some Boolean relationships have a familiar appearance, such as

$$A + 0 = A \qquad B \cdot 1 = B \qquad A \cdot (B + C) = A \cdot B + A \cdot C$$

But other Boolean equations look rather startling at first, $1 + 1 = 1$ being a case in point.

Rather than presenting a formal treatment of the subject, we'll focus attention on the differences between Boolean and ordinary algebra. To that end, Table 14.2−1 lists selected Boolean theorems that will be needed

Table 14.2−1
Boolean theorems.

1	$X \cdot X = X$	AND theorems
2	$X \cdot \overline{X} = 0$	
3	$X + 1 = 1$	OR theorems
4	$X + X = X$	
5	$X + \overline{X} = 1$	
6	$X + XY = X$	Absorption theorems
7	$X(X + Y) = X$	
8	$XY + X\overline{Y} = X$	
9	$(X + Y)(X + \overline{Y}) = X$	
10	$\overline{X \cdot Y} = \overline{X} + \overline{Y}$	DeMorgan's theorems
11	$\overline{X + Y} = \overline{X} \cdot \overline{Y}$	

for our work here. The symbols X and Y are used in this table to emphasize the fact that all of these theorems remain valid when a variable is consistently replaced by its complement or by some function of other variables.

Since the logical AND acts like multiplication, we henceforth omit the dot symbol except where helpful for emphasis. In absence of parentheses, the AND operation always takes precedence over the OR operation—just like "multiplication before addition." In theorem 6, for instance, the expression $X + XY$ compactly represents $X + (X \cdot Y)$. Furthermore, any operation appearing under a negation bar is to be performed before taking the logical inverse, so $\overline{X + Y}$ in theorem 10 stands for $\overline{(X + Y)}$.

Theorems 1 through 5 state significant consequences of the AND and OR operations. These theorems, and any other valid Boolean relation, may be proved by constructing a truth table. To illustrate the procedure, the truth tables in Fig. 14.2–1 constitute proofs of theorems 2 and 11.

Figure 14.2–1

Truth table proofs:
(a) $X \cdot \overline{X} = 0$;
(b) $\overline{X + Y} = \overline{X} \cdot \overline{Y}$.

X	$X \cdot \overline{X}$
0	$0 \cdot 1 = 0$
1	$1 \cdot 0 = 0$

(a)

X	Y	$\overline{X + Y}$	$\overline{X} \cdot \overline{Y}$
0	0	$\overline{0 + 0} = \overline{0} = 1$	$\overline{0} \cdot \overline{0} = 1 \cdot 1 = 1$
0	1	$\overline{0 + 1} = \overline{1} = 0$	$\overline{0} \cdot \overline{1} = 1 \cdot 0 = 0$
1	0	$\overline{1 + 0} = \overline{1} = 0$	$\overline{1} \cdot \overline{0} = 0 \cdot 1 = 0$
1	1	$\overline{1 + 1} = \overline{1} = 0$	$\overline{1} \cdot \overline{1} = 0 \cdot 0 = 0$

(b)

Other Boolean relations can be derived from the AND and OR theorems together with algebraic factoring or expansion based on the usual commutative, associative, and distributive laws. As an example of this algebraic manipulation, we prove theorem 6 by writing

$$X + XY = X \cdot 1 + X \cdot Y$$

$$= X \cdot (1 + Y) \qquad \text{(Distributive law)}$$

$$= X \cdot 1 \qquad \text{(Theorem 3)}$$

$$= X$$

Theorems 1 and 6 then yield theorem 7 via the expansion $X(X + Y) = X \cdot X + X \cdot Y = X + XY = X$.

Since the value of $X + XY$ equals X, regardless of the value of Y, we say that the variable Y has become *absorbed* and does not influence the result. Theorems 8 and 9 demonstrate other absorption effects in Boolean algebra. Absorption is important because it often leads to simplified gate networks.

Finally, theorems 10 and 11 show what happens when we take the logical inverse after performing the AND or OR operation. These theorems are particular cases of the more general **DeMorgan's rule,** which states that

> The complement of any Boolean expression may be obtained by negating each variable individually, changing each AND operation to OR, and changing each OR operation to AND.

This procedure is easily carried out, provided that you follow the order of precedence and watch for "hidden" AND operations.

Example 14.2–1

Suppose the specification for a certain logic function leads to the expression

$$F = \overline{A\overline{B} + C} \cdot (\overline{A}C + B)$$

Direct implementation as a combinational network would require eight gates (3 AND, 2 OR, 3 NOT) and numerous internal connections. To reduce hardware costs and assembly time, we seek a simpler but equivalent function calling for fewer gates.

We start by applying DeMorgan's rule to the first term of the function, which becomes

$$\overline{A\overline{B} + C} = \overline{(A \cdot \overline{B}) + C} = (\overline{A} + \overline{\overline{B}}) \cdot \overline{C} = (\overline{A} + B)\overline{C}$$

since $\overline{\overline{B}} = B$. Algebraic manipulation now yields

$$F = (\overline{A} + B)\overline{C} \cdot (\overline{A}C + B) = (\overline{A} + B)(A C \overline{C} + B\overline{C})$$

$$= (\overline{A} + B)(A \cdot 0 + B\overline{C}) \qquad \text{(Theorem 2)}$$

$$= (\overline{A} + B)B\overline{C} = B(B + \overline{A}) \cdot \overline{C}$$

$$= B\overline{C} \qquad \text{(Theorem 7)}$$

We have thus eliminated the variable A and simplified the function to one that requires just two gates.

Exercise 14.2–1

(a) Construct a truth table proving theorem 10.
(b) Use theorem 2 to prove theorem 8.
(c) Apply DeMorgan's rule to find $\overline{F}$ when $F = \overline{A}(B + \overline{C}D)$.

Logic Network Design

Designing a logic network involves finding an appropriate logic function and then simplifying it, if possible, for economical implementation. Here, we present two systematic methods for logic network design based on the truth table of the desired function.

To introduce these methods, suppose we need a logic network with inputs A, B, and C that produces $F = B$ when $A = 0$ and $F = C$ when $A = 1$. The truth table for this function is given in Fig. 14.2–2a, where the lines have been numbered 0 through 7 according to the decimal equivalent of the binary input word $[ABC]$. We'll derive two different expressions relating F to A, B, and C by examining the lines of the truth table.

First, consider those lines where $F = 1$. On line 2, for instance, we find $F = 1$ when $[ABC] = 010$. This particular input word is uniquely identified by the function $m_2 = \overline{A} \cdot B \cdot \overline{C} = \overline{A}B\overline{C}$, which is called a **minterm** and has the property that $m_2 = 1$ if and only if $A = 0$, $B = 1$, and $C = 0$. To obtain

Figure 14.2–2
(a) Truth table with minterms. (b) AND-OR network for standard sum of products.
(c) Minimized network.

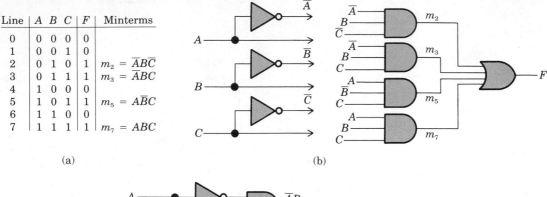

Line	A	B	C	F	Minterms
0	0	0	0	0	
1	0	0	1	0	
2	0	1	0	1	$m_2 = \overline{A}B\overline{C}$
3	0	1	1	1	$m_3 = \overline{A}BC$
4	1	0	0	0	
5	1	0	1	1	$m_5 = A\overline{B}C$
6	1	1	0	0	
7	1	1	1	1	$m_7 = ABC$

(a)

(b)

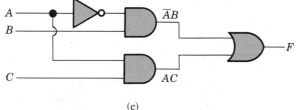

(c)

the minterm for any line, you simply

> Write the logical-AND "product" of all input variables and complement each variable whose value equals 0 on the line in question.

The minterms m_3, m_5, and m_7 likewise identify the other three input words that should produce $F = 1$.

Since we want $F = 1$ when any one of these minterms equals 1, our desired logic function is just the logical-OR "sum"

$$F = m_2 + m_3 + m_5 + m_7$$

$$= \overline{A}B\overline{C} + \overline{A}BC + A\overline{B}C + ABC \tag{1}$$

We call any function in this general form a **standard sum of products** (SSOP). The corresponding implementation diagrammed in Fig. 14.2–2b is known as an **AND-OR network.** The interconnections from the input lines have been omitted here so you can clearly see the one-to-one relationship between the minterms and the AND gates. Frequently, input words are stored in a register having both direct and negated outputs, so the inverters would be unnecessary.

Closer examination of Eq. (1) reveals that m_2 and m_3 are identical except that $\overline{C}$ appears in one and C in the other. This pattern leads to simplification via the absorption property $XY + X\overline{Y} = X$ in theorem 8.

Specifically, if we combine m_2 with m_3 and regroup variables, then we get

$$m_2 + m_3 = (\overline{A}B)\overline{C} + (\overline{A}B)C = \overline{A}B$$

Furthermore, m_5 and m_7 also differ only by the negation of one variable so

$$m_5 + m_7 = (AC)\overline{B} + (AC)B = AC$$

Now we can write

$$F = (m_2 + m_3) + (m_5 + m_7) = \overline{A}B + AC \qquad (2)$$

which is a *minimized* SOP function. The minimized AND-OR network in Fig. 14.2–2c requires four gates, in contrast to the eight gates needed in Fig. 14.2–2b.

As just illustrated, Boolean absorption produces desirable simplifications whenever an SSOP function contains minterm pairs that differ by the negation of one variable. Additional simplifications are possible if two reduced pairs again differ by the negation of one variable.

Next we derive an alternative expression for F by focusing on the lines of the truth table where $F = 0$. Our original truth table is repeated in Fig. 14.2–3a, and each line with $F = 0$ is now identified by a *negated minterm* called a **maxterm.** To obtain the maxterm for any line, you simply

Write the logical-OR "sum" of all input variables and negate each variable whose value equals 1 on the line in question.

Figure 14.2–3
(a) Truth table with maxterms. (b) OR-AND network for standard product of sums. (c) Minimized network.

Line	A	B	C	F	Maxterms
0	0	0	0	0	$M_0 = A + B + C$
1	0	0	1	0	$M_1 = A + B + \overline{C}$
2	0	1	0	1	
3	0	1	1	1	
4	1	0	0	0	$M_4 = \overline{A} + B + C$
5	1	0	1	1	
6	1	1	0	0	$M_6 = \overline{A} + \overline{B} + C$
7	1	1	1	1	

(a)

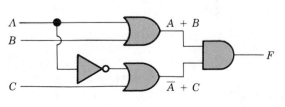

(c)

Being the complement of a minterm, a maxterm equals 0 when the inputs match the values listed on the corresponding line. For instance, the maxterm $M_4 = \overline{m_4} = \overline{A} + B + C$ has the property that $M_4 = 0$ if and only if $[ABC] = 100$.

Since we want $F = 0$ when any one of the listed maxterms equals 0, our alternative logic function is the logical-AND "product"

$$F = M_6 \cdot M_1 \cdot M_4 \cdot M_6$$
$$= (A + B + C)(A + B + \overline{C})(\overline{A} + B + C)(\overline{A} + \overline{B} + C) \tag{3}$$

We call any function in this general form a **standard product of sums** (SPOS). The corresponding implementation diagrammed in Fig. 14.2–3b is known as an **OR-AND network**, with a one-to-one relationship between the maxterms and the OR gates.

When an SPOS function contains pairs of maxterms that differ by the negation of one variable, it can be simplified using the absorption property $(X + Y)(X + \overline{Y}) = X$ from theorem 9. Equation (3) contains two such pairs, which simplify to

$$M_0 \cdot M_1 = [(A + B) + C][(A + B) + \overline{C}] = A + B$$
$$M_4 \cdot M_6 = [(\overline{A} + C) + B][(\overline{A} + C) + \overline{B}] = \overline{A} + C$$

Hence, the *minimized* POS function becomes

$$F = (M_0 \cdot M_1)(M_4 \cdot M_6) = (A + B)(\overline{A} + C) \tag{4}$$

Figure 14.2–3c diagrams the implementation.

It should be emphasized that the networks in Figs. 14.2–2 and 14.2–3 produce the same logic function F, but they do so in different ways. The SOP and POS networks happen to have the same number of gates because there are equal numbers of minterms and maxterms for this particular function.

Example 14.2–2

A certain electric power plant supplies four large loads, A, B, C, and D. The power demands are such that an auxiliary generator must be activated whenever any three loads are on at the same time, or when loads A and B are both on. We'll derive a logic function F to control the auxiliary generator $(0 = \text{OFF}, 1 = \text{ON})$.

The four-variable truth table for this problem is given in Fig. 14.2–4. Since there are fewer 1's than 0's in the output column, we choose to work with the SSOP function

$$F = m_7 + m_{11} + m_{12} + m_{13} + m_{14} + m_{15}$$

The four-variable minterms listed in the truth table are easily determined by the previously stated method, so we proceed to the task of minimizing our expression.

Figure 14.2–4
Truth table for
Example 14.2–2.

Line	A	B	C	D	F	Minterms
0	0	0	0	0	0	
1	0	0	0	1	0	
2	0	0	1	0	0	
3	0	0	1	1	0	
4	0	1	0	0	0	
5	0	1	0	1	0	
6	0	1	1	0	0	
7	0	1	1	1	1	$m_7 = \overline{A}BCD$
8	1	0	0	0	0	
9	1	0	0	1	0	
10	1	0	1	0	0	
11	1	0	1	1	1	$m_{11} = A\overline{B}CD$
12	1	1	0	0	1	$m_{12} = AB\overline{C}\,\overline{D}$
13	1	1	0	1	1	$m_{13} = AB\overline{C}D$
14	1	1	1	0	1	$m_{14} = ABC\overline{D}$
15	1	1	1	1	1	$m_{15} = ABCD$

A scan down the list of minterms reveals several pairs that differ by one variable for pairwise simplification, such as

$$m_{12} + m_{13} = AB\overline{C} \qquad m_{14} + m_{15} = ABC$$

We could also combine m_{15} with m_7 or m_{11} to get

$$m_7 + m_{15} = BCD \qquad m_{11} + m_{15} = ACD$$

These four pairs include all six minterms, and the Boolean property $X = X + X + X$ from theorem 4 allows us to use m_{15} three times for simplification purposes. Accordingly, we write

$$F = m_7 + m_{11} + m_{12} + m_{13} + m_{14} + (m_{15} + m_{15} + m_{15})$$

$$= (m_7 + m_{15}) + (m_{11} + m_{15}) + (m_{12} + m_{13}) + (m_{14} + m_{15})$$

$$= BCD + ACD + AB\overline{C} + ABC$$

But $AB\overline{C} + ABC = AB$, so the minimized SOP function becomes

$$F = BCD + ACD + AB$$

This function would be implemented as an AND-OR network with three AND gates.

Exercise 14.2–2

A logic network with three inputs produces $F = \overline{B \cdot C}$ when $A = 0$ and $F = \overline{B} + C$ when $A = 1$. Find the SSOP function and minimize it.

NAND and NOR Logic

Most logic gates are constructed with transistors at the output and, as a consequence, they have built-in inversion. Thus, an AND gate with inversion becomes a NOT-AND or **NAND gate,** while an OR gate with inversion becomes a NOT-OR or **NOR gate.** The inversion proves to be a blessing in disguise because, as we will soon see, any logic function can be implemented entirely with NAND gates or NOR gates.

The schematic symbol and truth table for a two-input NAND gate are given in Fig. 14.2–5a. The "bubble" at the output of the symbol stands for logical inversion after the AND operation. From DeMorgan's theorem, the NAND operation can be written in two ways:

$$F = \overline{A \cdot B} = \overline{A} + \overline{B} \tag{5}$$

The second expression here leads to an alternative symbol for the NAND gate given in Fig. 14.2–5b, where the input bubbles stand for inversion before the OR operation.

Figure 14.2–5
(a) NAND-gate symbol and truth table.
(b) Alternative symbol.

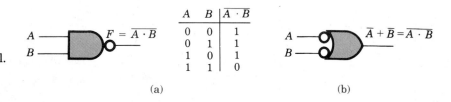

A	B	$\overline{A \cdot B}$
0	0	1
0	1	1
1	0	1
1	1	0

(a)   (b)

Now let the inputs of a NAND gate be tied together, so the output becomes $F = \overline{A \cdot A} = \overline{A}$—just like a NOT gate. For convenience, we'll represent this property by the NAND inverter in Fig. 14.2–6a. If we connect a NAND inverter at the *output* of a NAND gate, as in Fig. 14.2–6b, then we have the equivalent of an AND gate since

$$\overline{(\overline{A \cdot B})} = A \cdot B$$

On the other hand, if we connect NAND inverters at the *input* of a NAND gate, as in Fig. 14.2–6c, then we have the equivalent of an OR gate since

$$\overline{\overline{A} \cdot \overline{B}} = \overline{\overline{A}} + \overline{\overline{B}} = A + B$$

These results prove that the NAND gate is a *universal* gate, meaning that you can use NAND gates to build any desired combinational logic network.

Figure 14.2–6
(a) NAND inverter. (b) NAND implementation of AND. (c) NAND implementation of OR.

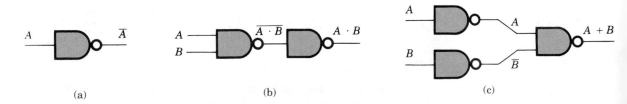

(a)   (b)   (c)

Furthermore, the design of NAND-gate networks involves no more labor than AND-OR network design. You simply

Start with the diagram of an AND-OR network for the desired
function and replace *all* gates with NAND gates.

To demonstrate the validity of this assertion, consider the SOP function
$F = AB + CD$ implemented by the AND-OR network in Fig. 14.2–7a. The
network in Fig. 14.2–7b produces exactly the same output since the inversion bubbles added at the outputs of the AND gates counteract the bubbles
inserted at the inputs of the OR gate. But, from Fig. 14.2–5b, an OR gate
with input bubbles performs the NAND operation. Hence, the final NAND
network in Fig. 14.2–7c is, indeed, equivalent to the original AND-OR network.

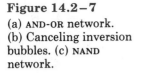

Figure 14.2–7
(a) AND-OR network.
(b) Canceling inversion
bubbles. (c) NAND
network.

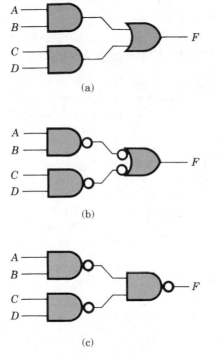

(a)

(b)

(c)

Another universal gate is the NOR gate, whose symbol and truth table
are given in Fig. 14.2–8a for the case of two inputs. The resulting output is

$$F = \overline{A + B} = \overline{A} \cdot \overline{B} \tag{6}$$

which justifies the alternative symbol in Fig. 14.2–8b.

A single-input NOR gate acts as an inverter, and the OR and AND operations can be obtained by combining NOR inverters with NOR gates. To
design a complete NOR network, you simply

Start with the diagram of an OR-AND network for the desired
function and replace *all* gates with NOR gates.

Figure 14.2–8
(a) NOR-gate symbol
and truth table.
(b) Alternative symbol.

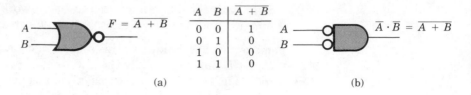

(a)

(b)

Thus, NOR networks would be used to implement functions expressed in POS form rather than SOP form.

Exercise 14.2–3

(a) Derive the NAND network in Fig. 14.2–7c by replacing the gates in Fig. 14.2–7a with NAND equivalents from Fig. 14.2–6.
(b) Draw NOR-gate networks that produce $F = A + B$ and $F = A \cdot B$.

Minimization with Karnaugh Maps[†]

The task of simplifying SSOP and SPOS expressions becomes easier with the help of a **Karnaugh map** (pronounced Car-no). This minimization technique is particularly effective for functions with three or four variables, and it also allows us to incorporate additional information known as *don't care conditions*.

Figure 14.2–9 shows the general structure of a three-variable map. (Temporarily ignore the curved lines here.) Each square cell represents a *minterm* and has been so marked. The labels along the edges make it simple to convert a minterm symbol to its expression, or vice versa. For instance, minterm m_6 appears at the intersection of the AB column and the $\overline{C}$ row, indicating that $m_6 = AB\overline{C}$. The peculiar arrangement of minterms is designed such that

> Minterms in adjacent cells always differ by the negation of *one* variable, which is therefore absorbed by the logical-OR combination of adjacent minterms.

This absorption property also holds for cells on opposite edges, so you should consider opposite edges of the map to be adjacent.

Figure 14.2–9
Karnaugh map for
three-variable
functions.

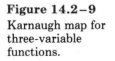
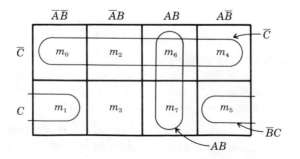

The combination of adjacent minterms is called a 2-cell **subcube.** Similarly, the combination of adjacent 2-cell subcubes forms a 4-cell subcube, which results in the absorption of two variables. To illustrate, the three closed curves in Fig. 14.2–9 identify subcubes corresponding to the absorption simplifications

$$m_6 + m_7 = AB\overline{C} + ABC = AB$$

$$m_1 + m_5 = \overline{A}\,\overline{B}C + A\overline{B}C = \overline{B}C$$

$$m_0 + m_2 + m_6 + m_4 = (\overline{A}\,\overline{B}\,\overline{C} + \overline{A}B\overline{C}) + (AB\overline{C} + A\overline{B}\,\overline{C})$$

$$= \overline{A}\,\overline{C} + A\overline{C} = \overline{C}$$

You can easily find these absorptions by inspection of the map — just eliminate any variable that appears with negation in half of the subcube cells and without negation in the other half.

To minimize a function F written as an SSOP expression, you start with a blank Karnaugh map and proceed as follows:

1. Map the function by putting 1's in the cells that represent the minterms of F.

2. Locate any isolated 1's and draw 1-cell subcubes around them.

3. Draw the fewest number of largest possible subcubes around the adjacent 1's until every 1 is "covered" by at least one subcube. Overlapping subcubes are permitted.

4. Write the simplified product term for each subcube, and form the logical-OR sum to obtain the minimized SOP function.

A minimized POS function is obtained in almost the same way, drawing upon the fact that $\overline{F} = 1$ when $F = 0$. Hence, you map the 0's of F and take the complement of the resulting SOP expression.

As an example of mapping technique, consider the three-variable SSOP function

$$F = m_2 + m_3 + m_7 = \overline{A}B\overline{C} + \overline{A}BC + ABC$$

The corresponding map of 1's in Fig. 14.2–10a is covered by two overlapping 2-cell subcubes, so the minimized SOP function is

$$F = \overline{A}B + BC$$

For the minimized POS function, the map in Fig. 14.2–10b has 0's where there previously were blanks. The 0's are covered by a 4-cell subcube and a 2-cell subcube, from which $\overline{F} = \overline{B} + A\overline{C}$. Complementing $\overline{F}$ then yields

$$F = \overline{\overline{B} + A\overline{C}} = B(\overline{A} + C)$$

This result clearly agrees with our SOP expression since $B(\overline{A} + C) = \overline{A}B + BC$.

Figure 14.2–10
Map for minimization of a: (a) SSOP function; (b) SPOS function.

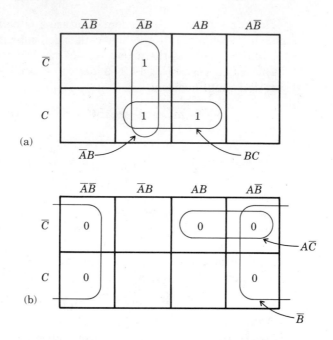

(a)

(b)

Minimizing a function of four variables requires the four-variable map structure in Fig. 14.2–11, which has the same adjacency properties as before. Thus, the logical-OR combination of minterms in a subcube again eliminates any variable that appears with complementation in half of the cells and without complementation in the other half. Illustrative 2-cell, 4-cell, and 8-cell subcubes are shown in the figure. The following example carries out a four-variable minimization.

Figure 14.2–11
Karnaugh map for four-variable functions.

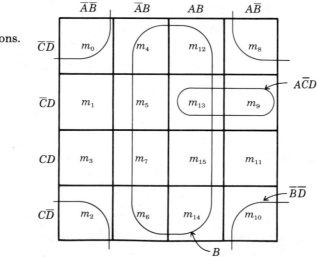

Example 14.2–3   Suppose a logic network is needed to convert decimal digits from the 8421 BCD code to the 5311 code, as listed back in Table 14.1–2. The truth table for this problem is given in Fig. 14.2–12a, where $[ABCD]$ stands for the input word and $[F_4 F_3 F_2 F_1]$ stands for the output word. There are only ten valid input words, representing the digits 0 through 9. However, the four-variable truth table still has 16 lines.

Figure 14.2–12
Truth table and Karnaugh map with don't-care states.

Digit	A B C D	Line	F_4 F_3 F_2 F_1
	8421 Code		5311 Code
0	0 0 0 0	0	0 0 0 0
1	0 0 0 1	1	0 0 0 1
2	0 0 1 0	2	0 0 1 1
3	0 0 1 1	3	0 1 0 0
4	0 1 0 0	4	0 1 0 1
5	0 1 0 1	5	1 0 0 0
6	0 1 1 0	6	1 0 0 1
7	0 1 1 1	7	1 0 1 1
8	1 0 0 0	8	1 1 0 0
9	1 0 0 1	9	1 1 0 1
	1 0 1 0	10	× × × ×
	1 0 1 1	11	× × × ×
	1 1 0 0	12	× × × ×
	1 1 0 1	13	× × × ×
	1 1 1 0	14	× × × ×
	1 1 1 1	15	× × × ×

(a)

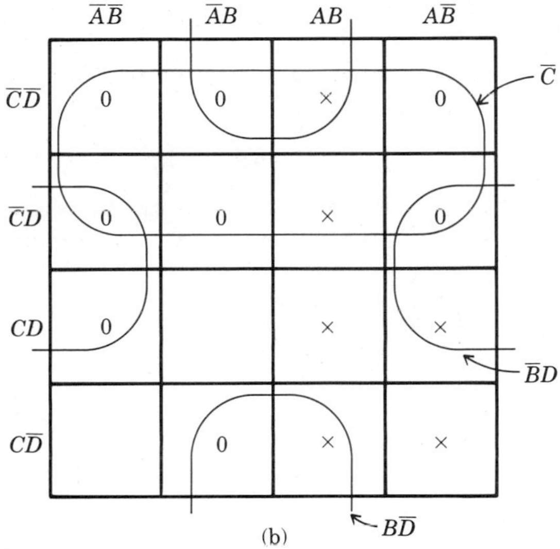

(b)

Since the last six input words are invalid and should never occur, we put ×'s under F_4, F_3, F_2, and F_1 in these lines. The ×'s are called **don't-care states** because we don't care what the output would be when the corresponding input combination never occurs. Accordingly, for purposes of minimization, we are free to interpret any × as either a 0 or a 1. We'll obtain a minimized POS function for output bit F_2 taking advantage of the don't-care conditions.

The Karnaugh map for F_2 is constructed in Fig. 14.2–12b, including all the 0's and ×'s. If we take the ×'s in the top two rows to be 0's, then we can form an 8-cell subcube absorbing A, B, and D. The remaining 0's are covered by two 4-cell subcubes, as shown. Hence, our result becomes

$$F_2 = \overline{\overline{C} + B\overline{D} + \overline{B}D} = C(\overline{B} + D)(B + \overline{D})$$

Adding any other subcubes to the map would increase the number of terms in F_2.

Exercise 14.2–4   Consider the four-variable function

$$F = m_0 + m_2 + m_4 + m_5 + m_6 + m_7 + m_{10}$$

Use a map in the form of Fig. 14.2–11 to find the minimized SOP and POS expressions.

14.3 STANDARD LOGIC MODULES

Certain gate configurations occur time and again in digital system design. Responding to this need, manufacturers have produced several versatile logic networks in the form of standard integrated-circuit modules. This section surveys the properties and typical applications of some popular combinational logic modules, including exclusive-OR gates, decoders, multiplexing units, and read-only memories.

Exclusive-OR Gates

The **exclusive-OR (XOR)** gate is a two-input module whose output equals 1 when just one of the two inputs equals 1. We write this operation as

$$F = A \oplus B \tag{1a}$$

where $A \oplus B$ stands for "A or B but not both." The XOR gate symbol is given in Fig. 14.3–1a. The accompanying truth table shows that

$$0 \oplus 0 = 0 \qquad 0 \oplus 1 = 1 \qquad 1 \oplus 0 = 1 \qquad 1 \oplus 1 = 0 \tag{1b}$$

which differs from the OR operation in that $1 \oplus 1 = 0$.

Figure 14.3–1

XOR gate: (a) symbol; (b) truth table; (c) NOR-gate network.

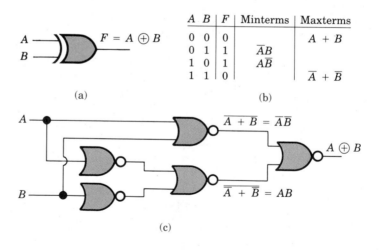

A	B	F	Minterms	Maxterms
0	0	0		$A + B$
0	1	1	$\overline{A}B$	
1	0	1	$A\overline{B}$	
1	1	0		$\overline{A} + \overline{B}$

(a)   (b)

(c)

The truth table also shows that building an XOR module requires several basic gates, since

$$F = \overline{A}B + A\overline{B} = (A + B)(\overline{A} + \overline{B}) \tag{2}$$

Figure 14.3–1c diagrams the NOR-gate network obtained directly from the

SPOS expression in Eq. (2). This particular implementation allows us to get AB and $\overline{A}\,\overline{B}$ as additional outputs, if desired.

Although an XOR gate has only two inputs, the XOR operation is associative and commutative. It then follows that

An XOR sum of three or more variables equals 1 if and only if the input word contains an *odd* number of 1's.

The XOR network in Fig. 14.3–2 exploits this property to generate the parity bit P needed to convert an 8421 BCD code word to an $8421P$ code word. A similar network could detect errors in $8421P$ code words.

Figure 14.3 – 2
Parity bit generator.

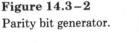

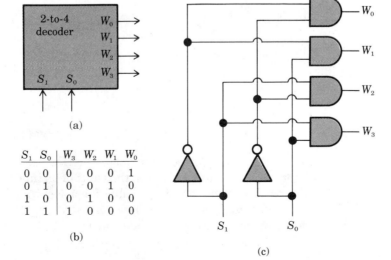

Exercise 14.3 – 1   Let an inverter be connected at the output of an XOR gate to get $F = \overline{A \oplus B}$. Explain why the resulting module is called a *coincidence* gate.

Decoders and Demultiplexers

Decoders belong to the general family of *multiple-output* logic networks, having $k \geq 2$ outputs. However, only one of the outputs is activated to the 1 state at any particular instant, while the remaining $k - 1$ outputs stay in the unactivated 0 state. An n-bit **control word** selects the output to be

Figure 14.3 – 3
(a) 2-to-4 decoder.
(b) Truth table.
(c) Gate diagram.

S_1	S_0	W_3	W_2	W_1	W_0
0	0	0	0	0	1
0	1	0	0	1	0
1	0	0	1	0	0
1	1	1	0	0	0

(b)

(c)

activated, and the number of selectable outputs is limited to $k \leq 2^n$ since there are 2^n different control words. We call such a network an **n-to-k decoder.**

As a specific example, consider the 2-to-4 decoder represented by Fig. 14.3–3a. This module accepts a 2-bit control word $[S_1 S_0]$ and has $k = 4$ outputs labeled W_0 through W_3. The truth table in Fig. 14.3–3b emphasizes the fact that all outputs equal 0 except the one selected by $[S_1 S_0]$. Upon examining the output columns of the truth table, we find that each output function corresponds to a different control-word *minterm*. Hence,

$$W_0 = \bar{S}_1 \bar{S}_0 \qquad W_1 = \bar{S}_1 S_0 \qquad W_2 = S_1 \bar{S}_0 \qquad W_3 = S_1 S_0 \qquad (3)$$

which leads directly to the gate diagram in Fig. 14.3–3c. The truth table and gate diagram for decoders with $n > 2$ follow the same pattern, and you simply omit or ignore some of the AND gates when there are fewer than 2^n outputs.

To illustrate *decoding* in action, Fig. 14.3–4 shows a 4-to-16 decoder arranged for BCD-to-decimal decoding. An 8421 BCD code word applied at the control terminals selects one of ten decimal digit outputs, D_0 through D_9. Thus, for instance, the input word $[S_3 S_2 S_1 S_0] = 0001$ activates output W_1 corresponding to digit D_1. An invalid BCD code word is detected by the OR gate connected to outputs W_{10} through W_{15}. In general, any 4-bit code can be decoded by appropriately assigning the outputs of a 4-to-16 decoder.

Figure 14.3–4
BCD-to-decimal decoding with a 4-to-16 decoder.

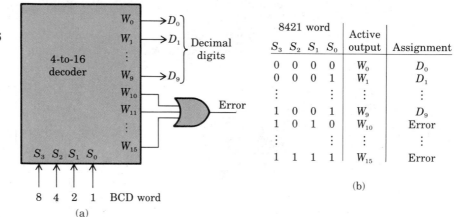

8421 word				Active output	Assignment
S_3	S_2	S_1	S_0		
0	0	0	0	W_0	D_0
0	0	0	1	W_1	D_1
⋮			⋮	⋮	⋮
1	0	0	1	W_9	D_9
1	0	1	0	W_{10}	Error
⋮			⋮	⋮	⋮
1	1	1	1	W_{15}	Error

(b)

A **demultiplexer,** abbreviated DMUX, is a decoder augmented such that the active output takes on the value of an input signal. Figure 14.3–5a shows the symbol and truth table for a 4-output DMUX with input X, control word $[S_1 S_0]$, and outputs labeled Y_0, Y_1, Y_2, and Y_3. Each output function can be written in the form

$$Y_i = W_i \cdot X \qquad i = 0, 1, \; \ldots \qquad (4)$$

Figure 14.3–5
(a) Demultiplexer
with four outputs.
(b) Equivalent selector
switch.

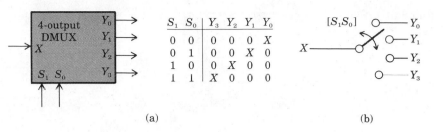

(a)

(b)

where W_i denotes the ith control-word minterm, as in Eq. (3). Accordingly, the gate network would look like Fig. 14.3–3c with the additional input X at each AND gate.

Conceptually, a 4-output DMUX acts like the selector switch in Fig. 14.3–5b, where $[S_1 S_0]$ determines which output terminal is connected to the input terminal. Similar switch models hold for demultiplexers with 2^n outputs. Thus, the n-bit control word of a DMUX allows you to steer a signal X to one of 2^n different destinations.

Standard decoder and demultiplexer modules are available with $n = 2$, 3, or 4. These modules are usually fabricated with NAND rather than AND gates, so the outputs actually become negated. Some modules include an additional DISABLE input, which, when activated, puts *all* outputs in the same logic state.

Exercise 14.3–2   Modify the output assignments in Fig. 14.3–4 to decode the Excess-3 code listed in Table 14.1–2.

Multiplexers

Multiplexers are *multiple-input* logic networks that perform the opposite function of a demultiplexer. A multiplexer or MUX has a single output that takes on the value of one of 2^n input signals, as selected by an n-bit control word. Consequently, multiplexers are also called **data selectors.**

Figure 14.3–6 shows the symbol, truth table, and equivalent selector switch for a 4-input MUX with inputs labeled X_0 through X_3. Since control-word minterms select the input, and since all minterms equal zero except one corresponding to the selected input, we can write the output as

$$Y = W_0 X_0 + W_1 X_1 + W_2 X_2 + W_3 X_3 \tag{5a}$$

$$= \overline{S}_1 \overline{S}_0 X_0 + \overline{S}_1 S_0 X_1 + S_1 \overline{S}_0 X_2 + S_1 S_0 X_3 \tag{5b}$$

If $[S_1 S_0] = 10$, for example, then $Y = 0 \cdot X_0 + 0 \cdot X_1 + 1 \cdot X_2 + 0 \cdot X_3 = X_2$. The gate diagram in Fig. 14.3–6c implements Eq. (5).

Standard multiplexer modules are available with $n = 2$, 3, or 4. The corresponding properties of 8-input and 16-input multiplexers follow by generalization of the 4-input case. Like decoders and demultiplexers, the

Figure 14.3–6
(a) Multiplexer with four inputs.
(b) Equivalent selector switch. (c) Gate diagram.

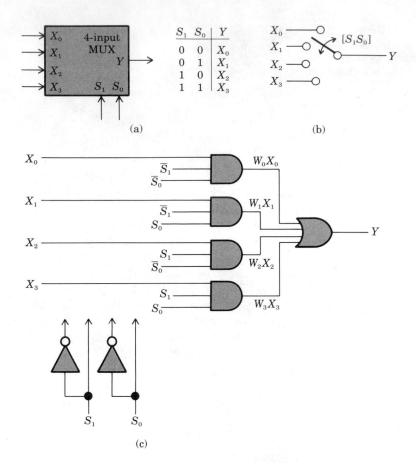

(a)

S_1	S_0	Y
0	0	X_0
0	1	X_1
1	0	X_2
1	1	X_3

(b)

(c)

output of a standard multiplexer is usually negated, and a DISABLE input may also be provided.

When operated as a data selector, a MUX allows you to pick out one of 2^n input signals for processing or display. Connecting a MUX to a DMUX creates a **time-division multiplexing system** (TDM) for bit-by-bit transmission of several digital signals over one data wire.

Additionally, a MUX with 2^n inputs can be programmed to serve as a **function generator,** producing any desired function of $n - 1$ variables. Using a standard MUX module for logic function generation saves both design time and assembly costs. The procedure is illustrated in the following example.

Example 14.3–1
Three-variable function generator.

Any function of three variables A, B, and C, can be generated using a 4-input MUX with $S_1 = A$ and $S_0 = B$. The MUX is *programmed* by assigning values to X_0, X_1, X_2, and X_3 such that Y equals the desired function F. Thus, to construct the programming table, we must determine the value of F for each of the four possible values of the control word $[S_1 S_0] = [AB]$.

Suppose, for instance, that the function happens to be given in the SSOP form

$$F = \overline{A}\,\overline{B}\,\overline{C} + \overline{A}\,\overline{B}C + \overline{A}\,B\overline{C} + A\overline{B}\,\overline{C} + A\overline{B}C$$

Setting $A = B = 0$ yields $F - \overline{C} + C = 1$, so we want $X_0 = 1$ since $Y = X_0$ when $S_1 = S_0 = 0$. Then setting $A = 0$ and $B = 1$ yields $F = \overline{C}$, so we want $X_1 = \overline{C}$. Continuing this process for $[AB] = 10$ and $[AB] = 11$, we get the complete programming table in Fig. 14.3–7a. The resulting programmed MUX is diagrammed in Fig. 14.3–7b. The 0 and 1 states needed at the inputs would be obtained by connections to appropriate low and high voltages.

Figure 14.3–7

Function generator: (a) programming table; (b) MUX diagram.

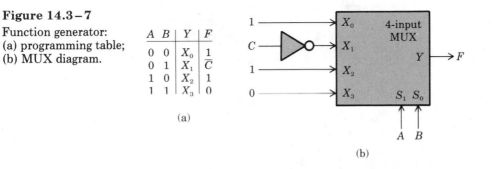

A	B	Y	F
0	0	X_0	1
0	1	X_1	$\overline{C}$
1	0	X_2	1
1	1	X_3	0

(a)

(b)

Although we worked here with an SSOP expression, any complete definition of the desired function is sufficient for programing purposes. Furthermore, our method readily extends to programming an 8-input MUX for any function of four variables or a 16-input MUX for any function of five variables.

Exercise 14.3–3   Construct the truth table for $F = A \oplus B \oplus C$ and program a 4-input MUX to generate this function. (Save your truth table for Exercise 14.3–4.)

Read-Only Memories

A **read-only memory** or **ROM** essentially consists of 0's and 1's stored in a two-dimensional array or matrix. The designation *read-only* means that the bits can be "read" at the output, but they cannot be changed. Furthermore, the bit pattern is **nonvolatile** and remains fixed in place even when the power goes off. This permanence and nonvolatility distinguishes read-only memories from the temporary storage units we'll discuss in Chapter 15.

To introduce ROM structure, consider the diode matrix in Fig. 14.3–8a. Each horizontal **word line** connects to each vertical **output line** through a diode in series with a **fusible link.** However, some of the links have been intentionally "blown" to break the connection. Thus, if we use a

Figure 14.3–8
Read-only memory
using diode matrix.

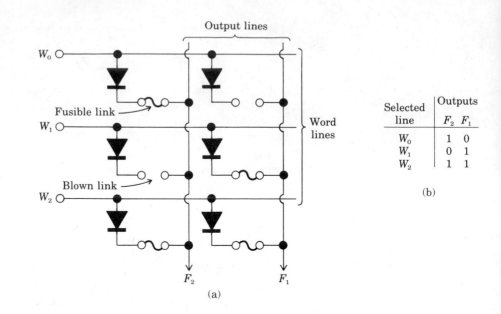

(a)

| Selected | Outputs |
line	F_2 F_1
W_0	1 0
W_1	0 1
W_2	1 1

(b)

decoder to select the top word line, then W_0 goes high and causes output F_2
to go high via the connecting diode, but the unconnected output F_1 remains
low. The table in Fig. 14.3–8b lists all the output combinations for this
partial segment of a ROM.

Figure 14.3–9a represents a small but complete ROM with an input
decoder, four word lines, and three output lines. The solid dots in the
matrix stand for connected diodes, while the open circles stand for broken
connections. The inputs S_1 and S_0 select a word line which, in turn, determines the values of the three outputs F_2, F_1, and F_0. The corresponding
truth table is given in Fig. 14.3–9b. This table clearly displays the one-to-
one relationship between the output bit values and the matrix connection
pattern.

Figure 14.3–9
4-word × 3-bit ROM
with input decoder.

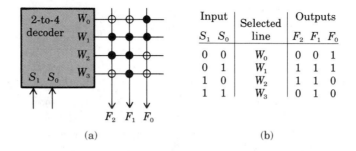

(a)

| Input | | Selected | Outputs | | |
S_1 S_0		line	F_2	F_1	F_0
0 0		W_0	0	0	1
0 1		W_1	1	1	1
1 0		W_2	1	1	0
1 1		W_3	0	1	0

(b)

Viewed as a *memory device*, our ROM stores three data bits on each of
the four word lines. A particular data word $[F_2F_1F_0]$ is selected by $[S_1S_0]$
and becomes available for reading on the output lines. Hence, this unit
would be called a 4-word × 3-bit read-only memory.

But our ROM can also be viewed as a *function generator,* since $W_0 - W_4$ are minterms of $[S_1 S_0]$. Thus, suppose that we let $S_1 = A$ and $S_0 = B$, where A and B are input variables. Referring to the column for F_2 in Fig. 14.3–9b, we find that

$$F_2 = W_1 + W_2 = \overline{A}B + A\overline{B} = A \oplus B$$

so F_2 happens to be the exclusive-OR function. Similarly,

$$F_1 = W_1 + W_2 + W_3 = A + B \qquad F_0 = W_0 + W_1 = \overline{A}$$

Other functions of A and B could be generated by additional output lines appropriately connected to the word lines.

Larger ROMs are available as integrated-circuit modules, complete with decoders. A typical unit might have nine inputs, $2^9 = 512$ word lines, eight outputs, and a matrix storing $512 \times 8 = 4{,}096$ bits. ROM bit patterns are preprogrammed by manufacturers for common functions such as code converters or character generators. Specialized applications require a **programmable ROM (PROM)** whose links can be blown selectively by the user with the help of a PROM programmer. An **erasable PROM (EPROM)** allows the user to restore all links, thereby erasing the program, so the EPROM can be reprogrammed if desired.

Finally we should mention the **programmable logic array (PLA),** which is similar to a PROM with n inputs but has fewer than 2^n word lines. Correspondingly, the input decoder is replaced by a decoding matrix so the user can program the input connections to the word lines as well as the output connections. To illustrate the advantage of a PLA, suppose that you need to generate eight different functions that involve only 48 minterms of 16 input variables. This task can be carried out using a $16 \times 48 \times 8$ PLA that stores just $48 \times 8 = 384$ bits, whereas a PROM with 16 inputs would have $2^{16} = 65{,}536$ word lines and an array of $65{,}536 \times 8 = 524{,}288$ bits. Thus, PLA implementation reduces size, cost, and programming time.

Example 14.3–2
Seven-segment display decoder.

Consider a calculator that performs computations in 8421 BCD and presents results as decimal digits on the familiar 7-segment LED display depicted by Fig. 14.3–10a. The **display decoder** receives a BCD word $[B_3 B_2 B_1 B_0]$ and generates outputs that light the appropriate segments of the display.

Figure 14.3–10b gives the truth table for the output functions $Y_a - Y_f$ such that $Y_a = 1$ turns on segment a, etc. This truth table leads directly to the display decoder implemented by a ROM, as diagrammed in Fig. 14.3–10c. The 16×7 ROM has four inputs and $2^4 = 16$ word lines, of which only the first 10 are needed here. (Alternatively, a $4 \times 10 \times 7$ PLA might be used, but the savings would be small.)

Display decoders are also available as standard IC modules. Since there is a large market for these modules, they actually consist of "hard-wired" gates rather than programmed arrays.

Figure 14.3–10
(a) 7-segment display.
(b) Truth table for BCD decoding.
(c) ROM display decoder.

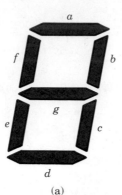

(a)

B_3	B_2	B_1	B_0	Digit	Y_a	Y_b	Y_c	Y_d	Y_e	Y_f	Y_g
		Input					Outputs				
0	0	0	0	0	1	1	1	1	1	1	0
0	0	0	1	1	0	1	1	0	0	0	0
0	0	1	0	2	1	1	0	1	1	0	1
0	0	1	1	3	1	1	1	1	0	0	1
0	1	0	0	4	0	1	1	0	0	1	1
0	1	0	1	5	1	0	1	1	0	1	1
0	1	1	0	6	1	0	1	1	1	1	1
0	1	1	1	7	1	1	1	0	0	0	0
1	0	0	0	8	1	1	1	1	1	1	1
1	0	0	1	9	1	1	1	1	0	1	1

(b)

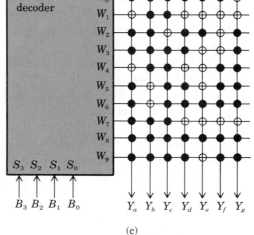

(c)

Exercise 14.3–4   Diagram an 8×3 ROM programmed to generate $F_2 = A \oplus B \oplus C$, $F_1 = (A \oplus B) \cdot C$, and $F_0 = AB$.

14.4 IC LOGIC GATES AND FAMILIES[†]

The building blocks of digital systems are fabricated in various integrated-circuit forms known as **logic families** or **technologies.** This section starts with the general properties of IC logic gates and families. Then we'll examine typical gate circuits from the widely used TTL, ECL, and CMOS families.

Properties of IC Logic Gates

Ideally, a logic gate acts like a perfect switch with two distinct states. But IC logic gates consist of transistors, and their behavior departs from ideal switching in various respects that will be considered here. For simplicity, we'll focus on the properties of inverters since they have just one input. Of course, similar properties would be exhibited by multi-input NAND or NOR gates when the inputs change one at a time.

Figure 14.4–1a represents a typical transfer curve for an inverter whose output voltage v_o goes from the HIGH state to the LOW state as the input voltage v_i increases. Although such transfer curves are informative, they fail to provide a complete picture because the high and low levels depend upon loading, temperature, and other factors, and they also vary somewhat from gate to gate. Consequently, manufacturers usually specify *worst-case values* shown on the generalized level diagram in Fig. 14.4–1b. The voltage V_{OH} stands for the minimum value of the high output level, and V_{IL} is the corresponding maximum input value. Conversely, V_{OL} stands for the maximum value of the low output level, and V_{IH} is the corresponding minimum input value. Thus, the manufacturer guarantees that $v_o \geq V_{OH}$ for any $v_i \leq V_{IL}$ and that $v_o \leq V_{OL}$ for any $v_i \geq V_{IH}$.

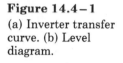

Figure 14.4–1
(a) Inverter transfer curve. (b) Level diagram.

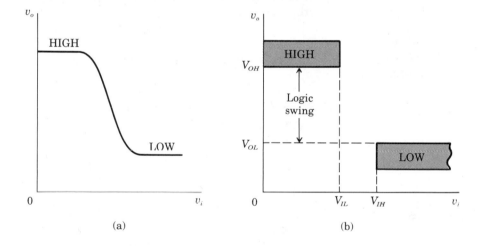

(a)   (b)

Given the worst-case values of a particular logic family, you can determine the **logic swing** which equals the difference $V_{OH} - V_{IH}$. You can also determine the high-level and low-level **noise margins** defined by

$$NM_H \triangleq V_{OH} - V_{OL} \qquad NM_L \triangleq V_{IL} - V_{OL} \tag{1}$$

The significance of noise margin is illustrated in Fig. 14.4–2 where noise $n(t)$ contaminates the output of gate 1 so the input to gate 2 becomes $v_{i2} = v_{o1} + n(t)$. Extreme noise peaks will then cause errors to appear at the output of gate 2. However, we see from the accompanying level diagram that the noise has no adverse effect if $n(t) < NM_L$ when $v_{o1} \leq V_{OL}$ and if $n(t) > -NM_H$ when $v_{o1} \geq V_{OH}$. Reliable operation therefore requires noise margins larger than the noise peaks.

Logic swing and noise margins are reduced by loading effects when the output of one gate drives the inputs of several gates — a situation known as

Figure 14.4−2
Illustration of noise margin.

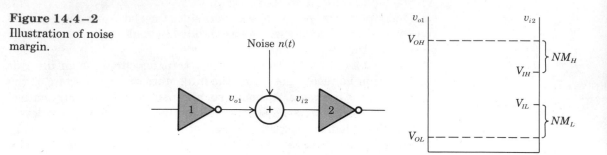

fan-out. To explain fan-out loading, we'll use the circuit in Fig. 14.4−3a as a simplified model of the output portion of an IC gate. The two switches with series resistance represent transistors activated by input circuitry not shown. Such a gate is said to have **active pull-up** and **active pull-down** in the sense that opening T_L and closing T_H "pulls up" the output to the HIGH state, whereas opening T_H and closing T_L "pulls down" the output to the LOW state.

Figure 14.4−3
(a) Gate model with active pull-up and pull-down. (b) HIGH-state fan-out.
(c) LOW-state fan-out.

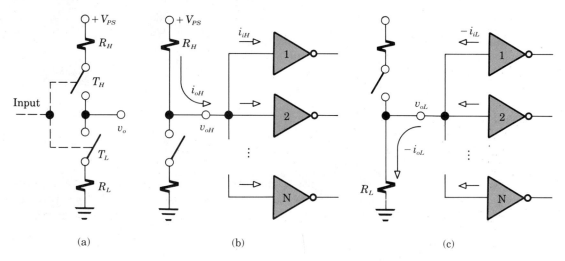

Figure 14.4−3b shows our gate model in the HIGH state (T_H closed) and loaded by fan-out to N gates. Each driven gate draws input current i_{iH}, so the driving gate must become a *current source* to supply the output current $i_{oH} = N i_{iH}$. Hence, $v_{oH} = V_{PS} - R_H(N i_{iH})$ and the high level decreases with increasing fan-out. When the driving gate is in the LOW state (T_L closed) as shown in Fig. 14.4−3c, it must become a *current sink* to accept current going from the driven gates to ground. Now i_{iL} and i_{oL} are negative quanti-

ties, so $v_{oL} = R_L(-i_{oL}) = R_L N(-i_{iL})$ and the low level increases with increasing fan-out. Manufacturers specify maximum values of input and output currents for the HIGH and LOW states, and they also indicate the maximum allowable fan-out for $v_{oH} \geq V_{OH}$ and $v_{oL} \leq V_{OL}$.

Some IC gates with active pull-up and pull-down can be modified to incorporate a special feature called **tristate output.** Referring to Fig. 14.4–3a, suppose an auxiliary input opens both T_H and T_L. We then have a *high-impedance state* that disconnects the gate from anything attached to its output terminal. Figure 14.4–4a gives the symbol and truth table for a tristate inverter whose high-Z state is activated by the DISABLE input D, regardless of the value of the direct input A. A typical application is portrayed by Fig. 14.4–4b, where several input and output units are connected via tristate inverters to a wire serving as the **data bus.** If all DISABLEs are activated except D_1 and D_3, for instance, then data travels along the bus from input unit 1 to output unit 3 without affecting the other units.

Figure 14.4–4

(a) Inverter with tristate output. (b) Connections to a data bus.

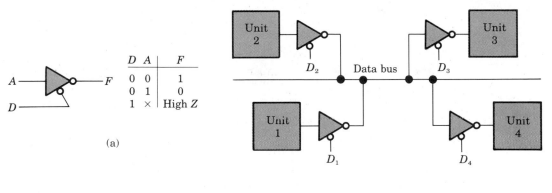

D	A	F
0	0	1
0	1	0
1	×	High Z

(a)

(b)

Another special feature known as **wired logic** is a way of obtaining "no-cost" gates by tying output terminals together. As an example of wired logic, suppose an IC gate still functions properly when the pull-up transistor is omitted and replaced by an external **pull-up resistor.** If we connect the terminals of two such gates to the same pull-up resistor, as diagrammed in Fig. 14.4–5a, then the resulting output Y goes high only when both pull-down transistor switches are open — corresponding to $A = 0$ and $B = 0$. Thus, $Y = \overline{A} \cdot \overline{B}$ and we have built the wired-AND gate symbolized by Fig. 14.4–5b. Similarly, a wired-OR gate results when a pull-down resistor replaces the pull-down transistors. In either case, however, the individual outputs A and B no longer exist.

Finally, we should give some attention to the *dynamic* behavior of IC gates. Every gate circuit includes unwanted but unavoidable energy stor-

Figure 14.4–5

Wired-AND gate:
(a) circuit diagram;
(b) symbol.

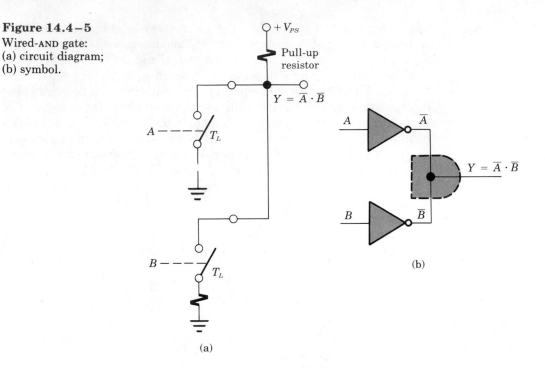

(a)

(b)

age, primarily due to stray capacitance and stored charge carriers in semi-conductor devices. As a result, the transition from one output state to the other always requires a certain amount of time. Moreover, there will always be some time delay before a change at the input produces a change at the output.

Figure 14.4–6 shows illustrative input and output waveforms for an inverter with a pulsed input. The parameter labeled t_{PHL} is the high-to-low propagation time measured between the 50% points of the leading edge of the input and output waveforms. Similarly, t_{PLH} is the low-to-high propagation time. The average of these two is called the **propagation delay,**

Figure 14.4–6

Inverter waveforms
showing propagation
delay.

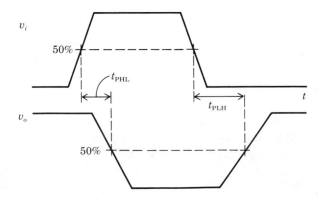

defined by

$$t_{PD} \triangleq \tfrac{1}{2}\,(t_{PHL} + t_{PLH}) \tag{2}$$

Typical values of t_{PD} range from about 100 ns down to 1 ns. Energy storage also limits the rate at which the output can be switched periodically between the HIGH and LOW states. The **maximum switching frequency** f_{max} is inversely proportional to t_{PD} and ranges from 10 MHz to 500 MHz.

Rapid switching requires relatively large currents to charge or discharge stray capacitance in a short time. Gates designed for high-speed operation must therefore carry larger currents and dissipate more power than "slower" gates. Power dissipation is obviously an important consideration because excessive heating may damage the IC chip. The trade-off between speed and power dissipation is brought out by the **delay-power product**

$$DPP \triangleq t_{PD}P_{dis} \tag{3}$$

This product serves as a figure of merit for comparing logic families. It has the units of energy and is measured in picojoules.

Logic Families

A logic family consists of an assortment of mutually compatible gates, logic modules, registers, flip-flops, and other functional units. Thus, you can build almost any desired logic network simply by connecting together appropriate units from one family, provided that you observe the fan-out limits. Most logic families also include special units for interfacing with peripheral devices (keyboards, displays, etc.) or with members of another logic family.

Compatability within a logic family is achieved through the replication of a basic gate circuit in numerous combinations and with various modifications to implement different functions. Since the circuitry for one gate takes up very little space, a single integrated-circuit chip has room for many gates. The equivalent number of gates on a logic chip varies from less than a dozen to more than 10,000, according to the **scale of integration** — small-scale integration (SSI), medium-scale integration (MSI), large-

Table 14.4–1

Categorization of IC logic chips.

Category	Number of Gates	Typical Functions
SSI	≤ 11	Individual gates, flip-flops, AND-OR networks.
MSI	12 to 99	Decoders, multiplexers, registers, counters, adders.
LSI	100 to 9,999	Microprocessors, small memories, analog/digital converters, special-purpose units.
VLSI	$\geq 10,000$	Microcomputers, large memories, signal processors.

scale integration (LSI), or very-large-scale integration (VLSI). These categories are defined in Table 14.4–1, which includes typical logic functions offered in each category.

SSI chips contain only a few gates, but the gates can be interconnected on the chip in various ways or they can have separate input and output leads. Consequently, as many as 75 different SSI units may be available in one logic family. SSI chips are mounted in packages with 14 or 16 external pins, two pins being reserved for the power-supply and ground connections. To illustrate, Fig. 14.4–7 shows the logic diagram and pin assignments for two representative SSI units.

Figure 14.4–7
Diagrams of SSI units.
(a) Triple three-input NAND gates.
(b) Four-wide 3-2-2-3-input AND-OR-INVERT gate.

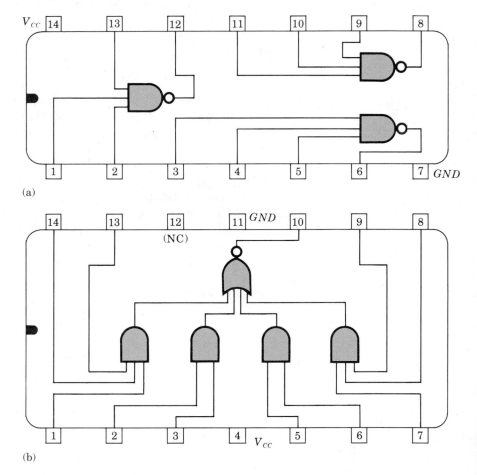

Although most logic functions can be implemented in any logic family, the distinctive properties of logic families depend upon the **technology** used to fabricate the basic gate circuit. Table 14.4–2 gives a chronological listing of the various technologies that have been developed for logic circuits. RTL, DCTL, and DTL existed as discrete-component technologies

Table 14.4–2

Logic-circuit technologies.

Technology	Description	Comment
RTL	Resistor-transistor logic	Obsolete.
DCTL	Direct-coupled transistor logic	Evolved into MOS and I²L.
DTL	Diode-transistor logic	Evolved into TTL.
HTL	High-threshold logic	DCTL modified for noisy environments.
TTL	Transistor-transistor logic	General purpose; several subfamilies.
ECL	Emitter-coupled logic	Highest switching speed.
PMOS NMOS	p-channel/n-channel MOSFET logic	Highest density for VLSI systems.
CMOS	Complementary-symmetry MOSFET logic	General purpose; low power dissipation.
I²L	Integrated-injection logic	Special purpose VLSI.

prior to the invention of integrated circuits, but they have subsequently become obsolete or evolved into newer forms. HTL, MOS, and I²L are technologies designed for special applications rather than general purposes. Examples of DTL and NMOS gates were given in Section 10.2. We therefore limit further consideration to TTL, ECL, and CMOS, the three most widely used logic families.

For comparison purposes, Tables 14.4–3 and 14.4–4 summarize the salient features and typical performance properties of the three major logic families. It should be emphasized that we are dealing here with *typical* properties rather than exact values. More complete information about any IC logic line is contained in **data books** published by the manufacturer.

Table 14.4–3

Properties of TTL, CMOS, and ECL.

	TTL	ECL	CMOS
Basic gate circuit	NAND	NOR/OR	NAND or NOR
Wired logic	AND	OR	AND
Supply voltage	5.0 V	−5.2 V	3–18 V
Logic swing	2.2–3.3 V	0.8 V	$\approx V_{DD}$
Noise margin	0.3–0.7 V	0.3 V	1–2.5 V
Fan-out	10	20	50
Static gate power	1–20 mW	25–40 mW	0.01 μW

Table 14.4–4

Dynamic performance of logic families.

Family/Series		t_{PD} (ns)	P_{dis} (mW)	DPP (pJ)	f_{max} (MHz)
TTL	74	10	10	100	35
	74S	3	20	60	125
	74ALS	4	1	4	70
ECL	100K	0.75	40	30	500
CMOS	4000	100	0.1	10	10
	HC	8	0.2	1.6	40

TTL is a bipolar (BJT) technology whose versatility made it the initial industry standard for SSI and MSI functions. Various TTL modifications over the years have led to several distinct **subfamilies** or **series.** Although the TTL subfamilies differ in dynamic performance, the family as a whole is characterized by moderate noise margin, power dissipation, and speed.

ECL is another bipolar technology, but one specifically designed to minimize propagation delay. However, the high-speed design results in large power dissipation and small noise margins. Consequently, ECL is used primarily for large computers, high-frequency counters, and other applications where speed is the foremost concern.

CMOS gates employ MOSFETS rather than BJTs to achieve larger noise margin and fan-out, with very low static power dissipation. The small size of MOSFETs also allows the high packing density needed for VLSI circuits. Additionally, unlike TTL and ECL, some CMOS units operate from any supply voltage within the 3–18 V range. These properties make CMOS ideal for battery-powered systems such as calculators, watches, and portable digital instruments.

The major disadvantage of CMOS is the long propagation delay that characterized early versions. Subsequent modifications have improved CMOS performance to the point that it has now largely supplanted I²L in LSI and VLSI applications, and it has become more popular than TTL for many SSI and MSI applications. Furthermore, CMOS-TTL interfaces permit the design of optimized systems using both technologies — CMOS subsystems for power conservation, and TTL for subsystems that require higher speed.

IC Gate Circuits

We conclude this chapter with a brief discussion of TTL, ECL, and CMOS gate circuits. An examination of these circuits will further clarify the differences between the three major logic families.

Figure 14.4–8a diagrams a simplified TTL NAND gate with two inputs. The inputs are applied to a **multi-emitter transistor** T_1, which can have as many as eight emitters. The configuration of T_1 is such that it functions like a cluster of *pn*-junction diodes, one from the base to each emitter and one from base to collector. This circuit is therefore equivalent to the diode-transistor NAND gate back in Fig. 10.2–3 (without D_2). However, multi-emitter transistors take up less space on an IC chip than individual diodes would.

To bring out the diode operation of T_1, Fig. 14.4–8b shows the conditions when both inputs are high compared to v_{B1}. The emitter diodes are reverse biased and draw very little current, but current i_{B1} flows through the collector of T_1 into the base of T_3. Thus, $v_{B1} = 2V_\gamma \approx 1.4$ V and $i_{B1} = (V_{CC} - 2V_\gamma)/R_B$. Assuming that this base current saturates T_3, the low output will be

$$v_{oL} = V_{sat} \approx 0.2 \text{ V}$$

Figure 14.4–8
(a) Simplified TTL
NAND gate.
(b) Conditions for low
output. (c) Conditions
for high output.

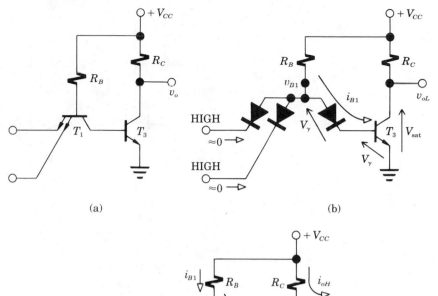

(a)

(b)

(c)

The output goes high when either input is driven by the low output of
another gate, as shown in Fig. 14.4–8c. Here the driving gate sinks current
from the forward-biased emitter diode of T_1 so $v_{B1} = V_{sat} + V_\gamma \approx 0.9$ V —
too low to allow current through the collector of T_1 into the base of T_3.
With T_3 now cut off, the high output is

$$v_{oH} = V_{CC} - R_C i_{oH}$$

whose value depends upon the output loading.

   Loading effects are greatly reduced by the standard Series 74 TTL
circuit diagrammed in Fig. 14.4–9. Transistors T_3 and T_4 constitute a
totem-pole output circuit, with T_2 being the driver stage. When all
inputs are high, i_{B1} saturates T_2 which forces T_3 into saturation, so $v_{oL} =
V_{sat}$ as before. But when any input goes low, T_2 and T_3 are cut off and

Figure 14.4–9
Series 74 TTL NAND
gate.

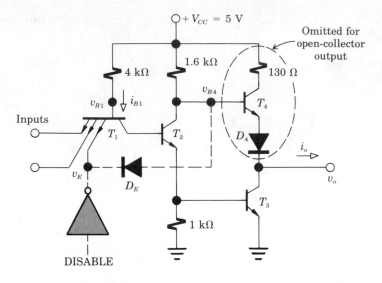

$v_{B4} \approx V_{CC} = 5$ V. Thus, after accounting for the voltage drops across D_4 and the base-emitter junction of T_4, the high output becomes

$$v_{oH} = v_{B4} - 2V_\gamma \approx 3.6 \text{ V}$$

Transistor T_4 remains active to keep v_{oH} constant while supplying current i_{oH} drawn by the load. Note that this design reduces the potential logic swing, since v_{oH} must be less than the 5-V supply.

The dashed lines in Fig. 14.4–9 indicate two variations of the standard TTL gate. An **open-collector gate** omits the upper half of the totem pole starting at the collector of T_3. This open-collector output permits *wired-AND logic* using external pull-up resistors. For *tristate* TTL, the circuit includes diode D_E and an inverter connected as shown to an additional emitter of T_1. When the DISABLE input goes high, v_E and v_{B4} are pulled low to cut off both T_3 and T_4, thereby establishing the high-impedance output state.

The switching speed of standard TTL is limited by charge storage in saturated BJTs. All of the improved TTL subfamilies feature nonsaturating **Schottky transistors** along with other modifications to increase switching speed and decrease the delay-power product. The subfamily series and names are:

Series 74S	Schottky TTL
Series 74LS	Low-power Schottky TTL
Series 74AS	Advanced Schottky TTL
Series 74ALS	Advanced Low-power Schottky TTL

The dynamic performance of Series 74S and 74ALS was compared with standard TTL in Table 14.4–4. Further details can be found in texts on digital electronics.

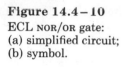

Figure 14.4–10
ECL NOR/OR gate:
(a) simplified circuit;
(b) symbol.

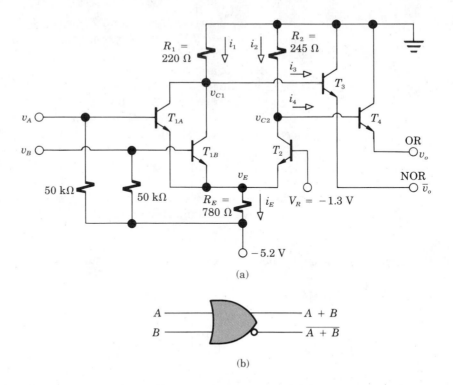

(a)

(b)

Emitter-coupled logic achieves even higher switching speed by reducing the logic swing together with avoiding transistor saturation. Figure 14.4–10a gives a simplified diagram of the basic ECL gate circuit for the case of two inputs. When appropriately terminated, the open emitters of T_3 and T_4 provide complementary NOR and OR outputs as symbolized by Fig. 14.4–10b. Consistent with high-speed design, complementary outputs eliminate the need for additional inverters and their accompanying delay. External signal connections must be made via transmission lines to preserve pulse shapes during rapid switching.

Logic levels in the ECL circuit are referenced to the node at the highest potential. Grounding this node prevents extraneous voltage fluctuations, so a negative supply voltage (-5.2 V) must be used. The reference voltage V_R is derived from another BJT circuit included on the chip but omitted in Fig. 14.4–10a.

The input transistors T_{1A} and T_{1B} work with T_2 in a *current-switching mode,* as follows. When both inputs are low, the input transistors are cut off but V_R biases T_2 in the active state. Thus, $i_1 = i_3 \approx 0$ and $v_{C1} \approx 0$, while $v_E = -V_R - V_\gamma \approx -2$ V, $i_2 \approx i_E = v_E/R_E \approx 4$ mA, and $v_{C2} \approx -1$ V. The levels at the output emitters are then

$$v_{oL} = v_{C2} - V_\gamma \approx -1.7 \text{ V} \qquad \bar{v}_{oH} = v_{C1} - V_\gamma \approx -0.7 \text{ V}$$

When either input goes high, the corresponding input transistor becomes active and v_E rises slightly to cut off T_2, so $i_2 = i_4 \approx 0$ whereas $i_1 \approx i_E$. If the high input is -0.7 V, then the resulting outputs are $v_{oH} \approx -0.7$ V and $\bar{v}_{oL} \approx -1.7$ V.

CMOS logic circuits differ radically from BJT circuits. The basic CMOS building block is the inverter given in Fig. 14.4–11 along with its voltage transfer curve. The n-channel and p-channel MOSFETs have the same threshold voltage $V_T < V_{DD}/2$, and both gate terminals are connected to the input. A low input at $v_{iL} < V_T$ turns T_n off and T_p on, since $v_{GS} = v_{iL} < V_T$ and $v_{SG} = V_{DD} - v_{iL} > V_T$. Conversely, a high input at $v_{iH} > V_{DD} - V_T$ turns T_n on and T_p off. Hence, this circuit operates like the two-switch gate model back in Fig. 14.4–3a, except for the transition region where $V_T \le v_i \le V_{DD} - V_T$. The supply voltage V_{DD} can be any value above $2V_T$ and below the MOSFET breakdown point.

Figure 14.4–11
CMOS inverter: (a) circuit; (b) transfer curve.

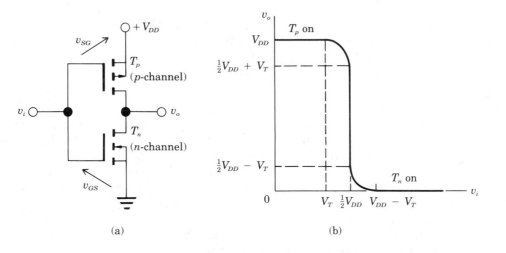

(a)   (b)

Tristate output is implemented in CMOS using transmission gates of the type described in Section 10.2. Other CMOS gates are made by augmenting the basic inverter with additional MOSFETs in series and parallel. For instance, Fig. 14.4–12 shows a CMOS NAND gate whose output goes low only when both inputs go high — so T_{nA} and T_{nB} are on while T_{pA} and T_{pB} are off. There are no power-dissipating resistors here, and no current paths through the MOSFETs from V_{DD} to ground. Hence, the power required by this circuit is less than 1 μW under static conditions. Furthermore, since the MOSFETs draw negligible input current, the static fan-out can be very large.

But power dissipation goes up dramatically under dynamic conditions because the MOSFETs must then carry current to charge or discharge

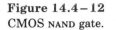

Figure 14.4 – 12
CMOS NAND gate.

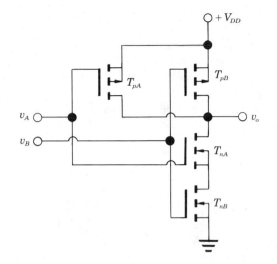

stray load capacitance. **Buffered CMOS,** designated B-series, provides more output current and achieves faster switching by incorporating additional output inverters made with larger, higher-current MOSFETs. Such circuits usually include buffering input inverters. Thus, a B-series CMOS NOR gate actually consists of a NAND gate with input and output buffers, as symbolized by Fig. 14.4 – 13. Buffering also permits wired logic in CMOS since the *p*-channel MOSFET can be omitted from the output inverter, leaving an **open-drain output** for connection to an external pull-up resistor.

Figure 14.4 – 13
Logic diagram of buffered CMOS NOR gate.

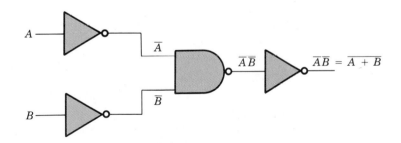

The HC-series CMOS is fabricated with polysilicon in place of metal for the MOSFET gate structure. This modification, together with buffering and smaller feature size, increases switching speed and reduces the delay-power product. Many HC-series units implement the same functions and have the same pin configurations as TTL units, thereby facilitating the conversion of TTL designs to CMOS.

PROBLEMS

14.1–1 A logic network in an airplane actuates the autopilot ($F = 1$) when the radio signal is on ($S = 1$) and the trouble indicator is off ($T = 0$), and the command is given by the pilot ($P = 1$) or by the copilot ($C = 1$). Write the logic expression for F and draw the gate diagram.

14.1–2 A logic network in a ventilation system actuates the main blower ($F = 1$) when the thermostat calls for heating ($H = 1$) or cooling ($C = 1$), or when vapors are detected ($V = 1$) and the auxiliary blower is off ($B = 0$). Write the logic expression for F and draw the gate diagram.

14.1–3 A logic network in a traffic signal actuates the trouble indicator ($F = 1$) when the green lights are on in both directions ($G_1 = G_2 = 1$) or when the red lights are off in both directions ($R_1 = R_2 = 0$). Write the logic expression for F and draw the gate diagram.

14.1–4 A logic network in an automobile actuates a buzzer when the ignition key is on ($K = 1$) and the driver's seat belt is not buckled ($D = 0$) or the passenger's seat belt is not buckled ($P = 0$) and a pressure switch indicates that that seat is occupied ($S = 1$). Write the logic expression for F and draw the gate diagram.

14.1–5 Convert 165.3_{10} to a binary word with six bits after the binary point. Then express the truncated result in octal and hexadecimal notation, and use base-8 expansion to find the decimal value of the truncation error.

14.1–6 Do Problem 14.1–5 for 240.9_{10}.

14.1–7 Convert 43.04_{10} to a binary word with eight bits after the binary point. Then express the truncated result in octal and hexadecimal notation, and use base-16 expansion to find the decimal value of the truncation error.

14.1–8 Do Problem 14.1–7 for 30.85_{10}.

14.1–9 Direct and reverse double-dabbel are easily modified to convert decimal numbers to any other base N. You simply divide or multiply by N, rather than 2, so the remainders or carries will be integers ranging from 0 to $N - 1$. Using this method, convert 48.1_{10} to a base-3 number with eight digits.

14.1–10 Using the method in Problem 14.1–9, convert 404.18_{10} to a base-5 number with eight digits.

14.1–11 Using the method in Problem 14.1–9, convert 420.9_{10} to a base-12 number with six digits. Let $10_{10} = A_{12}$ and $11_{10} = B_{12}$.

14.1–12 Write an expression for the smallest nonzero decimal quantity that can be represented by a binary number with m bits after the binary point. Then find the minimum value of m such that the decimal-to-binary truncation error does not exceed 0.0005_{10}.

14.1–13 A certain computer stores floating-point numbers in a form equivalent to $x_{10} = y \times 16^{(z-64)}$, where y is represented by a 24-bit binary fraction and z is represented by a 7-bit binary integer. Estimate the largest possible value of x_{10}.

14.1–14 Estimate the smallest nonzero value of x_{10} for the floating-point number in Problem 14.1–13.

14.1–15 Consider a 4-bit BCD code with weighting factors 4, 2, 2, 1. Construct the code table to show that some decimal digits can be represented by more than one code word.

14.1–16 Do Problem 14.1–15 with weighting factors 3, 3, 2, 1.

14.1–17 Construct the code table for a 4-bit BCD code with weighting factors 8, 4, −2, −1.

14.2–1 Use algebraic manipulation and theorems to simplify the function $F = (A + \bar{B})(A + C\bar{D})$. Then apply DeMorgan's rule to your result to find $\bar{F}$.

14.2–2 Do Problem 14.2–1 for $F = AB\bar{C} + BD(A\bar{C} + \bar{A}D) + \bar{A}\bar{B}D$.

14.2–3 Use algebraic manipulation and theorems to simplify the function $F = (A + B\bar{C})(A\bar{B} + C)\bar{D}$. Factor any common terms in your result and apply DeMorgan's rule to find $\bar{F}$.

14.2–4 Do Problem 14.2–3 for $F = \bar{A}B(C + D) + B(\bar{A}\bar{C} + \bar{B}D + C\bar{D})$.

14.2–5 A logic network has inputs A, B, and C. It produces $F = \bar{C}$ when $A = 0$, and $F = \bar{B}$ when $A = 1$. Construct the truth table for F, identify the minterms and maxterms, and find the minimized SOP and POS expressions.

14.2–6 Do Problem 14.2–5 for a network that produces $F = B + C$ when $A = 0$, and $F = \overline{BC}$ when $A = 1$.

14.2–7 Do Problem 14.2–5 for a network that produces $F = B + C$ when $A = 0$, and $F = BC$ when $A = 1$.

14.2–8 Farmer A has a bag of grain B, a chicken C, and a dog D. He does chores in a barn with two rooms, labeled 0 and 1, and he does not want to move B, C, and D each time he goes between rooms. However, C will eat B if left alone together, and D will eat C if left alone together. The farmer needs a logic network that produces $F = 1$ when an intended move would result in something being eaten. The inputs to the network are four switches such that $A = 1$ means that the farmer is in room 1, etc. Construct the truth table and obtain the minimized SOP function.

14.2–9 A logic network in an industrial process controller has inputs from four binary sensors, A, B, C, and D. The output must stop the process ($F = 0$) when $A = 0$ and $C = D = 1$, regardless of B, or when $A = 1$, $B + C = 1$, and $D = 0$. Construct the truth table and obtain the minimized POS function.

14.2–10 A majority logic network with inputs A, B, C, and D produces $F = 1$ whenever three or more inputs equal 1. Construct the truth table and obtain the minimized SOP function.

14.2–11 Show that $\overline{A(\overline{BC})} \neq \overline{(\overline{AB})C}$, so the NAND operation is not associative.

14.2–12 Show that $\overline{A + (\overline{B + C}} \neq \overline{(\overline{A + B})} + C$, so the NOR operation is not associative.

14.2–13 Verify algebraically that the output in Fig. 14.2–2c remains the same when all gates are replaced by NAND gates.

14.2–14 Verify algebraically that the output in Fig. 14.2–3c remains the same when all gates are replaced by NOR gates.

14.2–15 Suppose you only have available two NAND gates and two NOR gates, each with two inputs. Devise a network to implement the function $F = \overline{A}\overline{\overline{B}}CD$.

14.2–16 Do Problem 14.2–15 for $F = \overline{A + B + \overline{C} + \overline{D}}$.

14.2–17 Do Problem 14.2–15 for $F = \overline{\overline{A} + \overline{B} + \overline{CD}}$.

14.2–18 Do Problem 14.2–15 for $F = \overline{(A + \overline{B})\overline{C}\overline{D}}$.

14.2–19 A logic network with three inputs produces $F = 1$ whenever there is an odd number of 1's at the input. Find the minterms and maxterms, and use a Karnaugh map to show that the SSOP and SPOS functions cannot be simplified.

14.2–20 Using a Karnaugh map, obtain the minimized SOP function for a logic network with inputs A, B, and C that produces $F = B$ when $A = 0$ and $F = B + C$ when $A = 1$.

14.2–21 A logic network with inputs A, B, and C produces $F = 1$ when two and only two of the inputs equal 1. Use a Karnaugh map to obtain the minimized POS function.

14.2–22 A logic network tests the input word $[ABCD]$ and produces $F = 1$ if the word does not belong to the 5311 code listed in Table 14.1–2. Use a Karnaugh map to obtain the minimized SOP function.

14.2–23 Partner A owns 40% of a company, while partners B, C, and D each own 20%. They need a logic network with four input switches for YES/NO votes that produces $F = 1$ when a majority of the ownership votes YES. Let $A = 1$ stand for a YES vote by partner A, etc. Obtain the minimized SOP function using a Karnaugh map.

14.2–24 Using a Karnaugh map, obtain the minimized SOP function for the network in Problem 14.2–9.

14.2–25 A logic network with inputs A, B, C, and D produces $F = 1$ when $A = B$ and $C = D$ at the same time. Use a Karnaugh map to obtain the minimized POS function.

14.2–26 Using a Karnaugh map, obtain the minimized POS function for the network in Problem 14.2–23.

14.3–1 Show from Eq. (2) that $X \oplus 0 = X$, $X \oplus 1 = \overline{X}$, $X \oplus X = 0$, and $X \oplus \overline{X} = 1$.

14.3–2 Using Eq. (2), show that $\overline{X \oplus Y} = X \oplus \overline{Y} = \overline{X} \oplus Y$.

14.3–3 Use Eq. (2) and the property in Problem 14.3–2 to expand $(X \oplus Y) \oplus Z$ and $X \oplus (Y \oplus Z)$ in SOP form. Compare your results to verify that the XOR operation is associative.

14.3–4 Diagram a NAND-gate network that produces $F = A \oplus B$. What additional outputs would be available?

14.3–5 An equality detector has two 3-bit input words, $[A_1 B_1 C_1]$ and $[A_2 B_2 C_2]$, and produces $F = 1$ when $A_1 = A_2$, $B_1 = B_2$, and $C_1 = C_2$. Devise a network consisting of XOR and NOR gates for this task.

14.3–6 A **Gray code** has the property that successsive code words differ by exactly one bit. The table in Fig. P14.3–6 compares the 3-bit Gray code words $[A'B'C']$ with the standard binary code words $[ABC]$. Note that $A' = A$, whereas B' and C' can be viewed as functions of A, B, and C. Use a truth table to obtain minimized SOP expressions for B' and C'. Then devise a network of XOR gates that has inputs A, B, and C and produces the outputs A', B', and C'.

14.3–7 Develop the programming table for $X_0 – X_3$ so a 4-input MUX with $[S_1 S_0] = [AB]$ generates the function $F = B + \overline{C}$ when $A = 0$ and $F = BC$ when $A = 1$.

14.3–8 Do Problem 14.3–7 for $F = A(B \oplus C)$.

Figure P14.3–6

Binary			Gray		
A	B	C	A'	B'	C'
0	0	0	0	0	0
0	0	1	0	0	1
0	1	0	0	1	1
0	1	1	0	1	0
1	0	0	1	1	0
1	0	1	1	1	1
1	1	0	1	0	1
1	1	1	1	0	0

14.3–9 Do Problem 14.3–7 for $F = (A + \overline{C})(\overline{B} + C)$.

14.3–10 Develop the programming table for X_0–X_7 so an 8-input MUX with $[S_2 S_1 S_0] = [ABC]$ generates the function $F = \overline{A}CD + \overline{A}BC + A\overline{CD}$.

14.3–11 Do Problem 14.3–10 for $F = (\overline{A} \oplus B) + (C \oplus D)$.

14.3–12 Do Problem 14.3–10 for $F = (A + \overline{B} + \overline{D})(B + \overline{C} + D)(\overline{A} + \overline{B} + C)$.

14.3–13 An 8×3 PROM with input decoder is to be programmed to convert binary code words to the Gray code described in Problem 14.3–6. Using the link notation W_i–F_j, list the links that must be blown to get $F_1 = B'$ and $F_0 = C'$ when $[S_2 S_1 S_0] = [ABC]$.

14.3–14 An 8×3 PROM with input decoder is to be programmed to convert the Gray code described in Problem 14.3–6 to standard binary code words. Using the link notation W_i–F_j, list the links that must be blown to get $F_1 = B$ and $F_0 = C$ when $[S_2 S_1 S_0] = [A'B'C']$.

14.3–15 An Excess-3 BCD code word $[ABCD]$ is to be converted to an 8421 code word $[A'B'C'D']$. Table 14.1–2 shows that $D' = \overline{D}$, so a NOT gate will convert the LSB. The other bits can be converted by a 16×4 PROM. Using the link notation W_i–F_j, list the links that must be blown to get $F_2 = A'$, $F_1 = B'$, and $F_0 = C'$ when $[S_3 S_2 S_1 S_0] = [ABCD]$. Also list the word lines not activated.

14.3–16 A 5311 BCD code word $[ABCD]$ is to be converted to an Excess-3 code word $[A'B'C'D']$. Table 14.1–2 shows that $A' = A$, so the MSB requires no conversion. The other bits can be converted by a 16×4 PROM. Using the link notation W_i–F_j, list the links that must be blown to get $F_2 = B'$, $F_1 = C'$, and $F_0 = D'$ when $[S_3 S_2 S_1 S_0] = [ABCD]$. Also list the word lines not activated.

Chapter
15 Digital Systems

Culminating our study of information processing, we come at last to systems that exploit the capabilities and advantages of digital electronics. We'll consider systems that are entirely digital, and hybrid systems that include both analog and digital sections. Hybrid instrumentation systems have become increasingly important as advances in digital technology — especially the microprocessor — make it possible to replace analog hardware with digital ICs.

The first section of this chapter introduces clocked flip-flops that perform the sequential logic needed for digital counters and registers. The second section discusses digital instrumentation systems, emphasizing the tasks of analog-to-digital and digital-to-analog conversion. The remaining two sections concern digital computation, starting with arithmetic networks and going on to microprocessors and computers.

Objectives

After studying this chapter and working the exercises, you should be able to do each of the following:

- Write sequential equations for clocked SR, D, T, and JK flip-flops (Section 15.1).

- Analyze a counter or shift register by sketching the timing diagram (Section 15.1).

- Write an expression relating the analog output to the digital input of a DAC (Section 15.2).

- Calculate the resolution of an ADC and identify the differences between parallel-comparator, dual-slope, and successive-approximation converters (Section 15.2).

567

- Diagram a half adder, full adder, and parallel binary adder (Section 15.3).

- Carry out binary addition and subtraction using 2's-complement arithmetic (Section 15.3).

- Show how a binary adder is modified to perform subtraction or multiplication (Section 15.3).

- Draw the diagram of a bus-organized computer, explain the significance of CPU word length, and describe the operating principles of a microprocessor (Section 15.4). †

15.1 FLIP-FLOPS, COUNTERS, AND REGISTERS

Digital systems often perform cumulative operations on information bits strung out over time. Such tasks involve **sequential logic** whose output values depend on *past* input values, whereas the outputs of a combinational network would depend entirely on current input values. Sequential operation thus requires *memory* in the form of storage units that temporarily hold digital information for subsequent processing. Furthermore, in contrast to the permanence of a read-only memory, the storage units for sequential logic must have *read/write* capability so the contents can be updated at any time by appropriate input signals.

This section starts with the basic 1-bit read/write memory cell called a latch. Then we'll construct various types of flip-flops by combining latches with gates to incorporate a clock signal for timing purposes. Finally, we'll show how clocked flip-flops are interconnected in counters, registers, and related sequential-logic modules.

Latches and Flip-Flops

Figure 15.1–1a represents a **SET-RESET *(SR)* latch.** This bistable memory cell holds one logic variable Q, but both Q and its complement $\overline{Q}$ may be

Figure 15.1–1
SR latch: (a) symbol; (b) truth table; (c) transition table.

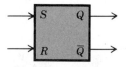

S_n	R_n	Q_{n+1}	$\overline{Q}_{n+1}$	
0	0	Q_n	$\overline{Q}_n$	(No change)
0	1	0	1	(RESET)
1	0	1	0	(SET)
1	1	—	—	(Not allowed)

S_n	R_n	Q_n	Q_{n+1}	
0	0	0	0	(stays RESET)
0	0	1	1	(stays SET)
0	1	0	0	(stays RESET)
0	1	1	0	(RESET)
1	0	0	1	(SET)
1	0	1	1	(stays SET)
1	1	0	—	$\left(\begin{array}{c}\text{Not}\\ \text{allowed}\end{array}\right)$
1	1	1	—	

(a)   (b)   (c)

available as outputs. The value of Q is established by applying input *pulses*, as follows:

- A pulse at the S input ($S = 1$) produces the **SET state** in which $Q = 1$ and $\overline{Q} = 0$.

- A pulse at the R input ($R = 1$) produces the **RESET state** in which $Q = 0$ and $\overline{Q} = 1$.

Obviously, the latch cannot be in both states at the same time, so simultaneous set and reset pulses are not allowed.

The bistable property of a latch means that it stays in either state after an input pulse ends. Having been put in the SET state, for instance, a latch remains set indefinitely until a reset pulse comes along — regardless of any intervening pulses at the S input.

We describe the behavior of an SR latch by writing Q_n to denote its *present state*, just before an input goes to some new value S_n or R_n. The resulting *next state* Q_{n+1} may or may not differ from Q_n, according to the *sequential equation*

$$Q_{n+1} = S_n + \overline{R}_n \cdot Q_n \tag{1a}$$

This equation is subject to the restriction

$$[S_n R_n] \neq 11 \tag{1b}$$

since $[S_n R_n] = 11$ represents the forbidden case of simultaneous input pulses.

The truth table in Fig. 15.1–1b follows directly from Eq. (1). Note on the first line that Q_{n+1} equals Q_n when $[S_n R_n] = 00$, indicating that no change of state occurs during the interval between input pulses. This memory effect is brought out more clearly by the **transition table** in Fig. 15.1–1c. Here we treat Q_n like another input whose value determines Q_{n+1} if both S_n and R_n go to logic 0.

Figure 15.1–2 shows two ways of building an SR latch using *cross-coupled* NAND gates or NOR gates. The outputs of the lower gates have been labeled P, rather than $\overline{Q}$, for purposes of analysis. To confirm that Fig. 15.1–2a satisfies the set/reset conditions, suppose that both inputs remain constant. We can then drop the time-sequence subscripts and write the static logic equations

$$Q = \overline{\overline{S} \cdot P} = S + \overline{P} \qquad P = \overline{\overline{R} \cdot Q} = R + \overline{Q} \tag{2}$$

Hence, the input combination $[SR] = 10$ produces the SET state with $Q = 1 + \overline{P} = 1$ and $P = 0 + \overline{Q} = \overline{Q} = 0$, while $[SR] = 01$ produces the RESET state with $P = 1 + \overline{Q} = 1$ and $Q = 0 + \overline{P} = \overline{P} = 1$. We also see from Eq. (2) that the cross-coupled feedback sustains either state in absence of input pulses, since $Q = 0 + \overline{P} = \overline{P}$ and $P = 0 + \overline{Q} = \overline{Q}$ when $[SR] = 00$. The analysis of Fig. 15.1–2b goes along similar lines.

Figure 15.1–2

Implementations of an SR latch.

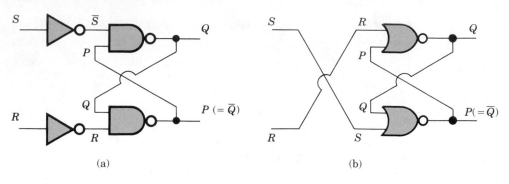

(a)   (b)

A **clocked SR flip-flop** differs from a latch in that output transitions are controlled by a timing waveform called the **clock signal.** Consider, for example, the gate configuration diagrammed in Fig. 15.1–3a, where Ck stands for the clock input. Setting $Ck = 0$ prevents set/reset pulses from reaching the cross-coupled latch gates, thereby disabling the S and R inputs. The S and R inputs are enabled when $Ck = 1$. Usually, the clock signal is a periodic pulse train like Fig. 15.1–3b, so set/reset transitions can occur at the output only during clock pulses.

Figure 15.1–3

(a) Clocked SR flip-flop. (b) Clock waveform.

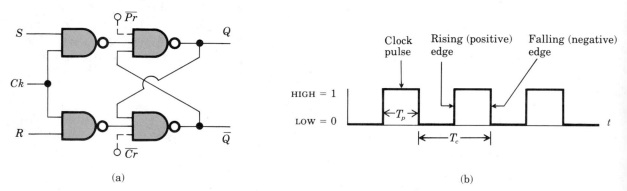

(a)   (b)

Equation (1) and our previous latch tables still hold for the clocked SR flip-flop. Now, however, S_n, R_n, and Q_n refer to values during the nth clock pulse, and the resulting next state Q_{n+1} lasts until the next clock pulse. Any S or R input pulses applied between clock pulses have no effect at the output.

A clocked flip-flop may also be equipped with separate inputs for **direct preset** (Pr) and/or **direct clear** (Cr), indicated by the dashed lines in Fig. 15.1–3a. These direct inputs make it possible to establish an appropriate initial state while $Ck = 0$. The negation overbars reflect the fact that $\overline{Pr} = 0$ has the same effect as $S = 1$ and $\overline{Cr} = 0$ has the same effect as $R = 1$. Setting $\overline{Pr} = \overline{Cr} = 1$ deactivates the direct inputs.

Many sequential systems operate in a **synchronous mode,** with all flip-flops controlled by a master clock. Synchronous operation facilitates the overall timing and coordination needed to accommodate short and long processing tasks. It also reduces the hazard posed by stray noise pulses and "glitches" that occasionally appear in digital systems. Unwanted input pulses have potentially long-lasting and sometimes devastating consequences if they cause erroneous flip-flop transitions. But synchronous operation eliminates erroneous transitions between clock pulses, when the data signals are supposed to be inactive.

Precise synchronization can be achieved using a clock whose pulse duration T_p is much less than the clock cycle T_c. Alternatively, and more commonly, a square-wave clock signal having $T_p = T_c/2$ is used in conjunction with **edge-triggered flip-flops.** To explain this technique, suppose we modify the flip-flop back in Fig. 15.1–3a by connecting the clock through a trigger device. If the trigger generates very short pulses only when the clock goes from high to low, then the S and R inputs will be disabled except for brief instants at the **falling** or **negative edge** of each clock pulse. All output transitions thereby become synchronized with negative clock edges. Likewise, if the trigger generates enabling pulses only when the clock goes from low to high, then all output transitions would be synchronized with **rising** or **positive edges.**

Henceforth, we'll assume that clocked flip-flops trigger on negative edges unless otherwise stated. We'll represent an edge-triggered SR flip-flop by Fig. 15.1–4a, where the negation bubble and triangle at the clock input denote negative-edge triggering. (The negation bubble would be

Figure 15.1–4

Negative-edge-triggered SR flip-flop: (a) symbol; (b) waveforms.

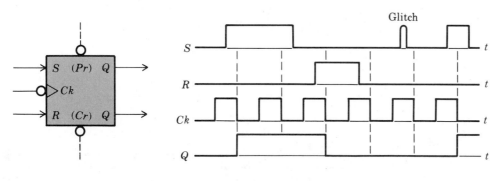

(a) (b)

omitted in the case of positive-edge triggering.) The bubbles at the direct inputs indicate that they act like $\overline{S}$ and $\overline{R}$, if the flip-flop is so equipped. Direct inputs are said to be **asynchronous** because they can produce output transitions at any time, irrespective of the clock.

Figure 15.1–4b depicts illustrative waveforms for an SR flip-flop with negative-edge triggering. You should study these waveforms thoroughly, particularly with respect to the following properties:

- An output transition initiated by a SET or RESET pulse always lines up with the falling edge of a clock pulse.

- No subsequent output change occurs when a SET or RESET pulse lasts longer than one clock cycle, or when successive pulses appear at the same input.

- An input pulse or glitch has no effect if it does not overlap a triggering instant.

Similar output signals are obtained from flip-flops built in the **master-slave configuration,** a configuration that samples the inputs while the clock is high and produces output transitions when the clock goes low (see Problem 15.1–1).

A flip-flop property omitted from Fig. 15.1–4b is the **propagation delay** resulting from internal energy storage. As a consequence of this delay, output transitions actually lag behind the triggering instants by a small time interval t_{PD}. However, typical values of t_{PD} are much less than a microsecond, so the flip-flop delay is usually small compared to the clock cycle — except for operation at very high clock frequencies.

Now suppose we drive the R input of a clocked SR flip-flop from an inverter connected to the S input, per the diagram in Fig. 15.1–5a. Such an

Figure 15.1–5
D-type flip-flop:
(a) implementation
with SR flip-flop;
(b) symbol;
(c) waveforms.

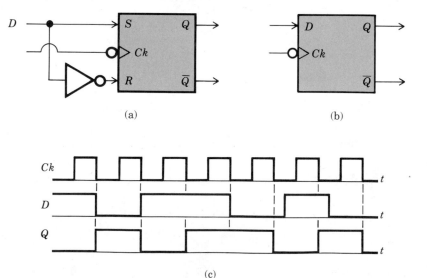

(a)   (b)

(c)

arrangement eliminates the forbidden input combination $[S_n R_n] = 11$, but it also eliminates $[S_n R_n] = 00$ and makes Q_{n+1} completely independent of Q_n. We call this device a **D-type flip-flop,** and we represent it by Fig. 15.1–5b. Setting $S_n = \overline{R}_n = D_n$ in Eq. (1a) then yields the simple sequential equation

$$Q_{n+1} = D_n \tag{3}$$

Thus, the output Q always takes on the previous value of the input D. The waveforms in Fig. 15.1–5c also reveal that a synchronized input transition appears at the output exactly one clock cycle later.

Another special clocked device is the **T-type flip-flop** symbolized by Fig. 15.1–6a. Sequential operation with input T yields

$$Q_{n+1} = T_n \overline{Q}_n + \overline{T}_n Q_n = \begin{cases} Q_n & T_n = 0 \\ \overline{Q}_n & T_n = 1 \end{cases} \tag{4}$$

The distinctive feature of this flip-flop is its **toggle mode,** obtained when we keep $T = 1$. The outputs Q and $\overline{Q}$ then switch alternately between high and low at the end of each clock pulse, as plotted in Fig. 15.1–6b. Observe that the Q and $\overline{Q}$ waveforms look like another clock signal with the pulse rate divided by two. Also notice that a T flip-flop in the toggle mode could be used as a single-input latch, since successive pulses applied to the Ck terminal will alternately set and reset the flip-flop.

Figure 15.1–6

T-type flip-flop: (a) symbol; (b) toggle-mode waveforms with $T = 1$.

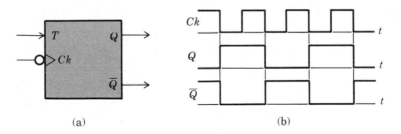

(a)   (b)

Lastly, we come to the versatile **JK flip-flop** represented by Fig. 15.1–7a. The equation for sequential operation with inputs J and K is

$$Q_{n+1} = J_n \overline{Q}_n + \overline{K}_n Q_n \tag{5}$$

You can confirm from Eq. (5) that $Q_{n+1} = Q_n$ when $[J_n K_n] = 00$, $Q_{n+1} = 0$ when $[J_n K_n] = 01$, and $Q_{n+1} = 1$ when $[J_n K_n] = 10$—identical to an SR

Figure 15.1–7

JK flip-flop: (a) symbol; (b) truth table.

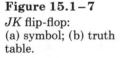

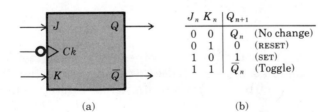

J_n	K_n	Q_{n+1}	
0	0	Q_n	(No change)
0	1	0	(RESET)
1	0	1	(SET)
1	1	$\overline{Q}_n$	(Toggle)

(a)   (b)

flip-flop with $S = J$ and $R = K$. But unlike an SR flip-flop, the input combination $[J_n K_n] = 11$ is allowed and it produces the *toggled* output $Q_{n+1} = \overline{Q}_n$. Thus, a JK flip-flop becomes a T flip-flop when $J = K = T$. The truth table in Fig. 15.1–7b summarizes the properties of a JK flip-flop.

Commercially available IC flip-flops include JK and D types in various packaging arrangements, with or without direct inputs. Edge-triggered units may have negative or positive triggering, but the output transitions of master-slave units usually occur at negative clock edges. Clocked SR flip-flops and T flip-flops are not offered as separate products because a JK flip-flop readily substitutes for either of them.

Example 15.1–1   Suppose that a digital system is needed to generate the output signal Y plotted in Fig. 15.1–8a, given a square-wave clock and unsynchronized *START* and *STOP* pulses at the input. The desired output is a pulse train with pulse duration $T_c/2$ and period $2T_c$. The first output pulse begins at the negative clock edge just after the *START* pulse arrives, and a STOP pulse immediately produces $Y = 0$ which then remains unchanged until the next *START* pulse.

Figure 15.1–8
Waveforms and system diagram for Example 15.1–1.

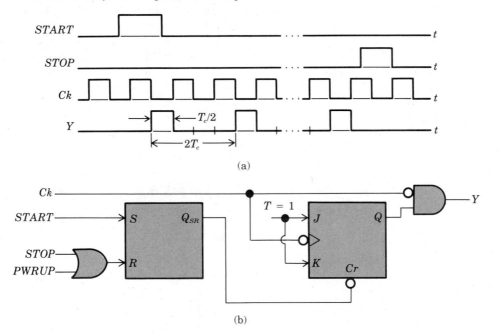

(a)

(b)

Since the Y pulses have period $2T_c$, a toggle-mode flip-flop seems ideal for this application. Indeed, a careful examination of the waveforms in Fig. 15.1–6b reveals that we can generate Y by forming the logical-AND product

$Y = \overline{Ck} \cdot Q$. (A quick sketch of Ck, Q, $\overline{Ck}$, and $Ck \cdot Q$ should help you see this.) The corresponding implementation is diagrammed in Fig. 15.1–8b using a JK flip-flop with $J = K = T = 1$.

The diagram also includes an SR latch connected to the direct clear for the $START/STOP$ function. A $START$ pulse puts the latch in its set state, so $\overline{Cr} = Q_{SR} = 1$ and the flip-flop begins to toggle. Toggling ends when a $STOP$ pulse resets the latch and we have $\overline{Cr} = Q_{SR} = 0$, which holds the flip-flop in its reset state ($Q = 0$) until the next $START$ pulse.

But we must ensure that the latch and flip-flop initially have $Q_{SR} = Q = 0$ before the first $START$ pulse. Initialization can be accomplished by generating an additional pulse $PWRUP = 1$ when the system is energized. Applying the $PWRUP$ pulse to the R input via an OR gate then resets both the latch and the flip-flop. Similar provisions are needed in most sequential systems because we cannot otherwise guarantee correct initial states for the latches and flip-flops.

Exercise 15.1–1   Write static logic equations for the latch in Fig. 15.1–2b. Then confirm the RESET, SET, and memory conditions by letting $[SR] = 01$, 10, and 00.

Exercise 15.1–2   Redraw the Q waveform in Fig. 15.1–4b for the case of: (a) positive-edge triggering; (b) a simple clocked flip-flop like Fig. 15.1–3a.

Counters

A counter is a sequential system that has *N possible states,* each state being associated with a number. Counting takes place by causing the system to step through its states in numerical order in response to successive input pulses. After N pulses, the counter reverts to its initial state and the counting cycle may repeat. We refer to N as the counter's **modulus.**

To illustrate these concepts, let Fig. 15.1–9a represent a sequential system containing four flip-flops. Each flip-flop output Q has two possible values, and the four outputs taken together form a 4-bit binary word $[Q_3 Q_2 Q_1 Q_0]$ corresponding to the state of the circuit. (Don't confuse the positional notation Q_0, Q_1, . . . with our sequential notation Q_n and Q_{n+1}.) The internal connections, yet to be described, are arranged to achieve the following action: A pulse applied at the RESET terminal resets all flip-flops and establishes the initial state $[Q_3 Q_2 Q_1 Q_0] = 0000$. The first input pulse produces $[Q_3 Q_2 Q_1 Q_0] = 0001$, and each subsequent pulse increases the count in standard binary order, as tabulated in Fig. 15.1–9b. The initial state then reappears after the 16th input pulse, so we have a counter with modulus $N = 16 = 2^4$.

Figure 15.1–9c shows the **timing diagram** for the 4-bit binary counter. A periodic train of input pulses has been assumed here, but the timing relations would be the same in the case of nonperiodic input pulses. We have also assumed that all flip-flop transitions occur at falling edges of

Figure 15.1-9

(a) Four-bit binary counter. (b) Counting sequence. (c) Timing diagram.

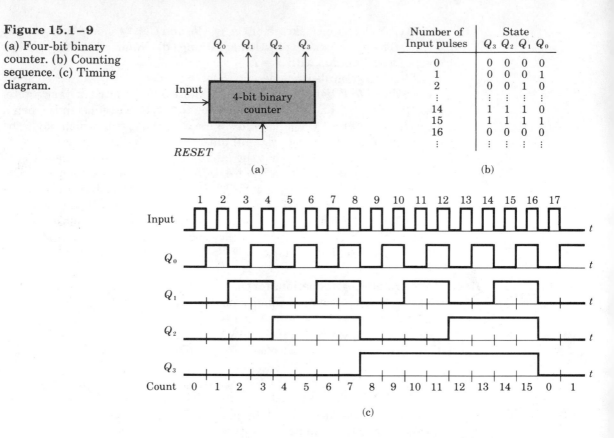

(a)

(b)

(c)

the input pulses, consistent with negative-edge triggering. The state $[Q_3Q_2Q_1Q_0] = 0001$ thus appears immediately after the end of the first input pulse and it lasts until the end of the second input pulse.

The waveforms in our timing diagram bring out the fact that a counter fundamentally acts as a *divider*, since $N = 16$ input pulses produce just one complete pulse at the Q_3 output. Hence, when the number of input pulses arriving in a certain time interval is $N_{in} = 16 \times m$, the resulting number of output pulses at Q_3 will be $N_{out} = m = N_{in}/16$.

If desired, we can detect any particular counting state using a simple **decoder.** For instance, an AND gate with inputs $\overline{Q}_3$, $\overline{Q}_2$, Q_1, and Q_0 would produce $\overline{Q}_3 \cdot \overline{Q}_2 \cdot Q_1 \cdot Q_0 = 1$ when $[Q_3Q_2Q_1Q_0] = 0011 = 3_{10}$, which occurs between the 3rd and 4th input pulse. Decoding gates are sometimes used to build **digital waveform generators** of the type illustrated in Fig. 15.1-10, where negation bubbles represent inverters at the AND-gate inputs. The AND-OR network here detects the 3rd count and the 8th through 11th count via the minimized sum-of-products expression

$$Y = \overline{Q}_3\overline{Q}_2Q_1Q_0 + Q_3\overline{Q}_2$$

When counting continues beyond $N = 16$ clock pulses, the counter resets and the output waveform Y repeats itself periodically.

Figure 15.1–10
Digital waveform generator.

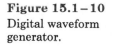

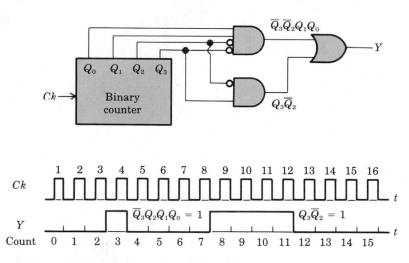

Decoding gates are also used to build counters with other values of N. In particular, suppose we need a **decade counter** which has $N = 10$. A 4-bit binary counter is easily adapted to that task by forcing it back to the initial state after 9 counts. Accordingly, we must decode the 10th state $[Q_3 Q_2 Q_1 Q_0] = 1010$ to generate a pulse that resets the counter. Figure 15.1–11 shows the modified counter and the relevant portion of the timing diagram. Note that the 10th state appears momentarily after the 9th count, until the flip-flops become reset by the *CARRY* pulse from the decoding gate.

Figure 15.1–11a also suggests how you connect two or more decade units in cascade to obtain a multiple-decade counter. Each decade unit counts *CARRY* pulses from the preceding decade, so the 2nd decade counts

Figure 15.1–11
(a) Decade counter.
(b) Timing diagram and *CARRY* pulse.

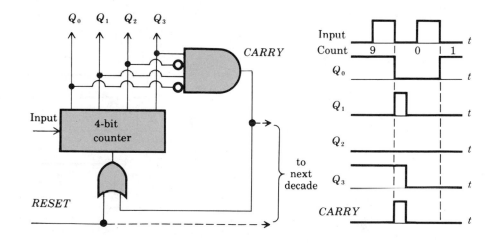

(a)   (b)

every 10th input pulse, the 3rd decade counts every 100th input pulse, etc. Thus, a cascade of M decades has modulus $N = 10 \times 10 \times \cdots = 10^M$. The M 4-bit output words then constitute the *BCD code* for the value of the count at any particular time.

Let's generalize our results so far by considering a counter composed of n flip-flops. If sequential input pulses advance the n-bit output word in standard binary order, then the counter could have any modulus satisfying the condition

$$N \leq 2^n \tag{6}$$

Values of N less than 2^n require auxiliary decoding gates to reset the counter from the Nth state. Similarly, a cascade of M n-bit counters has

$$N = N_1 \times N_2 \times \cdots \times N_M \leq (2^n)^M \tag{7}$$

For instance, to get $N = 3600$ using 4-bit counters, we must cascade at least $M = 3$ units since $(2^4)^3 = 4096 > 3600$ but $(2^4)^2 = 256 < 3600$.

As for the internal circuitry of an n-bit counter, Fig. 15.1–12a diagrams a simple implementation using a chain of T (or JK) flip-flops with $T = 1$. The pulses being counted are applied to the clock terminal of FF_0 whose output Q_0 toggles after each input pulse and completes a full HIGH/

Figure 15.1–12
(a) Ripple counter.
(b) Waveforms with propagation delay.

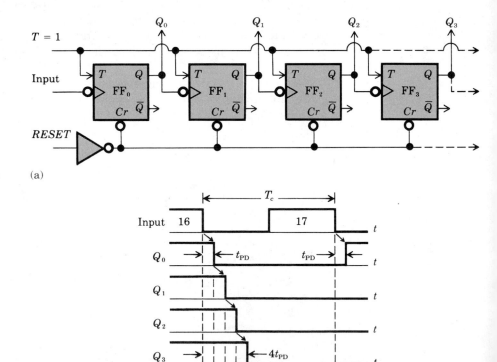

LOW cycle for every *two* input pulses (see Fig. 15.1–6b). Since Q_0 goes to the input of FF_1, Q_1 completes a full cycle for every *four* input pulses, and so on along the chain. Therefore, a chain of $n = 4$ flip-flops can, indeed, produce the counting sequence and timing diagram in Fig. 15.1–9. The extension to other values of n is straightforward.

The system in Fig. 15.1–12a is called a **ripple counter** because the 16th (Nth) pulse produces a downward transition at Q_0, which triggers Q_1, which triggers Q_2, which triggers Q_3 — like falling dominos rippling from left to right. But the initial state does not reappear instantaneously, for each flip-flop introduces delay t_{PD} whenever it changes state. Figure 15.1–12b gives a time expansion of the events between pulses 16 and 17 including propagation delay, and we see that the initial state has been delayed by $4t_{PD}$ after the trailing edge of the 16th pulse. Also note that these waveforms pass through the state $[Q_3 Q_2 Q_1 Q_0] = 1100$, so a decoding circuit with $Y = Q_3 \cdot Q_2 \cdot \overline{Q}_1 \cdot \overline{Q}_0$ might produce a brief spurious output.

The potential decoding errors of a ripple counter are alleviated by the **synchronous counter** in Fig. 15.1–13. Here, the flip-flops operate in parallel, the input pulses being applied simultaneously to all clock terminals. Thus, all output transitions take place at essentially the same time. The other connections and the AND gates are arranged such that Q_1 toggles after $Q_0 = 1$, Q_2 toggles only after $Q_1 \cdot Q_0 = 1$, etc., which yields the desired waveforms and counting sequence. Despite its name, a synchronous counter could be used to count nonperiodic pulses.

Figure 15.1–13
Synchronous counter.

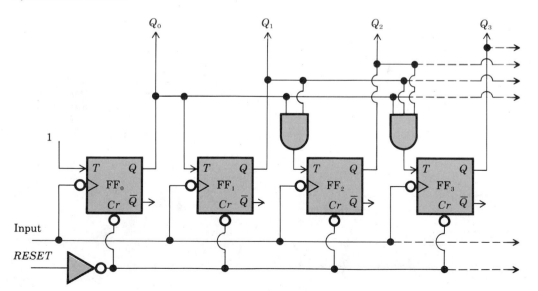

With the help of some additional gates, a ripple or synchronous counter becomes a reversible **up/down counter.** Up/down counters feature a control terminal that allows us to *reverse the counting sequence* when desired. Timing considerations generally favor synchronous counters for up/down counting.

An entirely different principle characterizes the **Johnson counter** in Fig. 15.1–14, which consists of D flip-flops arranged in a ring. The input pulses are applied to all clock terminals, like a synchronous counter, but the complemented output $\overline{Q}_4$ from the last stage goes back to the D terminal of the first stage. (Hence, this circuit is also known as a **twisted ring** or **switched-tail ring counter.**) To understand the accompanying wave-

Figure 15.1–14
(a) Johnson counter. (b) Timing diagram. (c) Counting sequence.

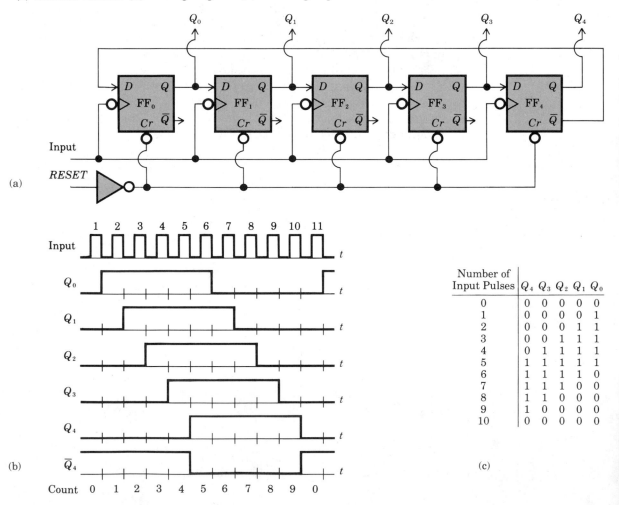

Number of Input Pulses	Q_4	Q_3	Q_2	Q_1	Q_0
0	0	0	0	0	0
1	0	0	0	0	1
2	0	0	0	1	1
3	0	0	1	1	1
4	0	1	1	1	1
5	1	1	1	1	1
6	1	1	1	1	0
7	1	1	1	0	0
8	1	1	0	0	0
9	1	0	0	0	0
10	0	0	0	0	0

(c)

forms and counting sequence, you must recall that the output of a D flip-flop equals the previous state at the D terminal. If all flip-flops are initially reset, then $\bar{Q}_4 = 1$ and the first input pulse produces $Q_1 = 1$, which in turn produces $Q_2 = 1$ after the second pulse, and so forth until $Q_0 = Q_1 = Q_2 = Q_3 = Q_4 = 1$ after the 5th pulse. Now $\bar{Q}_4 = 0$, so $Q_0 = 0$ after the 6th pulse, and the counter arrives back at the initial state after the 10th pulse.

The five-stage Johnson counter is called a **divide-by-10 counter** because the waveforms complete one full cycle for every 10 input pulses. However, this counter requires one more flip-flop than a 4-bit decade counter, and it counts in a nonstandard sequence. In exchange for these disadvantages, it has the advantage of synchronous flip-flop transitions implemented without additional AND gates. The five-stage ring configuration also forms the basis of a handy **divide-by-N counter** that allows you to select any modulus in the range $2 \leq N \leq 10$ using, at most, one auxiliary gate.

A representative listing of standard IC counters consist of items such as

- 4-bit binary and decade counters (ripple or synchronous)
- 4-bit synchronous up/down counters (binary or decade)
- 12-bit ripple counters
- Johnson counters with decoded decimal outputs
- Divide-by-N ring counters

Special features may include: *CARRY* outputs for cascading, *PRESET* inputs for establishing an arbitrary initial state, and built-in decoders for seven-segment displays.

Exercise 15.1–3   A counter with a 60-Hz input produces output pulses once per minute, once per hour, once per day, and once per week. (a) What is the minimum number of flip-flops needed? (b) How many divide-by-N counters are needed, if $N \leq 10$?

Exercise 15.1–4   Construct the timing diagram and counting sequence to verify that the ripple counter in Fig. 15.1–12a counts *down* from 1111 to 0000 if the $\bar{Q}$ terminals are connected to the succeeding clock terminals, still taking $[Q_3 Q_2 Q_1 Q_0]$ as the output.

Registers

A register is a group of flip-flops designed for *temporary storage* of binary data, one bit per flip-flop. D-type flip-flops are particularly well suited to this purpose — so much so that they are also known as *data* flip-flops.

Consider, to begin with, the 4-bit register in Fig. 15.1–15, where the binary data word $[B_3 B_2 B_1 B_0]$ might come from a combinational network. We store this data word by applying a *LOAD* pulse to the clock terminals so that $Q_0 = B_0$, $Q_1 = B_1$, $Q_2 = B_2$, and $Q_3 = B_3$ after the pulse ends. The data word now remains at the Q terminals for as long as desired, barring a power-supply failure, and the complemented word appears at the $\overline{Q}$ terminals (not shown). Later on, a new input word can be "written over" the current one by applying another *LOAD* pulse. We describe this register's operation as **parallel-input/parallel-output** since all input bits are stored simultaneously and, having been stored, are available simultaneously at the output terminals.

Figure 15.1–15
Parallel-input/parallel-output register.

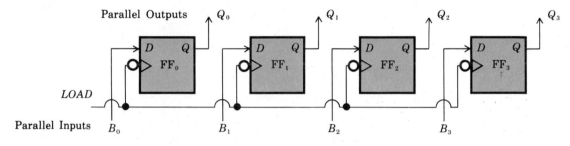

But suppose a data word arrives in **serial** form, one bit at a time. The storage of serial data requires a **shift register** like Fig. 15.1–16a, where bits shift from one flip-flop to the next after each clock pulse. The incoming bits are fed into the register synchronously with the clock pulses by setting $SHIFT = 1$. We then set $SHIFT = 0$ to disable the clock input, and the entire 4-bit word is available at the parallel output terminals. The waveforms in Fig. 15.1–16b illustrate this **serial-to-parallel conversion.** Note that the first flip-flop cleans up any *distortion* present in the incoming waveform — due, perhaps, to the imperfections of a data transmission system. If, at a later time, we wish to convert the stored word back to serial form, then we simply let $SHIFT = 1$ again so the bits appear sequentially at the serial output terminal Q_3.

With the addition of parallel inputs, a shift register becomes capable of **parallel-to-serial conversion** as well as serial-to-parallel conversion. Figure 15.1–17 diagrams an implementation of this feature using 2-input multiplexers similar to the 4-input MUX back in Fig. 14.3–6. Setting $LOAD = 1$ connects the parallel inputs to the D terminals, and storage is triggered by the next clock pulse. Setting $LOAD = 0$ and $SHIFT = 0$ then keeps the data word available at the parallel outputs. Serial operation begins when $LOAD = 0$ and $SHIFT = 1$. A **universal register** would have the further capability of *bidirectional* shifting, left to right or right to

Figure 15.1–16
(a) Shift register. (b) Waveforms with input $[B_3 B_2 B_1 B_0] = 1011$.

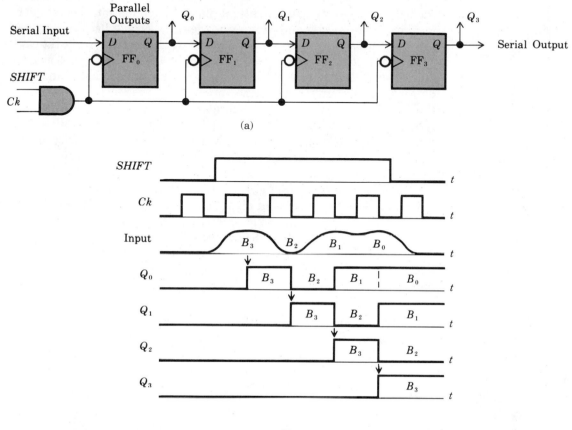

(a)

(b)

Figure 15.1–17
Shift register with serial and parallel inputs.

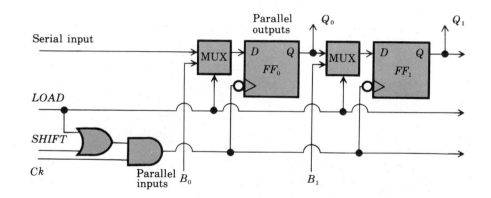

left. Usually, a master *CLEAR* terminal is also provided to reset all flip-flops.

While we're on the subject of shift registers, we should mention the related concept of **recirculating memory.** Recirculation is obtained by loading a shift register and connecting the serial output back to the serial input, so the data bits continually cycle through the register while the clock ticks. This scheme is essential for a *dynamic* register consisting of **charge-coupled devices** (CCDs) or other memory cells that cannot hold data bits indefinitely. Dynamic memory cells require much less space than flip-flops on an IC chip, which facilitates construction of very long registers. For instance, you can buy an 18-pin unit that contains nine CCD registers, each 1024 bits long, and provides a total storage capacity of 9216 data bits. (Other memory units will be considered in Section 15.4)

Example 15.1–2
Frequency counter

The illustrative system in Fig. 15.1–18a includes a flip-flip, counter, divider, and register to carry out digital measurement of the frequency of a sinusoidal input. The sinusoid is applied to a trigger circuit that generates one pulse per cycle, say at each positive-going zero-crossing of the waveform. The *COUNT* signal gates the trigger pulses to a counter for exactly one second, so the decoded displayed count equals the input frequency measured in hertz.

Figure 15.1–18
(a) Frequency counter. (b) Timing diagram.

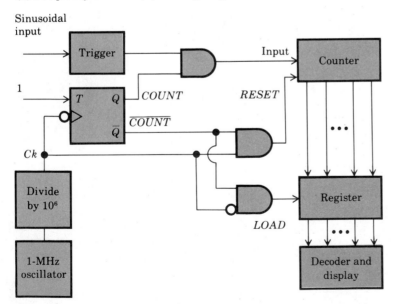
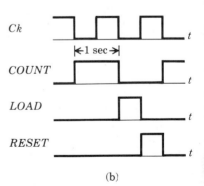

(a)

The *COUNT* signal is derived from a highly stable 1-MHz oscillator (usually quartz-crystal controlled) whose output passes through a six-decade divider to become a one-cycle-per-second clock waveform *Ck*. Connecting *Ck* to a flip-flop with $T = 1$ produces a *COUNT* waveform that is HIGH for one second and LOW for the next second, as shown in Fig. 15.1–18b. (If we bypass three decades of the divider, so it divides by 10^3, then *COUNT* will be HIGH for one millisecond and the displayed value will equal the input frequency measured in kilohertz.)

Counting ceases when *COUNT* goes LOW, and a *LOAD* pulse transfers the count bits in parallel to a storage register. The register's parallel outputs, in turn, go to the decoder/display unit. Another pulse now resets the counter, and a new counting cycle begins when *COUNT* goes HIGH. Meanwhile, the register continues to hold the previous result. The *LOAD* and *RESET* waveforms are generated from $\overline{COUNT}$, *Ck*, and $\overline{Ck}$ using AND gates.

15.2 DIGITAL INSTRUMENTATION

An instrumentation system extracts desired information about some physical quantity. Although the quantity of interest is usually *analog* in nature, *digital* instrumentation often provides the best way of processing and display. For one reason, digital systems have memory capability that would be difficult to implement with analog hardware. For another, the operating characteristics of a programmable digital processor are readily altered by changing the software — a simpler task than modifying analog hardware.

However, digital processing of physical quantities requires converting data from analog to digital form. Furthermore, some of the digitally processed results may be needed as analog output signals. Consequently, modern instrumentation systems usually have the *hybrid* structure represented by Fig. 15.2–1, with both analog and digital units. These two units are interfaced with an **analog-to-digital converter (ADC)** and a **digital-to-analog converter (DAC)**.

Figure 15.2–1
Hybrid instrumentation system.

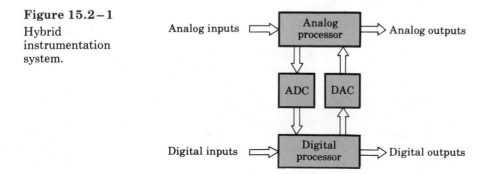

The analog portion of a hybrid instrumentation system carries out simple operations, but the sophisticated computations and decision-making are left to the digital processor. This unit receives digitized information from the ADC and, perhaps, direct digital inputs. Some of the resulting digital outputs might go directly to display and control devices, while others are converted to analog signals by the DAC.

Since DACs and ADCs play key roles in digital instrumentation, most of this section is devoted to the study of digital-to-analog and analog-to-digital conversion. The section then concludes with an examination of digital processing in the context of a typical data acquisition system.

Digital-to-Analog Conversion

A digital-to-analog converter accepts a digital input word and generates an analog output quantity, usually a voltage. To start with a simple case, let the 4-bit word $[B_3 B_2 B_1 B_0]$ be in natural binary code representing the decimal integer

$$x = B_3 \times 2^3 + B_2 \times 2^2 + B_1 \times 2^1 + B_0 \times 2^0$$

The analog output voltage should then be proportional to x, say in the form

$$v_q = (8B_3 + 4B_2 + 2B_1 + B_0)\Delta v$$

The subscript q emphasizes the fact that v_q is a *quantized* voltage. The proportionality constant Δv is called the **step size** because v_q increases by Δv when the least-significant bit B_0 changes from 0 to 1. Thus, the possible values of v_q are $0, \Delta v, 2\Delta v, \ldots, 15\Delta v$.

Generalizing from 4 bits to n bits, let the input $[B_{n-1} \ldots B_1 B_0]$ again be a natural binary code word. Digital-to-analog conversion with step size Δv now yields the output voltage

$$v_q = (2^{n-1}B_{n-1} + \cdots + 2B_1 + B_0)\Delta v \tag{1}$$

The largest output value is then

$$v_{q-\text{max}} = (2^n - 1)\Delta v \tag{2}$$

This value occurs when $B_{n-1} = \cdots = B_1 = B_0 = 1$, corresponding to the decimal integer $x_{\text{max}} = 2^n - 1$.

Figure 15.2–2 diagrams a **weighted-resistance DAC** that implements Eq. (1) using an op-amp circuit to form a weighted sum of voltages. Weighting factors proportional to 2^i are established by a bank of n input resistors with

$$R_i = R_0/2^i \qquad i = 0, 1, \ldots, n-1 \tag{3}$$

Voltages proportional to the input bits are obtained by applying the input word to a bank of n switches connected to a reference voltage $- V_{\text{ref}}$, so the ith switch output is an off/on voltage

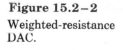

Figure 15.2–2

Weighted-resistance DAC.

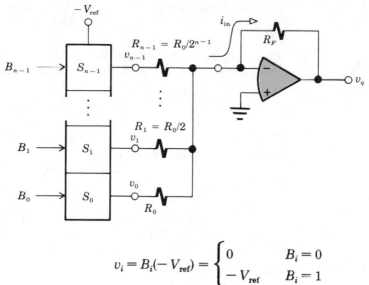

$$v_i = B_i(-V_{ref}) = \begin{cases} 0 & B_i = 0 \\ -V_{ref} & B_i = 1 \end{cases}$$

Since the input of the op-amp acts like a virtual short, the total current resulting from the v_i voltages will be

$$i_{in} = \frac{v_{n-1}}{R_{n-1}} + \cdots + \frac{v_1}{R_1} + \frac{v_0}{R_0}$$

$$= \frac{2^{n-1}}{R_0} v_{n-1} + \cdots + \frac{2^1}{R_0} v_1 + \frac{2^0}{R_0} v_0$$

and the final output voltage is

$$v_q = -R_F i_{in} = (2^{n-1}B_{n-1} + \cdots + 2B_1 + B_0)\left(\frac{R_F}{R_0} V_{ref}\right)$$

Hence, v_q has the form of Eq. (1) with step size

$$\Delta v = \frac{R_F}{R_0} V_{ref} \qquad (4)$$

which can be adjusted for the desired value.

By way of illustration, suppose you want to build a 4-bit DAC with $\Delta v = 1$ V and $V_{ref} = 5$ V. Suppose further that the input resistors must satisfy $R_i \geq 1$ kΩ to prevent excessive loading of the reference voltage source. Equation (3) indicates that the smallest R_i corresponds to $i = n - 1 = 3$, so you might take $R_3 = R_0/2^3 = 1$ kΩ and $R_0 = 2^3 \times 1$ kΩ $= 8$ kΩ. The other input resistors are then $R_1 = R_0/2 = 4$ kΩ and $R_2 = R_0/4 = 2$ kΩ, and Eq. (4) yields the feedback resistor $R_F = (\Delta v/V_{ref})R_0 = (1/5) \times 8$ kΩ $= 1.6$ kΩ.

With larger values of n, the weighted-resistance DAC requires a wide range of resistance values that makes accuracy a significant problem. An

Figure 15.2–3
R-2R ladder for DAC.

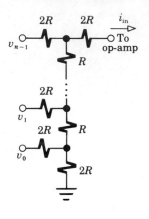

improved design for $n \geq 8$ employs the resistor network in Fig. 15.2–3, known as an **R-2R ladder.** This ladder network has just two resistance values, and it is easily fabricated in IC form with a high degree of precision. When an *R-2R* ladder replaces the input resistors in Fig. 15.2–2, the output is still given by Eq. (1) but the step size becomes

$$\Delta v = \frac{R_F}{2^n \times 3R} V_{\text{ref}} \tag{5}$$

The derivation of Eq. (5) involves a simple exercise in Thévenin/Norton manipulations (see Problem 15.2–11).

Sometimes we need a **bipolar DAC** capable of both positive and negative output voltages. Correspondingly, the digital code must provide information about the algebraic sign of the output. Figure 15.2–4 shows one way of modifying a DAC circuit for the case of **sign-magnitude coding.** Here, the **sign bit** B_* controls a polarity switch at the output, while the remaining bits represent the magnitude in natural binary form. The upper op-amp produces a positive voltage $v_{qa} = -R_F i_{\text{in}}$, the same as before. But the lower op-amp produces a *negative* voltage $v_{qb} = -R_b i_R = -v_{qa}$ since $i_R = v_{qa}/R_a$ and $R_b = R_a$. Switching between the two op-amp terminals therefore yields $v_q = 0, \pm\Delta v, \pm 2\Delta v, \ldots$, in accordance with the sign and magnitude bits.

Figure 15.2–4
Bipolar DAC for sign-magnitude coding.

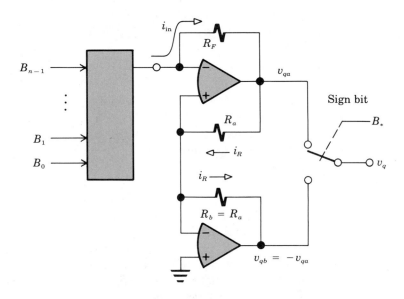

Exercise 15.2–1   Consider a weighted-resistance DAC with $n = 8$, $R_7 = 1$ kΩ, and $v_{q-\max} = V_{\text{ref}} = 10$ V. Calculate Δv and find the required values for R_0 and R_F.

Analog-to-Digital Conversion

We have seen how a DAC produces an analog output in one-to-one relationship with a digital input. The operation of an analog-to-digital converter (ADC) is less direct because the analog input quantity may take on any value within some range, whereas the digital output is limited to a finite number of different words. Analog-to-digital conversion therefore inherently involves round-off or quantizing error.

Suppose, for example, that we have an analog voltage v_a known to be in the range $0–8$ V. To represent an arbitrary value of v_a with a 2-bit word, we might use the scheme in Fig. 15.2–5a where the 8-V range has been divided into four equal intervals. A quantized value v_q is defined at the midpoint of each interval, and a code word $[B_1 B_0]$ is assigned to represent each value of v_q. Any value of v_a falling between 2.0 and 4.0 V would then be represented by $v_q = 3$ V and encoded as $[B_1 B_0] = 01$. Such digitization always introduces some **quantizing error** $v_a - v_q$ which, in this case, has extreme values of ± 1.0 V.

Figure 15.2–5
Quantizing and encoding an analog voltage.

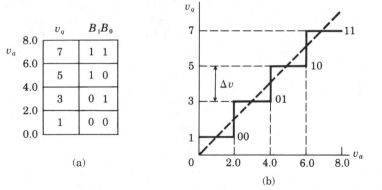

(a)

(b)

The plot of v_q versus v_a in Fig. 15.2–5b brings out the fact that an ADC makes a "staircase" approximation of the analog quantity. Here, the staircase starts at $v_q = 1$ V and has steps of height $\Delta v = 2$ V occurring at the threshold voltages $v_a = 2.0$, 4.0, and 6.0 V. Alternatively, the bottom step might be set at $v_q = 0$ V, if desired, which increases the maximum quantizing error to 2.0 V. Regardless of the position of the bottom step, a staircase having more steps of smaller height would improve the approximation. But each step must be represented by a unique code word, so better accuracy requires longer code words.

To generalize the foregoing observations, consider an ADC that accepts a unipolar analog voltage in the range $0 \le v_a \le V_R$ and produces n-bit output words. Since there are 2^n different n-bit words, the analog range V_R can be divided into 2^n equal intervals with step size

$$\Delta v = \frac{V_R}{2^n} \qquad (6)$$

The maximum value of $|v_a - v_q|$ will be $\Delta v/2$ or Δv, depending on the position of the bottom step. The ratio $\Delta v/V_R$ is called the **resolution**, usually expressed in percent as

$$\frac{\Delta v}{V_R} \times 100 = \frac{100}{2^n}\% \qquad (7)$$

Thus, for instance, an 8-bit ADC has a resolution of $100/2^8 \approx 0.4\%$. Equations (6) and (7) also hold for a bipolar ADC with analog voltage range $-V_R/2 \le v_a \le V_R/2$.

Conceptually, the most straightforward analog-to-digital circuit is a **parallel-comparator ADC,** as shown in Fig. 15.2–6a for the unipolar case with $n = 2$. The input v_a is applied simultaneously to $2^n - 1 = 3$ comparators. The comparators compare v_a with fixed threshold voltages established by $2^n = 4$ equal resistors connected in series with the range reference voltage V_R. Hence, each comparator output Q goes either high or low depending on whether v_a falls below or above the respective threshold value. An encoder then translates the characteristic pattern $[Q_3 Q_2 Q_1]$ into the output word $[B_1 B_0]$ according to the table in Fig. 15.2–6b.

Figure 15.2–6
(a) Parallel-comparator ADC. (b) Encoding table.

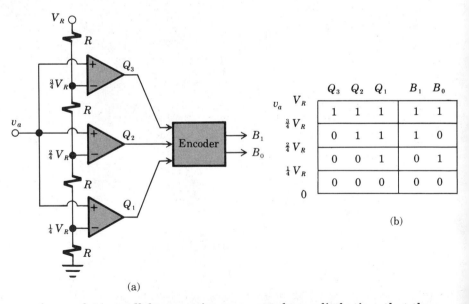

		Q_3	Q_2	Q_1	B_1	B_0
v_a	V_R	1	1	1	1	1
	$\frac{3}{4}V_R$	0	1	1	1	0
	$\frac{2}{4}V_R$	0	0	1	0	1
	$\frac{1}{4}V_R$	0	0	0	0	0
	0					

(b)

(a)

A complete parallel conversion process takes so little time that these ADCs are commonly known as **flash converters.** However, the quick response causes problems if rapid fluctuations of v_a change the output word before it has been completely processed by the instrumentation system. Consequently, flash converters are usually equipped with a **sample-and-**

hold (S/H) circuit like Fig. 15.2–7. A positive *SAMPLE* pulse turns on the FET switch, allowing the capacitor to charge up to the sample value v_{as}. The switch then becomes an open circuit, and the capacitor keeps v_{as} constant until the next *SAMPLE* pulse.

Figure 15.2–7
Sample-and-hold circuit.

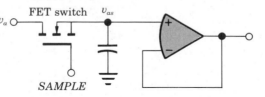

The major liability of parallel conversion is that a separate comparator is needed for each of the $2^n - 1$ threshold levels. Increasing the output word to 8 bits to get 0.4% resolution would therefore require $2^8 - 1 = 255$ comparators! Flash converters available as integrated circuits typically have 63 comparators, 6-bit outputs, and 1.6% resolution, but they can work at conversion rates up to 100 MHz. Such units are best suited to applications calling for very high speed rather than great precision.

Better accuracy at slower speeds is achieved by the family of ADCs that employ a *single* comparator with a *variable threshold voltage*. One of the most popular members of this family is the **dual-slope converter,** whose block diagram appears in Fig. 15.2–8a. This ingenious system uses an op-amp integrator to generate the threshold voltage $v_r(t)$, and it includes an *n*-bit counter that eventually holds the output word. A digital clock and a controller coordinate the conversion operations. We'll explain those operations by walking through one conversion cycle and examining the waveform $v_r(t)$ in Fig. 15.2–8b. We assume that the input v_a is positive, constant, and smaller than the range reference voltage V_R.

The conversion cycle begins at $t = t_1$, in synchronism with a clock pulse, when the controller resets the counter and connects switch S to v_a. The capacitor is initially uncharged, so the positive value of v_a applied to the integrator causes $v_r(t)$ to ramp down with slope $-v_a/RC$. Hence, the comparator output Q immediately goes high and the AND gate passes clock pulses to the counter. The counter overflows after 2^n clock pulses at time $t_2 = t_1 + 2^n T_c$, T_c being the clock period. Upon receiving the *OVERFLOW* signal from the counter, the controller switches S to the fixed voltage $-V_R$. Now $v_r(t)$ ramps back up with slope $+V_R/RC$ until time $t = t_3$ when $v_r(t_3) = 0^+$ and Q goes low. This event stops the counting, and the controller completes the cycle by putting S in the grounded position.

Since the counter resets at t_2, it records M counts during the interval $t_2 \leq t \leq t_3$. The final code word $[B_{n-1} \cdots B_1 B_0]$ therefore represents the quantized voltage

$$v_q = M \frac{V_R}{2^n} = M \, \Delta v \tag{8}$$

Figure 15.2–8

(a) Dual-slope converter. (b) Integrated waveform.

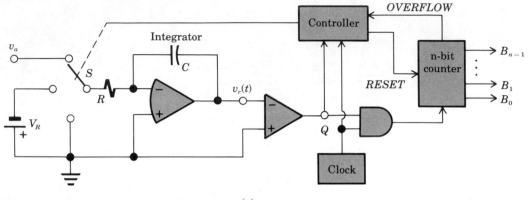

(a)

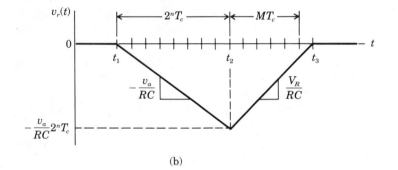

(b)

To relate v_q to v_a, we observe from the slopes in Fig. 15.2–8b that

$$\frac{V_R}{RC}(t_3 - t_2) = -v_r(t_2) = \frac{v_a}{RC} 2^n T_c$$

We also observe, in general, that $t_3 - t_2$ equals MT_c plus some fraction of a clock period, so

$$MT_c \le (t_3 - t_2) < (M+1)T_c$$

These two expressions put bounds on $v_q = M\,\Delta v$ given by

$$v_a - \Delta v < v_q \le v_a \qquad\qquad \textbf{(9)}$$

Thus, the quantizing error $v_a - v_q$ always falls within the range $0 \le (v_a - v_q) < \Delta v$.

The resolution of a dual-slope ADC depends only on the size of the counter, since $\Delta v / V_R = 1/2^n$. However, the time required for a complete conversion cycle is $(t_3 - t_1) \le 2 \times 2^n T_c$, so the maximum conversion rate in repetitive operation is

$$f_{\text{conv}} = \frac{1}{2^{n+1}T_c} = \frac{f_{\text{clock}}}{2^{n+1}} \tag{10}$$

which decreases as n increases. If $f_{\text{clock}} = 10$ kHz, for instance, then we can get 4-bit resolution at $f_{\text{conv}} \approx 300$ Hz or more accurate 8-bit resolution at the lower rate $f_{\text{conv}} \approx 20$ Hz. The dual-slope principle thereby allows the designer to *exchange speed for accuracy,* or vice versa.

Dual-slope converters have special advantages for **digital voltmeters** (DVM) and related instrumentation that must perform reliably in the face of interference and temperature variations. When v_a has been contaminated with time-varying interference and an S/H circuit is not used at the input, v_q becomes the quantized representation of the *average value*

$$V_a = \frac{1}{2^n T_c} \int_{t_1}^{t_1 + 2^n T_c} v_a(t) \, dt$$

Hence, AC interference is averaged out by taking $2^n T_c$ to be an integer multiple of the AC period. When the temperature changes, the only temperature-sensitive parameter in Eq. (8) is the reference voltage V_R, which can be stabilized by a temperature-compensated Zener diode. Further refinements for a DVM would include polarity reversal to allow negative values of v_a, an input attenuator for range selection, and a modified counter with direct BCD output display of three or more decimal digits.

Some applications require faster rates than a dual-slope ADC and better accuracy than a flash converter. These requirements are best satisfied by the sophisticated **successive-approximation converter** diagrammed in Fig. 15.2–9a. The n-bit register drives a DAC to generate the quantized threshold voltage v_{qr}, whose value is updated after each clock pulse. Successive values of v_{qr} are compared with the current analog sample value v_{as} held by the S/H circuit.

The controller executes the successive-approximation algorithm synchronized with the clock as follows:

1. Reset the register, issue a SAMPLE pulse, and let $k = n - 1$.
2. Try $B_k = 1$. If Q goes high, then $v_a > v_{qr}$ so keep $B_k = 1$. If Q goes low, then $v_a < v_{qr}$ so set $B_k = 0$ at the next clock pulse.
3. Repeat the second step for $k = n - 2, \ldots, 1$, and 0.
4. Issue the READ signal and go back to the first step.

Figure 15.2–9b shows the resulting waveform v_{qr} and the register contents when $n = 4$ and v_{as} falls between $6 \, \Delta v$ and $7 \, \Delta v$.

Figure 15.2–9
(a) Successive-approximation converter. (b) Waveform and register contents.

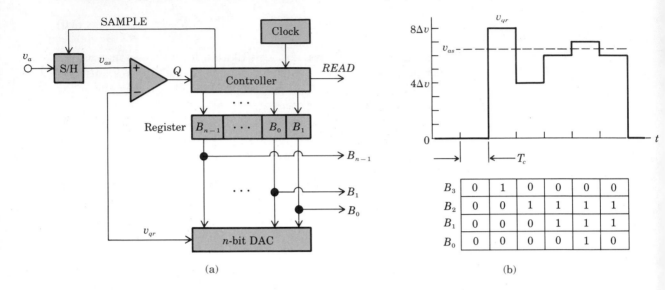

<table>
<tr><td>B_3</td><td>0</td><td>1</td><td>0</td><td>0</td><td>0</td><td>0</td></tr>
<tr><td>B_2</td><td>0</td><td>0</td><td>1</td><td>1</td><td>1</td><td>1</td></tr>
<tr><td>B_1</td><td>0</td><td>0</td><td>0</td><td>1</td><td>1</td><td>1</td></tr>
<tr><td>B_0</td><td>0</td><td>0</td><td>0</td><td>0</td><td>1</td><td>0</td></tr>
</table>

(a)   (b)

The algorithm takes one clock period per bit, plus one more at the beginning and at the end. Thus, the conversion rate of a successive-approximation ADC is

$$f_{\text{conv}} = \frac{1}{(n+2)T_c} = \frac{f_{\text{clock}}}{n+2} \qquad (11)$$

Commercial 10-bit models are available for rates up to 500 kHz. They combine the speed and accuracy needed in most digital systems that perform on-line signal processing.

Last, we should mention the **shaft encoder,** a low-speed device used in electromechanical systems to detect and digitize the angular position θ of a rotating shaft. Figure 15.2–10a depicts a simplified shaft encoding disk with $n = 3$ concentric bands divided into conducting and nonconducting segments. Fixed-position carbon brushes press on these bands and pick up a DC voltage from the conducting segments. The segments shown here have been arranged for the **Gray code** output tabulated in Fig.15.2–10b, which has the property that adjacent words differ by exactly one bit. Using a Gray code rather than a natural binary code minimizes errors resulting from nonzero brush width. Optical shaft encoders with photoelectric sensors have as many as $n = 18$ bands, for a resolution of about 0.0004%.

Exercise 15.2–2   Consider a 3-bit bipolar ADC with analog voltage range $-1.6 \text{ V} \le v_a \le +1.6 \text{ V}$.

Figure 15.2–10
(a) Shaft encoder disk.
(b) Angular codes.

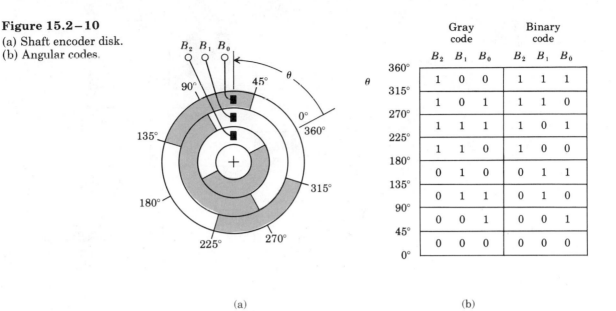

	Gray code			Binary code		
θ	B_2	B_1	B_0	B_2	B_1	B_0
360°						
	1	0	0	1	1	1
315°						
	1	0	1	1	1	0
270°						
	1	1	1	1	0	1
225°						
	1	1	0	1	0	0
180°						
	0	1	0	0	1	1
135°						
	0	1	1	0	1	0
90°						
	0	0	1	0	0	1
45°						
	0	0	0	0	0	0
0°						

(a)   (b)

(a) Calculate Δv, and plot v_q versus v_a taking $v_q = \Delta v/2$ when $v_a = 0^+$.
(b) Follow the pattern in Fig. 15.2–6b to construct the coding table for parallel-comparator implementation.

Exercise 15.2–3 The ADC for a certain application must have $f_{conv} = 8$ kHz with a resolution of about 0.1%. Find n and determine the corresponding clock frequency for dual-slope conversion and for successive-approximation conversion.

Data Acquisition and Digital Processing

Collecting and processing experimental data is a tiresome and error-prone chore when done by hand. (Remember your physics labs?) Fortunately, when suitable analog-signal transducers exist, the rest of the task can be turned over to a modern multichannel **data acquisition system** or **data logger** containing the functional elements shown in Fig. 15.2–11. Such a system might handle 40 or more input quantities, both analog and digital, with a scan rate of perhaps 20 channels per second and 0.01% ADC resolution. Besides collecting, displaying, and printing the data, this system might also carry out various numerical operations and deliver processed data to a computer terminal or peripheral recording device for further analysis.

Several of the blocks in Fig. 15.2–11 have been previously examined, and the internal workings of others will be discussed subsequently. Our purpose here is to present an overview of digital data processing by de-

Figure 15.2–11
Data acquisition
system.

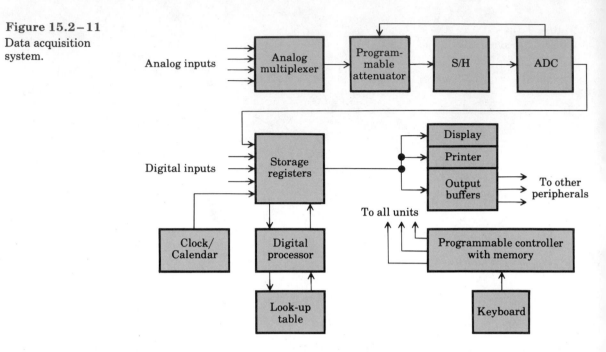

scribing typical operations on one of the analog inputs, say the temperature signal $x(t)$, generated by a thermal sensor.

The analog multiplexer periodically samples $x(t)$ and other analog inputs at a rate governed by the controller. The interleaved analog samples then pass through the **programmable attenuator** (see Problem 15.2–4) for digitizing by the ADC. Feedback from the ADC to the attenuator provides automatic ranging, and the digitized data and range information are loaded into storage registers. Other registers hold data obtained from direct digital inputs and time/date information provided by the clock/calendar. The remainder of the system consists essentially of a small digital computer that can be programmed from the keyboard. In fact, you can buy an expansion card that turns a personal computer into a data acquisition system.

The digital processor, in conjunction with the controller, might average the temperature data over a specified number of samples. At a predetermined time (or upon command by the operator), the processed values are displayed, printed, and transferred to peripheral devices. The output includes identifying labels for each data value, along with logging information from the clock/calendar and the keyboard entries. The entire cycle then automatically repeats every few seconds, or perhaps just once a week, according to the instructions stored in the controller.

Clearly the controller and digital processor constitute the "brains" of such a system, and we'll have more to say about the arithmetic and control circuits in Sections 15.3 and 15.4. Right now it's enough to state that the processor is capable of basic arithmetic calculations involving past and present data values and stored constants, and that the controller tells the processor what calculations to perform. Armed with this knowledge, we can explain how the system performs simple but illustrative digital-processing operations.

Suppose our temperature signal $x(t)$ has been sampled and digitized every T seconds to yield the sequence of data points $x_0, x_1, x_2, \ldots$, where x_k stands for the current sample value at $t = kT$. These values differ somewhat from $x(kT)$ because of quantizing, but the quantizing error will be negligible if the system has a high-resolution ADC. The main point here is that *numerical operations* by the digital processor can supplant the need for more cumbersome *analog processing*. It's a trivial matter, for example, to produce an output sequence $y_0, y_1, \ldots$ calculated from the formula

$$y_k = \tfrac{1}{10}\left(x_k + x_{k-1} + \cdots + x_{k-9}\right)$$

which corresponds to the *running average* of the past 10 input values. (You've probably written a similar computer program.) Other common operations include:

- *Scaling,* to put data values into appropriate units.
- *Editing,* to remove questionable values.
- *Linearizing,* to correct the effects of a nonlinear transducer.

Some operations might draw upon conversion factors stored in the look-up table.

In addition — and more significantly — the processor can simulate the action of a *filter,* in the sense that the output sequence corresponds to the sample values that would have been obtained had $x(t)$ been passed through an analog filter before sampling. The system then functions as a **digital filter.** For instance, suppose the controller is programmed to calculate

$$y_k = \frac{1}{\tau + T}\left(Tx_k + \tau y_{k-1}\right) \tag{12a}$$

where y_{k-1} is the previously computed value and the constant τ is large compared to T. This expression is called a **difference equation** because it involves $y_k - y_{k-1}$, as brought out by rewriting Eq. (12a) in the form

$$\tau\left(\frac{y_k - y_{k-1}}{T}\right) + y_k = x_k \tag{12b}$$

Furthermore, since $(y_k - y_{k-1})/T = \Delta y/\Delta t \approx dy/dt$, our difference equation approximates the *differential equation*

$$\tau\frac{dy}{dt} + y = x$$

But this differential equation describes a first-order lowpass filter with analog input $x(t)$, output $y(t)$, and cutoff frequency $\omega_{co} = 1/\tau$. Hence, the digital processor acts essentially like a lowpass filter.

Bandpass filters, notch filters, and lowpass filters with steeper cutoffs are simulated by higher-order difference equations that also include y_{k-2}, y_{k-3}, etc. The digital implementation still remains quite simple compared to the equivalent analog filter. Furthermore, the controller can be programmed to change the filter characteristics as the nature of the data changes, thereby creating an automatic **adaptive filter.** In general, digital processors are capable of highly sophisticated signal processing operations — operations that might be beyond the scope of practical analog hardware.

15.3 ARITHMETIC NETWORKS

Arithmetic networks carry out the numerical calculations in digital computers and instrumentation systems. Since there are literally hundreds of different units for various applications, this section focuses on those elementary networks that embody fundamental concepts of binary arithmetic. In particular, we'll start with *binary adders* and we'll go on to show how adders are augmented to perform *subtraction* or *multiplication.* We'll also examine a simplified version of the *arithmetic logic unit* found in microprocessors and computers.

Binary Addition

Suppose that two decimal numbers have been encoded as binary words and we want to generate another word corresponding to the sum of the decimal numbers. This is the task of a binary adder. We'll develop the method and implementation of binary addition here, initially assuming positive numbers and natural binary coding. Throughout, the addition operation will be denoted by writing *plus,* to avoid confusion with the logical-OR operation.

The building block for binary adders is the **half adder** (HA) symbolized by Fig. 15.3–1a. This unit adds two input bits, A and B, to produce a

Figure 15.3–1

Half adder: (a) symbol; (b) truth table; (c) gate diagram.

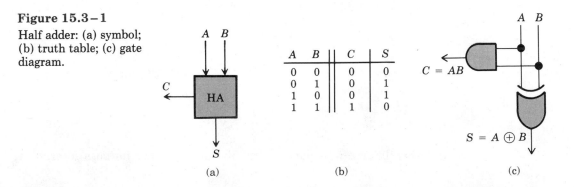

A	B ‖	C	S
0	0	0	0
0	1	0	1
1	0	0	1
1	1	1	0

$C = AB$

$S = A \oplus B$

(a)   (b)   (c)

sum bit S and a **carry bit** C. The outputs are related to the inputs in accordance with the simple rules of base-2 addition, as follows:

$$
\begin{array}{cccc}
0 & 0 & 1 & 1 \\
\text{plus } 0 & \text{plus } 1 & \text{plus } 0 & \text{plus } 1 \\
\hline
0\ 0 & 0\ 1 & 0\ 1 & 1\ 0
\end{array}
$$

These rules lead directly to the truth table in Fig. 15.3–1b, from which we obtain the logic expressions

$$C = AB \qquad S = \overline{A}B + A\overline{B} = A \oplus B \qquad\qquad (1)$$

Figure 15.3–1c shows the corresponding network consisting of an AND gate and an XOR (exclusive-OR) gate. (The NOR-gate network back in Fig. 14.3–1c may also serve as a half adder since it generates AB in the process of forming $A \oplus B$.)

Although a half adder suffices for one-bit words, the addition of n-bit words must take account of intermediate carries. Accordingly, the **full adder** (FA) represented by Fig. 15.3–2a adds an input carry C_{in} to A plus B to obtain the sum bit S and output carry C_{out}. A full adder thus performs the two additions written symbolically in Fig. 15.3–2b where we have introduced the **partial carry** G and the **partial sum** P defined by

$$G \triangleq AB \qquad P \triangleq A \oplus B \qquad\qquad (2)$$

We see from the symbolic addition that

$$S = P \oplus C_{\text{in}} = A \oplus B \oplus C_{\text{in}} \qquad\qquad (3a)$$

We also see that $C_{\text{out}} = 1$ occurs only when $A = B = 1$ or when $P = 1$ and $C_{\text{in}} = 1$, so

$$C_{\text{out}} = G + PC_{\text{in}} = AB + (A \oplus B)C_{\text{in}} \qquad\qquad (3b)$$

Hence, a full adder can be implemented using two half adders and an OR gate connected per Fig. 15.3–2c.

Figure 15.3–2
Full adder: (a) symbol;
(b) operations;
(c) implementation.

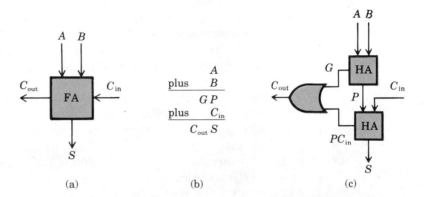

(a) (b) (c)

Now we have the necessary background to understand the complete n-bit adder in Fig. 15.3–3a. The input words to be added are denoted by

$[A_{n-1} \ldots A_1 A_0]$ and $[B_{n-1} \ldots B_1 B_0]$, and the output consists of the sum word $[S_{n-1} \ldots S_1 S_0]$ and the final carry C_n. The unit employs n full adders arranged such that $C_0 = 0$ and

$$S_k = A_k \oplus B_k \oplus C_k$$
$$C_{k+1} = A_k B_k + (A_k \oplus B_k) C_k \tag{4}$$

for $k = 0, 1, \ldots, n - 1$. Since $C_0 = 0$, the full adder for the least-significant bits A_0 and B_0 could be replaced by a half adder. However, we'll subsequently encounter practical uses for the external carry input.

Figure 15.3–3
(a) n-bit adder. (b) Addition of 4-bit words.

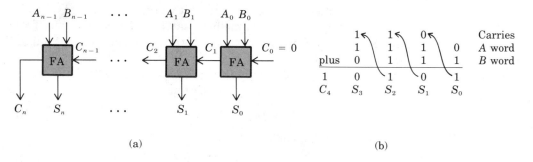

(a)   (b)

Figure 15.3–3b illustrates how a 4-bit adder performs the binary version of the decimal addition 14 plus 7 = 21. Note that we get $C_4 = 1$ in this case, meaning that the sum is actually the 5-bit word $10101 = 21_{10}$. In general, the addition of two n-bit words yields a result given by $[C_n S_{n-1} \ldots S_1 S_0]$, which consists of $n + 1$ bits. If a particular system is limited to n-bit words, then an **overflow** results whenever $C_n = 1$. Hence, the nth carry must be applied to an overflow-detection unit if it cannot be included in the output word.

The arithmetic network in Fig. 15.3–3a is called a **parallel adder** because all bits are added almost simultaneously. But each full adder introduces propagation delay, so some amount of time elapses while the carries travel from right to left — a phenomenon known as **carry ripple.** The more complicated **carry-look-ahead adder** reduces time delay and attains greater speed, typically taking less than 20 ns to add 16-bit words.

When speed is not a major concern, the **serial adder** in Fig. 15.3–4 offers the advantage of minimizing hardware since it requires just one full adder. The successive input bits A_k and B_k enter serially from shift registers, and the resulting sum bit S_k goes to an output shift register. Meanwhile, the output carry C_{k+1} is fed back and delayed one clock cycle by a D flip-flop to become the next input carry.

Figure 15.3–4
Serial adder.

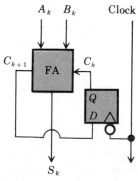

Another type of addition network is the **BCD adder,** which operates on decimal digits represented by 8421 BCD words. Recall from Table 14.1–2 that the 8421 code uses only the first 10 of the 16 possible 4-bit binary words, the remaining 6 words being invalid. Figure 15.3–5 shows the corresponding modifications of a parallel adder needed for one decade of BCD addition. The detection unit generates the next decade carry $C_{out} = 1$ whenever the binary sum exceeds $01001 = 9_{10}$. The correction unit then increases the binary sum by adding $0110 = 6_{10}$ to skip over the 6 invalid words. A cascade system that adds M decimal digits in BCD would contain M such units, with each output carry applied to the carry input of the next stage. Systems like this are commonly found in calculators and digital instruments having decimal displays.

Figure 15.3–5
BCD adder.

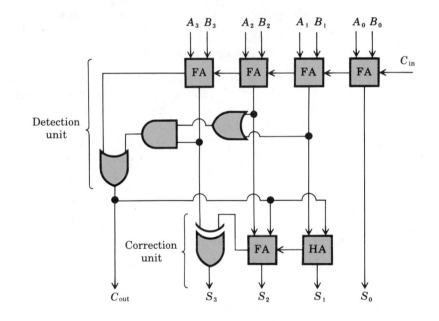

Exercise 15.3–1

Apply the rules of base-2 addition to construct the truth table for a full adder with inuts A, B, and C_{in}. Then show from your table that

$$S = \overline{A}\,\overline{B}C_{in} + \overline{A}B\overline{C}_{in} + A\overline{B}\,\overline{C}_{in} + ABC_{in}$$

$$C = AB + AC_{in} + BC_{in}$$

which are alternative versions of Eqs. (3a) and (3b).

Exercise 15.3–2

Let the BCD adder in Fig. 15.3–5 have $C_{in} = 0$, $[A_3 A_2 A_1 A_0] = 1000 = 8_{10}$, and $[B_3 B_2 B_1 B_0] = 0111 = 7_{10}$. Follow through all the internal operations to confirm that the output sum and carry represent the correct result 15_{10} in BCD.

Signed Numbers and Subtraction

Signed numbers and subtraction go hand in hand in base-2, just as they do in base-10. Thus, although binary subtraction networks may be used for special applications, general-purpose arithmetic networks carry out subtraction by *adding negative numbers.* Here we'll consider signed numbers and subtraction in the context of **2's-complement arithmetic,** the method employed by nearly all microprocessors and digital computers. BCD subtraction follows similar lines but is somewhat more complicated.

Positive and negative numbers are distinguished in 2's-complement arithmetic by a **leading sign bit,** 0 for positive and 1 for negative. The magnitude of a positive number is expressed in natural binary code, as usual, but the magnitude of a negative number is converted to its 2's complement. To form the **2's complement** of a binary word you just

Complement each bit, changing 0 to 1 or 1 to 0, and then add 1 to
the least significant bit of the complemented word.

For instance, the 2's complement of 1100 becomes 0011 plus $1 = 0100$. And since $1100 = 12_{10}$, the representation of -12_{10} with a 4-bit magnitude is 1*0100, where we've inserted an asterisk after the sign bit for clarity. (An arithmetic network designed to handle such words would, of course, automatically recognize the leading bit as the sign bit.)

The advantages of 2's-complement representation are brought out by studying the following examples:

$$5_{10} = 0*0101 \qquad 7_{10} = 0*0111 \qquad 12_{10} = 0*1100$$

$$-5_{10} = 1*1011 \qquad -7_{10} = 1*1001 \qquad -12_{10} = 1*0100$$

You can confirm that taking the 2's complement of an entire word, including the sign bit, is equivalent to changing the sign of the decimal number. Furthermore, adding a signed binary number to its 2's complement yields 10*0000, in agreement with $x_{10} - x_{10} = 0$ if we simply ignore the carry to the left of the sign bit. Now consider the decimal subtractions $12 - 7$ and $7 - 12$ written as the binary additions

$$
\begin{array}{rc@{\qquad}rc}
12 & 0*1100 & 7 & 0*0111 \\
- \ 7 \Rightarrow \text{plus} & 1*1001 & - \ 12 \Rightarrow \text{plus} & 1*0100 \\
\hline
5 & 10*0101 & - \ 5 & 01*1011 \\
\end{array}
$$

After dropping the extra carry, we again have the correct results in both cases.

The hardware implementation of 2's-complement arithmetic is illustrated by Fig. 15.3–6. This particular unit adds or subtracts signed binary numbers with 4-bit magnitudes and leading sign bits, the latter being identified by asterisk subscripts. Either of the input words, or both, may represent negative numbers. The control variable M_0 determines the mode of operation via

$$X_k = B_k \oplus M_0 = \begin{cases} B_k & M_0 = 0 \\ \overline{B_k} & M_0 = 1 \end{cases} \tag{5}$$

which holds for all bits in the X word.

Addition takes place when $M_0 = 0$, so the X word equals the B word and $C_0 = 0$. Hence, the network just adds the B word to the A word in the standard manner. Subtraction takes place when $M_0 = 1$, so the X word becomes the bit-by-bit complement of the B word. Injecting $C_0 = 1$ then completes the 2's complement of the B word. The network thus subtracts by adding the 2's-complemented B word to the A word.

Figure 15.3–6
2's-complement
adder/subtractor.

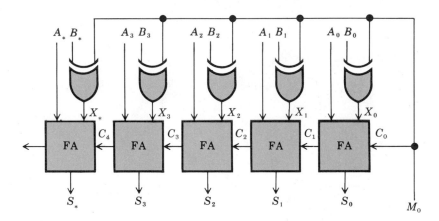

The output sign bit S_* conveys the correct sign of the result in either mode, provided that the magnitude of the sum remains within the allowed range. However, incorrect results may be caused by **magnitude overflow** when the X word has the same sign as the A word. The overflow condition appears as an erroneous sign bit, rather than an output carry, and it can be detected by a simple logic network.

Exercise 15.3–3

Let $M_0 = 1$ in Fig. 15.3–6, and let the B word represent 14_{10}. Find $[S_* S_3 S_2 S_1 S_0]$ when the A word represents 0_{10}, 15_{10}, 14_{10}, and -14_{10}. Comment on your results.

Arithmetic Logic Units

An arithmetic logic unit (ALU) consists of a binary adder/subtractor with auxiliary gates and additional control inputs. These inputs allow you to select *logic* operations as well as arithmetic operations. ALUs also include provision for making numerical comparisons, which further enhances their versatility.

Figure 15.3–7a diagrams one section of a simplified ALU. The control bit M_0 determines X_k as in Eq. (5). The half adders produce the partial carry $G_k = A_k X_k$, the partial sum $P_k = A_k \oplus X_k$, and the sum $S_k = A_k \oplus$

$X_k \oplus C_k$. An OR gate with inputs G_k and P_k generates the logic function $A_k + X_k$. These four functions are delivered to a MUX whose control bits M_1 and M_2 select the output function F_k. Hence, setting $[M_2 M_1 M_0] = 001$ gives $F_k = A_k \cdot \overline{B}_k$, while $[M_2 M_1 M_0] = 110$ gives $F_k = A_k \oplus B_k \oplus C_k$.

Figure 15.3–7
(a) Section of arithmetic logic unit. (b) Function table.

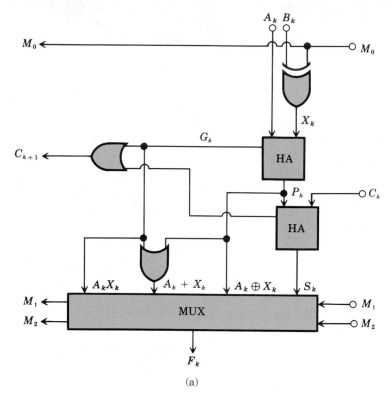

M_2	M_1	M_0	Function
0	0	0	$A \cdot B$
0	0	1	$A \cdot \overline{B}$
0	1	0	$A + B$
0	1	1	$A + \overline{B}$
1	0	0	$A \oplus B$
1	0	1	$A \oplus \overline{B}$
1	1	0	A plus B
1	1	1	A minus B

(a)

Our complete ALU would contain several sections like Fig. 15.3–7a, all under the control of $[M_2 M_1 M_0]$. The table in Fig. 15.3–7b lists the eight control words and the corresponding output functions. Besides addition and subtraction, we have six available logic functions that operate pairwise on the bits in the A and B words. Thus, for instance, the entry $A \cdot B$ means that $F_k = A_k \cdot B_k$ for each section.

Numerical comparisons can be made with the help of the additional OR gate in Fig. 15.3–8a, where $Z = F_* + \cdots + F_1 + F_0$. If we select the arithmetic function A *minus* B, and if the A word equals the B word, then all sections yield $F_k = 0$. The numerical equality $A = B$ is therefore indicated by $Z = 0$. Conversely, $Z = 1$ indicates the numerical inequality $A \neq B$, in which case the sign bit F_* distinguishes between $A > B$ and $A < B$. Figure 15.3–8b summarizes these tests.

Figure 15.3–8
OR gate for numerical comparisons.

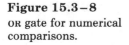

Z	F_∞	Meaning
0	0	$A = B$
1	0	$A > B$
1	1	$A < B$

(a)   (b)

Commercial arithmetic logic units offer all the features we've mentioned, and then some. Indeed, most ALUs have 5-bit control words that actuate 16 logic functions and 16 arithmetic functions in addition to numerical comparison.

Binary Multiplication

Multiplication of unsigned binary numbers is a simple task, involving only addition and position shifting. Consider, for instance, the multiplication of 4-bit words shown in Fig. 15.3–9a. The partial products are formed by multiplying the entire multiplicand by successive bits of the multiplier, which produces either all 0's or a copy of the multiplicand. Each partial product is shifted left to line up with its multiplier bit. Hence, the product of two 4-bit words is an 8-bit word.

Figure 15.3–9
(a) Binary multiplication.
(b) Shift-and-add method.

```
      1100   Multiplicand
      1011   Multiplier
      ----
      1100
      1100      Partial
      0000      products
      1100
    --------
    10000100  Product
```

```
              1100   Multiplicand
              1011   Multiplier
            ------
              0000
      plus    1100   1st partial product
            ------
              1100   1st partial sum
      plus    1100   2nd partial product
            --------
            100100   2nd partial sum
      plus    0000
            --------
           0100100
      plus   1100
          ---------
          10000100   Product
```

(a)   (b)

But this multiplication method does not lend itself immediately to hardware implementation because it requires adding all the partial products at once. A better approach from the hardware viewpoint is the **shift-and-add method** illustrated in Fig. 15.3–9a. Here, the 1st partial sum is obtained by adding the 1st partial product to 00000, the 2nd partial sum is obtained by adding the shifted 2nd partial product to the 1st partial sum, and so forth.

Figure 15.3–10 diagrams a system that uses the shift-and-add method to multiply $[A_3 A_2 A_1 A_0]$ by $[X_3 X_2 X_1 X_0]$. The system includes an 8-bit shift register with parallel inputs and outputs connected to a 4-bit adder. Shifting the register contents to the right relative to the adder is therefore equivalent to shifting partial products to the left. The operational sequence begins with register cells Q_7–Q_4 cleared and the multiplier load into cells Q_3–Q_0. Thus, the register initially holds

$$[Q_7 Q_6 Q_5 Q_4 Q_3 Q_2 Q_1 Q_0] = 0000 X_3 X_2 X_1 X_0$$

Subsequent shifts then yield $Q_0 = X_k$ with $k = 1$, 2, and 3.

Figure 15.3–10
System for shift-and-add multiplication.

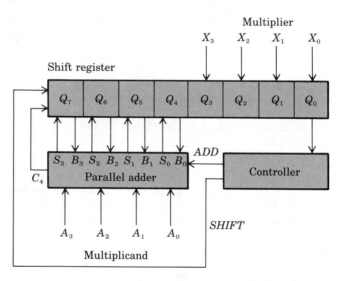

For each partial sum, the controller examines the multiplier bit X_k currently stored in the Q_0 cell. If $X_k = 0$, then the partial product is all 0's and no addition is required. Accordingly, the controller issues commands to:

> Shift the register contents right one cell
> Set $Q_7 = 0$

If $X_k = 1$, then the partial product equals the multiplicand which must be added to the current partial sum. Accordingly, the controller issues commands to:

> Add $[A_3 A_2 A_1 A_0]$ to $[Q_7 Q_6 Q_5 Q_4]$
> Load the result $[S_3 S_2 S_1 S_0]$ into $[Q_7 Q_6 Q_5 Q_4]$
> Shift the register contents right one cell
> Set $Q_7 = C_4$

After these steps have been performed four times, the entire shift register holds the final 8-bit product.

Modifications of Fig. 15.3–10 to handle signed numbers are relatively straightforward. The sign bits must be stored in separate cells, and the magnitudes of negative numbers represented in 2's-complement form must be converted to natural binary code. Additional logic gates are also needed to generate the sign of the product and to convert a negative result back to 2's-complement form if desired.

Obviously, binary multiplication implemented as sequential shift-and-add steps takes much more time than parallel addition or subtraction. The sequential steps for *binary division* are even more time consuming. In very high speed applications, computational time can be reduced using "hard-wired" gate networks designed specifically for multiplication or division. More commonly, however, multiplication and division are *software operations* performed by controlling a general-purpose unit similar to Fig. 15.3–10. As a matter of fact, if we replace the adder with an ALU, then we have most of the functional elements of the central processor in a digital computer.

Exercise 15.3–4 Taking $[A_3 A_2 A_1 A_0] = 1101$ and $[X_3 X_2 X_1 X_0] = 0110$ in Fig. 15.3–10, find the contents of the shift register after each step of the multiplication sequence. Check your final result.

15.4 COMPUTERS AND MICROPROCESSORS[†]

This closing section presents an overview of computers, primarily from the viewpoint of digital-systems engineering. Following an introduction to general concepts, we'll take a closer look at microprocessors and the relationship between hardware and software in microprocessor operations. Lastly, we'll examine the large-scale semiconductor memory units needed for computer systems.

Computer Concepts

Conceptually, a programmable digital computer consists of five functional sections represented as blocks in Fig. 15.4–1. Interactions between sections are indicated by three types of flow lines for data and instructions, control signals, and addresses.

The **input section** delivers **data** to be processed along with **instructions** that constitute the program for the task in question. The **storage section** holds the data and instructions encoded as binary words, and eventually receives results. The **control section** decodes instruction words, issues appropriate signals to the other sections, and supplies **addresses** that tell the storage section where to find data and where to put results. The **arithmetic logic section** (ALU) performs processing operations on stored data and transfers the results to storage. Finally, the **out-**

Figure 15.4–1
Functional sections of
a digital computer.

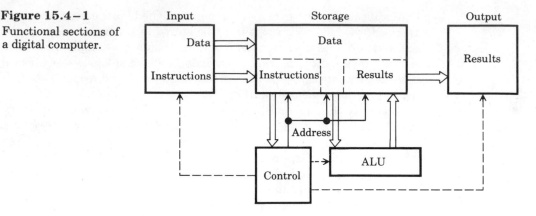

put section converts stored results to the type of presentation dictated by
the program.

Every computing system must implement each of the five conceptual
functions, but the physical layout usually takes a form like Fig. 15.4–2.
The features and purpose of each of the hardware units shown here are
summarized as follows:

- The **central processing unit** (CPU) is the heart of the computer,
  encompassing the controller, the system clock, the ALU, and several
  registers.

Figure 15.4–2
Bus-organized
computing system.

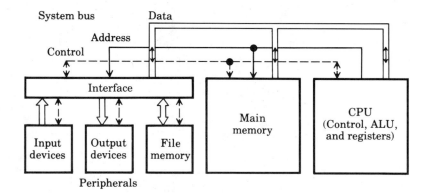
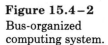

- The **main memory** provides "on-line" storage directly accessible to
  the CPU. It contains a permanent read-only section as well as thousands or even millions of semiconductor memory cells for temporary
  read/write storage.
- The **interface** serves as the buffer to external units called **peripherals.** Information and control signals pass through the interface via
  parallel or serial ports.

- The **file memory** consists of magnetic disk, drum, or tape drives. These "off-line" devices provide nonvolatile mass storage for data and programs, which are loaded to or from the main memory as needed.
- **Input/output devices** include units such as keyboards, light pens or "mice," video monitors, printers, plotters, telephone-line modems, and analog-to-digital and digital-to-analog converters.

Although not explicitly indicated in our diagram, some peripherals combine input and output in one unit — the telephone-line modem being an obvious example.

The system configuration in Fig. 15.4 – 2 is said to be **bus-organized** since all internal communications take place through a set of wires known as the **system bus.** Actually, the system bus consists of three separate wire groups for data, addressing, and control. The two-way **data bus** handles data and instructions. The one-way **address bus** carries address words from the CPU. The **control bus** conveys CPU commands to the other units, but a few of its wires send "interrupt" and "ready" signals generated by peripherals back to the CPU.

With some modifications, Fig. 15.4 – 2 could equally well represent a desktop personal computer or a giant mainframe computer system. The difference in performance between small and large computers depends largely upon two properties of the CPU: *execution speed,* and *word length.* The significance of execution speed should be self-evident, but the importance of word length deserves further clarification.

Suppose that the CPU in a small computer has a word length of eight bits (or one byte) for data and instructions. An 8-bit limit on *instruction words* poses no serious problem since it allows an instruction set of up to $2^8 = 256$ basic operations, and more complicated operations can be carried out as a sequence of steps. However, an 8-bit *data word* corresponds to only 0.4% resolution or two decimal digits of BCD data — clearly inadequate for most calculations. Greater numerical accuracy would then require *serial processing* of two or more words for each quantity. Serial data processing involves considerable overhead time for extra word transfers and carry-bit handling, thereby reducing the system's performance.

The length of *address words* also affects performance because each storage location must have a unique address. Thus, regardless of actual memory size, 8-bit address words would limit the usable memory to a paltry 256 locations. Serial addressing does not offer a viable alternative since it requires more hardware in the storage section and is even more time consuming than serial data processing. Consequently, the address bus in a small computer usually accomodates longer words than the data bus.

Mainframe systems and supercomputers work with 32- to 64-bit words for maximum speed and memory access. Separate hardware sections in the CPU are dedicated to specific tasks such as addition/subtraction, multiplication, and floating-point number conversion. This configuration permits high-speed *parallel processing* whereby several operations go on at once rather than one after the other. Additional processors in the interface section support a gamut of peripheral devices, time-shared terminals, and networking.

Most other computers, including personal computers, are built around a **microprocessor** — a single IC chip containing all the circuitry needed for the CPU. The term *micro* here refers to physical size, as distinguished from performance. Indeed, a 32-bit microprocessor with 24-bit addressing comes close to the performance of some mainframe CPUs. And parallel "number crunching" is also possible with the help of another chip called a **coprocessor.** Applications that require less computing power may be satisfied by the "computer on a chip" consisting of an 8- or 16-bit microprocessor plus on-board memory and interfaces. Mass production of these units has reduced their cost to the point that they can be built into products such as microwave ovens, video games, and automobile ignition systems. The versatile microprocessor therefore deserves further attention.

Microprocessors

Microprocessors come in several word lengths and have different capabilities, depending upon the number of registers, the amount of on-board memory, and the intended applications. Here we introduce general microprocessor characteristics by describing the internal structure and functions of a simplified hypothetical 8-bit unit. This treatment should provide sufficient background for you to pursue the details of a particular model. It will also shed some light on the workings of a CPU in a larger computer and the relationship between a computer program and machine operations.

The structure or **architecture** of our hypothetical microprocessor is given by the block diagram in Fig. 15.4–3. Three buffers interface with the system bus, and a two-way internal bus carries data and instructions between the arithmetic logic section on the left, the control section in the middle, and the register stack on the right. But it must be emphasized at the start that very few of the units have direct or "hard-wired" connections to other units. Instead, interconnections take the form of *gates* that the control section *enables* or *disables,* according to the program instructions. To put this another way — programming at the machine-language level boils down to *wiring with software.*

As to the specific units in Fig. 15.4–3, the arithmetic logic section consists of an ALU similar to Fig. 15.3–7, an **accumulator** (A), and a **status register.** One input word for the ALU always comes from the accumulator, which also holds the output upon completion of the operation. The status register contains condition-code bits or **flags** that indicate when the result of an ALU operation generated a carry ($C = 1$), or was equal to zero ($Z = 1$) or negative ($S = 1$). The control section includes an

instruction register (IR) to hold instruction words, a **decoder** that interprets those words, and a **control generator** that produces the corresponding control signals in synchronism with the clock. The register stack has four *general-purpose registers* (R_1–R_4) and a **selector** that permits data transfers to or from any register. There are also two special registers: a **program counter** (PC) to keep track of the program sequence, and a **memory address register** (MAR) that feeds the address bus.

Figure 15.4–3
Hypothetical
microprocessor.

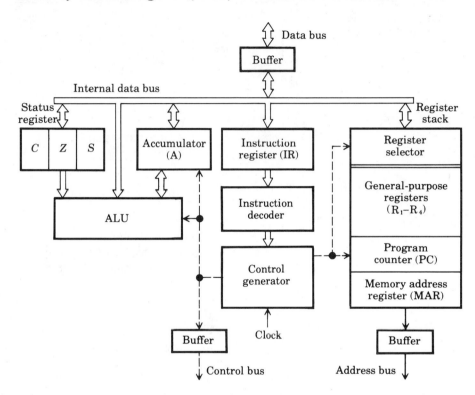

For simplicity's sake, we assume a memory capacity of just 256 locations so the MAR and PC will be 8-bit registers. All other registers also hold 8 bits, except the status register. An actual 8-bit microprocessor has additional status bits, other special registers, and a PC and MAR that each holds 16 bits.

Now consider a hypothetical instruction word such as 10110001. The first two bits might signify an ALU operation, the next three bits identify the operation to be addition, and the last three bits mean that the word to be added to the accumulator word resides in register R_1. When this instruction is transferred from memory to the instruction register, the decoder and control generator enable gates that yield the connections of Fig. 15.4–4. The ALU then adds the words in A and R_1, and the sum appears in A at the next clock pulse. Similarly, another instruction causes the A word to be copied into R_2, a "store"operation, or vice versa for a "load" operation. However, loading or storing a memory word requires *two* instructions, the first to identify the operation and the second to provide the memory address of the word in question. (A third instruction word would be needed to complete a 16-bit address.)

Figure 15.4–4
Connections produced by an instruction word.

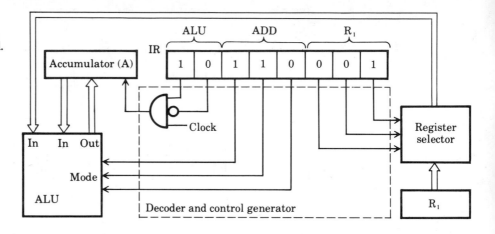

Table 15.4–1 lists the sequence of instructions in an illustrative program for our microprocessor. All addresses are given as decimal numbers, and instructions have been expressed in English. The actual machine language would, of course, consist entirely of binary digits. The purpose of this program is to sum N numbers stored in memory locations 051 to $051 + N - 1$ and place the result in location $051 + N$. The value of N is found in location 050. The program uses the general-purpose registers for the following variables: R_1 holds *SUM,* the sum in progress; R_2 holds *COUNT,* which counts down from N to zero; R_3 holds *NADR,* the address of the next number to be added to *SUM;* and R_4 holds *NUMB,* the number addressed in R_3. Instructions 000–005 initialize the program variables; instructions 006–015 constitute the summing loop; and instructions 016–017 place the result in memory.

With the help of the program description and the comments in the table, you should be able to follow the sequence of instructions. But bear in mind the distinction between the *address* of a memory location and the *contents* of that location! For example, instruction 002 calls for R_2 to be loaded with the contents of the location whose address is given in instruction 003, namely at location 050. We call this a **direct address** instruction, as contrasted with the **indirect address** sequence in instructions 006–007 where the address of the data to be loaded is found in R_3. (Some microprocessors have a special **index register** for the role played by R_3 in this program.) Instruction 004 illustrates **immediate addressing,** since the word to be loaded in R_3 is identical to the next instruction word.

Table 15.4–1
Microprocessor program.

Address	Instruction	Clock cycles	Comment
000	Clear A	4	} Puts $SUM = 0$ in R_1.
001	Store A in R_1	5	
002	Load R_2 from address	4	} Puts $COUNT = N$ in R_2.
003	050	3	
004	Load R_3 with	4	} Puts $NADR = 051$ in R_3.
005	051	3	
006	Load R_4 from address in	4	} Puts $NUMB$ in R_4.
007	R_3	6	
008	Load A from R_1	5	
009	Add A and R_4	5	} Adds $NUMB$ to SUM.
010	Store A in R_1	5	
011	Increment R_3	5	Adds 1 to $NADR$.
012	Decrement R_2	5	Subtracts 1 from $COUNT$.
013	Load A from R_2	5	
014	Jump if A $\neq$ 0 to address	4	} Loops back to instruction 006 if $COUNT \neq 0$.
015	006	4	
016	Store R_1 at address in	4	} Puts result in memory.
017	R_3	6	

Now we turn to the actual **execution** of this program. The instructions first must be stored in memory at the locations shown in Table 15.4–1, and the data must be stored in locations 050 to $051 + N$. Pressing the START button sets the program counter to 000 and initiates the sequence of operations listed in Table 15.4–2. At the first clock pulse, the control unit transfers the contents of the program counter to the memory address register and increments the program counter, as symbolized by PC → MAR and PC + 1 in the table. At the next clock pulse, the word addressed in the MAR is transferred to the instruction register, symbolized by (MAR) → IR, so the IR now holds the binary word meaning "Clear A." This instruction is decoded and carried out during the next two clock pulses. Complete execution of instruction word 000 requires a total of four *clock cycles*. Proceeding similarly, instruction 001 takes five clock cycles, while instructions 002–003 require seven cycles.

Note that the second cycle of instruction 003 is not a decode operation because the previously decoded word told the control unit to look for an address in the next word. By means such as this, the microprocessor distinguishes between instructions, data, and addresses, even though they are all just strings of binary digits stored at various memory locations. Also note how the "jump" in instructions 014–015 resets the PC to 006 if the status register has $Z = 0$, meaning that A $\neq$ 0.

Table 15.4–2 does not give the complete breakdown of all the operations in our simple program, but the clock cycles required for each instruction have been listed in Table 15.4–1. Since the loop is repeated N times,

Table 15.4–2
Microprocessor program execution.

Address	Clock cycle	Operation	PC	MAR	IR	A	R_1	R_2
					Register contents			
000	1	PC → MAR, PC + 1	001	000	—	—	—	—
	2	(MAR) → IR			"Clear A"			
	3	Decode						
	4	Execute CLEAR				0		
001	1	PC → MAR, PC + 1	002	001				
	2	(MAR) → IR			"Store A in R_2"			
	3	Decode						
	4	Select R_1						
	5	Execute STORE					0	
002	1	PC → MAR, PC + 1	003	002				
	2	(MAR) → IR			"Load R_2 from"			
	3	Decode						
	4	Select R_2						
003	1	PC → MAR, PC + 1	004	003				
	2	(MAR) → MAR		050				
	3	Execute LOAD						N
.	.	.	.	.	.	.	.	.
.	.	.	.	.	.	.	.	.
.	.	.	.	.	.	.	.	.
014	1	PC → MAR, PC + 1	015	014	—	COUNT	SUM	COUNT
	2	(MAR) → IR			"Jump if A = 0 to"			
	3	Decode						
	4	Select Z						
015	1	PC → MAR, PC + 1	016	015				
	2	(MAR) → MAR		006				
	3	Execute Z TEST						
	4	MAR → PC if Z = 0	(006)					
.	.	.	.	.	.	.	.	.
.	.	.	.	.	.	.	.	.
.	.	.	.	.	.	.	.	.

once for each number added, program execution takes a total of $33 + N \times 48$ clock cycles. Therefore, if the clock frequency is 1 MHz and $N = 200$, the program execution time will be $(33 + 200 \times 48)/1$ MHz $= 9633 \, \mu s \approx 10$ ms.

Because we have been dealing with a *hypothetical* microprocessor, the program instructions, execution sequence, and clock-cycle requirements differ in various respects from those of an *actual* microprocessor. Nonetheless, Tables 15.4–1 and 15.4–2 correctly imply that even a very simple program becomes quite detailed at the machine-language level. When you write a program in a higher-level language such as Pascal or FORTRAN, it is first processed by another program called a **compiler** that translates your program statements into a sequence of basic operations represented by binary code words. The compiler thereby spares you most of the tedious bookkeeping involved in wiring with software.

But like any other program, the compiler must be stored in memory, and it might use up so much space that there would be insufficient memory left for your encoded program. In that event, you must resort to **assembly language programming** along the lines of Table 15.4–1, which can be translated into machine language by a much shorter program called the **assembler.** We thus see that the software capability of a computer depends in part on the memory capacity. It's appropriate, then, that our discussion concludes with a brief consideration of memory units.

Memory Units

The main memory of a computer contains two different units, a **read-only memory (ROM)** and a **random-access memory (RAM).** The ROM *permanently* stores fixed programs and data such as assemblers, code converters, and look-up tables. The RAM *temporarily* stores data and instructions supplied by the user, aong with results generated by the CPU. A RAM must therefore consist of **read/write cells.** Furthermore, in contrast to the nonvolatile ROM, most RAMs are **volatile** devices whose contents vanish forever when the power goes off.

Both a ROM and a RAM permit direct access to any addressable location without passing through other locations. An important parameter in this regard is the **access time,** defined as the time required to carry out a read or write operation. Typical values range from about 400 ns down to 40 ns.

Two other important memory parameters are size and organization. The *size* of a memory simply equals the maximum number of stored bits. Memory *organization* refers to grouping the bits into words that constitute addressable locations. Suppose, for instance, a ROM holds a total of $8192 = 2^{13}$ bits, so individual addresses would have to be 13 bits long. But if 2^{13} bits are organized into 8-bit memory words or bytes, then we need just 10 address bits since there will be $2^{13}/8 = 2^{10} = 1024$ words. We say that such a ROM stores 1 K $\times$ 8 bits or 1 K-byte where, by convention, 1 K stands for 1024.

In principle, memory locations can be accessed using a decoder as previously depicted for small ROMs in Figs. 14.3–9c and 10c. However, each location then requires its own word line emanating from a decoder AND gate, so the decoder for our 1 K-byte ROM would have to contain 1024 gates. A more practical method is **two-dimensional addressing** illustrated in Fig. 15.4–5. Here, the 2^{13} bits are stored in a 64-word $\times$ 128-bit matrix, and the first 6 address bits are applied to a decoder having 64 gates and word lines. The word lines intersect 128 bit lines that are divided into 8 groups and connected to multiplexers with $128/8 = 16$ inputs. The remaining 4 address bits applied to the multiplexers select from the 16 available words the desired 8-bit word to be read at the output. Since each MUX contains 16 AND gates plus one OR gate, this addressing scheme reduces the gate count from 1024 to $64 + 8 \times (16 + 1) = 200$.

Figure 15.4–5
ROM with two-dimensional addressing.

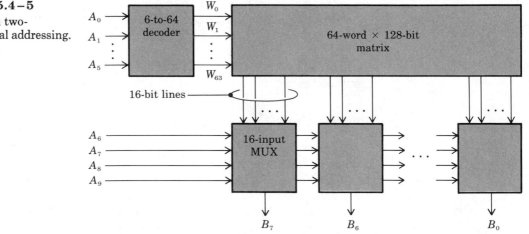

Efficient two-dimensional addressing is essential for integrated-circuit implementation of large read-only memories. Similar addressing methods are used for random-access memories, which often store many more bits than a ROM. We'll develop the organization of complete RAMs after examining a basic read/write memory cell.

Figure 15.4–6 diagrams a **static MOS cell** whose cross-coupled inverters form a bistable flip-flop. The cell is enabled by making X go HIGH so the FET switches conduct; this allows B_{in} to set or reset the flip-flop, producing $B_{out} = \overline{B}_{in}$. The switches become open circuits when X goes LOW, but the flip-flop retains its state for subsequent readout. Complete implementation of this cell requires six transistors, including two for each inverter. Using capacitance for information storage reduces the requirement to three transistors per cell and permits more memory per chip. Such cells are said to be **dynamic,** since they must be **refreshed** periodically to make up for charge leakage from the capacitor.

Figure 15.4–6
MOS memory cell.

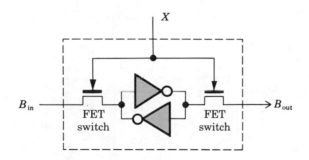

Now let's build a primitive 2×2 array or **memory plane** using static cells, as shown in Fig. 15.4–7. The control bits X_0 and X_1 each enable a *row* of cells, whereas Y_0 and Y_1 each enable a *column*. Thus, when X_0 and Y_1 are high, and X_1 and Y_0 are low, only the cell marked $X_0 Y_1$ will be enabled. If the **chip enable** (CE) is also high while the **read/write select** $R/\overline{W}$ is low,

Figure 15.4–7
Four-bit RAM.

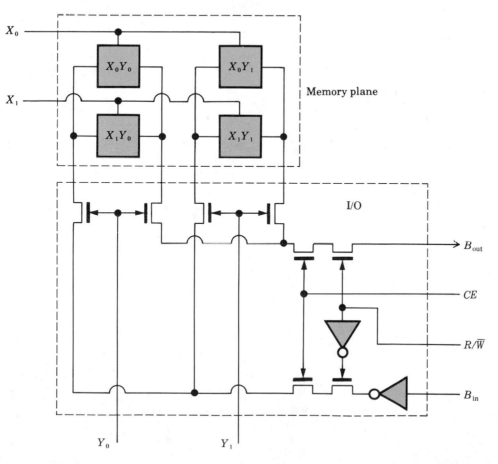

then an information bit can be written into cell $X_0 Y_1$ through the NOT gate that compensates for the cell's inversion. Similarly, the contents of $X_0 Y_1$ can be read by making $R/\overline{W}$ go high.

Figure 15.4–8
4 K × 1 bit RAM.

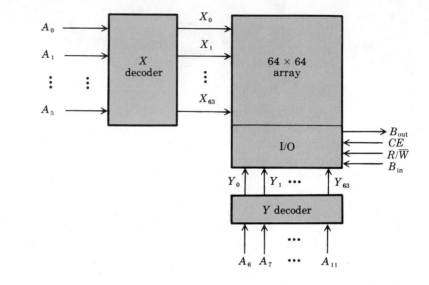

Figure 15.4–9
16 K × 8 bit RAM.

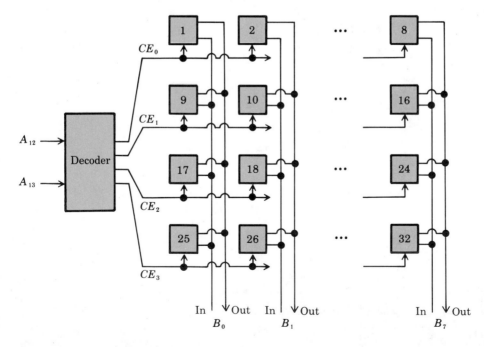

Expanding our array to 64×64 and adding decoders for the X and Y lines yields the $4 \text{ K} \times 1$-bit RAM in Fig. 15.4–8. A 12-bit address word selects any one of the $2^{12} = 4 \times 1024 = 4$ K locations, each holding 1 bit. For a $4 \text{ K} \times 8$-bit RAM, we could put 8 chips in a row and apply the same address word to all of them. Each chip then provides the location for one of the 8 bits from each of the 4 K words. This method is often used to expand the memory of a personal computer with $64 \text{ K} \times 1$-bit RAM chips.

Finally, Fig. 15.4–9 represents an array of 32 chips organized into a $16 \text{ K} \times 8$-bit RAM. All chips receive the first 12 address bits (not shown) while A_{12} and A_{13} are decoded to select a *row* of chips via the CE terminals. This arrangement stores a total of 2^{14} words $\times$ 8 bits per word $= 131{,}072$ bits (or 128 K-bits). The astonishing advances of IC technology have made it possible to put all the circuitry on just one chip!

PROBLEMS

15.1–1 Figure P15.1–1 is a **master-slave flip-flop** built with two clocked SR flip-flops like Fig. 15.1–3. Taking $Q_1 = Q_2 = 0$ initially, draw the S_2, R_2, and Q_2 waveforms produced by the S, R, and Ck signals in Fig. 15.1–4b. Comment on your results by comparing the Ck and Q_2 waveforms.

Figure P15.1–1

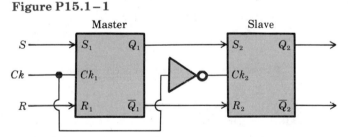

15.1–2 Do Problem 15.1–1 for the case where the inverter in Fig. P15.1–1 is moved to the Ck_1 input rather than the Ck_2 input.

15.1–3 Draw the Q waveform produced when the Ck and D signals in Fig. 15.1–5c are applied to a T flip-flop with $T = D$. Take $Q = 0$ initially.

15.1–4 Draw the Q waveform produced when the Ck and D signals in Fig. 15.1–5c are applied to a JK flip-flop with $J = D$ and $K = 1$. Take $Q = 0$ initially.

15.1–5 Draw the Q waveform produced when the Ck and D signals in Fig. 15.1–5c are applied to a JK flip-flop with $J = 1$ and $K = D$. Take $Q = 0$ initially.

15.1–6 By comparing Eqs. (3), (4), and (5), explain how a D flip-flop can be made to act like: (a) a T flip-flop in the toggle mode; (b) a JK flip-flop.

15.1–7 Suppose that auxiliary gates are connected to a clocked SR flip-flop to get $S = T\overline{Q}$ and $R = TQ$, where T is an external input. Construct the transition table to show that this arrangement acts as a T flip-flop.

15.1−8   Suppose that auxiliary gates are connected to a clocked SR flip-flop to get $S = J\overline{Q}$ and $R = KQ$, where J and K are external inputs. Construct the transition table to show that this arrangement acts as a JK flip-flop.

15.1−9   Inputs A and B are connected to a JK flip-flop through auxiliary gates. Construct the truth table showing Q_{n+1} as a function of A_n and B_n when:

(a) $J = A \oplus B$ and $K = \overline{B}$;

(b) $J = \overline{AB}$ and $K = A\overline{B}$.

Compare your results with Fig. 15.1−7b.

15.1−10   Consider a digital waveform generator consisting of a clock driving a 3-bit binary counter with a logic network connected to the outputs Q_2, Q_1, and Q_0. Obtain the minimized sum-of-products expression needed to produce waveform X in Fig. P15.1−10.

Figure P15.1−10

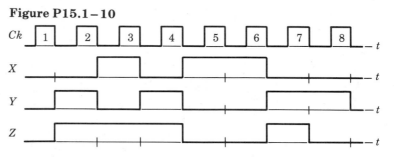

15.1−11   Do Problem 15.1−10 for waveform Y.

15.1−12   Do Problem 15.1−10 for waveform Z.

15.1−13   Using the fewest possible 4-bit binary counters, diagram a system that counts to 180 and resets.

15.1−14   Using the fewest possible 3-bit binary counters, diagram a system that counts to 100 and resets.

15.1−15   Diagram a system that uses 4-bit binary counters to produce a pulse at output X after every 8 input pulses and a pulse at output Y after every 400 input pulses.

15.1−16   Diagram a system using 6-bit binary counters to carry out the task in Exercise 15.1−3.

15.1−17   Suppose a 3-bit binary ripple counter is built with flip-flops that have propagation delay $t_{PD} = T_c/4$, where T_c is the period of the clock input. Construct the timing diagram and identify the false counts represented by $[Q_2 Q_1 Q_0]$.

15.1−18   A synchronous counter with $N = 3$ can be built using two JK flip-flops connected such that $K_0 = K_1 = 1$, $J_0 = \overline{Q}_1$, and $J_1 = Q_0$. The input is applied to both Ck terminals. Take $[Q_1 Q_0] = 00$ as the initial state and construct a table similar to Fig. 15.1−9b but also listing J_1 and J_0. Then draw the timing diagram.

15.1−19   A synchronous counter with $N = 3$ can be built using two T flip-flops connected such that $T_0 = \overline{Q}_1 + Q_0$ and $T_1 = Q_1 + Q_0$. The input is applied to both Ck terminals. Take $[Q_1 Q_0] = 00$ as the initial state and construct a table similar to Fig. 15.1−9b but also listing T_1 and T_0. Then draw the timing diagram.

15.1–20 If Q_4, rather than $\overline{Q}_4$, is fed back to FF$_0$ in Fig. 15.1–14a, then we have an *untwisted* ring counter.

  (a) Taking $[Q_4 Q_3 Q_2 Q_1 Q_0] = 10000$ as the initial state, draw the timing diagram to show that $N = 5$.

  (b) What would happen if the initial state is $[Q_4 Q_3 Q_2 Q_1 Q_0] = 00000$?

15.1–21 Figure P15.1–21 shows a shift register with feedback via a logic network to form a **binary sequence generator**. The initial state is set at $[Q_2 Q_1 Q_0] = 111$ before the clock goes on. Find the resulting output sequences available at Q_0, Q_1, Q_2, and D_0 with: (a) $D_0 = Q_1 \oplus Q_2$; (b) $D_0 = \overline{Q}_0 Q_1 Q_2$.

Figure P15.1–21

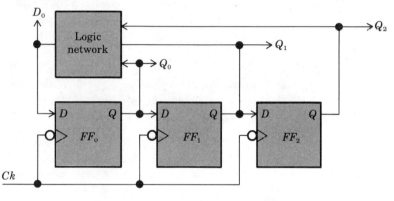

15.1–22 Do problem 15.1–21 with: (a) $D_0 = Q_0 \oplus Q_2$; (b) $D_0 = \overline{Q}_0 \overline{Q}_1 Q_2$.

15.1–23 Do Problem 15.1–21 with: (a) $D_0 = \overline{Q_1 Q_2}$; (b) $D_0 = \overline{Q}_0 Q_2$.

15.1–24 Diagram a system similar to Fig. 15.1–18 that measures the approximate time between a *START* pulse and a *STOP* pulse appearing at separate inputs. The system is to be initialized by a manual *RESET*, and the display should show the time in tenths of a second.

15.1–25 Diagram a system similar to Fig. 15.1–18 that measures the approximate time between a *START* pulse and a *STOP* pulse appearing at the same input. The system is to be initialized by a manual *RESET*, and the display should show the time in milliseconds.

15.1–26 A certain binary signal X consists of a series of pulse bursts having variable durations of 5–10 seconds. The time from the start of one burst to the start of the next is at least 20 seconds. Diagram a system similar to Fig. 15.1–18 that counts and displays the number of pulses in successive bursts. Use an auxiliary 4-bit counter with two decoding gates to generate the *LOAD* and *RESET* signals.

15.2–1 A certain weighted-resistance DAC with $R_0 = 32$ kΩ and $V_{ref} = 15$ V is to have $v_{q-max} = 12$ V and $\Delta v \le v_{q-max}/50$. Find the minimum required value of n and the corresponding values of Δv, R_{n-1}, and R_F.

15.2–2 Do Problem 15.2–1 with $\Delta v \le v_{q-max}/500$.

15.2–3 Do problem 15.2–1 with $\Delta v \le v_{q-max}/2000$.

15.2–4   Suppose V_{ref} in Fig. 15.2–2 is replaced by a time-varying analog voltage $v_a(t)$. We then have a **multiplying DAC** or **programmable attenuator** with output $v_q(t) = -Kv_a(t)$, where the multiplying factor K depends on $[B_{n-1} \ldots B_2 B_1 B_0]$ and could be controlled by computer. Taking $n = 4$, find R_F/R_0 such that $K_{max} = 1.5$. What are the other possible values of K?

15.2–5   Modify Eqs. (1) and (2) for a 4-bit weighted-resistance DAC when the input $[B_3 B_2 B_1 B_0]$ represents a decimal digit using a BCD code with weighting factors 5, 3, 1, 1. Then write i_{in} in terms of B_i, R_i, and V_{ref} to find the required resistance ratios R_3/R_0, R_2/R_0, and R_1/R_0.

15.2–6   Do Problem 15.2–5 for a BCD code with weighting factors 3, 3, 2, 1.

15.2–7   Suppose a decimal integer x in the range 0–99 is represented by two 8421 BCD code words, say $[B_3 B_2 B_1 B_0]$ for the tens place and $[A_3 A_2 A_1 A_0]$ for the ones place. Diagram a system using two 4-bit weighted-resistance DACs and another op-amp to get $v_{out} = 0.1x$ V.

15.2–8   Consider an n-bit DAC using an R-$2R$ ladder with $R = 5$ kΩ. Find n, v_{q-max}, V_{ref}, and R_F such that $\Delta v = 20$ mV, $v_{q-max} \approx 5$ V, and $R_F \approx 10$ kΩ. Take V_{ref} to be one of the following standard values: 5, 6, 8, 12, 15, 18, or 24 V.

15.2–9   Do Problem 15.2–8 with $\Delta v = 5$ mV and $v_{q-max} \approx 10$ V.

15.2–10   Do Problem 15.2–8 with $\Delta v = 2$ mV and $v_{q-max} \approx 16$ V.

15.2–11   Let the R-$2R$ ladder in Fig. 15.2–3 have $n = 3$ sections with input voltages v_0, v_1, and v_2. Starting at the lowest node, perform successive Thévenin/Norton conversions to show that the short-circuit current into the op-amp is

$$i_{in} = (4v_2 + 2v_1 + v_0)/(8 \times 3R)$$

Then extrapolate this expression for arbitrary n and derive Eq. (5).

15.2–12   Suppose v_a in Fig. 15.2–6 comes from a temperature sensor. Choose an appropriate value for V_R and add auxiliary logic to turn on a heater ($H = 1$) when $v_a < 6$ V and to sound an alarm ($X = 1$) when $v_a < 3$ V or $v_a > 9$ V.

15.2–13   Consider a 3-bit parallel-comparator ADC whose input v_a comes from a liquid level sensor. Choose an appropriate value for V_R and add auxiliary logic to actuate a fill valve ($F = 1$) when $v_a < 6$ V, actuate a drain valve ($D = 1$) when $v_a > 12$ V, and sound an alarm ($X = 1$) when $v_a < 4$ V or $v_a > 14$ V.

15.2–14   Careful examination of Fig. 15.2–6b reveals that $B_1 = Q_2$ and $B_0 = Q_3 + \overline{Q}_2 Q_1$. Construct a similar table for a 3-bit converter and obtain the expressions for B_2, B_1, and B_0 in terms of Q_1–Q_7.

15.2–15   Carry out the details leading to Eq. (9).

15.2–16   Redraw Fig. 15.2–9b for the case of $\Delta v = 0.5$ V and $v_a = 1.7$ V.

15.2–17   Redraw Fig. 15.2–9b for the case of $\Delta v = 0.2$ V and $v_a = 2.1$ V.

15.2–18   Redraw Fig. 15.2–9b for the case of $\Delta v = 0.25$ V and $v_a = 3.3$ V.

15.2–19   Consider a digital processor that edits questionable data in the input sequence x_k, $k = 0, 1, 2, \ldots$ Write a set of equations relating the edited output sequence y_k to x_k for the following algorithm: Keep any data point whose value falls within $\pm 60\%$ of the average of the previous two; otherwise, replace the data point by that average.

15.2–20 Do Problem 15.2–19 for the following algorithm: Keep any data point whose value falls within $\pm 60\%$ of the average of the previous and subsequent points; otherwise, replace the data point by that average. Do not use x_{k+1} to compute y_k. (Why?)

15.2–21 The differential equation of a first-order highpass filter is $\tau \, dy/dt + y = \tau \, dx/dt$. Obtain an expression similar to Eq. (12a) for the digital-filter implementation.

15.2–22 A second-order filter that passes bandwidth B centered at frequency ω_0 is described by the differential equation $d^2y/dt^2 + B \, dy/dt + \omega_0^2 y = B \, dx/dt$. Obtain an expression similar to Eq. (12a) for the digital-filter implementation using the approximation $(y_k - 2y_{k-1} + y_{k-2})/T^2 \approx d^2y/dt^2$.

15.2–23 Do Problem 15.2–22 for a second-order notch filter that rejects bandwidth B centered at frequency ω_0, and is described by the differential equation $d^2y/dt^2 + B \, dy/dt + \omega_0^2 y = d^2x/dt^2 + \omega_0^2 x$.

15.3–1 Draw a NOR-gate network for a full adder based on Fig. 14.3–1c.

15.3–2 Draw a NAND-gate network for a full adder based on the expressions in Exercise 15.3–1.

15.3–3 Diagram a system of half adders and full adders that produces the sum of three input words, $[X_2 X_1 X_0]$, $[Y_2 Y_1 Y_0]$, and $[Z_2 Z_1 Z_0]$.

15.3–4 Draw the block diagram of a system of BCD adders that produces the sum of three input words, $[X_3 X_2 X_1 X_0]$, $[Y_3 Y_2 Y_1 Y_0]$, and $[Z_3 Z_2 Z_1 Z_0]$.

15.3–5 A sequence of no more than ten decimal integers is to be summed. The integers are represented by BCD words available at the output terminals of a register like Fig. 15.1–15, and the next word appears immediately after each load pulse. Draw a block diagram of a system for this task using two BCD adders.

15.3–6 Let $[Y_3 Y_2 Y_1 Y_0]$ be a BCD word in either the 8421 or Excess-3 code from Table 14.1–2. Explain how you could use the adder-subtractor in Fig. 15.3–6 to convert from one type of code to the other.

15.3–7 The adder/subtractor in Fig. 15.3–6 can overflow only when the X and A words have the same sign, and the overflow results in an incorrect output sign. Use these properties to devise a logic function that produces $E = 1$ for overflow detection.

15.3–8 Overflow in Fig. 15.3–6 can be detected by examining the carries C_* and C_4. Determine the overflow indication by performing the calculations 9 plus 7, 9 plus 6, (-9) plus (-7), and (-9) plus (-8). Then devise a network to produce $E = 1$ when overflow occurs.

15.3–9 The **1's complement** of a signed binary integer is obtained simply by complementing all bits. Thus, for instance, -9 is represented by $0^*1\overline{0}\,\overline{0}\,\overline{1} = 1^*0110$. If a carry overflow is produced when two such numbers are summed, then it must be added to the least significant bit of the result. This **end-around carry** never changes the sign bit of the sum. Use 1's-complement arithmetic to calculate 7 plus (-12), 12 plus (-7), and 12 plus (-12). Explain your result in the last case.

15.3–10 Modify Fig. 15.3–6 for operation with 1's-complement numbers as described in Problem 15.3–9. Hint: You will need additional half adders.

15.3–11 Suppose a fourth mode bit M_3 is incorporated into the ALU in Fig. 15.3–7 such that A_k is replaced by $\overline{A}_k$ when $M_3 = 1$. What arithmetic functions are produced when $[M_3 M_2 M_1 M_0] = 1110$ and $[M_3 M_2 M_1 M_0] = 1111$?

15.3 – 12   Suppose the C_0 input for the ALU in Fig. 15.3–7 is separated from the M_0 input. What arithmetic functions are produced when $[M_2 M_1 M_0 C_0] = 1100$ and $[M_2 M_1 M_0 C_0] = 1110$?

15.3 – 13   Suppose the 2's-complement words $[Y_* Y_3 Y_2 Y_1 Y_0]$ and $[Z_* Z_3 Z_2 Z_1 Z_0]$ are to be multiplied by the system in Fig. 15.3–10. Devise an auxiliary system that produces the correct magnitude word $[A_3 A_2 A_1 A_0]$ from the Y word. Also devise a logic network to get the sign bit P_* of the product.

15.3 – 14   Set up the multiplication of $[A_1 A_0]$ times $[B_1 B_0]$ in symbolic form. Then devise a system using gates and half adders to produce the resulting product $[P_3 P_2 P_1 P_0]$.

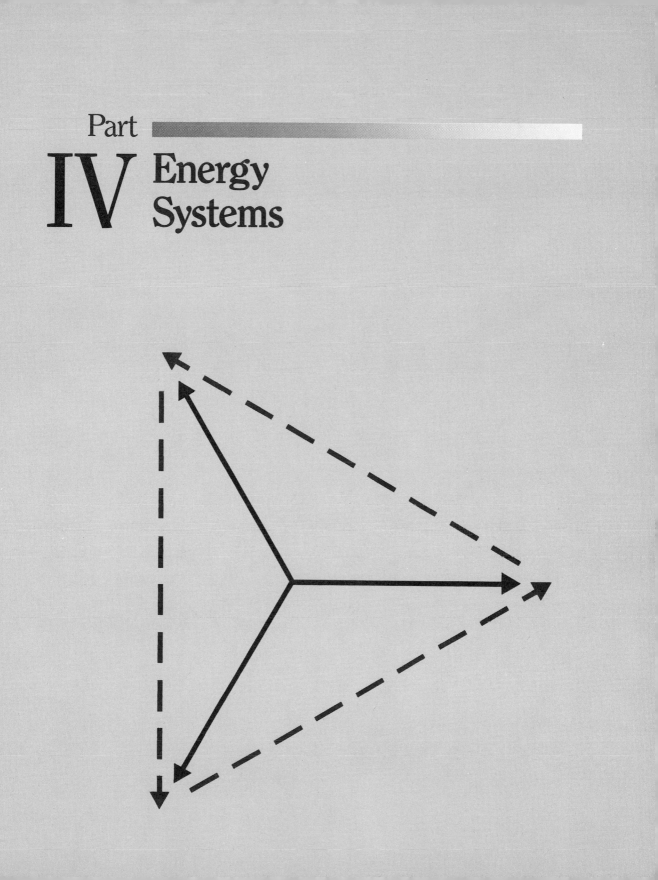

Chapter

16 AC Power Systems

Although energy appears in many different forms, the vast majority of all energy delivered from one point to another across the globe is handled by AC power systems. Appropriately, then, we begin our study of energy systems with an introduction to AC power engineering.

This chapter defines important concepts of AC power, analyzes three-phase circuits for power transfer, and describes representative power transmission and distribution systems. A major concern throughout our investigation is the crucial task of efficient power delivery from source to load. Various other practical considerations will also be discussed as we go along, including methods for AC power measurement.

Objectives

After studying this chapter and working the exercises, you should be able to do each of the following:

- Carry out AC power calculations, distinguishing between real, reactive, and apparent power (Section 16.1).

- Define power factor, and determine the element needed to correct the power factor of a given combination of loads (Section 16.1).

- State the advantages of three-phase systems for power transfer (Section 16.2).

- Find the voltages, currents, and power in a balanced three-phase circuit with a wye or delta load (Section 16.2).

- Explain how power can be measured in a single-phase or three-phase circuit (Sections 16.1 and 16.2).

- Identify the components of an AC power network, and explain why transformers play a critical role in power distribution (Section 16.3). †

16.1 AC POWER

Back in Section 6.3 we saw that instantaneous power in an AC circuit has an oscillatory behavior. The details and implications of that behavior are explored more fully here, starting with the concepts of real and reactive power. Then we'll define the apparent power and power factor of a load, and show how power-factor correction improves the efficiency of power transfer from an AC generator to an energy-consuming load. We'll also look at wattmeters for AC power measurements.

Real and Reactive Power

Two symbolic conventions commonly used in electric power engineering need to be stated at the outset of our study of AC power. These conventions simplify notation, and they will be employed throughout this chapter.

Our first convention reflects the fact that AC power systems operate at a *fixed frequency,* usually

$$\omega = 2\pi \times 60 \text{ Hz} \approx 377 \text{ rad/sec}$$

We may therefore ignore frequency-dependent effects and write any load impedance in the streamlined form

$$\underline{Z} = R + jX \tag{1a}$$

Here R and X stand for the AC resistance $R(\omega)$ and reactance $X(\omega)$, respectively, when evaluated at the specified frequency. We'll also drop the subscript on the impedance angle θ_z and let

$$\theta \triangleq \sphericalangle \underline{Z} = \arctan(X/|R|) \tag{1b}$$

Figure 16.1–1 shows our load model and the corresponding impedance triangle. This series model holds equally well for a parallel load or a combination of series and parallel elements, with R and X being interpreted as the *equivalent series values.* In particular, if you know the polar impedance

Figure 16.1–1
Equivalent load and impedance triangle.

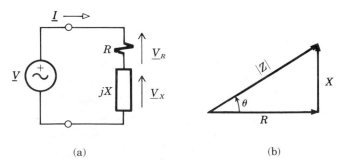

(a)   (b)

$\underline{Z} = |Z|\underline{/\theta}$ for any load, then

$$R = |Z|\cos\theta \qquad X = |Z|\sin\theta \qquad \text{(1c)}$$

as follows from Fig. 16.1–1b.

Our second convention combines phasor notation with the *root-mean-square values* from Eq. (17), Section 6.3. The terminal voltage and current phasors in Fig. 16.1–1a will be written in the usual form

$$\underline{V} = |V|\underline{/\theta_v} \qquad \underline{I} = |I|\underline{/\theta_i} \qquad \text{(2a)}$$

But rather than equating $|V|$ and $|I|$ to the peak sinusoidal values V_m and I_m, we'll express phasor magnitudes in *rms units* as

$$|V| \triangleq V_m/\sqrt{2} \qquad |I| \triangleq I_m/\sqrt{2} \qquad \text{(2b)}$$

Ohm's law for AC circuits still relates $\underline{V}$ and $\underline{I}$ by

$$\underline{V} = \underline{Z}\underline{I}$$

since $\sqrt{2}$ cancels out on both sides when we convert from peak to rms values. Thus, as usual,

$$|V| = |Z||I| \qquad \theta_v = \theta + \theta_i \qquad \text{(3)}$$

However, the phasors defined by Eq. (2) now represent the sinuosoidal waveforms

$$v(t) = \sqrt{2}|V|\cos(\omega t + \theta_v) \qquad i(t) = \sqrt{2}|I|\cos(\omega t + \theta_i) \qquad \text{(4)}$$

where $\sqrt{2}|V| = V_m$ and $\sqrt{2}|I| = I_m$.

Having stated our conventions, we're ready to consider the instantaneous power $p(t) = v(t)i(t)$ delivered to the load in Fig. 16.1–1a. First, we use Eq. (4) with $\theta_i = \theta_v - \theta$ and we apply standard trigonometric identities to obtain

$$p(t) = v(t)i(t) = 2|V||I|\cos(\omega t + \theta_v)\cos(\omega t + \theta_v - \theta)$$

$$= |V||I|\cos\theta\,[1 + \cos 2(\omega t + \theta_v)]$$

$$+ |V||I|\sin\theta\sin 2(\omega t + \theta_v)$$

Next, focusing on the quantities $|V||I|\cos\theta$ and $|V||I|\sin\theta$, we let

$$P \triangleq |V||I|\cos\theta \qquad Q \triangleq |V||I|\sin\theta$$

Now $p(t)$ can be written as

$$p(t) = p_R(t) + p_X(t) \qquad \text{(5a)}$$

where

$$p_R(t) = P[1 + \cos 2(\omega t + \theta_v)] \qquad p_X(t) = Q\sin 2(\omega t + \theta_v) \qquad \text{(5b)}$$

We have thereby decomposed the instantaneous power into two components, as illustrated in Fig. 16.1–2 with $\theta_v = 0$ and $\theta > 0$.

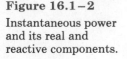

Figure 16.1–2

Instantaneous power
and its real and
reactive components.

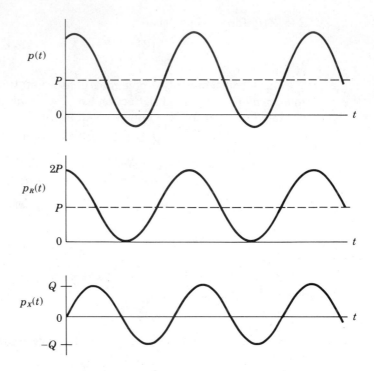

The component $p_R(t)$ oscillates between $2P$ and 0 and has the *average value* $P = |V||I| \cos \theta$. To interpret this quantity, we note from Fig. 16.1–1 that $\cos \theta = R/|Z|$, $|V|/|Z| = |I|$, and $R\underline{I} = \underline{V}_R$, so

$$P = |V||I| \cos \theta = |I|^2 R = |V_R|^2/R \qquad (6)$$

Hence, $p_R(t)$ accounts for the *power absorbed* by the load's resistance. Accordingly, we call P the **real power** measured in **watts** (W).

The other component $p_X(t)$ oscillates between $+Q$ and $-Q$ and has *zero average value*. Consequently, $p_X(t)$ does not contribute to power absorbed by the load. Instead, we note from Fig. 16.1–1 that $\sin \theta = X/|Z|$, $|V|/|Z| = |I|$, and $jX\underline{I} = \underline{V}_X$, so

$$Q = |V||I| \sin \theta = |I|^2 X = |V_X|^2/X \qquad (7)$$

Hence, $p_X(t)$ accounts for *energy exchange* between the source and the load's reactance. Accordingly, we call Q the **reactive power,** and we measure it in **volt-amperes reactive** (VAr) to emphasize that reactive power represents alternating energy storage rather than power transfer. (Be careful not to confuse the symbol Q used here with the quality factor of resonant circuits defined in Section 6.3.)

If the load happens to be entirely resistive, then $X = 0$, $Q = 0$, and $p(t) = p_R(t)$. But if the load happens to be entirely reactive, then $R = 0$,

$P = 0$, and $p(t) = p_X(t)$. Furthermore, in the case of a single inductor or capacitor, Eq. (7) becomes

$$Q_L = |I_L|^2 \omega L = |V_L|^2 / \omega L \qquad \text{(8a)}$$

$$Q_C = -|I_C|^2 / \omega C = -|V_C|^2 \omega C \qquad \text{(8b)}$$

The negative value of Q_C simply means that the waveform $p_X(t)$ would be inverted in Fig. 16.1–2. More generally, a load with inductive reactance $(X > 0)$ has $Q > 0$, while a load with capacitive reactance $(X < 0)$ has $Q < 0$. We therefore say that an inductive load "consumes" reactive power, while a capacitive load "produces" reactive power.

Since reactive power represents an oscillatory energy exchange, it does not contribute to power transfer. On the contrary, reactive power often reduces power-transfer efficiency by increasing the rms current required to deliver a specified average power from the source to the load. The increased current then results in power wasted as ohmic heat in the generator's internal resistance and the resistance of the wires connecting source to load. An example will help illustrate this effect.

Example 16.1–1 Figure 16.1–3 diagrams a parallel capacitive load connected to an AC voltage generator with $|V| = 600$ V (rms). The combined source and wire resistance are represented by $R_s = 8\ \Omega$. We want to find the resulting rms current, the real and reactive power, and the power-transfer efficiency.

Figure 16.1–3
Circuit for Example 16.1–1.

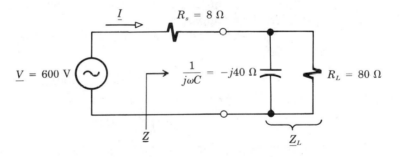

The load impedance is $\underline{Z}_L = 80\|(-j40) = 16 - j32$, so the source sees the total impedance

$$\underline{Z} = R_s + \underline{Z}_L = 24\ \Omega - j32\ \Omega = 40\ \Omega\ \underline{/53.1°}$$

Hence, $|I| = 600$ V$/40\ \Omega = 15$ A, and Eqs. (6) and (7) give

$$P = 15^2 \times 24 = 5400\ \text{W} = 5.4\ \text{kW}$$

$$Q = 15^2 \times (-32) = -7200\ \text{VAr} = -7.2\ \text{kVAr}$$

Since R_s dissipates $P_s = |I|^2 R_s = 1800$ W, the power actually delivered to the load resistance R_L is $P_L = P - P_s = 3600$ W. The **power-transfer**

efficiency is therefore

$$P_L/P = 3600/5400 \approx 67\%$$

a rather low efficiency.

But suppose the capacitive reactance could be removed from the load so that $|I| = 600 \text{ V}/(8 + 80) \ \Omega = 6.82$ A. The ohmic heat loss drops to $|I|^2 R_s \approx 372$ W while the load power increases slightly to $|I|^2 R_L \approx 3719$ W. The efficiency now becomes $P_L/P = 3719/(3719 + 372) \approx 91\%$, and we have a considerable improvement.

Exercise 16.1–1   Find $|I|, P, Q,$ and P_L/P for the circuit in Fig. 16.1–3 when an inductor with impedance $j50 \ \Omega$ is inserted in series with R_s.

Apparent Power and Power Factor

The product of the rms voltage and current at the terminals of a load is called the **apparent power,** denoted by $|V\|I|$ and expressed in **volt-amperes** (VA). This quantity is easily measured with the help of simple AC meters. In general, however, apparent power does *not* equal the actual power absorbed by the load. Instead, when we square and add Eqs. (6) and (7) we find that $|V\|I|$ is related to P and Q by

$$|V\|I| = \sqrt{P^2 + Q^2}$$

The form of this equation suggests a triangular relationship between real, reactive, and apparent power.

To develop the geometric picture, we use the complex conjugate of the current phasor

$$\underline{I}^* \triangleq |I|\underline{/-\theta_i} = |I|\underline{/\theta - \theta_v}$$

Then we define the **complex power**

$$\underline{S} \triangleq \underline{V}\underline{I}^* = |V\|I|\underline{/\theta} \tag{9a}$$

or, in rectangular form,

$$\underline{S} = |V\|I|\cos\theta + j|V\|I|\sin\theta = P + jQ \tag{9b}$$

Figure 16.1–4 shows the resulting complex-plane diagram known as the **power triangle,** a right triangle with hypotenuse

$$|S| = \sqrt{P^2 + Q^2} = |V\|I| \tag{10}$$

Figure 16.1–4
Power triangle.

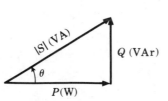

Q (VAr)

θ

P(W)

This triangle has exactly the same shape as the impedance triangle back in Fig. 16.1–1b since

$$\underline{S} = |I|^2 R + j|I|^2 X = |I|^2 \underline{Z}$$

as follows from Eq. (9b) with $P = |I|^2 R$ and $Q = |I|^2 X$.

Another significant property of complex power is the **conservation law,** which states that:

> When several loads are connected to the same source, the total complex power from the source equals the *sum* of the complex powers of the individual loads.

Thus, if two loads draw $\underline{S}_1 = P_1 + jQ_1$ and $\underline{S}_2 = P_2 + jQ_2$, respectively, then the source must supply

$$\underline{S} = \underline{S}_1 + \underline{S}_2 = (P_1 + P_2) + j(Q_1 + Q_2) \tag{11}$$

The conservation property, combined with the power triangle, eliminates the need for finding combined impedances and thereby simplifies many power calculations.

Loads that consume large amounts of power are usually characterized in terms of the **power factor,** defined by

$$\text{pf} \triangleq \frac{P}{|S|} = \cos \theta \tag{12}$$

Clearly, the power factor of a passive load always falls within the range $0 \le \text{pf} \le 1$ since $P \ge 0$ and $|S| \ge P$. Unity power factor simply means that $|S| = P$, which occurs when $Q = 0$ and $\theta = 0$. Thus, a load with pf = 1 has zero equivalent series reactance and draws the minimum source current $|I| = P/|V|$ for a fixed value of P. Otherwise, any load with $Q \ne 0$ draws $|I| = P/(|V| \times \text{pf}) > P/|V|$. In view of the increased current that must be supplied when pf < 1, electric utilities usually charge higher rates to large industrial consumers whose plants operate at low power factors.

Given the power rating and power factor of a load, you can easily compute the apparent power via $|S| = P \times \text{pf}$. Then you can compute the *magnitude* of the reactive power from Eq. (10) rewritten as

$$Q = \pm \sqrt{|S|^2 - P^2} = \pm |S|\sqrt{1 - \text{pf}^2} \tag{13}$$

However, the *sign* of Q depends upon the nature of the load's reactance. If the load is *inductive,* so $Q > 0$ and $\theta > 0$, then we say that it has a **lagging power factor** — meaning that the current phasor lags the voltage phasor. Conversely, if the load is *capacitive,* so $Q < 0$ and $\theta < 0$, then we say that it has a **leading power factor** — meaning that the current phasor leads the voltage phasor. Table 16.1–1 summarizes this power-factor terminology.

Many industrial loads include inductive motor windings and therefore have lagging power factors. Capacitors are often used in conjunction with such loads for the purpose of **power-factor correction.** An appropriate

Table 16.1–1

Power Factor	Conditions
Unity	$X = 0,\ \theta = 0,\ Q = 0$
Lagging	$X > 0,\ \theta > 0,\ Q > 0$
Leading	$X < 0,\ \theta < 0,\ Q < 0$

capacitor connected in parallel with an inductive load cancels out the reactive power and the combined load has pf = 1, thereby minimizing current drawn from the source.

Example 16.1–2

Figure 16.1–5a represents an industrial plant operating from a 500-V source at 60 Hz. The plant consists of two loads having

$$P_1 = 48 \text{ kW} \qquad \text{pf}_1 = 0.60 \text{ lagging}$$

$$P_2 = 24 \text{ kW} \qquad \text{pf}_2 = 0.96 \text{ leading}$$

We infer from this data that load #1 is highly inductive, while load #2 is slightly capacitive. The additional capacitor C has been included to correct the plant's overall power factor. We'll first use conservation of complex power and the power triangle to combine the two loads. Then we'll determine the value of C needed to get unity power factor.

Figure 16.1–5

Plant diagram and power triangle for Example 16.1–2.

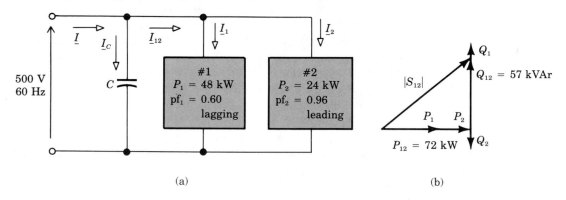

(a)   (b)

The apparent and reactive power for load #1 are found from Eqs. (12) and (13) to be

$$|S_1| = 48 \text{ kW}/0.6 = 80 \text{ kVA}$$

$$Q_1 = +\sqrt{80^2 - 48^2} = +64 \text{ kVAr}$$

where we've taken $Q_1 > 0$ because pf_1 is lagging. The rms current drawn by this load is

$$|I_1| = 80 \text{ kVA}/500 \text{ V} = 160 \text{ A}$$

Similarly, for load #2 we find that

$$|S_2| = 25 \text{ kVA} \qquad Q_2 = -7 \text{ kVAr} \qquad |I_2| = 50 \text{ A}$$

Now we can construct the combined power triangle shown in Fig. 16.1–5b with

$$P_{12} = P_1 + P_2 = 72 \text{ kW} \qquad Q_{12} = Q_1 + Q_2 = +57 \text{ kVAr}$$

Thus,

$$|S_{12}| = \sqrt{72^2 + 57^2} = 91.8 \text{ kVA}$$

$$|I_{12}| = 91.8 \text{ kVA}/500 \text{ V} = 184 \text{ A}$$

and $\text{pf}_{12} = 72/91.8 = 0.784$, which is still lagging since $Q_{12} > 0$.

Table 16.1–2 summarizes our results so far. You should note carefully those quantities that add directly (P and Q) and those that do not ($|S|$ and $|I|$). The last two lines of this table pertain to power-factor correction using an ideal lossless capacitor with $P_C = 0$ and $Q_C = -57$ kVAr. The corresponding value of C is calculated from Eq. (8b) as

$$C = -Q_C/\omega |V_C|^2 = 605 \ \mu\text{F}$$

where $|V_C| = |V| = 500$ V and $\omega = 377$ rad/sec.

Table 16.1–2

Load	P (kW)	Q (kVAr)	$|S|$ (kVA)	$|I|$ (A)
#1	48	+64	80	160
#2	24	−7	25	50
#1 and #2	72	+57	91.8	184
C	0	−57	57	11
Plant	72	0	72	144

The entire plant with its corrected power factor has

$$P = P_{12} + P_C = 72 \text{ kW} \qquad Q = Q_{12} + Q_C = 0$$

so $\text{pf} = 1$ and $|S| = P$. Accordingly, the rms current from the source will be

$$|I| = 72 \text{ kVA}/500 \text{ V} = 144 \text{ A}$$

which is less than the current for load #1 alone! In this situation, the source sees a plant that appears to be purely resistive and draws the minimum current $|I| = P/|V|$. But the plant actually contains inductive and capacitive reactances exchanging stored energy via "reactive currents." These reactive currents circulate entirely within the plant, accounting for the larger values of $|I_{12}|$ and $|I_1|$. A phasor diagram would confirm, of course, that $\underline{I}_1 + \underline{I}_2 = \underline{I}_{12}$ and $\underline{I}_{12} + \underline{I}_C = \underline{I}$.

Exercise 16.1–2   An AC motor operates from a 200-V source at 60 Hz. The motor draws $P = 8$ kW and $|I| = 85$ A. Calculate pf and Q. Then find the motor's equivalent impedance $\underline{Z}$ in rectangular form.

Exercise 16.1−3

An 8-Ω resistive heating element and a capacitor C are connected in parallel with the motor in Exercise 16.1−2. The entire plant has unity power factor. Make a table like Table 16.1−2 for this plant, and evaluate C.

Wattmeters

Electrical power can be measured using the meter diagrammed in Fig. 16.1−6. The heart of this instrument is an **electrodynamometer** movement with a current-sensing coil carrying $i_i(t)$ and a voltage-sensing coil carrying $i_v(t)$. We'll show in Section 17.4 that the steady-state deflection angle of the rotating pointer is proportional to the *average* of the *product* of the currents $i_v(t)$ and $i_i(t)$. Thus,

$$\gamma_{ss} = \frac{K_M}{T} \int_0^T i_v(t) i_i(t) \, dt \tag{14}$$

where K_M is a proportionality constant and T is the period of the current waveforms.

Figure 16.1−6
Wattmeter with electrodynamometer movement.

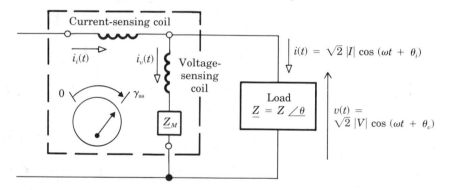

To measure the AC power P delivered to the load, the impedance $\underline{Z}_M$ in series with the voltage-sensing coil must be a large resistance $R_M \gg |Z|$ such that

$$i_v(t) = \frac{\sqrt{2}|V|}{R_M} \cos (\omega t + \theta_v) \qquad i_i(t) \approx i(t) = \sqrt{2}|I| \cos (\omega t + \theta_i)$$

The corresponding steady-state pointer angle becomes

$$\gamma_{ss} = \frac{2K_M|V||I|}{R_M T} \int_0^T \cos (\omega t + \theta_v) \cos (\omega t + \theta_i) \, dt$$

$$= \frac{K_M}{R_M} |V||I| \cos (\theta_v - \theta_i) = \frac{K_M}{R_M} P$$

since $\cos (\theta_v - \theta_i) = \cos \theta$ where $\theta = \angle \underline{Z}$. Thus, with appropriate dial calibration, we have an AC *wattmeter*.

Actually, the meter reading will differ slightly from the predicted value because $i_i(t) = i(t) + i_v(t) \neq i(t)$. If you require better accuracy, you simply disconnect the load and subtract the resulting no-load reading from the original value. More sophisticated instruments have a *compensating winding* that automatically cancels the no-load term.

Hereafter, we'll assume an ideal AC wattmeter represented symbolically by Fig. 16.1–7, where the current input terminal is labeled $\pm$. In terms of the voltage and current phasors, the meter reads

$$P = |V||I| \cos (\sphericalangle \underline{V} - \sphericalangle \underline{I}) = \mathrm{Re}\,[\underline{V}\underline{I}^*] \qquad (15)$$

This reading equals the power delivered through the meter since an ideal meter has zero voltage drop across the current-sensing coil and draws zero current through the voltage-sensing coil.

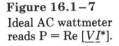

Figure 16.1–7
Ideal AC wattmeter
reads P = Re [$\underline{V}\underline{I}^*$].

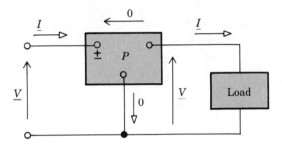

16.2 THREE-PHASE SYSTEMS

AC power has decided practical advantages over DC power in generation, transmission, and distribution. Its one major drawback is the oscillatory nature of the instantaneous power flow $p(t)$ in the *single-phase* circuits we have been discussing. However, a *three-phase* circuit can have *constant* instantaneous power, which eliminates the pulsating strain on generating and load equipment. As a further bonus for power transmission, a balanced three-phase system delivers more watts per kilogram of conductor than an equivalent single-phase system. For these reasons, almost all bulk electric power generation and consumption takes place in three-phase circuits. The properties of such circuits certainly deserve our attention.

Three-Phase Sources and Symmetrical Sets

The left-hand portion of Fig. 16.2–1a represents a three-phase generator. It consists of three AC voltage sources with equal amplitudes $\sqrt{2}V_p$, frequency ω, and relative phase shifts of 120°, so that

$$v_a = \sqrt{2}\,V_p \cos \omega t$$
$$v_b = \sqrt{2}\,V_p \cos (\omega t - 120°) \qquad (1)$$
$$v_c = \sqrt{2}\,V_p \cos (\omega t + 120°)$$

Figure 16.2–1
(a) Three-phase
generator with resistive
load. (b) Voltage
waveforms.

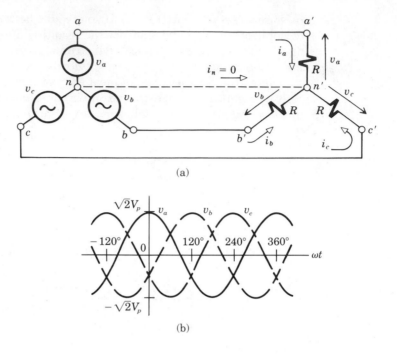

These three voltages are measured with respect to the common connection point n, called the **neutral.** Their waveforms are plotted in Fig. 16.2–1b versus ωt to emphasize the phase shifts.

Equation (1) defines a **symmetrical three-phase set** with **phase sequence** a-b-c, the latter referring to the time sequence of the waveform peaks. Other phase sequences, such as a-c-b, would have the same properties with the terminals relabeled. Thus, unless otherwise stated, the a-b-c sequence will be assumed throughout.

The generator has been connected to a load consisting of three equal resistances tied together at n', as shown to the right of Fig. 16.2–1a. Much can be gleaned from studying this elementary three-phase system. In particular, the load currents i_a, i_b, and i_c are directly proportional to their respective voltages, and their waveforms will have the same shapes. Hence, these currents also constitute a symmetrical three-phase set. Looking more carefully at Fig. 16.2–1b, we discover the important property

$$v_a + v_b + v_c = 0$$

which holds at each and every instant of time! Any symmetrical three-phase set has this property since it corresponds to the trigonometric identity

$$\cos \alpha + \cos (\alpha - 120°) + \cos (\alpha + 120°) = 0 \qquad \textbf{(2)}$$

for any angle α. Its impact here emerges when we note that the neutral current is $i_n = -(i_a + i_b + i_c) = -(v_a + v_b + v_c)/R = 0$. Accordingly, the wire connecting n to n' may just as well be omitted.

Next we examine the instantaneous powers $p_a = v_a^2/R = (2\,V_p^2/R) \cos^2 \omega t$, etc., and the total, $p = p_a + p_b + p_c$, delivered from the generator to the load. The pulsating waveforms p_a, p_b, and p_c are sketched in Fig. 16.2–2. Point-by-point summation then shows that $p = 3\,V_p^2/R$—a constant value for every instant of time! The mathematical proof is left for you as an exercise.

Figure 16.2–2
Three-phase power waveforms.

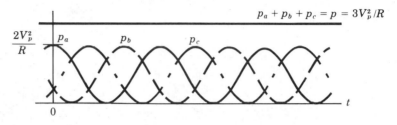

We are now in position to make two statements of profound significance regarding this three-phase system:

- It requires fewer conductors and less copper than three single-phase systems handling the same amount of power.

- The instantaneous power is constant, rather than pulsating as in a single-phase system.

The second property implies reduced mechanical strain at the generator and smoother power delivery to the load equipment—analogous to the performance of a six-cylinder engine as compared with a single-cylinder engine.

For a more detailed study of three-phase sources, consider the configuration in Fig. 16.2–3a—descriptively called a **wye-connected generator.** The neutral point n may or may not be externally available, or safety considerations may call for it to be grounded. The three sources here have been labeled with the phasors

$$\underline{V}_a = V_p\,\underline{/0^\circ} \qquad \underline{V}_b = V_p\,\underline{/-120^\circ} \qquad \underline{V}_c = V_p\,\underline{/+120^\circ} \qquad \textbf{(3)}$$

Figure 16.2–3b gives the phasor diagram. These voltages are known as the **phase voltages,** as distinguished from the **line voltages** $\underline{V}_{ab}$, $\underline{V}_{bc}$, and $\underline{V}_{ca}$. Three-phase generators are usually rated in terms of their line voltages, and the relationship between line and phase voltage is easily developed by phasor analysis.

Clearly, $\underline{V}_{ab} = \underline{V}_a - \underline{V}_b$ in Fig. 16.2–3a, and the phasor construction of Fig. 16.2–3c immediately reveals that $\sphericalangle\,\underline{V}_{ab} = 30^\circ$. Letting V_ℓ denote the rms value of the line voltage, so $V_\ell = |\underline{V}_{ab}|$, we see that $V_p^2 = (V_\ell/2)^2 +$

Figure 16.2–3
Wye-connected
generator and phasor
diagrams.

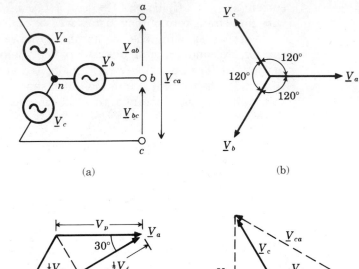

(a)   (b)

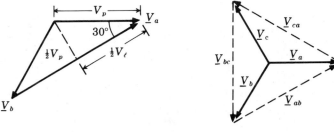

(c)   (d)

$(V_p/2)^2$. Solving for V_ℓ yields

$$V_\ell = \sqrt{3}\; V_p \tag{4}$$

Therefore, the rms line voltage equals $\sqrt{3} \approx 1.7$ times the rms phase voltage. By symmetry, the other line voltages have the same magnitude, but are
shifted $\pm 120°$ relative to $\underline{V}_{ab}$. Thus

$$\underline{V}_{ab} = V_\ell \;\underline{/30°} \qquad \underline{V}_{bc} = V_\ell \;\underline{/-90°} \qquad \underline{V}_{ca} = V_\ell \;\underline{/150°} \tag{5}$$

which is another symmetrical three-phase set in phasor notation.

Figure 16.2–3d depicts the relationship between these two sets and
brings out the fact that $\underline{V}_{ab} + \underline{V}_{bc} + \underline{V}_{ca} = 0$ — as it should according to
KVL. Moreover, by redrawing Fig. 16.2–3b with the vectors "tip-to-tail,"
we find that $\underline{V}_a + \underline{V}_b + \underline{V}_c = 0$, thus confirming the identity in Eq. (2).

Theoretically, the same line voltages could be produced by the **delta-
connected generator** in Fig. 16.2–4. The equivalent neutral point and
phase voltages are defined by the accompanying phasor diagram, where $\underline{V}_{ab}$
has been taken as the reference so $\underline{V}_a = (V_\ell/\sqrt{3})\,\underline{/-30°}$, etc. Delta-connected generators are seldom used in practice because any departure from
the balanced symmetrical-set condition would produce an unwanted internal current circulating around the delta loop.

Figure 16.2–4

Delta-connected
generator and phasor
diagram.

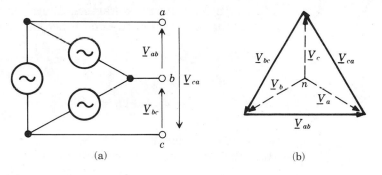

(a)　　　　　　(b)

Exercise 16.2–1　　Use Eq. (2), plus the additional relations $\cos^2 \beta = \frac{1}{2} + \frac{1}{2} \cos 2\beta$ and $\cos (\gamma \mp 240°) = \cos (\gamma \pm 120°)$ to prove that $p = 3 \, V_p^2/R$ in Fig. 16.2–2.

Balanced Loads

A three-phase load generally consists of both resistance and reactance, and may be wye- or delta-connected irrespective of the source configuration. The load is said to be *balanced* when the three branches have equal impedances, a matter of great practical importance.

Consider the **balanced wye load** in Fig. 16.2–5a, where $\underline{Z}_a = \underline{Z}_b = \underline{Z}_c = \underline{Z}_Y$ with $\underline{Z}_Y = |Z_Y|\underline{/\theta}$. The power factor is, of course, pf = $\cos \theta$. From symmetry, the potential at point n' will equal that at the generator's neutral, whether connected to it or not. Consequently, each impedance has a phase voltage across it, and the resulting line currents $\underline{I}_a$, $\underline{I}_b$, and $\underline{I}_c$ have the equal rms values

$$I_\ell = \frac{V_p}{|Z_Y|} = \frac{V_\ell}{\sqrt{3}|Z_Y|} \tag{6}$$

The phase angle of $\underline{I}_a$ is then $\sphericalangle \underline{V}_a - \sphericalangle \underline{Z}_Y = -\theta$, since $\underline{V}_a = \underline{V}_a/\underline{Z}_Y$. The other current angles differ by $\pm 120°$ as diagrammed in Fig. 16.2–5b.

Figure 16.2–5

(a) Balanced wye load,
$\underline{Z}_a = \underline{Z}_b = \underline{Z}_c = |Z_Y|\underline{/\theta}$.
(b) Phasor diagram.

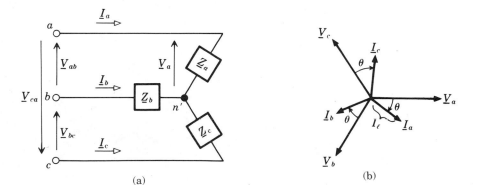

(a)　　　　　　(b)

Having determined the voltages and currents, we can use Eqs. (4) and (6) to find the real and reactive power *per phase,* namely,

$$P_p = V_p I_\ell \cos\theta = \frac{V_\ell I_\ell}{\sqrt{3}} \cos\theta = \frac{V_\ell^2}{3|Z_Y|} \cos\theta$$

$$Q_p = V_p I_\ell \sin\theta = \frac{V_\ell I_\ell}{\sqrt{3}} \sin\theta = \frac{V_\ell^2}{3|Z_Y|} \sin\theta$$

The total real and reactive powers delivered from the generator to the load are then exactly *three times the phase values,* so the total apparent power is

$$|S| = \sqrt{(3P_p)^2 + (3Q_p)^2} = 3\sqrt{P_p^2 + Q_p^2}$$

But since $\cos^2\theta + \sin^2\theta = 1$, we have

$$|S| = \sqrt{3}\, V_\ell I_\ell = \frac{V_\ell^2}{|Z_Y|} \tag{7a}$$

Thus, we can write the total real and reactive powers as

$$P = 3P_p = |S|\cos\theta \qquad Q = 3Q_p = |S|\sin\theta \tag{7b}$$

Furthermore, the *total instantaneous power is constant* at $p = P$, just like a balanced resistive load, because the reactive powers per phase constitute a symmetrical set with the property $p_{aX} + p_{bX} + p_{cX} = 0$.

We now turn to the **balanced delta load** configuration in Fig. 16.2–6, where $Z_{ab} = Z_{bc} = Z_{ca} = Z_\Delta$ and the line voltage appears across each impedance. Rather than starting again from scratch, we can apply our previous results to this case by calculating an **equivalent wye impedance** that draws the same line current as a balanced delta load. For this purpose we note in Fig. 16.2–6 that

$$I_a = I_{ab} - I_{ca} = (V_{ab} - V_{ca})/Z_\Delta$$

We also note that

$$V_{ab} - V_{ca} = (V_a - V_b) - (V_c - V_a) = 2V_a - V_b - V_c$$

$$= 3V_a - (V_a + V_b + V_c) = 3V_a$$

Figure 16.2–6
Balanced delta load, $Z_{ab} = Z_{bc} = Z_{ca} = |Z_\Delta|\,\underline{/\theta}$.

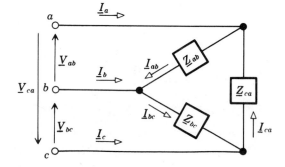

Thus, $\underline{I}_a = 3\underline{V}_a/\underline{Z}_\Delta = \underline{V}_a/(\underline{Z}_\Delta/3)$, and a balanced delta load is equivalent to a balanced wye load with

$$\underline{Z}_Y = \tfrac{1}{3}\underline{Z}_\Delta \tag{8}$$

Then, combining Eqs. (6) and (8), the rms line current becomes

$$I_\ell = 3\frac{V_P}{|Z_\Delta|} = \sqrt{3}\frac{V_\ell}{|Z_\Delta|} \tag{9}$$

This result should be contrasted with the rms delta current $|\underline{I}_{ab}| = |\underline{I}_{bc}| = |\underline{I}_{ca}| = V_\ell/|Z_\Delta|$. Equations (7a) and (7b) still hold as they stand, with $\theta = \angle\underline{Z}_\Delta$. Observe that a delta load draws three times the current and power of a wye load having the same branch impedances.

To summarize, we have seen that the symmetry of a balanced three-phase system leads to very simple expressions for voltage, current, and power, whether the load has a wye or delta configuration. This symmetry allows us to carry out analyses by focusing on just *one* of the three phases.

Example 16.2-1

Figure 16.2-7a depicts a three-phase source with $V_\ell = 45$ kV (*kilo*volts) connected to two balanced loads—a wye load with branch impedance $\underline{Z}_w = (10 + j20)$ Ω and a delta load with branch impedance $\underline{Z}_d = 50$ Ω. Each of the three connecting wires has 2 Ω resistance. Our task is to find the line current and the various powers. For this task, we'll take advantage of the symmetry and concentrate on one phase, say the phase that includes terminals a and a'.

Figure 16.2-7
(a) Three-phase system for Example 16.2-1.
(b) Equivalent circuit of one phase.

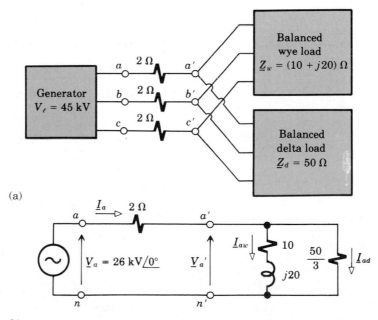

(a)

(b)

First, we recognize that the two loads are in *parallel*, in the sense of having the same voltages across their terminals. Second, we apply Eq. (8) to obtain the equivalent wye impedance of the delta load,

$$\underline{Z}_{Yd} = (50/3) \; \Omega$$

Third, we calculate from Eq. (4) the phase voltage at the source

$$V_p = 45 \text{ kV}/\sqrt{3} \approx 26 \text{ kV}$$

Now we can draw the equivalent circuit diagrammed in Fig. 16.2–7b, assuming a wye-connected source and taking $\underline{V}_a$ for the phase reference. The circuit also includes a resistanceless wire from n to n', reflecting the fact that these two points must be at the same potential in a balanced system. If you question the parallel load configuration in Fig. 16.2–7b, then you should go back to Fig. 16.2–7a and draw the complete wye load together with the complete wye equivalent for the delta load.

The work having progressed this far, the rest moves along rapidly. We combine the parallel elements to get

$$\underline{Z}_{a'n'} = (10 + j20) \| (50/3) = 10 + j5$$

and the total phase impedance seen by the generator is

$$\underline{Z}_{an} = 2 + \underline{Z}_{a'n'} = 12 + j5 = 13 \; \Omega \; \underline{/22.6°}$$

We thus obtain

$$\theta = \measuredangle \underline{Z}_{an} = 22.6° \qquad I_\ell = |\underline{I}_a| = 26 \text{ kV}/13 \; \Omega = 2 \text{ kA}$$

Then, since $V_\ell = 45$ kV, Eq. (7) yields

$$|S| = \sqrt{3} \times 45 \text{ kV} \times 2 \text{ kA} = 156 \text{ MVA}$$

$$P = |S| \cos 22.6° = 144 \text{ MW} \qquad Q = |S| \sin 22.6° = 60 \text{ MVAr}$$

Notice that powers come out in *mega* values when we deal with kilovolts and kiloamps, typical of very large three-phase systems.

As for the individual loads, we find that $\underline{V}_{a'} = \underline{V}_a - 2\underline{I}_a = 22.4$ kV $\underline{/-3.9°}$, $|I_{aw}| = |V_{a'}|/|10 + j20| = 1.00$ kA, and $|I_{ad}| = |V_{a'}|/(50/3) = 1.34$ kA. Direct power computations then yield

$$P_w = 3 \times 10|I_{aw}|^2 = 30 \text{ MW}$$

$$Q_w = 3 \times 20|I_{aw}|^2 = 60 \text{ MVAr}$$

$$P_d = 3 \times (50/3)|I_{ad}|^2 = 90 \text{ MW}$$

$$Q_d = 0$$

We also see that each of the three wires dissipates $2 \; \Omega \times |I_a|^2 = 8$ MW — clearly an intolerable situation.

Exercise 16.2–2 Find $\underline{I}_a$, $\underline{I}_b$, $\underline{I}_c$, $|S|$, P, and Q when a balanced delta load with $\underline{Z}_\Delta = (12 + j9)\ \Omega$ is supplied by a generator with $\underline{V}_{ab} = 10\ \text{kV}\ \underline{/0°}$. (Save your results for use in Exercise 16.2–3.)

Power Measurements

Be it balanced or unbalanced, a three-phase load consists of three branches capable of consuming power. It would therefore seem that *three* separate power measurements are needed to find the total power delivered to the load. Indeed, the four-wire system with an unbalanced wye load depicted by Fig. 16.2–8 requires three wattmeters as shown for power measurement. However, no current flows between n and n' in any other system configuration, and *two* power measurements turn out to be sufficient.

Figure 16.2–8
Four-wire system with three wattmeters.

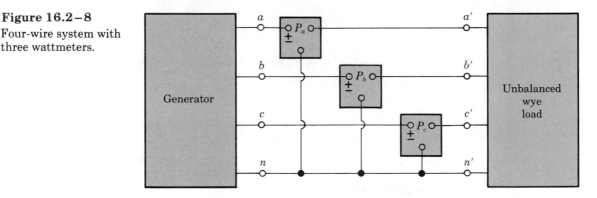

As a case in point, take a balanced or unbalanced delta load with two wattmeters arranged per Fig. 16.2–9. The complex power delivered to $\underline{Z}_{ab}$ is $\underline{S}_{ab} = \underline{V}_{ab}\underline{I}_{ab}^*$, and we can write the real power consumed by this branch as $P_{ab} = \text{Re}\,[\underline{S}_{ab}]$. Adding like expressions for P_{bc} and P_{ca} gives the total load power

$$P = \text{Re}\,[\underline{S}_{ab}] + \text{Re}\,[\underline{S}_{bc}] + \text{Re}\,[\underline{S}_{ca}] = \text{Re}\,[\underline{S}] \qquad \textbf{(10a)}$$

Figure 16.2–9
Delta load with two wattmeters.

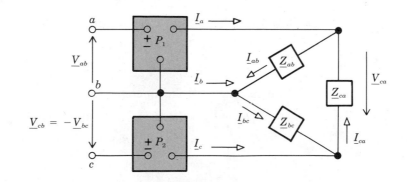

where, from the conservation property,

$$\underline{S} = \underline{V}_{ab}\underline{I}_{ab}^* + \underline{V}_{bc}\underline{I}_{bc}^* + \underline{V}_{ca}\underline{I}_{ca}^* \tag{10b}$$

We then draw upon Eq. (15), Section 16.1, to write the two wattmeter readings as

$$P_1 = \text{Re}\,[\underline{V}_{ab}\underline{I}_a^*] \qquad P_2 = \text{Re}\,[\underline{V}_{cb}\underline{I}_c^*] \tag{11}$$

in which $\underline{V}_{cb} = -\underline{V}_{bc}$. We'll prove that $P_1 + P_2 = P$, even though the individual readings have no particular physical significance.

Our proof starts by observing from Fig. 16.2–9 that $\underline{I}_{ab} = \underline{I}_a + \underline{I}_{ca}$ and $\underline{I}_{bc} = \underline{I}_{ca} - \underline{I}_c$. Substituting these relations in Eq. (10b) yields

$$\begin{aligned}
\underline{S} &= \underline{V}_{ab}(\underline{I}_a + \underline{I}_{ca})^* + \underline{V}_{bc}(\underline{I}_{ca} - \underline{I}_c)^* + \underline{V}_{ca}\underline{I}_{ca}^* \\
&= \underline{V}_{ab}\underline{I}_a^* - \underline{V}_{bc}\underline{I}_c^* + (\underline{V}_{ab} + \underline{V}_{bc} + \underline{V}_{ca})\underline{I}_{ca}^* \\
&= \underline{V}_{ab}\underline{I}_a^* + \underline{V}_{cb}\underline{I}_c^*
\end{aligned}$$

since $-\underline{V}_{bc} = \underline{V}_{cb}$ and $\underline{V}_{ab} + \underline{V}_{bc} + \underline{V}_{ca} = 0$ under any conditions. Now, using Eqs. (10a) and (11), we see that $P = \text{Re}\,[\underline{S}] = \text{Re}\,[\underline{V}_{ab}\underline{I}_a^* + \underline{V}_{cb}\underline{I}_c^*] = \text{Re}\,[\underline{V}_{ab}\underline{I}_a^*] + \text{Re}[\underline{V}_{cb}\underline{I}_c^*]$ and

$$P = P_1 + P_2 \tag{12}$$

Hence, the sum of the wattmeter readings equals the total power to a balanced or unbalanced delta load.

Equation (12) also holds for a balanced wye load and for an unbalanced wye load without a connection between n and n'. Furthermore, the two wattmeters may be rearranged in any configuration as long as each current-sensing coil carries a different line current and the reference ends of the voltage-sensing coils are connected to a line whose current does not pass through a current-sensing coil.

If the load happens to be *balanced*, then the two wattmeter readings also provide sufficient data to determine the magnitude of the total *reactive* power. For a balanced load with impedance angle θ, we know from Eq. (7b) that

$$Q = \sqrt{3}\,V_\ell I_\ell \sin\theta$$

Taking $\underline{V}_a$ as the phase reference, we also know that

$$\underline{V}_{ab} = V_\ell\,\underline{/30°} \qquad \underline{V}_{cb} = -\underline{V}_{bc} = V_\ell\,\underline{/90°}$$

$$\underline{I}_a^* = I_\ell\,\underline{/\theta} \qquad \underline{I}_c^* = I_\ell\,\underline{/\theta - 120°}$$

Then, using Eq. (11), the *difference* between the readings is

$$\begin{aligned}
|P_1 - P_2| &= |\text{Re}\,[V_\ell I_\ell\,\underline{/\theta + 30°}] - \text{Re}\,[V_\ell I_\ell\,\underline{/\theta - 30°}]| \\
&= V_\ell I_\ell |\cos(\theta + 30°) - \cos(\theta - 30°)| \\
&= V_\ell I_\ell |\sin\theta|
\end{aligned}$$

where we have invoked the trigonometric identity

$$\cos(\theta - 30°) - \cos(\theta + 30°) = 2\sin\theta\sin 30° = \sin\theta$$

Thus, the magnitude of the reactive power is given by

$$|Q| = \sqrt{3}|P_1 - P_2| \tag{13}$$

The total apparent power may now be calculated from $|S| = \sqrt{P^2 + Q^2}$.

Exercise 16.2−3   Let two power meters be inserted into the system in Exercise 16.2−2 such that $P_1 = \text{Re}\,[\underline{V}_{ac}\underline{I}_a^*]$ and $P_2 = \text{Re}\,[\underline{V}_{bc}\underline{I}_b^*]$. Evaluate P_1 and P_2, and verify that Eqs. (12) and (13) yield correct results.

Unbalanced Loads [†]

Much of the simplifying symmetry goes out the window when a three-phase load has *unequal* impedances. We must then resort to brute-force analysis involving all three branches. Moreover, an unbalanced wye load requires special attention regarding the presence or absence of a neutral connection, either directly or via grounds at generator and load.

Consider, on the one hand, the unbalanced *four-wire* system in Fig. 16.2−10a. The neutral connection forces the corresponding phase voltage to appear across each branch, so

$$\underline{I}_a = \underline{V}_a/\underline{Z}_a \qquad \underline{I}_b = \underline{V}_b/\underline{Z}_b \qquad \underline{I}_c = \underline{V}_c/\underline{Z}_c \tag{14}$$

The current through the fourth wire is then

$$\underline{I}_n = -(\underline{I}_a + \underline{I}_b + \underline{I}_c)$$

whereas a balanced system would have $\underline{I}_n = 0$.

On the other hand, the *three-wire* system in Fig. 16.2−10b has an ungrounded or "floating" neutral point n' at the load, thereby forcing

Figure 16.2−10

Unbalanced wye loads:
(a) four-wire system;
(b) three-wire system.

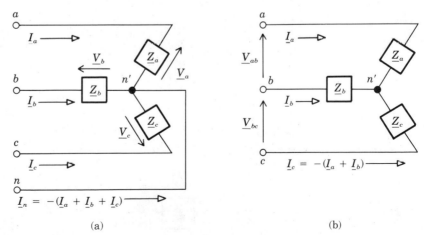

(a)   (b)

$\underline{I}_a + \underline{I}_b + \underline{I}_c = 0$. But we must now work with *line* voltages since phase voltages no longer appear across each branch. Utilizing $\underline{I}_c = -(\underline{I}_a + \underline{I}_b)$, we write two loop equations for $\underline{I}_a$ and $\underline{I}_b$ in terms of $\underline{V}_{ab}$ and $\underline{V}_{bc}$ to obtain

$$\underline{Z}_a \underline{I}_a - \underline{Z}_b \underline{I}_b = \underline{V}_{ab}$$

$$\underline{Z}_c \underline{I}_a + (\underline{Z}_b + \underline{Z}_c)\underline{I}_b = \underline{V}_{bc} \tag{15}$$

This pair of equations can be solved for $\underline{I}_a$ and $\underline{I}_b$ and thence $\underline{I}_c$. If desired, the voltage at the load's neutral relative to the generator's neutral is determined from any of the three phases via

$$\underline{V}_{n'n} = \underline{V}_a - \underline{Z}_a \underline{I}_a = \underline{V}_b - \underline{Z}_b \underline{I}_b = \underline{V}_c - \underline{Z}_c \underline{I}_c$$

A balanced system would, of course, have $\underline{V}_{n'n} = 0$.

Systems with unbalanced delta loads cannot have a fourth current path and the branch currents are easily calculated when the line voltages appear directly across the corresponding impedances. However, the situation becomes appreciably more complicated when generator and/or line

Figure 16.2–11

(a) Unbalanced delta load with series impedances.
(b) Equivalent wye load.

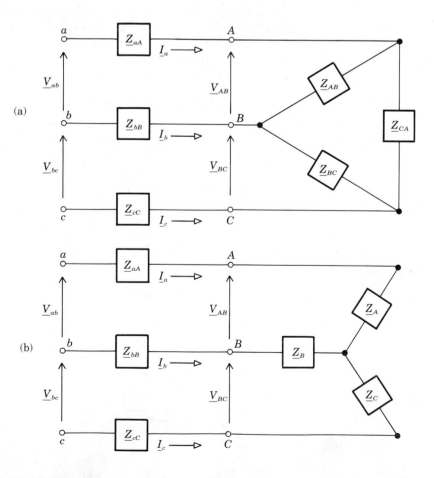

impedances cause an intervening voltage drop, as in Fig. 16.2–11a. For the study of such cases, the unbalanced delta load can be converted to an *equivalent wye load* shown in Fig. 16.2–11b. The resulting unbalanced three-wire system is then analyzed by applying Eq. (15) with $\underline{Z}_a = \underline{Z}_{aA} + \underline{Z}_A$, etc.

The elements of the equivalent unbalanced wye load in Fig. 16.2–11b are given by the **delta-to-wye transformation**

$$\underline{Z}_A = \frac{\underline{Z}_{AB}\underline{Z}_{CA}}{\underline{Z}_{ABC}} \qquad \underline{Z}_B = \frac{\underline{Z}_{BC}\underline{Z}_{AB}}{\underline{Z}_{ABC}} \qquad \underline{Z}_C = \frac{\underline{Z}_{CA}\underline{Z}_{BC}}{\underline{Z}_{ABC}} \qquad \textbf{(16a)}$$

where

$$\underline{Z}_{ABC} \triangleq \underline{Z}_{AB} + \underline{Z}_{BC} + \underline{Z}_{CA} \qquad \textbf{(16b)}$$

Note that $\underline{Z}_A$ equals the product of the two delta impedances connected to point A ($\underline{Z}_{AB}$ and $\underline{Z}_{CA}$) divided by the sum of the delta impedances — and likewise for $\underline{Z}_B$ and $\underline{Z}_C$. This transformation is derived by equating voltages and currents at terminals A, B, C, in both parts of Fig. 16.2–11. If the delta load happens to be balanced, so $\underline{Z}_{AB} = \underline{Z}_{BC} = \underline{Z}_{CA} = \underline{Z}_\Delta$, then Eq. (16a) reduces to Eq. (8) in the sense that $\underline{Z}_A = \underline{Z}_B = \underline{Z}_C = \underline{Z}_\Delta/3$.

Example 16.2–2

Suppose the system in Fig. 16.2–11a has the following parameter values:

$$V_\ell = 900 \text{ V} \qquad \underline{Z}_{aA} = \underline{Z}_{bB} = \underline{Z}_{cC} = 2.5 \ \Omega$$

$$\underline{Z}_{AB} = j15 \ \Omega \qquad \underline{Z}_{BC} = 20 \ \Omega \qquad \underline{Z}_{CA} = (20 + j15) \ \Omega$$

We want to find the line currents and the total real and reactive power supplied by the generator.

First, we apply the delta-to-wye transformation from Eq. (16a) with $\underline{Z}_{ADC} = 40 + j30$ to obtain

$$\underline{Z}_A = 3.6 + j4.8 \qquad \underline{Z}_B = 10 \qquad \underline{Z}_C = j7.5$$

Hence, the total equivalent wye impedances are

$$\underline{Z}_a = 6.1 + j4.8 \qquad \underline{Z}_b = 12.5 \qquad \underline{Z}_c = 2.5 + j7.5$$

where we've added the 2.5-Ω series resistance in each line.

Next, we take $\underline{V}_{ab}$ as the phase reference so $\underline{V}_{ab} = 900 \text{ V } \underline{/0°} = 900$, whereas $\underline{V}_{bc} = 900 \ \underline{/-120°} = -450 - j780$. Equation (15) then becomes

$$(6.1 + j4.8)\underline{I}_a \qquad - 12.5\underline{I}_b = 900$$

$$(2.5 + j7.5)\underline{I}_a + (15 + j7.5)\underline{I}_b = -450 - j780$$

Solving with the help of Cramer's rule yields

$$\underline{I}_a = 0.90 - j36.9 = 36.9 \text{ A } \underline{/-89°}$$

$$\underline{I}_b = -57.4 - j17.7 = 60.0 \text{ A } \underline{/-163°}$$

The remaining line current is

$$\underline{I}_c = 56.5 + j54.6 = 78.5 \text{ A } \underline{/44°}$$

since $\underline{I}_c = -(\underline{I}_a + \underline{I}_b)$.

Finally, we calculate the real and reactive powers directly as

$$P = |I_a|^2 R_a + |I_b|^2 R_b + |I_c|^2 R_c = 688 \text{ kW}$$

$$Q = |I_a|^2 X_a + |I_b|^2 X_b + |I_c|^2 X_c = 527 \text{ kVAr}$$

The impedance angle of the total load is not defined for the unbalanced case. Nonetheless, the equivalent power factor seen from the generator can be determined via

$$\text{pf} = P/|S| = P/\sqrt{P^2 + Q^2} = 0.79$$

a lagging power factor since $Q > 0$.

Exercise 16.2–4   Let the system in Fig. 16.2–11a have $\underline{V}_{ab} = 208$ V $\underline{/0°}$, $\underline{Z}_a = j10 \ \Omega$, $\underline{Z}_b = (6 + j8) \ \Omega$, and $\underline{Z}_c = 10 \ \Omega$. Find all four line currents. Then calculate P, Q, and pf.

16.3   POWER TRANSMISSION AND DISTRIBUTION †

This section traces the route of AC power from generating plants to consumers. We'll see how the limitations of power transmission lines are overcome through the use of very high voltages, generally obtained with the help of step-up transformers. Then we'll look at power networks, which include step-down transformers to distribute electric power to homes and industries.

Power Transmission Lines

Figure 16.3–1

Model for one phase of a power transmission line.

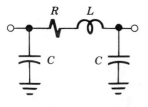

Most electric power is conveyed from the generating plants to consumers by AC three-phase overhead transmission lines. Such lines constitute *distributed circuits* in the sense that resistance, capacitance, and inductance are distributed continuously along their entire length. Energy flows in the form of a *traveling* wave moving with finite but very high velocity, nearly 3×10^8 m/sec.

Fortunately, most transmission lines are electrically "short" compared to the wavelength of the traveling wave. Each phase may then be approximated by the *lumped* circuit of Fig. 16.3–1, with parameter values proportional to line length. A representative line rated for 100 kV and 1 kA might have

$$R \approx 50 \text{ m}\Omega/\text{km} \qquad L \approx 1 \text{ mH/km} \qquad C \approx 10 \text{ nF/km}$$

This lumped model is valid for lines up to about 200-km long. It is further simplified by the fact that the capacitive current to ground is typically quite small compared to the total current, and that the inductive reactance is typically much greater than the resistance. Thus, we'll disregard the line capacitance entirely and take account of the resistance only when we compute the voltage drop and power dissipation on the line itself.

Consider, then, our simplified model for one phase with a generator at one end and a load at the other, as in Fig. 16.3–2. In a balanced system, each phase of the load consumes the complex power

$$\underline{S}_2 = \underline{V}_2\underline{I}^* = P_2 + jQ_2 \tag{1a}$$

and each phase of the generator produces

$$\underline{S}_1 = \underline{V}_1\underline{I}^* = P_1 + jQ_1 \tag{1b}$$

where $\underline{V}_1$ and $\underline{V}_2$ are phase voltages. These voltages differ by the voltage drop along the line, so

$$\underline{V}_1 = \underline{V}_2 + (R + jX)\underline{I}$$

Normally, the line impedance $R + jX$ is small enough that

$$|V_1| \approx |V_2| \approx V_p$$

where we interpret V_p as the system's average or *nominal phase voltage*. However, we cannot ignore the phase shift

$$\delta \triangleq \angle\underline{V}_1 - \angle\underline{V}_2 \tag{2}$$

which is called the **power angle.** The significance of the power angle is brought out in the following analysis.

Figure 16.3–2

Simplified per-phase model.

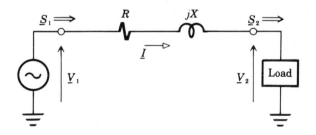

First, let $\underline{V}_2 = V_p\,\underline{/0°}$ be the reference phasor so that $\underline{V}_1 \approx V_p\,\underline{/\delta} = V_p e^{j\delta}$. Next, drawing upon the property $R \ll X$, we can write $\underline{I} \approx (\underline{V}_1 - \underline{V}_2)/jX$ and

$$\underline{I}^* \approx \frac{\underline{V}_1^* - \underline{V}_2^*}{-jX} = j\frac{V_p e^{-j\delta} - V_p}{X} = j\frac{V_p}{X}(e^{-j\delta} - 1) \tag{3}$$

Inserting $\underline{I}^*$ into Eq. (1) we obtain, after a few manipulations

$$P_1 \approx P_2 \approx \frac{V_p^2}{X} \sin \delta \tag{4}$$

$$Q_1 \approx Q_2 + 2\frac{V_p^2}{X}(1 - \cos \delta) \tag{5}$$

The total real and reactive powers are, of course, three times these values.

Equation (4) obviously lacks somewhat in accuracy, for P_1 must be slightly greater than P_2 to include power dissipation in the line resistance R. Nonetheless, it correctly shows that real power varies with the power angle and attains a maximum value

$$P_{\max} \approx \frac{V_p^2}{X} \tag{6}$$

which occurs when $\delta = 90°$ — so $\underline{V}_1$ is perpendicular to $\underline{V}_2$. We thus see that the maximum real power capacity of a transmission line is inversely proportional to its series reactance and proportional to the square of the nominal voltage.

Operating a line at close to its maximum real power capacity does not turn out to be an attractive proposition owing to the accompanying large values of *reactive* power. For, if $\delta = 90°$ then $\cos \delta = 0$ in Eq. (5) and $Q_1 \approx Q_2 + 2P_{\max}$. The motivation for keeping reactive power Q_1 at relatively low values is brought out by calculating the *power dissipation* along the line. Since $\underline{I}$ flows through the line resistance R in Fig. 16.3–2, the power dissipated is $P_R = R|I|^2$. But $|I| = |S_1|/|V_1| \approx |S_1|/V_p$, so

$$P_R \approx R\frac{|S_1|^2}{V_p^2} = \frac{R}{V_p^2}(P_1^2 + Q_1^2) \tag{7}$$

which means that power lost as line heating increases as the reactive power increases.

Another important observation from this result relates to the power transmission efficiency

$$\text{Eff} = \frac{P_2}{P_1} = 1 - \frac{P_R}{P_1} \tag{8}$$

Clearly, efficiency increases with V_p because P_R decreases inversely with V_p^2. The combined implications of Eqs. (4)–(8) explain why large amounts of power must be transmitted at *high voltages*. Invariably, doing so requires a step-up transformer between the generator and line, and a step-down transformer at the other end.

Example 16.3–1   A generator with phase voltage $V_s = 20$ kV is to be connected via a three-phase transmission line to a large plant. The transmission line has $R = 2.5\ \Omega$ and $X = 40\ \Omega$. The plant is a balanced load, corrected to unity power factor, requiring 4 kV phase voltage and 3 kA line current so $P_1 = 12$ MW per phase.

Clearly, we need a *step-down* transformer to get 4 kV at the load. But we also need a *step-up* transformer at the generator to get $P_{max} > 12$ MW which, from Eq. (6), requires

$$V_p \approx \sqrt{XP_{max}} > 22 \text{ kV}$$

Moreover, the larger the value of V_p, the better the efficiency will be.

Suppose, then, that we use the per-phase configuration diagrammed in Fig. 16.3–3, taking the load voltage as the phase reference. The 25:1 step-down transformer at the load means that $|V_2| = 25 \times 4$ kV $= 100$ kV and $|I| = 3$ kA/25 $= 0.12$ kA, both phasors having 0° angle. Accordingly,

$$\underline{V}_1 = (R + jX)\underline{I} + \underline{V}_2 = (100.3 + j4.8) \text{ kV} = 100.4 \text{ kV} \underline{/2.74°}$$

which calls for an input step-up transformer with the turns ratio $N = 100.4/20 = 5.02$. The power angle and nominal phase voltage are

$$\delta = \measuredangle \underline{V}_1 - \measuredangle \underline{V}_2 = 2.74° \qquad V_p = \tfrac{1}{2}(|V_1| + |V_2|) \approx 100.2 \text{ kV}$$

Direct power calculations yield

$$Q_1 = |I|^2 X + Q_2 = 0.576 \text{ MVAr} \qquad P_R = |I|^2 R = 0.036 \text{ MW}$$

so the efficiency of the proposed system is Eff $= 1 - (0.036/12) = 99.7\%$.

Figure 16.3–3

Circuit for Example 16.3–1.

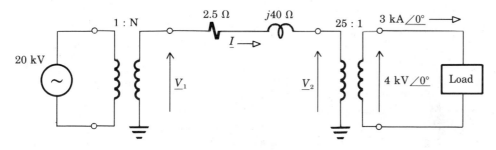

Exercise 16.3–1   Rework the calculations in Example 16.3–1 for the case of a 6:1 step-down transformer at the load.

Power Networks

An AC power network distributes electrical energy from sources to loads in a geographic region served by a particular electric utility company. Here we'll examine the salient features of an illustrative network whose schematic diagram is given in Fig. 16.3–4.

Figure 16.3−4

Schematic diagram of a
power network.

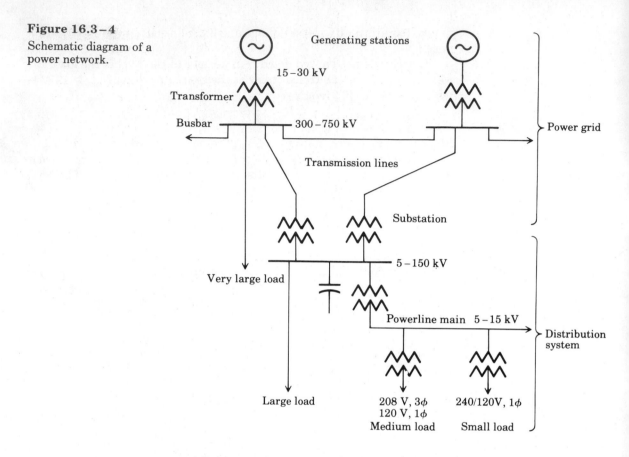

AC power originates at *generating stations,* where rotating machines
convert energy to electrical form. These machines function as three-phase
sources with line voltages ranging from 15 to 30 kV. (The operating principles of three-phase generators will be covered in section 18.2.) The line
voltage is then raised to 300 – 750 kV by step-up transformers, represented
by the double-sawtooth symbols in Fig. 16.3 – 4. Massive *busbars* serve as
the interconnecting nodes for the high-voltage transmission lines which,
together with the generating stations, constitute the **power grid.** Not
shown in the diagram, but essential nonetheless, are *circuit breakers* at the
end of each transmission line to protect the system from overload fault
conditions.

Power may flow along a transmission line in either direction between
nodes of the grid, in accordance with load demands. The direction and
amount of power flow is controlled by adjusting the relative phase angles of
the node voltages. To see how this control works, suppose that a generator
and a load are connected to each end of the transmission line back in Fig.
16.3 – 2. Equation (4) still applies, and it shows that P_1 is proportional to
$\sin \delta$. If $\angle \underline{V}_2 > \angle \underline{V}_1$, so $\delta < 0$, then P_1 is negative and power actually flows
in the reverse direction.

Although a very large load—a steel mill, for instance—might be fed directly from a power-grid node, most other loads are supplied through a lower-voltage **distribution system.** Distribution starts at *substations,* where step-down transformers reduce the incoming transmission-line voltage to 5–150 kV. Large loads and capacitance for power-factor correction are connected to the output busbar, along with another step-down transformer feeding the *powerline main* at 5–15 kV. The powerline main, in turn, supplies local transformers for medium and small loads—including your home.

Clearly, transformers play pivotal roles in AC power networks. We should therefore now give some attention to three-phase transformer configurations. (The general analysis of power transformers will be taken up in Section 17.2.)

A three-phase transformer actually consists of three pairs of windings, each pair having the same turns ratio N. For the step-up transformer at a generating station, the primary windings are delta-connected while the secondary windings are wye-connected, as depicted in Fig. 16.3–5. This wye-delta configuration provides an additional step-up factor of $\sqrt{3}$ over and above the turns ratio. Let V_ℓ be the generator's line voltage and take $\underline{V}_{ab}$ as the reference phasor so that

$$\underline{V}_{ab} = V_\ell \,\underline{/0°} \qquad \underline{V}_{bc} = V_\ell \,\underline{/-120°} \qquad \underline{V}_{ca} \,\underline{/+120°}$$

A simple phasor construction then yields

$$|\underline{V}_{a'b'}| = |N\underline{V}_{ab} - N\underline{V}_{bc}| = \sqrt{3}NV_\ell \qquad \textbf{(9)}$$

Hence, by symmetry, all three output line voltages have been stepped up by the factor $\sqrt{3}N$.

Figure 16.3–5
Three-phase step-up transformer.

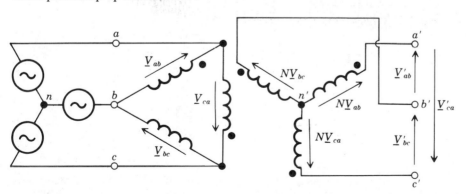

Three-phase step-down transformers may be delta-wye or wye-wye connected. Figure 16.3–6 shows a wye-wye configuration with *four* output terminals for the purpose of supplying both *three-phase* (3-ϕ) and *single-*

Figure 16.3–6
Three-phase step-down transformer with single-phase loads.

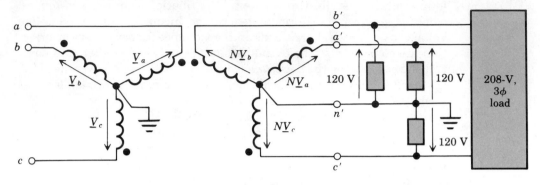

phase (1-ϕ) power to a medium load, such as an office building, store, or small factory. If V_p is the input phase voltage and $N = 120/V_p$, we then have 120-V, 1-ϕ AC between any of the "hot" output terminals (a', b', or c') and the grounded neutral n'. But we also have three-phase power with line-to-line voltage $\sqrt{3} \times 120 \approx 208$ V. The latter would be used for large heating elements or motors bigger than about five horsepower, whereas the former would supply lights and small machines.

Most residential service takes the form of single-phase, dual-voltage AC obtained from a transformer like that shown in Fig. 16.3–7. The single-phase primary is connected across any two lines, and the secondary has a grounded center tap. Thus, an appropriate turns ratio N will yield 240-V line-to-line at the output, with two 120-V terminals relative to ground — one of which has a 180° phase shift.

Figure 16.3–7
Step-down transformer for dual-voltage load.

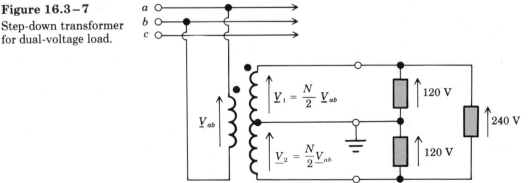

You may have noticed that the arrangement in Fig. 16.3–7 presents an *unbalanced* load to the three-phase network, and so does that of Fig. 16.3–6 unless the 120-V load elements happen to be equal. But with literally thousands of residential loads connected across each of the three line pairs, and with many medium loads also connected, the law of averages takes effect and the total composite load on the distribution system appears to be virtually balanced.

Exercise 16.3-2   Let the transformers in Figs. 16.3-6 and 16.3-7 be fed from the same three-phase main with $V_\ell = 7.6$ kV. Find the turns ratio required in each case to get the indicated output voltages.

PROBLEMS

16.1-1   Modify Eqs. (6) and (7) to express P and Q in terms of $|V|$, R, and X.

16.1-2   Obtain expressions for P and Q in terms of $|I|$ when R and jX in Fig. 16.1-1a are reconnected in parallel.

16.1-3   A certain load draws $P = 720$ W and $|I| = 10$ A at 120 V and 60 Hz. The load current decreases when the frequency increases. Find the element values to represent the load as: (a) two elements in parallel; (b) two elements in series.

16.1-4   Do Problem 16.1-3 for $P = 960$ W.

16.1-5   Do Problem 16.1-3 for the case when the load current increases as frequency increases.

16.1-6   An AC motor having impedance $\underline{Z}_M = (3.6 + j4.8)$ Ω is supplied from a 240-V, 60-Hz source.
(a) Calculate pf, $|I|$, P, and Q.
(b) The power factor is corrected to unity by connecting a capacitor in parallel with the motor. Find C and the power P and current $|I|$ supplied by the source.
(c) The power factor is corrected to unity by connecting a capacitor in series with the motor. Find C and the current $|I|$ and power P supplied by the source. Also calculate the voltage $|V_M|$ across the motor.

16.1-7   Do Problem 16.1-6 with $\underline{Z}_M = (3.0 + j5.2)$ Ω.

16.1-8   Do Problem 16.1-6 with $\underline{Z}_M = (1.68 + j5.76)$ Ω.

16.1-9   Two loads are connected in parallel with a 2000-V source. The individual power factors and currents are:

$$pf_1 = 0.5 \text{ lagging} \qquad |I_1| = 40 \text{ A}$$
$$pf_2 = 0.8 \text{ leading} \qquad |I_2| = 15 \text{ A}$$

(a) Find the total power and current from the source, and calculate the power factor of the combined loads.
(b) A capacitor is now added in parallel, increasing the power factor to 0.95 lagging. Find the current drawn from the source and the capacitor's reactive power.

16.1-10   Do problem 16.1-9 for the case of:

$$pf_1 = 0.75 \text{ leading} \qquad |I_1| = 20 \text{ A}$$
$$pf_2 = 0.28 \text{ lagging} \qquad |I_2| = 50 \text{ A}$$

16.1–11   Do problem 16.1–9 for the case of:

$$\text{pf}_1 = 0.7 \text{ leading} \qquad |I_1| = 20 \text{ A}$$

$$\text{pf}_2 = 0.4 \text{ lagging} \qquad |I_2| = 30 \text{ A}$$

16.1–12   An industrial dryer operates at 600 V and requires 50 A. The unit consists of a fan in parallel with a heater. The fan draws 20 kW and has a power factor of 0.8 lagging. Use a power triangle to find the resistance of the heater, assuming that it has unity power factor.

16.1–13   An industrial furnace operates at 1300 V and requires 100 A. The unit consists of a blower in parallel with a heater. The blower draws 50 kW and has a power factor of 0.707 lagging. Use a power triangle to find the resistance of the heater, assuming that it has unity power factor.

16.1–14   An inductive motor draws 24 kW and 40 A from a 1000-V, 60-Hz source. The total current drops to 25 A when a capacitor is connected in parallel with the motor. Show from a power triangle that there are two possible values of C, and find those values.

16.1–15   An inductive motor draws 16 kW and 68 A from a 500-V, 60-Hz source. The total current drops to 40 A when a capacitor is connected in parallel with the motor. Show from a power triangle that there are two possible values of C, and find those values.

16.1–16   A capacitor is to be connected in parallel with a motor that draws $P = 10$ kW and $Q = 6$ kVAr from a 240-V, 60-Hz source. Derive an expression for C in terms of the resulting power factor. Then find the cost of obtaining pf = 0.9, 0.95, and 1.0 when 240-V capacitors are priced at $0.25 per microfarad.

16.1–17   When the source voltage and frequency are known, the real and reactive power drawn by a load can be determined using an AC ammeter and a known capacitor. First the load current $|I_1|$ is measured without the capacitor. Then the total current $|I_2|$ is measured with the capacitor connected in parallel. By considering $|VI_1|^2 - |VI_2|^2$, derive an expression for Q in terms of known and measured quantities. Then derive an expression for P in terms of Q and other known quantities.

16.1–18   The wattmeter in Fig. 16.1–6 becomes a **varmeter** when $\underline{Z}_M = jX_M$, where X_M is an inductive reactance with $|X_M| \gg |Z|$. Show from Eq. (14) that $\gamma_{ss} \approx K_M Q / X_M$.

16.2–1   A balanced wye load is connected to a 60-Hz three-phase source with $\underline{V}_{ab} = 208$ V $\underline{/0°}$. The load has pf = 0.5 lagging, and each phase draws $P_p = 6$ kW.
(a) Calculate I_ℓ and find $\underline{Z}_Y$, $\underline{I}_a$, $\underline{I}_b$, and $\underline{I}_c$ in polar form.
(b) What value of C should be put in parallel with each load element to minimize the current from the source, and what is the resulting line current?

16.2–2   Do Problem 16.2–1 with pf = 0.8 lagging and $P_p = 7.2$ kW.

16.2–3   A balanced delta load is connected to a 60-Hz three-phase source with $\underline{V}_a = 120$ V $\underline{/0°}$. The load has pf = 0.6 lagging, and each phase load draws $P_p = 9$ kW.
(a) Calculate I_ℓ and find $\underline{Z}_\Delta$, $\underline{I}_{ab}$, $\underline{I}_{bc}$, and $\underline{I}_{ca}$ in polar form.
(b) What value of C should be put in parallel with each load element to minimize the current from the source, and what is the resulting line current?

16.2–4   Do Problem 16.2–3 with pf = 0.4 lagging and $P_p = 9.6$ kW.

16.2–5 Figure P16.2–5 represents a high-voltage transmission line connecting a balanced delta load to a three-phase generator with $\underline{V}_{ab} = 2600$ V $\underline{/0°}$. The transmission-line and load impedances are $\underline{Z}_{TL} = (0 + j6)$ Ω and $\underline{Z}_\Delta = (45 + j60)$ Ω. Use an equivalent circuit for one phase to find $\underline{I}_a$, $\underline{V}_{a'n'}$, and $\underline{V}_{a'b'}$. Also calculate the total real and reactive power supplied by the generator.

Figure P16.2–5

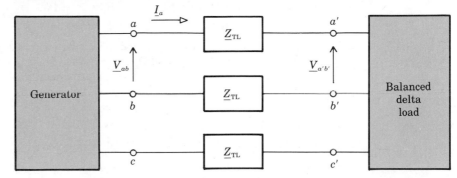

16.2–6 Do Problem 16.2–5 for the case of $\underline{Z}_{TL} = (4 + j0)$ Ω and $\underline{Z}_\Delta = (15 + j36)$ Ω.

16.2–7 Do Problem 16.2–5 for the case of $\underline{Z}_{TL} = (8 + j25)$ Ω and $\underline{Z}_\Delta = (48 + j90)$ Ω.

16.2–8 Two wattmeters are used to measure the properties of a balanced three-phase motor operating at $V_\ell = 1300$ V. The motor is known to be inductive, and the meters read $P_1 = 149$ kW and $P_2 = 59$ kW. Find P, Q, and I_ℓ. Then calculate $\underline{Z}_Y$ in rectangular form.

16.2–9 Do Problem 16.2–8 for the case of $P_1 = 114$ kW and $P_2 = 186$ kW.

16.2–10 When a load has a low power factor, one of the meters in Fig. 16.2–9 may have a *negative* reading. As an example, find P_1 and P_2 when $\underline{V}_{ab} = 208$ V $\underline{/0°}$ and the load is balanced, with $\underline{Z}_\Delta = (2.4 + j5.5)$ Ω. Then use direct power calculations to confirm that Eqs. (12) and (13) still hold.

16.2–11 Do Problem 16.2–10 with $\underline{Z}_\Delta = (2.1 + j7.2)$ Ω.

16.2–12 Suppose the load in Fig. 16.2–6 has developed an open-circuit fault, so $\underline{Z}_{ca} = \infty$ while $\underline{Z}_{ab} = \underline{Z}_{bc} = (5 + j12)$ Ω. Calculate $\underline{I}_a$, $\underline{I}_c$, P, and Q when $\underline{V}_{ab} = 208$ V $\underline{/0°}$. Then construct a phasor diagram to find $\underline{I}_b$.

16.2–13 Suppose the load in Fig. 16.2–10a has developed an open-circuit fault, so $\underline{Z}_c = \infty$ while $\underline{Z}_a = \underline{Z}_b = (4 + j3)$ Ω. Calculate $\underline{I}_a$, $\underline{I}_b$, P, and Q when $\underline{V}_{ab} = 208$ V $\underline{/0°}$. Then construct a phasor diagram to find the current $\underline{I}_n$ through the neutral line.

16.2–14 Suppose the load in Fig. 16.2–10b has developed an open-circuit fault, so $\underline{Z}_c = \infty$ while $\underline{Z}_a = \underline{Z}_b = (12 + j5)$ Ω. Calculate $\underline{I}_a$, $\underline{I}_b$, P, and Q when $\underline{V}_a = 120$ V $\underline{/0°}$. Then construct a phasor diagram to find the voltage $\underline{V}_{n'n}$ across the two neutral points.

16.2–15 Find $\underline{I}_a$, $\underline{I}_b$, $\underline{I}_c$, and P in Fig. 16.2–10b when $\underline{V}_{ab} = 240$ V $\underline{/0°}$, $\underline{Z}_a = \underline{Z}_b = 10$ Ω, and $\underline{Z}_c = 5$ Ω.

16.2–16 Find $\underline{I}_a$, $\underline{I}_b$, $\underline{I}_c$, P, and Q in Fig. 16.2–11a when $\underline{V}_{ab} = 600$ V $\underline{/0°}$, $\underline{Z}_{aA} = \underline{Z}_{bB} = \underline{Z}_{cC} = (1 + j0)$ Ω, $\underline{Z}_{AB} = (9 + j9)$ Ω, $\underline{Z}_{BC} = (9 + j0)$ Ω, and $\underline{Z}_{CA} = (9 - j9)$ Ω.

16.2–17   Find $\underline{I}_a$, $\underline{I}_b$, $\underline{I}_c$, P, and Q in Fig. 16.2–11a when $\underline{V}_{ab} = 300$ V $\underline{/0°}$, $\underline{Z}_{aA} = \underline{Z}_{bB} = \underline{Z}_{cC} = (0 + j1)$ Ω, $\underline{Z}_{AB} = \underline{Z}_{CA} = (4 + j0)$ Ω, and $\underline{Z}_{BC} = (8 + j0)$ Ω.

16.3–1   Consider a transmission line with $X = 250$ Ω and $R = 30$ Ω. The load draws $P_2 = 100$ MW and has power factor pf. Find δ when $V_p = 400$ kV. Then make a table comparing the estimated values of Q_1, P_R, and Eff for pf $= 1.0$ and for pf $= 0.8$ lagging.

16.3–2   Do Problem 16.3–1 for the case of $P_2 = 240$ MW.

16.3–3   Do Problem 16.3–1 for the case of $V_p = 300$ kV.

16.3–4   Consider a transmission line modeled in the form of Fig. 16.3–1 with $R = 2$ Ω, $\omega L = 10$ Ω, and $1/\omega C = 400$ Ω. In view of the capacitors, the input current $\underline{I}_1$ will differ from the load current $\underline{I}_2$. Use lumped-circuit analysis to find P_R, $\underline{V}_1$, δ, and $|I_1/I_2|$ when $\underline{V}_2 = 200$ kV $\underline{/0°}$ and the load draws $\underline{S}_2 = 240$ MW $+ j100$ MVAr. Then find P_2 and Q_2 by calculating $\underline{S}_1$.

16.3–5   Do Problem 16.3–4 for $\underline{S}_2 = 240$ MW $+ j0$.

16.3–6   Do Problem 16.3–4 for $\underline{S}_2 = 240$ MW $+ j200$ MVAr.

16.3–7   Derive Eqs. (4) and (5) from Eqs. (1) and (3).

16.3–8   Construct a phasor diagram to obtain $\underline{V}'_{ab}$ in Fig. 16.3–5 when $\underline{V}_{ab} = V_\ell$ $\underline{/0°}$ and the generator has an a-b-c phase sequence.

16.3–9   Do Problem 16.3–8 with an a-c-b phase sequence.

16.3–10   Suppose the transformer in Fig. 16.3–6 has $N = 1/20$ and the balanced three-phase load draws 100-A line current with pf $= 0.8$ lagging. Find the primary currents $|I_a|$, $|I_b|$, and $|I_c|$ when the single-phase impedances are $\underline{Z}_{a'n'} = (2 + j0)$ Ω, $\underline{Z}_{b'n'} = (3 + j0)$ Ω, and $\underline{Z}_{c'n'} = \infty$. Also calculate the ground current $|I_{n'n}|$.

16.3–11   Do Problem 16.3–10 with pf $= 0.6$ lagging, $\underline{Z}_{a'n'} = (2.6 + j1.5)$ Ω, $\underline{Z}_{b'n'} = \infty$, and $\underline{Z}_{c'n'} = (3 + j0)$ Ω.

Chapter

17 Magnetics, Induction, and Electromechanics

Electrical systems that generate, convert, or control large amounts of energy almost always involve devices whose operations depend on *magnetic* phenomena. Accordingly, we focus attention here on magnetics in relationship to electric power engineering.

We'll start with magnetic fields and flux, induction, magnetic circuits, and magnetic materials. Then we'll apply our results to two important practical topics: magnetic coupling in transformers, and energy conversion by electromechanical devices.

Objectives

After studying this chapter and working the exercises, you should be able to do each of the following:

- Identify the properties and units of magnetic flux and flux density (Section 17.1).

- State Faraday's law and its relationship to inductance (Section 17.1).

- Estimate the reluctance, flux, and energy stored in a simple magnetic circuit (Section 17.2).

- Describe the nonlinear effects and losses associated with a magnetic circuit (Section 17.2).

- Write the voltage-current equations for a two-winding transformer with mutual inductance (Section 17.3).

- Select an appropriate model to find the voltages and currents in a transformer circuit (Section 17.3).

- Calculate the magnetic force and induced voltage associated with a current-carrying conductor in a uniform magnetic field (Section 17.4).

- Analyze a simple electromechanical device with a translating or rotating coil or a moving-iron member (Section 17.4).

17.1 Magnetics and Induction

This section introduces the concepts of magnetic field and flux, giving particular attention to the case of current-carrying coils. We'll also state Faraday's law of induction, and we'll use it to show how a wire coil becomes an inductor with energy stored in a magnetic field.

Magnetic Field and Flux

A magnet exerts force on another nearby magnet without any mechanical connection between them. We visualize this force-at-a-distance effect in terms of a *magnetic field* in the space surrounding the magnet. Magnetic field is a vector quantity — like electrical and gravitational fields — and a complete investigation requires the mathematical tools of vector analysis. However, most engineering applications of magnetic fields involve configurations that can be treated in terms of a related scalar (nonvector) quantity called the *magnetic flux*. We will limit our consideration here to such cases.

Figure 17.1–1a qualitatively depicts the magnetic field around a simple bar magnet. Lines emanating from the magnet's north pole and returning to the south pole represent the direction of the field at various points in space. Thus, for instance, a compass needle or other small magnet placed somewhere in the field tends to become tangent to the field line at that point. The alignment force is strongest close to the magnet and decreases with distance, as represented by the spacing of the field lines.

Quantitatively, we describe the strength and direction of a magnetic field by a vector called the **flux density.** The flux-density magnitude B is proportional to the force exerted by the field on a moving charge. If a charge q moves with velocity component u perpendicular to the field, as in Fig. 17.1–1b, then the charge experiences a force of magnitude

$$f = Buq \tag{1}$$

This force is not in the same direction as the field but, instead, is mutually perpendicular to both u and B. Thus, the directions of u, B, and f are related to each other like the directions of a right-hand coordinate system.

The unit of flux density is the **tesla** (T) which, from Eq. (1), equals force in newtons divided by charge times velocity. Equivalently, and more conveniently for later work, we express this in terms of current in amperes by writing

$$1\ \text{T} = 1\ \text{N/A} \cdot \text{m}$$

for which we have noted that $1\ \text{C} \cdot \text{m/sec} = 1(\text{C/sec}) \cdot \text{m} = 1\ \text{A} \cdot \text{m}$.

Whenever flux density is uniform and perpendicular to some cross-sectional area A — as approximated at the ends of the bar magnet in Fig. 17.1–1a — we say that the total **flux** ϕ passing through that area is

$$\phi = BA \tag{2}$$

Figure 17.1–1 (a) Magnetic field. (b) Force on a moving charge.

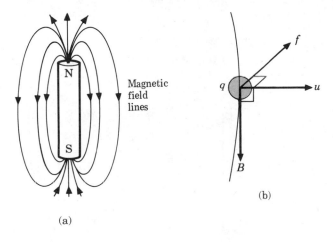

Magnetic field lines

(b)

(a)

Flux is measured in **webers** (Wb), where

$$1 \text{ Wb} = 1 \text{ T} \cdot \text{m}^2 = 1 \text{ N} \cdot \text{m/A}$$

Although flux is not a vector quantity, it can be either positive or negative depending on the direction of B.

Magnetic field and flux are created not only by permanent magnets but also by electrical currents. In particular, consider the long wire carrying current i in Fig. 17.1–2a. The flux density at any radial distance d from the axis of the wire is found to be

$$B = \frac{\mu i}{2\pi d} \tag{3}$$

The proportionality constant μ, called the permeability, will be discussed shortly.

The direction of B at the points shown in Fig. 17.1–2a is perpendicular out of or into the page, indicated by the symbols $\cdot$ and $\times$ to represent the "tip" and "tail feathers" of the vector arrow. It then follows from the circular symmetry that the field lines form closed circles around the wire, as seen in the endwise view of Fig. 17.1–2b where $\otimes$ symbolizes the current arrow going into the page. The clockwise rotational direction of B relative to i may be visualized with the help of the **right-hand rule:**

> Point your right thumb in the direction of the current and your fingers curl in the direction of the magnetic field.

The radially increasing spacing of the circles in Fig. 17.1–2b reflects the fact that flux density decreases as $1/d$.

Figure 17.1–2

Magnetic field around a current: (a) side view; (b) end view.

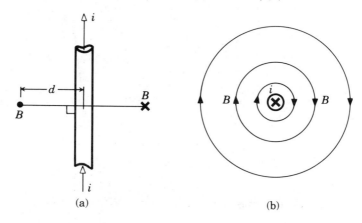

(a)   (b)

To obtain a more uniform field desired for many applications, the wire can be wound into a helical coil or **solenoid** of radius r and height h, as in Fig. 17.1–3a. If $h \gg r$, then the resulting interior flux density is

$$B = \frac{\mu N i}{h} \tag{4a}$$

where N is the number of turns in the coil. The total flux will then be

$$\phi = \frac{\mu N i \pi r^2}{h} \tag{4b}$$

Equation (4a) is derived by integrating the field contributions from each differential current element, and Eq. (4b) comes from $\phi = BA$ with cross-sectional area $A = \pi r^2$. Both of these expressions are valid anywhere inside the solenoid except close to the ends where the field gets somewhat weaker due to flux *leakage* between the turns. This leakage is shown in the axial field map, Fig. 17.1–3b.

The **permeability** μ in Eqs. (3) and (4) accounts for the magnetic property of the medium in question. The units are webers per ampere-

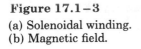

Figure 17.1−3
(a) Solenoidal winding.
(b) Magnetic field.

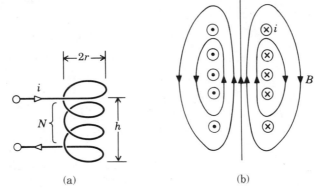

meter (Wb/A · m) or, equivalently, henrys per meter (H/m). The permeability of free space is

$$\mu_0 = 4\pi \times 10^{-7} \tag{5}$$

Many ordinary materials have $\mu \approx \mu_0$. **Ferromagnetic materials** such as iron and steel have much larger values of μ, usually expressed in terms of the **relative permeability**

$$\mu_r \triangleq \mu/\mu_0$$

Figure 17.1−4
Toroidal winding.

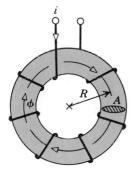

Typically, $\mu_r \approx 1000$ for a good magnetic material, but such materials are quite *nonlinear* in the sense that the value of μ varies with the flux density.

Returning to Eq. (4), we see that inserting a ferromagnetic core into a solenoid will substantially increase the internal magnetic field. Moreover, a high-permeability core reduces flux leakage between the turns of the coil, so that nearly all the field lines exit from the top of the core and reenter at the bottom. Under this condition the solenoid looks externally just like a permanent magnet.

Suppose, then, that the core is formed in the shape of a doughnut or **toroid** with the ends joined together as in Fig. 17.1−4. The magnetic field is thereby contained almost entirely within the core, and the flux becomes

$$\phi = \frac{\mu NiA}{\ell_\phi} \tag{6}$$

Here, $\ell_\phi = 2\pi R$ is the *average length* of the flux path through the core and A is the cross-sectional area (but not necessarily circular). High-μ toroids and similar configurations play important roles in devices that require a large confined flux.

Example 17.1−1

Consider a 400-turn coil wound on a toroidal iron core with average radius $R = 8$ cm $= 0.08$ m, cross-section radius $r = 2$ cm $= 0.02$ m, and relative permeability $\mu_r = 800$ so $\mu = 800\mu_0 \approx 10^{-3}$. The interior flux produced by $i = 1.0$ A will be

$$\phi = 10^{-3}\,\frac{\text{Wb}}{\text{A} \cdot \text{m}}\,\frac{400 \times 1.0\text{ A} \times \pi(0.02\text{ m})^2}{2\pi \times 0.08\text{ m}}$$
$$= 10^{-3}\text{ Wb} = 1.0\text{ mWb}$$

Hence, the flux density is

$$B = \frac{\phi}{A} = \frac{10^{-3}\text{ Wb}}{\pi(0.02\text{ m})^2} \approx 0.8\text{ T}$$

These values are typical for coils of moderate dimension with ferromagnetic cores. Had the core been of nonmagnetic materials, the current required for the same flux value would have been $800 \times 1.0 = 800\ A$!

Exercise 17.1–1   Suppose a solenoid has 10 turns per centimeter wrapped around a nonmagnetic core of radius 5 cm. What current is required to produce $\phi = 1$ mWb inside the solenoid? And what is the corresponding flux density? (Save your results for Exercise 17.1–3.)

Faraday's Law of Induction

We have seen that a current-carrying wire wound into a coil produces magnetic flux through the coil. Now consider the situation in Fig. 17.1–5, where flux ϕ from some external source passes through an N-turn coil. If the flux varies with time, then **Faraday's law** predicts an **induced voltage** across the terminals of the coil. This voltage, denoted e, is related to the rate of change of flux by

$$e = N\,\frac{d\phi}{dt} \qquad (7)$$

As a consequence of the induced voltage, a current i must flow through the coil when we connect it to a resistor or any other element closing the loop.

Figure 17.1–5
Voltage induced by time-varying flux.

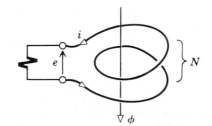

The polarity of induced voltage and the direction of the resulting current are governed by **Lenz's law,** an important companion to Faraday's law. Lenz's law tells us that:

Any induced quantity has a polarity or direction that tends to oppose the cause that induces it.

This law confirms that the polarity of e and direction of i are correctly shown in Fig. 17.1–5 when $d\phi/dt > 0$, for the induced current flow would produce flux through the coil in the upward direction (remember the right-hand rule) to oppose the increasing external flux in the downward direction. If you find this somewhat confusing, just keep in mind the principle that induction processes cannot be self-sustaining.

An interesting variation of Faraday's law occurs in the set-up portrayed by Fig. 17.1–6. A metal bar of length ℓ travels along conducting rails, cutting across a uniform magnetic field with flux density B perpendicular to the plane of motion. The bar, rails, and terminating element form a loop or coil with just one turn. The bar moves at constant velocity u and travels distance $u\,dt$ in time dt—during which time the flux encircled by the loop increases by the amount $d\phi = B\,dA = B\ell u\,dt$. Hence, the flux through the loop increases at the rate $d\phi/dt = B\ell u$ and there will be an induced voltage at the terminals given by

$$e = B\ell u \tag{8}$$

as follows from Eq. (7) with $N = 1$.

Figure 17.1–6
Voltage induced by a moving conductor.

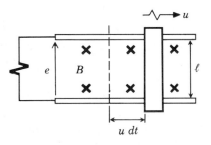

Equation (8) will be used extensively to analyze electromechanical transducers (Section 17.4) and electrical generators (Chapter 18). Any voltage induced by moving a conductor across a magnetic field is sometimes called an **electromotive force,** or **emf** for short.

Exercise 17.1–2   Since the number of turns N is a dimensionless quantity, Eq. (7) implies that the units of flux may be written as 1 Wb = 1 V · sec. Verify this expression by using the facts that force equals work per unit distance and voltage equals work per unit charge.

Inductance and Stored Energy

Faraday's law also holds for *inductors*. Applying a voltage v to a coil produces time-varying flux through it and establishes a magnetic field with stored energy.

Figure 17.1–7
Toroidal inductor.

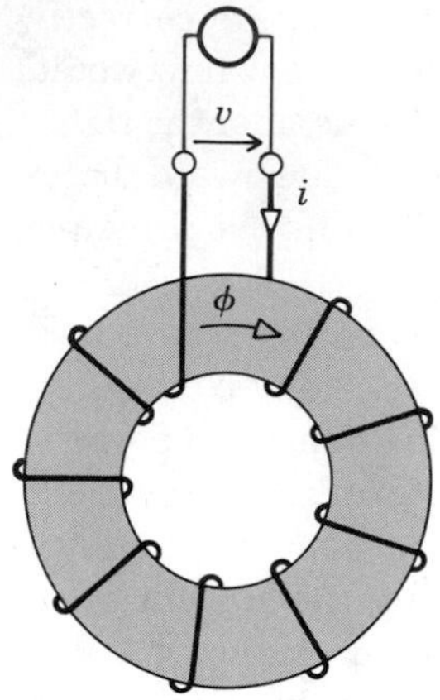

Take the toroid in Fig. 17.1–7 as a case in point. We know from Eq. (6) that the flux at any instant of time is $\phi = (\mu N A/\ell_\phi)i$. Therefore, the voltage required to produce a changing current and, as a result, a changing flux is

$$v = N\frac{d\varphi}{dt} = N\frac{\mu N A}{\ell_\phi}\frac{di}{dt}$$

This expression has the familiar form $v = L\,di/dt$ with the inductance L given by

$$L = \frac{\mu N^2 A}{\ell_\phi}$$

Alternatively, since $\mu N A/\ell_\phi = \phi/i$, we can write

$$L = \frac{N\phi}{i} \tag{9}$$

Equation (9) says that inductance equals the number of *flux linkages* ($N\phi$) *per unit current*.

In our initial discussion of inductance as a circuit element, we stated that there was stored energy in a magnetic field given by

$$w = \tfrac{1}{2}Li^2$$

Now we can relate the stored energy to the flux or flux density inside the toroidal coil. For this purpose, we use Eqs. (9) and (6) to write $Li^2 = (N\phi/i)i^2 = \phi Ni = BANi$ with $Ni = \phi\ell_\phi/\mu A = B\ell_\phi/\mu$. Thus,

$$w = \frac{\ell_\phi}{2\mu A}\phi^2 = \frac{A\ell_\phi}{2\mu}B^2 \tag{10}$$

where the product $A\ell_\phi$ equals the *volume* of the toroid. This magnetic stored energy comes, of course, from the electrical power $p = vi$ delivered to the coil. We can recover the energy by replacing the source with a load resistance, so that $\phi \to 0$ as $t \to \infty$ and the voltage induced by the decreasing flux delivers power to the load.

Despite their specialized derivations, the preceding results are universally valid. Indeed, Eq. (9) serves as the *definition* of inductance for any N-turn coil enclosing flux ϕ produced by current i. Similarly, Eq. (10) gives the magnetic energy stored by a uniform magnetic field B in any region with volume $A\ell_\phi$. However, care must be exercised in the case of a ferromagnetic material whose permeability μ is not constant.

Example 17.1–2

The 400-turn toroid in Example 17.1–1 was found to have $\phi = 1.0$ mWb when $i = 1.0$ A. The corresponding inductance is

$$L = 400 \times \frac{1.0\text{ mWb}}{1.0\text{ A}} = 400\text{ mH} = 0.4\text{ H}$$

The stored energy will be $w = 0.2$ J, calculated either from Eq. (10) or from $w = \frac{1}{2}Li^2$.

Exercise 17.1–3   Find the value of h such that the solenoid in Exercise 17.1–1 has $L = 10$ mH.

17.2 MAGNETIC CIRCUITS AND MATERIALS

Motors, generators, and other magnetic devices require intense magnetic fields. Sometimes a permanent magnet satisfies that need. More commonly, however, the field must be created by wrapping a current-carrying coil around a ferromagnetic core to form a magnetic circuit. For this reason, the study of magnetic circuits and ferromagnetic core materials is an important part of electric power engineering.

Magnetic Circuits

Consider the simple magnetic circuit of Fig. 17.2–1, where a high-permeability core channels the flux produced by an N-turn winding. Our understanding of such configurations is enhanced by viewing the flux ϕ in a magnetic circuit as analogous to the current in an electric circuit. This viewpoint reflects the property that flux, like current, always "travels" around a closed path. Then, since flux production depends on both the number of turns and the current through the winding, we define a **magnetomotive force (mmf)**

$$\mathcal{F} \triangleq Ni \qquad (1)$$

Magnetomotive force serves as the source in a magnetic circuit, similar to the source voltage (or electromotive force) in an electric circuit. We measure mmf in ampere-turns (A-t) to underscore the presence of N in Eq. (1).

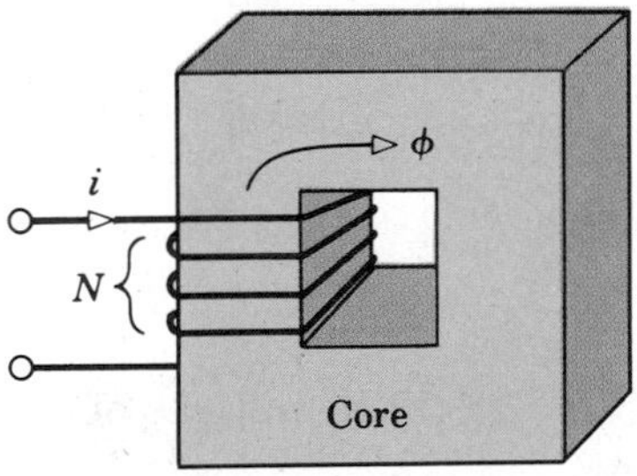

Figure 17.2–1
Magnetic circuit.

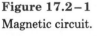

Completing the circuit analogy, we define the magnetic **reluctance** $\mathcal{R}$ of the flux path through the core such that

$$\mathcal{F} = \mathcal{R}\phi \qquad (2)$$

Reluctance, in ampere-turns per weber (A-t/Wb), is the magnetic equivalent of electrical resistance. Hence, $\mathcal{F} = \mathcal{R}\phi$ becomes the magnetic version of Ohms' law, $v = Ri$. Table 17.2–1 summarizes our analogy and includes some additional quantities.

Table 17.2–1

Magnetic Quantity	Symbol	Unit	Electric Analog
Flux	ϕ	Wb	Current
mmf	$\mathcal{F} = Ni$	A-t	Voltage (emf)
Reluctance	$\mathcal{R}$	$\dfrac{\text{A-t}}{\text{Wb}}$	Resistance
Permeability	μ	$\dfrac{\text{Wb}}{\text{A} \cdot \text{m}}$	Conductivity
Flux density	B	T	Current density
Magnetizing force	H	$\dfrac{\text{A-t}}{\text{m}}$	Field strength

Figure 17.2–2 illustrates the concept of an equivalent electrical circuit for the magnetic circuit of Fig. 17.2–1. The source $\mathcal{F}$ causes flux ϕ to "flow" around the closed loop while the core reluctance tends to resist it. Thus, ϕ equals $\mathcal{F}/\mathcal{R}$ and an "mmf drop" $\mathcal{R}\phi = \mathcal{F}$ exists across $\mathcal{R}$. In principle, magnetic-circuit analysis using $\mathcal{F} = \mathcal{R}\phi$ exactly mimics the analysis of resistive electrical circuits when we have linear (constant μ) core material. In practice, however, we encounter a major difficulty when it comes to calculating reluctances for specific cores.

Figure 17.2–2
Electrical analogy for a magnetic circuit.

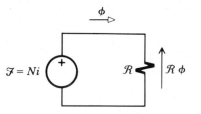

If the core happens to be a leakage-free toroid, then Eq. (6), Section 17.1, shows that

$$Ni = \frac{\ell_\phi}{\mu A}\,\phi$$

Thus, by comparison with the definitions of $\mathcal{F}$ and $\mathcal{R}$, the reluctance is

$$\mathcal{R} = \frac{\ell_\phi}{\mu A} \tag{3}$$

Based on this result, we can *estimate* the reluctance of other core configurations as follows:

1. Subdivide the core into sections having constant cross-sectional area A_k, permeability μ_k, and average flux path length ℓ_k.

2. Take the reluctance of each section to be

$$\mathcal{R}_k = \frac{\ell_k}{\mu_k A_k} \tag{4}$$

3. Combine reluctances in accordance with the circuit analogy.

Estimation yields fair results for high-μ cores, consistent with the fact that we seldom know permeability values to better than 5% accuracy. More precise results therefore require experimental measurements.

To put our estimation method to work, take the case of a high-μ core broken by a short **air gap** as shown in Fig. 17.2–3a. The same flux ϕ passes through both core and gap, so we have the *series* magnetic circuit diagrammed in Fig. 17.2–3b. The total reluctance thus becomes $\mathcal{R} = \mathcal{R}_c + \mathcal{R}_g$ and

$$\phi = \frac{\mathcal{F}}{\mathcal{R}} = \frac{Ni}{\mathcal{R}_c + \mathcal{R}_g}$$

It frequently turns out that $\mathcal{R}_c \ll \mathcal{R}_g$ and $\phi \approx Ni/\mathcal{R}_g$ because $\mathcal{R}_c$ and $\mathcal{R}_g$ are proportional to $1/\mu_c$ and $1/\mu_0$, respectively, and a ferromagnetic core has $\mu_c \gg \mu_0$.

Figure 17.2–3
Magnetic circuit with air gap.

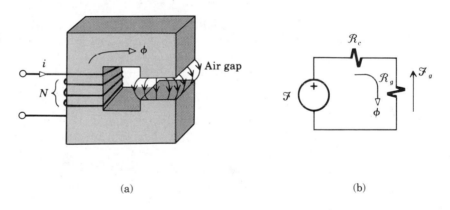

(a)   (b)

As illustrated in Fig. 17.2–3a, the magnetic field tends to "bulge out" across an air gap. This effect, known as **fringing,** increases the effective cross-sectional area of the gap and can be accounted for by the following rule of thumb:

> Add the gap length ℓ_g to both of the gap's transverse dimensions when calculating the area A_g to be used in Eq. (4).

You should take account of fringing, even in estimations, because the gap reluctance usually constitutes a major portion of the total reluctance.

Figure 17.2–4 depicts a more complicated core configuration with two windings. The accompanying equivalent circuit has two mmf sources and both series and parallel reluctances to represent the various flux paths. Note that the "polarity" of $\mathscr{F}_2$ opposes ϕ_2, as determined from the right-hand rule.

Figure 17.2–4

Magnetic circuit with two windings.

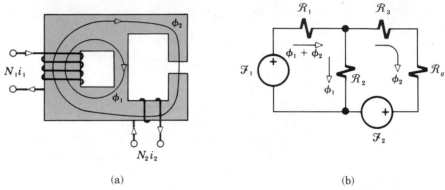

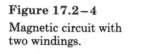

(a)   (b)

Finally, combining the definitions of reluctance $\mathscr{R}$ and inductance L, we obtain a simple and useful relationship between them,

$$L = \frac{N\phi}{i} = \frac{N}{i}\frac{\mathscr{F}}{\mathscr{R}} = \frac{N}{i}\frac{Ni}{\mathscr{R}} = \frac{N^2}{\mathscr{R}} \qquad (5)$$

Equation (5) holds only when the core has just one winding. With two or more windings, an interactive effect known as *mutual inductance* occurs. This interaction is implied in Faraday's law and forms the basis of transformer operation.

Example 17.2–1

Suppose the toroid described in Example 17.1–1 is cut radially and pried apart, creating an air gap with $\ell_g = 0.5$ cm. We'll use the electric circuit analog from Fig. 17.2–3 to find the resulting changes of the magnetic circuit.

First, adding ℓ_g to the gap's cross-sectional diameter to account for fringing, we get $A_g = \pi(2r + \ell_g)^2/4$ and

$$\mathscr{R}_g = \frac{\ell_g}{\mu_0 A_g} = \frac{0.005}{4\pi \times 10^{-7} \times \pi(0.045^2/4)} = 2.5 \times 10^6$$

Next, for the core's reluctance, we have $\ell_c = 2\pi R$, $A_c = \pi r^2$, and $\mu_c = 10^{-3}$, so

$$\mathscr{R}_c = \frac{2\pi \times 0.08}{10^{-3} \times \pi(0.02)^2} = 0.4 \times 10^6$$

Thus

$$\mathscr{R} = \mathscr{R}_c + \mathscr{R}_g = 2.9 \times 10^6 \text{ A-t/Wb}$$

and the gap contributes about 90% of the total reluctance.

To obtain the same flux value as before (1.0 mWb), we must have

$$\mathcal{F} = \mathcal{R}\phi = (2.9 \times 10^6 \text{ A-t/Wb})(10^{-3} \text{ Wb}) = 2900 \text{ A-t}$$

Since $N = 400$, the required current becomes

$$i = \frac{2900}{400} = 7.25 \text{ A}$$

whereas it took just 1.0 A to get $\phi = 1.0$ mWb without the air gap. The increased current results in more stored energy, which is confirmed from

$$L = \frac{N^2}{\mathcal{R}} = \frac{400^2}{2.9 \times 10^6} = 55 \text{ mH}$$

and

$$w = \tfrac{1}{2}Li^2 = \tfrac{1}{2}(55 \times 10^{-3}) \times 7.25^2 = 1.45 \text{ J}$$

Furthermore, most of the stored energy now resides in the air gap since the energy stored in the core is still 0.2 J, as we determined in Example 17.1–2.

Exercise 17.2–1 Suppose the core in Fig. 17.2–3a is 7 cm high and 7 cm wide, each leg is 2 cm wide and 2 cm deep, and the air gap is 0.5 cm high. Estimate $\mathcal{R}$ with $\mu_c = 0.001$, and find i such that $\phi = 0.5$ mWb when $N = 200$.

Magnetization and Hysteresis

Up to now we have ignored the complicated behavior of ferromagnetic core materials hidden behind the permeability μ. Such materials not only are nonlinear but also have a magnetic "memory," as exemplified by permanent magnets and magnetic recording media. We will not go into the underlying *domain theory* of magnetism here, but we will describe the external properties of magnetic materials. For that purpose we introduce the **magnetizing force** or **magnetic field intensity**

$$H \triangleq \frac{\mathcal{F}}{\ell_\phi} = \frac{Ni}{\ell_\phi} \tag{6}$$

where ℓ_ϕ, again, stands for flux path length. Magnetizing force H represents the *mmf per unit length* (A-t/m) and is analogous to electric field strength $\mathcal{E}$ — the emf per unit length.

Let's proceed by analogy, recalling that when an electric field exists in a medium with conductivity $\sigma = 1/\rho$ and cross-sectional area A, the resulting current is $i = \sigma \mathcal{E} A$. When a magnetizing force H exists in a toroidal core with permeability μ and cross-sectional area A, the resulting magnetic flux will be $\phi = \mu(Ni/\ell_\phi)A = \mu H A$. Permeability μ, therefore, is the "magnetic conductivity" of the material. Furthermore, dividing by the area converts flux ϕ to flux density B and yields the important relation

$$B = \mu H \tag{7}$$

Thus, a plot of B versus H — called the **normal magnetizing curve** — reveals whether the material in question has constant permeability $\mu = B/H$.

Figure 17.2–5 shows typical magnetizing curves for cast iron and sheet steel. Also shown, but barely visible, is the curve for nonmagnetic material, a straight line with very small slope $\mu \approx \mu_0$. The magnetizing curve for cast iron has much greater initial slope (implying $\mu_r = \mu/\mu_0 \gg 1$) that gradually decreases with increasing H. The curve for sheet steel rises even more sharply and then flattens off rather abruptly. (The slope actually approaches μ_0 as $H \rightarrow \infty$.) We therefore say that such materials become quickly *saturated*, since increasing H beyond the "knee" of the curve yields little further increase in B.

Clearly, these ferromagnetic materials are *nonlinear* in the sense that the value of μ depends on the magnetizing force and, in particular, de-

Figure 17.2–5
Typical magnetizing curves.

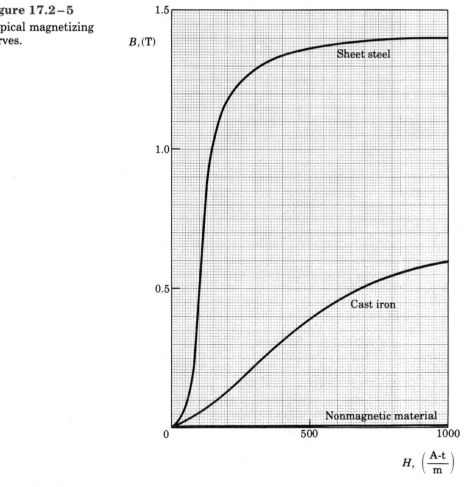

creases with increasing H. This nonlinearity means that flux ϕ is not directly proportional to current i and that, therefore, inductance $L = N\phi/i$ and reluctance $\mathcal{R} = \mathcal{F}/\phi$ are not truly constants. Fortunately, approximate calculations based on an average value of μ often prove satisfactory if the final result is below saturation.

When magnetizing force varies with time, we must also take account of the **hysteresis** effect as shown in Fig. 17.2–6. Here, the material is ini-

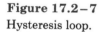

Figure 17.2–6
Hysteresis effect.

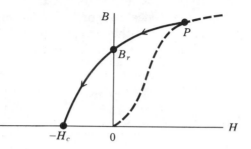

tially saturated at some point P on the B-H curve and the magnetizing force is removed ($H = 0$). But instead of going to zero flux density along the curve, B drops only slightly to a nonzero value B_r known as the **retentivity.** We have now created a *permanent magnet* that retains flux density $B = B_r$ without an external magnetizing force. This material can be demagnetized only when we apply a reverse magnetizing force of strength H_c, known as the **coercivity.** The special alloys used for such permanent magnets are called "hard" magnetic materials and typically have $B_r \approx 1$ T and $H_c \approx 10^4$ A-t/m. "Soft" magnetic materials, on the other hand, have relatively little hysteresis.

Figure 17.2–7
Hysteresis loop.

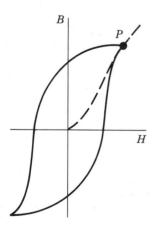

If magnetic material experiences an oscillating magnetizing force, then the value of B lags behind H and the B-H relationship takes the form of the **hysteresis loop** in Fig. 17.2–7. The extreme tip at point P falls on the normal magnetizing curve, indicated by the dashed line. Such a hysteresis loop is produced in a magnetic circuit, for instance, when AC winding current generates a sinusoidal $H(t)$ waveform. The corresponding flux waveform $\phi(t)$ and any induced voltage $e(t)$ will then be *distorted* sinusoids, which may cause problems in transformers and motors.

Hysteresis also prevents direct measurement of a material's normal magnetizing curve. Instead, we must measure the tips of several hysteresis loops of various sizes. The resulting B-H plot is then called an *average* magnetizing curve because the actual value of B for a given value of H may be either somewhat above or below the plotted points, depending upon the material's magnetization history.

Example 17.2–2

To illustrate the use of B-H curves for magnetic circuit design, consider a toroidal core with $\ell_c = 0.08\pi$ m and $A_c = \pi(0.02 \text{ m})^2$—the same dimensions as in Examples 17.1–1 and 17.2–1. The core could be made of either sheet steel or cast iron. We'll estimate the mmf needed to get $\phi = 0.6$ mWb, without and with a 0.5-cm air gap.

First, we calculate the required flux density $B = \phi/A_c \approx 0.48$ T. The curves in Fig. 17.2–5 then show that the magnetizing force must be $H \approx 100$ A-t/m for sheet steel or $H \approx 650$ A-t/m for cast iron. The corresponding core mmf is $\mathcal{F}_c = H\ell_c \approx 25$ or 160 A-t.

Now, accounting for the air gap, we write $\mathcal{F} = \mathcal{F}_c + \mathcal{F}_g$ where $\mathcal{F}_g = \mathcal{R}_g\phi = 2.5 \times 10^6 \times 0.6 \times 10^{-3} = 1500$ A-t. Thus, we would need $\mathcal{F} \approx 1525$ A-t with a sheet-steel core or $\mathcal{F} \approx 1660$ with a cast-iron core. Since the air gap dominates in either case, the inherent inaccuracies of the core calculations become relatively unimportant.

Exercise 17.2–2   Consider a magnetic circuit with $N = 60$ turns and a sheet-steel core having $\ell_\phi = 0.1\pi$ m and $A_c = \pi(0.025$ m$)^2$. Calculate H, B, ϕ, $\mathcal{R}$, and L when: (a) $i = 1$ A, and (b) $i = 2$ A.

Losses

A magnetic device always has losses in the sense that some fraction of the input energy will be converted to unwanted heat. The most obvious loss is ohmic heating in the windings resulting from the small but inevitable winding resistance R_w. Often referred to as *copper loss*, this $R_w i^2$ power dissipation occurs with either constant or time-varying current. Time-varying current also produces two other forms of heating in the *core* itself, due to hysteresis and eddy currents.

Hysteresis loss represents the energy required to go around the hysteresis loop (Fig. 17.2–7), when $H(t)$ has a cyclical time variation that alternately magnetizes and demagnetizes the core. For a sinusoidal variation at frequency $f = \omega/2\pi$, the average hysteresis power loss is given by the empirical formula

$$P_h = k_h f B_m^n A_c \ell_c \tag{8}$$

where B_m is the maximum flux density and k_h and n are characteristics of the core. The value of n, called the *Steinmetz exponent*, ranges from about 1.5 to 2.0, with 1.6 being a typical value.

Eddy-current loss comes from localized currents induced in the core by a time-varying flux. Illustrating this phenomenon, Fig. 17.2–8a shows $\phi(t) = B(t)A_c$ passing through the cross-section of a magnetic core. Since the iron or steel core material is a conductor, we have a conducting path encircling the flux (similar to that of Fig. 17.1–5). The corresponding induced voltage produces an eddy current i_e around the path, and the current, in turn, causes ohmic heating. From Faraday's law, it can be shown that a sinusoidal flux variation produces an average eddy-current power loss

$$P_e = k_e \frac{(A_c f B_m)^2}{R_e} \tag{9}$$

where R_e is the equivalent core resistance and k_e another core constant. Unless special steps are taken, P_e may account for well over half of the total input power of a magnetic circuit.

Fortunately, we can minimize eddy-current losses simply by constructing the core from thin sheets or **laminations** parallel to the flux, with insulating varnish between them, as in Fig. 17.2–8b. If there are m laminations, then each one has a smaller area $A' = A_c/m$ and a larger resistance $R'_e \approx m R_e$; therefore, the total loss will be reduced by a factor of about $1/m^2$.

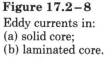

Figure 17.2–8
Eddy currents in:
(a) solid core;
(b) laminated core.

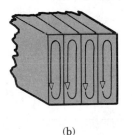

(a)

(b)

Values of the core constants k_h and k_e are not usually available. Rather, they are inferred from power and efficiency measurements. The efficiency of a magnetic device intended for power transfer (a transformer, for example) typically exceeds 95%, including core and copper losses.

17.3 MUTUAL INDUCTANCE AND POWER TRANSFORMERS

Having completed the necessary background, we are ready to treat magnetic coupling in transformers. We'll start with the concept of mutual inductance as the circuit manifestation of magnetic coupling. Then we'll develop circuit models and analysis techniques for two-winding transformers in general and power transformers in particular.

Magnetic Coupling and Mutual Inductance

Figure 17.3–1a depicts two coils wound on the same core, a **primary winding** with N_1 turns and a **secondary winding** with N_2 turns. The core material may or may not be ferromagnetic, but it is assumed to be operating over an essentially linear region of its magnetization curve. We'll demonstrate the existence and nature of the magnetic coupling between the coils by deriving the relationships for their terminal voltages and currents.

Figure 17.3–1

(a) Magnetic core with two windings and flux leakage. (b) Equivalent circuit diagram.

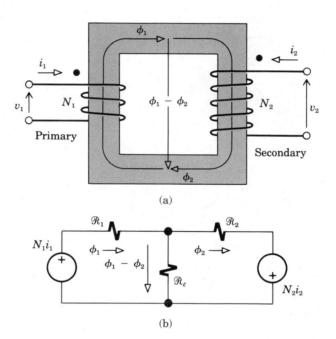

(a)

(b)

Since either coil might serve as the input of a transformer, we'll take the reference direction for both currents into the upper terminals marked with dots. Current i_1 entering the dotted terminal of the primary establishes the mmf $\mathcal{F}_1 = N_1 i_1$ which, by the right-hand rule, tends to drive flux in the clockwise direction around the core. Likewise, current i_2 establishes $\mathcal{F}_2 = N_2 i_2$, which also tends to drive flux in the clockwise direction. In general, then, the dots mean that:

> Current entering the dotted end of one winding produces flux in the same direction as the flux produced by current entering the dotted end of the other winding.

We'll subsequently show that these polarity dots agree with our previous convention for the ideal transformer.

An ideal core would confine the flux completely within it. In reality, however, some *leakage flux* escapes from the core so the total flux ϕ_1 linking the primary differs from the total flux ϕ_2 linking the secondary. The diagram for the magnetic circuit therefore takes the form of Fig. 17.3–1b, where $\mathcal{R}_\ell$ represents the reluctance of the leakage path. Loop analysis for ϕ_1 and ϕ_2 in this diagram yields the pair of equations

$$(\mathcal{R}_1 + \mathcal{R}_\ell)\phi_1 - \mathcal{R}_\ell\phi_2 = N_1 i_1$$

$$-\mathcal{R}_\ell\phi_1 + (\mathcal{R}_2 + \mathcal{R}_\ell)\phi_2 = N_2 i_2$$

Simultaneous solution then gives

$$\phi_1 = \frac{N_1 i_1}{\mathcal{R}_{11}} + \frac{N_2 i_2}{\mathcal{R}_M} \qquad \phi_2 = \frac{N_1 i_1}{\mathcal{R}_M} + \frac{N_2 i_2}{\mathcal{R}_{22}} \qquad \textbf{(1a)}$$

where we have defined

$$\mathcal{R}_{11} = \mathcal{R}_1 + \mathcal{R}_2\|\mathcal{R}_\ell \qquad \mathcal{R}_{22} = \mathcal{R}_2 + \mathcal{R}_1\|\mathcal{R}_\ell$$

$$\mathcal{R}_M = \mathcal{R}_1 + \mathcal{R}_2 + (\mathcal{R}_1\mathcal{R}_2/\mathcal{R}_\ell) \qquad \textbf{(1b)}$$

The quantities $\mathcal{R}_{11}$ and $\mathcal{R}_{22}$ are just the equivalent reluctances seen by the mmfs $\mathcal{F}_1$ and $\mathcal{F}_2$, respectively. The mutual reluctance $\mathcal{R}_M$ accounts for the fact that $\mathcal{F}_1$ contributes to ϕ_2 and $\mathcal{F}_2$ contributes to ϕ_1.

Next we draw upon Faraday's law and Eq. (1a) to obtain expressions for the terminal voltages, namely

$$v_1 = N_1 \frac{d\phi_1}{dt} = \frac{N_1^2}{\mathcal{R}_{11}} \frac{di_1}{dt} + \frac{N_1 N_2}{\mathcal{R}_M} \frac{di_2}{dt}$$

$$v_2 = N_2 \frac{d\phi_2}{dt} = \frac{N_1 N_2}{\mathcal{R}_M} \frac{di_1}{dt} + \frac{N_2^2}{\mathcal{R}_{22}} \frac{di_2}{dt}$$

These two equations can be cleaned up and made more understandable by introducing the quantities

$$L_1 \triangleq \frac{N_1^2}{\mathcal{R}_{11}} \qquad L_2 \triangleq \frac{N_2^2}{\mathcal{R}_{22}} \qquad M \triangleq \frac{N_1 N_2}{\mathcal{R}_M} \qquad \textbf{(2)}$$

Now we can write

$$v_1 = L_1 \frac{di_1}{dt} + M \frac{di_2}{dt} \qquad v_2 = M \frac{di_1}{dt} + L_2 \frac{di_2}{dt} \qquad \textbf{(3)}$$

Notice that v_1 and v_2 depend on both currents i_1 and i_2 because ϕ_1 and ϕ_2 in Eq. (1a) depend on both i_1 and i_2.

To interpret the symbols L_1 and M, let the secondary be open-circuited. Then $i_2 = 0$, $di_2/dt = 0$, and Eq. (3) becomes

$$v_1 = L_1\, di_1/dt \qquad v_2 = M\, di_1/dt$$

Hence, L_1 corresponds to the inductance of the primary by itself, called the **self-inductance,** while M represents the *magnetic coupling* from primary to secondary. This coupling arises because part of the flux ϕ_1 produced by $N_1 i_1$ links the secondary and induces the open-circuit emf $M\, di_1/dt$ when i_1 varies with time. Conversely, if $di_1/dt = 0$, then

$$v_1 = M\, di_2/dt \qquad v_2 = L_2\, di_2/dt$$

Hence, L_2 is the self-inductance of the secondary, and M now represents magnetic coupling from secondary to primary. Since the coupling is the same in either direction, M is named the **mutual inductance** between the coils.

Refering to Eq. (2), we see that the self-inductances L_1 and L_2 are proportional to N_1^2 and N_2^2, respectively, while the mutual inductance is proportional to $N_1 N_2$. Accordingly, M is related to L_1 and L_2 by

$$M = k\sqrt{L_1 L_2} \tag{4a}$$

where, from Eqs. (1b) and (2),

$$k = \frac{\mathcal{R}_\ell}{\sqrt{(\mathcal{R}_1 + \mathcal{R}_\ell)(\mathcal{R}_2 + \mathcal{R}_\ell)}} \le 1 \tag{4b}$$

We call k the **coupling coefficient.** Maximum or *unity coupling* ($k = 1$) occurs when $\mathcal{R}_\ell = \infty$, so there is no leakage and all of the flux links both coils. The mutual inductance then has the largest possible value $M_{\max} = \sqrt{L_1 L_2}$. Otherwise, leakage reduces the amount of flux linking both coils, resulting in $k < 1$ and $M < \sqrt{L_1 L_2}$.

Finally, consider the coil configuration in Fig. 17.3–2 where the dot at the lower end of the secondary indicates that it has been wound in the opposite sense compared to Fig. 17.3–1a. Reversing the sense of this wind-

Figure 17.3–2
Magnetic circuit with reversed secondary winding.

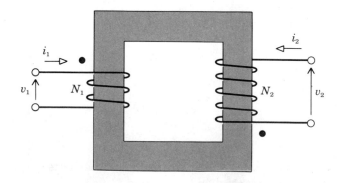

ing reverses the polarity of the mmf $\mathcal{F}_2 = N_2 i_2$, so the terminal voltages and currents are now related by

$$v_1 = L_1 \frac{di_1}{dt} - M \frac{di_2}{dt} \qquad v_2 = -M \frac{di_1}{dt} + L_2 \frac{di_2}{dt} \qquad (5)$$

which differ from Eq. (3) in that M has been replaced by $-M$. Thus, reversing one winding changes the sign of the magnetic coupling in both directions.

Circuit Analysis with Mutual Inductance

We have seen that mutual inductance is associated with two coils that always have self-inductance. Those properties are represented in circuit diagrams by the schematic symbols shown in parts (a) and (c) of Fig. 17.3–3. These symbols may stand for a two-winding transformer with mutual inductance coupling— as distinguished from an ideal transformer —or they may stand for any two coils coupled by a magnetic field. In any case, the magnetic coupling is assumed to be linear.

Figure 17.3–3
Symbols and tee equivalent models for self and mutual inductance.

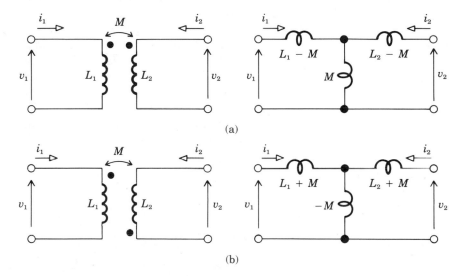

(a)

(b)

The symbol in Fig. 17.3–3a means that the mutual-inductance terms have the same sign as the self-inductance terms, and the terminal voltages and currents are related by

$$v_1 = L_1 \frac{di_1}{dt} + M \frac{di_2}{dt} \qquad v_2 = M \frac{di_1}{dt} + L_2 \frac{di_2}{dt}$$

But exactly the same relationships also hold for the three-inductor network in Fig. 17.3–3b, where the total current down through the inductor labeled M will be $i_1 + i_2$. Thus,

$$v_1 = (L_1 - M)\frac{di_1}{dt} + M\frac{d}{dt}(i_1 + i_2) = L_1\frac{di_1}{dt} + M\frac{di_2}{dt}$$

$$v_2 = (L_2 - M)\frac{di_2}{dt} + M\frac{d}{dt}(i_1 + i_2) = M\frac{di_1}{dt} + L_2\frac{di_2}{dt}$$

Accordingly, Fig. 17.3–3b is known as the **tee equivalent model** for mutual inductance when it has the same sign as the self-inductance terms.

The schematic symbol in Fig. 17.3–3c is used when the mutual inductance opposes the self-inductance terms, indicated by the dot location at the lower end of the secondary. The voltage-current relationships are then given by Eq. (5), and Fig. 17.3–3d is the corresponding tee equivalent model. This model also follows from Fig. 17.3–3b with M replaced by $-M$.

The tee equivalent models explicitly incorporate the terminal voltage-current relationships, thereby expediting circuit analysis with mutual inductance. Furthermore, having identified the appropriate model based on the dot convention, we are free to take any convenient reference polarities for the voltages and currents. However, these models involve artificial inductors, sometimes negative, and they imply the existence of three energy-storing elements, even though a two-winding device has only two magnetic fields for energy storage. You should therefore bear in mind that the models correctly describe terminal relationships, not internal behavior.

Tee models have another potential drawback, namely the direct electrical connection across the bottom. That connection is eliminated by inserting a fictitious ideal transformer to obtain the alternative model in Fig. 17.3–4a, which holds when the mutual inductance has the same sign

Figure 17.3–4

Tee models with an ideal transformer:
(a) arbitrary N_0;
(b) simplification with $N_0 = M/L_1$.

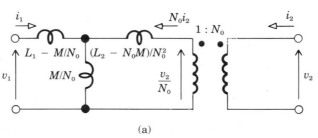

(a)

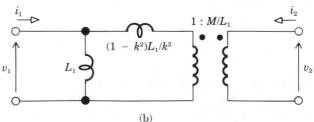

(b)

as the self-inductance terms. (Changing the sign of M would take care of the other case.) You can easily confirm the validity of this model with any value of the ideal transformer's turns ratio N_0. Taking $N_0 = M/L_1$ then reduces the model to two nonnegative inductors, as shown in Fig. 17.3–4b. Either of these models may be used when an external current-carrying element connects the lower terminals of the primary and secondary, whereas the original tee models would short-circuit such an element.

More commonly, we are concerned with circuits like Fig. 17.3–5a, where a transformer couples a time-varying source to some load. We'll analyze this important configuration taking the secondary current i_2 directed toward the load, as shown, and assuming exponential time variation. Figure 17.3–5b gives the corresponding diagram with the tee model from Fig. 17.3–3b and an arbitrary load impedance $Z(s)$. Our diagram also includes the case of AC excitation, of course, simply by changing to phasor notation and setting $s = j\omega$.

Figure 17.3–5
(a) Transformer coupling source to load.
(b) Equivalent circuit with tee model.

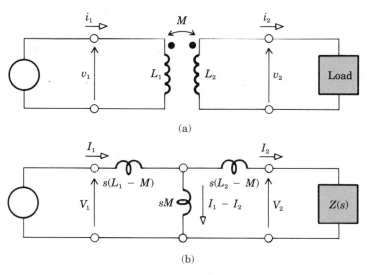

(a)

(b)

From KVL applied around the left and right loops we get the pair of equations

$$s(L_1 - M)I_1 + sM(I_1 - I_2) = sL_1I_1 - sMI_2 = V_1$$

$$sM(I_1 - I_2) - s(L_2 - M)I_2 - Z(s)I_2 = sMI_1 - [sL_2 + Z(s)]I_2 = 0$$

Solving for I_1 and I_2 and using $V_2 = Z(s)I_2$ then yields the voltage ratio and input impedance in the form

$$\frac{V_2}{V_1} = \frac{sMZ(s)}{s^2(L_1L_2 - M^2) + sL_1Z(s)} \tag{6a}$$

$$\frac{V_1}{I_1} = \frac{s^2(L_1L_2 - M^2) + sL_1Z(s)}{sL_2 + Z(s)} \tag{6b}$$

These expressions agree with the fact that the transformer has two energy-storing fields since s^2 is the highest power of s here, apart from $Z(s)$. Also note that $(L_1 L_2 - M^2)$ cannot be negative because $M = k\sqrt{L_1 L_2}$ with $k \le 1$.

A transformer with *unity coupling* confines the energy storage to just one field, and the term $s^2(L_1 L_2 - M^2)$ vanishes when $k = 1$. Equation (6a) then reduces to

$$\frac{V_2}{V_1} = \frac{M}{L_1} = \sqrt{\frac{L_2}{L_1}} = \frac{N_2}{N_1} = N \qquad \textbf{(7a)}$$

where we have used Eqs. (1b) and (2) with $\mathcal{R}_\ell = \infty$ to equate the voltage ratio to $N = N_2/N_1$. This result suggests that a unity-coupled transformer acts like an *ideal* transformer with turns ratio $N = N_2/N_1$. But the input impedance differs from that of an ideal transformer since, with $k = 1$, Eq. (6b) becomes

$$\frac{V_1}{I_1} = \frac{sL_1 Z(s)}{sL_2 + Z(s)} = \frac{L_1}{L_2} \frac{Z(s)}{1 + Z(s)/sL_2} \qquad \textbf{(7b)}$$

However, if $Z(s)$ is small compared to sL_2, then $1 + Z(s)/sL_2 \approx 1$ and

$$\frac{V_1}{I_1} \approx \frac{L_1}{L_2} Z(s) = \frac{N_1^2}{N_2^2} Z(s) = \frac{Z(s)}{N^2}$$

which we recognize as load impedance reflected by an ideal transformer. Moreover, the current ratio will be

$$\frac{I_2}{I_1} = \frac{V_2/Z(s)}{I_1} = \frac{1}{Z(s)} \frac{V_1}{I_1} \frac{V_2}{V_1} \approx \frac{1}{N}$$

just as in the case of an ideal transformer. We have thus seen that a transformer with mutual inductance behaves essentially like an ideal transformer with the same turns ratio, provided that $k \approx 1$ and $|Z(s)| \ll |sL_2|$.

Example 17.3–1

Let the circuit in Fig. 17.3–5 have an AC source at $\omega = 100$ and load resistor $R = 60\ \Omega$. Let the transformer have $L_1 = 1$ H, $L_2 = 4$ H, and $k = 0.95$, so $M = 0.95\sqrt{1 \times 4} = 1.9$ H. Setting $s = j\omega$ and $Z(s) = R$ in Eqs. (6a) and (6b), and inserting the numerical values yields the phasor relations

$$\frac{V_2}{V_1} = \frac{j\omega M R}{-\omega^2(L_1 L_2 - M^2) + j\omega L_1 R} = 1.59 \underline{/-33°}$$

$$\frac{V_1}{I_1} = \frac{-\omega^2(L_1 L_2 - M^2) + j\omega L_1 R}{j\omega L_2 + R} = 17.7\ \Omega\ \underline{/41.5°}$$

For comparison, an ideal transformer with $N = N_2/N_1 = \sqrt{L_1/L_2} = 2$ would have $\underline{V_2}/\underline{V_1} = N = 2$ and $\underline{V_1}/\underline{I_1} = R/N^2 = 15\ \Omega$

Exercise 17.3–1   Show that the terminal voltages and currents in Fig. 17.3–4a are related by Eq. (3).

Exercise 17.3–2   Apply impedance analysis to the tee model in Fig. 17.3–3b to find the equivalent inductance seen looking into the primary when the secondary is short circuited. Express your result in terms of L_1 and k, and investigate the extreme cases $k = 0$ and $k = 1$.

Power Transformers [†]

Transformers designed for AC power applications invariably have laminated steel or iron cores with coils wrapped one over the other, which all but eliminates flux leakage. Consequently, we can take $k \approx 1$ and $\phi_2 \approx \phi_1$. Furthermore, if the core material operates over an essentially linear region of its magnetization curve, then

$$v_1 = N_1 \, d\phi_1/dt \qquad v_2 \approx N_2 \, d\phi_1/dt = (N_2/N_1)v_1$$

But the linear assumption deserves closer attention, as does the matter of internal power dissipation. Both of these topics will be pursued here.

To investigate linearity, let the primary have an applied AC voltage with rms value $|V_1|$, so $v_1 = \sqrt{2}|V_1| \cos(\omega t + \theta_1)$. The resulting flux under steady-state linear conditions is

$$\phi_1 = \frac{1}{N_1} \int v_1 \, dt = \frac{\sqrt{2}|V_1|}{\omega} \sin(\omega t + \theta_1)$$

This flux produces the oscillating flux density $B = \phi_1/A$, where A is the cross-sectional area of the core. Reasonably linear operation requires the peak value of B to be less than the nominal saturation level B_{max} on the magnetization curve (see Fig. 17.2–5). Otherwise, core saturation and hysteresis will cause severe waveform distortion. Hence, to ensure that $|\phi_1| < \phi_{max} = B_{max} A$, the applied rms voltage must be limited by

$$|V_1| < \frac{\omega N_1 \phi_{max}}{\sqrt{2}} = 4.44 f N_1 B_{max} A \qquad \textbf{(8)}$$

in which we have set $\omega = 2\pi f$ to yield a convenient design expression. The secondary voltage is similarly limited.

When Eq. (8) holds we could, in principle, represent a power transformer by any of our previous models. However, since large amounts of power are involved, we must also consider unavoidable losses within the transformer. For that purpose, we'll employ the modified AC model in Fig. 17.3–6, which is based on Fig. 17.3–4b with $k \approx 1$, so $M/L_1 = k\sqrt{L_2/L_1} \approx N_2/N_1 = N$. The two inductors appear here as the **magnetizing reactance** $X_m = \omega L_1$ and the much smaller **leakage reactance** $X_\ell = \omega(1 - k^2)L_1/k^2$. The two resistors have been added to represent internal power dissipation. A large shunt resistance R_c accounts for all the *core losses* from eddy currents and hysteresis, while a small series resistance R_w accounts for the *copper losses* in both windings.

Figure 17.3–6

AC model of a power transformer.

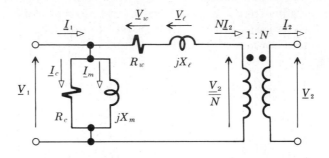

Of course variables such as $\underline{I}_c$ and $\underline{V}_w$ associated with the internal modeling elements do not exist as distinct physical quantities. Nonetheless, they can be used to calculate external variables such as $\underline{I}_1$ and $\underline{V}_2$. By way of illustration, Fig. 17.3–7 shows the phasor construction for $\underline{V}_1$ and $\underline{I}_1$ when we take $\underline{V}_2$ as the phase reference and the load has a lagging power factor, so $\underline{I}_2$ lags $\underline{V}_2$.

Figure 17.3–7

Phasor diagram for a power transformer when $\underline{I}_2$ lags $\underline{V}_2$.

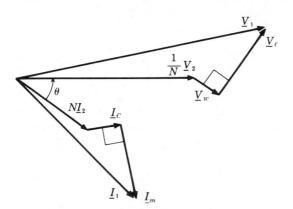

Although there are several other ways to represent the behavior of power transformers, the model in Fig. 17.3–6 readily lends itself to parameter measurements — a necessary and important task because manufacturers seldom provide all the information needed to analyze the performance of a power transformer. The data listed on the nameplate of a transformer usually consists of the voltage ratings and the rated (or maximum) apparent power at the specified frequency. From these we can calculate the turns ratio and the current ratings. For instance, the nameplate legend "60 Hz, 30 kVA, 1200/240 V" means that $|V_1|_{\max} = 1200$ V, $|V_2|_{\max} = 240$ V, $N = 240/1200 = 1/5$, $|I_1|_{\max} = 30$ kVA/1200 V $= 25$ A, and $|I_2|_{\max} = 30$ kVA/240 V $= 125$ A. (We might also turn this transformer around and operate it in the step-up mode with $N = 5$, $|V_1|_{max} =$ 240 V, etc.) Given the nameplate information, the values of R_c, X_m, R_w, and X_ℓ can be determined experimentally using a wattmeter, rms ammeter (AM), and rms voltmeter (VM) in the following procedure.

First, as diagrammed in Fig. 17.3–8a, we *open-circuit* the secondary and adjust the primary voltage at or below its rated value. The real power P_{oc} displayed on the wattmeter is dissipated entirely by R_c, while X_m absorbs all the reactive power $Q_{oc} = \sqrt{|V_1|^2|I_1|^2 - P_{oc}^2}$. Hence,

$$R_c = \frac{|V_1|^2}{P_{oc}} \qquad X_m = \frac{|V_1|^2}{Q_{oc}} \qquad \text{(9a)}$$

(Measuring the open-circuit secondary voltage here would also give us the turns ratio via $N = |V_2|/|V_1|$.) Next, we *short-circuit* the secondary and adjust the primary current at or below its rated value to measure P_{sc} and find Q_{sc}, as represented by Fig. 17.3–8b. Since almost all of I_1 flows through the small series impedances, we now have

$$R_w \approx \frac{P_{sc}}{|I_1|^2} \qquad X_\ell \approx \frac{Q_{sc}}{|I_1|^2} \qquad \text{(9b)}$$

If the calculated values do not support this approximation, then more accurate results must be obtained from the additional measurement of the short-circuit secondary current.

Figure 17.3–8

Power transformer measurements: (a) open circuit; (b) short circuit.

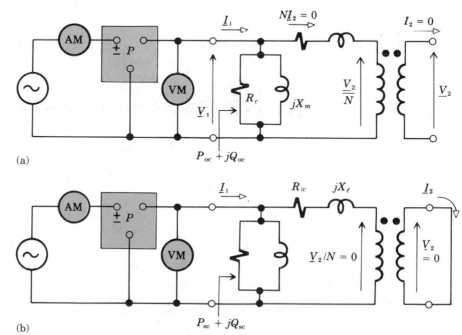

(a)

(b)

Example 17.3−2

Measurements on a certain step-up transformer with $N = 5$ yield, via Eq. (9),

$$R_c = 40\ \Omega \qquad X_m = 24\ \Omega \qquad R_w \approx 0.05\ \Omega \qquad X_\ell \approx 0.3\ \Omega$$

These values justify the approximation in the short-circuit calculations since $|R_w + jX_\ell| \approx 0.6\ \Omega$ whereas $|(R_c \| X_m)| = 20.5\ \Omega \gg |R_w + jX_\ell|$. We'll investigate the performance of this transformer when it couples a 120-V AC source to a 30-Ω resistive load, assuming that the linearity condition in Eq. (8) is satisfied. For comparison purposes, we note in advance that an ideal transformer would have $V_2 = 5 \times 120 = 600$ V, delivering $P_L = 600^2/30 = 12$ kW to the load.

Figure 17.3−9 shows the equivalent circuit diagram incorporating our model with the load resistance relected into the primary as $30\ \Omega/N^2 = 1.2\ \Omega$. Routine analysis then gives

$$I_c = \frac{120}{40} = 3\ \text{A} \qquad I_m = \frac{120}{j24} = -j5\ \text{A}$$

$$5I_2 = \frac{120}{1.25 + j0.3} = 90.8 - j21.8 = 93.4\ \text{A}\ \underline{/-13.5^\circ}$$

$$I_1 = I_c + I_m + 5I_2 = 93.8 - j26.8 = 97.6\ \text{A}\ \underline{/-15.9^\circ}$$

Thus, the load has $I_2 = 18.68\ \text{A}\ \underline{/-13.5^\circ}$, $V_2 = 30\ I_2 = 560.4\ \text{V}\ \underline{/-13.5^\circ}$, and $P_L = 10.47$ kW — all values that are somewhat lower than those of the ideal transformer because of the losses and reactances in the real transformer.

Figure 17.3−9
Circuit diagram for
Example 17.3−2.

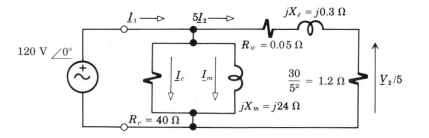

Going to the primary side, we note that I_1 lags V_1 by $\theta = 15.9^\circ$. The apparent, real, and reactive powers from the source are thus

$$|S| = 120\ \text{V} \times 97.6\ \text{A} = 11.71\ \text{kVA}$$

$$P = |S|\cos\theta = 11.26\ \text{kW}$$

$$Q = |S|\sin\theta = 3.21\ \text{kVAr}$$

This means that the transformer dissipates $P - P_L = 800$ W and yields a respectable power-transfer efficiency of $P_L/P \approx 93\%$.

Exercise 17.3−3   Find all the internal currents and the real and reactive powers for the transformer model in Example 17.3−2 when it has a short-circuited secondary and $\underline{V}_1 = 60$ V $\underline{/0°}$. What values of R_w and X_ℓ would be predicted by Eq. (9b)?

17.4 ELECTROMECHANICAL TRANSDUCERS

The concepts developed in the first two sections of this chapter are combined here to derive the operating principles of common electromechanical transducers. These devices convert electrical energy to mechanical work, or vice versa, via interaction taking place in a magnetic field. We should therefore perhaps refer to them as electro-*magneto*-mechanical transducers, a more descriptive but tongue-twisting name.

Our study begins with the general inter-relationship between magnetic force and induced voltage found whenever a current-carrying conductor cuts flux lines. Then we use our results to analyze transducers that feature wire coils moving in a uniform field, distinguishing between those with translational motion and those with rotational motion. We'll also examine devices with moving-iron members rather than moving coils.

Magnetic Force and Induced Voltage

Recall from Faraday's law that if a conductor of length ℓ moves at velocity u across a uniform B field (as in Fig. 17.1−6), then the field induces a voltage given by

$$e = B\ell u \tag{1}$$

This equation, known as the **"Blu" law,** embodies one of the two basic principles underlying the behavior of most electromechanical devices. The other basic principle pertains to the magnetic force experienced by the conductor, for magnetic force and induced voltage always go hand in hand when a current-carrying conductor cuts flux lines.

To develop the expression for magnetic force, consider the situation in Fig. 17.4−1 where the conductor carries current $i = dq/dt$ in a direction perpendicular to the B field. Let dq be the amount of charge in the differential length $d\ell$ at any instant, and let $u_q = d\ell/dt$ be the average charge velocity. Since the field would exert force $f = Bu_q q$ on an isolated moving charge q, the differential force df on element $d\ell$ will be

$$df = Bu_q \, dq = B\frac{d\ell}{dt} \, dq = B\frac{dq}{dt} \, d\ell = Bi \, d\ell$$

Integrating over the entire length ℓ then yields

$$f = B\ell i \tag{2}$$

which we call the **"Bli" law.** The direction of f is mutually perpendicular to i and B, as shown.

Figure 17.4–1

Magnetic force on a current-carrying wire.

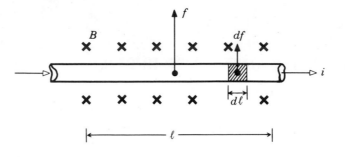

We can put this magnetic force to work in the simple system depicted by Fig. 17.4–2a. The conductor is free to slide along rails that supply current from an electrical source, and the resulting force pushes the conductor to the right. (The rails may also experience magnetic force, but they are presumed to be fixed in place.) A pulley arrangement allows the moving conductor to lift a weight at constant velocity u, producing the mechanical output power

$$p_m = fu = B\ell iu$$

This system therefore acts as a primitive **motor,** converting electrical energy to mechanical work.

Figure 17.4–2

(a) Primitive motor.
(b) Primitive generator

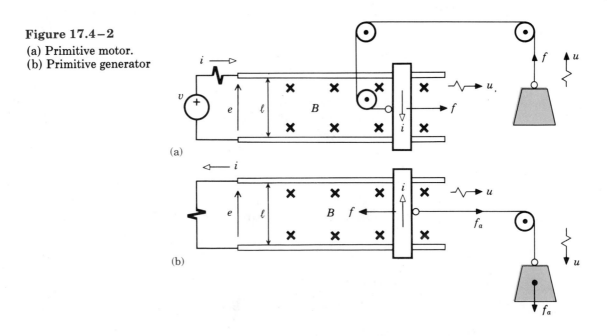

The electrical source supplies power $p_e = vi$ and there are no other energy sources here. (In particular, energy is not drawn from the magnetic field, although the small field produced by the current loop does slightly alter the value of B near the conductors.) Hence, we must have $p_e \geq p_m$ or $vi \geq B\ell iu$, so the required source voltage is

$$v \geq B\ell u$$

The equality holds in absence of ohmic heating, friction, and other losses within the system. The condition $v \geq B\ell u$ agrees with the "Blu" law since the source voltage must counterbalance the induced emf $e = B\ell u$ which appears across the moving conductor and, by Lenz's law, opposes the current i.

Our moving-conductor system becomes a primitive **generator** when modified as shown in Fig. 17.4–2b. Here, an applied force f_a pulls the conductor to the right at velocity u, so the magnetic field induces $e = B\ell u$ across the rails. Current then flows through the electrical load, thereby converting mechanical input power $p_m = f_a u$ to electrical output power $p_e = ei = B\ell ui$. Clearly, we must have $p_m \geq p_e$, so the required applied force is

$$f_a \geq B\ell i$$

This condition agrees with the "Bli" law since the applied force must counterbalance the opposing force $f = B\ell i$ produced by the magnetic field when current flows upwards through the moving conductor.

Any device capable of electromechanical energy conversion in either direction is called a **bilateral transducer.** In the case of moving-conductor devices, the conversion medium is the interaction between a magnetic field and the moving conductor. Regardless of the conversion direction, there will always be *both* a magnetic force $f = B\ell i$ and an induced voltage $e = B\ell u$. We obtain motor action by applying a voltage $v \geq B\ell u$, which overcomes the induced voltage; this permits current flow from the electrical source, and interaction with the magnetic field produces mechanical output power (Fig. 17.4–2a). Conversely, we obtain generator action by applying a force $f_a \geq B\ell i$, which overcomes the magnetic force; this permits conductor motion whose interaction with the magnetic field produces electrical output power (Fig. 17.4–2b). Most motors, generators, and various other electromechanical devices are based on this two-way interaction.

Example 17.4–1

Suppose the motor system in Fig. 17.4–2a has $B = 1$ T and $\ell = 0.5$ m and we want to lift a 100-kg mass at a rate of 1.5 m/sec (about 5 km/hr). The minimum required source voltage is $v = B\ell u = 0.75$ V, a surprisingly small value. But the magnetic force must raise the mass against the gravitational force Mg, so $f = 100$ kg $\times$ 9.8 m/sec $= 980$ N, and the source must provide a very large current, $i = f/B\ell = 1960$ A. Clearly, a gear-reduction system allowing faster conductor motion would yield more practical values in this case. For instance, with 200-to-1 reduction we could use $v = 150$ V and $i = 9.8$ A.

Exercise 17.4–1   Determine the mass M and its velocity u_M, such that the generator system of Fig. 17.4–2b produces 120 V across a 20-Ω resistance when $B = 1$ T and $\ell = 0.5$ m. Assume a gear arrangement with $u = 10\ u_M$.

Translating-Coil Devices

Many electromechanical transducers have moving conductors wound into a coil to multiply the magnetic force and induced voltage by the number of turns. Here we'll examine devices in which the coil is arranged for translational motion.

Figure 17.4–3 sketches the cross-sectional view and end view of a generic translating-coil device. It contains a cylindrical coil wrapped around a sleeve to allow back-and-forth motion perpendicular to a radial magnetic field in the air gap of a permanent magnet. A spring attached to the sleeve establishes the rest position and restrains lateral movement. Such a device becomes an *acoustic transducer,* for instance, when the sleeve is connected to the diaphragm of a microphone or loudspeaker — in which case the diaphragm usually serves as the spring.

Figure 17.4–3
Translating-coil device:
(a) cross section;
(b) end view.

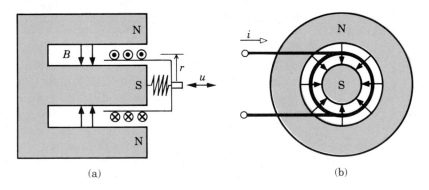

(a)   (b)

Let the coil have N turns, each with circumference $c = 2\pi r$, for a total moving conductor length $\ell = Nc$. Since every differential element of the moving conductor is perpendicular to both the radial B field and the velocity direction u, the magnetic force will be $f = B\ell i = BNci$ and it is accompanied by the induced voltage $e = B\ell u = BNcu$. We'll write these expressions in the compact form

$$f = \alpha i \qquad e = \alpha u \tag{3a}$$

where we have introduced

$$\alpha \triangleq 2\pi BNr = BNc \tag{3b}$$

The parameter α stands for the magnetic coupling coefficient in a translating-coil device.

When the device performs electrical-to-mechanical energy conversion (as in a loudspeaker), its behavior can be modeled by Fig. 17.4–4a. The electrical circuit on the left consists of an external voltage source v, the inductance L and resistance R associated with the coil, and a controlled voltage source to represent the induced emf $e = \alpha u$. The mechanical diagram on the right shows the magnetic force $f = \alpha i$ acting on a mass M representing the coil and everything that moves with it. The coil has displacement x relative to its rest position, and it moves with velocity $u = dx/dt$. Motion is opposed by the mechanical load force f_a, by the spring's restoring force f_K, and by a viscous friction drag force f_D. The spring force is proportional to displacement while the drag force is proportional to velocity, so $f_K = Kx$ and $f_D = Du$.

Figure 17.4–4

Translating-coil energy conversion: (a) electrical to mechanical; (b) mechanical to electrical.

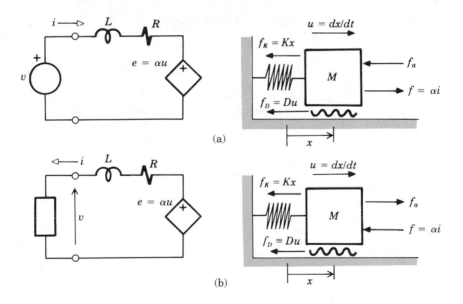

In like fashion, Fig. 17.4–4b models the behavior when a translating-coil device performs mechanical-to-electrical energy conversion (as in a microphone). Now f_a is an external applied force, and the electrical output is delivered to some load. Energy flows here from the mechanical input to the electrical output, whereas it flows from the electrical input to the mechanical output in Fig. 17.4–4a. But both devices are third-order systems that store energy in the inductor, spring, and moving mass, and both dissipate a portion of the input energy as ohmic and frictional heat.

To analyze Fig. 17.4–4a, we first apply Kirchhoff's voltage law around the electrical loop to get

$$v = L(di/dt) + Ri + e$$

Second, we sum mechanical forces according to Newton's third law, noting that $du/dt = d^2x/dt^2$ is the acceleration of the mass, so

$$M(du/dt) = f - (f_a + Du + Kx)$$

Substituting $e = \alpha u$ and $f = \alpha i$ and rearranging then yields

$$L\frac{di}{dt} + Ri = v - \alpha u \qquad M\frac{du}{dt} + Du + Kx = \alpha i - f_a \qquad \textbf{(4)}$$

Thus, we have a pair of differential equations coupled by the terms αu and αi.

Our next step involves extending the concept of *impedance* to include mechanical as well as electrical elements. We do this by assuming all variables to be exponential time functions, which reduces the differential equations to algebraic equations. Taking this approach to Eq. (4), we let $i = Ie^{st}$, $v = Ve^{st}$, $f_a = F_a e^{st}$, and $x = Xe^{st}$ so $u = dx/dt = sXe^{st} = Ue^{st}$. The resulting set of impedance equations becomes

$$(sL + R)I = V - \alpha U \qquad U = sX$$
$$(sM + D)U + KX = \alpha I - F_a \qquad \textbf{(5)}$$

where we've canceled the common factor e^{st}. The analysis of Fig. 17.4–4b follows identical lines and leads to similar equations.

Having obtained the impedance equations, they can be manipulated to get an algebraic expression for the variable of interest. The steady-state or natural response is then found using the methods previously developed in Chapters 6 and 7.

Example 17.4–2
Loudspeaker Frequency Response.

Suppose Fig. 17.4–4a represents a loudspeaker with input voltage v. The acoustical output of a loudspeaker is proportional to the diaphragm's velocity u, and the acoustical load force is also proportional to u. We want to find the AC steady-state relationship between v and u given that

$$f_a = 30u$$

and given the element values

$$L \approx 0 \qquad R = 8 \qquad \alpha = 16$$
$$M = 0.01 \qquad D = 2 \qquad K = 10{,}000$$

all in SI units.

After substituting $F_a = 30U$ into Eq. (5), along with the given numerical values, we have

$$8I = V - 16U \qquad U = sX$$
$$(0.01s + 2)U + 10{,}000X = 16I - 30U$$

Routine algebra then yields the transfer function

$$H(s) = \frac{U}{V} = \frac{200s}{s^2 + 6400s + 10^6} = \frac{200s}{(s + 160)(s + 6240)}$$

The presence of s^2 signifies a second-order system, since the inductance is negligible.

Now we set $s = j\omega$, corresponding to AC steady-state conditions. The resulting frequency response becomes

$$H(j\omega) = 0.032 \frac{j(\omega/160)}{1 + j(\omega/160)} \frac{1}{1 + j(\omega/6240)}$$

Comparing this expression with Eq. (9), Section 7.1, reveals that the loudspeaker acts as a *bandpass filter* with lower cutoff frequency $f_\ell = 160/2\pi \approx 25$ Hz and upper cutoff frequency $f_u = 6240/2\pi \approx 1$ kHz. This would be a suitable "woofer" in a high-fidelity system that also has a "tweeter" to handle frequencies above 1 kHz.

Exercise 17.4–2   Write the differential equations for Fig. 17.4–4b and derive the impedance equations

$$(sL + R)I = \alpha U - V \qquad U = sX$$
$$(sM + D)U + KX = F_a - \alpha I$$

Rotating-Coil Devices

Rotating-coil devices can be divided into two major categories according to the amount of rotation allowed. Here we'll examine devices whose coils are restricted to rotation angles less than 180°, as typically found in the movements of electrical meters. Our results provide background for the next chapter where we'll study generators and motors whose coils have continuous rotation.

Perhaps the simplest rotating-coil device is the common **d'Arsonval meter movement** illustrated in Fig. 17.4–5a. This transducer consists of a flat coil connected to an electrical source (not shown). The coil is suspended in a magnetic field so that magnetic force creates torque and rotates the coil about its vertical axis. Opposing torque from a pair of spiral springs results in an angular deflection γ indicated by the pointer attached to the coil assembly. Other torques may also oppose the magnetic torque.

The coil is more or less rectangular, as sketched in Fig. 17.4–5b, having length ℓ, width $2r$, and area $A = 2r\ell$. Its movement is restricted to the air gap region shown close up in Fig. 17.4–5c, where a uniform radial B field has been established between a permanent magnet and a stationary iron cylinder. With current i through the coil, each axial length of conductor experiences force $f = B\ell i$ at radius r from the center. Since there are $2N$ axial conductors in an N-turn coil, the total magnetic torque T becomes

$$T = 2N \times f \times r = 2NB\ell ir$$

Figure 17.4–5
(a) d'Arsonval meter
movement. (b) Coil
detail. (c) Air-gap detail.

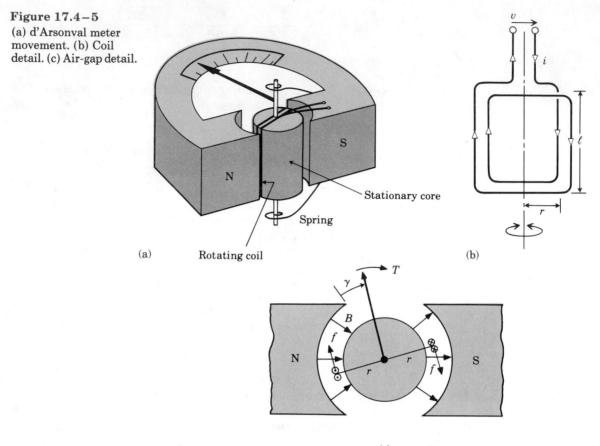

(a)   Rotating coil

Stationary core

Spring

(b)

(c)

If the coil rotates at some angular velocity $\omega = d\gamma/dt$, then each axial
conductor cuts through the B field with tangential velocity $u = r\omega$ and the
total induced emf is

$$e = 2N \times B\ell u = 2NB\ell r\omega$$

We'll write these expressions in the form

$$T = \beta i \qquad e = \beta\omega \tag{6a}$$

where

$$\beta \triangleq 2NB\ell r = BNA \tag{6b}$$

so β is the rotational coupling coefficient. See Eq. (3) for comparison with a
translating-coil device.

Figure 17.4–6 gives the model of the d'Arsonval movement and similar
rotating-coil devices with an electrical input source. The electrical circuit
should be self-explanatory. The mechanical diagram shows the magnetic

torque T acting on the rotational moment of inertia J representing all the rotating mass. Rotation is opposed by the spring torque $T_K = K\gamma$, a viscous friction drag torque $T_D = D\omega$, and any external applied torque T_a. Routine analysis gives the coupled differential equations

$$L\frac{di}{dt} + Ri = v - \beta\omega \qquad J\frac{d\omega}{dt} + D\omega + K\gamma = \beta i - T_a \qquad \textbf{(7a)}$$

which are directly comparable to Eq. (4). Letting $\omega = \Omega e^{st}$, $\gamma = \Gamma e^{st}$, etc., then yields the corresponding impedance equations

$$(sL + R)I = V - \beta\Omega \qquad \Omega = s\Gamma$$
$$(sJ + D)\Omega + K\Gamma = \beta I - T_a \qquad \textbf{(7b)}$$

The modeling and analysis of a rotating-coil transducer with mechanical input goes along similar lines.

Figure 17.4–6
Rotating-coil device with electrical input.

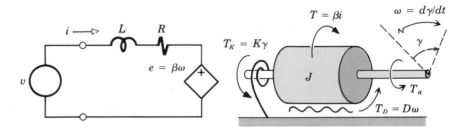

Now let a *constant* voltage $v = V_s$ be applied to a d'Arsonval meter with no external torque T_a. The coil begins to rotate and the meter eventually reaches mechanical and electrical equilibrium at some steady-state deflection angle γ_{ss}. The DC steady-state condition is found by setting $s = 0$ in Eq. (7b) along with $V = V_s$ and $I = I_s = V_s/R$, from which we obtain

$$\gamma_{ss} = \frac{\beta}{KR}V_s = \frac{\beta}{K}I_s \qquad \textbf{(8)}$$

Accordingly, we can use the d'Arsonval movement in a DC *voltmeter* or *ammeter*. The maximum allowable deflection is typically $90–120°$, limited by the uniform field region of the air gap. However, external circuitry described in Section 3.4 permits the measurement of current or voltage values larger than those predicted by Eq. (8).

Angular deflections greater than $180°$ are prevented by mechanical stops in a d'Arsonval meter. Otherwise, the magnetic torque would reverse its direction because the conductors would be in a B field pointed the opposite way. (You can see this effect in Fig. 17.4–5c by visualizing the coil position with $\gamma > 180°$.) Such *field and torque reversals* are inherent in all rotating-coil devices, and they present a major complication in the design of DC machinery. On the other hand, field reversals lend themselves quite naturally to the production of *alternating* voltages in an AC generator. Further discussion of these matters is taken up in Chapter 18.

Another type of meter movement is the **electrodynamometer** diagrammed in Fig. 17.4–7. Instead of a permanent magnet, the magnetic field is supplied by a solenoidal coil divided into two parts as shown. This coil is wound on a nonmagnetic core to avoid magnetic nonlinearity, so B and β are proportional to the current i_v. When the electrodynamometer is used for power measurements, as discussed in Section 16.1, the fixed solenoidal coil is the voltage-sensing coil while the rotating coil is the current-sensing coil. Each coil bears the label $\pm$ at one end to indicate that proper "up-scale" deflection will be obtained when the currents i_v and i_i simultaneously enter or leave the respective marked terminals.

Figure 17.4–7
Electrodynamometer movement.

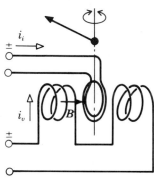

If both coil currents are constant, say $i_v = I_v$ and $i_i = I_i$, then the steady-state deflection angle is

$$\gamma_{\text{ss}} = \frac{K_B}{K} I_v I_i$$

which follows from Eq. (8) with $\beta = K_B I_v$. For the more general case of periodic time-varying currents $i_v(t)$ and $i_i(t)$, mechanical inertia provides an averaging effect and the steady-state deflection angle becomes

$$\gamma_{\text{ss}} = \frac{K_M}{T} \int_0^T i_v(t) i_i(t) \, dt \tag{9}$$

where K_M absorbs all the proportionality constants.

Example 17.4–3

To illustrate the dynamic behavior of a rotating-coil device, let's investigate the transient that occurs before a d'Arsonval movement reaches the steady-state condition in Eq. (8). We'll make the simplifying but reasonable assumption that the coil has negligible inductance and inertia, so the system reduces to first-order with energy stored only in the springs.

Since $v(t) = V_s$ starting at $t = 0$, the resulting deflection $\gamma(t)$ is just the step-response of a first-order system and has the form

$$\gamma(t) = \gamma_{ss}(1 - e^{-t/\tau}) = \frac{\beta V_s}{KR}(1 - e^{-t/\tau})$$

Hence, we only need to find the time constant τ associated with the natural response. For this purpose we start with Eq. (7a) and set $L = J = 0$, $v = 0$, and $T_a = 0$ to get

$$Ri = -\beta\omega \qquad D\omega + K\gamma = \beta i$$

Eliminating i and inserting $\omega = d\gamma/dt$ then yields

$$\frac{d\gamma}{dt} = -\frac{K}{D + \beta^2/R}\gamma$$

so the time constant is

$$\tau = \frac{D + \beta^2/R}{K} = \frac{RD + \beta^2}{RK}$$

which follows by comparison with Eq. (8), Section 5.2.

Note that both τ and γ_{ss} increase with the coupling coefficient β. This means that we have a trade-off between rapid response (small τ) and large steady-state deflection γ_{ss}, as illustrated in Fig. 17.4–8. Accordingly, meter design involves a compromise to achieve acceptable response time and deflection sensitivity.

Figure 17.4–8
Step response of a
d'Arsonval movement.

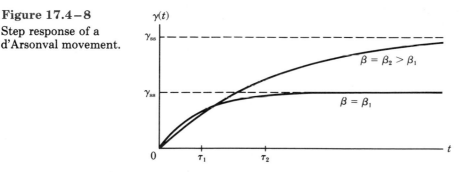

Exercise 17.4–3   Consider a d'Arsonval meter with $B = 0.5$ T, $K = 10^{-8}$ N · m/rad, $r = 5$ mm, and $\ell = 15$ mm. Find the number of coil turns N needed to get $90°$ full-scale deflection when $I_s = 10\ \mu A$. Then calculate the corresponding full scale voltage if the coil wire has a resistance of 4000 Ω/m.

Moving-Iron Devices †

In addition to force on current-carrying conductors, a magnetic field also acts on all nearby magnetic materials and tends to draw them into the densest part of the field — as demonstrated when a magnet picks up loose pieces of iron. Such forces exist at every air gap in a magnetic circuit, and are utilized by a variety of practical devices with movable iron members that convert the magnetic attraction into mechanical work. Because this gap force differs in nature from the induced force previously discussed, we must first develop a general expression for it that can be applied to specific configurations.

Figure 17.4–9 shows a portion of a magnetic circuit containing two plane-parallel iron faces with area A_g separated by a gap of length ℓ_g. The B field in the gap creates a force f_g that pulls the movable member toward the fixed piece and, consequently, alters the circuit's properties. For the purpose of analysis, we limit the movement to a very small *virtual displacement* dx and consider the resulting *virtual work* $dw_m = f_g\,dx$. If dw represents all other changes in energy caused by the displacement dx, then conservation of energy requires that $f_g\,dx + dw = 0$ or

$$f_g = -\frac{dw}{dx}$$

Assuming B remains constant, dw consists entirely of the reduced stored energy in the air gap, namely

$$dw = -\frac{B^2 A_g}{2\mu_0}\,dx$$

This follows from Eq. (10), Section 17.1, since the volume of the gap has decreased by an amount $A_g\,dx$ (ignoring fringing). Therefore,

$$f_g = \frac{B^2 A_g}{2\mu_0} = \frac{\phi^2}{2\mu_0 A_g} \qquad (10)$$

and the gap force is proportional to the square of the flux $\phi = BA_g$. This means that f_g is always inward, regardless of the direction of the flux.

An alternative expression for f_g is obtained by writing the stored energy in terms of the *gap reluctance* $\mathcal{R}_g = \ell_g/\mu_0 A_g$. Thus, $w = B^2 A_g \ell_g/2\mu_0 = \phi^2 \ell_g/2\mu_0 A_g = \tfrac{1}{2}\phi^2 \mathcal{R}_g$ and

$$f_g = -\tfrac{1}{2}\phi^2 \frac{d\mathcal{R}_g}{dx} \qquad (11)$$

Figure 17.4–9
Magnetic force at an air gap.

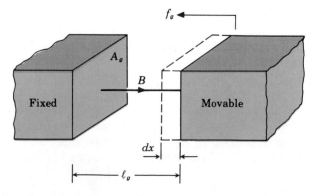

The minus sign agrees with the fact that $\mathcal{R}_g$ decreases as the gap gets smaller, so $d\mathcal{R}_g/dx$ is a negative quantity. Equation (11) has particular value for studying moving-iron devices when the core reluctance and magnetic nonlinearities are negligible.

As a case in point, consider the magnetically operated switch or **relay** illustrated in Fig. 17.4–10. Switching contacts for an external circuit are attached to a fixed core and to a hinged armature held open by the spring a distance ℓ_0 in absence of magnetic excitation. Current i through the operating coil then creates a flux and downward force, drawing the contacts together. We'll calculate f_g for a fixed coil current and arbitrary gap spacing $\ell_g = \ell_0 - x$, subject to the reasonable condition that the core has small reluctance compared to the gap.

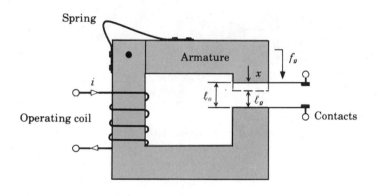

Figure 17.4–10
Magnetic relay.

The gap reluctance is $\mathcal{R}_g = \ell_g/\mu_0 A_g = (\ell_0 - x)/\mu_0 A_g$ so $d\mathcal{R}_g/dx = -1/\mu_0 A_g$ independent of the spacing. However, the flux depends on $\mathcal{R}_g$ and, therefore, on ℓ_g, since $\phi \approx \mathcal{F}/\mathcal{R}_g = \mu_0 NiA_g/\ell_g$. Substituting in Eq. (11) yields

$$f_g = \tfrac{1}{2}\left(\frac{\mu_0 NiA_g}{\ell_g}\right)^2\left(-\frac{1}{\mu_0 A_g}\right) = \frac{\mu_0(Ni)^2 A_g}{2\ell_g^2} \tag{12}$$

which clearly brings out that f_g varies inversely with ℓ_g^2, whereas the opposing spring force would be relatively constant. Based on this result, we would expect the *pull-in current* required to close the relay, starting with $\ell_g = \ell_0$, will be greater than the *hold-in current* when the relay is closed and $\ell_g < \ell_0$. Also note that an AC coil current can be used since f_g is proportional to i^2.

Figure 17.4–11 illustrates another type of moving-iron device, called a *magnetic actuator*. The magnetic field of the actuator draws an iron plunger into the air gap and operates a valve or similar mechanical device. The plunger slides against brass (nonmagnetic) spacers of thickness ℓ_g, so that the combined series reluctance of the upper and lower spacers is $\mathcal{R}_g = 2(\ell_g/\mu_0 bx)$, assuming all the flux passes through the area $A = bx$ corresponding to that portion of the plunger within the gap. Thus,

$dR_g/dx = -2\ell_g/\mu_0 bx^2$, $\phi = \mu_0 Nibx/2\ell_g$, and we have

$$f_g = \frac{\mu_0(Ni)^2 b}{4\ell_g} \tag{13}$$

Note that f_g is independent of the plunger position x, a sometimes desirable feature. In practice, this device usually has a cylindrical configuration with a solenoidal coil around the plunger, thus the more common name is **solenoid valve** or simply **solenoid.**

Figure 17.4–11
Magnetic actuator or solenoid.

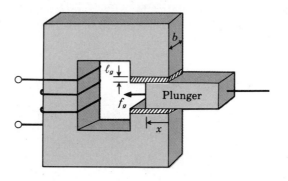

Example 17.4–4
Electromagnet

Figure 17.4–12 shows the cross-sectional dimensions of a large disk-shaped **electromagnet** of the type used to lift heavy pieces of iron or steel — a junked automobile, for instance. The inner and outer pole faces have areas A_1 and A_2, each equaling 0.283 m², and are protected by a brass sheet with $\ell_g = 2$ mm. Maximum force is attained when the load completely covers both faces, so that $\phi = Ni/R_g$ where

$$R_g = \frac{\ell_g}{\mu_0 A_1} + \frac{\ell_g}{\mu_0 A_2}$$

Equation (12) then gives the lifting force as

$$f_g = \frac{\phi^2}{2\mu_0 A_1} + \frac{\phi^2}{2\mu_0 A_2} = 0.0222(Ni)^2$$

Lifting a 1000-kg mass requires $f_g \geq 1000$ kg $\times$ 9.8 m/sec = 9800 N, or an mmf of

$$Ni \geq \sqrt{\frac{9800}{0.0222}} = 664 \text{ A-t}$$

Of course a larger mmf would be needed in practice to compensate for imperfect contact with the load.

Exercise 17.4–4

Calculate the pull-in and hold-in mmf values for the relay in Fig. 17.4–10 when $\ell_g = 8$ mm and 3 mm, respectively, and $A_g = 1$ cm². The opposing spring force is essentially constant at $f_K = 10^{-6}$ N.

Figure 17.4–12
Electromagnet.

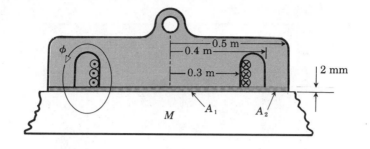

Figure 17.4–12
Electromagnet.

PROBLEMS

17.1–1   The **mixed English system** of units expresses length in inches and flux in kilolines, where 1 kiloline $= 10^{-5}$ Wb. Convert $B = 1$ T to English units.

17.1–2   Calculate the value of μ_0 in the English units described in Problem 17.1–1.

17.1–3   Suppose the toroid in Fig. 17.1–4 has $Ni = 100$ A, $\mu_r = 1200$, and a circular cross section. Calculate B and find the maximum possible value of ϕ when $R = 5$ cm.

17.1–4   Do Problem 17.1–3 with $R = 3$ cm.

17.1–5   Do Problem 17.1–3 with $R = 9$ cm.

17.1–6   Suppose $B = 0.01$ T in Fig. 17.1–6 and the rolling bar has 1-mm radius and carries $i = 15$ A in the upward direction. Find the net flux density at the right and left of the bar.

17.1–7   Do Problem 17.1–5 with $i = 50$ A.

17.1–8   The Earth's magnetic field strength is about 50 μT. Estimate the maximum voltage induced on an airplane with a 100-foot wingspan traveling at 600 miles per hour.

17.1–9   A small flat coil with N turns and area A has its plane aligned with the axis of a conductor carrying current i. The coil moves away from the conductor at constant velocity u, so the distance between them is $d = ut$. Derive an expression for the voltage induced in the coil.

17.1–10   Obtain an expression for the maximum value of L when the toroid in Fig. 17.1–4 has a circular cross section and fixed values of R and N.

17.1–11   A 3-mH inductor is to be built in the toroidal form of Fig. 17.1–4 with $R = 2$ cm and circular cross section of radius $r = 1$ cm. Find the required number of turns and estimate the total length of wire in centimeters for $\mu_r = 1$ and $\mu_r = 1000$.

17.1–12   Suppose the wire used for the coil in Fig. 17.1–3 has resistance R_0 ohms per meter. Assume $N \gg h/r$ and show that the total winding resistance is related to the inductance by $R_w \approx R_0 \sqrt{4\pi h L/\mu}$.

17.2–1   The core in Fig. P17.2–1 is 5-cm thick, and the other dimensions are shown in centimeters. The core material is linear, with $\mu = 0.005$. Draw the equivalent circuit, estimate the reluctances, and calculate the flux density in the middle leg when a coil having $Ni = 80$ A-t is around the left leg.

Figure P17.2–1

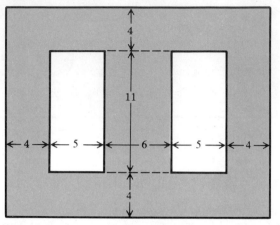

17.2–2   Do Problem 17.2–1 for the case where the middle leg, between the dashed lines, is made of a different material with $\mu = 0.001$.

17.2–3   A 200-turn coil is wrapped around the middle leg of the core described in Problem 17.2–1. The middle leg, between the dashed lines, is made of a different material with $\mu = 0.002$. Draw the equivalent circuit, estimate the reluctances, and determine the current required to get $B = 0.4$ T in each of the outer legs.

17.2–4   The core described in Problem 17.2–1 has a 0.2-cm air gap cut across the middle leg. There are 50-turn coils carrying current i on each of the outer legs. Estimate the value of i required to get $B = 0.05$ T in the gap.

17.2–5   The core described in Problem 17.2–1 has a coil carrying 0.5 A on the middle leg, and 0.3-cm air gaps cut across both of the outer legs. Estimate the number of turns in the coil such that $B = 0.1$ T in the gaps.

17.2–6   The core described in Problem 17.2–1 has identical coils carrying 3 A on each of the outer legs, and 0.2-cm air gaps have been cut across these legs. Estimate the total number of coil turns needed to get $B = 0.3$ T in the gaps.

17.2–7   Use Eq. (7) and Fig. 17.2–5 to evaluate $\mu_r = \mu/\mu_0$ for cast iron at $H = 100, 400,$ and 1000 A-t/m.

17.2–8   Use Eq. (7) and Fig. 17.2–5 to evaluate $\mu_r = \mu/\mu_0$ for sheet steel at $H = 100, 200,$ and 1000 A-t/m.

17.2–9   If a coil has a nonlinear core and carries $i = I_0 + \Delta I \cos \omega t$ with $\Delta I \ll I_0$, then the flux will be $\phi \approx \phi_0 + \Delta\phi \cos \omega t$ and the equivalent inductance is $L \approx N \Delta\phi/\Delta I$. Using Fig. 17.2–5, calculate L for a 200-turn coil when $I_0 = 0.1$ A, $\Delta I = 0.01$ A, and the core is cast iron with area $A = 10^{-4}$ m^2 and length $\ell = 0.025$ m. Compare your result with the value of $L_0 = N\phi_0/I_0$.

17.2–10   Do Problem 17.2–9 for the case of a sheet-steel core with $A = 10^{-4}$ m^2 and $\ell = 0.04$ m.

17.2–11   A magnetic circuit with $N = 50$ and $\ell_c = 0.5$ m is found to have an AC hysteresis loss of 10 W when the peak current is $I_m = 2$ A. Take $n = 1.5$ in Eq. (8) to estimate P_h for $I_m = 0.5$ A and 8 A, assuming a cast-iron core.

17.2–12   Do Problem 17.2–11 assuming a sheet-steel core.

17.2–13   AC measurements with constant voltage amplitude show that the combined hysteresis and eddy-current loss of a certain magnetic circuit is 10.5 W at $f = 50$ Hz and 13.2 W at $f = 60$ Hz. What would be the loss if the frequency is increased to 400 Hz?

17.3–1   Use the tee equivalent model to obtain an expression in terms of i_1 and i_2 for the total energy stored by the transformer in Fig. 17.3–3a.

17.3–2   Use the tee equivalent model to obtain an expression in terms of i_1 and i_2 for the total energy stored by the transformer in Fig. 17.3–3c.

17.3–3   Redraw Fig. P17.3–3 using the tee equivalent model to find L_{eq} for the terminal relationship $v = L_{eq}\, di/dt$.

Figure P17.3–3

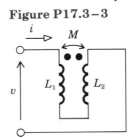

17.3–4   Do Problem 17.3–3 with the dot moved to the lower end of L_2.

17.3–5   Derive Eqs. (6a) and (6b) by solving the given loop equations for I_1 and I_2.

17.3–6   Let the transformer in Fig. 17.3–5 have $L_1 = 8$ H, $L_2 = 0.32$ H, and $M = 1.6$ H.
(a) Evaluate the coupling coefficient.
(b) Find the phasors V_2, I_2, and I_1 when $v_1 = 120 \cos 100t$ V and the load is $R = 24\ \Omega$.
(c) Repeat (b) for an ideal transformer with $N = \sqrt{L_2/L_1}$.

17.3–7   Do Problem 17.3–6 with $L_1 = 0.01$ H, $L_2 = 0.16$ H, and $M = 0.03$ H.

17.3–8   Do Problem 17.3–6 with $L_1 = 2$ H, $L_2 = 0.5$ H, and $M = 0.8$ H.

17.3–9   The circuit in Fig. P17.3–9 has $L_1 = 0.6$ H, $L_2 = M = 0.4$ H, and $R_2 = 4\ \Omega$. Draw the AC impedance diagram using the tee equivalent model. Then solve for the phasor currents I_1 and I_2.

Figure P17.3–9

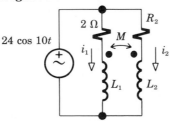

17.3–10   Do Problem 17.3–9 with $L_1 = 0.8$ H, $L_2 = 4.8$ H, $M = 1.6$ H, and $R_2 = 0$.

17.3–11   Do Problem 17.3–9 with $L_1 = 0.9$ H, $L_2 = 0.6$ H, $M = 0.7$ H, and $R_2 = 1\ \Omega$.

17.3–12   Figure P17.3–12 is called a **step-up autotransformer** when the input is $v_1 = v_{ab}$ and the output is $v_2 = v_{cb}$ across a load connected to terminals c and b.

   (a) Using the tee equivalent model, draw the impedance diagram with load $Z(s)$. Then solve loop equations for the current through $Z(s)$ to obtain an expression for the voltage ratio V_2/V_1.

   (b) Now let $k = 1$ and show that $V_2/V_1 = 1 + N$ where $N = \sqrt{L_2/L_1}$.

Figure P17.3–12

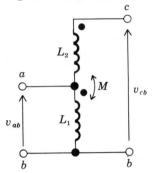

17.3–13   Figure P17.3–12 is called a **step-down autotransformer** when the input is $v_1 = v_{cb}$ and the output is $v_2 = v_{ab}$ across a load connected to terminals a and b.

   (a) Using the tee equivalent model, draw the impedance diagram with load $Z(s)$. Then solve loop equations for the current through $Z(s)$ to obtain an expression for the voltage ratio V_2/V_1.

   (b) Now let $k = 1$ and show that $V_2/V_1 = 1/(1 + N)$ where $N = \sqrt{L_1/L_2}$.

17.3–14   Calculate the parameter values for the power transformer model in Fig. 17.3–6, given the following measurement data.
Open-circuited secondary:

$$|V_1| = 120 \text{ V} \qquad |I_1| = 0.5 \text{ A} \qquad P = 36 \text{ W} \qquad |V_2| = 600 \text{ V}$$

Short-circuited secondary:

$$|V_1| = 60 \text{ V} \qquad |I_1| = 10 \text{ A} \qquad P = 240 \text{ W}$$

17.3–15   Calculate the parameter values for the power transformer model in Fig. 17.3–6, given the following measurement data.
Open-circuited secondary:

$$|V_1| = 240 \text{ V} \qquad |I_1| = 5.2 \text{ A} \qquad P = 480 \text{ W} \qquad |V_2| = 60 \text{ V}$$

Short-circuited secondary:

$$|V_1| = 50 \text{ V} \qquad |I_1| = 25 \text{ A} \qquad P = 750 \text{ W}$$

17.3–16   Derive more accurate expressions for R_w and X_ℓ when another ammeter is used to measure the short-circuit current $|I_2|$ in Fig. 17.3–8b. Assume that the open-circuit measurements are performed first.

17.3–17   Find $\underline{V}_2, \underline{I}_1, P, Q,$ and the efficiency P_L/P for a power transformer system with:

$$R_c = 400 \text{ }\Omega \qquad X_m = 300 \text{ }\Omega \qquad R_w = 8 \text{ }\Omega \qquad X_\ell = 64 \text{ }\Omega$$

$$N = 1/4 \qquad \underline{V}_1 = 1200 \text{ V } \underline{/0^\circ} \qquad \underline{Z}_L = (2.5 + j0) \text{ }\Omega$$

17.3–18   Do Problem 17.3–17 with:

$$R_c = 72 \ \Omega \qquad X_m = 45 \ \Omega \qquad R_w = 0.7 \ \Omega \qquad X_\ell = 1.0 \ \Omega$$
$$N = 5 \qquad \underline{V}_1 = 360 \ \text{V} \ \underline{/0°} \qquad \underline{Z}_L = (120 + j35) \ \Omega$$

17.3–19   Do Problem 17.3–17 with:

$$R_c = 900 \ \Omega \qquad X_m = 300 \ \Omega \qquad R_w = 4 \ \Omega \qquad X_\ell = 10 \ \Omega$$
$$N = 1/2 \qquad \underline{V}_1 = 1800 \ \text{V} \ \underline{/0°} \qquad \underline{Z}_L = (12 + j5) \ \Omega$$

17.3–20   Suppose a source voltage $\underline{V}_s$ with series resistance R_s is connected to the primary in Fig. 17.3–6. Derive approximate expressions for the open-circuit voltage and Thévenin equivalent impedance seen looking back into the secondary when $R_s \ll R_c$, $R_w \ll R_c$, and $X_\ell \ll X_m$.

17.4–1   Two parallel busbars in a power plant are 10-m long and 1-m apart. Find the force on each bar when they both carry 5,000 A.

17.4–2   The two blades of a large knife switch are 10-cm long and 4-cm apart. Find the force on each blade when they both carry 80 A.

17.4–3   Suppose the system in Fig. 17.4–2a is intended to lift a 50-kg mass at the rate $u_M - 2$ m/sec, and the pulleys are arranged such that the bar velocity is $u = 100u_M$. Calculate the required current i and the source voltage v when $B = 0.4$ T, $\ell = 0.25$ m, and the total electrical resistance is $R = 1 \ \Omega$. Assume that there are no mechanical losses.

17.4–4   Do Problem 17.4–3 including a frictional drag force $f_D = 0.6$ N on the bar.

17.4–5   A certain translating-coil device has $B = 0.5$ T, $N = 100$ turns, and $c = 0.2$ m. The device can be modeled by Fig. 17.4–4 with $R = 5 \ \Omega$, $D = 10$ N $\cdot$ s/m, and $L = M = K \approx 0$.

(a) What value of voltage v holds the coil stationary when $f_a = 60$ N is applied in the $-x$ direction?

(b) What is the resulting velocity u when the terminals of the coil are shorted and $f_a = 60$ N is applied in the $+x$ direction?

17.4–6   Suppose the device in Problem 17.4–5 is used as a velocity sensor, with $R_L = 20 \ \Omega$ connected to the coil's terminals. Obtain an expression for v in terms of u, and calculate the efficiency p_e/p_m.

17.4–7   Let the system in Fig. 17.4–4a have $R = 16 \ \Omega$, $D = 4$ N $\cdot$ s/m, $f_a = 60u$, and $L = K = M \approx 0$. Find the value of α that maximizes the ratio u/v. Then calculate the resulting efficiency p_m/p_e.

17.4–8   Consider a translating-coil device described by Eq. (4) with $L = M = f_a = 0$, $R = 100$, $D = 16$, and $\alpha = 20$, all in SI units.

(a) Obtain the differential equation relating x to v.

(b) Find $x(t)$ for $t > 0$ when $v = 0$ for $t > 0$ and the device is in the steady state with $v = 40$ V for $t < 0$.

17.4–9   Consider a vibration table described by Eq. (5) with $K = F_a = 0$, $L = 0.1$, $R = 5$, $M = 1$, $D = 20$, and $\alpha = 30$, all in SI units. Find the AC input impedance $Z(j\omega) = \underline{V}/\underline{I}$.

17.4–10   Find $H(s) = U/V$ when the loudspeaker in Example 17.4–2 has $M = 0$ and $L =$

0.025 H, all other parameters being the same. Is this system underdamped or overdamped?

17.4−11   A certain microphone is described by the equations in Exercise 17.4−2 with $V = R_L I$, where R_L is a load resistance. The parameter values in SI units are $M = L \approx 0$, $R = 50$, $R_L = 250$, $D = 0.1$, and $K = 60$.

(a) Find $H(s) = I/F_a$ in terms of α.

(b) What value of α is needed so that $H(j\omega)$ has a lower cutoff frequency at 50 Hz?

17.4−12   Consider a d'Arsonval meter movement described by Eq. (7b) with $L = T_a = 0$. Obtain an expression for $H(s) = \Gamma/V$. Then determine the condition on D for critical damping.

17.4−13   Calculate the force across the air gap when the magnetic circuit in Example 17.2−1 has $i = 2$ A.

17.4−14   Suppose the field in Fig. 17.4−9 is produced by a coil with constant mmf $\mathcal{F} = Ni$, inductance $L = N^2/\mathcal{R}$, and $\mathcal{R} \approx \mathcal{R}_g$. Show that Eq. (11) becomes $f_g \approx \frac{1}{2}i^2 \, dL/dx$.

17.4−15   Rewrite Eq. (13) in terms of the flux density B in the core. If saturation results in $B \leq 1.4$ T and if we must keep $\ell_g \leq b/10$ to minimize fringing, then what is the minimum value of b needed to get $f_g = 50$ N?

17.4−16   An **electrostatic transducer** is created by letting $B = 0$ in Fig. 17.4−9 and applying voltage v across the gap. We then have *capacitance* C with electric field $\mathcal{E} = v/\ell_g$, neglecting fringing.

(a) Show that the stored energy is $w = \frac{1}{2}\epsilon_0 \mathcal{E}^2 A_g \ell_g$. Then apply the virtual-work method to obtain the gap force $f_g = \frac{1}{2}\epsilon_0 \mathcal{E}^2 A_g$.

(b) Calculate the maximum value of f_g/A_g for an electrostatic transducer, given that an air gap has $\epsilon_0 = 10^{-9}/36\pi$ and $\mathcal{E} \leq 3 \times 10^6$. Compare with the maximum value of f_g/A_g for a moving-iron transducer with $B \leq 1.5$ T.

Chapter

18 Rotating Machines

The vast bulk of "man-made" electrical energy comes from rotating generators driven by mechanical prime movers such as steam or hydro turbines. And, after transmission, a sizable portion of the generated output is converted back to mechanical energy by electrical motors. It seems fitting, therefore, to close our study of electric energy with a look at the rotating machines typically found at either end of a system.

Following an introductory overview of rotating-machine concepts, we examine the three-phase synchronous machine operated as a generator or a motor. Then we consider the workhorse induction motor — both three-phase and single-phase — and other AC motors. A concluding section deals with DC machines. Our emphasis throughout will be on broad operating principles and characteristics, with only brief mention of some of the second-order effects.

Objectives

After studying this chapter and working the exercises, you should be able to do each of the following:

- Explain how rotating fields and unidirectional torque are produced in single-phase and three-phase AC machines (Sections 18.1 and 18.2).

- Construct the phasor diagram and analyze a synchronous machine operating as a generator, motor, or capacitor (Section 18.2).

- State the differences between synchronous and induction motors in terms of equivalent circuits, structural features, and torque-speed curves (Sections 18.2 and 18.3).

- Explain the role of the commutator in a DC machine, and draw the equivalent circuit and typical performance curve for a DC motor or generator (Section 18.4).

709

- Compare methods for controlling the speed of an AC or DC motor (Sections 18.3 and 18.4).

18.1 ROTATING-MACHINE CONCEPTS

From a giant three-phase generator producing hundreds of megawatts to a tiny battery-powered motor driving a portable tape recorder, nearly every rotating machine operates on the same basic principles. We discuss those principles here, starting with the "Blu" and "Bli" laws as they apply to an elementary machine. We then introduce the concept of rotating fields to explain how an AC machine develops unidirectional torque. Attention is also given to a number of practical matters, including structural features, losses and efficiency, and machine capacity and performance curves.

An Elementary Machine

Our prior study of rotating electromechanical transducers began with the assumption that the "working" coil or **armature** rotated in a fixed, uniform magnetic field. But a more common and more practical arrangement for continuous rotation is to rotate the field and keep the armature fixed, permitting direct (nonrotating) electrical connections to the coil.

Figure 18.1–1a illustrates a simplified generator with this configuration. The armature coil has N turns fastened in slots along the inner face of a stationary frame, called the **stator,** having length ℓ and radius r. A **rotor** — assumed for the moment to be a permanent magnet — revolves within the stator at the mechanical angular frequency ω_m, driven by a prime mover coupled to its shaft. The north and south poles of the rotor alternately sweep past the armature conductors, inducing a time-varying emf just as though the conductors moved through a fixed but nonuniform field. (The spinning rotor also creates a time-varying flux through the stator, which should be laminated to minimize the resulting eddy currents.) Since the field seen by the armature conductors reverses direction periodically as the rotor poles pass by, the induced emf $e(t)$ alternates in polarity. Hence, we have an **AC generator** or **alternator.**

For sinusoidal generation, the field in the air gap between the rotor and stator must vary sinusoidally with time. This requirement is met by shaping the rotor to obtain a *sinusoidal field pattern* in space around it. With reference to Fig. 18.1–1b, let B_r be the outward radial flux density at distance r from the center of the rotor, and let γ_r be an arbitrary angle measured clockwise from the rotor's north pole. The desired spatial variation of B_r is

$$B_r = B_R \cos \gamma_r \qquad (1)$$

where B_R is the maximum density at the north pole. Setting $\gamma_r = 180°$ then

Figure 18.1–1
(a) Elementary
generator. (b) Rotor
field. (c) Angular
coordinates.

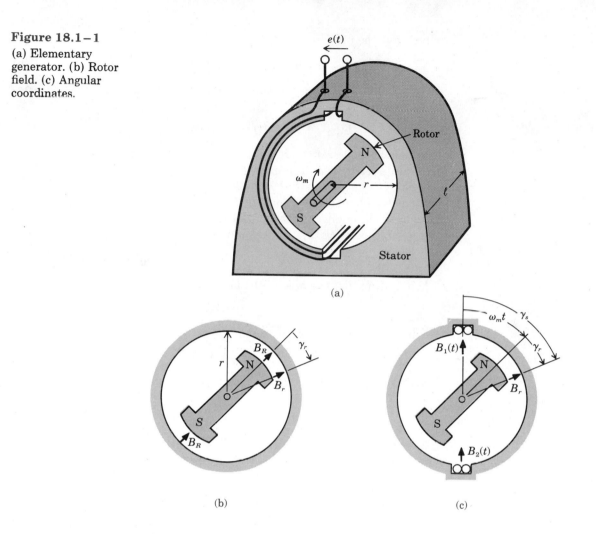

$e(t)$

Rotor

ω_m

r

N

S

ℓ

Stator

(a)

B_R

γ_r

r

N

B_r

S

B_R

(b)

$\omega_m t$ γ_s

$B_1(t)$

γ_r

N

B_r

S

$B_2(t)$

(c)

gives the inward-directed field $B_r = -B_R$ at the south pole, while $B_r = 0$ at $\gamma_r = \pm 90°$.

We account for rotor motion using the angular coordinates shown in Fig. 18.1–1c, where γ_s denotes an arbitrary angle measured clockwise from the top of the stator. If the instantaneous position of the rotor's axis is at the angle $\omega_m t$, then $\gamma_r = \gamma_s - \omega_m t$ and Eq. (1) becomes

$$B_r = B_R \cos (\gamma_s - \omega_m t)$$

This expression describes a sinusoidal field rotating clockwise (with the rotor) and having its north pole at $\gamma_s = 0$ when $t = 0$. To emphasize the time and space variation of B_r, we introduce the *rotating field vector*

$$\underline{B_R} \triangleq B_R \underline{/\gamma_s - \omega_m t} = B_R e^{j(\gamma_s - \omega_m t)} \qquad \textbf{(2a)}$$

which has the property that

$$B_r = \text{Re}[\underline{B_R}] = B_R \cos (\gamma_s - \omega_m t) \qquad \textbf{(2b)}$$

Setting $\gamma_s = 0$ then gives the time-varying outward field at the upper conductors as

$$B_1(t) = B_R \cos \omega_m t \qquad \textbf{(3)}$$

The field at the lower conductors varies in exactly the same manner but is inward directed, so $B_2(t) = B_1(t)$.

Now we can apply the "Blu" law to calculate the emf induced by the rotor field. There are N armature conductors of length ℓ in each slot, and they have the equivalent tangential velocity $u = r\omega_m$ relative to the moving rotor. Summing the voltages induced by $B_1(t)$ and $B_2(t)$ yields

$$e(t) = NB_1(t)\ell r\omega_m + NB_2(t)\ell r\omega_m = 2NB_R \ell r\omega_m \cos \omega_m t$$

Then, noting that the *area* of the armature coil is

$$A = 2\ell r$$

we arrive at the compact expression

$$e(t) = NB_R A\omega_m \cos \omega_m t \qquad \textbf{(4)}$$

This result shows that $e(t)$ has the desired AC time variation with the rms value

$$|E| = NB_R A\omega_m / \sqrt{2}$$

The electrical frequency in radians per second equals the mechanical rotation frequency ω_m.

When we connect a load to the terminals of the alternator, an AC current $i(t)$ flows through it. If the load has impedance angle θ, then $i(t)$ lags $e(t)$ by θ and

$$i(t) = \sqrt{2}\,|I| \cos (\omega_m t - \theta)$$

with $|I|$ being the rms current. But $i(t)$ also flows through the armature conductors, so the "Bli" law says that each conductor feels a clockwise tangential force f depicted in Fig. 18.1–2a. Since the stator is immobile, the rotor experiences a counterclockwise **reaction torque**

$$T(t) = N \times B_1(t)\ell i(t) \times r + N \times B_2(t)\ell i(t) \times r$$
$$= \sqrt{2}\,NB_R A\,|I| \cos \omega_m t \cos (\omega_m t - \theta)$$
$$= \frac{NB_R A\,|I|}{\sqrt{2}} [\cos \theta + \cos (2\omega_m t - \theta)] \qquad \textbf{(5)}$$

where we have inserted the identity for the product of cosines.

Figure 18.1–2b plots the time variations of $B_1(t)$, $i(t)$, and $T(t)$ versus $\omega_m t$. Although the reaction torque pulsates, it has the nonzero average

Figure 18.1–2

(a) Magnetic force and reaction torque.
(b) Waveforms.

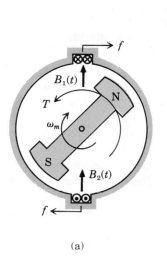

(a)

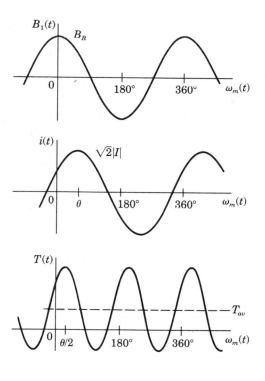

(b)

value

$$T_{av} = \frac{NB_R A \, |I|}{\sqrt{2}} \cos \theta$$

Since the torque from the prime mover must counteract $T(t)$, the mechanical input power is $p_m(t) = T(t)\omega_m$ in absence of mechanical losses. Furthermore, in absence of electrical losses, $p_m(t)$ must equal the electrical output power $p(t) = e(t)i(t)$. Hence, the variation of $T(t)$ has the same shape as the instantaneous AC power $p(t)$ previously sketched in Fig. 16.1–2.

It should be evident that our elementary machine is a *bilateral* device and, consequently, could also be operated as an *AC motor*. In particular, an AC voltage $v(t) \geq e(t)$ applied at the terminals forces stator current $i(t)$ in the opposite direction, and the resulting clockwise reaction torque sustains rotor motion. Such a motor is said to be **synchronous** because the mechanical rotation frequency equals the electrical frequency of the AC source.

Exercise 18.1–1   The rotor of a portable AC generator rotates at $\omega_m = 2\pi \times 60$ Hz and has $B_R = 0.5$ T, $r = 5$ cm, and $\ell = 10$ cm. How many armature conductors are needed so that $|E| = 120$ V?

Rotating Fields and Poles

To gain further insight on the operation of a synchronous machine, and to prepare for subsequent consideration of other machines, we next examine the interaction between the rotor's field and the magnetic field established by the armature current. Obviously, the rotor field rotates, being bound to the mechanical movement of the rotor. Not so obvious is the fact that the immobile armature coil also has rotating fields.

Figure 18.1–3a gives the reference armature-current direction for clockwise generator operation and the corresponding field pattern based on the right-hand rule (see Fig. 17.1–3b). The stator thus has an apparent south pole at $\gamma_s = -90°$ and a north pole directly opposite. If we place the permanent-magnet rotor in this field, as in Fig. 18.1–3b, then its north pole will be drawn towards the south pole of the stator until the two magnetic axes become aligned. The magnetic alignment torque T is proportional to $\sin \delta$, where δ is the angle between the magnetic axes, and the direction of T agrees with the reaction torque in Fig. 18.1–2a.

Figure 18.1–3

(a) Stator field.
(b) Angle δ between magnetic axes.

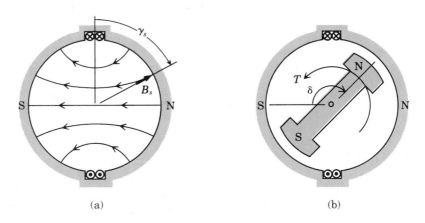

(a)   (b)

But we have not yet included the time variation of the armature current $i(t)$. For that purpose, we ignore any magnetic nonlinearity and assume that the outward radial stator field in the air gap is

$$B_s = -k_s i(t) \sin \gamma_s = k_s i(t) \cos (\gamma_s + 90°)$$

where the constant k_s depends on the armature turns N and the reluctance of the flux path through stator, air gap, and rotor. Inserting $i(t) = \sqrt{2}\,|I| \cos (\omega_m t - \theta)$ and expanding yields

$$B_s = \frac{k_s |I|}{\sqrt{2}} \cos [\gamma_s + 90° - (\omega_m t - \theta)]$$

$$+ \frac{k_s |I|}{\sqrt{2}} \cos [\gamma_s + 90° + (\omega_m t - \theta)] \tag{6}$$

A comparison of Eqs. (6) and (2b) now leads to the interpretation that B_s consists of *two rotating fields* moving in opposite directions. Accordingly, we write B_s in vector terms as

$$B_s = \text{Re}[\underline{B}_S] = \text{Re}[\underline{B}_{S+} + \underline{B}_{S-}]$$

where

$$\underline{B}_{S+} = B_0 \; \underline{/\gamma_s + 90° - (\omega_m t - \theta)}$$
$$\underline{B}_{S-} = B_0 \; \underline{/\gamma_s + 90° + (\omega_m t - \theta)}$$

(7)

with

$$B_0 = k_s |I| / \sqrt{2}$$

The vector $\underline{B}_{S+}$ rotates in the positive or clockwise direction, starting from $\gamma_s = -90°$ when $\omega_m t = \theta$, while $\underline{B}_{S-}$ rotates in the negative or counterclockwise direction. Their rotation is created by the stator current oscillations at frequency ω_m, not by any physical movement. In fact, the resultant stator field vector $\underline{B}_S = \underline{B}_{S+} + \underline{B}_{S-}$ constructed in Fig. 18.1–4a always remains along the horizontal axis of the armature coil.

Figure 18.1–4

Magnetic field vectors: (a) stator; (b) rotor and stator.

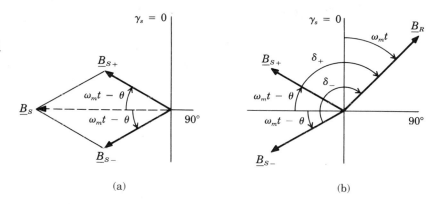

(a)　　　　　　(b)

Adding the rotor field vector $\underline{B}_R$ to our picture gives the diagram in Fig. 18.1–4b, which explains the torque variations $T(t)$. Specifically, the clockwise stator vector $\underline{B}_{S+}$ continually lags $\underline{B}_R$ by the angle $\delta_+ = 90° - \theta$ and tries to pull the rotor back in alignment with it. This effect accounts for the nonzero average value of $T(t)$ proportional to $\sin \delta_+ = \cos \theta$ in Eq. (5). But $\underline{B}_{S-}$ revolves past the rotor and alternately tugs it backward and forward. The instantaneous angle between $\underline{B}_{S-}$ and $\underline{B}_R$ is $\delta_- = 90° + 2\omega_m t - \theta$, accounting for the alternating torque component proportional to $\sin \delta_- = \cos (2\omega_m t - \theta)$.

For motor operation with a clockwise revolving rotor, the direction of the armature current must be reversed. Hence, the stator field vectors are also reversed from their positions in Fig. 18.1–4. Then $\underline{B}_{S+}$ leads $\underline{B}_R$ by a constant angle δ_+, and the resulting magnetic alignment torque pulls the rotor along in synchronous rotation. But the developed motor torque still pulsates as $\underline{B}_{S-}$ revolves past $\underline{B}_R$.

Next, we turn our attention to the structural modifications needed for a more realistic implementation of our elementary machine, operating as a motor or a generator. In practice, the rotor would almost certainly be an electromagnet instead of a permanent magnet, and it might have more than two poles. For instance, Fig. 18.1–5A illustrates a *4-pole* rotor with current-carrying **field windings** around each pole. The windings are series connected and brought out to a pair of **slip rings** upon which press carbon blocks known as **brushes.** This arrangement permits a rotating electrical connection for the **field current** I_f that comes from a small DC source called an **exciter.** The corresponding armature winding would have two equally spaced coils, as shown in Fig. 18.1–5b along with the apparent stator poles and typical flux paths through the rotor and stator. By tracing such paths, you can convince yourself that there must always be an *even number of poles* and one armature coil for each pair of rotor poles.

Figure 18.1–5

Four-pole machine:
(a) rotor with field windings and slip rings;
(b) stator.

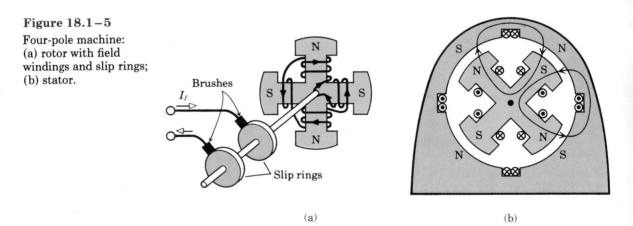

(a)   (b)

The advantage of multiple pairs of poles is *lower mechanical speed* for a given electrical frequency. For if p is the number of poles, then one full revolution of a p-pole rotor produces $p/2$ complete cycles of induced emf in any armature conductor. Therefore, the electrical frequency will be $\omega = (p/2)\,\omega_m$. If we measure electrical frequency in hertz, $f = \omega/2\pi$, and mechanical speed in **revolutions per minute** (rpm) denoted by n_s, then

$$n_s = \frac{60}{2\pi}\,\omega_m = \frac{60}{\pi}\frac{\omega}{p} = 120\,\frac{f}{p} \qquad \textbf{(8)}$$

The subscript s emphasizes the fact that we are dealing with *synchronous* speed. To get $f = 60$ Hz, a hydro turbine or other low-speed prime mover might drive an 18-pole alternator at $n_s = 7200/18 = 400$ rpm, whereas a high-speed steam turbine unit would have $p = 2$ or 4 and $n_s = 3600$ or 1800 rpm.

In the high-speed case, the protruding or **salient rotor poles** of Fig. 18.1–5 could be subjected to extreme mechanical stress. That stress is minimized in a cylindrical rotor structure that has **nonsalient poles** formed by distributing the field winding around the rotor. This configuration is illustrated in Fig. 18.1–6a with $p = 2$. A distributed winding is also shown for the armature coil. Most machines, in fact, have *distributed* rather than *concentrated* armature windings because the former makes better use of materials. Figure 18.1–6b shows a simplified winding pattern for one coil spanning an angle of $2 \times 360°/p$ in a p-pole machine. Because of the spatial distribution of conductors, the emf is somewhat reduced by a **winding factor** $k_w \approx 0.82$–0.96 compared to the emf induced in a concentrated coil.

Figure 18.1–6

Distributed windings: (a) two-pole machine with nonsalient rotor poles; (b) winding pattern for p poles.

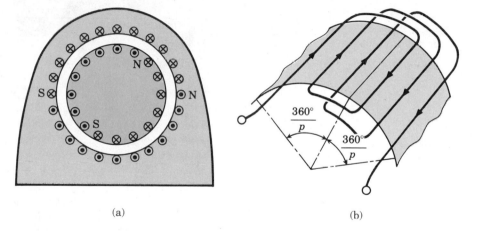

(a)   (b)

Exercise 18.1–2

Consider a generator with a resistive load, so $\theta = 0$. Redraw Fig. 18.1–4b for the three cases $\omega_m t = 0°$, $45°$, and $90°$. Then evaluate the resulting torque from $T(t) = T_0(\sin \delta_+ + \sin \delta_-)$.

Machine Performance Characteristics

Although the "Blu" and "Bli" laws describe the interaction across the air gap, a number of additional factors enter the picture when we talk about the external characteristics of an actual machine. Here we briefly consider some of the major factors and their influence on machine performance.

Obviously, a real machine always has internal losses that result in less than 100% energy-conversion efficiency. Take the synchronous motor, for example. The total input power will consist of AC power to the armature plus the smaller DC power for the rotor field, from which we must deduct **copper losses** caused by ohmic heating in the winding resistances and brush-contact loss at the slip rings. There will also be *core losses* due to hysteresis and eddy currents (primarily confined to the iron through which passes time-varying flux), and *mechanical losses* due to brush and bearing friction and air drag or windage. Core and mechanical losses are usually considered together under the general heading of *no-load rotational loss* simply because it's easier to measure their joint effect. An additional term called the *stray load loss* arises from nonuniform current and field distributions, and adds to the total **rotational losses.**

Figure 18.1–7 diagrams the power flow from electrical input to mechanical "shaft" output. Similar diagrams apply for other types of motors. For generators, the direction of power flows from the mechanical input to the electrical output. Typical copper and rotational losses fall in the neighborhoods of 2–10% and 1–15%, respectively, for a net conversion efficiency of 75–97%.

Figure 18.1–7
Power flow and losses
for a typical motor.

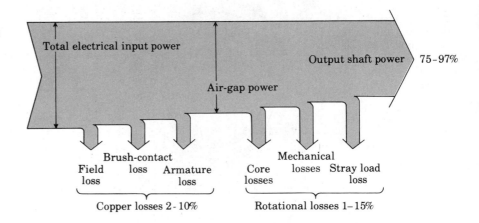

Internal losses lead to *heating,* which must be considered in the design of any sizable machine. Excessive heating breaks down the winding insulation and eventually leads to failure. The National Electrical Manufacturers Association (NEMA) has therefore established standard *ratings* of allowable temperature rise for various classes of insulation. NEMA has also standardized many frame sizes. Special ventilation and cooling techniques may be needed to put a large-capacity machine in the smallest possible frame.

Our power-flow diagram (Fig. 18.1–7) identifies a quantity P_{ag} known as the **air-gap power,** which represents the rate of energy conversion in the air gap exclusive of any losses. This quantity is of fundamental importance in determining a machine's capacity. A general expression for P_{ag} that applies to most machines can be written as

$$P_{ag} = kN_a I_a B_f An \qquad \textbf{(9)}$$

Here N_a is the number of armature conductors carrying the DC or rms AC current I_a, B_f is the maximum air-gap flux density per field pole, $A = 2\ell r$ is the axial area of the rotor turning at n rpm, and k is a constant whose value depends on the individual machine. For a generator, P_{ag} equals the average electrical power in watts (or kilowatts or megawatts) before deducting internal copper losses. For a motor, P_{ag} equals the mechanical power before deducting rotational losses. Mechanical power is often expressed in **horsepower** (hp), defined by

$$1 \text{ hp} = 746 \text{ W} \tag{10}$$

Horsepower, of course, is not an SI unit.

Some important conclusions about machine capacity can be inferred from Eq. (9). In particular, we see that power increases with the armature mmf $N_a I_a$—which also implies increasing copper losses and ohmic heating. A high-capacity motor or generator may therefore require armature conductors of large cross-sectional area, special high-temperature insulation, and perhaps an elaborate cooling system. Power also increases with both flux density B_f and physical size corresponding to the area A, neither of which can be made arbitrarily large. Magnetic saturation limits the former (as seen in Fig. 17.2–5), while mechanical and economic factors and available space limit the latter. Lastly, P_{ag} increases directly with rotor speed n, suggesting the potential need in high-capacity machines for high-speed bearings, auxiliary gear boxes, and similar mechanical provisions.

But we seldom, if ever, expect a machine to operate at its maximum capacity. Instead, it should provide the power demanded of it by the particular load in a given application. As a matter of fact, the intersection of the load characteristics and machine characteristics determines the actual operating point. For example, the terminal voltage and output current of a generator might be related by the characteristic curve sketched in Fig. 18.1–8. The dashed curve in the figure represents a load characteristic, not necessarily linear, that happens to intersect the generator's curve at the **rated** or **full-load** current I_{FL} and voltage V_{FL}. The corresponding full-load power $V_{FL} I_{FL}$ presumably would be less than the machine's maximum generating capacity to allow for occasional overload surges of brief duration.

Usually we want a voltage source to be "stiff," in the sense of maintaining constant voltage despite load-current variations. Loading effects,

Figure 18.1–8

Typical generator and load characteristics.

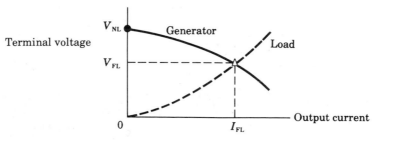

however, often result in a full-load voltage less than the **no-load** (open-circuit) voltage V_{NL}. As a measure of generator performance, we speak of the **voltage regulation** defined by

$$VR = \frac{V_{NL} - V_{FL}}{V_{FL}} \tag{11}$$

usually expressed in percent. All practical generators include an automatic *regulator* that adjusts the field excitation to keep voltage constant, as illustrated by Fig. 18.1–9.

Figure 18.1–9
Voltage regulation by field excitation.

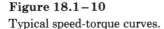

The performance characteristics of a motor are described by its **speed-torque** (or torque-speed) **curve.** The synchronous motor, of course, is a constant-speed machine, but the more common *induction motor* has a speed-torque curve similar to the one shown in Fig. 18.1–10a. Since the full-load speed n_{FL} is less than the no-load speed n_{NL}, we define the **speed regulation**

$$SR = \frac{n_{NL} - n_{FL}}{n_{FL}} \tag{12}$$

Figure 18.1–10
Typical speed-torque curves.

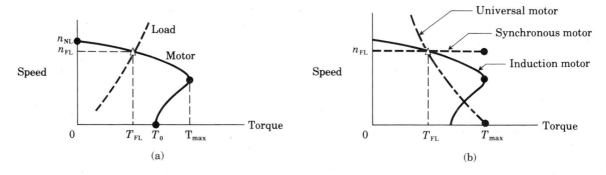

analogous to voltage regulation. The rated (full-load) horsepower may be only 30–50% of the maximum capacity, but there is a maximum or breakdown torque that cannot be exceeded without stalling the motor. The stalled torque or starting torque T_0 (at zero speed) will be greater than T_{FL} if the motor has been designed to come up to speed under loaded conditions.

Different motor applications have grossly different speed and torque requirements. For instance, think of the various requirements of an automobile starter, elevator drive, and magnetic-tape transport. Numerous types of motors have been developed to match various load categories — constant speed, adjustable speed, high starting torque, and so forth. Figure 18.1–10b illustrates the speed-torque characteristics of a hypothetical synchronous motor and a *universal motor* that have the same full-load and T_{max} values as the induction motor in part *a*. The curve for the universal motor approximates a "constant-horsepower" characteristic with high starting torque and low torque at high speed. (The details behind these curves will be discussed in the remaining sections of this chapter.)

Exercise 18.1–3 A certain 6-pole synchronous motor operates at $f = 60$ Hz and develops 100 hp at the shaft. The copper and rotational losses are each 10%. Calculate the electrical input power, the air-gap power, and the output torque, all expressed in SI units.

18.2 SYNCHRONOUS MACHINES

The elementary machine introduced in the previous section was found to be capable of operating as an AC generator or motor, but had the disadvantage of pulsating torque and instantaneous power. For improved performance, we turn to the balanced three-phase synchronous machine, in which three sets of rotating stator fields combine to produce constant torque and power. Similar results are attained by any *polyphase* machine with two or more symmetrical phase windings on the stator.

Three-Phase Generators

All bulk AC power fed into a power grid comes from three-phase synchronous generators. These machines have the same general structure as a single-phase alternator except that there are *three* sets of armature coils per pair of rotor poles. The coils for successive phases have separate terminals and equal angular spacing of $120°/(p/2)$ around the stator, so the equivalent electrical spacing is $120°$. Figure 18.2–1a shows the conductor positions in a 2-pole machine having one turn per coil. In general, there would be $N/(p/2)$ turns per coil, for a total of $6N$ uniformly distributed conductors.

Whether of cylindrical or salient-pole design, the rotor has field windings (not shown) that carry a DC current I_f producing an approximately sinusoidal angular field pattern with maximum value B_R. The rotor turns

Figure 18.2–1

(a) Simplified 2-pole three-phase machine. (b) Waveforms of open-circuit emf.

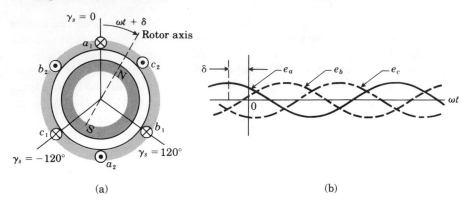

(a)   (b)

at the synchronous speed $n_s = 60\ \omega_m/2\pi = 120f/p$, and has an arbitrary initial angle δ at $t = 0$. As the rotor field sweeps past successive coils, it induces the open-circuit emf's

$$e_a = \sqrt{2}|E| \cos (\omega t + \delta)$$
$$e_b = \sqrt{2}|E| \cos (\omega t + \delta + 120°) \qquad \textbf{(1)}$$
$$e_c = \sqrt{2}|E| \cos (\omega t + \delta - 120°)$$

with $\omega = 2\pi f$ being the electrical angular frequency. You should have no trouble understanding the phase shifts here by comparing the waveforms in Fig. 18.2–1b with the angular positions of the rotor and stator coils.

Drawing upon Eq. (4), Section 18.1, we find that the rms voltage per phase is

$$|E| = \frac{\pi}{60\sqrt{2}}\ k_w\ NB_R An_s \qquad \textbf{(2a)}$$

where we have inserted the winding factor k_w along with $\omega_m = 2\pi n_s/60$. It can also be shown that the *flux per field pole* is $\phi = B_R A/p$ so

$$|E| = \sqrt{2}\pi k_w fN\phi = 4.44\ k_w fN\phi \qquad \textbf{(2b)}$$

where $f = pn_s/120$.

Now let the armature coils be connected in a wye configuration to a balanced three-phase load with lagging power factor pf $= \cos\theta$, as represented by Fig. 18.2–2a. The equivalent circuit for phase a (armature coil a_1-a_2) is diagrammed in Fig. 18.2–2b, where δ has been chosen such that the terminal voltage phasor $\underline{V}_a = |V|\ \underline{/0°}$ is the phase reference and leads the line current phasor $\underline{I}_a = |I|\ \underline{/-\theta}$. The terminal voltage differs from the open-circuit emf phasor $\underline{E}_a = |E|\ \underline{/\delta}$ due to an internal voltage drop across the **synchronous reactance** X_s representing the inductance of the armature winding. (There would also be some winding resistance, but its effect

Figure 18.2−2
(a) Three-phase
generator with
balanced load.
(b) Equivalent circuit
of one phase.

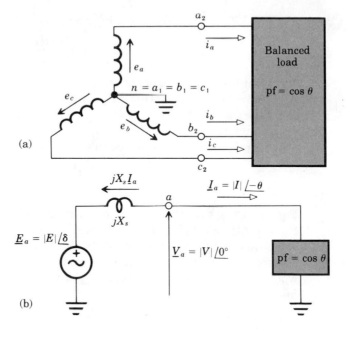

Figure 18.2−3
Phasor diagram for one
phase.

is negligible compared to X_s in a large machine and the generator's Thévenin equivalent impedance essentially equals jX_s.) Identical diagrams hold for the other two phases with an added phase shift of $\pm 120°$.

We'll subsequently show that synchronous reactance arises as a natural consequence of the internal rotating fields. Here, we proceed with the analysis of the generator's equivalent circuit using the phasor diagram in Fig. 18.2−3, which displays the relationship

$$\underline{V} = \underline{E} - jX_s\underline{I} \tag{3}$$

The phase subscript a has been omitted from $\underline{V}_a$, $\underline{E}_a$, and $\underline{I}_a$ since the diagram equally well represents any of the three phases. The phasor $jX_s\underline{I}$ equals the voltage drop across the synchronous reactance and is perpendicular to $\underline{I}$, so $jX_s\underline{I}$ leads $\underline{V}$ by $90° - \theta$. Hence, from the similar triangles, we see that $X_s|I| \cos \theta = |E| \sin \delta$. The average power P_p consumed by each

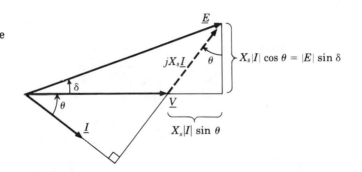

phase of the load can now be expressed as

$$P_p = |V||I| \cos \theta = \frac{|V||E|}{X_s} \sin \delta \qquad \textbf{(4)}$$

The total load power is, of course, $P = 3P_p$.

Observe from Eq. (4) that P_p is proportional to $\sin \delta$. Consequently, we call δ the **power angle**. If $|V|$ and $|E|$ have fixed values, then δ varies in accordance with the power drawn by the load. But there exists an inherent upper limit or **pullout power** per phase, occurring at $\delta = 90°$ and given by

$$P_{\text{max}} = \frac{|V||E|}{X_s} \qquad \textbf{(5)}$$

Figure 18.2–4 emphasizes this limitation in a plot of P_p versus δ based on Eq. (4). Also plotted is the curve for a salient-pole rotor, which provides somewhat higher pullout power at $\delta < 90°$ as a result of *reluctance torque* to be described in Section 18.3. In any case, the power rating of a generator would be substantially less than P_{max} to accommodate occasional overloads.

Figure 18.2–4
Average power versus power angle.

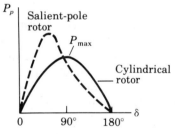

For all load variations below P_{max}, a generator should maintain constant terminal voltage $|V|$. This voltage regulation is accomplished by adjusting the DC field current I_f to increase or decrease B_R which, in turn, increases or decreases the emf per Eq. (2a). A plot of $|E|$ versus I_f therefore takes the shape of a *magnetization curve*, like Fig. 17.2–5, and magnetic saturation establishes an upper bound on $|E|$ for a given machine.

A large generator usually supplies its own field current from a **brushless exciter** consisting of an auxiliary rotor and stator. The auxiliary rotor, mounted on the same shaft as the main rotor, acts as a small AC generator whose current is rectified by a diode bridge and fed into the main rotor winding. The rectified current I_f is controlled by adjusting the external voltage applied to the auxiliary stator. This clever scheme eliminates the need for rotor slip rings and facilitates automatic voltage regulation.

Example 18.2–1   A certain large generator has $X_s = 25$ Ω, and automatic voltage regulation keeps the terminal voltage fixed at $|V| = 15$ kV. Magnetic saturation limits the emf to $|E| \leq 20$ kV, so Eq. (5) gives the absolute maximum load power per phase as

$$P_{\text{max}} = (15 \text{ kV} \times 20 \text{ kV})/25 \ \Omega = 12 \text{ MW}$$

We'll calculate the values of $|E|$ and δ when the load draws $P_p = 2.4$ MW and has pf $= 0.8$ lagging, corresponding to $\theta = 36.9°$.

We first determine the line current $|I| = P_p/(|V| \cos \theta) = 200$ A, which results in the internal voltage drop $X_s|I| = 5$ kV. The phasor emf is then found from Fig. 18.2–3 to be

$$E = 15 + 5 \sin \theta + j5 \cos \theta = 18 + j4 = 18.4 \text{ kV} \ \underline{/12.5°}$$

Hence, $|E| = 18.4$ kV and $\delta = 12.5°$. The combination of the synchronous reactance and the inductive load requires $|E|$ to be about 23% greater than $|V|$.

Now suppose that the power factor at the terminals is corrected to unity, so $\theta = 0°$. We then have $|I| = 160$ A, $X_s|I| = 4$ kV, and $E = 15 + j4 = 15.5$ kV $\underline{/14.9°}$. Note that the power angle δ increases to maintain P_p with the smaller value of $|E|$.

Exercise 18.2–1 Redraw Fig. 18.2–3 for the generator in Example 18.2–1 when $P_p = 2.4$ MW and $\theta = -36.9°$, so I leads V by $-\theta$. Then calculate E to show that $|E| < |V|$.

Three-Phase Fields

We now investigate the magnetic field in a three-phase machine. By doing so, we'll see why synchronous reactance appears in the equivalent circuit. Additionally, we'll gain valuable insight on the operation of three-phase generators and motors.

Recall from our previous study of an elementary machine that a stator coil carrying AC current creates a magnetic field represented by two rotating vectors. Since a three-phase machine has three such coils for each pair of poles, the total stator field will consist of *six rotating vectors*. To find those vectors, we must take account of the phase angles of the line currents as well as the spatial orientation of the coils. We'll perform this analysis on a generator supplying a balanced load with pf $= \cos \theta$, as in Fig. 18.2–2a. For clarity's sake, we'll continue to work with the 2-pole machine having stator coils arranged per Fig. 18.2–1a.

Coil a lies in the vertical plane, just like the single coil back in Fig. 18.1–3a, and it carries the current $i_a = \sqrt{2}|I| \cos (\omega t - \theta)$ where $\omega = \omega_m$ since $p = 2$. Thus, from Eq. (7), Section 18.1, we obtain the field vectors

$$B_{a+} = B_0 \underline{/\gamma_s + 90° - (\omega t - \theta)}$$

$$B_{a-} = B_0 \underline{/\gamma_s + 90° + (\omega t - \theta)}$$

which form the angles $\pm(\omega t - \theta)$ relative to the coil axis at $\gamma_s = -90°$. The axis of coil b lies along $\gamma_s = 120° - 90° = 30°$, and i_b has the instantaneous angle $\omega t - \theta - 120° = \omega t - \theta + 240°$, so

$$\underline{B}_{b+} = B_0 \underline{/\gamma_s - 30° - (\omega t - \theta + 240°)}$$

$$\underline{B}_{b-} = B_0 \underline{/\gamma_s - 30° + (\omega t - \theta + 240°)}$$

Similar reasoning for coil c yields

$$\underline{B}_{c+} = B_0 \underline{/\gamma_s - 150° - (\omega t - \theta + 120°)}$$

$$\underline{B}_{c-} = B_0 \underline{/\gamma_s - 150° + (\omega t - \theta + 120°)}$$

Superimposing the six stator vectors then yields the diagram in Fig. 18.2–5a.

Figure 18.2–5
(a) Three-phase stator field vectors. (b) Resultant stator field and rotor field vectors.

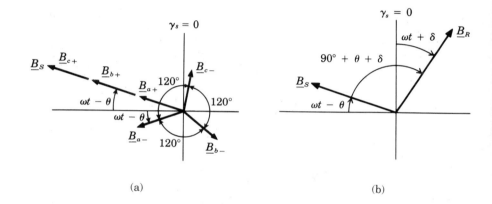

(a)　　　　　　　　　　(b)

This diagram reveals that the clockwise components always line up and *reinforce,* whereas the counterclockwise components form a *symmetrical three-phase set* and mutually cancel out. Therefore, a balanced three-phase stator field is completely represented by *one rotating vector*

$$\underline{B}_S = \underline{B}_{a+} + \underline{B}_{b+} + \underline{B}_{c+} = 3B_0 \underline{/\gamma_s + 90° - (\omega t - \theta)} \tag{6}$$

Figure 18.2–5b shows $\underline{B}_S$ in relation to the rotor field vector

$$\underline{B}_R = B_R \underline{/\gamma_s - (\omega t + \delta)} \tag{7}$$

whose instantaneous angle $\omega t + \delta$ comes from Fig. 18.2–1a.

We now see that the angle between $\underline{B}_S$ and $\underline{B}_R$ equals $90° + \theta + \delta$, independent of time. Accordingly, the magnetic reaction torque exerted by the stator on the rotor will be *constant,* rather than pulsating, and the instantaneous power demanded from the prime mover likewise will be constant. This observation agrees with our conclusion in Section 16.2 that a balanced three-phase system has *constant instantaneous power* $p(t) = 3P_p$.

Figure 18.2–5b also explains the *synchronous reactance* via the following argument where we take advantage of symmetry to focus on phase a.

The rotor field acting alone induces the open-circuit emf $\underline{E}_a = |E|\,\underline{/\delta}$, independent of the load current. But the rotating stator field also sweeps past coil a and creates an additional emf, say $\underline{E}_{as}$, so the terminal voltage becomes $\underline{V}_a = \underline{E}_a + \underline{E}_{as}$. (A single-phase machine would not have this additional emf because $\underline{B}_S$ always stays perpendicular to the stator conductors.) Since $\underline{B}_S$ is proportional to $|I|$ and lags $\underline{B}_R$ by $90° + \theta + \delta$, we can write

$$\underline{E}_{as} = X_s|I|\,\underline{/-90° - \theta} = -jX_s\underline{I}_a$$

where X_s is the proportionality constant and we have incorporated the current phasor $\underline{I}_a = |I|\,\underline{/-\theta}$. Thus,

$$\underline{V}_a = \underline{E}_a + \underline{E}_{as} = \underline{E}_a - jX_s\underline{I}_a$$

which justifies our equivalent circuit with synchronous reactance back in Fig. 18.2–2b.

Exercise 18.2–2 Derive the given expressions for $\underline{B}_{b+}$ and $\underline{B}_{b-}$ from $B_b = k_s i_b \cos(\gamma_s - 30°)$ by inserting $i_b = \sqrt{2}|I| \cos(\omega t - \theta + 240°)$ and expanding.

Synchronous Motors and Capacitors

When AC power is applied to the stator of our three-phase machine, it becomes a synchronous motor capable of driving a mechanical load at the synchronous speed n_s. Such motors are used in special applications where *constant speed* justifies the added expense of the DC exciter. A synchronous motor may also be used for *power-factor correction* — in which case it is sometimes called a synchronous capacitor.

Figure 18.2–6a gives the equivalent circuit for one phase of a synchronous motor with applied terminal voltage $\underline{V} = |V|\,\underline{/0°}$. We still have the synchronous reactance X_s and the induced emf $\underline{E} = |E|\,\underline{/\delta}$, with $|E|$ given by Eq. (2). However, $\underline{E}$ now acts as a "back emf" and opposes the stator current $\underline{I}$, so

$$\underline{V} = \underline{E} + jX_s\underline{I}$$

The resulting phasor diagram in Fig. 18.2–6b has $\underline{V}$ leading $\underline{E}$. Corre-

Figure 18.2–6
Synchronous motor:
(a) equivalent circuit of
one phase; (b) phasor
diagram when
underexcited.

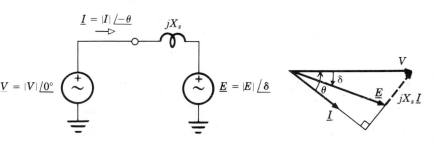

(a)　　　　　　　　(b)

spondingly, the stator field rotates ahead of the rotor field, pulling it along, and the power angle δ is *negative*. (Remember that positive phasor-diagram angles are measured in the counterclockwise sense.)

Negative power angle means that $(|V||E|/X_s)\sin\delta$ will be a negative quantity, reflecting the fact that the machine consumes power rather than producing it. By the same argument, a plot of motor power versus power angle takes the same shape as Fig. 18.2–4 with $-\delta$ replacing δ. Corresponding to P_{max} in Eq. (5), there is a *maximum motor torque* given by

$$T_{max} = \frac{3P_{max}}{\omega_m} = \frac{3p}{4\pi f}\frac{|V||E|}{X_s} \tag{8}$$

If the mechanical load torque exceeds this upper limit, then the rotor drops out of synchronism and grinds to a halt.

When $\underline{V}$ and $\underline{I}$ have the relationship in Fig. 18.2–6b, the motor acts like an inductive load with pf $= \cos\theta$ and is said to be **underexcited.** But we can correct the lagging power factor simply by increasing $|E|$, via the DC field current, until

$$|E| = \sqrt{|V|^2 + (X_s|I|)^2} \tag{9}$$

Then, as shown in Fig. 18.2–7a, $\underline{I}$ is colinear with $\underline{V}$ and $\theta = 0°$. A synchronous motor in this condition appears to be a purely resistive load with the desirable attribute of unity power factor.

Figure 18.2–7
Phasor diagrams:
(a) unity power factor;
(b) overexcited.

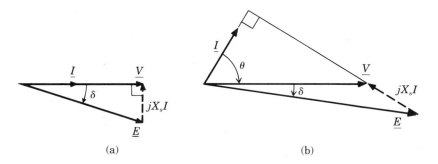

(a)   (b)

Increasing the excitation even further produces the **overexcited** condition diagrammed in Fig. 18.2–7b. Now $\underline{I}$ leads $\underline{V}$, so θ is negative and the motor has an equivalent *capacitive* reactance. In the extreme case where $\theta = -90°$ and $\delta \approx 0°$, the motor becomes a **synchronous capacitor** drawing $Q_p = -|V||I|$ and $P_p \approx 0$. Synchronous capacitors are used in industry to correct the lagging power factor of inductive loads when the size of a conventional capacitor would be impractically large.

Whether operated in the motor or capacitor mode, an additional mechanism is needed to bring the machine up to synchronous speed. For that purpose, the rotor usually has an auxiliary winding serving as an *induction motor* during start-up. This winding produces no torque at n_s, but it does help damp out unwanted mechanical "hunting."

Example 18.2–2   A certain synchronous motor with $X_s = 20\ \Omega$ is connected to a 4-kV, 60-Hz AC supply. We want to find the conditions for unity power factor when the motor drives an 800-hp load.

Assuming that the motor has negligible losses, both mechanical and electrical, the required power per phase is

$$P_p = \tfrac{1}{3} \times 800 \text{ hp} \times 746 \text{ W/hp} \approx 200 \text{ kW}$$

Thus, with unity power factor, $|I| = 200$ kW/4 kV $= 50$ A and $X_s|I| = 1$ kV. Figure 18.2–7a then shows that

$$\underline{E} = 4 - j1 = 4.12 \text{ kV } \underline{/-14°}$$

so $|E| = 4.12$ kV and $\delta = -14°$.

Exercise 18.2–3   Redraw Fig. 18.2–7b for the case of $\theta = -90°$ and $\delta = 0°$. Then find the equivalent capacitance per phase when the machine in Example 18.2–2 operates as a synchronous capacitor with $|E| = 7$ kV.

18.3 AC MOTORS

Routine applications — blowers, pumps, conveyors, and the like — seldom require the constant-speed property of a synchronous motor and do not warrant the expense. For these applications the preferred machine is almost always an *induction motor,* three-phase or single-phase depending on the power level. Indeed, the AC induction motor serves as a veritable rotating workhorse in home and industry. Its characteristics are detailed in this section, along with some special-purpose AC motors.

Three-Phase Induction Motors

Structurally, the simplest, most rugged, most reliable rotating machine is the three-phase *squirrel-cage induction motor.* The stator is identical to that of a synchronous machine with distributed windings, but the rotor winding consists of a "squirrel-cage" arrangement of conducting bars shorted together by two end rings. Figure 18.3–1a illustrates squirrel-cage construction, omitting the laminated rotor core that fills the remaining space. There are no rotating electrical connections nor a DC exciter supplying field current. Instead, the rotor field needed to develop mechanical torque comes from the stator field via induction.

For a qualitative description of this process, consider the simplified rotor in Fig. 18.3–1b and assume it to be initially at rest. The stator has an applied voltage of frequency f and its field $\underline{B}_s$ rotates at the synchronous speed $n_s = 120f/p$ rpm. As $\underline{B}_s$ sweeps past the rotor bars, it induces an emf which causes current to circulate in the rotor and thereby produces a rotor field $\underline{B}_R$. Torque develops from the interaction between $\underline{B}_s$ and $\underline{B}_R$, so the rotor begins to turn and eventually reaches a steady-state speed n.

Figure 18.3–1

Squirrel-cage induction motor: (a) rotor bars and end rings; (b) three-phase stator and rotor field vectors.

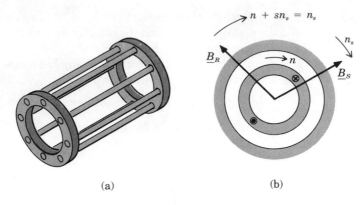

(a)   (b)

The rotor speed must be *less* than the synchronous speed, for there would be no induced emf if $n = n_s$ (why?), so no rotor current or field and no torque to sustain the rotation against a mechanical load. We therefore write

$$n = (1 - s)n_s$$

where s is called the **slip** and defined as

$$s \triangleq 1 - \frac{n}{n_s} \tag{1}$$

We also define the **slip speed**

$$sn_s = n_s - n$$

which equals the rotational rate of $\underline{B}_S$ relative to the rotor bars. The electrical frequency of the rotor's induced emf is then

$$f_r = \frac{p}{120} sn_s = sf \tag{2}$$

Accordingly, the resulting three-phase rotor field vector $\underline{B}_R$ rotates at sn_s rpm relative to the rotor. Since the rotor turns at n rpm, $\underline{B}_R$ has total speed $n + sn_s = n_s$ and stays in synchronism with $\underline{B}_S$. Therefore, the developed torque will be constant and free of pulsations, just as in a synchronous motor, even though the rotor turns at $n < n_s$.

For analytic purposes, we view the induction motor as a three-phase *transformer* with a fixed primary (the stator) and a rotating secondary (the rotor). We take account of the rotor's motion by observing that the secondary voltages and currents have electrical frequency $\omega_r = s\omega$ and that the induced secondary voltage will be proportional to s as well as to the effective turns ratio N. Figure 18.3–2a diagrams the resulting equivalent circuit for one phase, including winding resistance and reactance in both the primary and secondary. (Magnetizing inductance and core-loss resistance have been neglected.) The phasors $\underline{V}$ and $\underline{E}_1$ are the applied voltage and back emf, respectively. This circuit will lead to a quantitative expression for the developed torque after two manipulations.

Figure 18.3−2

Equivalent circuits for one phase of an induction motor.

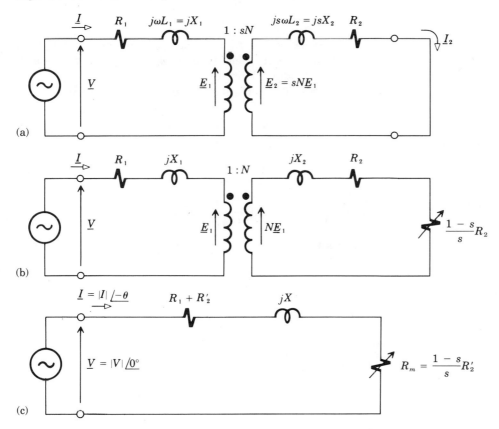

(a)

(b)

(c)

First, we apply KVL around the short-circuited secondary and divide by s to get

$$N\underline{E}_1 = \frac{1}{s}(jsX_2 + R_2)\underline{I}_2 = \left(jX_2 + R_2 + \frac{1-s}{s}R_2\right)\underline{I}_2$$

where R_2/s has been broken up into the sum $R_2 + (1-s)R_2/s$. Our expression for $N\underline{E}_1$ eliminates s from the secondary voltage and leads to the modified diagram in Fig. 18.3−2b. The constant resistance R_2 here represents power lost by ohmic heating in the rotor, while the speed-dependent resistance $(1-s)R_2/s$ represents electrical-to-mechanical energy conversion. Next, reflecting the secondary impedance into the primary yields the total motor impedance

$$\underline{Z} = R_1 + jX_1 + \frac{jX_2}{N^2} + \frac{R_2}{N^2} + \frac{1-s}{s}\frac{R_2}{N^2}$$

$$= (R_1 + R_2') + jX + R_m \tag{3a}$$

where $R_2' = R_2/N^2$, $X = X_1 + (X_2/N^2)$, and

$$R_m = \frac{1-s}{s} R_2' \tag{3b}$$

The resulting circuit model in Fig. 18.3–2c clearly brings out the fact that an induction motor acts as an *inductive* load whose AC resistance and power factor vary with the slip s. The model's parameter values R_1, R_2', and X can be determined by measurements similar to those used for a power transformer.

Now we calculate the torque from the electrical power delivered to the speed-dependent resistance R_m. Solving for $|I|$ in Fig. 18.3–2c yields

$$P_m = |I|^2 R_m = \frac{|V|^2 R_m}{(R_1 + R_2' + R_m)^2 + X^2}$$

$$= (1-s)|V|^2 \frac{sR_2'}{(sR_1 + R_2')^2 + s^2 X^2}$$

Since P_m represents the electromechanical power conversion per phase, the total developed torque will be

$$T = \frac{3P_m}{\omega_m} = \frac{3p|V|^2}{4\pi f} \frac{sR_2'}{(sR_1 + R_2')^2 + s^2 X^2} \tag{4}$$

in which we have expressed the rotor's mechanical angular velocity as $\omega_m = 2\pi n/60 = 4\pi(1-s)f/p$.

Figure 18.3–3 is a typical **torque-speed curve** obtained from Eq. (4) by plotting T versus slip s or rotor speed $n = (1-s)n_s$. This curve illustrates three significant properties:

- An induction motor develops torque at any speed less than the synchronous speed, but it is designed to operate near the rated speed n_{FL} somewhat less than n_s.

- The maximum torque T_{max} is developed below rated speed, allowing the motor to drive an overload $T > T_{FL}$ at reduced speed for short intervals of time.

- The starting torque T_0 at $n = 0$ (or $s = 1$) usually exceeds the rated torque, so the motor can start under load and come up to speed.

Figure 18.3–3
Torque-speed curve of an induction motor.

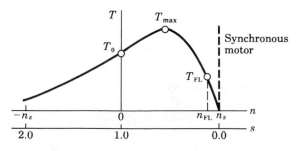

Contrast these induction-motor characteristics with those of synchronous motor, represented by the dashed line, which also has a maximum torque but operates only at n_s and has no starting torque of its own. On the other hand, the induction motor lacks the adjustment flexibility made possible by the variable excitation of a synchronous motor.

The maximum torque of an induction motor has the value

$$T_{max} = \frac{3p|V|^2}{8\pi f(R_1 + \sqrt{R_1^2 + X^2})} \tag{5a}$$

which occurs when

$$s = \frac{R_2'}{\sqrt{R_1^2 + X^2}} \tag{5b}$$

(The derivation of these expressions is outlined in Problem 18.3–14.) The rated torque normally occurs at very small slip, in the range $s = 0.01–0.05$. With $|s| \ll 1$, Eq. (4) becomes

$$T \approx \frac{3p|V|^2 s}{4\pi f R_2'} \tag{6}$$

This approximation holds whenever $sR_1 \ll R_2'$ and $sX \ll R_2'$.

Given the value of s, we can determine other characteristics by returning to the equivalent circuit in Fig. 18.3–2c. In particular, we see that each phase of the motor draws electrical power $P_p = |I|^2(R_1 + R_2' + R_m)$ to produce the mechanical power $P_m = |I|^2 R_m$. Hence, the electromechanical conversion efficiency is given by the resistance ratio

$$\text{Eff} = \frac{3P_m}{3P_p} = \frac{R_m}{R_1 + R_2' + R_m} \tag{7}$$

The efficiency varies with speed since $R_m = (1 - s)R_2'/s$.

Example 18.3–1

A certain 6-pole three-phase induction motor is designed for $T_{FL} = 85$ N · m with $|V| = 120$ V and $f = 60$ Hz. Electrical measurements yield the following parameter values:

$$R_1 = 0.30\ \Omega \qquad R_2' = 0.12\ \Omega \qquad X = 0.40\ \Omega$$

We'll predict the operating characteristics based on this information.

The synchronous speed is $n_s = 120f/p = 1200$ rpm, but the rated speed will be slightly less. Assuming Eq. (6) holds, we have

$$s_{FL} \approx \frac{4\pi f R_2'}{3p|V|^2} T_{FL} = 0.03$$

from which $s_{FL} R_1 = 0.009 \ll R_2'$ and $s_{FL} X = 0.012 \ll R_2'$, as required for the approximation. Thus, $n_{FL} = 0.97 n_s = 1164$ rpm and $\omega_m = 2\pi n_{FL}/60 = 122$ rad/sec, so the rated mechanical output power is $T_{FL}\omega_m = 10.4$ kW, or about 14 hp. Equation (7) then yields Eff $= 3.88/4.3 \approx 90\%$ at full load.

Next, inserting numerical values in Eq. (5), we find that $T_{max} = 215$ N $\cdot$ m $\approx 2.5T_{FL}$ at slip $s = 0.24$ or $n = 912$ rpm. The mechanical output power rises in this condition to almost twice the rated value. Finally, we set $s = 1$ in Eq. (4) to get the starting torque

$$T_0 = \frac{3p|V|^2}{4\pi f} \frac{R_2'}{(R_1 + R_2')^2 + X^2} = 123 \text{ N} \cdot \text{m}$$

confirming that $T_0 > T_{FL}$ for starting purposes.

Exercise 18.3 – 1   Use Fig. 18.2 – 3c and values given in Example 18.3 – 1 to calculate the rms current, power factor, and efficiency of the motor when $T = T_{max}$.

Starting and Speed Control

A large induction motor must have a **controller** designed to limit inrush current and excessive starting torque, to protect against overheating and, perhaps, to permit reversing direction. A schematic diagram of a simplified controller appears in Fig. 18.3 – 4, where the symbol —Ⓕ— stands for the operating coil of a magnetic relay (as in Fig. 17.4 – 10) that closes all pairs of contacts represented by —| |— . This controller functions as follows.

Pressing the forward button (FWD) energizes relay F and applies the line voltage to the motor through resistors R_S, whose voltage drops reduce

Figure 18.3 – 4
Controller for an
induction motor.

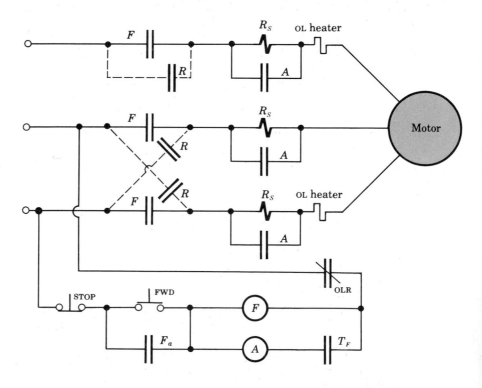

$|V|$ at the motor and thereby limit the starting torque and inrush current. Relay F also closes auxiliary contactor F_a so you can release the push button. After the motor comes up to the reduced-voltage speed, a time-delay contactor T_F closes and energizes relay A whose contacts then close and bypass the starting resistances. (In practice, this would be done in several steps using additional resistors, relays, and time-delay contactors.)

The normally closed contactor labeled OLR is part of an *overload relay* whose heater elements in series with the motor cause the relay to open in the event of prolonged or extreme overload conditions. Three other contacts labeled R will *reverse* the phase sequence applied to the motor and thus reverse the direction of rotation. For clarity, we have omitted the reverse button and the interlocking arrangement that opens the F contactors when the R contactors are closed.

But starting resistors do not provide effective nor efficient running-speed control. For applications requiring such speed control, the **wound-rotor motor** may be appropriate. Instead of short-circuited conducting bars, the rotor has a symmetrical three-phase winding with the same number of poles as the stator. The three phases are connected by slip rings to a set of external variable resistors (or three-phase **rheostat**) that increases the equivalent value of the secondary resistance.

Figure 18.3–5 gives the rotor diagram and the torque-speed curves for three values of resistance. The curves are obtained from Eq. (4) with the per-phase external resistance R_{ext} added to R_2'. The maximum torque remains unchanged, since it does not depend on R_2', but T_{max} now occurs at the increased slip

$$s = \frac{R_2' + R_{ext}}{\sqrt{R_1^2 + X^2}} \tag{8}$$

The rheostat, therefore, controls the starting torque and the running speed at rated torque. The price of this control, compared to the squirrel-cage motor, is not only its greater initial cost, but also its increased maintenance problems and its decreased efficiency, the latter attributable to power wasted as heat in the rheostat.

Figure 18.3–5
Wound-rotor motor:
(a) rotor diagram;
(b) torque-speed curves.

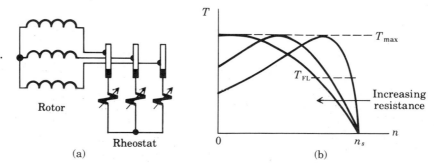

Recent advances in semiconductor power electronics have led to improved systems for adjusting and controlling the speed of an induction (or synchronous) motor. These systems employ controlled rectifiers and other circuitry to produce variable-frequency, variable-voltage three-phase AC power. A feedback path from a tachometer may be included to facilitate automatic speed control.

Exercise 18.3–2   Suppose the motor in Example 18.3–1 has a wound rotor.
(a) What external resistance (per phase) gives $T_{\max}$ at $s = 1$ to maximize starting torque?
(b) Find n_{FL} with this resistance in the rotor circuit.

Single-Phase Induction Motors

Single-phase induction motors are used in residential and other applications where three-phase AC is not available and the mechanical power requirement does not exceed about 5 hp. These motors have a squirrel-cage rotor but, of course, only one phase in the stator winding. They also must have special provision for starting purposes.

To explain the operating principles and starting problem of a single-phase induction motor, recall that the magnetic field of a single-phase stator consists of two components rotating at synchronous speed in opposite directions. These are represented in Fig. 18.3–6a by $\underline{B}_{S+}$ and $\underline{B}_{S-}$, and we assume the rotor to be turning at rate n in the clockwise direction. The interaction of $\underline{B}_{S+}$ with the rotor produces a torque-speed characteristic just like that of a three-phase motor (Fig. 18.3–3), and $\underline{B}_{S-}$ has the same effect but with the torque and speed in the opposite direction. The resulting average torque is then the sum plotted in Fig. 18.3–6b. Thus, although the motor can run in either direction, it has zero starting torque. When the motor is running in a given direction, the instantaneous torque consists of the average torque plus a pulsating component caused by the oppositely rotating stator field, similar to the conditions in a single-phase alternator. Of course, the torque *must pulsate* if the speed is constant, since a single-phase AC circuit delivers pulsating instantaneous power.

Figure 18.3–6
Single-phase induction motor: (a) stator field vectors; (b) torque-speed curve.

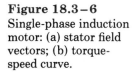
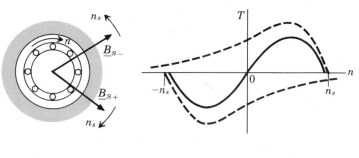

(a)   (b)

Starting torque can be obtained with the help of an auxiliary stator winding displaced by 90° (electrically) from the main winding, as represented in Fig. 18.3–7a. This smaller winding is designed so that its current is phase-shifted relative to the current through the main winding. Consequently, the forward rotating fields from both windings partially reinforce and produce net torque at zero rotor speed. A centrifugal switch mounted in the shaft disconnects the auxiliary winding after starting, and the machine normally runs on the main winding alone. Figure 18.3–7b shows a typical torque-speed curve of this machine, which is called a **split-phase motor.**

Figure 18.3–7

Split-phase motor:
(a) windings;
(b) torque-speed curve.

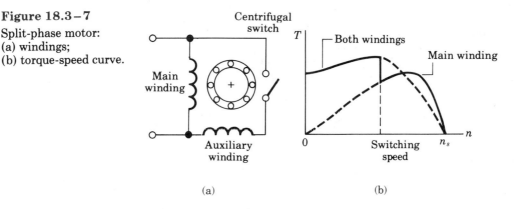

(a) (b)

Other variations of the split-phase strategy employ one or two capacitors to enhance the current phase shift and to improve running performance. Alternatively, very small motors (under $\frac{1}{20}$ hp) may have copper rings called **shading coils** around a portion of each of the two salient stator poles, depicted by Fig. 18.3–8. Induced currents in the shading coils distort

Figure 18.3–8
Shaded-pole motor.

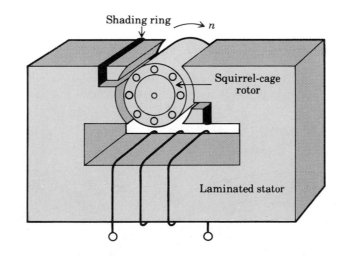

the field and produce starting and running torque in the direction from the unshaded to the shaded portion. The economical structure of this **shaded-pole motor** makes it a popular choice for hair dryers and similar small appliances.

Regardless of starting method, the running torque of a single-phase induction motor is proportional to the square of the line voltage. Crude but effective speed control is therefore possible by changing the line voltage, a technique illustrated in Fig. 18.3–9 with a typical load characteristic. This technique is implemented in low-power 2-speed or 3-speed devices, notably fans, through the simple expedient of inserting a voltage-dropping resistor.

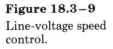

Figure 18.3–9
Line-voltage speed control.

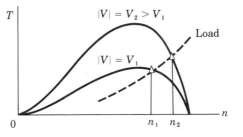

Special-Purpose AC Motors

Numerous special-purpose AC motors have been developed to fit particular needs. Three interesting types are described below. Another type, the important *universal motor,* will be covered in the next section.

Servomotors are *two-phase* induction motors designed for use in servomechanisms and automatic control systems. Fixed AC voltage of amplitude V_F is applied to one stator winding and an AC *control voltage* with variable amplitude V_C and 90° phase shift is applied to the other winding. Adjusting the control voltage yields a family of torque-speed curves like Fig. 18.3–10, where

$$T \approx k \left(\frac{V_C}{V_F} - \frac{n}{2n_s} \right) \tag{9}$$

which holds at speeds up to about 25% of n_s. This nearly linear relationship is valuable for effective performance of many control systems.

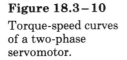

Figure 18.3–10
Torque-speed curves of a two-phase servomotor.

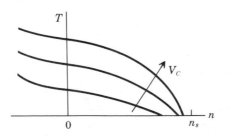

Figure 18.3–11
Reluctance motor.

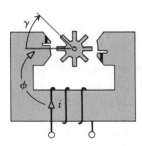

Hysteresis motors have a split-phase stator winding and a smooth cylindrical rotor of hard magnetic material. The rotating stator field magnetizes the rotor, but hysteresis causes the rotor magnetization to lag behind by a constant angle δ. The motor therefore develops constant torque at any speed up to n_s and will accelerate from standstill to steady-state synchronous rotation, provided that the load does not exceed the power rating — normally less than $\frac{1}{10}$ hp. Electric clocks, timing devices, and phonograph turntables are common applications of this simple synchronous motor.

Reluctance motors have a cogwheel-shaped rotor of soft magnetic material and, usually, a shaded-pole stator, as shown in Fig. 18.3–11. The rotor's shape causes the total reluctance of the flux path to vary with the rotor angle γ approximately as

$$\mathcal{R} = \mathcal{R}_0 - \Delta\mathcal{R}\cos p\gamma$$

where p is the number of rotor "teeth." What we have here is a *rotating-iron* device, and the rotational version of Eq. (11), Section 17.4, gives the *reluctance torque.*

$$T = -\tfrac{1}{2}\phi^2\frac{d\mathcal{R}}{d\gamma} = \tfrac{1}{2}\phi^2\,\Delta\mathcal{R}p\,\sin p\gamma \qquad \textbf{(10a)}$$

If flux ϕ is proportional to the AC current i, and if the instantaneous rotor angle is $\gamma = (2/p)(\omega t + \delta)$, then the torque consists of alternating components, plus a constant term

$$T_{\text{av}} = T_{\text{max}}\sin 2\delta \qquad \textbf{(10b)}$$

Thus, for $0 < \delta < 90°$, the rotor will turn at the synchronous speed $n_s = 120\,f/p$. The stator shading coil and a squirrel-cage winding on the rotor provide the initial torque needed to get things going. The advantage of the reluctance motor over the hysteresis motor is its lower mechanical speed since p can be a larger integer.

Reluctance-torque effects will occur in *any* AC machine with a *salient-pole* rotor. This observation, together with Eq. (10), accounts for the difference between the two curves of P_p versus δ for synchronous generators in Fig. 18.2–4.

Exercise 18.3–3   Carry out the steps between Eqs. (10a) and (10b), taking $i = \sqrt{2}\,|I|\cos\omega t$ and assuming $\mathcal{R}_0 \gg \Delta\mathcal{R}$ so $\phi \approx Ni/\mathcal{R}_0$.

18.4 DC MACHINES

AC motors are preferred for rotational drive units whenever possible, given their mechanical reliability and the ready access to AC power. But some applications require operating characteristics beyond the scope of AC motors with fixed source voltage and frequency. A DC motor may then be the machine of choice, especially for its ease of speed control. This section describes the general structure and characteristics of DC machines.

Commutation and Structural Features

DC machine operation depends upon a **commutator**, whose principles are illustrated by Fig. 18.4–1 for a simplified 2-pole machine. Here, a concentrated armature coil is on the rotor and the field windings are on the stator, just the opposite of a synchronous machine. The field windings consist of many turns of relatively thin wire, all connected in series and usually carrying a small current compared to the low-resistance, high-current armature. The salient stator poles are shaped to establish a fixed and nearly uniform radial field B_f in the air gap, a field that reverses direction from pole to pole. The potential torque reversals are canceled out by a *split-ring commutator* on the rotor shaft that reverses the direction of the DC armature current I_a when the plane of the coil crosses the field reversals.

Figure 18.4–1
(a) DC machine with split-ring commutator. (b) Waveforms.

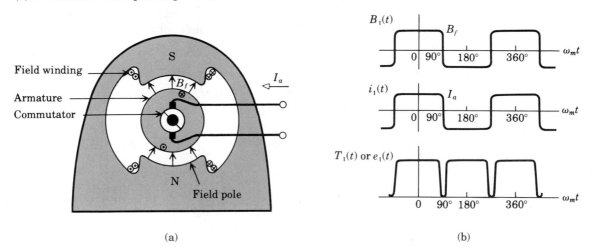

(a)   (b)

Now imagine traveling with one of the armature conductors while it rotates at frequency ω_m. The field we see varies as shown in Fig. 18.4–1b, but — thanks to the commutator — the current reverses direction simultaneously with the field reversals. Consequently, the torque $T_1(t)$ experienced by that conductor and the emf $e_1(t)$ induced across it stay almost constant, save for the dips at the instants of commutation. Similar curves hold for a p-pole machine when you substitute the equivalent electrical angle $(p/2)\omega_m t$ for the mechanical angle $\omega_m t$ in Fig. 18.4–1b.

Figure 18.4−2

(a) DC machine with multi-segmented commutator. (b) Induced emf.

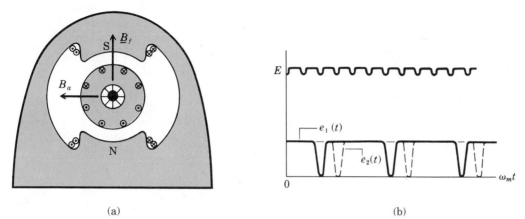

(a)

(b)

A *distributed* armature winding with a *multisegmented commutator* essentially eliminates the commutation dips and gives rise to the vector picture in Fig. 18.4−2a, where the armature field $\underline{B}_a$ remains fixed and perpendicular to the stator field $\underline{B}_f$. (The brushes are omitted from this drawing for clarity, as are the rather complicated connections to the commutator segments.) Since this picture applies at any mechanical speed, controlling the speed of a DC motor only requires adjusting the value of the armature current. Whether the machine functions as a motor or a generator, the total induced emf equals the sum of the contributions from the many series-connected armature wires and looks like Fig. 18.4−2b—a large constant voltage E with a small *ripple* due to commutation. A plot of the total magnetic torque would have the same shape.

Figure 18.4−3

Commutating and compensating windings.

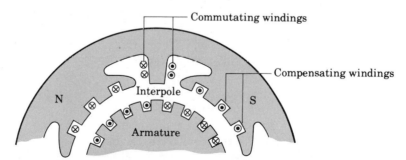

Large machines often have two other sets of windings illustrated in Fig. 18.4−3. **Commutating windings** on small **interpoles** midway between the field poles provide flux that assists commutation action and reduces sparking at the brushes caused by current reversals. (Recall that $v_L = L \, di/dt$, so sudden current reversal in an inductive armature winding generates a voltage pulse.) **Compensating** or **pole-face windings** carry current in the opposite direction of the adjacent armature conductors and thereby cancel the unwanted magnetic field caused by armature current that would otherwise distort the air-gap field — an effect known as *armature reaction*. Both of these auxiliary windings are connected in series with the armature.

Induced EMF and DC Generators

To calculate the value of E, let the armature have radius r and length ℓ, so the area of each pole is $A_p \approx 2\pi\ell r/p$ and the *flux per pole* will be

$$\phi = B_f A_p = B_f \frac{2\pi\ell r}{p}$$

assuming uniform air-gap flux density B_f. If there are N_a series-connected armature conductors between brushes, then armature rotation at constant angular velocity ω_m produces the open-circuit emf

$$E = N_a B_f \ell r \omega_m = (N_a p/2\pi)\phi\omega_m \tag{1}$$

Equivalently, in terms of the armature speed n, we have

$$E = k_a \phi n \tag{2a}$$

with

$$k_a = N_a p/60 \tag{2b}$$

since $n = 60\omega_m/2\pi$.

When current I_a flows in the armature, the resulting magnetic torque is given by

$$T = \frac{30}{\pi} k_a \phi I_a = \frac{N_a p}{2\pi} \phi I_a \tag{3}$$

Equation (3) follows from the fact that the mechanical power $T\omega_m$ must equal the electrical power EI_a.

The value of ϕ in the foregoing expressions depends on the field-winding current I_f in accordance with the magnetization properties of the complete magnetic circuit, Fig. 18.4–4 being a representative example. This has the same shape as a B-H curve (Fig. 17.2–5) since B is proportional to ϕ, and H is proportional to the field mmf $\mathcal{F} = N_f I_f$. For moderate values of field current we can use the linear approximation $\phi \approx N_f I_f/\mathcal{R}$ and write

$$E = k_f I_f n \qquad T = 30k_f I_f I_a/\pi \tag{4a}$$

where

$$k_f = k_a N_f/\mathcal{R} = pN_a N_f/60\,\mathcal{R} \tag{4b}$$

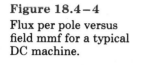

Figure 18.4–4

Flux per pole versus field mmf for a typical DC machine.

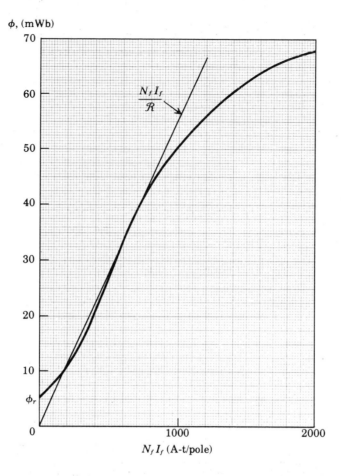

But saturation introduces nonlinearity at high field currents, while residual magnetism (retentivity) causes a residual flux ϕ_r when $I_f = 0$. Both effects turn out to have importance for certain types of DC machines.

Figure 18.4–5 gives schematic diagrams for the four common ways of connecting the armature and field. In the **separately-excited mode** (Fig. 18.4–5a), the field current comes from a separate source, whereas I_a and I_f are interdependent in the remaining three **self-excited modes**. With armature and field connected in **series** (Fig. 18.4–5b), I_f equals I_a, and the field winding must have lower resistance and fewer turns than the other configurations, unless most of I_a is bypassed through a small *diverter resistor* R_D. The parallel or **shunt** connection in Fig. 18.4–5c puts the same voltage across both windings so $I_f \ll I_a$ in view of the high field resistance. A variable resistor R_C called a *control rheostat* may be included to adjust I_f. The **compound** connection (Fig. 18.4–5d), has part of the field winding in series but mostly in parallel with the armature, so there are actually two

Figure 18.4 – 5
Armature and field
connections:
(a) separately excited;
(b) series; (c) shunt;
(d) compound.

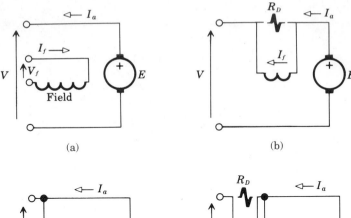

(a)   (b)

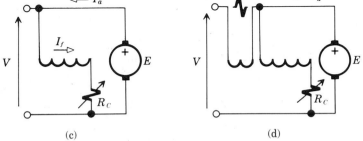

(c)   (d)

different field currents. The relative merits of these various connection schemes depend on the particular application.

In the case of generator operation, a mechanical prime mover (an AC induction motor, for instance) drives the armature at angular frequency ω_m and a DC voltage $V \leq E$ appears at the armature terminals. We usually want this terminal voltage to be sensibly independent of the output load current I. For a **separately-excited generator** having the equivalent circuit of Fig. 18.4 – 6a, we see that $I_a = I$ and

$$V = E - R_a I = k_a \phi n - R_a I \tag{5}$$

Figure 18.4 – 6
Separately-excited generator: (a) equivalent circuit; (b) V-I characteristic.

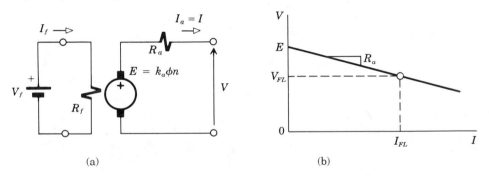

(a)   (b)

Thus the V-I characteristic is a straight line whose slope equals the negative of the small armature resistance R_a, as plotted in Fig. 18.4–6b.

The rated full-load voltage $V_{FL} = E - R_a I_{FL}$ will be somewhat less than the no-load voltage $V_{NL} = E$ when $I = 0$. The entire curve may be shifted up or down by changing the field voltage V_f, since ϕ depends on $I_f = V_f/R_f$. The total input power consists of $P_m = T\omega_m$, plus the small field power $P_f = V_f I_f$.

The self-excited **shunt generator** needs no separate field supply but its V-I curve in Fig. 18.4–7 falls off rapidly with increasing load, so a higher no-load voltage would be required for the same full-load voltage as that of a separately-excited generator. You can see why this happens by referring to Fig. 18.4–5c and noting that any decrease of the terminal voltage also decreases the field current I_f which, in turn, further reduces the terminal voltage. The opposite effect takes place in a **series generator** whose rising V-I curve (not shown) would have little practical value by itself. However, appropriate combination of the shunt and series characteristics in a **compound generator** yields a nearly constant V-I curve, with the series field providing increased flux when the shunt-field flux decreases. The windings can be proportioned such that $V_{FL} = V_{NL}$, in which case the generator is said to be *flat-compounded* and has the typical characteristics sketched in Fig. 18.4–7.

Figure 18.4–7

V-I characteristics of shunt and compound generators.

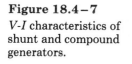

Example 18.4–1

A 4-pole shunt-connected generator is driven at 1200 rpm. The armature has total resistance $R_a = 0.5\ \Omega$ and $N_a = 100$ conductors between brushes. The field winding has total resistance $R_f = 40\ \Omega$ and $N_f = 180$ turns per pole. The magnetic characteristics are given by Fig. 18.4–4. We seek the rheostat resistance R_C such that $V = 300$ V when the load draws $I = 120$ A.

To begin the calculations, we note from Fig. 18.4–5c that $I = I_a - I_f$ so $I_a = I + I_f$ and

$$V = E - R_a(I + I_f)$$

Next, we use Eq. (2) to get

$$E = 8000\phi$$

where ϕ varies *nonlinearly* with I_f per Fig. 18.4–4. However, if we're lucky, the required flux may fall close to the linear approximation $\phi \approx N_f I_f / \mathcal{R}$ with $\mathcal{R} \approx 18{,}000$, and we would then have

$$\phi \approx 0.01 I_f$$

After substituting $E \approx 8000 \times 0.01 I_f$ and other numerical values, our V-I equation becomes

$$80 I_f - 0.5(120 + I_f) \approx 300$$

from which $I_f \approx 4.5$ A — so we get $E \approx 360$ V, $\phi \approx 45$ mWb, and $N_f I_f \approx 810$ A-t/pole.

Referring again to Fig. 18.4–4, we find that a flux of 45 mWb actually requires $N_f I_f \approx 840$ A-t/pole. Hence, the field current must be increased slightly to $I_f \approx 840/180 = 4.67$ A. Finally, since $I_f = V/(R_f + R_C)$ in Fig. 18.4–5c, the corresponding rheostat resistance is

$$R_C = (V/I_f) - R_f = 24.3 \ \Omega$$

However, in view of the approximations involved here, it would be more realistic to use a larger rheostat and adjust it as needed to get $V = 300$ V.

Exercise 18.4–1   Suppose the generator in Example 18.4–1 is series connected, without R_C or R_D, so $I = I_a = I_f$. Write V in terms of ϕ and I. Then use Fig. 18.4–4 to find V when $I = 3$ A and $I = 6$ A.

DC Motors and the Universal Motor

Like a DC generator, a DC motor may be self-excited or separately-excited. The latter offers a combination of good speed regulation and flexible control unmatched by any other type of motor. The series motor, however, has unique advantages as a universal motor. These properties, along with those of the shunt motor, will be explored below.

Consider the equivalent armature circuit of a separately-excited motor with applied voltage V in Fig. 18.4–8a. This differs from a generator in that I_a flows in the opposite direction and is opposed by the back emf $E = k_a \phi n$. Since $V = R_a I_a + E$, the rotor speed will be $n = (V - R_a I_a)/k_a \phi$. Substituting $I_a = \pi T / 30 k_a \phi$ from Eq. (3) yields the speed-torque relationship

$$n = n_{\mathrm{NL}} - bT \tag{6a}$$

where

$$n_{\mathrm{NL}} = \frac{V}{k_a \phi} = \frac{60 V}{N_a p \phi} \qquad b = \frac{\pi R_a}{30(k_a \phi)^2} = \frac{120 \pi R_a}{(N_a p \phi)^2} \tag{6b}$$

Figure 18.4–8

Equivalent armature circuits: (a) separately-excited motor; (b) shunt motor; (c) series (or universal) motor.

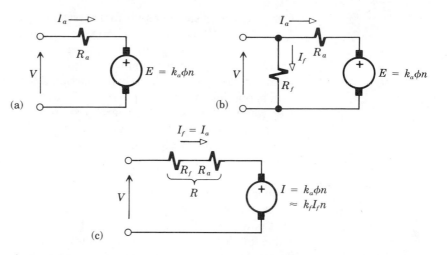

Figure 18.4–9

DC motor speed-torque curves.

Note that $n_{\mathrm{NL}} \rightarrow \infty$ when $\phi = 0$ and $V \neq 0$. Thus, to prevent potential runaway, the armature voltage should never be applied without field excitation.

Equation (6) defines the speed-torque curve plotted in Fig. 18.4–9 — a straight line starting from the no-load speed n_{NL} and decreasing with slope b. This line eventually intercepts the horizontal axis ($n = 0$) at the starting torque

$$T_0 = \frac{n_{\mathrm{NL}}}{b} = \frac{30 k_a \phi V}{\pi R_a} = \frac{N_a p \phi V}{2 \pi R_a}$$

The starting torque also equals the stall torque when the motor is subjected to excessive load.

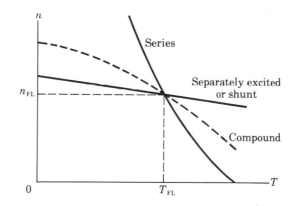

Both n_{NL} and b depend on the flux ϕ which, in turn, depends on the field current I_f. Usually, the flux is such that $b T_{\mathrm{FL}} \ll n_{\mathrm{NL}}$ so $n_{\mathrm{FL}} \approx n_{\mathrm{NL}}$ and we have a nearly *constant-speed* motor. The key to this behavior is the induced voltage $E = k_a \phi n$, which is just slightly less than V at full load.

Consequently, a small change in n corresponds to a large change in current $I_a = (V - E)/R_a$ and a large change in torque $T = (30/\pi)k_a\phi I_a$.

Closer examination of the expression for n_{NL} reveals that we can control the motor speed simply by adjusting the armature voltage V. Furthermore, since n_{NL} is directly proportional to V, smooth *reversal* of rotation is possible by reducing V to zero and then reversing its polarity. An electronic DC power supply performs these tasks quite well, although a three-phase rectifier is needed for power levels above about 5 kW. Prior to the development of high-power semiconductor devices, DC motors were supplied from a motor-generator (MG) set consisting of a DC generator driven by an induction motor. This arrangement, known as the *Ward-Leonard system,* also provides *regenerative braking* that slows down or stops the DC motor by reducing the generator's field current to the point that power flows back to the generator. Regenerative braking has special value for electric locomotives and similar applications where the coasting or braking motor actually returns power to the system.

Equation (6) and the corresponding speed-torque curve also hold for the **shunt motor** diagrammed in Fig. 18.4–8b. But now $I_f = V/R_f$, so ϕ depends on the applied voltage and rotation reversal via voltage polarity is not possible. By the same token, a control rheostat in series with the field winding provides speed control with a ratio of about 2:1 — as long as care is taken not to reduce the field too much.

The **series motor** in Fig. 18.4–8c turns out to have dramatically different characteristics because its flux varies with the armature current. For an analytic investigation we use the linear approximations of Eq. (4) and let $I = I_f = I_a$ and $R = R_f + R_a$. Thus,

$$V = RI + E \qquad E = k_f I n \tag{7a}$$

and the torque is proportional to current *squared*, namely,

$$T = 30\, k_f I^2/\pi \tag{7b}$$

Solving for n in terms of T yields

$$n = \sqrt{\frac{30}{\pi k_f}}\,\frac{V}{\sqrt{T}} - \frac{R}{k_f} \tag{8}$$

as plotted in Fig. 18.4–9.

We clearly see that the series motor is a *varying-speed* device that has large starting torque at $n = 0$ but tends to "run away" under no-load conditions, since $n \to \infty$ if $T \to 0$. This behavior would be suitable for a constant-power application that needs low torque at high speed and high torque at low speed. Other applications might call for the intermediate speed-torque characteristics of a **compound motor,** also plotted in Fig. 18.4–9.

A significant implication of Eq. (7) is that a series motor with *alternating* current $i = \sqrt{2}|I|\cos \omega t$ will develop pulsating but *unidirectional* torque $T(t) = (60\, k_f/\pi)|I|^2 \cos^2 \omega t$. It is therefore a **universal motor,** capable of operating with AC or DC excitation (providing that both stator and rotor are laminated). The typical speed-torque curves in Fig. 18.4–10 exhibit lower mechanical output power in the AC mode due to reactive voltage drops and magnetic saturation at the AC current peaks. Nonetheless, most universal motors are intended for use with an AC supply because the universal motor is a *high-speed* machine normally running around 5000–15,000 rpm. As such, it has a high ratio of output *power to weight* and best suits the needs of many hand tools and portable appliances — electric drills, vacuum cleaners, and so forth. Moreover, a simple thyristor circuit suffices for torque/speed control if required.

Figure 18.4–10

Speed-torque curves for a universal motor with DC or AC excitation.

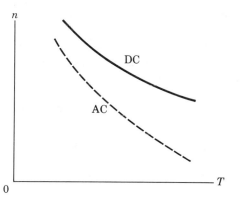

Example 18.4–2

Suppose the machine described in Example 18.4–1 is rated for $T_{FL} = 100$ N · m when operated as a shunt motor with $V = 200$ V and $R_c = 0$. The resulting field current will be $I_f = V/R_f = 5$ A, so $N_f I_f = 900$ and Fig. 18.4–4 shows that $\phi \approx 47$ mWb. After inserting numerical values, Eq. (6a) becomes

$$n = 638 - 0.533T$$

Hence, with $T = T_{FL}$, we find that

$$n_{FL} = 638 - 0.533 \times 100 \approx 585 \text{ rpm}$$

whereas $n_{NL} = 638$ rpm. The corresponding speed regulation of the motor is SR $= (638 - 585)/638 \approx 8\%$.

The mechanical power at full load is $P_m = T_{FL} \times (2\pi n_{FL}/60) = 6.13$ kW ≈ 8 hp. To find the electrical input power, we first use Eq. (2) to calculate the back emf $E = 183$ V. Then $I_a = (V - E)/R_a = 34$ A, so the armature draws $P_a = VI_a = 6.8$ kW. But the field winding also draws power $P_f = VI_f = 1$ kW, all of which is dissipated as ohmic heating. Thus, the motor's full-load efficiency is

$$\text{Eff} = \frac{P_m}{P_a + P_f} = \frac{6.13}{7.8} \approx 80\%$$

which neglects any mechanical losses.

Exercise 18.4-2   Consider a series motor running with $V = 120$ V, $R = 20$ Ω, and $k_f = 0.01$. Calculate the current, torque, and horsepower developed at 10,000 rpm. Also find the resulting efficiency in absence of mechanical losses.

PROBLEMS

18.1-1   Consider an elementary alternator having p poles. The radial field of the rotor is $B_r = B_R \cos \frac{1}{2} p \gamma_r$, with γ_r measured from the centerline of one of the north poles. Appropriately spaced stator conductors yield $e(t) = \frac{1}{2} p N B_R A \omega_m \cos \frac{1}{2} p \omega_m t$. Use a sketch like Fig. 18.1-1c to derive $e(t)$ when $p = 4$ and there are N stator conductors at $\gamma_s = 0°$, $90°$, $180°$, and $-90°$.

18.1-2   Do Problem 18.1-1 with $p = 6$ and N stator conductors at $\gamma_s = 0°$, $60°$, $120°$, $180°$, $-120°$, and $-60°$.

18.1-3   Suppose the stator in Fig. 18.1-1 has a primitive *distributed* winding. Using Eq. (1), obtain an expression for $e(t)$ when there are $N/2$ upper conductors at $\gamma_s = -45°$ and $45°$, and $N/2$ lower conductors at $\gamma_s = 135°$ and $-135°$. Then set $t = 0$ to find e_{max} and calculate the winding factor $k_w = e_{max}/N B_R A \omega_m$.

18.1-4   Do Problem 18.1-3 with $N/3$ upper conductors at $\gamma_s = -60°$, $0°$, and $60°$, and $N/3$ lower conductors at $\gamma_s = 120°$, $180°$, and $-120°$.

18.1-5   A certain 60-Hz synchronous machine is described by Eq. (9) with $k = 0.04$, $B_f = 0.5$ T, and $A = 0.1$ m². Its copper losses are 8% and rotational losses are 12%. If the machine has two poles and is operated as an alternator with $N_a I_a = 2500$ A-t, then what is the electrical output power?

18.1-6   Suppose the machine in Problem 18.1-5 has six poles and is operated as a motor. Calculate the armature mmf $N_a I_a$ needed to get 20 hp at the shaft.

18.1-7   Suppose the machine in Problem 18.1-5 has four poles and is operated as an alternator. Calculate the number of armature conductors N_a needed to deliver output current $I_a = 50$ A at 240 V with unity power factor.

18.1-8   A small 60-Hz synchronous machine is described by Eq. (9) with $k = 0.025$, $B_f = 1$ T, and $A = 0.05$ m². The machine delivers 600 W when operated as an alternator with $N_a I_a = 400$ A-t. Use efficiency considerations to determine the most likely number of poles the machine has.

18.1-9   Suppose the machine in Problem 18.1-8 develops 1 hp when operated as a motor with $N_a I_a = 700$ A-t. Use efficiency considerations to determine the most likely number of poles the machine has.

18.2-1   A three-phase generator with $X_s = 60$ Ω has automatic regulation that keeps the terminal voltage at $|V| = 3200$ V. Find $|E|$ and δ when the generator is connected to a balanced load drawing $P_p = 48$ kW at pf = 0.75 lagging.

18.2-2   Do Problem 18.2-1 with $P_p = 80$ kW at pf = 0.5 lagging.

18.2-3   An unregulated three-phase generator has $|E| = 600$ V and $X_s = 6$ Ω.
(a) Taking $|I| = 50$ A, calculate $|V|$, P_p, and δ for $\theta = 0°$, $30°$, and $-30°$.
(b) Sketch $|V|$ versus $|I|$ for $\theta = 0°$, $30°$, and $-30°$, including the points at $|I| = 0$ and $|V| = 0$.

18.2–4 Do Problem 18.2–3 with $|E| = 1200$ V and $X_s = 10\ \Omega$.

18.2–5 Do Problem 18.2–3 with $|E| = 450$ V and $X_s = 5\ \Omega$.

18.2–6 A balanced load is connected to a three-phase generator with $|V| = 1000$ V, $|E| = 1700$ V, and $X_s = 10\ \Omega$. Using phasor diagrams, find $|I|$ and δ when the load has pf $= 1.0$ and when it has pf $= 0.8$ lagging.

18.2–7 A balanced load is connected to a three-phase generator with $|V| = 2800$ V, $|E| = 4100$ V, and $X_s = 50\ \Omega$. Using phasor diagrams, find $|I|$ and δ when the load has pf $= 1.0$ and when it has pf $= 0.6$ lagging.

18.2–8 A lossless three-phase synchronous motor with $X_s = 6\ \Omega$ is operated at $|V| = 400$ V. If the motor has unity power factor and it develops 60 kW of mechanical output power, then what are the values of $|E|$ and δ?

18.2–9 Do Problem 18.2–8 for 90 kW of mechanical output power.

18.2–10 Consider a three-phase synchronous motor having $|E| = 1500$ V and $X_s = 50\ \Omega$. Evaluate P_{max} when $|V| = 2000$ V, and use a phasor diagram with $\delta = 90°$ to find the corresponding value of $|I|$. Then calculate θ and Q_p.

18.2–11 Do Problem 18.2–10 with $|V| = 800$ V.

18.2–12 An overexcited three-phase synchronous motor has $X_s = 18\ \Omega$ and is operated at $|V| = 900$ V. The motor draws $P_p = 30$ kW to drive a mechanical load, and it also draws $Q_p = -40$ kVAr to correct the power factor of other machines connected to the same line. Construct a phasor diagram to find $|E|$ and δ.

18.2–13 Do Problem 18.2–12 with $P_p = 15$ kW and $Q_p = -25$ kVAr.

18.3–1 An 8-pole induction motor operates at 5% slip with $f = 60$ Hz. Find the rotor speed, the slip speed, and the frequency of the rotor currents.

18.3–2 A 60-Hz induction motor operates at 585 rpm. Determine the number of poles, the slip, and the frequency of the rotor currents.

18.3–3 A 50-Hz induction motor operates at 720 rpm. Determine the number of poles, the slip, and the frequency of the rotor currents.

18.3–4 A 6-pole 60-Hz induction motor develops 50 hp at $n_{FL} = 1170$ rpm.
(a) Calculate s_{FL} and T_{FL}.
(b) Estimate n when $T = 1.2T_{FL}$ and $T = 0.8T_{FL}$.

18.3–5 Do Problem 18.3–4 for a 4-pole 50-Hz motor that develops 20 hp at $n_{FL} = 1455$ rpm.

18.3–6 A 2-pole three-phase induction motor operates at $f = 60$ Hz and $|V| = 240$ V. The motor has $R_1 = 0.6\ \Omega$, $R_2' = 0.2\ \Omega$, and $X = 1.2\ \Omega$.
(a) Calculate the torque, horsepower, and efficiency at $n = 3520$ rpm.
(b) Find $|I|$ and θ at $n = 0$ and 3520 rpm.

18.3–7 An 8-pole three-phase induction motor operates at $f = 60$ Hz and $|V| = 600$ V. The motor has $R_1 = 0.01\ \Omega$, $R_2' = 0.6\ \Omega$, and $X = 1.5\ \Omega$.
(a) Calculate the torque, horsepower, and efficiency at $n = 860$ rpm.
(b) Find $|I|$ and θ at $n = 0$ and 860 rpm.

18.3–8 Suppose the motor in Problem 18.3–6 drives a load that requires $T = n/50$. Assume small slip to determine the resulting operating speed.

18.3-9   Suppose the motor in Problem 18.3-7 drives a load that requires $T = n^2/8000$. Assume small slip to determine the resulting operating speed.

18.3-10   A 4-pole three-phase induction motor develops 50 hp at $n = 1750$ rpm. The source has $|V| = 400$ V and $f = 60$ Hz, and the motor draws $|I| = 31$ A with $\theta = 5°$. Estimate the values of R_2', X, and R_1.

18.3-11   A 6-pole three-phase induction motor develops 75 hp at $n = 965$ rpm. The source has $|V| = 800$ V and $f = 50$ Hz, and the motor draws $|I| = 19$ A with $\theta = 6°$. Estimate the values of R_2', X, and R_1.

18.3-12   Suppose the motor in Problem 18.3-6 has a wound rotor. Sketch the torque-speed curve for $R_{ext} = 0$ and $R_{ext} = 0.3$ Ω by finding the starting torque, the maximum torque, and the slip at maximum torque.

18.3-13   Suppose the motor in Problem 18.3-7 has a wound rotor. Sketch the torque-speed curve for $R_{ext} = 0$ and $R_{ext} = 0.4$ Ω by finding the starting torque, the maximum torque, and the slip at maximum torque.

18.3-14   Derive Eqs. (5a) and (5b) via the following method: Divide the numerator and denominator of Eq. (4) by s^2 and let $y = R_2'/s$; then solve $dT/dy = 0$ for y and use your result to obtain the expressions for s and T_{max}.

18.4-1   A certain 6-pole DC machine has $N_a = 40$, $R_a = 0.2$ Ω, $N_f = 200$, $R_f = 50$ Ω, and the field characteristics given in Fig. 18.4-4. Determine the flux ϕ and field voltage V_f needed to get $V = 200$ V when the machine is operated as a separately-excited generator with $n = 1400$ rpm and $I = 200$ A. Then calculate the generator's efficiency, assuming no mechanical losses.

18.4-2   Do Problem 18.4-1 with $n = 900$ rpm.

18.4-3   A certain 2-pole DC machine has $R_a = 1.2$ Ω, $N_a = 300$, $R_f = 40$ Ω, $N_f = 600$, and nearly constant reluctance $\mathcal{R} = 20{,}000$. The machine is operated as a separately-excited generator with $V_f = 120$ V and $n = 1000$ rpm. The terminal voltage is maintained at 600 V by adjusting a control rheostat in series with the field winding. Find the maximum and minimum values of R_C when the load current varies over the range 0-100 A.

18.4-4   Do Problem 18.4-4 with $\mathcal{R} = 15{,}000$ and $V_f = 75$ V.

18.4-5   Suppose the generator in Example 18.4-1 has $R_C = 24.3$ Ω, and the load draws $I = 0$. Obtain an expression for ϕ in terms of $N_f I_f$, and use a graphical construction on Fig. 18.4-4 to find ϕ. Then calculate the corresponding values of I_f, E, and V.

18.4-6   Do Problem 18.4-5 with $I = 40$ A.

18.4-7   Consider a shunt-connected generator with $\phi \approx k_f' \sqrt{I_f}$ and $R_a \ll R_f$. Show that the terminal current-voltage relationship is $I \approx (K\sqrt{V/R_f} - V)/R_a$ where $K = k_a k_f' n$.

18.4-8   Find the torque developed and armature current drawn by a separately-excited motor when $n = 0$ and $n = 1000$ rpm, given that $k_a = 4$, $\phi = 50$ mWb, $R_a = 2$ Ω, and $V = 300$ V.

18.4-9   A separately-excited motor rated for $T_{FL} = 100$ N · m has $n_{NL} = 1600$ rpm and $n_{FL} = 1200$ rpm when $V = 240$ V and $\phi = 30$ mWb. Find n_{NL} and n_{FL} when $V = 200$ V and $\phi = 20$ mWb.

18.4-10   What values of V and ϕ are needed so that the motor in Problem 18.4-9 has $n_{NL} = 1000$ rpm and $n_{FL} = 900$ rpm?

18.4−11   By finding dP_m/dn, show that a separately-excited motor develops maximum mechanical power P_m when $n = \frac{1}{2}n_{\text{NL}}$.

18.4−12   A certain shunt motor has $k_a = 5$, $R_a = 1\ \Omega$, and nearly constant reluctance so $\phi \approx KV$ with $K = 2 \times 10^{-4}$. The applied voltage is automatically adjusted to keep the speed at $n = 0.9n_{\text{NL}}$ even though the load torque varies from 20 to 50 N · m. What are the minimum and maximum values of V?

18.4−13   Do Problem 18.4−12 with $K = 10^{-4}$.

18.4−14   Suppose the machine in Problem 18.4−1 is operated as a shunt motor with $V = 250$ V. Find the speed, the shaft horsepower, and the efficiency when $T = 500$ N · m. Neglect mechanical losses.

18.4−15   Consider a series DC motor with $k_f = 0.01$, $R_f = 6\ \Omega$, $R_a = 2\ \Omega$, and $V = 300$ V. Calculate the speed, the shaft horsepower, and the efficiency when $T = 25$ N · m and when $T = 100$ N · m. Neglect mechanical losses.

18.4−16   Suppose the machine in Problem 18.4−3 is operated as a DC series motor with $V = 600$ V. Calculate the speed, the shaft horsepower, and the efficiency when $T = 10$ N · m and when $T = 100$ N · m. Neglect mechanical losses.

18.4−17   Consider a series motor with an applied AC voltage and no magnetic saturation. The phasors $\underline{V}$, $\underline{I}$, and $\underline{E}$ are then related by the circuit in Fig. 18.4−8c, except that an inductive impedance jX must be included in series with R. The average torque is $T = 30k_f|I|^2/\pi$, where $|I|$ is the rms current. Derive an expression similar to Eq. (8) for n in terms of k_f, T, R, X, and the rms voltage $|V|$.

Supplementary Reading

The textbooks listed below, by subject area, are recommended as supplements to this book. The symbol ‡ identifies more advanced treatments of particular subjects.

General Electrical Engineering

L. Bobrow, *Fundamentals of Electrical Engineering.* Holt, Rinehart, and Winston, New York, 1985.

A. Fitzgerald, D. Higginbotham, and A. Grabel, *Basic Electrical Engineering,* 5th edition. McGraw-Hill, New York, 1981.

C. Paul, S. Nasar, and L. Unnewehr, *Introduction to Electrical Engineering.* McGraw-Hill, New York, 1986.

S. Schwarz and W. Oldham, *Electrical Engineering: An Introduction.* Holt, Rinehart, and Winston, New York, 1984.

R. Smith, *Circuits, Devices, and Systems,* 3rd edition. Wiley, New York, 1976.

Circuit Analysis

L. Bobrow, *Elementary Linear Circuit Analysis,* 2nd edition. Holt, Rinehart, and Winston, New York, 1987.

D. Johnson, J. Hilburn, and J. Johnson, *Basic Electric Circuit Analysis,* 3rd edition. Prentice-Hall, Englewood Cliffs, 1986.

W. Hayt and J. Kemmerly, *Engineering Circuit Analysis,* 4th edition. McGraw-Hill, New York, 1986.

J. Nilsson, *Electric Circuits,* 2nd edition. Addison-Wesley, Reading, 1986.

R. Thomas and A. Rosa, *Circuits and Signals: An Introduction to Linear and Interface Circuits.* Wiley, New York, 1984.

System Dynamics and Control

C. Close and D. Frederick, *Modeling and Analysis of Dynamic Systems.* Houghton Mifflin, Boston, 1978.

R. Dorf, *Modern Control Systems*‡, 4th edition. Addison-Wesley, Reading, 1986.

D. Frederick and A. Carlson, *Linear Systems in Communication and Control*‡. Wiley, New York, 1971.

Operational Amplifiers

R. Irvine, *Operational Amplifier Characteristics and Applications,* 2nd edition. Prentice-Hall, Englewood Cliffs, 1987.

D. Johnson and V. Jayakumar, *Operational Amplifier Circuits: Design and Application,* Prentice-Hall, Englewood Cliffs, 1982.

Electronic Circuits

B. Dowding, *Principles of Electronics.* Prentice-Hall, Englewood Cliffs, 1988.

A. Malvino, *Electronic Principles,* 2nd edition. McGraw-Hill, New York, 1979.

J. Millman and A. Grabel, *Microelectronics‡,* 2nd edition. McGraw-Hill, New York, 1987.

A. Sedra and K. Smith, *Microelectronic Circuits‡,* 2nd edition. Holt, Rinehart, and Winston, New York, 1987.

R. Smith, *Electronics; Circuits and Devices,* 2nd edition. Wiley, New York, 1980.

Semiconductor Devices

S. Sze, *Semiconductor Devices: Physics and Technology‡.* Wiley, New York, 1985.

E. Yang, *Microelectronic Devices‡.* McGraw-Hill, New York, 1988.

Instrumentation and Measurements

G. Barney, *Intelligent Instrumentation: Microprocessor Applications in Measurement and Control.* Prentice-Hall, Englewood Cliffs, 1985.

W. Cooper and A. Helfrick, *Electronic Instrumentation and Measurement Techniques,* 3rd edition. Prentice-Hall, Englewood Cliffs, 1985.

S. Wolf, *Guide to Electronic Measurements and Laboratory Practices,* 2nd edition. Prentice-Hall, Englewood Cliffs, 1983.

Communication Systems

A. Carlson, *Communication Systems: An Introduction to Signals and Noise in Electrical Communication‡,* 3rd edition. McGraw-Hill, New York, 1986.

H. Stark, F. Tuteur, and J. Anderson, *Modern Electrical Communication; Analog, Digital and Optical Systems‡,* 2nd edition. Prentice-hall, Englewood Cliffs, 1988.

F. Stremler, *Introduction to Communication Systems,* 2nd edition. Addison-Wesley, Reading, 1982.

Digital Circuits and Systems

A. Dixon and J. Antonakos, *Digital Electronics with Microprocessor Applications.* Wiley, New York, 1987.

J. Gault and R. Pimmel, *Introduction to Microcomputer-Based Digital Systems.* McGraw-Hill, New York, 1982.

L. Jones, *Principles and Applications of Digital Electronics.* Macmillan, New York, 1986.

A. Malvino and D. Leach, *Digital Principles and Applications,* 3rd edition. McGraw-Hill, New York, 1981.

R. Sandige, *Digital Concepts Using Standard Integrated Circuits.* McGraw-Hill, New York, 1978.

Microprocessors and Computers

T. Bartee, *Digital Computer Fundamentals,* 5th edition. McGraw-Hill, New York, 1981.

W. Eccles, *Microprocessor Systems: A 16-Bit Approach.* Addison-Wesley, Reading, 1985.

G. Gibson and Y. Liu, *Microcomputers for Engineers and Scientists,* 2nd edition. Prentice-Hall, Englewood Cliffs, 1987.

H. Lam and J. O'Malley, *Fundamentals of Computer Engineering: Logic Design and Microprocessors.* Wiley, New York, 1988.

Electric Power Systems

O. Elgerd, *Basic Electric Power Engineering.* Addison-Wesley, Reading, 1977.

W. Stevenson, *Elements of Power System Analysis,* 4th edition. McGraw-Hill, New York, 1982.

Electric Machines

V. Del Toro, *Electric Machines and Power Systems.* Prentice-Hall, Englewood Cliffs, 1985.

L. Matsch, *Electromagnetic and Electromechanical Machines,* 2nd edition. Harper and Row, New York, 1977.

G. Slemon and A. Straughen, *Electric Machines.* Addison-Wesley, Reading, 1980.

Cramer's Rule

Cramer's rule provides an efficient organization for the work required to solve a set of simultaneous linear equations. We'll give explicit formulas here for the case of two and three unknowns. When there are more than three unknowns, the arithmetic becomes quite tedious and is best carried out by a computer or calculator program.

Consider a pair of linear equations with two unknowns, say y_1 and y_2, written in the standard form

$$a_1 y_1 + a_2 y_2 = x_1$$
$$b_1 y_1 + b_2 y_2 = x_2 \tag{1}$$

Cramer's rule gives the solution for the unknowns as

$$y_1 = D_1/D \qquad y_2 = D_2/D \tag{2}$$

where the Ds are *determinants*.

The denominator term D is the determinant of the 2×2 array of coefficients on the left side of Eq. (1), namely

$$D = \begin{vmatrix} a_1 & a_2 \\ b_1 & b_2 \end{vmatrix} = a_1 b_2 - a_2 b_1 \tag{3a}$$

The numerator determinants D_1 and D_2 are obtained by putting the x column on the right of Eq. (1) in place of the first or second column of the coefficient array. Thus,

$$D_1 = \begin{vmatrix} x_1 & a_2 \\ x_2 & b_2 \end{vmatrix} = x_1 b_2 - a_2 x_2$$
$$D_2 = \begin{vmatrix} a_1 & x_1 \\ b_1 & x_2 \end{vmatrix} = a_1 x_2 - x_1 b_1 \tag{3b}$$

Note in Eqs. (3a) and (3b) that you simply take the diagonal product from upper left to lower right and subtract the product from upper right to lower left.

Now consider the set of three simultaneous linear equations

$$a_1 y_1 + a_2 y_2 + a_3 y_3 = x_1$$
$$b_1 y_1 + b_2 y_2 + b_3 y_3 = x_2$$
$$c_1 y_1 + c_2 y_2 + c_3 y_3 = x_3 \tag{4}$$

Direct application of Cramer's rule gives the solution in the form

$$y_k = D_k/D \qquad k = 1, 2, 3 \tag{5}$$

where

$$D = \begin{vmatrix} a_1 & a_2 & a_3 \\ b_1 & b_2 & b_3 \\ c_1 & c_2 & c_3 \end{vmatrix} \tag{6}$$

The numerator determinants D_k are similarly defined with the x column in place of the kth column of the coefficient array.

The 3×3 determinant in Eq. (6) can be evaluated by *expansion*. In particular, expanding along the first row yields

$$\begin{vmatrix} a_1 & a_2 & a_3 \\ b_1 & b_2 & b_3 \\ c_1 & c_2 & c_3 \end{vmatrix} = a_1 \begin{vmatrix} b_2 & b_3 \\ c_2 & c_3 \end{vmatrix} - a_2 \begin{vmatrix} b_1 & b_3 \\ c_1 & c_3 \end{vmatrix} + a_3 \begin{vmatrix} b_1 & b_2 \\ c_1 & c_2 \end{vmatrix} \tag{7}$$

Each term here consists of a coefficient from the first row of the array multiplied by the determinant of the 2×2 array that results when you delete the row and column containing that coefficient. The 2×2 determinants are then computed as in Eq. (3a).

Expansion of a 3×3 determinant may be carried out along any row or column, and the resulting middle term is always subtracted from the other two terms. For instance, expanding D in Eq. (6) along the first column gives

$$\begin{vmatrix} a_1 & a_2 & a_3 \\ b_1 & b_2 & b_3 \\ c_1 & c_2 & c_3 \end{vmatrix} = a_1 \begin{vmatrix} b_2 & b_3 \\ c_2 & c_3 \end{vmatrix} - b_1 \begin{vmatrix} a_2 & a_3 \\ c_2 & c_3 \end{vmatrix} + c_1 \begin{vmatrix} a_2 & a_3 \\ b_2 & b_3 \end{vmatrix}$$

If one of the terms in the array equals zero, then you save some labor by expanding along the row or column containing that term.

As an alternative approach to the solution of three simultaneous equations, you can always use one of the equations to express an unknown in terms of the other two. Substitution into the remaining two equations thus reduces the problem to two equations with two unknowns. This method works best when you don't need the unknown being eliminated and one of the x terms equals zero to simplify the elimination process.

Example

To illustrate the calculations in Cramer's rule, let's use it to find the unknown y_1 from the set of three equations

$$7y_1 - 2y_2 \qquad\quad = 25$$
$$2y_1 + 3y_2 - 4y_3 = 0$$
$$-5y_2 + 8y_3 = 10$$

Since we only want y_1, and since $x_2 = 0$, we'll start with the elimination method of solution.

The first equation involves only y_1 and y_2. Hence, we eliminate y_3 from the third equation using the second equation rewritten as

$$y_3 = (2y_1 + 3y_2)/4$$

The reduced problem then becomes

$$7y_1 - 2y_2 = 25$$

$$4y_1 + y_2 = 10$$

Evaluating D and D_1 from Eq. (3) gives

$$D = \begin{vmatrix} 7 & -2 \\ 4 & 1 \end{vmatrix} = (7)(1) - (-2)(4) = 15$$

$$D_1 = \begin{vmatrix} 25 & -2 \\ 10 & 1 \end{vmatrix} = (25)(1) - (-2)(10) = 45$$

Therefore,

$$y_1 = 45/15 = 3$$

If desired, you can then find that $D_2 = -30$, so $y_2 = -30/15 = -2$ and $y_3 = [(2)(3) + (3)(-2)]/4 = 0$.

Of course, we could also have solved the complete set of three equations to find y_1. Expanding along the top row yields

$$D = \begin{vmatrix} 7 & -2 & 0 \\ 2 & 3 & -4 \\ 0 & -5 & 8 \end{vmatrix} = 7\begin{vmatrix} 3 & -4 \\ -5 & 8 \end{vmatrix} - (-2)\begin{vmatrix} 2 & -4 \\ 0 & 8 \end{vmatrix} + 0\begin{vmatrix} 2 & 3 \\ 0 & -5 \end{vmatrix}$$

$$= 7(24 - 20) + 2(16 - 0) + 0 = 60$$

$$D_1 = \begin{vmatrix} 25 & -2 & 0 \\ 0 & 3 & -4 \\ 10 & -5 & 8 \end{vmatrix} = 25(24 - 20) + 2(0 + 40) + 0 = 180$$

Hence, $y_1 = 180/60 = 3$, as before.

Tables of Mathematical Relations

The following tables summarize most of the mathematical relationships used in the text and in the problems.

Derivatives and Integrals

$$\frac{d}{dt} e^{st} = s e^{st} \qquad \int e^{st}\, dt = \frac{1}{s} e^{st}$$

$$\frac{d}{dt} \cos \omega t = -\omega \sin \omega t \qquad \int \cos \omega t\, dt = \frac{1}{\omega} \sin \omega t$$

$$\frac{d}{dt} \sin \omega t = \omega \cos \omega t \qquad \int \sin \omega t\, dt = -\frac{1}{\omega} \cos \omega t$$

Complex Numbers

$$\underline{A} = A_r + jA_i = |A|\underline{/\theta_A}$$

$$\mathrm{Re}[\underline{A}] = A_r = |A| \cos \theta_A$$

$$\mathrm{Im}[\underline{A}] = A_i = |A| \sin \theta_A$$

$$|A| = \sqrt{A_r^2 + A_i^2}$$

$$\measuredangle \underline{A} = \theta_A = \begin{cases} \arctan (A_i/A_r) & A_r \geq 0 \\ \pm 180° - \arctan (A_i/-A_r) & A_r < 0 \end{cases}$$

$$\underline{A}^* = A_r - jA_i = |A|\underline{/-\theta_A}$$

$$e^{\pm j\phi} = \cos \phi \pm j \sin \phi = 1\underline{/\pm\phi}$$

$$e^{j\phi} + e^{-j\phi} = 2 \cos \phi \qquad e^{j\phi} - e^{-j\phi} = j2 \sin \phi$$

Exponential and Logarithmic Functions

$$e^{\alpha} e^{\beta} = e^{(\alpha+\beta)} \qquad \frac{e^{\alpha}}{e^{\beta}} = e^{(\alpha-\beta)}$$

$$\log xy = \log x + \log y \qquad \log \frac{x}{y} = \log x - \log y$$

$$\log x^n = n \log x \qquad \log_a x = \frac{\log_b x}{\log_b a}$$

Trigonometric Functions

$$\cos \theta = \sin (\theta + 90°) \qquad \sin \theta = \cos (\theta - 90°)$$

$$\cos^2 \theta = \tfrac{1}{2}(1 + \cos 2\theta) \qquad \sin^2 \theta = \tfrac{1}{2}(1 - \cos 2\theta)$$

$$\sin (\alpha \pm \beta) = \sin \alpha \cos \beta \pm \cos \alpha \sin \beta$$

$$\cos (\alpha \pm \beta) = \cos \alpha \cos \beta \mp \sin \alpha \sin \beta$$

$$\sin \alpha \sin \beta = \tfrac{1}{2} \cos (\alpha - \beta) - \tfrac{1}{2} \cos (\alpha + \beta)$$

$$\cos \alpha \cos \beta = \tfrac{1}{2} \cos (\alpha - \beta) + \tfrac{1}{2} \cos (\alpha + \beta)$$

$$\sin \alpha \cos \beta = \tfrac{1}{2} \sin (\alpha - \beta) + \tfrac{1}{2} \sin (\alpha + \beta)$$

Series Expansions and Approximations

$$(1 + x)^n = 1 + nx + \frac{n(n - 1)}{2!} x^2 + \cdots$$

$$(1 + x)^{-n} = 1 - nx + \frac{n(n + 1)}{2!} x^2 - \cdots$$

$$e^x = 1 + x + \frac{1}{2!} x^2 + \cdots$$

$$\sin x = x - \frac{1}{3!} x^3 + \frac{1}{5!} x^5 - \cdots$$

$$\cos x = 1 - \frac{1}{2!} x^2 + \frac{1}{4!} x^4 - \cdots$$

Standard Component Values

Listed below are standard values of resistors, potentiometers, capacitors, and power-supply voltages for use in electronic circuit design problems. Actual values available from a particular manufacturer may be subject to variation.

Carbon Resistors

These resistors have tolerances of ±5%, ±10%, and ±20%, and power ratings of 1/8, 1/4, 1/2, 1, and 2 W. The range of resistance is:

$$1.0 \, \Omega - 20 \, \text{M}\Omega$$

Significant figures of standard values with ±5% tolerance are:

```
1.0   1.1   1.2   1.3   1.5   1.6   1.8   2.0
2.2   2.4   2.7   3.0   3.3   3.6   3.9   4.3
4.7   5.1   5.6   6.2   6.8   7.5   8.2   9.1
```

Potentiometers

Carbon potentiometers have power ratings of 1/8, 1/4, 1/2, 1, and 2 W. Typical standard values are:

```
250 Ω    500 Ω     1 kΩ      5 kΩ   10 kΩ
 25 kΩ    50 kΩ   100 kΩ   500 kΩ    1 MΩ
```

Ceramic and Polyester Capacitors

These general-purpose capacitors have voltage ratings from 25 to 1000 V. Typical ranges of capacitance are:

10 pF – 47 nF   ceramic disc capacitors
10 nF – 10 μF   metallized polyester capacitors

Significant figures of standard values with ±10% tolerance are:

```
1.0   1.2   1.5   1.8   2.2   2.7
3.3   3.9   4.7   5.6   6.8   8.2
```

Tantalum and Aluminum Capacitors

These polarized electrolytic capacitors have voltage ratings from 4 to 100 V. Typical ranges of capacitance are:

0.1 μF – 220 μF   tantalum electrolytic capacitors
0.1 μF – 22,000 μF   aluminum electrolytic capacitors

Significant figures of standard values with $\pm 20\%$ tolerance are:

1.0   1.5   2.2   3.3   4.7   6.8

Power-Supply Voltages

Standard DC voltages available from integrated-circuit regulators are:

5   6   8   12   15   18   24

Answers to Exercises

2.1–1 $i = 4.17$ A, $T = 10^5$ sec $= 27.8$ hours

2.1–2   15 mW

2.2–1 $i = 8$ A, $p = 1.6$ kW

2.3–1 $v = 15$ V, $p = 45$ mW

2.3–2 $R = 796$ Ω, $\Delta R = 1.59$ Ω

2.4–1   (a) $i = 4$ A; (b) $v_R = 0$; (c) $i = 2$ A; (d) $v_R = -15$ V

2.4–4 $v_1 = 14$ V, $i_2 = -3$ A

3.1–1   120 Ω, 120 W, 240 W, 360 W

3.1–2 $R_{eq} = 4$ Ω, $v_2 = 6$ V

3.1–3   For $R_L = 35$ Ω: $i = 3$ A, $v_L = 105$ V, $p_L = 315$ W, $p_s = 45$ W, Eff $= 87.5\%$

3.2–1 $R_{eq} = (1 + \beta)R$

3.2–2   10 kΩ

3.2–3 $v_{oc} = 60$ V, $i_{sc} = 5$ A, $R_o = 12$ Ω

3.2–4   2.5 mA

3.2–5 $i = 4 - 1.5 = 2.5$ mA

3.3–1 $v_1 = 9$ V, $v_2 = 10$ V

3.3–2 $i_1 = 0.5$ A, $i_2 = -1$ A

3.3–3   2 A

3.4–2 $R_v = 597$ Ω, $R_a \approx 0.075$ Ω

4.1–3 $A_i \approx 30$ dB, $G \approx 24$ dB

4.2–1 $v_{out}/v_{in} = A/(1 + A) = 0.99999$

4.2–2 $K = 10$, $R_F = 30$ kΩ, $R_2 = 3$ kΩ

5.1–1 $i = \begin{cases} 40 \text{ mA} & 0 < t < 50 \ \mu s \\ -20 \text{ mA} & 50 < t < 60 \ \mu s \end{cases}$

5.1–3 $i = 2 \cos 100t$ A, $v_L = -20 \sin 100t$ V, $v = 60 \sin 100t$ V

5.1–4 $I_L = 20$ A, $V_C = 120$ V, $w = 0.76$ J

5.2–1 $RC \, dv_C/dt + v_C = v$

5.2–2 $v(t) = 150t - 75 + 75e^{-2t}$

5.3–2 $v_R = \dfrac{R}{s^2 LCR + sL + R} V_0 e^{st}$

5.3–3 $H(s) = \dfrac{sL}{s^2 LC + sCR + 1}$

6.1-1 $t_0 = 0.01$

6.1-2 $\underline{V} = 24.5 + j8.1 = 25.8 \,\underline{/18.3°}$

6.1-3 $j\underline{A} = 5\,\underline{/36.9°}$, $\underline{A}\,\underline{A} = -7 - j24$, $1/\underline{A} = 0.12 + j0.16$

6.2-1 $\underline{v} = (16 + j30)e^{j30t} = 34e^{j(30t+61.9)}$

6.2-2
$$R(\omega) = \frac{R^2\omega L}{R^2 + (\omega L)^2} \qquad X(\omega) = \frac{R(\omega L)^2}{R^2 + (\omega L)^2}$$

6.2-3 $\underline{Z} = 24 + j7\ \Omega$, $i(t) = 4\cos(800t - 16.3°)$

6.2-4 $\underline{V_b}/\underline{V} = 1.19\,\underline{/-26.6°}$

6.3-2 $\omega_0 = 10^5$, $Q = 4$, $I_m = 41.2$ mA

6.3-3 $\underline{Z} = 5 + j5$, $P = 40$ W, $V_m = 20\sqrt{2}$ V

6.3-4 $v = 20\sqrt{2}\cos(100t + 45°) + 6\sqrt{10}\cos(300t - 161.5°)$

6.4-1 $i_1 = (N_2/N_1)i_2 + (N_3/N_1)i_3$

6.4-2 $I_2 = 2\,\underline{/-16.3°}$, $\underline{V_1} = 15\,\underline{/-53.2°}$

6.4-3 $R = 70.7\ \Omega$, $C = 1.41\ \mu$F, $P = 0.707$ W

7.1-2 $R_a \approx 1.6$ kΩ, $R_b \approx 320\ \Omega$

7.1-3 $Q = 42.8$, $R = 28.8$ kΩ, $C = 22.1$ pF

7.2-1
$$H(j\omega) = 10\,\frac{j\omega}{100 + j\omega}\,\frac{100}{100 + j\omega}, \quad H_{\max} = 14\text{ dB}$$

7.2-2
$$H(j\omega) = -1\,\frac{50 + j\omega}{50}\,\frac{400}{400 + j\omega}, \quad H_{\max} = 18\text{ dB}$$

7.3-1 $\tau = 40\ \mu$s, $v(t) = 48\,e^{-t/\tau}$ for $t \geq 0$

7.3-2 $i(0^-) = i(0^+) = 3$ mA, $i(\infty) = -5$ mA, $v(0^-) = 0$, $v(0^+) = -24$ V, $v(\infty) = 0$

7.3-3   (a) $v(D) \approx 1.8$ V, (b) $C \leq 0.02\ \mu$F

7.4-1   (b) $i(t) = e^{-100t} - 0.5\,e^{-200t}$

7.4-2   Example 7.4-1: $R > 50\ \Omega$; Exercise 7.4-1: $R < 200\ \Omega$

7.4-4 $v_L(t) = (\sqrt{10}/3)V_s\,e^{-100t}\cos(300t + 18.4°)$

8.1-2 $v_s = -0.0482$ V

8.1-3 $I_0 = 10\ \mu$A, $I_d = 64.8\ \mu$A, $i = 54.8\ \mu$A

8.2-1 $v_Q \approx 2.6$ V, $i_Q \approx 0.14$ A

8.2-2 $V_\gamma \approx 2$ V, $R_f \approx 6\ \Omega$, $i \approx 1/3$ A

8.2-3 $v_{\text{out}} = 20.19$ V and $i_Z = 48$ mA when $i_{\text{out}} = 0$; $v_{\text{out}} = 20.00$ V and $i_Z = 0$ when $i_{\text{out}} = 50$ mA

8.2-4 $i = 0$ for $v < -12$ V, $i = 0.5v + 1$ for $v > 3$ V

9.1-1 $A_v \approx -15$

9.1-3 $i_D = 0$ and $v_{DS} = 20$ V; $i_D = 0.5$ mA and $v_{DS} = 17.5$ V; $i_D = 3.85$ mA and $v_{DS} = 0.77$ V

9.1-4 $V_T = 3$ V, $I_{DSS} = 1.8$ mA

9.2-2 $v_{GS} = -4$ V, $R_K = 2$ kΩ, $v_{DS} = 13$ V

9.3-1 $A_v \approx -8.4$, $A_i \approx 84$

9.3-2 $i_C = 3$ mA and $v_{CE} = 6$ V; $i_C = 5.9$ mA and $v_{CE} = 0.2$ V

9.3-3 $i_E = 15.9$ mA

9.3-4 $i_B > 0.131$ mA, $R_B < 71$ kΩ

10.1−1 $R_1 = R_2 = 1$ kΩ
10.1−2 $v_{out} \approx 3.91$ V
10.1−3 $V_C \leq 3.64$ V, $V_{max} \leq 2.36$ V
10.2−1   (a) $v_A < 1.4$ V, $v_1 = 4.61$ V, $\beta_2 > 12.3$; (b) $v_A > 6.2$ V
10.2−2 $R_{DS} = 16$ kΩ, $v_{out} = 0.32$ V
10.2−3 $v_Q = 8.5$ V
10.3−1 $R_F = 24$ kΩ, $V_B = 3.125$ V
10.3−2   (a) $v_{out} = 0$ for $0.916RC < t < D_{in} + 0.916RC$; (b) $v_x < V_2$ so v_{out} stays at V_{DD}
10.3−3 $v_x(t) = V_{DD} + (V_1 - 2V_{DD})e^{-(t-t_2)/\tau}$

11.1−1 $A_v = -6$ for $n = 1$; $A_v = -600$ for $n = 3$
11.1−3 $R_D = 2.7$ kΩ, $R_K = 1.3$ kΩ, $R_1 = 1.8$ MΩ, $R_2 = 2.4$ MΩ
11.2−1
$$A_v \approx \frac{R_i^2 \mu^2}{(R_s + R_i)(R_i + R_c)} \frac{R_L}{R_C + R_L}$$
11.2−2 0.961 mA $\leq I_C \leq 1.026$ mA
11.2−3 $R_C = 2.7$ kΩ, $R_E = 1.2$ kΩ, $R_1 = 75$ kΩ, $R_2 = 36$ kΩ
11.3−1 $B = -0.04, -22.2 \leq A_f \leq -16.7$
11.3−2 $A = -g_m(R_D \| R_L)$, $B = -R_{K1}/(R_D \| R_L)$
11.3−3 0.727 with follower, 0.0385 without follower
11.3−4 $R_{i1} = 106$ kΩ, $\mu_1 = 0.997$, $R_{o1} = 0.0712$ kΩ, $A_v = -191$
11.4−2 $C_o = 0.47$ μF, $C_i = 1.5$ μF, $C_E = 220$ μF
11.4−3 $f_u \approx \omega_1/2\pi \approx 4$ MHz, $f_{max} \approx 22$ MHz, $|A_v|\omega_u = 2.3 \times 10^8$
11.5−1   (a) $\mu_d = 100$, $\mu_{cm} = -0.002$, CMRR ≈ 94 dB;
             (b) $v_{out} = 2 \cos \omega_d t - 0.002 \cos \omega_{cm} t + 0.1$
11.5−2 $R = 4.3$ kΩ, gain ≈ 33.7

12.1−1 $V_0 = 32.9$ V, $I_0 = 0.548$ A, $P_L \leq 0.312$ W, Eff $\approx 4.7\%$
12.1−2 $N = 2.24$, $I_C \approx 0.5$ A, $V_{CC} = 22$ V, $R_3 = 240$ Ω
12.1−3   For push-pull amplifier: $V_{CC} = 20$ V, $P_{CC} = 32$ W, $P_{D_{max}} \geq 5$ W, $i_{C_{max}} \geq 3$ A,
             $v_{CE_{max}} \geq 40$ V
12.2−1 $V_r = 6.93$ V, $V_{DC} = 48.5$ V, $i_{pk} = 12.5$ A
12.2−3 $14.73 \leq V_{DC} \leq 15.05$, $25 \leq I_Z \leq 188$ mA, $5.15 \leq V_{CE} \leq 14.0$ V
12.2−4
$$\frac{I_r}{I_{DC}} \approx \frac{4}{3\sqrt{1 + (2\omega L/R_L)^2}}$$
12.3−1 $T_A = -52°$C, $T_C = 188°$C
12.3−2
$$P_Q = \frac{T_{J_{max}} - T_A}{T_{J_{max}} - T_{CO}} P_{DO}$$

13.1−2   Triangular wave: $W = 5$ kHz; sawtooth wave: $W = 25$ kHz
13.1−3 $y(t) = 3 \sin 2\pi t - \sin 2\pi 3t$
13.1−4 $|H_{eq}(f)| = \sqrt{1 + (f/f_{co})^2}$, $\theta_{eq}(f) = +\arctan(f/f_{co})$
13.2−1   Frequencies: 0, 50, 200, 250, 950, 1000, 1200, 1250
13.2−3   (a) $B \geq 12$ kHz; (b) $B \geq 36$ kHz
13.3−1   (a) $v_L = 5 + 6 \cos(2\pi 60t + 90°)$ mV (b) $v_L = 5 + 0.006 \cos(2\pi 60t + 90°)$ mV

13.3−2 $N_{\text{out}} = 80$ pW, $n_{\text{rms}} = 2$ mV

13.3−3 $P_{\text{out}}/N_{\text{out}} = 5 \times 10^4$ when $T_A = 0$; $P_{\text{out}}/N_{\text{out}} \geq 10^5$ would require $T_A \leq -0.5T_0$

13.4−1 $G = 2 \times 10^7$, $G_1 = 40$, $G_2 = 1000$, $G_3 = 500$

13.4−2   Radio: 66 dB; transmission line: 120 dB

13.4−4   1.5%

14.1−1 $F = A + (B \cdot \overline{C})$

14.1−3   (a) 1,048,575; (b) 99,999; (c) 9,999

14.2−1   (c) $\overline{F} = A + \overline{B}(C + \overline{D})$

14.2−2 $F = (\overline{B} + \overline{C})(\overline{A} + \overline{C})(\overline{A} + \overline{B})$

14.2−4 $F = \overline{A}\,\overline{D} + \overline{A}B + \overline{B}CD = (B + \overline{D})(\overline{A} + \overline{B})(\overline{A} + C)$

14.3−1 $F = 1$ only when $A = B$

14.3−2 $D_i = W_{i+3}$ for $i = 0, 1, \ldots, 9$, Error $= W_0 + W_1 + W_2 + W_{13} + W_{14} + W_{15}$

14.3−3 $S_1 = A$, $S_0 = B$, $X_0 = X_3 = C$, $X_1 = X_2 = \overline{C}$

14.3−4 $[S_2 S_1 S_0] = [ABC]$, $F_2 = W_1 + W_2 + W_4 + W_7$, $F_1 = W_3 + W_5$, $F_0 = W_6 + W_7$

15.1−1 $Q = \overline{R + P} = \overline{R} \cdot \overline{P}$, $P = \overline{S + Q} = \overline{S} \cdot \overline{Q}$

15.1−3   (a) $6 + 6 + 5 + 3 = 20$; (b) $2 + 2 + 2 + 1 = 7$

15.2−1 $\Delta v = 0.0392$ V, $R_0 = 128$ kΩ, $R_F = 0.502$ kΩ

15.2−2   (a) $\Delta v = 0.4$ V; (b) 7 comparators

15.2−3 $n = 10$, $f_{\text{clock}} = 16.384$ MHz for dual-slope, $f_{\text{clock}} = 96$ kHz for successive-approximation

15.3−2 $C_{\text{out}} = 1$, $[S_3 S_2 S_1 S_0] = 0101$

15.3−3 $[1{*}0010] = -14_{10}$, $[0{*}0001] = 1_{10}$, $[0{*}0000] = 0_{10}$, $[0{*}0100] = 4_{10}$ (overflow)

16.1−1 $|I| = 20$ A, $P = 9600$ W, $Q = +7290$ VAr, $P_L/P \approx 67\%$

16.1−2   pf $= 0.471$ lagging, $Q = +15$ kVAr, $\underline{Z} = 1.11 + j2.35$ Ω

16.1−3   For plant: $P = 13$ kW, $|I| = 65$ A, $C = 995$ μF

16.2−2 $\underline{I}_a = (2/\sqrt{3})$ kA $\underline{/-66.9°}$, $P = 16$ MW, $Q = 12$ MVAr

16.2−3 $P_1 = 11.46$ MW, $P_2 = 4.53$ MW

16.2−4 $\underline{I}_a = 12$ A $\underline{/-120°}$, $\underline{I}_n = 18.17$ A $\underline{/-20.3°}$, $P = 2304$ W, $Q = 2592$ VAr, pf $= 0.664$ lagging

16.3−1 $\underline{V}_2 = 32.2$ kV $\underline{/38°}$, $N_1 = 1.61$, $V_p = 28.1$ kV, $Q_1 = 10$ MVAr, $P_R = 0.625$ MW, Eff $= 94.8\%$

16.3−2   0.0273, 0.0316

17.1−1   101 A, 0.127 T

17.1−3   1.01 m

17.2−1 $\mathcal{R} \approx 69 \times 10^5$, $i = 17.3$ A

17.2−2   (a) $H = 191$ A-t/m, $\phi = 2.3$ mWb, $L = 138$ mH; (b) $H = 382$ A-t/m, $\phi = 2.6$ mWb, $L = 78$ mH

17.3−2 $L_{\text{eq}} = L_1(1 - k^2)$

17.3−3 $\underline{I}_1 = 200$ A $\underline{/-80.2°}$, $R_w = 0.509$ Ω, $R = 0.296$ Ω

17.4−1 $M = 3.06$ kg, $u_M = 24$ m/sec

17.4−3 $N = 21$, $V_s = 42$ mV

17.4−4   1.01 A-t, 0.38 A-t

18.1–1 $N = 90$

18.1–2 $T(t) = 2T_0,\ T_0,\ 0$

18.1–3 93.3 kW, 83.9 kW, 594 N · m

18.2–1 $\underline{E} = 11.7$ kV $\underline{/20°}$

18.2–3 $|I| = 150$ A, $C \approx 100\ \mu$F

18.3–1 $|I| = 152$ A, pf = 0.86 lagging, Eff = 56%

18.3–2 (a) 0.38 Ω; (b) 1051 rpm

18.4–1 $V = 8{,}000\phi - 40.5I = \begin{cases} 118.5\ \text{V} & I = 3\ \text{A} \\ 181\ \text{V} & I = 6\ \text{A} \end{cases}$

18.4–2 $I = 1$ A, $P_m = 0.134$ hp, Eff = 83.3%

Index